Dragons and Tigers

A Geography of South, East and Southeast Asia

Third Edition

Dragons and Tigers

A Geography of South, East and Southeast Asia

Third Edition

Barbara Weightman
California State University, Fullerton

John Wiley & Sons, Inc.

VP AND PUBLISHER	*Jay O'Callaghan*
ACQUISITION EDITOR	*Ryan Flahive*
ASSOCIATE EDITOR	*Veronica Armour*
EDITORIAL ASSISTANT	*Darnell Sessoms*
MARKETING MANAGER	*Margaret Barrett*
MARKETING ASSISTANT	*Susan Matulewicz*
PRODUCTION MANAGER	*Janis Soo*
ASSISTANT PRODUCTION EDITOR	*Elaine S. Chew*
PHOTO DEPARTMENT MANAGER	*Hilary Newman*
PHOTO EDITOR	*Kathleen Pepper*
COVER PHOTOS	*(insets) © Barbara A.Weightman; (background, globe) © Tom Van Sant/Photo Researchers, Inc.*
COVER DESIGNER	*Hotfusion Pte Ltd*

This book was set in 9/12 Point Roman by Aptara, Inc. and printed and bound by Hamilton Printing Company. The cover was printed by Hamilton Printing Company.

This book is printed on acid free paper. ♾

Founded in 1807, John Wiley & Sons, Inc. has been a valued source of knowledge and understanding for more than 200 years, helping people around the world meet their needs and fulfill their aspirations. Our company is built on a foundation of principles that include responsibility to the communities we serve and where we live and work. In 2008, we launched a Corporate Citizenship Initiative, a global effort to address the environmental, social, economic, and ethical challenges we face in our business. Among the issues we are addressing are carbon impact, paper specifications and procurement, ethical conduct within our business and among our vendors, and community and charitable support. For more information, please visit our website: www.wiley.com/go/citizenship.

Library of Congress Cataloging-in-Publication Data:

Weightman, Barbara A.
Dragons and tigers: a geography of South, East, and Southeast Asia / Barbara A. Weightman. —3rd ed.
p. cm.
Includes index.
ISBN 978-0-470-87628-2 (pbk.)
1. South Asia—Geography 2. East Asia—Geography. 3. Southeast Asia—Geography. I. Title.
DS335.W37 2010
915—dc22

2010050386

Printed in the United States of America
10 9 8 7 6 5 4 3 2

I dedicate this work to

Louise Henriette Scheffer
(1932–2010)
Outstanding Teacher and Mentor

Preface

I have designed this book to meet the need for a comprehensive text on the geography of Asia. Knowledge of Asian landscapes is fundamental to academic and practical learning in the twenty-first century. Virtually half of the world's people live in Asia, 40 percent in China and India alone. With some of the fastest growing economies in the world, Asian nations, notably of the Pacific Rim, are critical components of North American and other countries' economic and political strategies. Australia, Canada, and the United States are key destinations for Asian migrants whose presence intensifies networks of trans-Pacific interrelationships. Some Asian nations are global leaders in environmental damage, and some are riddled with internecine strife. I think that learning about this dynamic and multicultural region is imperative.

Dragons and Tigers explores and illustrates conditions, events, problems, and trends of both larger regions and individual nations. I use a cross-disciplinary approach to discuss evolving physical and cultural landscapes. Nature-society interrelationships provide the foundation for my emphasis on social, economic, political, and environmental problems.

There is much new material in this 3rd edition. I have added a chapter on food, agriculture and food security. I have also addressed many additional gender issues. There are more "close up" boxes along with a variety of new maps, tables and photos. I have provided pertinent websites and current and multidisciplinary bibliographies at the end of each chapter. A revised glossary of terms is provided at the end of the text. I have tried to write a book rich and fascinating in content. My hope is that both instructors and students will want to read it.

Acknowledgments

I owe a debt of gratitude to Claudia Lowe, Eva Voigt and the late Louise Scheffer for their never-ending encouragement and support. I also gratefully acknowledge the contributions of those who reviewed various segments of the book. They are:

Christopher Airriess, *Ball State University*
Patrick Buckley, *Western Washington University*
Meera Chatterjee, *University of Akron*
Roman Cybriwsky, *Temple University*
George Demko, *Dartmouth College*
Christine Drake, *Old Dominion University*
Cindy Fan, *UCLA*
Philippe Foret, *University of Oklahoma*
Cub Kahn, *Oregon State University*
Maz Lu, *Kansas State University*
Kin Ma, *Grand Valley State University*
Melinda Meade, *University of North Carolina, Chapel Hill*
Norman Moline, *Augustana College*
David Nemeth, *University of Toledo*
Mary Lee Nolan, *Oregon State University*
Bimal K. Paul, *Kansas State University*
Yda Schreuder, *University of Delaware*
Richard Ulack, *University of Kentucky*
Jean-Paul Rodrigue, *Hofstra University*
Gil Schmidt, *University of Northern Colorado*
Wei Tu, *Georgia Southern University*
Susan M. Walcott, *University of North Carolina, Greensboro*

Special thanks to Harm de Blij and Peter O. Muller for allowing me such generous use of their maps. The people at John Wiley & Sons, Inc. also deserve credit. Jay O'Callaghan, Executive Publisher; Ryan Flahive, Executive Editor; Veronica Armour, Associate Editor; Kathleen Pepper, Photo Editor; Elaine Chew, Production Editor and Helen Walden, copy editor.

Contents

Chapter 5

Agriculture, Food, and Food Security 119

Chapter 6

South Asia: Creating Dilemmas of Diversity 150

Chapter 7

South Asia: Pakistan and the Himalayan States 180

Chapter 8

Chapter 9

Chapter 16

Chapter 1

The Big Picture: Major Influences

"The idea of tolerance and mutual concession is based on admitting the compatibility of many different philosophical views of the world."

HAJIME NAKAMURA (1964)

Introduction

Welcome to South, East, and Southeast Asia, a region of unparalleled diversity that is home to more than half of humankind. More than 3.8 billion of the world's 6.8 billion people inhabit a land area of about 8 million square miles (20,761,000 sq km). Here arose some of the world's most admirable and powerful civilizations. Here, too, are the world's highest mountains and several of its longest rivers. Vast, densely populated plains contrast with expanses of sparsely populated mountains, deserts, and tropical forests. Hundreds of ethnic groups, religions, and languages blend in a kaleidoscopic array of lifestyles and cultural landscapes. Democracies exist alongside socialist and militarist states. The region contains some of the globe's poorest countries as well as the ascendant **dragon and tiger economies** to which we are increasingly connected.

Think of the number of Japanese, Korean, or Taiwanese cars and electronic gadgets that you have. Check your clothing labels; where were they assembled? China? Thailand? Indonesia? Have you ever called an appliance repair service or checked on the balance of a credit card only to find yourself talking to someone in India? Do you eat Chinese, Indian, or Malaysian food? Is there a Buddhist or Hindu temple, or Muslim mosque near your community? We are all linked in some fashion to South, East, and Southeast Asia.

Diversity makes the pursuit of multiple themes necessary for exploring and appreciating this part of the world. However, in light of contemporary geographic thought, some themes will be more apparent than others. An overarching theme will be **landscape**, a concept that goes far beyond that which we observe. It encompasses our entire sensory experience; it is the content of our lived world. Individuals or groups do not function independently of other people or the forces of nature. Moreover, political, economic, social, and environmental systems are increasingly intertwined from local to global scales. Under this rubric, we will investigate both cultural landscapes and physical landscapes, in the context of their interrelationships.

In this text, we will encourage a strong commitment to understanding social and economic inequality. We will look at the pros and cons of capitalism—an economic system that is desired by most people but that expresses itself in growing inequities among various sectors of Asian populations.

We will stress the **social construction** of landscapes as well as **gender** and **quality of life** issues. How have internal and external forces affected ways of life? How well are people's needs being met? Who determines what these needs are? We will also focus on **environmental transformation.** How have local and externally imposed values, attitudes, policies, and practices altered physical, biological, and zoological realms? We will discover how political institutions mediate the interaction between humans and environments. In addition, we will explore the meanings and processes of development in the context of **globalization** and **world systems theory**. Is the world becoming increasingly similar? Or is it becoming more dissimilar? How global is the global economy? Is it becoming more fragmentary with the formation of competitive regional organizations?

Dragons and Tigers

Dragons and tigers represent the dynamic forces of power and change. They are symbols long associated with South, East, and Southeast Asia. In China's *I Ching*, or *Book of Changes*, the light and dark aspects of *yin* and *yang* are often expressed as dragon and tiger. In Daoism, *yang* or goodness is controlled by a green dragon; *yin* or evil is controlled by a white tiger. Many a sage sought the dragon–tiger elixir as a key to immortality. The Chinese Emperor wore *dragon robes* and sat on the *dragon throne*. The dragon is used as a metaphor for mountains and rivers—to cut a pass or install water-control works is to "tame the dragon." Bhutan's national flag incorporates a large black and white dragon on a background of red and yellow. Dragons bring luck and protection. They continue to be incorporated as architectural detail in temples, hotels, houses, and many other structures.

In the ancient *Sanscrit* language, "Naga" is a word that represents a being or deity in the form of a large snake, often with multiple heads. "Naga" can also mean "dragon" and is an important symbol in both Buddhism and Hinduism (Figure 1-1). In Indian myth, nagas are subterranean serpent-beings under the sea. Varuna, the ancient, religious Vedic god of storms, is King of the Nagas. Nagas are enemies of Garuda—a large, eagle-like creature popular in Indonesia. In Hindu myth the important god Vishnu is said to sleep on a serpent in the cosmic waters. Shiva, another important deity, wears a snake around his neck. Among the Dravidian population of southern India, if a cobra is accidently killed, it is cremated like a person according to Hindu ritual. There are also many temples where snakes are worshipped, especially in Kerala state in India's southwest. Kashmir's earliest inhabitants are believed to be nagas. Here, naga also means "spring" and many temples of worship were built by springs. Moreover, nagas remain associated with water by both Hindus and Muslims (the majority population in Kashmir).

Figure 1-1

This stone sculpture is one of many forms of a naga. Here, the beast acts as a guardian figure at a Laotian, Buddhist temple.
Photograph courtesy of B. A. Weightman.

The tiger, aside from being a real, although increasingly rare, animal, is fabled in stories throughout these realms. In Japan, tigers represented the warrior *samurai*. In Korea, the tiger is a protective figure, watching over the land (Figure 1-2). Indonesia is rich in tiger lore with legends of tiger spirits affecting families, and villages. In many parts of Asia, tiger teeth and claws are said to induce prowess and potency. In South Asia, the tiger, along with the elephant, is a national symbol of India, and tiger images appear on seals used by early Indus Valley civilizations more than 4,000 years ago. Betrothed couples in India wear red and yellow shawls that symbolize the tiger's stripes of fertility. East Asian mythology recognizes five types of tiger.

Five Types of Tiger.

1. White = West, Earth and afterlife
2. Blue = East, fertility and vegetation
3. Red = South and fire
4. Black = North and winter
5. Yellow = Sun and center of the universe

Figure 1-2

I took the picture of this large painting that was hanging on the wall of a hotel in Busan (Pusan), Korea. This is the "White Tiger of the West," one of four beasts that protect the cardinal directions in China, Japan, Vietnam, and Korea. Known as Baekyo, it is a prominent creature in Korea's creation myth. South Korea is one of the economic tigers of Asia. Photograph courtesy of B. A. Weightman.

Japan, Singapore, South Korea, and Taiwan have all been referred to as *dragon economies* or *economic tigers,* although today they are described as developed economies. India is struggling for "tigerhood." Although Vietnam's economy is growing and standards of living are rising rapidly, it is still called "Asia's baby tiger."

These beasts are frequently seen in service and product names. Dragonair flies out of Hong Kong. Nine Dragon Beer is sold in China. Singapore's Tiger Beer is exported around the world. Tiger Balm ointment is a trusted remedy for headaches and insect bites. Finally, there is a "dragon, phoenix, and tiger soup." The dragon is snake, the phoenix is chicken, and the tiger is cat.

Dragons and tigers have pervaded Asian landscapes in both subtle and obvious ways. That is why they have been given a place in the title of this book.

Snake Charming

Snakes also represent dragons. While snake charming probably originated in India as a component of traditional, medicinal practices, it is also found in Pakistan, Bangladesh, Sri Lanka, and Thailand. The snake, commonly a cobra, is placed in a basket, where it appears to be hypnotized and sways to the sound of a musical instrument such as a flute. In fact, snakes are deaf and more likely follow the flute as the charmer moves it back and forth. The charmer usually sits just out of range of the snake's reach. Sometimes the snake's fangs and/or venom glands are removed.

Snake charming is a hereditary profession with sons learning from their fathers. Most practitioners lead a migratory existence travelling among town and village markets and festivals. Of course, they are paid for their entertainment.

Although there are an estimated 800,000 charmers in India today, their numbers are shrinking dramatically. There are several reasons for this. Urbanization and deforestation has dramatically reduced the snake population. Further, the Indian government has all but outlawed the practice in line with the views of animal rights activists. Thanks to cable television and nature programs, many people are losing their fear and revulsion of snakes and understand that the snakes are not really hypnotized. As a result of these changes, snake charmers either hide away their activities in remote villages or give up the occupation and turn to begging, scavenging, or poorly-paid day-labor work in the cities.

But the snake charmers are not going away easily. As they are denied their means of earning a living, many are becoming angry and have turned to protesting. In 2004, a large group, waving their snakes, stormed the capital buildings in the state of Orissa (eastern India). An estimated 5,000 snake charmers poured into Kolkhata (Calcutta) in the northeast state of West Bengal to argue their plight.

Now the Indian government and rights groups are somewhat sympathetic to the snake charmers and are looking for ways to improve their lives. One suggestion is to train them to be snake caretakers and educators. Another is to focus on their music and have them join the ranks of street musicians. The government is also allowing a limited number of charmers to practice at popular tourist sites (Figure 1-3).

Figure 1-3

In India, snake charming is a dying profession in light of government strictures and perhaps better opportunities in other low-end occupations. However, snake charmers are allowed to operate in certain tourist areas such as here in New Delhi. Photograph courtesy of B. A. Weightman.

DEFINING "ASIA"

Asia is the world's largest continent, stretching from the Pacific Ocean to the Mediterranean Sea. As of 2010, it incorporates 54 countries—far too great a scope for coverage in one textbook. Asia's expanse and complexity require its division into more manageable and meaningful areas of study. For the sake of convenience, we will use the term "Asia" in this book while recognizing that we are not covering the entire continent. This text will concentrate on three geographic realms of Asia: South, East, and Southeast Asia (Figure 1-4).

According to geographer Harm de Blij (2009), **geographic realms** have three sets of spatial criteria. First, physical and cultural characteristics are the largest units into which we can divide the **ecumene** (i.e., the inhabited world). Second, geographic realms are founded on similarities of **functional interactions** between people and their natural environments. These are expressed in farmscapes, transportation networks, cityscapes, environmental damage, and countless other landscape features. Third, geographic realms focus on and incorporate the world's major population clusters. South, East, and Southeast Asia encompass the two largest clusters in India and China and the fourth largest in Indonesia. You will learn about the growth and distribution of these populations in a later chapter.

South Asia comprises India, Pakistan, and Bangladesh; the Himalayan states of Nepal and Bhutan; and the island nations of Sri Lanka and the Maldives. East Asia includes China, Mongolia, North and South Korea, Taiwan, and Japan. Southeast Asia encompasses the mainland countries of Myanmar, Thailand, Lao (Laos), Cambodia, and Vietnam, in addition to the maritime states of Malaysia, Singapore, Brunei, Indonesia, Timor-Leste, and the Philippines.

"ASIA" AS MYTH

These countries and their respective realms are parts of a larger entity called *Asia*. While we might think of Asia as a definable geographic region, actually it is more of an idea than a reality. More importantly, the evolution of the *idea of Asia* in Europe and the West shaped the nature and tenor of perceptions and interactions between Asian and non-Asian cultures throughout history. To understand this, we must explore what sometimes is called the **myth of Asia.**

There is no cultural or historical entity that equals Asia. It is more of a literary and psychological construct imposed from the outside than a geographical reality. Designations of "the Orient" and "the Far East" are also associated with the idea of Asia. Where do these terms come from, and what do they mean? What is Asia, and how did it come to be? How can it be a myth if it actually exists?

Asia, in fact, was defined by Europeans. From the time of ancient Greece, Europeans visualized the world as being divided into two or three continental parts: Asia, Europe, and sometimes Africa. Religiously inspired ***mappae mundi*** (maps of the world), portraying the Earth as flat, often depicted Asia as half the known world! This *understood world* could also be divided into east and west: **Orient** meaning *east,* derived from sunrise, and **Occident**, meaning *west,* from sunset. For Europeans, *Orient* meant east and Asia. However, the initially blurry concept of Asia was filled in over time and concurrently expanded spatially.

CONCEPTUALIZING ASIA

The term *Asia* came from the Greek word for sunrise and originally referred to the Anatolian peninsula (Turkey) in the seventh century BC, a region later known as Asia Minor. By the fifth century BC, Asia had acquired

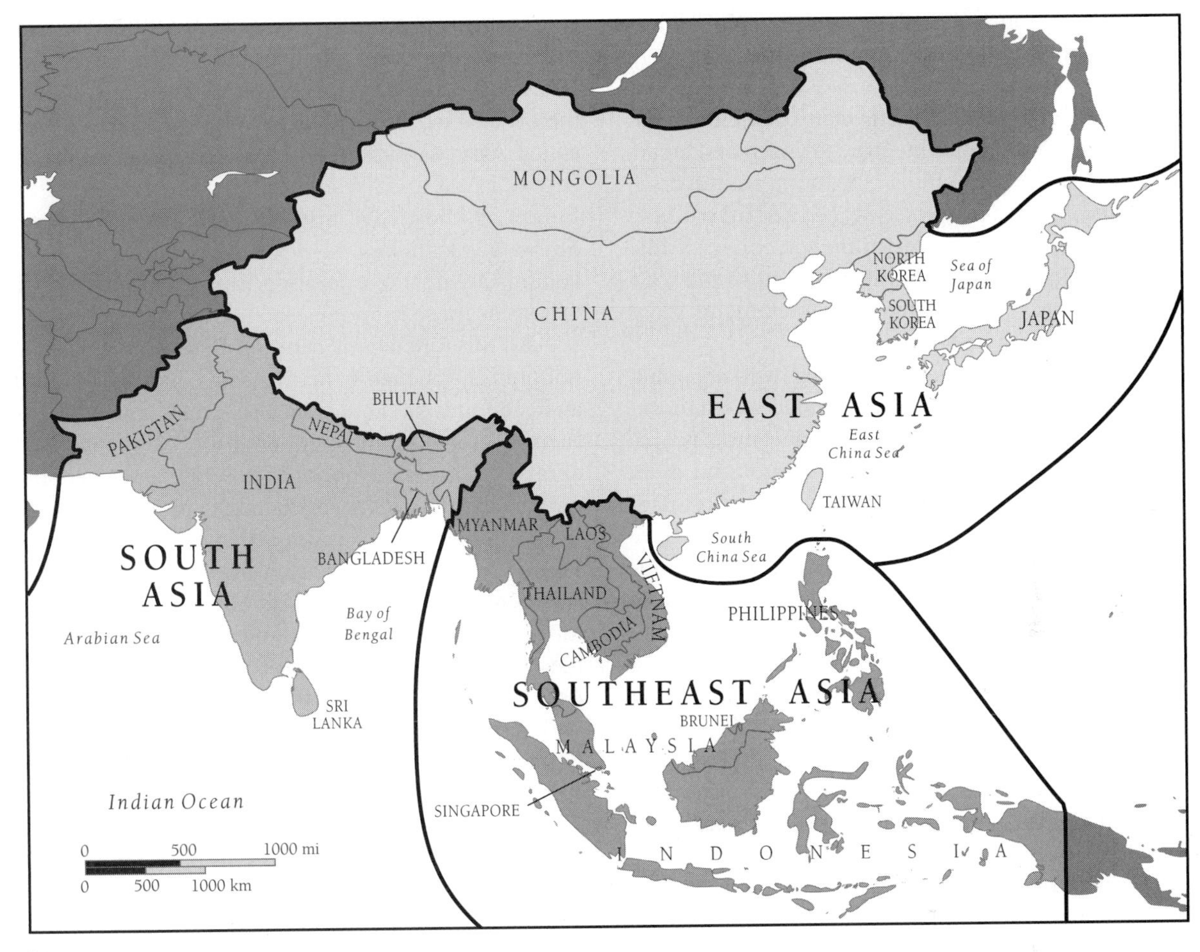

Figure 1-4

These three realms are based on similarities of functional interrelationships between people and their natural environments.

continental meaning. Throughout this evolutionary process, comparisons and contrasts were drawn between east and west.

Centuries later, Karl Marx (1818–1883) drew distinctions between Asia and Western Europe. He subscribed to the concept of "An Asiatic Mode of Production." In this exploitative, economic system, the urban rich and powerful sucked all profits from an undifferentiated rural peasantry laboring in virtual enslavement.

Meanwhile, Asia's colonizers had developed their own frameworks of thought. You will learn more details in later chapters. A general model is provided by geographer James L. Blaut (1994), who refers to a **colonizer's model** of the world: civilization and progress diffuse from the European center to the "culturally barren" periphery. This is an interesting perception considering that Asian civilizations were much older and more advanced and complex than those of the colonizers.

Two things are important here. First, since classical times, Western writers have commonly contrasted their own civilizations with those of Asia. Asia has been perceived as a single entity, its internal diversity homogenized in false unity, its characteristics and institutions portrayed as alien and inferior to those in Occidental realms. The term "inscrutable" was often used to describe the frustrated Western perception of Asia. Recent research on the brain indicates that Asians and Europeans may be "wired" to see the world differently. As the British writer Rudyard Kipling (1865–1936) said in the first line of his poem "The Ballad of East and West": "Oh, East is East, and West is West, and never the twain shall meet."

Second, **environmental determinism** colored historic and geographical accounts well into the twentieth century. Ellen Churchill Semple (1863–1932), an American geographer who taught at the University of Chicago and Clark University, was complimentary to China and Japan, yet described the people of India as ignorant and superstitious. She saw tropical peoples as servile and inferior due to the high temperatures in which they lived. According to Semple, anyone who went to tropical areas would soon be reduced to the lethargic state characteristic of those residing in these debilitating regions.

Derogatory assessments not only generated **xenophobia** (fear of foreigners) but also contributed to **ethnocentrism** and the **we–they syndrome.** Ethnocentric people judge foreign cultures by their own standards and often speak in *we* (better) versus *they* (not as good) terms. Economically developed, democratic Western nations have viewed themselves as superior to less developed, politically alternate Eastern ones. As you will see, this notion of cultural superiority continues to pervade East–West relations. Furthermore, the history of cartography reveals a long-established practice of centering the world on Europe. This promoted a **Eurocentric** worldview, the antithesis of China's **Sinocentric** (China-centered) view.

Recently, I purchased a school notebook printed in Vietnam. On the back cover is a Eurocentered world map captioned by the words, "The World of [the] Future." What sort of maps are in your classrooms and academic books? Examine some world geography, world history, or world civilization textbooks and note how many pages are devoted to Eastern versus Western civilizations. What does all this say about Western perceptions of Asia? And what does it say about European influence in Asia?

European notions of Asia were also colored by religious beliefs. After all, the Garden of Eden was there, the Three Wise Men came from the east, and the sun, the light of salvation, rises in the east. Prester (Presbyter) John, a wise Christian king, was believed to have ruled somewhere in Asia. In the Middle Ages, countless pilgrimages were launched to find his kingdom. In that vein of thought, the Orient was a source of light and goodness and, conversely, the Occident was dark and evil.

Beyond these scripturally-inspired conceptions, Asia was portrayed as a mysterious place of dog-headed men and other freakish beings, as well as the heathen tribes of Gog and Magog, poised to pounce upon the unwary. More significantly, the Orient was thought to possess unbounded wealth in spices, jewels, incense, and other exotica—an idea underpinning numerous land and sea explorations.

It is important to recognize that this collective, conflicting imagery of an ill-defined Asia derived from both real and imagined experiences with a region extending eastward from Egypt into Ukraine and the Russian grasslands and southward into India. At the time, China was merely a fabled land on the periphery of this vast expanse called Asia. As exploration fostered increasing contacts between East and West, more places were added to the European knowledge map. Asia, as a conceived region, ultimately stretched eastward and southward into the Pacific Ocean, incorporating Japan and such far-flung islands as those in Indonesia and the Philippines.

By the nineteenth century, European powers were not only entrenched in many Asian countries for political and economic gain but were also struggling with defining their own boundaries. A series of wars, the rise and demise of assorted political powers, and ever-shifting national boundaries necessitated the revision of European frontiers. This was accomplished at the **Congress of Vienna** in 1815. The division between Europe and Asia was demarcated from Russia's Ural Mountains, through the Caspian and Black seas, and along the eastern end of the Mediterranean Sea to the African continent. Today, Asia is recognized as part of the geographical landmass known as **Eurasia.** And so, there is a continental Asia in a Eurasian context. However, cultural, political, economic, and environmental regions, as we know them today, are no more coincident with these boundaries than with spatial constructs of earlier times. This is why it is preferable to consider smaller, more cohesive realms. Although South, East, and Southeast Asia are the regions designated for discussion, I will use the term Asia when referring to these collectively, or the Asian continent in general.

Reading Place Names

There are many ways to spell **toponyms** or place names, especially in the case of Asia. This is, in part, because of varying transliterations of regional scripts into the Roman alphabet. Some of the spellings for the Chinese capital city of Beijing are: Peiping, Peiching, and Peking. Westerners' inability to pronounce indigenous place names led to alternate spellings. Singapore is the English version of Singha Pura that was formerly known as Temasek. Different ethnic groups may have their own versions of toponyms. For example, Urumchi in northwestern China is a Uyghur name. The Chinese call it Wu-lu-mu-chi. Sometimes places disappear and reappear as something else. For instance, East Pakistan became

Bangladesh in 1971. Furthermore, place names are frequently changed. Tokyo was once known as Edo. Others are newer, as with the restoration of traditional names. Bombay is now called Mumbai. Be forewarned: If you send a letter to "Bombay," it will likely be returned to you marked "No such place."

Rangoon, Burma, is currently Yangon, Myanmar. However, in deference to the majority of Burmese who reject their country's military regime and still use the names "Rangoon" and "Burma," we will use Yangon (Rangoon) and Myanmar (Burma) in this textbook.

When speaking of a place in a historical context, we will spell it as it was referred to at that time, followed by the current spelling in parentheses—for example, "In old Siam (Thailand). . . ." In modern context, current names will stand alone—for example, "Chennai" instead of Madras. This might seem confusing, but, as you will find out, different maps and atlases employ different spellings. More importantly, residents of places frequently continue to use the name that has been in use the longest, or the one that they have a preference for. For example, in the 1990s, Calcutta was renamed Kolkata, but many people continue to use the name Calcutta. Although Saigon was renamed Ho Chi Minh City by the communists after the Vietnam War, most people still call it Saigon.

Religion and Belief Systems

Now let us turn to an aspect of culture of paramount significance: **religion**. Religion, or a set of sacred beliefs, pervades everyone's life, whether consciously or unconsciously. Aside from being evident in the cultural landscape in the form of sacred structures and other phenomena, religion functions in a variety of ways in society. Religions define male and female roles; affect political systems and policies; influence agricultural and dietary practices; shape environmental values; impact economic development; and, frequently, are used by politicians to create dissension and strife.

Hinduism, Buddhism, and Islam are the major religions in Asia, but Christianity and other belief systems such as Sikhism, Confucianism, Daoism, and animism are important in specific regional settings. At this juncture, I will discuss the origin, **diffusion** or spread, and key tenets of four religions in Asia: Hinduism, Buddhism, Christianity, and Islam. Others will be discussed in their regional contexts in subsequent chapters.

All four religions have their **hearth areas**, or places of origin, in continental Asia. Each of these diffused by one or more means: conquest, word of mouth (contact diffusion), migration of believers (relocation diffusion), and the work of missionaries (proselytization). Merchants and traders played significant roles in spreading religion, especially in the case of Buddhism and Islam (see Figures 1-5 and 1-6).

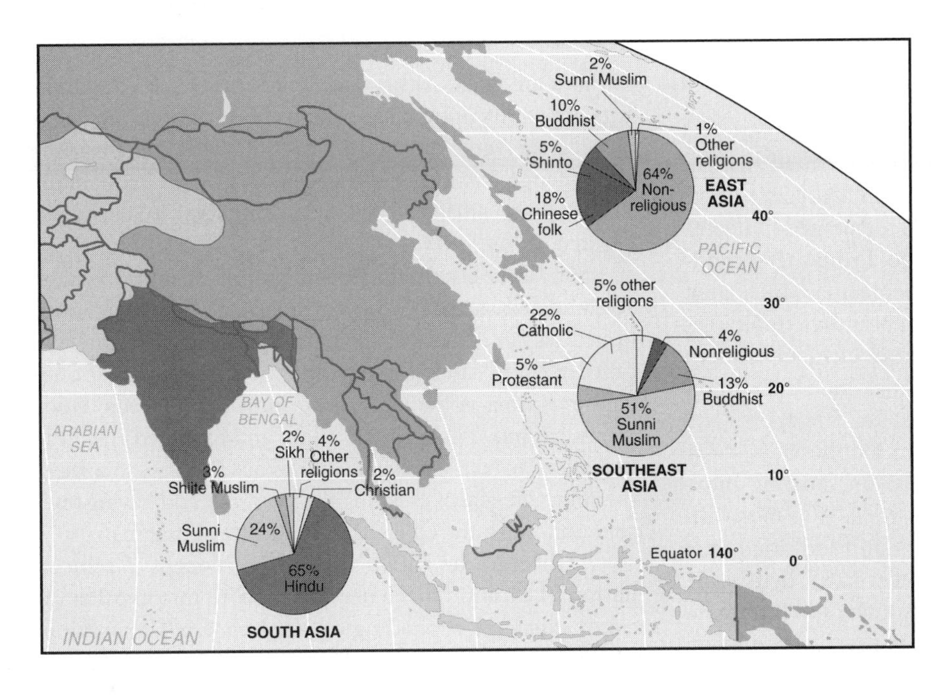

Figure 1-5

Distribution of major religions in East, South, and Southeast Asia. James M. Rubenstein, *Human Geography*, 9th edition, 2008 pp. 184-5. Upper Saddle River, NJ: Pearson Prentice-Hall.

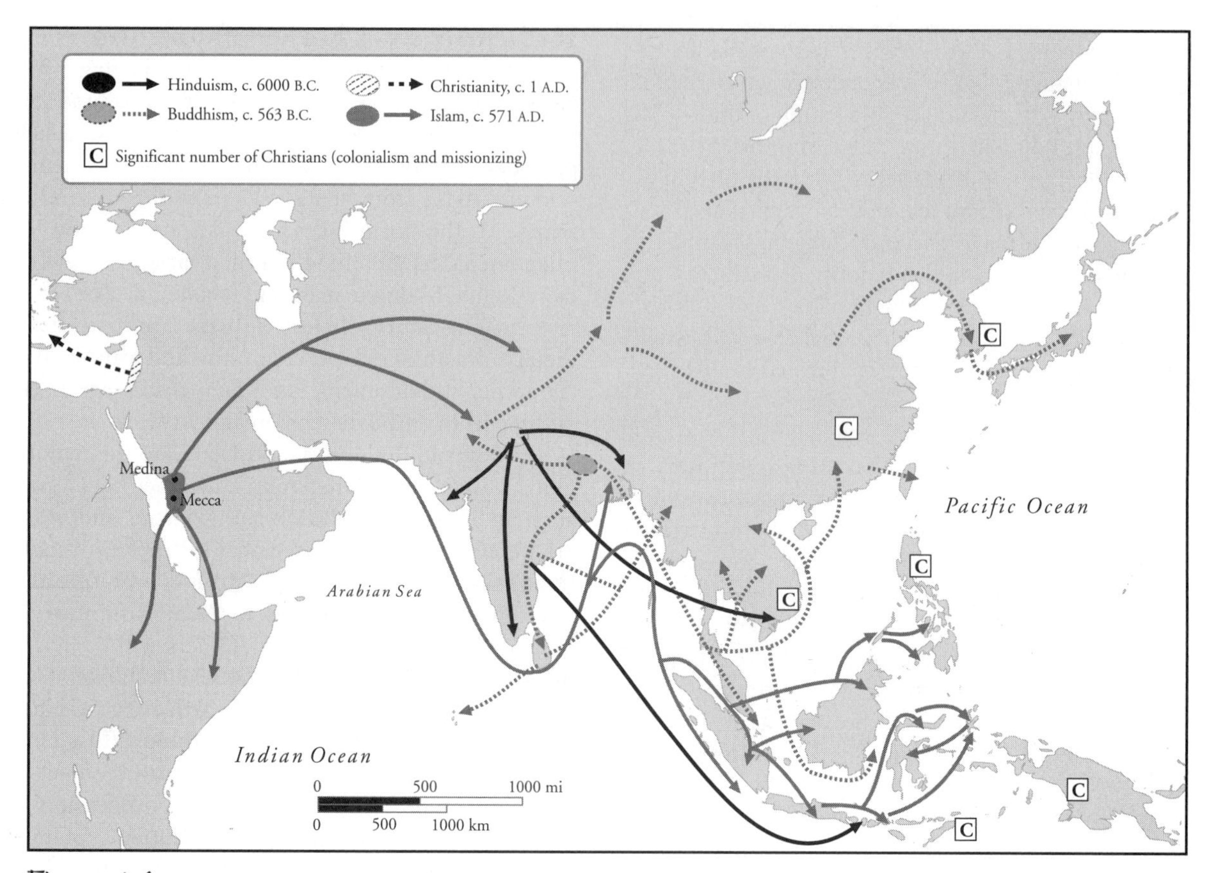

Figure 1-6

Origin and diffusion of major religions. Religions spread from hearth areas by word of mouth, migration, and proselytization.

HINDUISM

Hinduism is the oldest of the world's major religions. Its origins reach beyond the limits of documented history. Without an apparent founder, but with a substratum of belief from Indus Valley civilization, it emerged in northwestern India at least 6,000 years ago. From there, it diffused southward across the Indian peninsula. Much later, it spread eastward with the diffusion of Indian influences as early as the third century AD and with subsequent Indian migrations through the colonial era. Hindu communities still thrive in Nepal, Malaysia, Singapore, and on the islands of Bali and Lombok in Indonesia.

Hinduism is a **polytheistic** religion, hosting more than 333,000 (meaning countless) gods. The most important of these, in terms of daily worship and rituals, are *Shiva* and *Vishnu* (Figure 1-7). However, all deities are manifestations of one ultimate reality known as **Brahman.** There are, in essence, as many ways to practice Hinduism as there are gods. Although there is no set creed, adherents may draw upon an enormous body of literature. Earliest are the **Vedas**, from the Sanskrit word *veda*, meaning "knowledge." Another critical work is the **Upanishads.** This means "sitting down near" one's spiritual advisor or **guru**.

Hindus believe in the **transmigration of souls** based on accumulated **karma**. All beings possess a soul and are positioned within a hierarchy. Where one is, on what might be called a status ladder, is conditioned by the state of one's karma. Karma is the sum of an individual's deeds. Good deeds can move one up the ladder and bad deeds result in demotion. In fact, it may take numerous lifetimes to achieve the common goal of escaping the hierarchy and having one's soul unite with the One Ultimate Reality of Brahman.

Critical to the Hindu faith is the concept of **reincarnation:** the notion that a person is reborn into another cycle of existence depending on the karma of the previous

Figure 1-7

This Indian man is a "saddhu"—a holy man who has divested himself of worldly possessions to seek the One Ultimate Reality. His trident indicates that he is a devotee of Shiva. Saddhus migrate among sacred sites and live off donations from the public. As a true saddhu (not a beggar), this man asked for nothing. He was merely present. Yes, I did make a donation. Photograph courtesy of B. A. Weightman.

life. If a woman deliberately harms or kills another person or an animal such as a lizard, she might be reborn as that lizard in her next life. To quote a 1960s saying: "Keep your karma clean!"

Hinduism is entrenched in all aspects of Indian culture. We will learn more about its impact on society and development in the chapters on South Asia.

BUDDHISM

Buddhism also originated in northern India, in part as a reaction to the excesses of Hinduism prevalent by the sixth century BC. From here, it diffused with missionary monks and traders via trade routes into China, and from there, across the Korean peninsula to Japan. It also spread southward and became established in Sri Lanka, and from there it spread seaward to Southeast Asia.

Adapted to local circumstances, Buddhism diversified and evolved into many schools of thought. An extensive body of literature was created, and many temples and monasteries accumulated rich collections of sacred books, tablets, and scrolls.

Buddhism and its various schools evolved from the teachings of Prince Siddhartha Gautama, born around 563 BC in what today is Nepal. After years in search of an explanation of life's harsh realities, he awakened to a profound conclusion: The cause of suffering is desire. This realization was the key to his enlightenment as the **Buddha.**

That the world is filled with suffering and that suffering is caused by desire became the first two of the **Four Noble Truths.** The third is that it is possible to get rid of suffering by overcoming desire. The fourth Noble Truth tells one how to achieve this goal by following the **Eightfold Path.** By choosing this "right path" of behavior (e.g., right knowledge, aspiration, and livelihood), one can achieve ***nirvana***. As inscribed on the wall of a temple I visited in Myanmar: "*Nibbana* is not situated in any *Bhumis* [stages of being] nor is it heaven. It is a state which is dependent upon ourselves." Nirvana is an indescribable state of desirelessness and peacefulness. It is absolute bliss.

Buddha advocated withdrawal from the material world of desire into a contemplative mode, making meditation a key part of Buddhist practice. He also emphasized being generous, kind, and vowing not to harm people or animals. Consequently, vegetarianism is widely practiced in Buddhism.

As Buddhism diversified with expansion, two main schools of thought developed. **Mahayana Buddhism** is practiced in China, Korea, Japan, and, because of Chinese influence, in Vietnam. This school holds that salvation can be achieved through the intercession of divine beings. Mahayana Buddhists believe the Buddha to be their saviour but may commune with other deities as well.

One of the most popular of these divine beings is the **Bodhisattva** *Avalokitesvara* (in Sanskrit, the lord who looks in every direction). A Bodhisattva is one who postpones entry into nirvana in order to save other beings. *Avalokitesvara* is most popular in his female form and is enshrined in virtually all Buddhist temples. She is known as *Kuan-Yin* in China, *Kannon* in Japan, and *Kwanum* in Korea.

Theravada Buddhism developed in Sri Lanka as the monastic expression of Buddha's teachings. It places

Figure 1-8

It was sunrise in the mountain town of Mamyo, Myanmar (Burma), where I watched Buddhist monks walk past devotees who put rice and other food into their "begging bowls." This scene is ubiquitous throughout Theravada Buddhist communities. Photograph courtesy of B. A. Weightman.

responsibility on the individual for salvation through the accumulation of merit by good behavior and religious activity. Monastic orders, or *sangha*, are very important, and monks and nuns are highly respected. It is Theravada Buddhism that diffused to Southeast Asia, where it is very much a part of the cultural landscapes of Cambodia, Laos, Myanmar, and Thailand (Figure 1-8).

What happened to Buddhism in India? It rose to its zenith under Emperor Asoka in the third century BC. But Hinduism eventually reasserted itself, and Buddhist adherents became a rarity. Aside from a few practitioners and scattered temple relics, Buddhism has faded from the Indian landscape. However, in its Himalayan refuge situation, Bhutan remains a Buddhist kingdom. Buddhism is also important in Nepal, even though the Nepalese people are predominantly Hindu.

Buddhist temples and other landscape elements are increasingly common in Europe, North America, and Australia. This is a result of relocation diffusion as Chinese, Laotians, Vietnamese, and others continue to migrate to other parts of the world.

CONFUCIANISM

According to most scholars, Confucianism is more of a philosophy or a code of behavior than a religion. Founded in China by **Kung fu-tzu (Confucius)** around 500 BC, it stresses that humans are social beings obligated to others in a hierarchical order. In general, rulers, elders, and males hold superior positions in a properly ordered society. Confucian ideals have been most influential in East Asia as well as in Vietnam.

Although Confucianism may not be defined as a religion by some, there are many Confucian temples. These are particularly popular with students who make offerings and pray for success in their education.

Initially, Chinese communists castigated Confucianism as a negative and demeaning holdover from the archaic past. However, in recent years the government has recognized the belief system as having similarities with socialist ideals. Both emphasize education as a remoulding process to encourage group responsibility, social harmony, and service to the government of the people.

Today, South Korea is regarded as Asia's most Confucian country. There is a great respect for elders, and family responsibility is a very serious requirement. Even today, young people consider open defiance of their father as tantamount to a sin. We will learn more about Confucianism in Chapter 10.

DAOISM

Daoism is based on the teaching of the Chinese *Tao Te Ching* written in the sixth century BC. Its philosophical aspect (*Dao-chia*) is drawn from the writings of Lao-Tzu and Chuang-tzu and other early mystics. Daoism rejects competition, rank, and luxury. Lao-Tzu said that everyone wants to be at the top of the tree but if this actually happened, the tree would collapse. Thus he stresses non-action, non-control, and non-interference as means to achieve the unknowable *Dao—the Way*. The best way to survive is as a simple-thinking farmer, close to the earth—at one with nature.

Religious Daoism (*Dao-chiao*) emphasizes religious rituals aimed at achieving immortality In Daoism, there are many deities and spirits that are manifestations of the Dao. People seek happiness and prosperity by appeasing these forces often at Daoist temples.

Under communism in China, Daoism was renounced as superstitious and fatalistic. Practices were forbidden and temples destroyed However, with a more liberal government in the past two decades, Daoist temples are reappearing on the landscape and people once again are exhibiting Daoist pictures and objects in their homes.

Even now in the twenty-first century, Daoist philosophy remains an underpinning of thought and action for many Asians. It is also very important in the practice of traditional medicine. We will revisit Daoism in greater depth in Chapter 10.

CHRISTIANITY

The spread of Christianity from its hearth region in Israel/Palestine is closely tied to European imperialism. Carried by colonizers, administrators, soldiers, sailors, traders, and missionaries, Christian doctrine arrived in Asia especially after the thirteenth century. It flourished in context of colonization that began in earnest in the sixteenth century, but competition with already established religions diminished its impact. Although Christians can be found throughout Asia, the most significant concentrations are in Timor-Leste, the Philippines, eastern Indonesia, South Korea, and Vietnam.

Christianity is a **monotheistic** faith proclaiming the idea of a single *God*. A Jewish man known as **Jesus Christ** is believed to be the Son of God, the Messiah that had been promised roughly 1,500 years earlier. The Messiah would restore the relationship between God and humankind. The teachings of Jesus and his disciples spread, as the Word of God, via contact conversion. Numerous versions of the **Bible** offer both a history of Judaism and early Christianity in addition to the teachings of Jesus and his followers. Many Jews and Christians believe that the Bible is the Word of God. Missionizing continues to be a powerful force in the diffusion of Christianity (Figure 1-9).

Christians believe in salvation through faith in Jesus Christ and adherence to the teachings of the Bible. Historic circumstance has seen the continual division of Christianity. Two major branches, Roman Catholicism and Protestantism, with its multiple denominations, have made the greatest headway in Asia. Christianity has grown to the point that it is self-perpetuating. Jesus' birthday (Christmas) is celebrated by close to 300 million Christians in Asia with people of other faiths often joining in the celebrations.

Figure 1-9

I came across this Pentecostal mission in the remote Dayak community of Mencimai in Kalimantan on the island of Borneo. Most of the people of this region follow folk belief systems or are Muslim. Even Christian converts retain elements of their traditional beliefs. There are not many converts in Muslim Indonesia. The church is built on stilts because of the tropical rains and surface flooding.
Photograph courtesy of B. A. Weightman.

Saint Thomas came to the Malabar Coast of India (southwest) in 52 AD. Many converts were gained from the lowest classes of society attracted by human equality offered by Christians. When the Portuguese arrived in the sixteenth century, they passed anti-Hindu laws, forced conversions, and even persecuted some converts for not having enough fervor in their beliefs and practices. Many were tortured then burned at the stake.

The British arrived with their missionaries in the seventeenth century. To attract more Christian converts, they built schools and handed out clothes and food to the poor. Christianity remains a minority faith in India, but even today, many Hindu Indians attend Christian schools, especially Roman Catholic ones for girls. The largest populations of Indian Christians are found in Mumbai (Bombay), the former Portuguese colony of Goa on the west coast, Kerala and Tamil Nadu states in the south, and the northeastern tribal states of Manipur and Mizoram.

One famous Christian in India is **Mother Teresa** (1910–1997). Mother Teresa founded the Missionaries of Charity organization in Kolkhata (Calcutta) where she and her helpers tended to the diseased, dying, and destitute for decades. Her organization now has more than 500 branches worldwide responsible for caring for 90,000 lepers and feeding 500,000 people a year. Mother Teresa was given many honors and was awarded the Nobel Peace Prize in 1979.

The Portuguese imposed Christianity in their colony of Timor (now Timor-Leste), and the Philippines, which Is 80 percent Roman Catholic, adopted the faith when colonized by Spain. Estimates of the size of Korea's Christian population range from 28 to 49 percent. Moreover, Korea has the second largest Christian missionary force in the world after the United States. There is only a small minority of Christians in Japan, which is a largely secular country.

The Roman Catholic Jesuit Order brought Christianity to Vietnam in the sixteenth century. As Communist forces advanced from the north during the Vietnam War, Christians fled southward to escape persecution. Many *Montagnards*, tribal people of the western mountains who aided U.S. forces in the war, are Christians. Although the Communist regime proclaims freedom of religion, *Montagnards* continue to be tortured and imprisoned as "Agents of America."

Under the communists who took over China in 1949, the faith was suppressed and followers persecuted. For decades, Christians worshipped in "underground churches" commonly known as "house churches" to evade arrest for "disturbing public order." In recent years, the government has eased its pressure and in 2007, President Hu Jin-tao told Politburo leaders that, "The knowledge and strength of religious people must be mustered to build a prosperous society." The government estimates 21 million Roman Catholics and Protestants in the country. Outside estimates put the number at 70 million. Even greater numbers continue to worship in house churches.

ISLAM

Islam, which translates as *submission to God,* is the youngest of the major religions. It originated around the subsequently holy cities of **Mecca (Makkah)** and **Medina (al Madinah)** in Arabia in the sixth century AD. Islam swept east across Asia with successive waves of conquerors, traders, and missionaries. It replaced earlier religions in many areas. For example, Buddhism and Hinduism were once important in Indonesia but were virtually eradicated under the fervor of Islam. Islam is the dominant religion in Pakistan, Bangladesh, Malaysia, Indonesia, and Brunei. It is also a potent force in several other regions such as northern India and the southern Philippines. You will learn more about Islam in its regional contexts in later chapters.

Coming from the same core area, Islam has links to both Christianity and Judaism. It, too, is a monotheistic faith. Following a series of divine revelations, **Muhammad** (571–632) began preaching in Mecca and later Medina after being forced to flee there by his enemies. His teachings about a new way of life provided welcome structure in what was then a chaotic Arab world, and converts quickly accumulated. **Muslims** (those who submit to God) formed armies to invade and conquer the lands of **infidels**, or non-believers. Clashes with other established faiths resulted in the death of millions on all sides. Unfortunately, religious conflicts associated with Islam versus other belief systems still exist.

Muslims believe in one God, **Allah**, and recognize Muhammad as His prophet. Their house of worship is called a **mosque** (Figure 1-10). They also view Jesus Christ as a prophet. Like the Jews, they avoid pork, perceiving it as unclean. To Muslims, the will of Allah is absolute. The Word of Allah is contained in the **Koran**, or *Qur'an,* which is Arabic for "recitation." Pivotal to being a good Muslim are the **Five Pillars of Islam.** These are: acknowledgment of Allah as God and Muhammad as His Prophet; prayer five times a day wherever one is (Friday is "mosque-day"); a month of daytime fasting (known as **Ramadan**); the giving of alms; and at least one pilgrimage to Mecca. Held in the last month of the lunar calendar, this yearly event is known as the ***hajj*** and involves the convergence of millions of pilgrims in the holy cities of Mecca and Medina.

Dissension over succession to the Caliphate (ruling) authority eventually caused division in Islam, resulting in two major branches and several other sects. The majority of Muslims are orthodox Sunnis. The breakaway Shiites account for less than 15 percent. Shi'ism (*Shiah*) became rooted in Persia (Iran) in the sixteenth century and subsequently spread via Iraq and Afghanistan to Pakistan. It has even influenced Sunni Islam in India. Shiites are generally more fervent in their beliefs and more emotional in their practices. To them, religious

Figure 1-10

This magnificent mosque is at Shah Alam, the new capital of Selangor Province west of Kuala Lumpur, Malaysia. Named for the Sultan Salahuddin Abdul Aziz Masjid, it was built in 1988 and holds 16,000 people. The mosque is oriented to face the holy city of Mecca. Loudspeakers on the minarets announce the call to prayer five times a day. Photograph courtesy of B. A. Weightman.

leaders, or ***Imams***, are infallible, as are their interpretations of the Koran and other religious literature. Shiites believe that Imams should rule while Sunnis believe that religious leaders should advise rulers.

Islam defines certain life-roles, modes of dress and social behaviour. But, the degree of strictness varies greatly from place to place and has different impacts on societies, especially on the lives of women. Islam is also an influential force in governments of countries like Malaysia and Pakistan, where Muslims make up the majority population. However, Islam does not have a significant influence in Indonesia's government even though the country has the world's largest Muslim population.

Islamism

In recent years there has been a resurgence of religious fundamentalism in the Islamic world. Muslims refer to this phenomenon as religious revivalism. Many Muslims disapprove of what they perceive as the erosion of traditional values in the wake of Westernization and secularization. They see Western culture as extravagant and immoral, and Godless governments as blasphemous. Both private property and profit-making are incompatible with basic values promoted by the Koran. As long as economic conditions were improving, dissatisfaction remained muted. However, as populations grew, poverty increased, and unemployment rose, fundamental Islamic ways become more appealing. Hatred of the West was epitomized by the devastation of New York's World Trade Center on September 11, 2001, by Islamic terrorists. This and other atrocities are linked to al-Qaeda, which is dedicated to punishing the perceived enemies of Islam.

Many al-Qaeda members are educated in Wahabi Islam that promotes a literal interpretation of the Koran and *hadiths*—the sayings of Muhammad. Wahabism originated in Saudi Arabia in the eighteenth century and is generally considered a sect of Sunni Islam. It preaches against moral decline and influences of the West. Osama bin Laden is a Saudi Wahabi.

Wahabis are very active missionaries, and the Saudi Arabian government funds thousands of religious schools around the world. These schools are called madrassas, which is Arabic for "school." Most madrassas are apolitical and offer free education to the poorer segments of society. However, there are some that are extremely radical and focus almost entirely on extremist, religious doctrine. Girls are indoctrinated in separate schools where they must be fully covered and even wear gloves so that not an inch of flesh will be exposed.

Acts of terrorism have become all too common around the world, including in several Asian countries. Pakistan is embroiled in fighting Islamic terrorists along its frontier with Afghanistan. It is also attempting to thwart fundamentalism within its own government. Bombings in Pakistan are weekly, if not daily occurrences. More will be said on this issue in Chapter 6.

Terrorists blew up two nightclubs in Bali, Indonesia, in 2003 killing more than 200 people, mostly Australian tourists. More recent bombing of major hotels have occurred in the Indonesian capital of Jakarta. Thailand is facing insurgency in its predominantly Muslim southern provinces. And the Philippines is trying to cope with Islamic militants in its southern island of Mindanao. More recently there has been a spate of bombings in India and Pakistan. Mumbai (Bombay) was targeted with 13 bombs in 1993 and 7 explosions in 2006 whereby hundreds were killed and wounded. These acts of terrorism have been attributed to fundamentalist Islamic groups.

Many Muslims are calling for the imposition of Islamic, religious law called **Sharia**. Sharia deals with everything from politics and economics to social issues; it defines every aspect of a Muslim's life. There is great variation in interpretation of Islamic law—from extreme conservatism to relative liberalism. However, fundamental Islamists believe that Sharia should be imposed on all Muslims and non-Muslims alike. Some of the more extreme interpretations include: flogging or even execution (perhaps stoning to death) for adultery, blasphemy (speaking negatively about Islam), and homosexuality.

Sharia has many more strictures for women than it does for men. For example, women are admonished to be fully covered (even to wearing gloves in some regions) and to never go outside without being accompanied by a male relative. A woman can inherit money or property, but her share is only half that of her brothers. To prove rape, a woman must produce four male, Muslim witnesses. Also, a man can divorce his wife by telling her "*talaq*" ("I divorce you") three times. In 2003, Malaysia passed a law that permits a man to divorce his wife via text-message. You will learn more about the impact of conservative Islam in Chapter 4.

Although heinous acts have been fostered by some Islamic fundamentalists, it is important to note that not all Muslims are fundamentalists and not all fundamentalists are terrorists. Nevertheless, it is true that increasing numbers of Muslims are calling for a return to the precepts of the Koran as interpreted by religious leaders.

FAITH, EXPLORATION, AND TRADE

Trade had long existed between Asia and Europe, dating back to the Roman Empire. However, Arab Muslims laid the basis for a world trade system in the Mediterranean and Indian Ocean regions. As Islam is a way of life, there is no separation between religion and commerce. In fact, Islam is an ethical faith, providing even the rules of commerce. Believers who worked with each other understood rules of fair-trading practices. Non-believers who wanted to enter Muslim trading networks found conversion the easiest avenue. Each trading community served as a vehicle for the expansion of Islam.

In the eighth century, Arab Muslims conquered the Sind region of the Indus River delta. Until then, Buddhists had dominated trade, while Hindus focused on agriculture. Buddhist traders found it expedient to convert to Islam. Eventually, Islamic trading posts were established along India's west coast from Cambay to Calicut on the Malabar Coast. When the Muslim, Mughal Empire in India expanded to the Bay of Bengal (1574–1592), access was gained to both the Coromandel coast and adjacent coasts of Southeast Asia. Muslims thereby achieved hegemony over the region's sea trade.

Prior to the appearance of Arab traders, China had been using South Asian merchants to carry their cargo. Steel had been made in China since the eighth century and was in great demand for weapons and farm implements. As a result of their overland and sea expansion, Muslims became active in the China trade and remained instrumental well into the twelfth century. Muslim trading communities existed in several Chinese cities such as Canton (Guangzhou) on the southern coast (Figure 1-11).

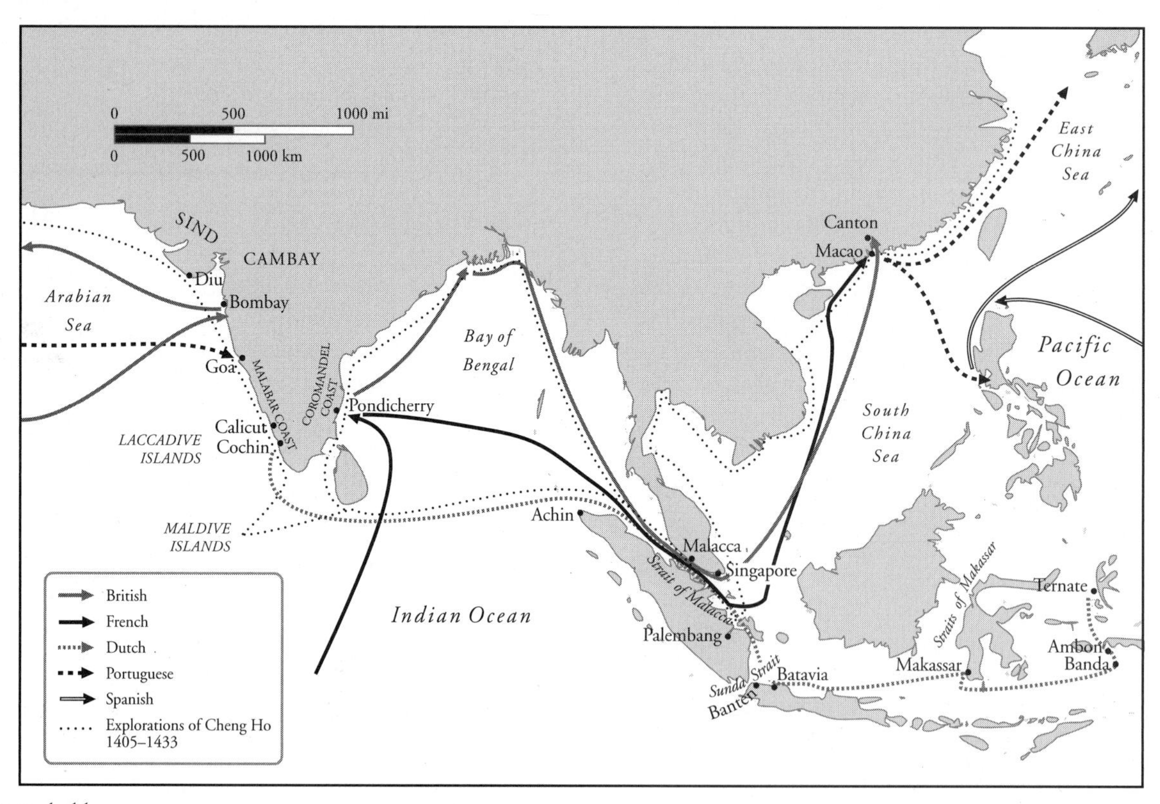

Figure 1-11

Trading centers and sea routes. Note that the Chinese explorer Cheng Ho navigated these waters long before the Europeans. The first European trading port was established by the Portuguese at Malacca (Melaka) in 1511.

Figure 1-12
I had always wondered what mace looked like in the nutmeg plant. Finally, I got a chance to see it in a market in Penang, Malaysia. Mace is the reddish stringy stuff covering the thin shell of the nut that is found in the pod. It's amazing to think that spices such as nutmeg and mace changed the face of the world.
Photograph courtesy of B. A. Weightman.

An international world system of commerce was now functioning via a network of trading ports and shipping lanes. By the fifteenth century, Muslim enclaves dotted the northern coasts of Java and Sumatra. Muslim-controlled Banten (Bantam) was also the focal point of Chinese trade in Southeast Asia. Makassar in Celebes (Sulawesi) was the key to the Spice Islands (Maluku). Achin (Aceh), Sumatra, was situated at the northern entry to the Straits of Malacca. Of great importance was the founding of a transhipment point or **entrepot** at Malacca (Melaka). Indians, Javanese, Chinese, and others migrated to these nodes of perceived opportunity where many of their descendants remain to this day.

Why this much interest? What was to be gained? Quite simply: riches. Demand was soon created for sandalwood and other aromatic woods; teak and other hardwoods; silk and porcelain; slaves and exotic animals; and countless other products. But pepper, along with cinnamon, cloves, nutmeg, and a host of other aromatic and edible spices became the most desired (Figure 1-12). Arabs conjured tales of gross birds harvesting cinnamon from ferocious bat- and snake-guarded swamps to dissuade others, especially the Europeans, from discovering their spice sources.

Neither Arabs nor Europeans were first in the spice trade. Indigenous states and empires were trading such commodities long before outsiders arrived. The Roman historian Pliny told of Indonesians in outrigger canoes driven by the prevailing winds unloading cargoes of cinnamon and cassia on the east coast of Africa in the first century AD. The great sea-borne empire of Srivijaya, centered in current-day Palembang near the east coast of Sumatra, controlled the Malacca and Sunda straits dominating east–west sea routes in the seventh century.

China was also involved in ocean trading. The talented Admiral Cheng Ho, in a series of voyages from 1405 to 1433, touched base at virtually every port in the region, from southern China to the east coast of Africa. He discovered that Chinese merchants were already well established in many of these places. Europeans truly were latecomers to this dynamic, lucrative world.

Europeans: Exploration, Commerce, and Conquest

By the late 1400s and early 1500s, Europeans became familiar with what Arabs had long been aware of: the annual reversal of winds associated with Asia's monsoon climatic system (see Chapter 2). "Monsoon" comes from *mausim,* Arabic for "season." By sailing west along the coast of the Atlantic Ocean and then south, around Africa's Cape of Good Hope and following the prevailing winds across the Indian Ocean between June and September, and doing the reverse from December to March, sailing vessels could make the round trip in less than a year.

This was the **Age of Exploration** for Europeans on a zealous mission to convert the heathen to Christianity. But the real driving force was economic and thus similar to the aims of earlier Islamic traders. And so the Europeans ventured across the globe in quest of spices, gold, and other riches. In 1488, Bartolomeu Dias of Portugal rounded the Cape of Good Hope and sailed as far as the Persian Gulf. Upon returning home he announced that the sea route to India was open! Then in 1498, Vasco da Gama sailed as far as Calicut on India's Malabar Coast and later entered Lisbon's harbor with the first cargo brought from the "Far East" to Europe by sea. Ferdinand Magellan pioneered the westward route around South America's Cape Horn in 1519. Reaching the Philippines in 1522, he initiated Spain's entry into the region. Holland's (The Netherlands') 350-year involvement in the East Indies (Indonesia) commenced with the 1595 voyage of Cornelius de Houtman, who returned to the Netherlands with three shiploads of pepper and nutmeg.

England's Sir Francis Drake obtained a cargo of spices, plus booty plundered from a Portuguese ship, in his circumnavigation of the world (1577–1580). This accomplishment, in concert with Britain's defeat of the Spanish

Armada in 1588, was a pivotal event in the sanctioning of further British expeditions. The race for territory was on, and it was inevitable that swords would cross.

Territorial gain and sea-lane control became paramount concerns. The importance of control over the Strait of Malacca cannot be underestimated. A plaque in the garden of the sultan's former palace in Malacca reads: "Whoever is lord of Malacca has his hand on the throat of Venice." The Italian city-state of Venice dominated international trade from the 1200s to the mid-1400s.

The Portuguese established trading ports from Africa to China, and by 1520 they dominated the seas. Intolerant of Islam and greedy for profit, they tried to force out the Muslims who remained along the coasts of Indonesia's islands and used Islam as a rallying point against the European intruders. The Portuguese moved into Cochin, Diu, Goa, and created a new port at Bombay (Mumbai) in India. They took the critical port of Malacca in 1511, and acquired control of Macao in 1557 (Figure 1-13). The Portuguese kept Ternate in the Spice Islands until 1570, when it was wrested away by the Dutch. However, the Portuguese remained in Timor-Leste until 1975.

Both the Dutch and the British were anxious to oust the Portuguese from South and Southeast Asia. Backed by stockholders, with monopoly privileges and the right to raise armies and navies, British and Dutch trading companies aggressively intruded into the area. **Factories**, or trading posts, were quickly set up throughout the region. The Dutch took Ternate in 1607 and founded Batavia on Java in 1619. Malacca fell into their hands in 1641, and Banten (by then British) was acquired in 1682. When the Dutch attacked the Johor port in 1784, the British responded.

Figure 1-13

This building reflects both Dutch and Malay architecture. Eventually it was used by the British as the exclusive Melaka Club. Now it is an Independence Memorial for Malaysia. Photograph courtesy of B. A. Weightman.

Britain had initially concentrated on India but became intrigued with the economic potential of Southeast Asia, where they vied with the Dutch. They seized Malacca in 1796 and gained the island of Singapore in 1819. France entered the picture and established the first of its settlements at Pondicherry, India, in 1674. However, the French were more interested in their African colonies and were not as aggressive as their British and Dutch contemporaries. And so the era of European colonialism was launched.

Colonialism and Imperialism

With colonization rampant in the fifteenth and sixteenth centuries, an extensive web of ports and shipping lanes fostered the evolution of an Atlantic economy as well. Silver and other resources from the New World became integral to economic exchange between Atlantic and Pacific realms. A world economy was emerging.

The number of imperial powers increased from two (Spain and Portugal) in the sixteenth century to five in the seventeenth century with the addition of Britain, France, and Holland. Collections of colonies worldwide were absorbed into the web of economic interchange. The only place not colonized in Southeast Asia was Siam (Thailand). It became a **buffer state** between British and French interests in mainland Southeast Asia.

Competition was fierce during this mercantilist period, during which raw materials were brought to the colonial power, where factories produced manufactured goods to be shipped back to the colonies. Under **mercantilism**, colonies were not allowed to manufacture such goods. Neither were they allowed to trade with other than their own colonial overlord. The Seven Years' War (1756–1763) between Britain and France ended French ambitions in India and North America and allowed Britain to become the world's leading power. With colonies around the globe, it was said that the sun never set on the British Empire.

Britain expanded further into India, and France reached into Indochina. Holland tightened its grip on the Dutch East Indies (Indonesia). In order to control their monopoly on the spice trade and to jack up prices, the Dutch began destroying all the spice plants on some of the islands. Fearing instant poverty, the islanders resisted, with terrible consequences. For instance, almost the entire population of the Banda Islands was exterminated. By attacking and blockading ships owned by indigenous peoples and the Chinese, the Dutch managed to impoverish other islands as well.

In 1869, the British- and French-owned Suez Canal was opened, ushering in the **Age of Imperialism** (1870–1914). Reducing the sea journey to Asia by thousands of miles had the effect of intensifying interactions between Europe and Asia. With increased territorial gains came friction. Rivalries, once confined to the European continent, spilled over into the colonies. Territories were exchanged between the British and the Dutch and between the British and the French with no consultation with the indigenous people.

Those colonized did not accept their inequitable situation without protest. Resistance movements were countered by imperialist wars of oppression. In England these were known as Queen Victoria's Little Wars. In British India, French Indochina, and Dutch Indonesia, at least 70,000 died in battle before the turn of the nineteenth century. Suppression of resistance in Burma alone (1823–1826) resulted in 15,000 lives lost.

World War I (1914–1918) was a cataclysmic event in history. It spearheaded a new era for the colonial world. The Treaty of Versailles in 1919 spurred nationalistic claims on the part of the colonies, and reassigned them to the victorious Western powers. But things could not be as they were; the so-called *Belle Epoch* was over.

The Colonial City

Colonial cities were intended to serve the interests of colonial powers. They were guarded outposts that functioned as military, economic, and administrative centers. Architecture was grand, meant to impress the masses with the power and glory of the empire (Figure 1-15). Symbols of colonial rule were expressed in the form of architectural styles, statues, and monumental buildings with Western crests. Tudor-style apartment buildings still exist in the former British Quarter of Shanghai, and Dutch-style churches can be found in Jakarta.

The colonial language was employed for naming places and streets and was the language of rule for administrative and educational purposes. Street signs such as Orchard Road and Rue Louis Pasteur remain in many former English and French colonial cities, and these languages continue to be spoken by educated elites.

To preserve race and class distinctions, separation was imposed on both people and landscape. Europeans built their own residential quarters, segregated from the indigenous people by physical barriers such as rail lines or waterways, or by military cantonments. Locals could enter the foreign quarters only with permission, and customarily did so in a servant capacity. Many sources refer to signs posted in China's and India's city parks stating, "No Chinese (Indians) Allowed."

Some cities were designed with a standardized plan such as set down in Spain's 1573 *Laws of the Indies*. This called for a grid-style street layout, centered on a main plaza surrounded by government buildings and a church. Similar but smaller plazas, built as the grid expanded, were intended to ensure the dissemination of order and godliness. This influence is particularly evident in the Philippines. Britain and France also liked the grid system because it represented order and regularity. Additionally, straight streets facilitated speedy access to trouble spots.

The colonial-style city differed widely from the traditional city with its narrow, winding streets and bustling crowds. The colonial city had one focal point, but the traditional city had several, each focused on a government, religious, or market center. To the occupiers, the planned colonial city symbolized order within chaos.

As colonial power strengthened, city architecture and all that it stood for became even more imposing. Local people were relegated even further into the socioeconomic background. In Singapore and Kuala Lumpur, the British, in concert with Chinese and Indian migrants, dominated practically all activity. Indigenous Malays faded from the scene. This was also true in the Dutch East Indies. To overcome such injustices, Malaysia had special policies favoring Malays for several years.

Imperialist Japan and the Rise of Nationalism

After centuries of feudalistic rule under powerful military leaders, or shoguns, the Emperor Meiji was restored to the throne in 1868. Young reformers were determined to bring Japan into the Industrial Era. The Industrial Revolution had been initiated around 1750 in England and

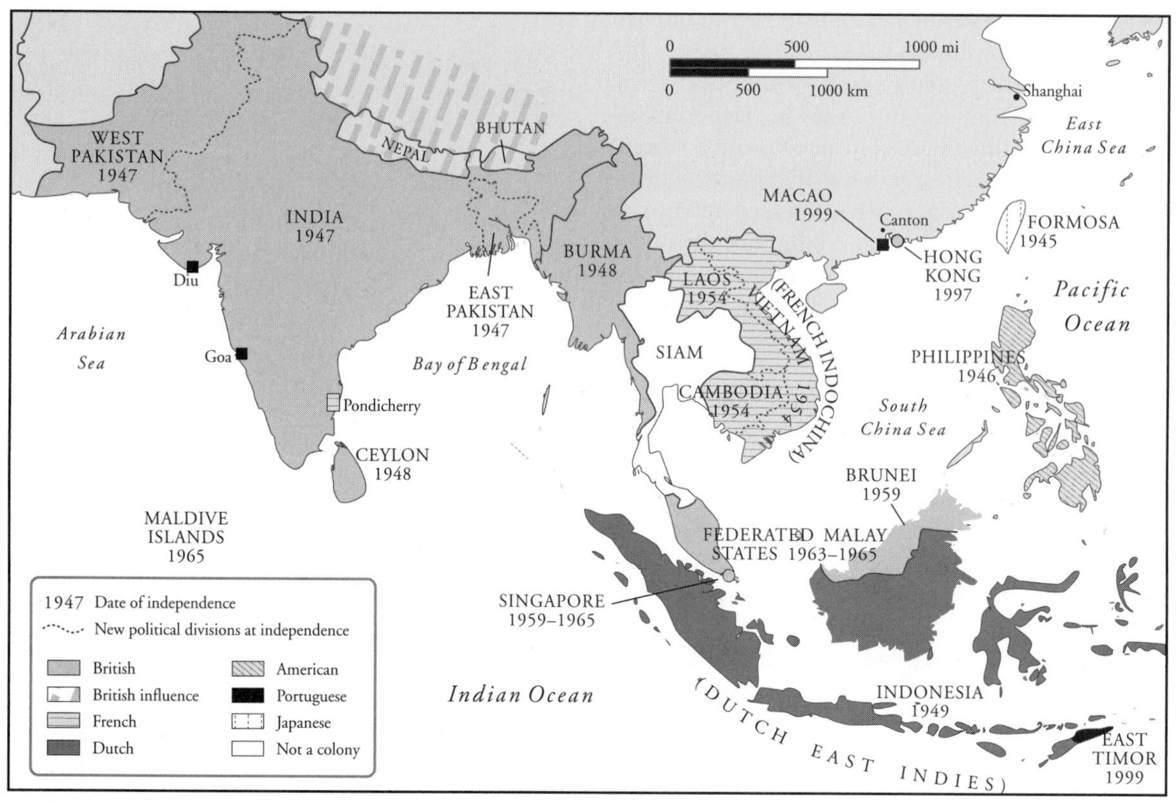

Figure 1-14

Most of Asia was either colonized or strongly influenced by foreign powers, with the exception of Thailand, which served as a buffer between English and French interests in Southeast Asia. The Philippines was a Spanish colony from 1565 to 1898 and an American one thereafter. Formosa (Taiwan) was a Japanese colony from 1895 to 1945. Although China was never a colony, many places such as Canton (Guangzhou) and Shanghai had foreign concessions.

was in full swing elsewhere. Some saw potential for the island nation as the "Britain of the Pacific." Why could Japan not have a colonial empire? After all, it needed raw materials for industrialization and space for expansion. Japan flexed its muscles, attacking and defeating both China and Russia at the turn of the century. China's island of Taiwan (1895) and footholds on the mainland were the rewards of victory. Japan's empire was taking shape. Korea was annexed in 1910. Japan's territorial ambitions were fervently pursued in China, and by the mid-1930s, Japan dominated a vast empire (Figure 1-14).

Japan's ambition was to establish a **Greater East Asia Co-Prosperity Sphere (GEACPS)**, incorporating Asians minus the European imperialists. As it defeated both indigenous and colonial forces, it propagated the slogan "Asia for the Asians." Its grandiose plans were foiled when it drew the United States into the Pacific theater of war by bombing Pearl Harbor, Hawaii, on December 7, 1941. By 1942, Japan controlled all of the northern Pacific west of the international date line. Japan then dominated 100 million people!

After its defeat in 1945, Japan was relieved of all its colonies, but the GEACPS had left a legacy: Japan had been the vanguard of Asian nationalism. "Asia for the Asians" became a cornerstone of independence struggles. But new socioeconomic-political theories were on the horizon to shape the entire process.

Decolonization and Independence

European imperialism began to crumble under the weight of independence movements before and after the first World War. After World War II (1939–1945), it crashed. During the process of **decolonization**, independence was both negotiated and fought for. Countless lives were sacrificed. France, which was the most resistant to giving up its colonies, lost 95,000 persons between

1946 and 1954, trying to keep its hold on Indochina. However, this significant loss was overshadowed by the deaths of well over half a million Vietnamese.

Asian independence leaders, schooled in the ideals of communism or democracy, were determined to build their own independent states characterized by dignity, equality, democracy, and self-determination. China's Sun Yat-Sen (Sun Zhong-shian) (1866–1925) was regarded as the first great Asian nationalist. Although China as a whole was not a colony, some coastal areas were colonized, and foreign powers strongly influenced its government. The name Sun Yat-sen, along with others such as Ho Chi-Minh (1890–1969) in Vietnam and Sukarno (1901–1970) in Indonesia, soon became familiar to newsreaders and listeners. Such leaders instilled a new sense of national consciousness among their respective peoples.

Marxism and Leninism were attractive to some independence leaders because they provided explanations for what was wrong with a world under imperialism and capitalism, told how to get rid of the old order, and offered a model to put in its place. Marx based his theories on the plight of an urbanizing and industrializing Europe. Lenin recognized the potential power of both urban workers and the rural peasantry. Asians witnessed the success of the Russian Revolution of 1917, when the imperialistic Tsarist regime was overthrown and large estates were confiscated by the state in an attempt to equalize the use of land. Lenin's calls for revolution hit home with anti-imperialist, pro-nationalist groups. It is important to understand that communism was not the driving force behind independence uprisings in Asia. Nationalism and patriotism were.

Social movements of unprecedented scale and devastation brought independence, although it was achieved sporadically (Figure 1-15). Each country has its own story to be spelled out in later pages. A patchwork quilt of political systems emerged: democratic, communist,

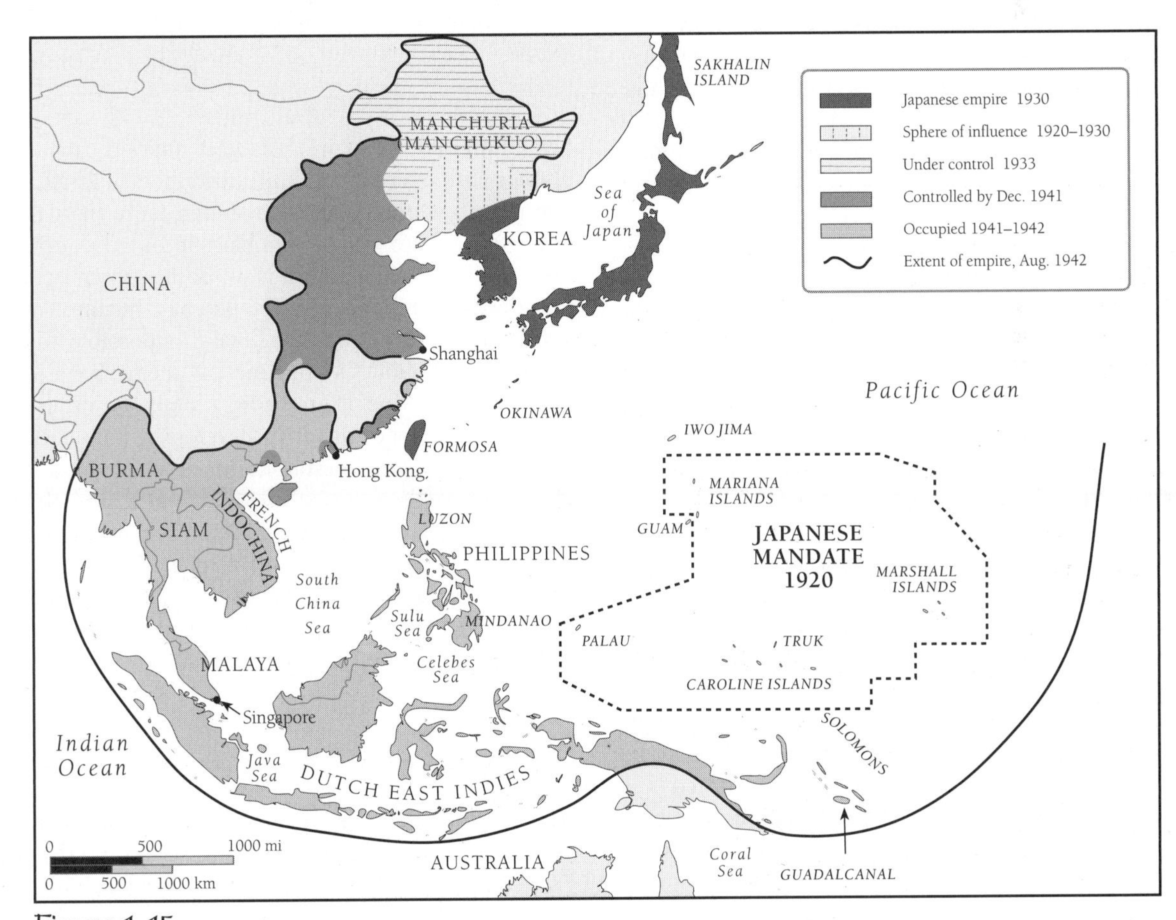

Figure 1–15

Japan's "Greater East Asia Co-Prosperity Sphere." Japan won Formosa from China as a prize of war in 1895. It annexed Korea in 1910 and established Manchukuo in 1932. It pressed for its goal to become "Britain of the Pacific" until its defeat by Allied Forces in 1945.

militarist. Whatever the end result, as the Chinese leader Mao Zedong (Mao Tse-tung) noted in 1957, "The east wind prevails over the west wind."

Prevalence of east-wind political, economic, and social systems did not bring stability, though. Communism with its massification, lack of incentives, and mismanagement was unable to achieve its ideals. It eventually collapsed in the Soviet Union in 1991. This momentous event, coupled with new economic realities of global proportions, spearheaded profound changes in Asia and the world.

A New Global Order

TIME-SPACE CONVERGENCE

Time-space convergence is an integral part of our changing world; more space can be covered in a shorter time span. In 1850, undersea communication cables were laid around the world. In that same year, an American-built clipper ship, the *Oriental,* made a record journey from Hong Kong to London in 97 days. In 1935, a freighter from Los Angeles, struggling in a winter storm, took 19 days to reach Tokyo. Sea-mail took weeks as well. Today, we can e-mail and conduct business transactions across the globe in mere seconds.

The vast, intertwined, and overlapping channels of communication produced by flows of technology, hardware, and software disseminated by transnational or multinational corporations, organizations, and agencies are called **technoscapes.** Dramatic changes in technology have spurred far-reaching alterations in both national and international political, economic, and social systems. This restructuring is manifesting profound changes for people and the environment (Figure 1-16).

Figure 1-16

I was on a river trip in Borneo and was amazed at the number of satellite dishes marking river and even interior communities. This scene is from Samarinda, a port on the Mahakam River in Indonesia. Televisions are often shared among houses, and American programming such as MTV is available. Note the stilt houses that allow for tidal changes and rising waters in the wet monsoon. Photograph courtesy of B. A. Weightman.

DISAPPEARING DIGITAL DIVIDE?

Researchers still talk about the **digital divide**. This idea purports that possession and use of electronic devices such as cell phones occurs primarily in rich countries that are mostly in the northern hemisphere. But the digital divide is narrowing at an unprecedented pace. In 2000, developing (poorer) countries accounted for around one-quarter of the world's 700 million or so mobile phones. By the beginning of 2009, they accounted for three-quarters of the world's 4 billion mobile phones. As some people may own several phones and others hook into the system with SIM cards, the actual number of phone owners is around 3.6 billion.

China is the world's largest market for mobile phones with over 700 million subscribers in 2009. Yet, in one month in that year, India added more than 15 million users. There, cell towers have sprouted in urban areas and along major roads. New, home-grown operators in India and China have developed new business models and industry structures that make it possible to make affordable cell phones for low-income people. Some aspects of the "Indian Model" are being adopted by other countries, rich and poor alike. In developing countries, data services such as mobile-phone-based advice on a variety of subjects such as agriculture and health are helping to improve the lives of millions. This extraordinary explosion of both technology and information is the most dominant force in the global system of production. Jeffrey Sachs, a development expert at Columbia University, calls the cell phone, "the single most transformative tool for development."

GLOBALIZATION

Globalization refers to the spread of international capitalism, business, investment, and ideas around the globe. Proponents claim that an expanding global economy will bring increased prosperity to the entire world. Supporters advocate the elimination of trade barriers such as tariffs so that developing countries can compete in the global market, thereby reducing their dependency on developed ones.

Multinational or transnational enterprises play a vital role in the globalization process. These corporations control financial, manufacturing, and distribution systems. Conglomerates with headquarters in North America, Europe, and Japan are rapidly expanding their footprints in Asia. Hong Kong is the Asian headquarters for America's Walmart, the single largest importer of Chinese goods. Walmart, the world's largest corporation, has more than 600 stores in China and Japan and opened its first outlet in India in 2009. Carrefour of France, the world's largest hypermarket chain, opened a store in Taiwan in 1989. Now the company has at least 90 hypermarkets in China and Indonesia. Monsanto, an American agricultural conglomerate, operates in 12 Asian countries. The Finnish company Nokia has its regional headquarters in Singapore and manufactures mobile phones and multimedia products in South Korea and China. Japan's advanced technology firm Fujitsu operates in at least 11 Asian countries and has connections with Germany's electronics firm Siemens, established in 14 Asian nations from India to the Philippines. More than 60 foreign IT and telecommunication firms are currently operating in Pakistan.

Transnationals are also emerging in other parts of Asia. South Korea's Hyundai completed the construction of an automobile assembly plant in Montgomery, Alabama, in 2005 and Kia opened one in Troup County, Georgia, in 2006. Tata, India's largest conglomerate, manufactures everything from cars to electronics. It has recently opened metalworks and automobile components plants in China. Moreover, India now hosts research and development firms for over 100 international companies. China hosts more than 750 such centers. Although China invests and builds primarily in other parts of Asia, it is now heavily involved in Latin American and African countries. China's presence in Africa will be discussed in Chapter 10.

Another feature of globalization is **offshoring**. Offshoring involves the substitution of less costly foreign labor for domestic labor. It began with the establishment of *call centers* in developing countries. For example, an American dialing a toll-free number causes a phone to ring in India, or a German sending an e-mail to Microsoft receives a reply from China.

India offers examples of how offshoring has expanded in scope in the twenty-first century. Sophisticated computer programs, once written in California's Silicon Valley, are coded in Bengaluru (Bangalore). Medical x-rays, previously interpreted by doctors in Germany, are now being analyzed by medics in New Delhi. Bank clerks are crunching numbers in Mumbai (Bombay) and sending them electronically to New York. And material for animated movies is being created in Hyderabad, not Hollywood.

Many Asian countries have jumped onto the offshoring bandwagon over the past decade. For instance, more than 250 foreign companies such as Shell, Cisco, Ericson, and HSBC have located In Malaysia to provide offshoring services. The Philippines is host to more than 800 call centers with multiple functions such as telemarketing, debt collection, mail ordering, and other customer services. This "Sunshine Industry" is one of the fastest growing sectors of the Philippines' economy. Read more about this phenomenon in the regional chapters.

WORLD SYSTEMS THEORY

Immanuel Wallerstein, an American sociologist, has proposed a **world systems theory** in relation to the process of globalization. He points to a dichotomy of capital and labor as competing agents reap an endless accumulation of capital. Corporations seek to maximize profits at the expense of worker safety, fair compensation, environmental protection, and the integrity and sovereignty of governments. Further, he argues that there is no Third World; there is only ONE world connected by a complex network of economic interchange relationships—that is, a "world economy" in a "world system."

According to Wallerstein, this globalization process, which began in the sixteenth century, has evolved into a lasting division of the world into **core regions, semi-peripheral regions**, and **peripheral regions.** Clearly, rich countries such as the United States, Canada, Australia, and Japan comprise the core, and poor ones such as Laos, Cambodia, and Nepal sit in the periphery. The core has high levels of technological development and manufactures complex products. Peripheral regions supply raw materials and cheap labor for the expansion of the core.

Economic interchange between core and periphery takes place on unequal terms. Poor countries are forced to sell their products at low prices to the core but must pay high prices for manufactured goods from the core. Semi-peripheral regions act like a core to the periphery and as a periphery to the core regions. At the end of the twentieth century, such countries as China, India, Brazil, and Mexico had moved from the peripheral to the semi-peripheral realm. We will revisit cores and peripheries and examine many aspects of development and its inequities in Chapter 4.

ANTI-GLOBALIZATION MOVEMENT

Wallerstein supports the growing worldwide **anti-globalization movement**, which argues that corporations seek to maximize profits at the expense of fair compensation, work-safety conditions and standards, environmental conservation, and the integrity of governments' independence and sovereignty. Protestors stress that globalization will accentuate inequality not only between rich and poor countries but also within individual countries. Poor farmers and other powerless people see globalization as a vehicle to keep them in a position of poverty and subservience. Although it is generally true that global economic integration allows economies of poor countries to expand, national incomes to rise, and new middle classes to emerge, this wealth is unevenly distributed. Significant monetary gains are enjoyed primarily by the upper echelons of society while millions remain destitute.

To some, Walmart personifies the wonders of a global marketplace that delivers low-cost goods to grateful consumers. But to many champions of the poor, Walmart is an arrogant, union-busting employer that denies its employees decent wages and benefits such as health insurance. Critics accuse the company of exploiting hundreds of thousands of anonymous poor workers who slave for long hours in "sweatshops," especially in Asian countries like China, Bangladesh, and the Philippines. To some of the cyber-critics who inhabit the "blogosphere," Walmart is a "hated capitalistic Satan."

Another target of the anti-globalization movement is the **World Trade Organization (WTO)**. The WTO, with more than 150 member states, sets rules for economic activities such as foreign investment and trade. Member countries are expected to reduce or eliminate barriers to trade. Poor countries are supposed to benefit from this new order. However, this is not always the case.

When the Philippines joined the WTO, its government assumed that world markets would be open to cheaper Filipino farm produce. When demand increased, more farm jobs would be created and wages would rise. However, this did not happen. Filipino farmers found themselves competing against American and European farmers whose governments subsidize their costs of production and export. Simultaneously, subsidized U.S. corn appeared in Filipino markets with prices cheaper than for corn grown in the Philippines. Consequently, several hundreds of thousands of agricultural jobs were lost and wages declined. Obviously, the WTO is not popular in the Philippines.

But globalization is only part of the picture. A countermovement is **regionalism.** Regional organizations, which include diverse nonprofit organizations, human-rights groups, political groups, and even governments, are concerned with such things as social, cultural, environmental, and economic issues. Although globalism and regionalism appear contradictory, they are integrated phenomena operating at varying spatial scales and settings. Regionalism is a response to loss of self-determination through incorporation into multinational behemoths.

Regional trading blocks, such as the European Union (EU) or the North American Free Trade Association (NAFTA), regulate economic interchange in the interests of reducing foreign competition and to maintain the flexibility of extensive markets. The **Asia Pacific Economic Cooperation (APEC)** is a similar organization.

APEC was organized in 1989 by then Australian Prime Minister Bob Hawke. This organization relies on ad hoc task forces and groups of business people, economists, and government officials to chart and achieve its goals. APEC's main goal is to achieve free trade in the Asia-Pacific region by 2020. The original membership of 12 has increased to 21, including East and Southeast Asian nations along with six non-Asian nations such as Canada, Mexico, the United States, and Australia. In other words, APEC is a Pacific Rim phenomenon.

The **Pacific Rim** is an economic concept of the late twentieth century. It refers to all the countries situated around the Pacific Ocean including those of North and South America, East and Southeast Asia, and Australia and New Zealand. **Asia-Pacific** is another, broader term. These countries' primary trading partners are one another.

While regional organizations set new paradigms for international negotiations, they are frequently criticized as being insensitive to concerns such as air pollution, toxic waste, environmental health, labor exploitation, and cultural survival.

THE DUAL IMPACT OF "AMERICANIZATION"

As I enjoyed a Coke on a beach in Vietnam, I was struck by the size of the canvas sign being used to block out the hot rays of the sun. There was the Marlboro Man riding into the sunset, lasso in hand. A tin Coca-Cola sign hung crookedly on the wooden, drink and snack serving bar. There are Coca-Cola museums in China, Japan, and the Philippines, and Taiwan boasts the world's largest inflatable Coke bottle.

Wahaha is China's largest beverage producer because it markets its products in rural areas where most Chinese still live. It has recently come up with a bottled drink for rural consumers who cannot afford real Coca-Cola. This spinoff is called "Future Cola."

American fast-food outlets are becoming common features of urban landscapes. For example, there is a Starbucks coffee shop on the Great Wall of China. As of 2008 there were 1,000 McDonalds restaurants in China. McDonalds has signed a deal with Sinopec, the huge gas station chain, to build drive-through restaurants along China's new highways.

In India, call center workers are trained to imitate American accents. In order to get it right, they watch American soap operas and game shows. During working lunches in offshoring firms, they are as likely to order from Domino's or Pizza Hut as from restaurants serving Indian food. Even India's poor are targeted by some products. For example, a cup of Pepsi can be bought for six rupees, about a penny, in rural areas.

Vietnamese versions of rap music boom in taxis and from music shops in Hanoi and Saigon. Japan is famous for its American-style game shows. You can still find pictures of Mohammed Ali and Rambo in village houses and markets in Asia. On a visit to a small village in northern Laos, an elderly man offered me a swig of homemade rice wine out of a Jack Daniel's bottle. Once I witnessed a group of elderly women squatting around a TV on the floor of a communal longhouse in a remote region of Indonesian Borneo. They were laughing, pointing, and shaking their heads at the apparent silliness in front of them. It was MTV.

These are all examples of American **popular culture**, which is spreading into the remotest corners of Asia. What impact does this trend have on lives and landscapes?

Although many, especially young people, have fallen in love with American-style fast food, clothing styles, music, and stars, there are others who are displeased with what they perceive as negative aspects of American culture. In places such as Taiwan, Malaysia, and Thailand, some people think that American popular culture is detrimental to traditional values, especially those concerning youth behavior and familial relationships and responsibilities. In India, cooked-chicken sellers protested against Kentucky Fried Chicken for ruining their businesses. Elsewhere, small, family-run shops are being replaced by franchise chains such as 7-Eleven, and homemade ice cream parlors by such franchises as Baskin-Robbins. Avon, Amway, and other direct marketers are changing long-held relationships between customers and local merchants.

Americanization is controversial. Although it is true that such enterprises provide employment, where do most of the profits go? Televisions are becoming ubiquitous in Asia. Although television can be a way of teaching and promoting public health or family planning, it also increases expectations and promotes consumerism.

Figure 1-17

This Kentucky Fried Chicken on the island of Java in Indonesia shows cultural integration with the inclusion of Ayam Goreng*—fried chicken with specific Indonesian spices, commonly in peanut sauce.* Photograph courtesy of B. A. Weightman.

This may be good for business, but is it good for society in general? How does this medium and American programming such as CNN, MTV, and various comedy, crime, and drama series influence perception and behavior? In South Korea and China, leaders are concerned about rising levels of obesity related to the increased availability of junk food. Malaysians and Taiwanese worry about inappropriate youth behavior. As young people move away from their families to pursue careers, older people worry about whether traditional networks of care will be available for them in their old age.

Conservatives, especially in Muslim populations, view popular culture as contributing to secularism and liberalism. While modernization and change often have positive consequences, many are convinced that their own national cultures should transcend such potentially "harmful" Western elements.

THE RISE OF ASIAN "POP-CULTURE"

Disillusionment with Americanization along with the expansion of communication systems within Asia have intensified the popularity of indigenous modes of popular culture. Two countries stand out in terms of their influence: Japan and Korea. Japanese movies are well attended, and many Japanese movie stars are icons in the region. Japanese *manga* comic books are read by millions of young people. The animated characters "Hello Kitty," "Ampan Man," and "Poke'mon" can be seen on T-shirts, backpacks, and a myriad of other items in shops around the world, especially in Europe.

Many fashion magazines in Hong Kong are Japanese, some translated into Cantonese. However, interest in Japanese popular culture is subsiding under a recent wave of Korean influences.

Korean popular culture has invaded the rest of Asia since public media began paying attention to it in the late 1990s. This phenomenon is known as **Korean Wave III**.

It all began with the showing of a Korean drama called *Star in My Heart* in China, Taiwan, and elsewhere in Asia. Before seeing the movie, many Asians perceived Korea to be rough, violent, and lacking material and cultural refinement. However, the drama was set in the context of impressive skyscrapers and high fashion. And the star—Ahn Jae-wook—was attractive to Chinese and Taiwanese women. *Shiri*, a film that explored a romance between North Korean and South Korean spies, offered a peek into the estranged relationship between those two countries. Suddenly Korea became very interesting.

Now, Korean movies, soap operas, Internet games, fashion, and music are all the rage in Asia. Music groups such as H.O.T. (disbanded 2002), Baby Vox, and CLON have adopted Western pop music and given it an Asian twist (Figure 1-18).

Korean media is regarded as "fresh and trendy." Perhaps more importantly, it not only contains Asian values and sentiments (often Confucian) but also assures Asian identity. Korean Wave III demonstrates than cultural influences are geo-cultural, not simply transnational.

Figure 1-18

Korean Wave III has arrived in Asia. ©KIM JAE-HWAN/AFP/Getty Images, Inc.

Recommended Web Sites

www.aag.org
Association of American Geographers homepage.

www.About.com:Geography
Learn how to become a geographer. Country information; historical geography and exploration; physical and cultural geography; geographic games and quizzes; downloadable outline maps.

www.amergeog.org/
American Geographical Society home page. Information on publications such as the *Geographical Review* and *Focus*.

www.atlapedia.com
Excellent downloadable physical and political maps.

www.atlas.mapquest.com/atlas
Downloadable maps.

www.cia.gov/library/publications/theworld-factbook.
The Central Intelligence Agency's World Factbook with information and statistics.

www.hrw.org
An organization dedicated to protecting the human rights of people around the world.

http://www.ifg.org/
An alliance of scholars and activists who are critical of current trends of globalization.

www.internetworldstats.com/stats2.htm
Information on Internet users around the world.

www.nationalgeographic.com/
Interactive maps and videos and excellent photos and fascinating articles about the world.

http://placesonline.org.
Outstanding site with links to Web sites around the world. Sponsored by the Association of American Geographers.

www.un.org/en/index.shtml
United Nations links to many sites with information about human development around the world.

www.en.wikipedia.org/wiki/Main_Page

While you can look up just about anything on this site, facts and figures may not be reliable because anyone can contribute information. Check the references at the end of each discussion. There is a good discussion of Immanuel Wallerstein and his theories on this site.

www.worldatlas.com/aatlas/world.htm

An online atlas with facts, figures, flags and maps for continents, countries, islands, cities, and every other type of territory on the planet.

www.worldbank.org

Information on WB projects and views on development.

www.wto.org

Maps and statistics on world trade. Student site.

Bibliography Chapter 1: The Big Picture: Major Influences

n.a. 2007. "All Shapes and Sizes: Asia's Financial Centres Reflect Its Vast Geography and Divergent Economies." *The Economist*, September 15:10–13.

Blaut, J. 1994. *The Colonizers Model of the World: Geographic Diffusionism and Eurocentric History.* New York: Guilford.

Brierley, Joanna Hall. 1994. *Spices: The Story of Indonesian Spice Trade.* Kuala Lumpur: Oxford Edition. New York: Wiley.

Craig, Timothy J., and Richard King, eds. 2002. *Global Goes Local: Popular Culture in Asia.* Toronto: University of Toronto Press.

Craig, Timothy J., ed. 2000. *Japan Pop: Inside the World of Japanese Popular Culture.* Toronto: University of Toronto Press.

Engammare, Valerie, and Jean-Pierre Lehmann. 2007. "Can Asia Avert a Globalization Crisis?" *Far Eastern Economic Review* 170/2:7–10.

Georgacas, Demetrius J. 1969. "The Name Asia for the Continent: Its History and Origin." *Names* 17:32–37.

Johnston, R. J., Peter J. Taylor, and Michael J. Watts, eds. 2004. *Geographies of Global Change.* Oxford: Blackwell.

Kitigawa, Joseph M. 1989. *The Religious Traditions of Asia.* New York: Macmillan.

Leong, Anthony. 2002. *Korean Cinema: The New Hong Kong.* Victoria, Canada: Trafford Publishing.

Levathes, Louise. 1994. *When China Ruled the Seas.* New York: Oxford.

Lewis, Martin W., and Karen E. Wigen. 1997. *The Myth of Continents: A Critique of Metageography.* Berkeley: University of California.

Meredith, Robyn. 2005. "The Next Wave of Offshoring." *Far Eastern Economic Review* 168/3:19–24.

Nakamura, Hajime. 1997. *Ways of Thinking of Eastern Peoples: India, China, Tibet, Japan.* London: Kegan Paul.

Risso, Patricia. 1995. *Merchants and Faith: Muslim Culture and Commerce in the Pacific Ocean.* Boulder, Colo.: Westview.

Rushford, Greg. 2007. "Is Wal-Mart Good For Asia?" *Far Eastern Economic Review* 170/10: 26.

Sung, Sang-yeon. 2008. "The High Tide of the Korean Wave III: Why Do Asian Fans Prefer Korean PopCulture?" *The Korea Herald*, February 4.

The Economist. 2009. "The More the Merrier: India and China Are Creating Millions of Entrepreneurs." *The Economist*, March 14:13–14.

The Economist. 2009. "The Next Great Wall: Buy Local Campaigns Raise Protectionist Barriers in Asia." *The Economist*, March 14:67–68.

The Economist. 2009. "The Noodle Bowl: Why Trade Agreements Are All The Rage in Asia." *The Economist*, September 5:48.

Wallerstein, Immanuel. 2004. *World-Systems Analysis: An Introduction.* Durham, N.C.: Duke University Press.

Chapter 2

Environments and People

"Those who would take over the earth/ And shape it to their will/ Never, I notice, succeed."

TAO TE CHING (CA 3000 BC)

Major Topographical Features

South, East, and Southeast Asia exhibit a grand and complex display of mountains, plateaus, drainage basins, and river valleys (Figure 2-1). All of these have been modified and have influenced people's actions and interactions since the beginning of human time.

The foundation for Asia's topography (landforms) is an extensive zone of ancient rocks that, in early geologic time, were folded, faulted, and intruded on countless occasions. Ongoing hydrological (water) and aeolian (wind) erosion and deposition have combined to generate Asia's physical geography as we know it today (Figure 2-2). Unlike North America and Europe, Asia was not covered with massive ice sheets, with the exception of high mountain systems such as the Himalayas. In fact, glaciation continues to play a critical role in sculpting mountainous regions.

The most imposing topographic features are the Himalayan ranges and adjacent Tibetan Plateau. The Greater or highest Himalayas extend 1,500 miles (2,415 km) from west to east with an impressive number of the world's highest peaks (check your atlas for locations). No fewer than 15 peaks top 25,000 feet (7,000 m). Annapurna at 26,535 feet (8,090 m), Dhaulagiri at 26,689 feet (8,137 m), Kanchenjunga at 28,208 feet (8,598 m), and Everest (*Qomolangma*) the highest at 29,028 feet (8,848 m) are the best known mountains. Imagine a fractured rock wall around 4 miles (6 km) high and 200 miles (322 km) across. This is the majesty of the Himalayas.

Although cut by numerous passes, the Himalayan ranges have posed a formidable barrier between Indian and Chinese realms, thereby demarcating racial, linguistic, and other cultural differences. The mountains themselves form a region of **cultural convergence** where lifeways and belief systems blend in splendid landscapes (Figure 2-3).

The history of the Himalayas began 200 million years ago with the process of continental drift when the landmass known as **Pangaea** broke into pieces. At that time, **Gondwanaland** (Africa) and what we know now as India proceeded toward another, larger landmass called **Laurasia**. When the collision occurred 45 million years ago, mountains emerged in a colossal upheaval. The **Tethys Sea** vanished, consumed in the process. When trekking near the Tibetan border, I was delighted to find two **ammonites**: flat, spiral fossil shells, evidence of life in the ancient watery world of the Tethys Sea.

The Himalayan ranges are a product of **plate tectonics**. The theory of plate tectonics postulates that the upper portion of the lithosphere (the outer part of the solid earth composed of rock) comprises a patchwork of rigid plates. These are rooted in the underlying, more plastic upper layer of the Earth's mantle. Movement along the fracture zones between the plates is responsible for a legion of topographic features and seismic phenomena.

Where plates diverge or move apart, molten material from the Earth's core wells up into the space. Where plates converge or come together, spectacular mountains, pyroclastic events, and deep ocean trenches appear.

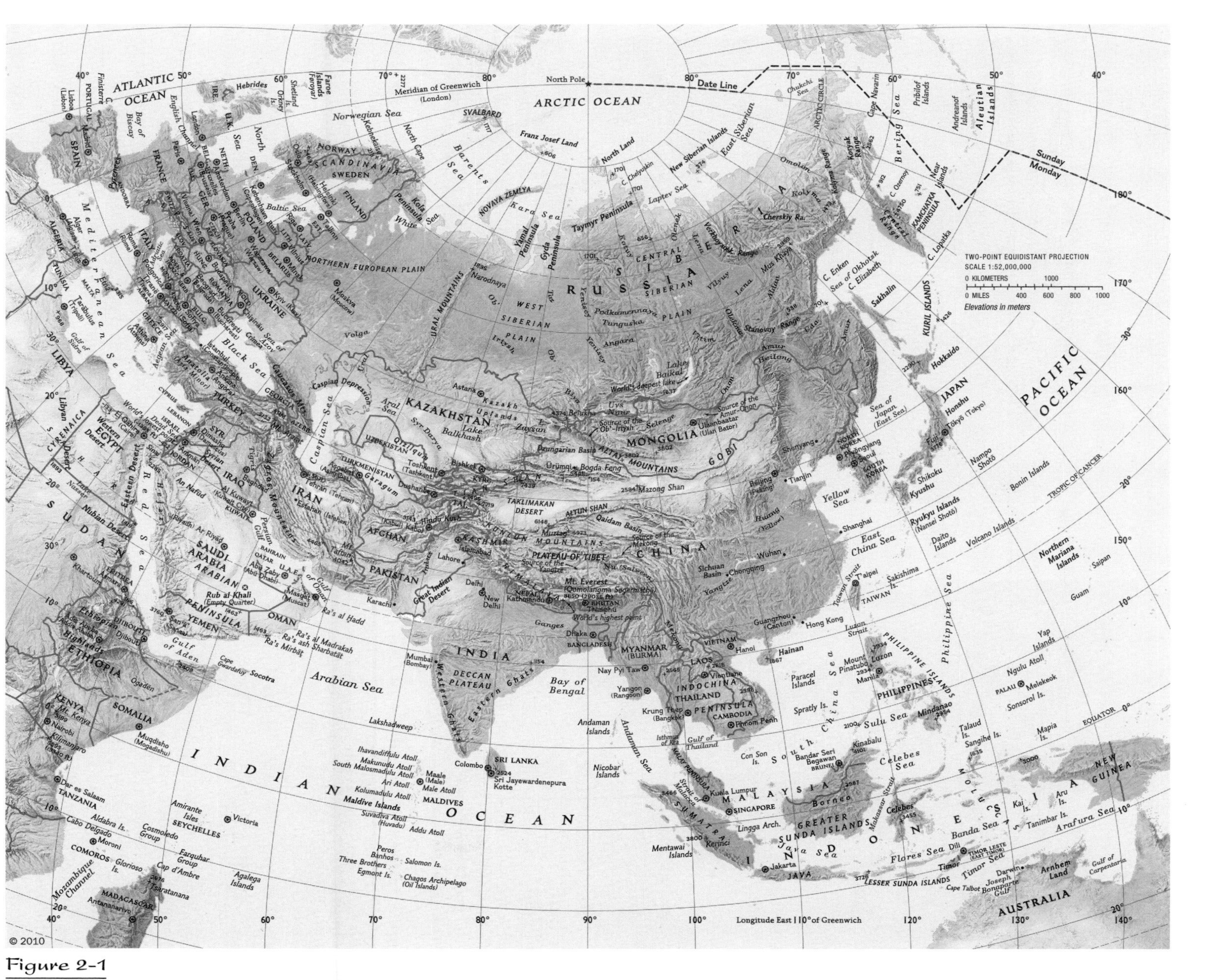

Figure 2-1

Physical geography of Asia. Note the extreme ruggedness of many parts of the continent and the limited amount of lowlands. Most populations live in lowland areas. ©NG Maps.

Figure 2-2

The Spiti Valley, in the Indian Himalayas, clearly illustrates the impact of water, wind, and ice in physical landscape formation. A small community with its barley fields clings to the base of the alluvial fan. The Spiti River is a tributary of the Indus River system. Photograph courtesy of B. A. Weightman.

Figure 2-3

Many Tibetans live in the Himalayas of India's Himachal Pradesh. This gompa (monastery), at almost 13,000 feet (3,890 m), was built in the sixteenth century. Photograph courtesy of B. A. Weightman.

Did you know that Asia is the world's most disaster-prone region?

The Pacific Ring of Fire

Half of all major natural disasters are in the Asia-Pacific region. Because of the abundance and intensity of seismic activity, the region is often referred to as the *Pacific Ring of Fire* (Figure 2-4).

Volcanic eruptions and earthquakes have wreaked havoc on humanity for millennia. The most powerful eruption in recorded history occurred on the Indonesian island of Sumbawa in 1815 when Mount Tambora (9,348 ft, 2,850 m) blew up 100 billion tons of volcanic rock and debris. The death toll was estimated at 92,000. China's Tangshan earthquake in 1976 killed more than 240,000 people, and a 1993 quake in western India's Maharashtra state took 12,000 lives. The Philippines' Mount Pinatubo eruption in 1991 and Japan's Kobe quake in 1995 registered relatively low death tolls but caused considerable damage. In 2001, a 6.9 quake in Gujarat, India, killed more than 20,000 people. At least 87,476 people died in 2008 in a 7.9 earthquake in China's Sichuan Province. And even more recently, on September 30, 2009 a 7.6 quake with its epicenter in the sea west of Padang in Indonesia's West Sumatra Province resulted in the deaths of some 1,115 people and caused immeasurable property damage.

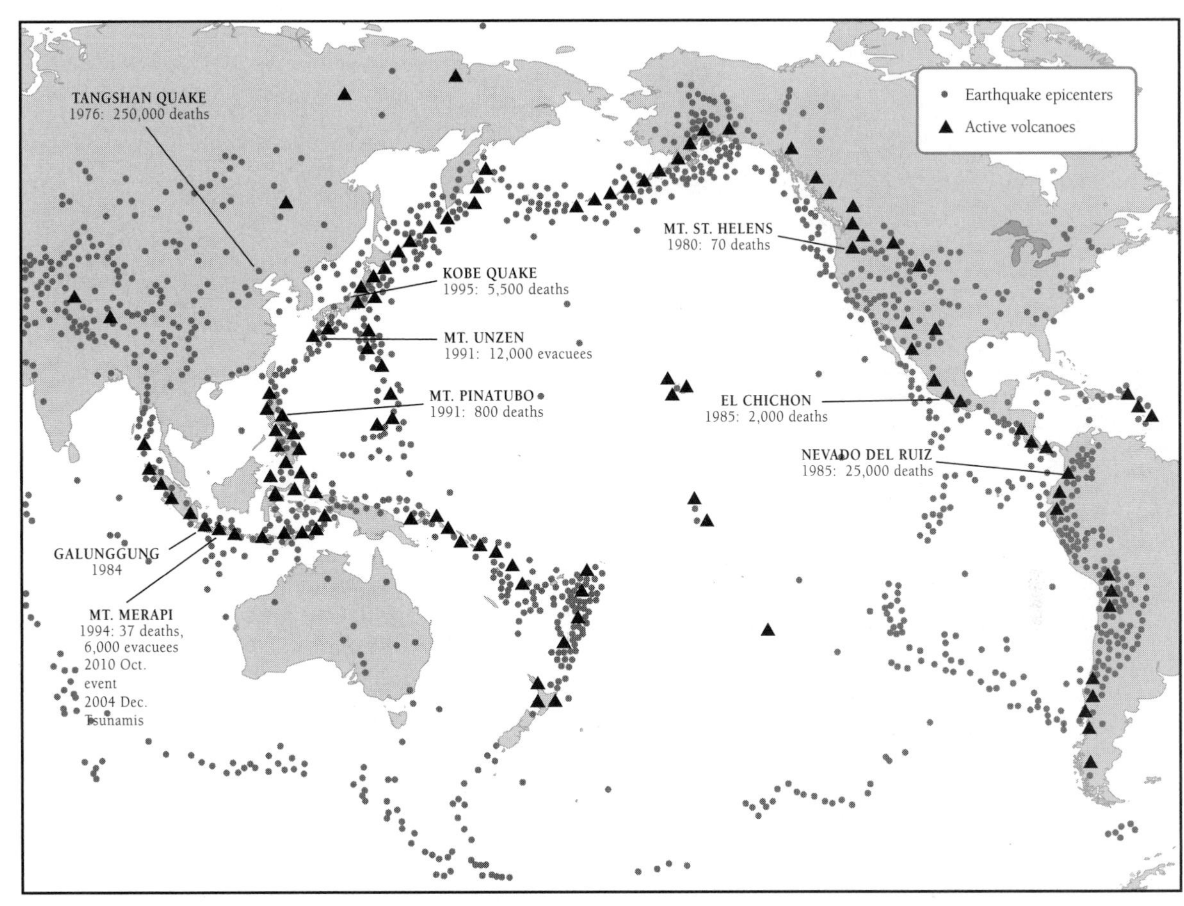

Figure 2-4

The Pacific Ring of Fire. Do you live in one of these major fault zones? How would you prepare for a seismic event?

Tsunamis, or seismic sea waves, can be even more damaging than the seismic event. The speed of a tsunami is proportional to the depth of the water; the deeper the water, the faster the wave. As the water approaches a shallow area, the fast-moving wave is slowed, causing the water to pile up on itself. Tsunamis have caused the deaths of hundreds of thousands of coastal and island people. For example, in 1976, a tsunami drowned 8,000 people in the Philippines' Moro Bay.

On the morning of December 26, 2004, a massive 9.0 earthquake struck off the western coast of Sumatra, Indonesia, setting off ocean shock waves that were felt more than 3,000 miles away on the coast of East Africa. The earthquake occurred when stress built up between the Burma and Indian plates, causing several hundred miles of the Burma plate to snap. This forced a massive displacement of water in the Indian Ocean. The resulting waves spread in all directions, moving as fast as 500 miles per hour. In the deep ocean, the waves may be imperceptible, but they slow down and gain height as they hit shallower water near shore. The worst-hit places were Aceh province on the island of Sumatra, where more than 80,000 were killed, and Sri Lanka, where more than 29,000 died. Thailand lost at least 5,000 people. In addition, thousands of people disappeared, their bodies swept out to sea. The tsunami also devastated infrastructures and ruined regional economies (Figure 2-5).

Although the outpouring of international aid was unprecedented for this disaster, the most affected areas such as Aceh province are not expected to recover for over a decade. And even as the recovery

Figure 2-5

Devastation from the 2004 tsunami in Aceh, Sumatra, Indonesia. ©JEWEL SAMAD/AFP/Getty Images, Inc.

efforts proceeded, another earthquake hit on March 28, 2005, also of great magnitude. (Refer to Chapter 15 for an update on progress and change in Aceh.)

In studying the December earthquake, scientists revised its scale from magnitude 9 to 9.3. Although this does not seem to be a large increase, 9.3 represents a 2 $^1/_2$ increase in scale. The December earthquake is now the second largest ever recorded in over 100 years of record keeping. The seafloor rupture of the earthquake was 750 miles, also the largest ever recorded, and three times larger than originally thought.

The Pacific Ring of Fire also includes the world's deepest ocean trench: the Mariana Trench, which descends 36,198 feet (11,033 m) below sea level. Another is the Indian Ocean's Java Trench at 23,376 feet (7,125 m) below sea level.

NORTH OF THE HIMALAYAS

If you were to cross the Himalayas northward you would be standing at an elevation of approximately 15,000 feet (4,500 m) at the edge of the stupendous Tibetan Plateau. You would be gasping for air and gaping in awe at more than 400,000 square miles (1,036,001 km^2) of rugged basins and ranges. Many of the rock-strewn basins contain permanent lakes. Mountain passes permit relatively easy access from one basin to another, but many of these passes are 17,000 feet (5,100 m) and higher. Here is where South Asia's four great rivers originate: the Sutlej, Indus, Ganges, and Brahmaputra (Tsang Po). China's major rivers, the Chang Jiang or Yangzi and the Huang He (Yellow River), also originate here. The northern perimeter of the Tibetan Plateau is bounded by other massive ranges such as the Kunlun and Karakoram. The 28,250 feet (8,611 m) peak K2 rises in the Karakoram.

Northward again are the arid basins and mountains of northwestern China. The Tarim Basin is largest in expanse and the Turpan (Turfan) Basin lowest in elevation. The Heavenly Mountains—the Tien Shan—stand as one of this region's major divides. Although the Tarim and Turpan and related basins are not important in terms of population density, they are of critical importance economically and politically.

EAST AND SOUTHEAST OF THE TIEN SHAN

Eastward is a mosaic of rugged mountains and plateaus, along with plains and more basins. The Qinling Range, Loess Plateau, North China Plain, and Sichuan Basin are only four of many significant elements of China's physical landscape. These regions are dominated by the mighty Huang He and the Chang Jiang or Yangzi. Millions of people make a livelihood in these and other river valleys, basins, and plains. Even further east are the craggy Korean peninsula and Japanese islands.

SOUTH AND SOUTHEAST OF THE HIMALAYAS

South of the Himalayas, the North Indian or Indo-Gangetic Plain sweeps from Pakistan, across India, to Bangladesh. This subsiding trough of 300,000 square miles (777,001 km^2) is filled with fertile alluvial soils deposited by the river systems of the Indus, Ganges, and

Figure 2-6

This is the always busy Chao Praya river in Bangkok, Thailand. Photograph courtesy of B. A. Weightman.

Brahmaputra. These water-deposited sediments, in concert with extensive river networks and relatively flat land, have helped make the Indo-Gangetic Plain one of the most densely populated regions on Earth.

South of the Indo-Gangetic Plain rises the Central Indian Plateau and the dissected Deccan Plateau. Covering most of peninsular India, the Deccan is bordered by narrow coastal plains along the western Malabar and eastern Coromandel coasts. The tear-shaped island of Sri Lanka lies about 20 miles (32 km) across a narrow strait to the southeast of the Coromandel coast.

Mainland Southeast Asia is a washboard of mountains and **fluvial** (river-related) landforms. Mountain ranges, with elevations as high as 18,000 feet (5,400 m), are generally aligned in a north-to-south direction. One exception is the S-shaped Annamite Cordillera that follows practically the entire coast of Vietnam from southern China. Mountain chains are arranged closely in the north with virtual slots of lowland in between. Further south, the slots widen into V-shaped valleys and eventually into open plains, basins, and deltas.

Mountains have created cultural barriers for the inhabitants of interstitial, riverine zones and they have encouraged ethno-linguistic diversity as each of hundreds of different culture groups maintained its own way of life in relative isolation from others for centuries.

Rivers are of paramount importance in mainland Southeast Asia. The Ayeyarwady (Irrawaddy), Chao Praya, Mekong, and Red River valleys have their headwaters on the Tibetan Plateau. So tortuous is the surrounding mountain terrain that the source of the Mekong was not found until 1994. River valleys foster support of millions of inhabitants and provide excellent sites for the growth and development of major cities, including most capitals (Figure 2-6).

Peninsular Southeast Asia extends 900 miles (1,449 km) from the Gulf of Thailand to the island of Singapore. The narrowest portion, the Isthmus of Kra, is roughly 40 miles (64 km) across. It widens to the south to form the Malay Peninsula.

ISLAND REALMS

South, East, and Southeast Asia are ringed by islands. Some, such as Sri Lanka and Singapore, are politically independent states. Others are part of island chains called **archipelagoes**. Japan, the Philippines, and Indonesia are all archipelagoes. Each comprises several large islands and hundreds, or even thousands, of smaller ones. Japan is dominated by the large island of Honshu, but there are three other sizable islands (Hokkaido, Kyushu, and Shikoku) and hundreds of lesser ones. The Philippines incorporates more than 7,000 islands along with the larger islands of Luzon and Mindanao. However, the biggest is not necessarily the most important. For instance, the Indonesian archipelago totals more than 17,000 islands, dominated in size by Sumatra, Sulawesi (Celebes), and Kalimantan (Borneo). However, Indonesians and their political and economic centers are most heavily concentrated on the much smaller island of Java.

New Guinea, the world's second largest island (after Greenland), is politically divided between

Papua New Guinea in the east and West Papua in the west. West Papua is part of Indonesia. This island is the focus of one of Indonesia's many political and environmental controversies (Chapter 15).

There are more island groups in the Indian Ocean. Strung between Myanmar and Sumatra is a 600 mile (966 km) chain of more than 200 islands. These are known as the Andaman and Nicobar Islands and belong to India. The majority are merely exposed tops of mountains protruding from the ocean floor. Southwest of India lie the Maldives, 115 square miles (298 km^2) of low-lying islands, banks, and reefs. Only about 10 percent of the roughly 2,000 of these are inhabited.

LIMESTONE AND CORAL ENVIRONMENTS

Chemically weathered limestone is the essence of **karst topography**. Fractured limestone formations, with at least 80 percent calcium carbonate, are subject to solution processes. Subsurface drainage channels are formed as water seeps through joints in the otherwise impermeable limestone. Chemical weathering takes place, resulting in pitted and lumpy surface forms and labyrinths of caves below (Figure 2-7). Karst regions in Asia are found in northwestern India, southern China, and in various parts of Southeast Asia. An especially beautiful area of karst features is Halong Bay in North Vietnam. Karst formations are popular tourist attractions, drawing sightseers, photographers, rock climbers, and spelunkers from around the world.

Are you aware that the world's best dive sites are in the Asia-Pacific region?

Coral Reefs

Thriving in warm, tropical waters, coralline formations cover an estimated 109,769 square miles (284,300 km^2) of oceans, occur in more than 100 countries, and comprise roughly a third of tropical coastlines. It is estimated that more than half of the world's coral reefs are in the Indian Ocean, occurring along continental shelves and around islands.

Biological process is responsible for the formation of coral reefs. A coral is a small marine animal related to other marine invertebrates such as anemones and jellyfish. Corals excrete calcium carbonate from the lower half of their polyp bodies, forming a hard, calcified skeleton. Reefs are formed by colonies of corals that live in a symbiotic relationship with algae. Over centuries, skeletons accumulate on raised submarine features. This biological formation makes up a coral reef.

Coral reefs, the most biologically diverse oceanic ecosystem, are the marine equivalent of rain forests, the richest ecosystem on Earth. As many of 100,000 reef species have been named thus

Figure 2-7

This is an example of tower "karst" topography in Guilin, China. Photograph courtesy of B. A. Weightman.

far. Scientists estimate that there could be as many as 1 to 3 million as yet unidentified.

An estimated 4,000 to 4,500 fish species inhabit the world's coral reefs—more than a quarter of all fish species. Sea turtles, marine mammals, and certain seabirds are also associated with coral environments. Their biological and economic importance cannot be underestimated.

Indonesian waters contain some of the world's most spectacular coral reefs that support more than 5,000 species of fish, 30 species of sea snakes, and 7 varieties of turtles, among countless other creatures. However, reef populations have been significantly reduced. Look at Sulawesi, where scientists believe that tropical marine life began. Each year, thousands of pounds of pear oysters, mother-of-pearl shells, and *trepang*—sea cucumbers—are exported. Shells such as the helmet conch and giant clams are sold as well.

Recent research off the coast of Indonesia's Papua Province (western New Guinea) found more than 50 species that are likely new to science, including 24 fish and 20 corals. Among the fish discovered were two species of bottom-dwelling sharks that use their pectoral fins to "walk" across the seafloor. Scientists are now working with the Indonesian government to protect this and other reef areas from commercial fishing and destructive fishing practices. Globally, reef fisheries provide food and livelihood for tens of millions of people in the tropics and subtropics, who harvest a large diversity of reef species. For example, in the Philippines, some 209 species are taken from one single reef area. A booming commercial fishing industry is also taking its toll on coral reef communities. This trade supplies export markets, the restaurant and hotel industries, and the live-fish trade of Southeast Asia.

Coral reefs protect coasts from wave erosion and contribute to the process of sandy beach formation. Anecdotal evidence and satellite imagery suggest that reefs provided vital protection from the impacts of the 2004 Indian Ocean tsunami. In Sri Lanka some of the most severe damage occurred along coastlines where reefs had been reduced by extensive coral mining.

Failing to recognize the intrinsic value of coral reefs, people are destroying and endangering them in a variety of ways. Reefs are mined for construction materials such as lime. Corals and tropical fish are harvested for the aquarium trade. Dredging for sand for cement has damaged them and disrupted the marine community. Land reclamation projects and fish ponds have eliminated them. Reefs are ruined by heavy oil spills and the dumping of industrial waste. Speedboats, mechanical fishing devices, and fish blasting have been extremely detrimental. Fish blasters use explosives to kill large numbers of fish easily.

In 2004, it was estimated that 20 percent of the world's coral reefs have been destroyed. The greatest immediate threats to reefs are overfishing and pollution from poor land management practices. Some 50 reef species are now listed as "threatened." Intensified urbanization and agriculture increase the runoff of sediments and nutrients to reefs, thereby smothering corals by reducing light penetration and/or oxygen levels. Another Indonesian study, in an area that was subject to such pollution stresses, demonstrated a 30 to 60 percent reduction in species diversity.

Since the 1970s, many nations have recognized the significance as well as the severity of loss of their coral reefs and have taken steps to protect them. Action has also been spurred by the potential for tourism dollars generated by dive sites and ancillary facilities.

A significant focal area for action is the **Coral Triangle**, referred to as the "Amazon of the Sea". This is an extensive region of coral reefs that fringes six countries in Asia: the Philippines, Sarawak and Sabah (parts of Malaysia on the northern part of Borneo), east and central Indonesia, Timor-Leste, Papua New Guinea, and the Solomon Islands. Home to one-third of the globe's reefs and over 3,000 types of fish, this giant ocean community is also the spawning grounds for the world's largest tuna fishery. More than 120 million people depend on these resources for their livelihoods. A plan of action is currently being drafted to deal with overfishing, blast fishing (using bombs to kill or stun fish) and pollution runoff. Several international and local agencies such as the Nature Conservancy, USAID, and the Asian Development Bank are involved. Nevertheless, abuses continue and now include unlimited and uncontrolled tourist activities. Unfortunately, coral reefs continue to be threatened around the world. As they are part of a larger marine ecosystem, destruction of reef formations is detrimental to many other life-forms as well.

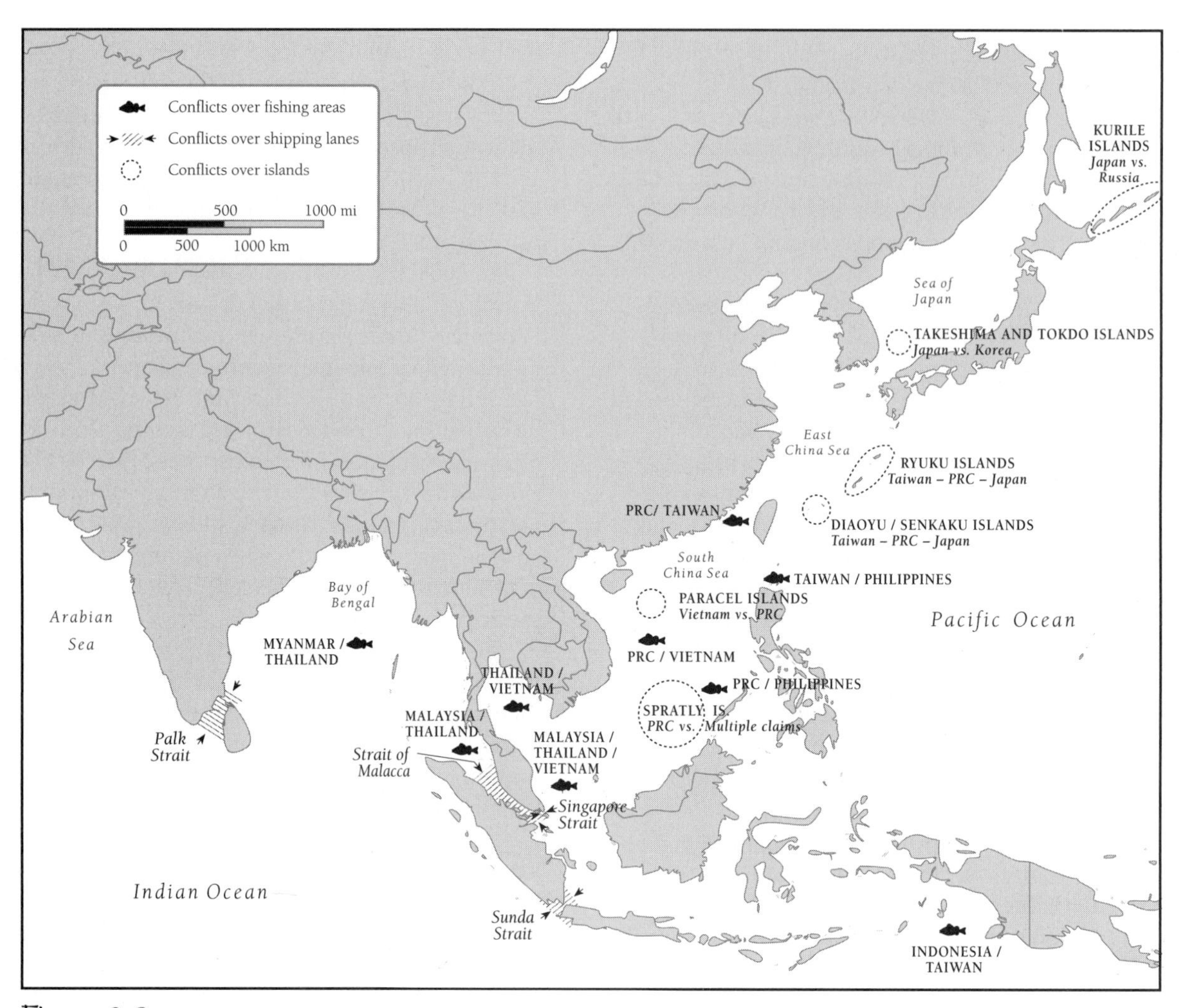

Figure 2-8

Conflicts at sea. Conflicts arise over ocean territories such as straits and islands, and over rights to resources such as fish and minerals.

Physical geography can also affect political geography. We know that rivers, mountain ridges, and other physical features are often used to demarcate political boundaries. Although some are disputed, most of these natural boundaries are stable. But bodies of water are problematic to define, presenting many opportunities for territorial disagreements (Figure 2-8). Many of these disputes will be discussed in the regional chapters.

Ocean Disputes and Conflicts

Oceans and seas are problematic because it is difficult to demarcate, in any practical sense, who owns which territory or mineral rights, or even who owns fish. In 1982, the United Nations issued the Law of the Sea Treaty, which allows each coastal state to claim a 200-nautical-mile **Exclusive Economic Zone (EEZ)** in which it controls resource rights. Countries are authorized to sell or lease rights in the EEZ to multinational corporations and can control fishing and pollution. However, problems arise when fishers stray over boundaries or pollution comes from another country's EEZ.

Designation of EEZs has not solved all problems. The 200-nautical-mile EEZ does not work where distance between states is less than 400 nautical miles. To solve this dilemma, a median line is used. This line divides the waters equidistantly between the opposing countries. For example, Malaysia and Indonesia share the Strait of Malacca, whose width ranges from 8 to

126 nautical miles. The EEZ of each country ranges from 4 to 63 nautical miles. Nevertheless, disputes are common in such narrow passages partly because of the multitude of islands found in these waterways.

The case of the **Timor Gap** provides an example. In 1989, Australia and Indonesia struck a deal to explore and exploit the seabed between them. Here is thought to be one of the world's largest reserves of hydrocarbons. This perceived breakthrough followed years of controversy. In 1972, Australia and Indonesia agreed to formalize their sea boundaries along the deep Timor Trough. However, Timor-Leste was still controlled by Portugal. Both Australia and Portugal made conflicting claims to the seabed, issuing overlapping exploration licenses to oil companies. This disputed section became known as the Timor Gap. After Timor-Leste became independent in 2002, it signed a deal with Australia allowing Timor-Leste 90 percent of royalties from the shared zone of exploration.

Another point of controversy is the amount of territory that islands can claim. A small island can claim rights to fish and minerals such as manganese and oil over a disproportionate sea area. This makes possession of islands, no matter how small, critical. Jurisdiction over islands fuels tensions and conflicts in the Indian and Pacific Oceans.

The Law of the Sea Treaty also guarantees ships innocent passage while passing through the waters of one country to the next. Passages such as the Strait of Malacca are vital to world shipping. To close these narrow waterways would be a relatively simple task in a time of conflict.

There are those who proclaim oceans and seas to be humanity's common heritage, a concept introduced by the Dutch as *mare liberum* in the sixteenth century. Unfortunately, this ideal is impractical in a commercial world, and the "scramble for the seas" has intensified.

Monsoon Asia

South, East, and Southeast Asia have been termed **Monsoon Asia** by geographers and other scientists. This is because these realms are affected by the **monsoon**—intense, seasonally shifting wind systems that involve an annual cycle of wet and dry seasons. Beyond equatorial regions, summers are wet and winters are dry (Figure 2-9).

Equatorial regions experience year-round rainfall, although there are "wet" and "less wet" periods. This depends on the position of the sun, prevailing winds at the time, and topography. As wind systems shift north and south, different islands and even parts of islands receive more or less precipitation. Examine both January and July wind and precipitation patterns on a more detailed map in an atlas. This information is critical when planning a trip. You probably do not want to arrive in Bali in the wet period.

Another feature of equatorial regions is convectional rainfall. The combination of solar energy and abundant moisture availability in oceans and biotic communities causes rapid evaporation. Warm, tropical air rises, cools, and condenses. The inability of cool air to hold as much moisture as warm air produces rain. This convectional rainfall occurs nearly every day in the hotter, wetter months. In the drier months when evaporation rates are lower, several days might pass without a deluge.

Prior to rain, the humidity builds to an intolerable level. There is so much moisture in the air that 78°F (26°C) feels like 100°F (38°C). As the day wears on, storm clouds fill the sky, and in late afternoon, torrential rains wash the landscape and relieve the humidity. This is one reason why people work their farms in the early morning. Midday is a time to seek relief from the sun's intensity.

Latitudinal position is also significant, affecting both moisture availability and temperatures. In general, moisture decreases and temperature ranges increase inland and away from the equator. For example, average yearly precipitation at Guangzhou, on China's southern coast, is about 27 inches (686 mm). At Hohhot, capital of Inner Mongolia in China's interior, it is 17 inches (432 mm). Guangzhou's yearly temperature range is 56°F to 83°F (13°C to 28°C). Hohhot's is 8°F to 71°F (–13°C to 22°C).

Both rainfall and temperatures are critical criteria for agriculture because they determine the need for irrigation and the length of growing season. A **growing season** is the number of consecutive, frost-free days in a year.

What causes the monsoon seasons? The size of the Asian landmass or **continentality**, combined with its proximity to the Indian Ocean, are keys to explaining variations in air pressure and related upper-air circulation.

Because it is such a large landmass, extreme measures of air pressure form over the land in winter and summer: As the earth gets colder in the winter, cold, heavier air forms a high-pressure zone over the landmass. In summer, as the land gets warmer, the air gets warmer too and as warm air rises, a lower air pressure zone is generated. Typically, winds blow from high to low pressure zones. In general, winter brings cold, dry air from the continental interior and summer brings moist, warm air from the surrounding ocean. Water bodies have modifying influences

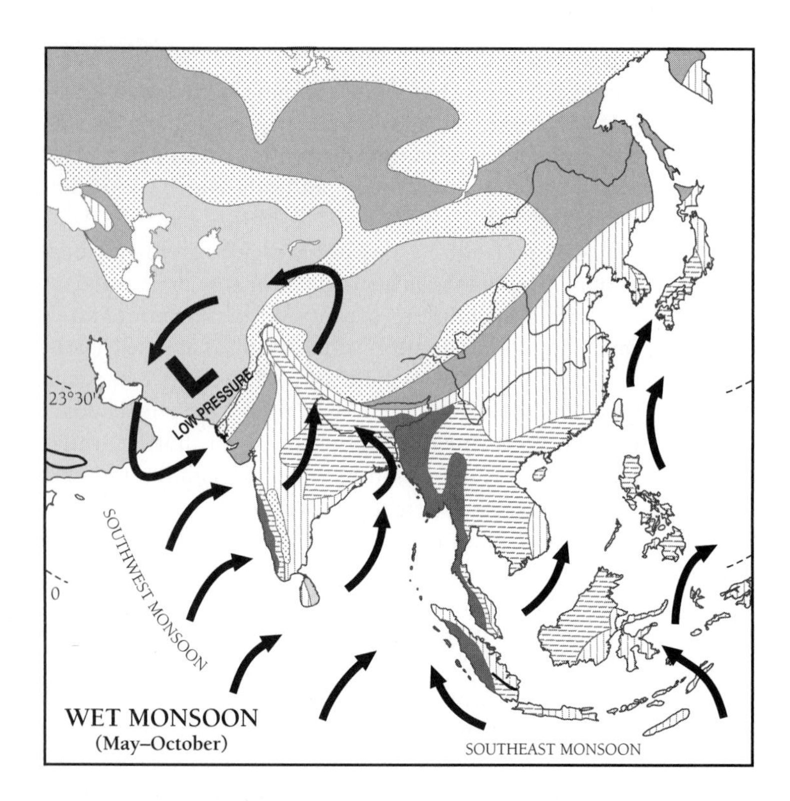

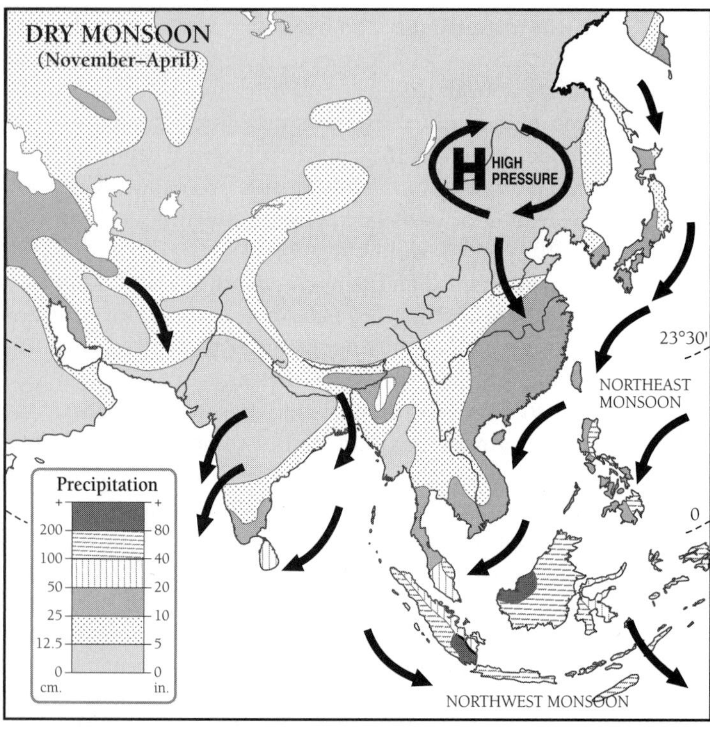

Figure 2-9

Monsoon seasons. Remember that winds blow from high to low (pressure) and are named by the direction from which they came. All of Asia is influenced by the monsoon.

on climate. In other words, if you live near a coast, winters will be somewhat warmer and summers will be somewhat cooler. If you live in the interior of a large continent you will probably be either freezing or frying depending, of course, on elevation and latitude.

Now, for a more scientific description. The interior of Asia is far removed from modifying maritime influences. Therefore, in winter, it is dominated by a high-pressure anticyclone while the central area of the Indian Ocean is dominated by the equatorial low-pressure trough known as the **Intertropical Convergence Zone (ITCZ).** Winds blow from zones of high pressure to zones of relatively lower pressure. Consequently, cold, dry winds blow from the continental interior to the periphery, dessicating (drying) the landscape in the process. Temperatures average between 60° and 68°F (16° and 20°C) at lower elevations.

Cold increases with elevation. The normal lapse rate is 3.5°F per 1,000 feet (6.4°C per 1,000 m). The impact of cold is worsened with the **wind-chill factor,** whereby body heat is lost more rapidly with stronger winds. Further, because of diminished atmospheric density at higher altitudes, the air's ability to absorb and radiate heat is reduced. Exposure to the sun's ultraviolet energy is intensified. While high elevations can produce oxygen deprivation and sunburn, exposure to extreme cold can induce **hypothermia**—frostbite with subsequent loss of body parts, or freezing to death.

Extremely cold temperatures, while found in the mountains and near the poles, can be experienced in lowland areas as well. One January, I was in a New Delhi train station. The platforms were overflowing with hundreds of homeless individuals and families huddled under blankets seeking escape from the bitter cold. A newspaper reported that 17 people had frozen to death in the streets overnight.

With desiccating winds and the coming of summer, the period of March to June becomes very hot. The winds begin to shift, and people pray and prepare for the coming of the wet monsoon. Ideally, the wet season occurs between June and September, but one sure thing about the monsoon is its unpredictability. If it is early, seeds can be washed away; if it is late, crops can wither and die. This explains the many ceremonies and festivals at this time of year to propitiate various gods and spirits associated with these essential rains.

At the time of the wet monsoon, continentality causes a low-pressure zone to develop in the interior. The ITCZ shifts northward toward the Tropic of Cancer. Solar radiation heats the landmass, leaving the surrounding waters relatively cooler. A subtropical high-pressure cell emerges over the Indian Ocean. As hot air sweeps outward across the ocean's surface, it quickly evaporates water. This huge, thunderous, sodden air mass moves into the lower-pressure ITCZ, bringing voluminous rains to the area south of the Himalayas. Precipitation is highest in coastal areas and on the windward slopes of mountains. As air masses move inland, available moisture diminishes and less precipitation is experienced.

Mountains and other areas of raised relief are highly influential as **topographic barriers.** The Himalayas, for example, provide a physical barrier to incoming air masses, channeling them westward across the Indo-Gangetic Plain. As moisture-loaded air meets this topographic barrier, it is forced to rise, cool, and condense, providing **orographic precipitation.** As a result, the windward slopes of the Himalayas receive abundant rainfall. The subsequently dry air masses flow over the crest and down the leeward slope, warming and drying the landscape in its path. Consequently, protected leeward slopes, or **rainshadow** areas, are comparatively dry.

Cherrapunji, India, situated at 4,309 feet (1,313 m) in the foothills of the Himalayas (just north of Bangladesh), is the record holder for most precipitation in a single year. Toronto, Canada, averages 30.1 inches (76 cm); Atlanta, Georgia, 50.4 inches (128 cm); and Singapore 89.3 inches (227 cm). In contrast, Cherrapunji averages 37.5 feet (1,143 cm) of precipitation a year. Such torrential rains, sweeping against the mountains from the Bay of Bengal, have caused the deaths of hundreds of thousands of people.

Tropical Cyclones

Another messenger of destruction is the tropical cyclone, termed a **typhoon** in the Pacific. Tropical cyclones typically develop between the latitudes of 8° and 15°, when sea temperatures rise above 80°F (28°C). Beginning as a weak, low-pressure cell, the typhoon funnels as a deep, circular low. Moving with prevailing wind systems, it intensifies further and delivers violent storms. Depending on the location, typhoons generally strike between May and September.

Here are some details on Asia's more recent cyclone disasters.

- **1991, Bangladesh and India:** 138.000 deaths; 2 million displaced
- **1997, Vietnam (Typhoon Tina):** 4,000 deaths—mostly fishermen; 3,000 boats destroyed
- **1999, Orissa State in Eastern India:** estimated 10,000 deaths

- **2006, China (Typhoon Saomai):** official count 458 deaths; unofficial count 1,000+ deaths; 1.7 million displaced
- **2008, Myanmar (Burma) (Cyclone Nargis):** 146,000+ deaths; thousands missing; military junta withholds information
- **2009, Philipppines (Tropical Storm Ketsana):** 500 deaths
- **2009, Taiwan (Typhoon Morakot):** 461+ deaths; 192+ missing; floods and mudslides

Regional Climatic Patterns

Although the monsoon affects the entire Asian region, smaller regions have their own climate characteristics (Figure 2-10). If you are interested in more detail regarding temperature ranges and precipitation amounts, be sure to consult your atlas. I will simplify here.

Climates in South, East, and Southeast Asia can be generalized into several categories:

- **Subarctic:** cold, dry winters and cool summers
- **Steppe:** semi-arid, cool to cold and with short grasslands
- **Hot Desert:** arid with hot summers and winters
- **Cold Desert:** arid with hot summers and cold winters
- **Highlands:** mountains and plateaus with temperatures falling with increasing elevation
- **Humid Continental:** cold winters and cool summers
- **Humid Continental:** cool winters and hot summers
- **Humid Subtropical:** cool to hot with year-round precipitation

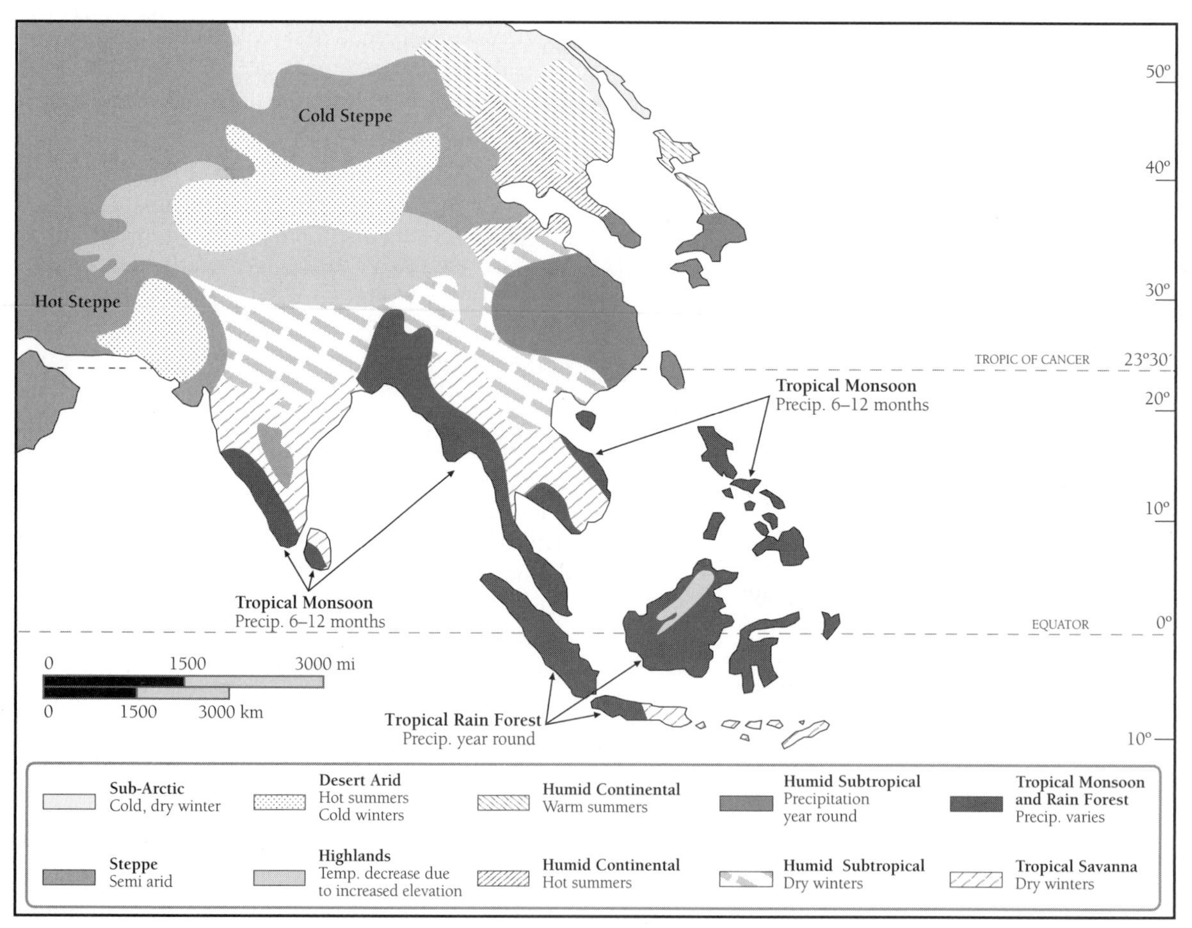

Figure 2-10

Climatic zones. Climate is a year-round phenomenon that varies with latitude, elevation, prevailing winds, and distance from moderating effects of water.

- **Humid Subtropical:** warm to hot with dry winters
- **Tropical:** variable precipitation throughout the year
- **Tropical Monsoon:** wet summers and dry winters

To understand this pattern, you can ask the following questions. How near or far from the equator is a place? Does it experience continental extremes or moderating maritime influences? What is the elevation? Is the place influenced by the presence of a topographic barrier? Answering these questions will help you understand regional similarities and differences in climate.

Global Warming and Climate Change

We have all heard about **global warming** and **climate change** as being perhaps **the** major issues of our time. However, most of us have little understanding of what is actually happening. In order to get a grasp on the processes involved, we need to investigate both the **atmosphere** and the **greenhouse effect**.

THE ATMOSPHERE

All life on Earth depends on oxygen in the atmosphere. Most of Earth's atmosphere (90%) lies within 12.4 miles (20 km) of the surface. The atmosphere provides thermal insulation, which prevents extreme changes in temperature. Unequal heating of Earth's atmosphere and surface creates long-term climate and short-term weather patterns. Variations in heating and resultant pressure differences generate winds that help drive ocean currents. The atmosphere also transfers large quantities of heat from equatorial to polar latitudes.

The atmosphere is composed primarily of nitrogen gas (N_2) (78.08%) and oxygen (O_2) (20.94%) by volume. A variety of other compounds, such as water vapor (H_2O), carbon dioxide (CO_2) and ozone (O_2) make up the remaining .98 percent. While oxygen levels appear to be stable at this time, CO_2 levels are not. The present concentration of CO_2 in the atmosphere is higher than it has been for at least 400,000 years! Since the amount of CO_2 in the atmosphere is very small, its concentration can be changed easily by the addition of more of it from a number of sources.

Scientists have concluded that humans had altered the makeup of the Earth's atmosphere even before the Industrial Revolution (1750) by clear-cutting forests in Europe, China, and the Middle East. In the process, carbon from the vegetation is converted to CO_2 and released into the atmosphere. Soils, devoid of their biotic cover, emit CO_2 at rates greater than ever before. According to the 2007 report of the International Panel on Climate Change (IPCC), the measure of **greenhouse gases** in the atmosphere has increased significantly. CO_2 levels have risen 30 percent, methane gas (CH_4) content has more than doubled, and nitrous oxide (NO_2) concentrations have increased by at least 15 percent.

CARBON DIOXIDE AND THE GREENHOUSE EFFECT

Greenhouse gases let short-wavelength radiation from the sun pass through the atmosphere but they absorb long-wave radiation emitted by the Earth's surface, thereby heating the atmosphere. Increase in greenhouse gases is caused by such factors as:

- Exponential growth of global population with increasing consumption. Until recently, high consumption rates have been assigned to industrialized nations, mostly in the northern hemisphere. Now, rapid consumption growth is taking place in China (1.34 billion people) and India (1.2 billion). If per capita commercial energy trends continue unabated, the typical Chinese will out-consume the typical American before 2040, with Indians surpassing Americans by 2080.
- Growing need for energy to support such activities as heating and cooling homes, cooking food, watching TV, and surfing the Internet.
- Increasing use of automobiles and other machines burning fossil fuels.

Fossil fuel deposits, including coal, oil, and natural gas, took tens of millions of years to form, but we are burning them up at an unprecedented rate. Fossil fuel consumption's key by-product is CO_2. Because of carbon emissions, Earth atmosphere's CO_2 levels will at least double pre-1860 amounts by 2150 if present trends continue.

The United States accounts for more than 20 percent of carbon emissions from fossil fuel burning. While emission rates continue to increase, the increments of increase are lessening as we take action to improve the environment.

Currently, however, the most significant increases in carbon emissions are taking place in Asia. China's emissions rose by 9.1 percent in 2005 and now experts recognize that it is emitting more CO_2 into the atmosphere

Table 2-1 CO_2 Emissions: Selected Country Rankings 2006 (Rankings out of 210 countries)

Rank	Country	Annual CO_2 Emissions (in 1,000s of metric tons)	% of Total Global Emissions
1.	China	6,103,493	21.5
2.	United States	5,752,289	20.2
3.	Russia	1,564,669	5.5
4.	India	1,510,351	5.3
5.	Japan	1,293,409	4.6
8.	Canada	544,680	1.9
9.	South Korea	475,247	1.7
16.	Australia	372,013	1.3
19.	Indonesia	333,483	1.2
42.	North Korea	79,111	0.3
126.	Nepal	3,241	<0.1
189.	Timor-Leste	176	<0.1

than the United States. Researchers at the University of California, Berkeley now calculate that China's emissions have increased 11 percent a year from 2004 to 2010 (Figure 2-13). Information from the Netherlands Environmental Assessment Agency Report of 2007 details the amount of CO_2 emissions from 299 countries and ranks them in terms of the percentage of their global contribution. Table 2-1 shows some of the report's country rankings in terms of carbon emissions in 2006.

METHANE GAS AND THE GREENHOUSE EFFECT

Methane (CH_4) is another greenhouse gas that is on the increase. Methane comes from such things as leaking gas pipelines, coal mines, and municipal landfills (trash). Agricultural practices including the use of fertilizers, wetlands and flooded farmlands, such as rice paddies, along with the belching of cows and other ruminants also produce methane gas. Although we hear a lot about CO_2, we don't hear much about the impact of methane gas. Pound for pound, methane in the atmosphere traps 25 times more of the sun's heat than CO_2 does.

Recent research demonstrates that global warming is causing melting of ice and permanently frozen ground in the Arctic. Permanently frozen ground is called **permafrost**. Permafrost, which averages 80 feet (24.4 m) in thickness, covers 8.8 million square miles (22,792,000 km^2) of the northern hemisphere. In summer, when the sun is in the northern hemisphere, there is a surface thaw of a few feet and thousands of lakes form (these freeze up again in winter). Permafrost is full of dead plant and animal material that has been locked in cold storage for thousands of years. When thawed, this matter emits methane gas. Because of global warming, the Arctic region is heating up twice as quickly as the rest of the globe.

Even a modest thaw of the permafrost that lies beneath these ephemeral lakes could trigger a vicious cycle: warming thaws permafrost, which creates lakes, which thaw permafrost and free more gas, which intensifies warming, thereby creating more lakes, and so on. Researchers are considering these processes as the reasons why atmospheric methane concentrations shot up in 2007 and have stayed high ever since. Other signs indicate that permafrost is melting under the Arctic Ocean floor, thereby loosening the cap on large pockets of methane gas stored deeper down.

Atmospheric concentrations are sampled on a daily basis at dozens of sites worldwide. By plugging these measurements into global climatic models, scientists have concluded that methane gas is responsible for one-third of the current warming trend.

PHYSICAL IMPACTS

Analysis of tree rings, soil cores, and historical records in the northern hemisphere indicate that temperature increases of the twentieth century were the largest of any century of the past 1,000 years. The 1990s were the warmest decade and 1998 the warmest year thus far. The 2000s are predicted to shatter previous global warming records. The global average surface temperature is projected to rise from 34.5° to 42.4° Farenheit (1.4°–5.8°C) from 1990 to 2100. Humidity and precipitation will also increase. The real and potential consequences of climate change are dire.

Climate-related disasters are often perceived as natural events, but as we now know, humans have had a heavy hand in their creation. Such disasters include those caused by heat waves or cold snaps, droughts, floods, landslides, avalanches, wildfires, hurricanes, cyclones, typhoons, tornadoes, or winter storms. Warming sea temperatures can generate stronger storms.

Another serious effect of global warming is the rise of sea levels that threaten low-lying areas, especially during storm events. Consider the fact that about 17 million Bangladeshis live less than 3.05 feet (1 m) above sea level. Flooding is a regular occurrence in the wet monsoon, but floodwater levels are getting higher and affecting more land area. Most of the islands of the Maldives lie even lower. There, the government is building sea walls and planting mangroves to quell the surging water. Moreover, of the 33 cities projected to have at least 8 million residents by 2015, some 21, such as Bangkok and Mumbai, have coastal or near-coastal locations and therefore will have to contend with the impacts of rising sea levels. Many of these cities, such as Bangkok, Mumbai, and Manila, are in Asia. IPCC predictions of rising sea levels for the twenty-first century range from a low of 7 to 15 inches (18–38 cm) to a high of 10 to 23 inches (26–59 cm).

GLOOM AND DOOM?

Does all this scary data mean that we have nothing to look forward to but devastation and disaster? Fortunately, steps are being taken by several governments to slow global warming and mitigate its effects. For instance, under what are known as **carbon offset programs**, governments agree to invest in actions and projects that will reduce or even eliminate greenhouse gas emissions. These programs might include the closing of coal-fired energy plants, harvesting energy from wind machines and solar panels, recycling trash, producing "smart" vehicles, or giving cattle garlic or other natural substances to reduce their flatulence.

Positive steps are also being taken in Asian countries. For example, Japan, Taiwan, and China are now the world's leading producers of solar panels. The Indian Green Building Council, with more than 500 representatives throughout the country, is promoting the construction of "green buildings." These structures are energy efficient, use less water, and generate less waste. However, as of 2009, only 400 had been built, mostly in Mumbai.

Malaysia is one of the most active countries in terms of developing green technology. Aware that the airline industry is responsible for at least two percent of carbon emissions, Malaysian Airlines is committed to flying more direct routes (with fewer fuel-consuming stops and starts) at more economical speeds. The government also has an intensive and wide-ranging reforestation program.

THE HUMAN FACTOR

Quality of life in various climate zones with their associated environmental shortcomings depends very much on what coping mechanisms are available. Flood control works, irrigation systems, water storage facilities, healthy work animals or functional machinery, electricity, sealed (paved) roads, insurance against crop loss, and emergency food and medical supplies are only a few.

Millions of people in Asia have minimal to no means to cope with disaster. An erratic monsoon, a drought, a tropical cyclone, or a lengthy cold spell can mean ruination and total despair for those living on the edge of survival. Here is an example. In 1998, rains and pests destroyed the crops in a small area in southeastern India. Farmers unknowingly had purchased illegally diluted fertilizers and pesticides. Deeply in debt and sensing nothing but hopelessness, hundreds of farmers ingested their remaining chemicals. They believed their deaths would bring a government stipend to their wives, thereby saving their land from seizure by the moneylenders. Government payment was sporadic and insufficient, and many women and their children are now landless and destitute.

India experienced a 23 percent shortfall of rain during the 2009 four-month monsoon season. As a result, rice, sugar cane, and groundnut (peanuts) crop yields are down. These not only are export crops but also are important in national dietary regimens. Further, reservoirs have not collected enough water for irrigation throughout the dry season.

People will not necessarily starve, because the government will distribute food from its substantial food stores. However, this will not be simple because at least half of India's 600,000 villages are not even accessible by road. The greater danger is that prices of food will rise due to lack of availability. Price rises might also be experienced by countries that import Indian food because there will be less of it.

The very poorest of the India's 235 million farmers, already malnourished, will certainly be at even greater risk for health problems and even death. In previous years of food shortages, there have been "rice riots" with resultant personal injuries and property destruction. These events harm local communities and place an even greater strain on newly required government resources.

Those of us living in a high-tech world probably find it difficult to grasp these circumstances and levels of desperation. But for millions in Asia and elsewhere, desperation is still a fact of life.

Natural Vegetation

Natural vegetation implies undisturbed, mature plant communities. Major terrestrial ecosystems are called **biomes**. Biomes are characterized by particular climates, soils, plants, and animals. Climate is the most important factor in the generation of these biotic communities. Earth biomes are usually named for their dominant vegetation. Because of thousands of years of occupation, an undisturbed biome is a rarity. However, examination of natural, model biomes helps us to understand original vegetation and reveals the measure of human impacts. The following biomes are coincident with the climate zones of the regions of Asia under discussion in this book.

TROPICAL RAINFORESTS

Tropical rain forests are most prominent in Southeast Asia, where equatorial conditions provide plentiful year-round moisture and consistently warm temperatures. Here is Earth's most diverse biome. Of 250,000 flowering plant species in the world, 170,000 are tropical. Of these, 40,000 are in Asia, with 25,000 in Malaysia and Indonesia. Western Europe, west of Russia and north of the Alps, has 50 indigenous species. Eastern North America has 171. In contrast, one study in Indonesia revealed more than 170 different plant species in a single acre (about the size of a football field). Another 123 acre research plot on the Malay Peninsula exhibited 830 indigenous species.

Think of rain forests as having layers within which every plant competes for sunlight and moisture. Each layer has a different microclimate, with the upper layer having the most light, heat, and wind. The forest's biomass is in the **canopy**, high above the forest floor. Leaves appear on tree trunks at approximately 15 feet (5 m) above ground, and this begins the canopy that reaches upward to 100 feet (30 m) or more. Poking out of the canopy are the tallest trees, called **emergents. Dipterocarps**, with their two-winged seeds, are said to be the "apex of evolution." Of grand stature, they dominate the upper canopy and house a wealth of flora and fauna.

The middle level of the canopy is densest (Figure 2-11). It has a spectacular array of mammals, birds, and other species. **Lianas** (woody climbers and vines) loop around and between trees. These are the bane of foresters because they can so entangle a tree that even when the tree is completely cut down, it remains dangling, tied to adjacent trees. Some lianas are 8 inches (20 cm) in diameter. Orchids, ferns, and other **epiphytes** inhabit this zone as well. Epiphytes are physically, but not nutritionally, supported by other plant life.

Figure 2-11

Tropical rain forest. Where light penetrates, vegetation can be very dense. Note the lianas. Dense forest like this is often called "jungle." Photograph courtesy of B. A. Weightman.

Beneath this exceptionally rich environment lies a more humid, darker understory and the forest floor. Ferns, bamboo, shrubs, and herbs shade a relatively open floor that is strewn with decomposing plant and animal litter and fungi. This rather thin layer provides some nutrients to vegetation. However, most tropical soils are deceivingly infertile, leached of their nutrients by copious rainfall and high temperatures. At lower layers, aluminum and iron oxides form rock-like layers called **laterite.** Since antiquity, laterite has been cut and used as building blocks for temples and other structures. Even giant trees have shallow roots to tap the litter layer for food. Some of these have supportive bases called **buttresses** (Figure 2-12).

Figure 2-12

Here is a dramatic example of a buttressed tree in the tropical rain forest. Since soil fertility is at the surface, trees develop sprawling surface rather than tap-root systems. Some large species of trees develop blade-like supports like these. Photograph courtesy of B. A. Weightman.

Where light penetrates, dense growth can be found at floor level. Dense vegetation, or jungle, exists along rivers or in areas that have been cleared and abandoned. These seemingly impenetrable areas belie the more open rain forest in their vicinity.

Humans have decimated Earth's rain forests by clearing vast areas for settlement and such activities as cattle rearing and plantation agriculture (e.g., palm oil). Millions of acres have been removed for timber and firewood. Worldwide, 65,000 square miles (169,000 km^2) of rain forest, an area nearly the size of Wisconsin, is lost. Logging, plantation agriculture, and their impacts will be discussed further in the regional chapters.

TROPICAL MONSOON FORESTS

Found primarily in mainland Southeast Asia and the Indian subcontinent, the monsoon forest is a transitional biome between tropical rain forests and grasslands. Unlike tropical rain forests, monsoon forests must adapt to distinct wet and dry seasons. Many trees are semi-deciduous; they lose part of their foliage in the dry months (Figure 2-13).

Tropical monsoon forests lack a continuous canopy. Their trees average only 50 feet (15 m) in height. Often cut for firewood, they make poor lumber for construction. Teak, a highly valued hardwood for carpentry, is an exception. Trees also produce waxes, gums, resins, and lacquer. Monsoon forests possess a rich collection of wildlife including tigers, sloth bears, and the one-horned rhinocerous.

Do your shoes have rubber soles? Do you use a palm oil-based soap? Have you used lacquer or latex paint? Have you used bamboo skewers or wooden chopsticks? Is there a coconut fiber mat at your front door? Bought any bamboo products lately? How much of the tropical forest are you using?

Riches of the Tropical Forests

Tropical forests are rife with desirable products such as fruit, tubers, plants, and other food crops; rattans and fibers; oils and resins, in addition to wood. Traditional ways of gathering forest products have little impact compared to modern logging activity. Logging is a prime reason why more than 40 percent of Asia's

Figure 2-13

This monsoon forest lines the Rapti River in southern Nepal. It is winter, dry season, and many of the trees have lost all or at least some of their leaves to compensate for the water shortage. Most of these trees are semi-deciduous. The Rapti is part of the Ganges River system and consequently is a sacred river. Photograph courtesy of B. A. Weightman.

tropical forests have been destroyed. Foresters seek certain types of trees to cut for local use and export. I will describe two of these here: **teak** (*tectona grandis*) and **meranti** (*shorea spp*).

Teak is regarded as a fine, medium-weight hardwood. It is noted for its strength, stability, and durability. Termite resistant, it is valued for constructing housing and furniture. It used to be in great demand for ships and is still used in the construction of yachts. Teak is native to India, Vietnam, Burma, and Thailand, but is grown commercially on Java and in Sri Lanka. Logging for export has devastated teak forests in Myanmar and teak forests have already been eliminated from Thailand.

There are several species of *meranti*, an equatorial forest wood that is called *seraya* in Malaysia. These are giant, buttressed trees achieving heights of over 200 feet (61 m). Meranti are among the most important export woods of Asia. Logs are sent to Australia, Europe, and Japan for use in construction and in the manufacture of plywood.

Rattans and fibers are also in demand. Used for furniture, matting, and basketry, these are collected and assembled for local use and export. Some of the commercial rattans are climbing palms that wind their way around the forest for lengths of 250 feet (76 m) and more. Rattans have been known to grow to 600 feet (183 m). The main suppliers are the Malay Peninsula, Indonesia, and China. Kapok, a waterproof fiber harvested from ripe tree pods, and coconut fiber called coir, serve multiple uses.

Bamboo is among the oldest materials used by humanity. Seeds and shoots are edible, as are the worms that live inside (Figure 2-14). Water pipes, containers, spears, house frames, and construction scaffolding are only some of its uses. Bamboo is a grass related to corn and wheat. There are at least 1,400 species of the grass, and it grows around the world in most environments. However, commercial species are tropical, and 60 percent of these are in Asia.

A woody perennial, it can grow to 60 feet (18 m). Some fast-growing varieties can grow a foot in one day. *Moso* bamboo, commonly cultivated in China, is the fastest growing plant on Earth. This variety can shoot up as much as three feet in a single day and reaches heights of 75 feet (23 m).

While bamboo has long had a variety of uses in Asia, today, industrialized countries have found many new uses for it. Bamboo flooring is growing in popularity and it is even being used for basketball courts. Kitchen cabinets, furniture, elegant boxes, computer cases, skateboards, and clothing among other products are being made from bamboo.

Bamboo cultivation has boomed in Asia over the past two decades, especially in China. Chinese bamboo generally grows on small, family-owned plots. Some of these plots have been farmed for centuries. For a number of years, Chinese farmers cleared swaths of natural forest in order to grow their ever-more profitable bamboo. Fortunately,

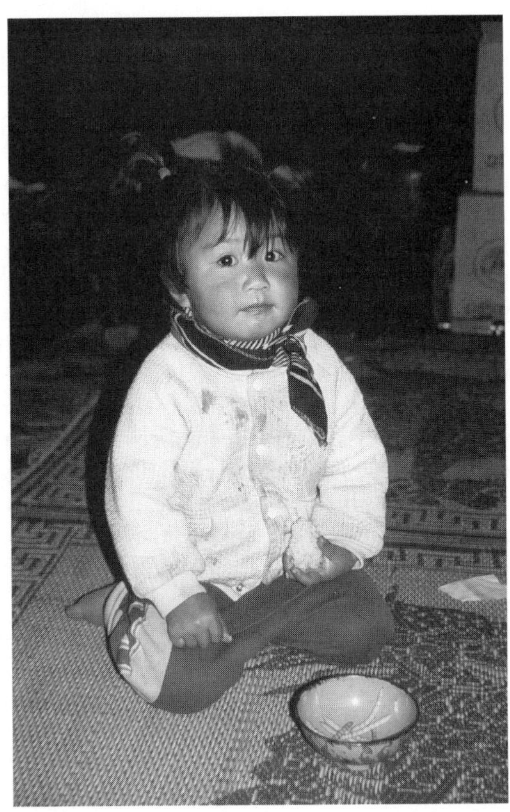

Figure 2-14

I took this photo in a small village in northern Laos where I was offered live bamboo worms to eat. Of course, to us they are quite revolting, but this little girl was eating them like candy.
Photograph courtesy of B. A. Weightman.

China changed its forest policy in the late 1990s and this practice has become less common.

Proponents of the bamboo industry claim that the plants capture CO_2 and require very little fertilizer or pesticides and virtually no irrigation water. Further, heavy machinery is not needed to cut it, and its root systems help to prevent soil erosion. Consequently, it is marketed as a "green" product. However, since most consumers live thousands of miles away from the producers, transshipment of bamboo and its products does leave a sizeable carbon footprint. Also, large quantities of water and noxious chemicals are used in the processing of bamboo into cloth and other materials.

Tropical forests are also rich in essential oils. Vietnam, for instance, produces around 400 of these, including anise (black licorice or anisette). In India, Sri Lanka, and Timor-Leste, aromatic sandalwood has a recorded history of 4,000 years and is one of the oldest known perfumery ingredients. Sandalwood oil is commercially distilled from trees at least 30 years old. Extracted from plant leaves, patchouli oil was used to scent the carpets and rugs of India. Now cultivated commercially, especially in Malaysia, it is a key ingredient in French perfumes.

Resins, known as *copals* and *dammars* (the Malay word for resin), are exudates; they ooze out of trees as a result of injury. These can be collected in soft or hard form. Old, fossilized, hard lumps, buried beneath the ground for ages, are highly prized. Copals are used in paints and resins. Softer than copal and soluble, dammars are used for such things as varnishes and wallpaper finishes. Gum resins and gum turpentine are drawn from pine forests in the Himalayan foothills and elsewhere.

Forest products are primary sources of pharmaceuticals. Strychnine is but one example. Indigenous to the monsoon forests of western, coastal India, it is still widely employed there and in Southeast Asia as a stimulant, tonic, or emetic. Strychnine is also used for spear and arrow hunting and to poison vermin. You have probably read about it in mystery novels as a means of murder. Prepared from the seed of the snakewood tree, large doses of strychnine are definitely deadly.

MANGROVES

Mangroves are littoral (coastal) plant formations of tropical and subtropical regions where trees and shrubs have adapted to inundated and saline conditions (Figure 2-15). They grow in intertidal zones between land and sea. At least 50 million years old, they originated in Malaysia and diffused around the globe. A total of 69 mangrove species have been documented worldwide, with the highest diversity occurring in Southeast Asia. Mangroves are capable of rapid growth. A yearly growth of 540 feet (164 m) has been recorded in Java. Palembang, Sumatra, a coastal port in the thirteenth century, is now 31 miles (50 km) inland.

The world's largest remaining mangrove concentration is the **Sunderbans** of Bangladesh in the Bay of Bengal, and the Bengal tiger makes its home in this forested area. Mangroves comprise habitats for hundreds of living organisms from mammals, birds, crustaceans, and fish, to algae, fungi, and bacteria. Four hundred types of fish can be found in the Sunderbans alone. This entire community is called **mangal**.

Figure 2-15

Mangroves line the banks of the Mekong river delta in Vietnam. Photograph courtesy of B. A. Weightman.

Mangal environments are significant for several reasons. Capable of generating over three tons of organic matter per acre per year, mangroves build bottom sediment with their multiple roots. Animals such as water buffalo graze on them. They help to maintain water quality in coastal zones by trapping sediments, organic materials and nutrients—an activity that can help the functioning of nearby coral reefs. They also provide nutrients for the myriad fish and other species that inhabit their waters.

Mangrove realms are vital for the livelihood of millions. Their wood is used in construction and provides a rich charcoal. Sugar from the *nypah* palm can be turned into alcohol and machinery fuel. Prawns, crabs, fish, and other delicacies that live among mangrove roots are harvested for food. An estimated 60 percent of India's commercially important coastal fish are directly associated with mangrove habitats. The vegetation stabilizes shorelines and decreases coastal erosion.

Large-scale mangrove destruction is a relatively recent phenomenon. Unfortunately, mangroves are disappearing rapidly with increased population, new settlements, urban expansion, industrialization, commercial fishing and fish-farming (e.g., shrimp), and tourism. In fact, half of the world's mangroves are gone. Mangroves are often used as waste dumps and ruined by factory effluents and oil spills. International companies cut mangroves for pulp and particleboard. They are cleared for the development of tourist complexes. Grandiose damming and water-diversion schemes alter the fresh–salt water balance essential to the survival of both mangroves and the creatures within. Massive losses of mangrove forests have also resulted in the release of large quantities of stored carbon, contributing to human-induced global warming and climate change.

In parts of Bangladesh, Sri Lanka, Thailand, Vietnam, and other countries where dense mangroves still exist, the impacts of the 2004 tsunami were significantly mitigated. Since governments have become aware of the role of mangroves in saving lives and property, they are supporting project initiatives to replant them along their coastlines. However, it has been a learning process.

Several projects have been controlled by international organizations such as the World Bank. Despite heavy funding, these operations have been largely unsuccessful. For example, large-scale replanting schemes in the Philippines have had a survival rate of only 10 to 30 percent. The mangroves were installed in relatively unpopulated areas to avoid property ownership issues. Consequently, there were no people to maintain them. More significantly, lack of eco-knowledge often resulted in the wrong species being planted—species that could not withstand severe wave action, for example. Moreover, these were "top-down" efforts; the wishes and knowledge of local people were given little, if any, consideration. Now, mangrove-replanting projects in the Philippines are low-budget and locally led. The current survival rate is 97 percent. National governments are coming to realize that when communities have shared interests in actions that will affect their lives, success is more likely.

A Cambodian mangrove project, supported by the International Development Research Center of Canada, is based on the inputs of local villagers and fishermen. Incentives such as "rice for labor" are offered for adults, and educational materials such as books, pencils, and rulers are given to teachers and students in exchange for their assistance.

TROPICAL SAVANNA

Tropical savannas are transitional zones between monsoon forests and even drier regimes. They are often described as *tropical wet and dry regimes* because they experience distinct wet and dry seasons. Tropical savannas have an average yearly rainfall of 30 to 50 inches (76–127 cm), most of which falls in summer. A mere 2 to 4 inches (5.08–19.16 cm) falls in the four, dry months of winter. Savanna lands do not get particularly cold or hot. Monthly temperatures average around 64°F (18°C).

Vegetation is typically **xerophytic**, or drought-resistant. Savannas reveal trees, shrubs, and grasses in various combinations depending on moisture and temperature. When it is very dry, grasses become brittle and shrivel up, and trees drop their leaves. In some places, there are intermittent streams that only appear on the surface in the wet season. As these evaporate in the dry months, fish burrow into the remaining mud to survive the drought.

Spontaneous or set fires function to maintain the savanna. Early in the dry season (winter), "cool" fires stimulate new growth. Late-occurring "hot" fires rage out of control, killing trees, seeds, and every living thing. However, in the fire aftermath, soils are more fertile and better drained, and savanna lands are widely used for grazing and crops such as grains and groundnuts (peanuts). However, crops rarely survive without irrigation in dry periods. Consequently, governments such as those of India, Pakistan, and Thailand, are developing large-scale irrigation schemes in their savanna regions.

DESERTS

Xerophytic and often thorny shrubs and succulents are found in deserts, which are defined as regions receiving less than 10 inches (254 mm) of precipitation a year. Some plants bloom only sporadically, waiting perhaps years for rain. These are called **ephemerals.** Great swathes of sandy or stony desert are devoid of vegetation.

Cold Deserts

Montane deserts occur at very high elevations (above 10,000 feet; 3,000 m) and are exceedingly dry. These can be found on the Tibetan Plateau and in the region north of the Himalayas. (Refer to Figures 2-2 and 2-3). These areas are deep in the continental interior and behind formidable mountains, out of reach of most moisture. They are normally cold all year and often experience snow in winter. The Tibetan Plateau is the largest and highest plateau in the world and covers an area three times the size of Texas (950,000 mi^2; 2.5 million km^2).

The Gobi, which means "waterless plain" in Mongolian, is situated in Mongolia and northern China. Covering an area of 500,000 square miles (1,300,000 km^2), it is one of the largest deserts on Earth and is the least populated region outside of the polar ice caps. Winters can be extremely cold and summers unbearably hot. Because of sparse vegetation and openness, heat loss is rapid and nights are relatively cooler. When I spent a couple of weeks camping in Mongolia, I wore a T-shirt during the day and a heavy, wool jacket at night.

Wild, two-humped, Bactrian camels, that can survive on the limited supplies of salt water in the sesert, are critically endangered. Only about 950 remain in the Gobi. Also, the desert is encroaching on surrounding grasslands. Overgrazing and the increasing use of trucks are destroying the delicate grasses, which are soon subsumed by blowing sand.

The **Taklimakan** covers most of the **Tarim Basin** in northwestern China. With an area of 100,000 square miles (270,000 km^2), it is China's largest and most arid desert. It is also one of the world's biggest shifting-sand deserts. Sand dunes cam rise as high as 1,000 feet (300 m). Sandy areas are called ***shamo***. There are also vast, windswept expanses of nothing but stones and pebbles (Figure 2-16). **Diurnal temperatures** (day to night) can vary by as much as 300s. When I travelled across the Taklimakan one summer, in a very rickety car with an inebriated driver and no air-conditioning, the daytime temperatures were above 120°F (49°C). Winter temperatures can dive below minus 15°F (–26°C). Local wisdom describes this desert as "the one you go into but never come out!"

Hot Desert

The **Great Indian** or **Thar Desert** of northwestern India and southeastern Pakistan is the size of Louisiana and North Carolina combined—that is, 92,000 square miles (238,700 km^2). Most of the Thar is in the Indian State of Rajasthan. It was here that India detonated its first nuclear device in 1974. This arid environment gets less than 1 inch (2.5 cm) of rain a year. In a particularly wet year, it might get almost 2 inches (5 cm). Most (90%) of the precipitation arrives from July to September from the wet Monsoon, which delivers torrential rains elsewhere. However, rainfall here is erratic and unpredictable. Average annual temperatures range from 75° to 79° Farenheit (24°–26°C) but it can get much cooler in the winter.

Figure 2-16

Large expanses of the Taklimakan desert is made up of sand dunes (shamo) *and stony areas* (gebi).
Photograph courtesy of B. A. Weightman.

The vegetation is xerophytic with grasses, shrubs, and some trees. Plants have wide-spreading, surficial roots in order to suck up any available moisture. Leaves only appear after rain. The desert is inhabited by a variety of wildlife such as the desert fox and the short-tailed, tufted-ear caracal—a wild cat. These animals eat birds and rodents.

The Great Indian Desert is subject to severe soil erosion from strong wind action. Shifting dunes bury roads and rail lines. In order to achieve some measure of control over wind erosion, trees are being planted as windbreaks. Most of the trees, such as eucalypts and acacias, are imported from other countries such as Israel, Chile, and Australia. Overgrazing of sparse grasses, and low-yield agriculture add to the problem. In fact, 33 percent of crops fail because of wind, sand-blasting, and lack of water.

STEPPE

In the plateau regions beyond the Tibetan Plateau, such as China's Nei Mongol (Inner Mongolia), there are vast areas covered with short grasses (Figure 2-17). Referred to as **steppes**, these grasslands are the traditional home of Mongol nomads who move their animal herds from one pasture site to another in seasonal cycles. Steppes are China's most important grazing lands, but as increasing numbers of ethnic Chinese are moving into the region, they are being used increasingly for grain cultivation.

MID-LATITUDE FORESTS

Many parts of China, Japan, and Korea possess broadleaf, **deciduous forests** characteristic of middle-latitude

Figure 2-17

Here in Inner Mongolia, China, cattle graze on short grass steppe. Note the temporary holding pen in the background. Pens are often made of stones that can be found on the steppe. After a few days the cattle will be moved to a fresh grazing area.
Photograph courtesy of B. A. Weightman.

Figure 2-18

High in the Indian Himalayas—15,000 feet (4,500 m)—a horse grazes on vegetation far above the tree line. Rich in grasses and flowers in summer, this terrain is covered with ice and snow in the winter months. Photograph courtesy of B. A. Weightman.

locations. Deciduous trees drop their leaves in the cold season, using winter dormancy to survive until the warm days of spring. Oak, birch, and elm are three of the more familiar varieties. Needle trees such as pine, fir, and spruce exist in cooler northern regions and at higher elevations. There are nearly 2,000 woody plants in China's forests alone. These forests are important for both wood and non-wood products. Ginkgo, chestnuts, walnuts, tallow oil, tung oil, and birch juice are some of these. The vast majority of mid-latitude forests have been eliminated or damaged for the benefit of large human populations in these parts of the world.

MOUNTAIN VEGETATION

Mountain plant communities vary vertically because of climatic change with elevation. This is known as **altitudinal zonation** of vegetation. Everything from tropical to mixed deciduous forests, to needle-leaf forests and grasslands exist with increases in altitude. Forested areas give way to alpine meadows, stunted trees, and the **tree line**, where neither growing season nor soil depth will support tree life. Here the sparse, stunted trees give way to the low-lying flora, mosses, and lichens of subarctic **tundra** (Figure 2-18). Even higher are barren rock and ice. As with other biomes, population pressure is having deleterious impacts on mountain environments, especially in the Himalayas.

Wealth of Wildlife Lost

The existence of two broad zoogeographical divisions—Gondwanan and Laurasian—can be explained by plate tectonics. About 15 million years ago, the continental plate of Antarctica, Australia, and New Guinea split apart, leaving Antarctica behind. The Australia–New Guinea section collided with southeastern Laurasia, thereby creating the Malaysian and Indonesian archipelagoes. The collision took place just west of Sulawesi. This undersea fracture marks the division between Laurasian and Gondwanan faunal types. As mentioned earlier in this chapter, India, once part of Gondwanaland (Africa), crashed into Laurasia as well. In the process, India served as a raft carrying Gondwanan fauna to Laurasia.

Western and eastern **Malesia** (Malaysia and Indonesia) have strikingly different animal communities. Plant differences are not as distinct because plants more readily disperse across water. A combination of Gondwanan and Laurasian fauna inhabits the western region, while Laurasian live in the east. Here, animals such as the tree kangaroo and pygmy glider have similarities with Australian species.

The British naturalist Alfred Russel Wallace explored both South American and Southeast Asian rain forests. During his stint in Asia (1854–1862), Wallace discovered these two distinct zoogeographic regimes. He described their fauna as having "endless eccentricities of form and extreme richness of colour." The line of demarcation between the two zones is known as the **Wallace Line**.

Thousands of wildlife species live in Asia, many not yet identified. Many are **endemic**—they occur in a given area only. Having a restricted range, these are easily wiped out. Indonesia, the Philippines, China, India, and Japan have the most endemic species of flora and fauna, and these are seriously threatened.

Since international focus is usually on tropical regions, other climes are often ignored or forgotten. Think about this: Half of Japan's native mammals are in danger of being eliminated across the archipelago. Two kinds of wolf and three types of bat are now extinct. Another 22 species, including the Japanese river otter, have not been seen in the wild for a decade. More than 2,000 species of terrestrial vertebrates live in China. Many are at risk. Endemic species such as the lesser and giant panda, the golden monkey, and the Yangzi alligator are seriously endangered. Przewalski's horse, the Earth's last truly wild horse, has not been seen in the wild since 1966.

The diverse environments of South Asia provide habitats for a plethora of animals, birds, and other living things. However, Asian elephants, one-horned rhinoceros, sloth bear, and other species have sharply reduced populations (Figure 2-19). This is attributable to human pressure for land, modification or destruction of habitats, and poaching. Asian lions, once widespread and avidly sought by hunters, are now confined to India's Gir National Park in the western state of Gujarat.

Some Indian wildlife have adapted to living in human settlements. Bats, birds, mongooses, and monkeys subsist on the refuse produced by the inhabitants of cities, towns, and villages. The common langur and rhesus macaque are the most prominent, especially around temples. Here they can feed off food offerings and neither Hindus nor Buddhists will harm them.

Figure 2-19

I spotted this elephant getting a drink of water in Kerala state in southern India. The elephant had just completed a day's work hauling logs in the forests of the Western Ghats. Elephants, as work animals, are rapidly being replaced by vehicles that are responsible for much greater environmental damage and pollution.
Photograph courtesy of B. A. Weightman.

Indonesia is the most mammal-rich, with close to 700 species of higher vertebrates, but it has more threatened species of mammals and birds than any other country. Nevertheless, in its forests can be found wildlife inhabiting every layer: from proboscis monkeys (Figure 2-20), gibbons, orangutans, flying foxes (bats), gliding snakes, and frogs in the canopy; to tree kangaroos and other arboreals of the understory; to clouded leopards, tigers, and other cats of the lower understory and floor; to tapirs, rhinos, and wild pigs, foraging on the floor. Birds and insects are in countless varieties throughout. Unfortunately, many of these species are disappearing as forests are cut down. Increasing numbers of individuals and groups are exploiting forest habitats for commercial gain (Figure 2-20). Logging and plantation agriculture (e.g., palm oil) are the two major contributors to forest destruction and resultant loss of wildlife habitats.

Table 2-2 shows that the numbers and percentages of endangered and vulnerable species in Asia continue to rise. The International Union for the Conservation of Nature (IUCN) is the world's most comprehensive inventory of the conservation status of plant and animals. A threatened species is likely to become vulnerable or endangered if actions are not taken. A species that is vulnerable is likely to become endangered, and if it is endangered, it is likely to become extinct. Some species have reached the critical stage where they are almost extinct. Note that this list does not include threatened birds and fish of which there are many hundreds.

Figure 2-20

I discovered this poor creature in a small park/zoo in Banjarmasin, Kalimantan, Indonesia. It is an illegally-caged proboscis monkey, known for its large nose. Given the rapid rate of deforestation in this region, the proboscis monkey is a threatened species. In another tiny cage was a baby orangutan that never stopped crying. Rescue efforts cannot keep up with the illegal poaching of wild animals. Photograph courtesy of B. A. Weightman.

Table 2-2 2004 IUCN Red List of Threatened, Endangered, and Vulnerable Animal Species in East, South, and Southeast Asia (If you don't recognize some of these animals, look them up on the Internet, where you will find a description and often a photograph)

Country	Number of Threatened Species on Red List	Percent Threat-ened 1996	Percent Threat-ened 2000	Percent Endan-gered 2000	Examples	Percent Vulner-able 2000	Examples
Bangladesh	109	12	24	10	Asian Elephant Tiger Ganges dolphin	18	Black bear Clouded leopard Eurasian otter
Bhutan	99	20	23	10	*Pygmy dog Snow leopard Red panda Golden monkey	15	Assamese macaque Sloth bear Sikkim rat
Brunei	157	6	8	5	Orangutan Banteng Otter civet	8	Proboscis monkey Marbled cat Dugong
Burma (Myanmar)	251	12	15	13	*Javan rhinoceros Elephant Blue whale	27	Malayan porcupine Eld's deer Sperm whale
Cambodia	117	18	22	9	Tiger Asian elephant *Sumatran rhinoceros	20	Golden cat Malay tapir Pygmy loris
China	394	?	20	34	*Gansu shrew Giant panda *Przewalski's horse *Wild Bactrian camel	48	Wild yak Siberian musk deer Chinese mountain cat Stump-tail macaque

(continued)

Table 2-2 (*continued*)

Country	Number of Threatened Species on Red List	Percent Threatened 1996	Percent Threatened 2000	Percent Endangered 2000	Examples	Percent Vulnerable 2000	Examples
India	316	24	28	37	Pygmy hog Spotted civet Malabar langur Asian elephant	56	Black buck Four-horned antelope Gaur Kashmir cave bat
Indonesia	316	25	32	52	*Javan rhinoceros *Orangutan *Silvery gibbon *Alpine wooly rat *Tiger	103	Javan langur Bay cat Babirusa Tree kangaroo Dugong
Japan	132	22	30	24	*Bonin fruit bat Amami rabbit Stellar's sea lion Ryukyu flying fox	16	Asian black bear Harbor porpoise Northern fur seal White whale
Laos	172	17	19	12	*Javan rhinoceros Asian elephant Black gibbon *Tiger	25	Fishing cat Smooth-coated otter Clouded leopard
Malaysia	286	15	18	21	Tiger Asian elephant Borneo orangutan Proboscis monkey	33	
Mongolia	134	9	9	3	Snow leopard *Wild Bactrian camel Long-eared jerboas	10	Asian wild as Siberian musk deer Wolverine
Nepal	167	17	18	11	*Pygmy hog Tibetan antelope Wild water buffalo One-horned rhinoceros	22	Sloth bear Gaur Himalayan taur Marbled cat
North Korea	?	?	?	6	Northern right whale Blue whale Tiger	9	Eurasian otter Siberian musk deer Harbor porpoise Sperm whale

(*continued*)

Table 2-2 *(continued)*

Country	Number of Threatened Species on Red List	Percent Threatened 1996	Percent Threatened 2000	Percent Endangered 2000	Examples	Percent Vulnerable 2000	Examples
Pakistan	151	9	14	7	Snow leopard Wooly flying squirrel Markhar	14	Black buck Moufflon (Urial) Wild goat Fishing cat
Philippines	153	32	33	20	*Visayan wart pig *Cloud rat Calamian deer Visayan spotted deer	30	Samar squirrel Palawan fruit bat Flying lemur Phillippines warty pig
South Korea	49	12	29	5	Blue whale Fin whale Sei whale Northern right whale	9	Long-tailed goral Dhole Asian black bear Sperm whale
Sri Lanka	88	16	24	5	*Nillu rat Asian elephant Jungle shrew Blue whale	9	Rusty spotted cat Slow loris Toque macaque Sloth bear
Taiwan	63	16	22	—		14	Eurasian otter Taiwan vole Harbor porpoise

* critical

THE ANIMAL TRADE

Unfortunately, too many people hunt and trade in exotic live beasts, their skins, horns, feathers, claws, and other body parts. According to the Wildlife Conservation Society's (WCS) 2006 report, illegal trade in wildlife is a $6 billion a year industry.

Live Animals

Nearly 99,000 primates were imported into the United States between 1995 and 2002. The favorite acquisitions are various species of monkeys and baby orangutans. Orangutans have even been carried into the country in hand luggage! Typically the animal's parents are poisoned or shot and the baby is captured for shipment. Americans also import birds such as parrots and cockatoos, large snakes, turtles, and other exotic creatures. The WCS believes that the death rate for transported birds is as high as 80 percent.

The World Wildlife Federation describes the world's largest illicit animal market in Jakarta, Indonesia. At the Pramuka Market, one can purchase, orangutans, gibbons, eagles, parrots, and even a Komodo dragon (lizard) among hundreds of other species. Many of these beasts are taken or sent by air to Singapore, Malaysia, Thailand, and the Middle East.

India's Wildlife Crime Database, maintained by the Wildlife Protection Society of India, established some estimates of illegally traded animals. These include 846 tigers, 3,140 leopards, and 585 otters (skins) that were poached between 1994 and 2008. From 2000 to 2008, 320 elephants were killed.

Animal Byproducts

A bigger part of the animal trade is in animal parts, especially among Asian countries. Many animal parts are used in traditional medicine but others are collected for souvenirs or to wear. Ten tons of tiger bones (from 500 to 1,000 tigers) were sold between 1995 and 2002. Bear bladders and tiger penises are very much in demand as well. In India such unusual items as mongoose hair, tiger whiskers, and bear bile are garnered for the international market for traditional medicine. Stuffed snakes, inscribed turtle shells, and carved elephant tusks are also very popular around the world.

Every year, at least $300 million worth of wildlife products enter the United States. Elephant skin boots; crocodile shoes and belts (some of this trade is legal); matted butterflies (stuck by pins into a display box); snake skins; and even bear paws are only a sampling of illegally imported items.

In my travels throughout Asia, I have visited (reluctantly) a number of markets selling illegally acquired animals and their by-products. I saw a bear for sale in Inner Mongolia (China) that was in a cage so small that it could barely turn around. In one small town, there were wolf skins hanging on a board in an alley and a woman wanted to know if I wanted to buy a custom-made pair of wolf-skin boots. In a shop in Vietnam, I spied bottles of "cobra tonic" for virility. In each large, green, beer-type bottle was a whole cobra floating in liquid.

In Guangzhou, China, I went to a market where there were otters, spotted cats, and other animals crammed into tiny cages, without food or water, waiting to be sold, I presume for food. I also went to a "Snake Restaurant" where they not only cooked snakes but also kept them for their gall bladders. A man took the snake out of its cage and put one foot on its tail and the other on its neck. Then, he grabbed the middle of the snake and, with what looked like a scalpel, slit it open and took out its little gall bladder. The organ was then put into a jar with some kind of preservative in it and placed on a shelf in the window for the public to see. (No, I didn't eat snake.) Clearly, these are images that have stuck in my mind.

Future Prospects

Prospects for the protection of wildlife are bleak. Although there are international agreements, and national laws against the illegal trade of animals and their parts, there is too much money at stake, especially in poorer countries. The Indian mafia is integrally involved in this sad, illegal activity. However, it is not only India; crime syndicates worldwide are making fortunes in the "industry." The worst of the action is in Southeast Asia. Confiscation, fines, and arrests are limited. Borders are porous and many poorly paid officials and police are all too ready to take bribes and look the other way.

Cambodia has a total of only 31 trained officials to monitor and make arrests of those who are poaching such animals as tigers, pythons, and pangolins. Cambodian wildlife is increasingly in demand as "stocks run dry in Vietnam." The wildlife trade in Vietnam was an estimated $66.5 million in 2002. Now, Vietnamese traders go to the largely unguarded border and contract with Cambodians to get the animals for them.

Those individuals and organizations that are working to save and protect wildlife are pessimistic. There is too much money to be made by unscrupulous people. A point to note: After the drug and arms trades, wildlife poaching is considered to be the most lucrative crime on the planet!

The removal of wildlife negatively impacts **biodiversity**. Biodiversity refers to the range of variation of living organisms and is usually defined in terms of genes, species, and ecosystems. It is generally accepted that we should strive to preserve biodiversity for several reasons: present and potential uses of biological resources especially in medicine; to maintain a biosphere supportive of healthy human life; and for its own sake, based on a concern for the well-being of all living things.

TRADITIONAL MEDICINE

Recent findings reveal that tropical medicinal plants are more commonly found in "disturbed habitats"—where hunting and gathering or slash-and-burn agriculture is practiced, and that the virgin forest is quite devoid of them. However, the knowledge of these plants and their uses is dying as traditional populations modernize and become more urbanized.

Does this mean that we should save the people and disturbed habitats and not worry about pristine rain forest? Remember that there are unknown, potential uses for rain forest flora and that this forest, with its own complex of ecosystems, is integral to the larger biosphere in which we live.

Some environmentalists speak of **environmental criticality** when referring to environmental degradation and associated socioeconomic deterioration. Removal of forest, disturbed or not, is removal of habitat for all living creatures. Ways of life are destroyed and associated knowledge is lost. Thoughtless development reduces biodiversity of flora, fauna, and humans. Quality of life for all is surely related to the health of habitat.

Do the medicines you use derive from tropical plants or animals?

Plants, Animals, and Medicine

The United States imports more than $20 million worth of rain forest plants a year. In fact, 40 percent of medications sold in North America derive from plants. These are either harvested from forests by local people or are commercially grown. Cinchona trees, originally from South America, are grown in India to produce the anti-malarial drug quinine. Malaria is carried by mosquitos in tropical and subtropical regions. But new, quinine-resistant mosquitos have evolved, prompting researchers to investigate anti-malarial potential of new plants.

Both modern and traditional medicines impact biotic communities. In India, for example, 2,500 plants are either used by traditional healers or exported. Nepal exports 90 percent of its 700 medicinal plants and herbs. These countries, along with Sri Lanka, are noted for traditional *Ayurvedic* medical practice. Grounded in Hinduism, Ayurvedic healing relies on natural products.

Traditional medicine from India, China, and elsewhere entails an enormous *pharmacopia*, some of which has proved invaluable to modern, medical practitioners (Figure 2-21). For example, reserpine, the source of modern tranquilizers, is extracted from the serpentine root shrub. The National Cancer Institute and other organizations, intent on finding cures for cancer and HIV-related diseases, are actively collecting tropical leaves, twigs, roots, and rhizomes from both forests and traditional healers. Thousands of samples have been collected, yet these account for only seven to eight percent of the thousands of flowering plants in Asia. Many of these are vanishing in the wake of development.

Also useful in medicine are animal products. Snake venom derivatives can relieve phlebitis. They also can duplicate symptoms of muscular dystrophy and myasthenia gravis, enabling scientists to understand these diseases better.

Animal products are widely used in traditional practice, but Western scientists put little faith in the curative powers of most of them. A more important point is that the majority of Asian people believe in their efficacy. And, through experiments with placebos, scientists have shown that belief in a medicine's effectiveness can bring about actual improvement.

Figure 2-21

This is a traditional medicine pharmacy in South Korea, where animal and plant products are essential. Here you can see deer antlers and packages of ginseng roots. Photograph courtesy of B. A. Weightman.

Faith in the mysterious workings of faunal cures drives the demand for animal parts. In China, powdered deer antlers are employed to counter aging, and ground, dried gecko is taken for respiratory problems. In Vietnam, boiled monkey bones are believed to improve circulation. Indonesia's trade in *jammu* or traditional medicinals involves millions of dollars a year. While many, better-educated urbanites are turning to Western imports, increasingly available in markets and pharmacies, rural and older populations continue to rely on long-trusted remedies.

One of the most exploited animals is the rhinoceros. Only a few hundred survive in India, Nepal, and Indonesia. A mere 70 remain on the island Java. Rhino parts are craved for their imagined powers as analgesics and aphrodisiacs. Its droppings are said to cure leech bites, and its rawhide ameliorates skin afflictions. Its "horn" (actually hard, agglutinated hair) guarantees the seller instant wealth. Unfortunately, rhino horn remains in demand even with public education and legal restraints on its trade.

THINKING ABOUT THE ENVIRONMENT

Environment is a multidimensional concept, and definitions range from anthropocentric to bioethical: the environment exists to support humanity versus it has value in its own right. The former leans toward the idea that there is a sharp division between people and nature, the latter toward the notion that human and natural worlds are entwined in an all-encompassing life-world. Those who view environment in terms of its use value tend to think that problems result from mismanagement and that better management of resources will fix things. The bioethical perspective is that nonhuman phenomena have intrinsic moral significance apart from their utility. Therefore, humans have a moral responsibility to care for and protect all aspects of nature. Between these "use it" and "save it" perspectives lie numerous other interpretations.

There is a common perception that Asian culture groups, at least historically, lived closer to and had more regard for nature. Nature must be important because both East and Southeast Asians employ natural metaphors regarding social position and morality. For example, reason (*li*) is fundamental to the history of Chinese philosophy. Originally, the character for *li* meant "well-distributed veins on minerals or precious stones." Later, it came to mean "principle" and then, "universal principle." Besides, nature is intrinsic to religions and belief systems. For instance, numerous gods in Hinduism such as Ganesh (an elephant) or Hanuman (a monkey) are animals. The popular god Vishnu rides on a bull. However, there is no single Asian concept of nature.

Did you know that almost-extinct tigers are descendants of the extinct saber-toothed tigers of Siberia?

Fate of the Tiger

At the turn of the twentieth century, an estimated 100,000 tigers roamed from Siberia to Indonesia. In the dawn of the twenty-first century, there are perhaps 2,500 remaining. Three subspecies have become extinct in the past three years. Big game hunting in the colonial era and loss of natural habitat are partly responsible for this catastrophe. Another important cause is population pressure and need for settlement space. However, the major culprit is poaching—the illegal slaughter of these magnificent beasts for money. To cite one Indian conservationist: "Every tiger that walks into the forest is a cash register."

From the tip of its nose to the tip of its tail, virtually every piece of a tiger is useful in traditional Chinese and other Eastern medical practices. Here are only a few examples.

- **Claws** cure insomnia
- **Teeth** treat fever
- **Brains**, when mixed with oil and rubbed on the body, cure acne and laziness
- **Eyeballs** are for malaria and epilepsy
- **Whiskers** prevent toothache and give the user courage
- **Tails** sooth skin diseases
- **Penises** promote virility
- **Bones** cure rheumatism and paralysis

A dead tiger can fetch as much as US$40,000 for its various body parts in the retail markets of East Asia.

In addition to these "medical" uses, there is a huge demand for tiger skins, claws, and teeth for ornamental purposes. An Indian tiger skin can bring as much as US$15,000 in Arab nations. Tibet has become a virtual shopping mall for tiger products. Chinese (not Tibetan) vendors in Lhasa hawk dozens of pelts in the back rooms of their shops.

"Acquiring" a tiger is a horrible affair (for the tiger). According to an account from the Wildlife Protection Society of India, the killer sets a few metal traps near a watering hole in a wildlife park where tigers are supposed to be protected. Once a thirsty tiger accidently steps into the trap, the jagged metal teeth clamp onto its paw. The tiger howls and the noise can easily alert a park ranger who may be quite far away. In order to silence the beast so that the ranger can't locate it, the poacher rams a spear down its throat and rips out its vocal cords. Then, at his leisure, he can poison, electrocute, or shoot the cat.

What is being done to save the tiger? "Project Tiger," launched 30 years ago, is the most high-profile conservation program in the world. It and other programs such the Convention on International Trade in Endangered Species Act (CITES), are admittedly on the verge of failure as tiger populations continue to shrink.

India is home to about half the world's wild tigers. Skins are trafficked via Nepal and Tibet to China and through Burma to other lucrative markets in East Asia. The rural population, deriving no benefit from their proximity to the big cats, considers them as dangerous vermin. After all, about 300 rural dwellers in India lose their lives to leopard, elephant, and tiger attacks each year. Moreover, cats carry off and kill valuable livestock for food.

Poachers take advantage of the disagreements between farmers and conservationists and can pay villagers small sums, perhaps US$25, for their assistance in tracking the beasts. If and when the poachers are caught, the tiger skins and body parts are confiscated by the government and the perpetrator is arrested. However, given the degree of government corruption in India, the tiger products frequently end up back on the market, and this time officials and/or police reap the profits. In 2008, 411 cases were filed against illegal hunters and poachers for killing tigers. Not a single one of these secured a conviction.

China, which joined CITES in 1981, banned domestic trade in tiger products in 1993. It also illegalized the use of tiger bone in traditional pharmacopeias. However, growing affluence in China has only increased demand.

Recognizing that the tiger is a renewable resource, entrepreneurs have taken to farming them. Tigers breed very easily in captivity; an estimated 5,000 have been bred in this manner. However, these animals will not be returned to the wild; they will be slaughtered for their products. It should be noted that no tiger breeding program has ever successfully returned a tiger into its natural habitat; not trained to survive in the wild by its mother, it soon dies. In the wild, tiger cubs spend at least three years with their mothers learning the ways of their world. On tiger farms, cubs are taken away from their mothers at three months; they are not supposed to learn anything. "Farmers" stockpile skins to make them scarcer and therefore more expensive. They also have created a market for wine made from the bones.

China has identified a couple of original habitats for the south China tiger, one of the most endangered species, and is conducting a bold experiment in re-wilding and reintroduction. It has even sent tigers to large conservation farms in South Africa where they can learn survival skills. Efforts are also being made to educate and integrate local villagers into ecotourism.

Although culture is frequently seen in opposition to natural forces, these remain connected in a holistic, contextual existence. While Westerners assume a direct connection between people's perception of nature and their treatment of it, this is not necessarily the case in Asia, where contextualism and pragmatism are far more important. Nature can be worshipped and harmed simultaneously; one can pay homage to the tree prior to felling it.

Certainly, Asian perceptions of nature have not stood in the way of its destruction. Yes, Western impacts, foreign companies, and foreign investment all play their roles. However, political and economic elites, the formulators of national policies, facilitate, contribute to, and benefit financially from environmentally-damaging development schemes. With all the rhetoric about saving endangered beasts and habitats, the juggernaut of development continues to affect the health and well-being of nature and humanity alike.

Current thinking does not dichotomize use or development from environment. Since humanity cannot survive without natural resources and these cannot exist if depleted or destroyed, development and environment are inexorably linked. Sustainable development is essential in light of human impacts on air, water, flora, fauna, and other natural systems. It is critical for the survival and well-being of future generations.

Conceptually, sustainable development sounds like the right thing to do. But putting these ideas into practice unearths such questions as these: What is the precise meaning of sustainable, and who sets the standards? Who or what should be sustained? Is a geomorphic phenomenon such as a prehistoric cave entitled to the same consideration as a tiger population? Are animals more important than fish or plants?

In the upcoming chapters, you will read more about environmental problems and efforts to contain them. Keep in mind that because of the ephemeral characteristic of nature, local problems become regional ones, and regional problems become global ones. Keep in mind, also, that it is the poor who suffer most directly from environmental degradation. Unlike the majority of those living in the industrialized nations of the northern hemisphere, the majority of Asians live close to the land, without the mediation of air conditioners, water purifiers, and so forth. Geographer W. M. Adams (1995) says, "Sustainable development is synthetic, and constructed within the confines of current convention in northern-dominated debate about development. . . ." While debate lingers among policymakers, the full force of modern development continues to transform whatever is left of the natural world. In Chapter 3, we will examine the dynamics of population, agriculture and food supply, and related social inequity.

Recommended Web Sites

www.bing.com/images/search?q=global%20warming&FORM=IGRE#
Wonderful photos relevant to global warming. Maps and diagrams.

www.bing.com/images/search?q=tropical+forest&FORM=IGRE#
Thousands of magnificent photos of tropical rain forests, tropical dry forests, and forest fires. Scientific and educational diagrams and other types of illustrations.

www.bing.com/images/search?q=mangroves&FORM=IGRE#
Excellent photos, maps, diagrams, and information about mangroves.

www.blueplanetbiomes.org/rainforest.htm
Excellent description of rainforests, climate, plants, and animals.

www.climatesscience.gov
Information on climate change from NASA, etc.

http://earth.google.google.com
Explore the Earth with aerial photos.

www.eoearth.org
Photos and interactive features about deserts including the Gobi, Taklimakan, and Tibetan Plateau.

www.livescience.com
Animals, health, culture, and environment.

www.NationalGeographic.com
Articles about physical and human environments of the world.

http://oceanworld.tamu.edu/students/coral/index/html
A good discussion of coral reefs.

www.panda.org/
All about the panda and China's efforts to save it.

www.ScientificAmerican.com/bbg/60-second-extinction
Countdown of extinction of species and efforts to save them.

www.wcme.org.uk/data/database/r/-anml-combo.html
IUCN Red List of endangered and vulnerable wildlife.

www.worldwildlife.org/
Information and photos about the predicament of wildlife. Includes information on the impact of environmental conditions such as pollution and climate change.

www.wspa-international.org
World map of animal welfare, photos and short articles about the plight of wildlife around the world. Read about the fate of the Mekong River dolphin.

Bibliography Chapter 2: Environments and People

Adams, W.M. 1995. "Sustainable Development?" In *Geographies of Global Change*, eds. R.J. Johnston. Peter J. Taylor, and Michael J. Watts, pp. 354–373. Oxford: Blackwell.

ADB. 2009. "The Economics of Climate Change in Southeast Asia." *Southeast Asia: A Regional Review*. Asian Development Bank Report.

Allsopp, Michelle et.al. 2007. "Oceans in Peril: Protecting Marine Diversity." *Worldwatch Report 174*. Washington D.C.: World Watch Institute.

ASEAN. 2009. *Manifesto on Combating Wildlife Crime in Asia*. Thailand: Ministry of Natural Resources.

Balik, Michael J., Elaine Elizabetsky, Sarah A. Laird eds. 1996. *Medicinal Resources of Tropical Forests*. New York: Columbia University.

Bruun, Ole, and Arne Kalland. 1995. *Asian Perceptions of Nature*. Richmond, Surrey, G.B.: Curzon.

Durst, Patrick B., Ward Ulrich, and M. Kashio eds. 1993. *Non-Wood Forest Products of Asia*. New Delhi: Oxford and IBH.

Hayden, Thomas. 2010. *National Geographic: State of the Earth 2010*.

Inman, Mason. 2009. "China CO2 Emissions Growing Faster Than Anticipated." *National Geographic News* October 12.

n.a. 2007. *Intergovernmental Panel on Climate Change. 4th Report.*

International Union for the Conservation of Nature. 2007. *Red List of Threatened Animals.*

McConnell, Robert L, and Daniel C. Abel. 2009. *Environmental Issues 3rd Edition.* Upper Saddle River, N.J: Pearson Prentice Hall.

McNeely, Jeffrey A., and Paul S. Sochaczewski. 1998. *Soul of the Tiger.* New York: Doubleday.

Mackay, Richard. 2002. *The Penguin Atlas of Endangered Species.* New York: Penguin.

Ministry of the Environment. 2006. *Final Report—Mangrove Action Project—Cambodia.*

Mitra, Barun S. 2005 "How the Market Can Save the Tiger." *Far Eastern Economic Review* 168/6: 44–47.

Mitra, Barun. 2007. "China's Market Plan to Save the Tiger." *Far Eastern Economic Review* 170/5: 50–52.

n.a. 2009. "East Asia: Saving the 'Amazon of the Seas'." *IRIN United Nations Humanitarian News and Analysis October 19.*

n.a. 2007. Netherlands Environmental Assessment Agency. *Report on CO2 Emissions.*

Nijhuis, Michelle. 2009. "Bamboo Boom." *Scientific American: Earth 3.0* 19/2: 60–65.

Peissel, Michel. 1997. *The Last Barbarians: The Discovery of the Source of the Mekong in Tibet.* New York: Holt.

Saghai, Bittu. 1997. "Tigerland." *New Internationalist* 288: 24–25.

Scott, Geoffrey A. 1981. "Mangroves and Man in the Malay Archipelago." *Sea Frontiers* 27: 258–266.

Siegel, Benjamin. 2006. "How to Kill the Tiger." *The Economist* July 31: 50–51.

Simpson, Sarah. 2009. "The Peril Below the Ice." *Scientific American Earth 3.0* 19/2: 30–37.

Starke, Linda ed. 2007. *Vital Signs 2007-2008.* The Worldwatch Institute. New York: W.W. Norton.

Stone, David. 1994. *Biodiversity of Indonesia.* Singapore: Archipelago.

Tsui, Bonnie. 2007. "Saving Coral Reefs Becomes a Tourism Priority." *New York Times* June 24.

Vesilind, P.J. 1984. "Monsoons." *National Geographic* 166: 712–747.

Ward, Geoffrey C. 1997. "Making Room for Wild Tigers." *National Geographic* 192: 2–45.

Chapter 3

Population, Gender, and Disparity

Somehow or other man has dominated women from ages past, and so woman has developed an inferiority complex. She has believed in the truth of man's invested teaching that woman is inferior to man. But the seers among men have recognized her equal status.

(MOHANDAS "MAHATMA" GANDHI) 1869–1948

Population Growth and Dispersal

Eight thousand years ago, a mere 5 to 10 million people inhabited the Earth. Today, the Earth's population is more than 6.8 billion. Most of this growth has taken place since 1830, when the world population first reached 1 billion. In 2009, South, East, and Southeast Asians accounted for more than 3.7 billion, or about 54 percent of the global population (Figure 3-1). How and why did this happen? What are the ramifications?

Estimates of early Asian populations indicate that they were relatively small, with sporadic periods of increase and decrease. By the mid-nineteenth century, population numbers had jumped significantly. However, growth was not an even progression. Before the advent of modern medicine, population numbers fluctuated across the globe with environmental disasters, war, famine, and disease. For example, in 1279, the Mongol conquerors of China exterminated some 35 million peasants. A plague pandemic from 1347 to 1351 killed 75 million of the then 300 million Eurasians.

Although worldwide high-mortality disasters have been largely quelled, regional catastrophes continue to make dents in population totals. For instance, 17,000 died in Bombay from plague in the late nineteenth century. A famine and government policies killed at least 30 million Chinese from 1959 to 1961. A million Bangladeshis perished in a hurricane in 1970. And, in one of the worst disasters in human history, at least two million Cambodians were exterminated by government policy from 1975 to 1978! Nevertheless, overall numbers remain sizable and provide large **base populations** to generate notable increments of increase even though population growth rates have been declining in recent years.

Figure 3-2 shows the distribution of Asian populations. What accounts for this pattern? A combination of historical and geographical factors provides the answer. Asian populations originated in and expanded from a series of core areas over a period of more than 4,000 years. However, where they cluster today reflects the availability of resources such as water supplies, fertile soils, and minerals.

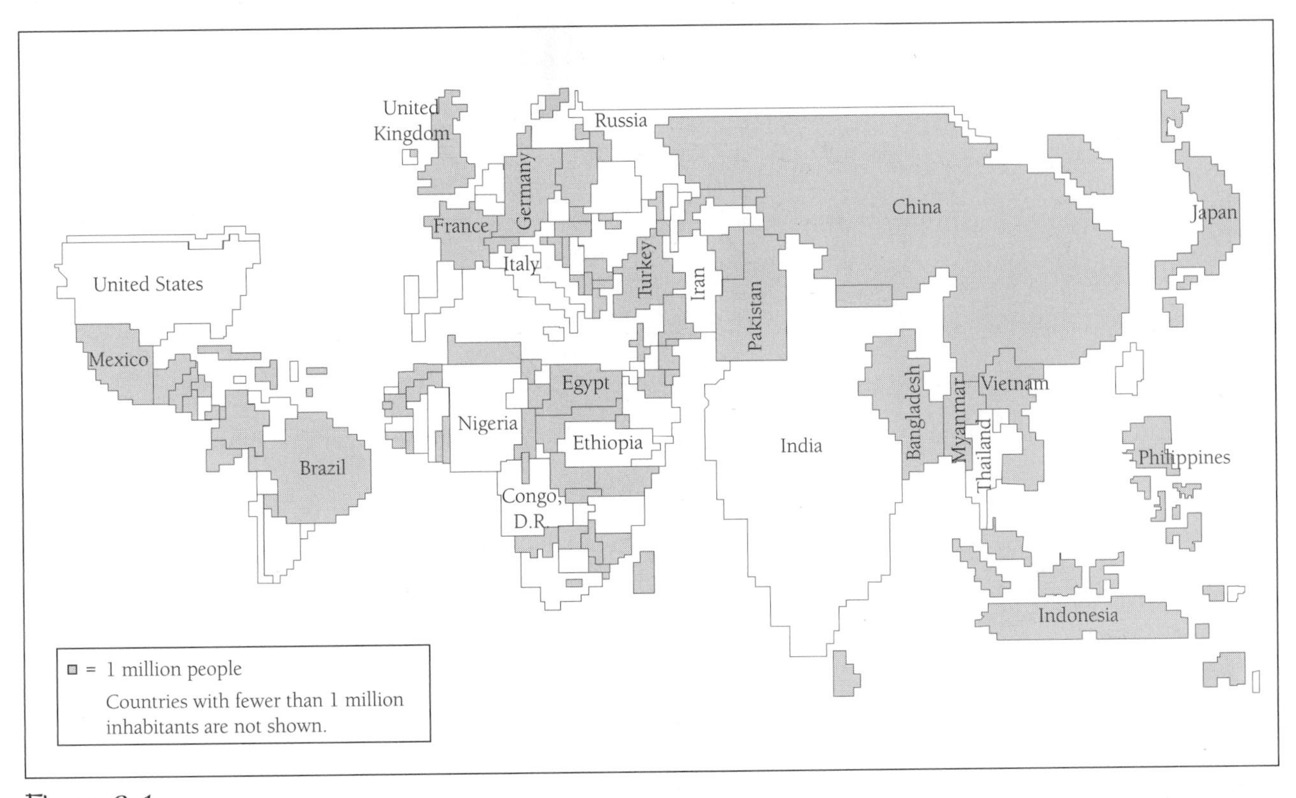

Figure 3-1

This map is called a "cartogram." It shows the relative size of countries around the world according to their population size. Compare Asia's size to that of other countries. James M. Rubenstein, *Introduction to Cultural Geography,* 9th edition, 2008 p. 47. Upper Saddle River, NJ: Prentice-Hall.

CORE AREAS OF GROWTH AND DIFFUSION

Around 3000 BC, there were two major clusters of population in East and South Asia: the North China Plain and the Indus River valley of northwestern India (now Pakistan) (Figure 3-3). These regions possessed agricultural systems productive enough to permit growth and stimulate expansion. A good measure of political control, social organization, and diversification of labor in rich environments allowed the development and maintenance of irrigation projects, as well as effective storage and distribution of agricultural surpluses. Population growth at this time derived from agrarian expansion.

Other smaller clusters existed around Japan's Inland Sea, in southern Korea, the Red River basin of northern Vietnam, Java, the upper Ganges River valley, and elsewhere in South Asia. By the year 1 AD regional populations had accumulated in these places.

By the fifth century, secondary population concentrations emerged in association with the upper Mekong River delta, Vietnam's central coast, central Myanmar, China's Szechuan basin, and southern India.

From the eleventh to the fourteenth centuries, periods of effective political control, accompanied by improvements in agricultural methods, fostered expansion into China's central Yangzi River region, Cambodia's Tonle Sap basin, the wet zone of Sri Lanka, and the lower Ganges River valley.

Growth burgeoned in the nineteenth century under the impact of European influences. European technology, coastally-oriented trade, and imposed political structures combined to affect both population increases and change in distributions. New flood-control technology with large-scale drainage canals facilitated the development of river deltas. Ports and their hinterlands became focal points of development

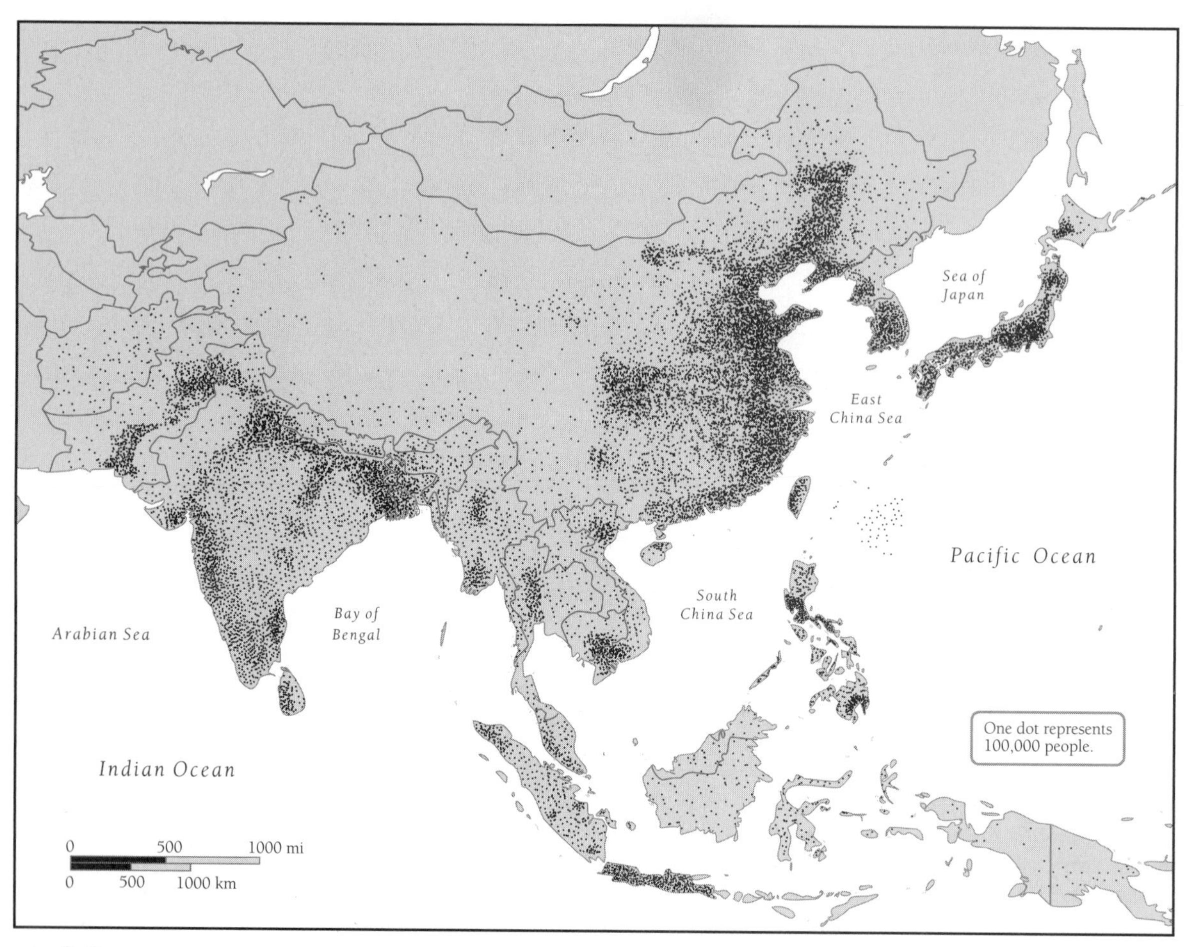

Figure 3-2

Population distribution. Compare this map with the one of physical features. Note the relationship between population, plains and basins, and the significance of rivers. *World Regional Geography: A Development Approach*, 8th edition, © 2004. Upper Saddle River, NJ: Prentice-Hall. Reprinted by permission.

and important migration destinations. Most of Asia's dominant cities grew dramatically during this period. Tokyo, Hong Kong, Shanghai, Calcutta (Kolkata), Bombay (Mumbai), Bangkok, Singapore, and Jakarta are some of these.

Core areas were critical to population growth and distribution in Asia. In fact, every country except for Laos is based on one or more core areas. However, there is more to the story. The landscape of population, like any other landscape, is socially constructed; landscapes are made by people in their particular life-world context. Social organization is especially significant in the development of Asian landscapes.

SIGNIFICANCE OF THE GROUP

Population growth and expansion were founded on intensive agriculture based on human labor. This is an undertaking requiring cooperative land management and production systems. Cooperation is encouraged because in Asia, the family is central and the group is more important than the individual. Established settlements were (and continue to be) **nucleated**; dwellings were clustered in villages Figure 3-4).

People and animals function in daily and seasonal routines of movement to and from their village and surrounding farmland. Cooperation is particularly important

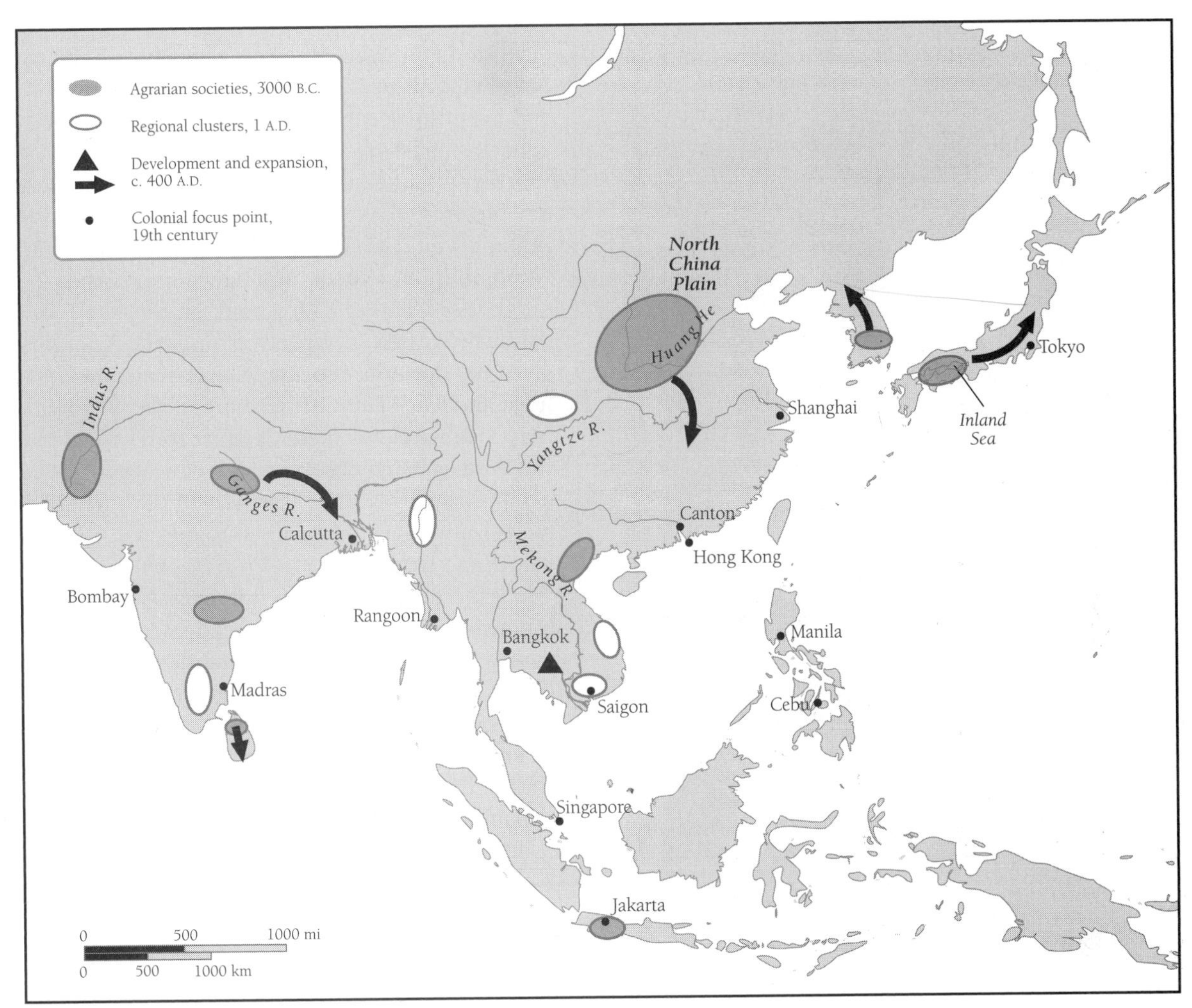

Figure 3-3

Core areas and diffusion of development. Note how alluvial lowlands fostered settlement as well as the importance of rivers and coasts in colonial settlement.

at planting and harvest times and in the maintenance of irrigation and flood-control systems, but this can only work in a situation of social and political stability. Consequently, long periods of stability coincided with population growth. Long-term chaos brought about population decline.

Thinking and Acting Collectively

In contrast to many societies where the individual is the locus of responsibility and action, Asian societies are characterized by collectivism. From birth, Asians are imbued with group consciousness extending far beyond familial and national levels to otherworldly existences. Identity derives from sets of expectations and responsibilities covering all aspects of life from emotional to economic. Long-established relationships among individuals, families, clans, work groups, and so forth are regarded as both mutually beneficial and necessary to nourish and perpetuate family or organizational lineage. Thought and action revolve around "we," and collective agreement supersedes individual will.

The existence of different standards for different people is generally understood and accepted. The

Figure 3-4

This aerial view over Pakistan shows two nucleated villages. People work the fields surrounding the villages. Much of this land is fragmented, meaning farmers do not own contiguous fields. Many farmers are tenant farmers working for landlords. Linear settlement patterns can be seen along roads and irrigation channels. Photograph courtesy of B. A. Weightman.

fact that royalty or elite groups exhibit abundant wealth, conspicuous consumption, or questionable behavior is tolerated as long as duties are performed correctly and moral responsibilities are met in ordering and providing for society at large. Authoritarian structure and order are considered prudent.

Religious beliefs frequently support a stratified status quo. Hierarchical concepts of order in the sacred realm bolster hierarchical relationships and consequential order in the secular world. *Carpe diem* (in Latin, "seize the day") and "you only live once" make little sense in context of profoundly enduring and seemingly endless continuities of collective existence.

If you ranked the five factors that shape everyday life, from the traditional Asian perspective, the ranking would be: fate, luck, ***feng-shui*** (in China; *vashtu* in India), virtue, and education, in that order. (See Chapter 10 for more on *feng-shui*). In contrast, Western perspectives rank education first, give only lip service to virtue, and denigrate fate, luck, and *feng-shui*. Furthermore, it seems ironic that Asian students these days are stereotyped in the West as being obsessed with education. To the extent that this is true, it demonstrates how rapidly globalization has undermined traditional Asian values and worldviews.

It is true that Western influences/philosophies espousing individual rights, freedom, and social equity are poking holes in the fabric of group identity in Asia. Change is most evident in places where values of individual achievement and entrepreneurial success are becoming prominent. Nevertheless, change does not mean erasure of tradition. Asians remain grounded, to varying degrees, in traditional collectivist consciousness.

POPULATION INCREASE AND EXPANSION

Population numbers and distribution were tied to agricultural productivity. In stable and prosperous times, villages expanded and even coalesced. New farmlands were opened up and people moved into them. Where equal or multiple-share inheritance was practiced, lands continued to be subdivided into ever-smaller plots. As populations grew and good lands filled, marginal lands became sought after. In-migration of one population forced the out-migration of another. Southeast Asia received countless non-Chinese ethnic groups as well as the Chinese in context of China's southward expansion. Repeated invasions of Aryans and later Mongols into northwestern India drove then-existing populations eastward and southward there as well. Population migrations did not end in ancient history; they are ongoing. As we will see later, Chinese movement into Manchuria is a fairly recent phenomenon, and the Chinese continue to migrate into their northern, western, and southwestern frontier regions. Vietnamese migration down the

Annamite coast continues to press southward into the lower Mekong delta. Population processes and patterns are constantly changing in light of social and environmental circumstances.

THE SIGNIFICANCE OF CHILDREN

The key dynamic of population growth is the fact that people have children. This reality draws much comment and criticism, especially when the child-bearers appear not to be able to support their offspring. The necessity for having children does not rest solely on the concept of societal reproduction. Children are the core of Asian family-group culture. Children are essential for the respect and continuity of family lineage. With strong bonds between people and their lands, children (usually sons) receive and are entrusted with guardianship of family property through generations. Children sustain the group, becoming a source of pride. Men and women are truly worthy when they produce children, more so with sons. Not having children is inconceivable to the majority of Asians.

There are also practical reasons for having children. In traditional societies, children contribute their services and labor very early. Four-year-olds look after two-year-olds. Seven-year-olds work at farm tasks or help in the family business. Children make important contributions toward their own and their families' survival. They are also expected to take care of their parents. Both boys and girls expect to look after their aged parents or parents-in-law, especially where there are no old-age insurance programs or care facilities (as in most places).

But why do people have so many children? History demonstrates that life is uncertain. Natural disasters can strike at any time. Famine can quickly decimate a population. North Korea suffered a severe famine in 1998, although the extent of the famine's impact remains disputed. War has been all too common in the twentieth century, and millions continue to experience fighting as a way of life. Health care is available only to some and limited for just about everybody. In other words, if a family has only one or two children, the chances are not good that any will survive to carry out family responsibilities.

In the War Museum of Hanoi, there is a picture of a woman who lost nine sons in wars fought in Vietnam from the 1950s to the 1970s. Who will care for her in her old age? People cannot be expected to have fewer children if many are needed to ensure an ongoing supply for labor, income, military duties, parental care, and the like. However, quality of life improvements can induce reduction in family size. Provision of family planning has also helped to reduce family size.

REGIONAL DYNAMICS

Table 3-1 lists statistical data used in traditional geodemographic analysis. Traditional assessment of population trends is based on information about growth and change, density and distribution, and population composition. This information is gathered by various organizations from official sources based on censuses in individual countries.

Data are not necessarily reliable. Counts frequently are not accurate, in part because not all people are easily accessed. Moreover, people may lie about the number of children they have or the number of individuals in their household to avoid taxes, prohibitions on family size, or residency rules. Governments might skew figures for economic and political purposes. Another important point about statistics is that there are vast regional differences within countries. For example, areas that are not near centers of development, such as major cities, typically have higher birth and death rates. It is important to look at these figures, not as absolute facts, but rather as *indicators* of conditions and trends.

GROWTH AND CHANGE

To understand population growth and change, we look at several statistical measures.

- **Birth Rate (BR):** Number of births per 1,000 people in a population per year.
- **Death Rate (DR):** Number of deaths per 1,000 people in a population per year.
- **Natural Increase Rate (NIR):** Births per 1.000 minus deaths per 1,000 per year, usually expressed in percent. For example, Laos: BR = 28/1,000 minus DR = 7/1,000 = an NIC of 21/1,000 = 2.1%.
- **Total Fertility Rate (TFR):** The average number of children a woman would have, assuming BRs remain constant throughout her childbearing years (usually considered to be ages 15–49).

Variation is a key word for understanding population dynamics in Asia. Table 3-1 demonstrates that some countries have relatively high birth rates (BRs). Nepal,

Table 3-1 Demographics

Country or Region	Population (millions) Mid-2009	Birth Rate Per 1,000	Death Rate Per 1,000	Nat. Inc. Rate %	Total Fertility Rate	Percent Change By 2050	Percent <15/>65 Years Old
Australia	21.9	14	7	0.7	2.0	55	19/30
Canada	33.7	11	7	0.4	1.6	24	17/14
U.K.	61.8	13	9	0.4	1.9	24	18/16
U.S.A.	306.8	14	8	0.6	2.1	43	20/13
South Asia							
Bangladesh	162.2	23	7	1.6	2.5	37	32/4
Bhutan	0.7	25	8	1.7	3.1	46	32/5
India	1,171.0	23	7	1.6	3.1	49	32/5
Maldives	0.3	22	4	1.8	2.3	51	30/5
Nepal	27.5	29	9	2.1	3.1	67	37/4
Pakistan	180.8	30	7	2.3	4.0	85	38/4
Sri Lanka	20.5	19	7	1.2	2.4	24	26/7
East Asia							
China	1,331.4	12	7	0.5	1.6	8	19/8
Hong Kong SAR*	7.0	11	6	0.5	1.1	25	13/13
Macao SAR*	0.6	9	3	0.5	1.0	53	13/7
Japan	127.6	9	9	−0.0	1.4	–25	13/23
North Korea	22.7	16	10	0.5	2.0	01	22/9
South Korea	48.7	9	5	0.4	1.2	–13	17/10
Mongolia	2.7	24	6	1.8	2.6	48	33/4
Taiwan	23.1	8	6	0.2	1.0	–7.0	17/10
Southeast Asia							
Brunei	0.4	16	3	1.3	1.7	64	26/4
Cambodia	14.8	25	8	1.7	3.0	61	35/3
Indonesia	243.3	21	6	1.5	2.5	41	29/6
Laos	6.3	28	7	2.1	3.5	70	39/4
Malaysia	28.3	21	5	1.6	2.6	43	32/4
Myanmar (Burma)	50.0	21	10	1.1	2.3	27	27/5
Philippines	92.2	26	5	2.1	3.3	63	35/4
Singapore	5.1	10	4	0.6	1.3	10	18/9
Thailand	67.8	15	9	0.6	1.8	8	22/7
Timor-Leste	1.1	40	9	3.1	6.5	184	45/3
Vietnam	87.3	17	5	1.2	2.1	29	26/7

* Special Autonomous Regions of China
Source: Population Data Sheet 2009. Population Reference Bureau.

Pakistan, Mongolia, Cambodia, Laos, and Timor-Leste are among these. Some, such as Myanmar and North Korea, have higher death rates (DRs) than other countries. The difference between these two measures is the natural increase rate (NIR). While Timor-Leste has a high BR (40/1,000), it has a much lower DR (9/1,000). This results in a very high NIR (3.1%).

Total fertility rates (TFRs) have fallen quite rapidly over the past 20 years. In Asian countries (excluding the Middle East), women had an average of 5.82 children between 1950 and 1955. By 1980–1985, the TFR had dropped to 3.54. From 2000 to 2005, the TFR had fallen to 2.39. To illustrate further, Bangladesh's TFR was 7 in 1975 and is now 2.5.

What does this mean in terms of population growth? To answer that question we have the notion of **replacement fertility rate (RFR)**. Replacement fertility rate means that a population is growing just enough to replace itself in the future. In developed countries the RFR is about 2.1. This assumes that two parents will have two children to replace them after they die. The extra 0.1 is meant to account for the facts that many children will never reach adulthood and couples might choose not to have any children.

Look at the Percent Change by 2050 column. This indicates the anticipated percentage increase or decrease of a population from 2009 to 2050 given the current NIR. Pakistan's population, with its NIR of 2.3 percent and TFR of 4.0, is expected to increase 85 percent. This means that Pakistan's population will grow from its current 180.8 million to 335.2 million by 2050. Notice that Japan is predicted to have 25 percent fewer people in 2050. Having both low BR and DR, Japan's NIR of –0.0 percent and TFR of –25 are insufficient to replace its population. Does this mean that the Japanese will ultimately disappear? Read more on this topic in Chapter 12.

It is important to note that these are "snapshot" projections of current conditions. In fact, long-term trends show an overall decline in NIRs. For instance, Bangladesh's NIRs dropped from 2.2 percent in 2003 to 1.6 percent in 2009, and South Korea's from 0.7 to 0.4 percent. More education for women, greater government support for family planning, increasing urbanization, and improving economic status all contribute to the reduction of population growth. Nevertheless, while women are bearing fewer children than they did in the past (TFRs), countries with large percentages of youth will have more women to have children as time ensues. Consequently, population growth is expected to continue.

Although India and China do not stand out in terms of NIRs and TFRs, they are notable in terms of their current population totals and projections for the future. Because of the size of these base populations, even a small percentage increase adds many people to the total. So, while India's NIR is only 1.6 percent, if current growth rates continue, it will add another 577 million people by 2050. China, with its low NIR of 0.5 percent, nevertheless will grow by nearly 106 million people by 2050. We will learn more about India's and China's population issues separately in the regional chapters.

DENSITY AND DISTRIBUTION

Density and distribution data measure spatial characteristics of populations. Refer back to Figure 3-2 and note that some areas are densely populated while others are sparsely populated. Environmental and historic circumstances contribute to these patterns. People tend to avoid extreme environments such as deserts and rugged, high-elevation terrain. Access to resources is crucial to survival. For example, arable land is found on plains and along waterways, and this is where most populations reside. Past circumstances have led to a phenomenon known as **historic inertia**—people migrate to and continue to live in places where other people are already established and opportunities are perceived to exist. Consequently, densely populated areas most likely will continue to be densely populated areas.

Density figures are misleading. **Arithmetic density** indicates how many people exist per square mile (km) of land in a particular country. The Philippines, for instance, had a density of 160 persons per square kilometer in 1980. Now its density is 307 per square kilometer. However, people are not evenly distributed throughout the islands. The northern island of Luzon is home to millions more than the south of Mindanao, and numerous islands have no inhabitants. Moreover, there are parts of metropolitan Manila that have population densities 100 times those of many outlying regions. Clearly, arithmetic densities reveal nothing about actual spatial distribution of people.

An alternative to arithmetic density is **nutritional density**, which relates numbers of people to amount of cultivated land. Nutritional density assumes a direct relationship between the number of people and land productivity potential. This information is interesting but still misleading because it assumes that each person has the same amount of land and equal access to food supply. Of course, we know that this is not the case.

Density data provide only crude measures. Laos is a relatively sparsely populated country yet it is regarded as one of the poorest in Asia. Taiwan, one of the most densely populated, is seen as one of the richest. Densely populated Singapore, also one of the richest, has virtually no arable land. On the other hand, some say that Bangladesh is one of the most densely populated countries in the world and it suffers from extreme poverty. Density does not account for variation in human or land productivity such as differences in dietary requirements, or quality of available diet, food imports, or unequal access to food supplies.

POPULATION COMPOSITION

Population composition data reveal the sex and age structure of a particular population. **Population pyramids** are visual representations of relationships between males and females, and among different *age cohorts* (Figure 3-5). An age cohort represents different age groups usually in five-year segments. For example, persons between the ages of 50 and 55 would be an age cohort. In population pyramids, age cohorts are divided by gender. Some pyramids show percentages of males and females in each cohort. The pyramids indicate numbers of people (in thousands) that are easier to understand.

Assuming that a baby stays alive for 60 years or more, it will "move upward" through the pyramid as the years go by. For instance, a newborn (0 years old) in 2009 is in the 0–5 year cohort. By 2020, it will be in the 15–20 year cohort. For the sake of discussion, if all the babies stay alive and reach the age of 60, the whole giant 0–5 cohort that you see now will be in the 60–65 age cohort in 2070. Of course, the probability of this happening is remote.

Most populations for Asian countries have wide bases, indicating high NIRs and youthful populations. Considering that it is female cohorts ages 15–45 that are potential child-bearers, this age group will increase dramatically as younger cohorts age. For example, a five-year-old girl in 2000 will be a potential mother in 2010 (or perhaps even earlier).

Timor-Leste has a very small population of only 1.1 million. However, its NIR is 3.1 percent and its TFR is 6.5. This means that if these rates continue, the country's population will multiply 184 percent by 2050! If you examine the pyramid again, you can see a sudden rise in births in the 1990s. This reflects the ending of years of violence beginning in 1976 when Indonesia forcibly took over the country and initiated a campaign of "pacification" that lasted for two decades. An estimated 100,000 to 250,000 Timorese were killed. Obviously, this situation negatively affected population growth.

Japan stands in sharp contrast to Timor-Leste because its population is clearly shrinking dramatically. As you can also see by the pyramid, there are many more people over 35, especially in the over-65 category, in Japan than there are in Timor-Leste. Look for more on Japan's population in Chapter 12.

FAMILY PLANNING

Virtually all South Asian, East Asian, and Southeast Asian countries have population planning programs, although these are unevenly accessible and acceptable

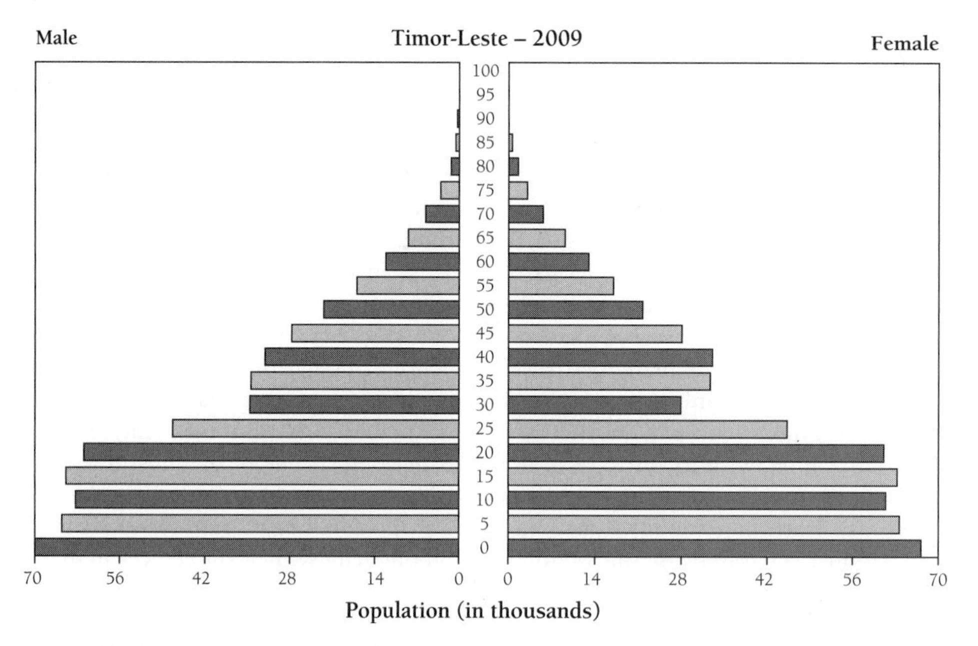

Source: U.S. Census Bureau, International Data Base.

Figure 3-5

Compare these population pyramids. Timor-Leste's broad-based pyramid indicates a high rate of natural increase until the 1990s when the natural increase rate started to decline. The narrowing base of Japan's pyramid shows that its population is declining. The narrowing base of Vietnam's pyramid reflects the success of its family planning programs.

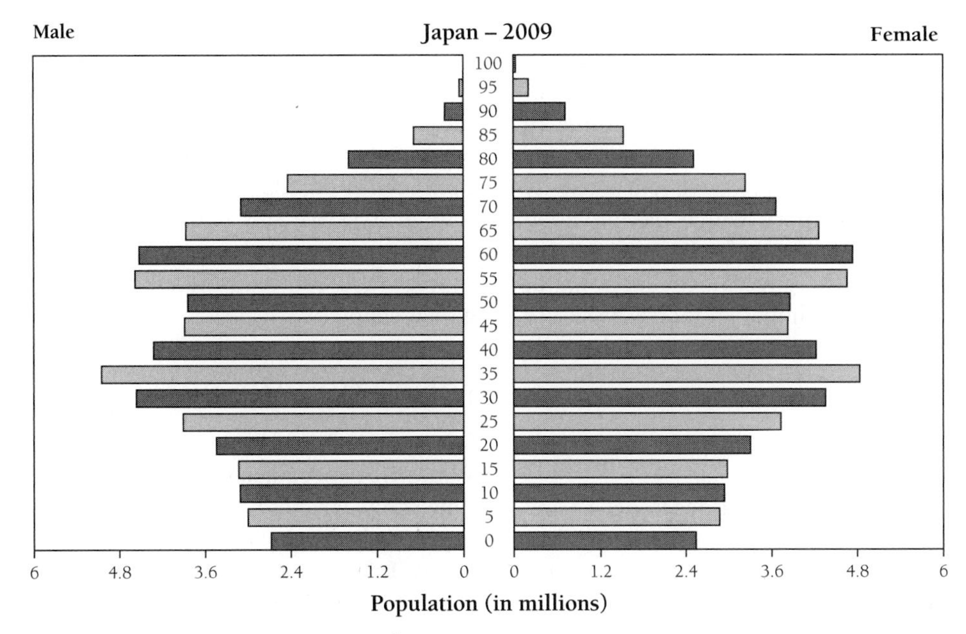

Source: U.S. Census Bureau, International Data Base.

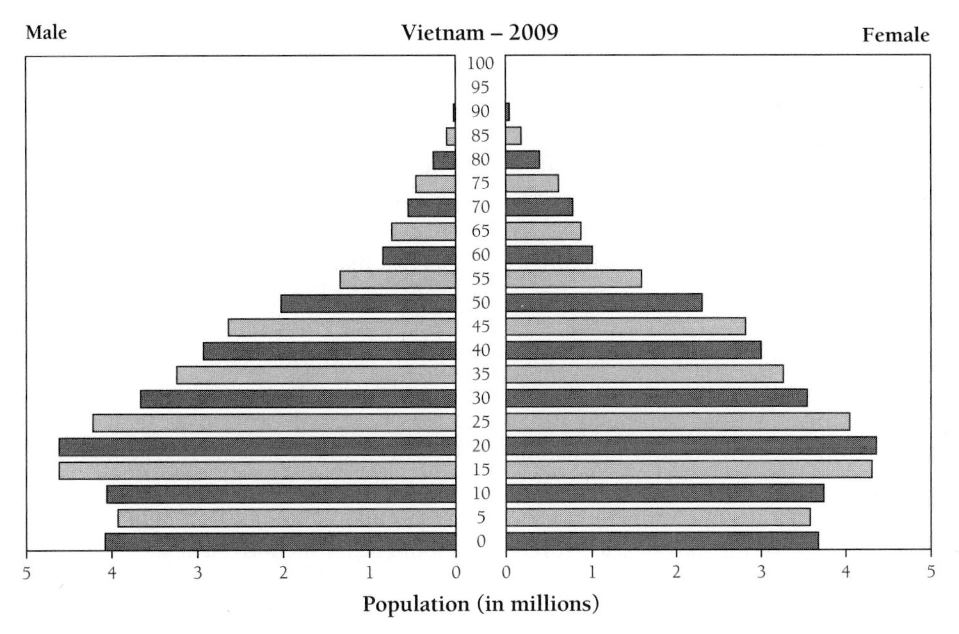

Source: U.S. Census Bureau, International Data Base.

Figure 3-5 (continued)

across the region. Less educated and more tradition-bound societies are less likely to want to reduce their family size.

You can see the impact of family planning in the Vietnam pyramid where younger age cohorts are getting smaller. Population reduction programs have been promoted in Vietnam since the 1960s. The country has high rates of free abortions. However, contraception use has increased from 68 percent in 1996 to 78 percent in 2006, although rates are lower in many more traditional rural areas. The most common form of contraception is the IUD (inter-uterine device) although the use of birth-control pills and condoms is growing. Pills are unpopular due to misinformation about their side effects, especially the perceived high risk of cancer. Gender issues and cultural barriers continue to limit the use of condoms.

Vietnam's population programs have relied heavily on international donor support, which has provided 84 percent of the budget especially for IUDs. Since this support ended in 2009, the government faces major challenges for the future. It hopes to increase the use of pills and condoms through education and media campaigns.

DEPENDENCY RATIOS (DR)

Table 3-1 lists percentages of populations under 15 and over 65 years old. These age groups are perceived as not working and therefore as being "dependent" for survival on the middle-age groups, who are assumed to be working. A **dependency ratio (DR)** is easily determined. Just add the percentages of population under 15 and over 64 years old; divide the total by the percentage of population from 15 to 64 years old; then multiply this figure by 100 to get the result. For example, Pakistan has an under-15 population of 42 percent and an over-65 population of 4 percent for a total of 46 percent "dependents." This leaves 34 percent of the population as the "producers" and supporters. Consequently Pakistan's DR is 82 percent. In contrast, Singapore, with its highly educated and richer society, had a DR of 37 percent.

Don't be fooled by dependency ratios, because they are often based on faulty assumptions especially concerning developing countries. In Asian and many other societies, significant numbers of young and old people work and contribute to their own survival, as well as that of their extended family. Dependency ratios do not account for this fact. Nevertheless, they are at least somewhat indicative of problems governments around the world are facing.

Japan is clearly an unusual case. It is the only country in the world to have a higher percentage of its population over 65 than below 15. This bodes dire consequences not only for providing services to the elderly also for having enough people of employment age to work in the country's vast industrial sector in the near future. In contrast, most Asian countries face the opposite challenge of trying to provide services, such as health, education, and employment, to their predominantly young populations.

Basic population statistics should be interpreted as estimates of numerical change. They reveal nothing tangible about the well-being of people. On the other hand, projections and related measures are used to illustrate perceived population problems such as "overpopulation." Since population "control" measures are ubiquitous in Asia, we can assume that governments believe they have too many people relative to available resources. You can read more about this idea in the regional chapters.

POPULATION AND QUALITY OF LIFE

Selected quality of life statistics are shown in Table 3-2. While **life expectancies** at birth (LE) have improved over time, only Japanese, Singaporeans, and Taiwanese can expect to live as long as people in Canada or the United States. **Infant mortality rates (IMR)** in the first year of life and child mortality rates (CMR) in the first five years of life are excellent indicators of living conditions. Pakistan, Myanmar (Burma), Laos, and Timor-Leste, among the poorest, exhibit very high IMRs.

LIFE EXPECTANCY (LE)

Life expectancy figures indicate how long one can expect to live from birth. In most societies, women outlive men. LEs are quite low in many Asian countries. As of 2009, a male child born in Myanmar (Burma) can look forward to a life span of 59 years and a female to 63 years. In Taiwan, which is much more developed and richer, LEs are significantly higher.

The percentage of children under five years of age suffering from malnutrition, as represented by being underweight or under-height, is very high for many countries. You will notice that in six Asian countries, 40 percent or more children are underweight and/or shorter than they should be for their age. These children will most likely not live as long as their healthier counterparts.

Poor countries experience high **infant mortality rates (IMRs)**. This population measure refers to the number of babies that perish before they are one year old per 1,000 live births in a year. Typically, these data are derived from officially registered births. However, many births, especially in squatter settlements, urban slums and remote rural areas, are not registered so the data are not always accurate.

Life expectancy and infant mortality rates are obviously related to an array of factors. Poor diet, disease—especially diarrhea for newborns—lack of medical

Table 3-2 Quality of Life Indicators for Infants and Children (2009)

Country or Region	- Life Expectancy at Birth Male/Female	Infant Mortality Rate Age 0–1 per 1,000 Live Births	Children Malnourished In 2004, %	Children under 5 Under weight, %	Children under 5 Under-Height, %
United States	75/78	7.0	—	—	—
Canada	78/83	5.0	—	—	—
United Kingdom	77/82	7.0	—	—	—
South Asia					
Bangladesh	64/66	48.0	27	2005: 39.2	47.8
Bhutan	67/68	40.0	n.d.	1999: 14.1	47.7
India	63/65	55.0	21	2006: 43.5	47.9
Maldives	72/74	10.0	14	2001: 25.7	31.9
Nepal	63/64	48.0	15	2006: 38.8	49.3
Pakistan	66/67	67.0	23	2001: 31.3	41.5
Sri Lanka	67/75	15.0	21	2000: 22.8	18.4
East Asia					
China	71/75	21.0	9	2002: 6.8	21.8
Hong Kong SAR*	79/86	1.8	Incl. with China	—	—
Macao SAR*	79/84	3.0	Incl.with China	—	—
Japan	79/86	2.6	—	—	n.d.
North Korea	61/66	54.0	32	—	
South Korea	76/83	4.0	5	—	44.7
Mongolia	62/69	41.0	29	2004: 4.8	23.5
Taiwan	75/82	4.7	—	—	—
Southeast Asia					
Brunei	75/80	7.0	5	—	—
Cambodia	59/63	62.0	26	2006: 28.4	43.7
Indonesia	69/73	34.0	17	2004: 19.4	28.6
Laos	63/66	64.0	19	2000: 36.4	48.2
Malaysia	72/77	9.0	5		—
Myanmar (Burma)	59/63	75.0	n.d.	2003: 29.6	40.6
Philippines	66/72	23.0	16	2003: 20.7	33.8
Singapore	78/83	2.1	—	2000: 03.3	33.8
Thailand	66/72	7.1	17	2006: 07.0	15.7
Timor-Leste	60/62	67.0	22	2002: 40.6	55.7
Vietnam	72/76	15.0	14	n.d.	35.8

* Special Autonomous Regions of China
— Negligible
n.d. No recent data available
Sources: *Population Data Sheet 2009*. Population Reference Bureau; World Health Organization (2009). *United Nations Development Program Report (2009)*.

care, lack of sanitation and access to clean water, and frequency of war and natural disasters are only some of these.

Access to health care and other services is very important for the survival potential of children. Hospitals may be nonexistent and doctors may be few. In India for example, there are a mere 60 physicians for every 100,000 people; in the United States there are 256. Clinics, particularly in remote areas, are likely to be poorly supplied. I have visited many clinics and hospitals in various parts of Asia and can assure you that conditions in most are grim.

Personal Experiences with Health Care in Asia

Once I went to a clinic in Simla (a popular Indian tourist destination) because I had a bad cut on my knee. The female doctor had completed medical school in New Delhi where she was well trained. However, she had very few supplies in her one-room clinic. There were some tongue depressors, a jar of cotton balls, a role of gauze, a large bottle of iodine closed with a cork, and little else. Red, rubber hot-water bottles were hanging on a wall. I have no idea what these were used for. Thankfully, she was able to tend to my knee.

Doctors may be poorly trained and are not necessarily fully licensed physicians. They may be little more than medical aides with a few weeks' training. I met a young "doctor" in Xian, China, who told me he was a cardiologist. When I inquired about his training, he said that he had gone to "cardiology college" for four years after finishing high school.

At Number Four People's Hospital, a "nurse" wanted to give me an injection because I had a sinus infection. The room was filthy, and a "doctor" squatted on the top of a dirty, rough, wooden table where he smoked a hand-rolled cigarette. There was a large, deep sink that was covered with blood and who knows what other stains. There was a grimy mop standing up in the corner of the sink. The nurse rummaged around in a drawer and came up with a big, rusty-looking syringe that was unpackaged and clearly not sterilized. The needle looked like something from the Dark Ages—very long and so thick that I could look up the hole at the end of it. She smiled broadly and said: "I give you needle; I very good; I practice on orange." I said *hsieh-hsieh* (thank you) and made a hasty exit.

China has recently built over 2,000 modern, state-of-the-art hospitals, but these are situated in large cities. Thus they could serve the local residents as well as country folk who came to town to work or visit. However, it's not working like that. Tests and treatments are costly, and medical insurance no longer exists for most people. Keep in mind that the majority of Chinese still live in rural areas, earn little more than $US 2.00 a day, and have no social safety nets other than their families.

On another occasion, I was on my way to Laos from Thailand and I heard from an international health-care volunteer that her small hospital in the northern tribal region had virtually no supplies. I was going into the country by land and knew that the border guards were instructed not to let foreigners bring any goods in. I purchased rubber gloves, bandages, Dettol (a disinfectant), and syringes in Bangkok. The needles were easy to get because they are distributed freely to drug addicts in an effort to reduce the occurrence of AIDS. Since you can only get one a day, I had to visit several dispensaries. I packed the supplies in my backpack under my camera equipment and put a US$20 on the top (which is a lot of money in Laos). Needless to say, I crossed the border with ease; the US$20 disappeared and I was able to deliver the goods to the hospital.

ARE THERE TOO MANY PEOPLE ON EARTH?

Thomas Malthus published his *Essay on the Principle of Population* in 1798. In it he forecast that population growth would soon outstrip the world's food supply. Malthusian theory was touted into the 1990s and then fell out of favor as demographers and others recognized that overall fertility rates were falling in the context of development and increasing affluence. Predictions are that the global fertility rate will dip below the global replacement rate by 2020.

Neo-Malthusians claim that while fertility rates are important, it is *absolute numbers* that count. They note that another 2.4 billion people will be added to the global population over the next 40 years. If, because of high fertility rates in earlier generations, there is a bulge of women of childbearing years, more children will be born overall even though individual women are having fewer births. Families will be smaller but there will be more of them. At current rates, the global population will be over 9 billion in 2050.

Today, the critical issue related to population increase is environmental damage. As development helps hundreds of millions of people to escape grinding poverty, it also entrenches them more deeply in the capitalist economy. As we saw in Chapter 2, modern consumption patterns are devastating the global environment. At this point, a poor

African or Asian generates a mere 0.1 ton of CO_2 a year, compared to 20 tons for each American. Further, those areas of the planet with the fastest growing populations are those most vulnerable to climate change and global warming—water shortages, mass migration, and food shortages. Neo-Malthusians warn that, if the poor copy the patterns of wealth creation and consumption exhibited by other affluent societies, there will be grim consequences for the planet.

Patriarchy and Gender Issues

Patriarchy refers to societies in which family unit structures focus on males who exert authority over the rest of the family members. In patriarchal systems, men also dominate social, economic, and political realms. Unfortunately, the majority of women in Asia are forced into subordinate roles and statutes embedded in a wide array of male-dominated cultures. The power of patriarchy varies across countries, regions, and social classes. Generally, patriarchy is more deeply embedded in more traditional, rural areas.

Gender is a socially constructed cultural institution. Gender roles are expected and assumed in cultural contexts, and society is organized around gender. Institutional behaviors and relationships are structured according to understood relative positioning and roles of men and women. In most cultures, gender inequity is the norm. Mencius (ca 372–289 BC), an influential Chinese philosopher and principle interpreter of Confucianism, outlined "The Three Subordinates:" A woman was to be subordinate to her father in youth, her husband in maturity, and her son in old age.

To understand gender inequity is to appreciate **power relationships**. Men and women are accorded unequal roles in virtually every society. This means that women do not have equal power to access education, earn comparable wages, make economic or political decisions, or even control their own destiny regarding inheritance, marriage, childbearing, or career. Women seldom have power over land or resources, or equal, if any, participation in public decision making. (See Table 3-4 below for exceptions in political leadership.)

In patriarchal societies, there is a widespread belief, held by both sexes, that men are smarter and more productive individuals. Families prefer to have boys because of their greater earning capacity. Girls start to work at a young age and learn their upcoming role as mother by tending younger siblings. Furthermore, females have less access to food and usually are the last to eat.

EXAMPLES FROM SOUTH ASIA

According to a recent UNICEF report about women in Muslim Bangladesh, women are frequently exposed to violence in their own homes. They are subject to mental abuse and physical torture by their husbands and in-laws.

Most marriages are arranged, often with girls as young as 12 being forced to wed older men. Child marriage is practiced by both Hindus and Muslims and is most prevalent in rural areas where women have little if any education. Betrothals may be arranged between an 8-year-old girl and a 10-year-old boy. Once the girl begins menstruation, she moves to her husband's home and is expected to obey her mother-in-law and get on with the business of having sons.

There are roughly 3,000 divorces a year. All an Islamic man has to do is say "I divorce thee" three times and the marriage is over. Fathers are recognized as legal guardians of the children and that often leads to mistreatment of his children by their stepmother. No child can be admitted to school without her father's name on the application form.

In Pakistan and northern India, where women's roles are also sharply defined, ***purdah*** is practiced, especially in the countryside. *Purdah* involves the concealment of women from nonrelated males. This means that women remain in the home, which is usually divided into male and female spheres. If a woman must go outside, she is to be fully covered and accompanied by a male relative.

The situation in northern India, where both Muslim and Hindu communities exist, is an interesting one. Low-caste (low socioeconomic ranking) Hindu women are not put into *purdah* and are free to go out as they wish. However, they are usually subjected to sexual harassment by higher-caste men. If these low-status groups are able to improve their economic standing, they begin to practice *purdah*. Increased wealth means that female labor is no longer needed outside the house and she can now be kept "ritually pure."

In many parts of India, a girl is considered nothing more than a liability. A village proverb holds that, "When you raise a daughter, you are watering another man's plant." This is said because she will be married off and so will not contribute to her parents' welfare. And, to their chagrin, her marriage family will have to support her. She will go into her husband's family compound as a lowly source of domestic labor. She will only gain status when she produces male children. Power and mobility will ensue when she enters the mother-in-law phase of her life.

If a woman's husband dies, no matter what the circumstances, she is held in disgrace. A widow is scorned and often kicked out of her house. The best scenario is that she is bound to a life of drudgery under the rule of her mother-in-law. Widows seldom remarry.

DOWRY

In Hindu marriages, the girl's family is expected to pay a **dowry** to the groom's family. This may include large sums of cash and material goods or perhaps land and cattle. This practice originated as a means of transferring wealth between high-status families. It has been outlawed by the Indian government but apparently is more common than ever before. Increasing family affluence and education for males have reinforced the dowry custom. Young men see themselves as more valuable and therefore deserving of higher dowries.

Many families cannot make their promised dowry payments because of declining economic circumstances. This can lead to "bride-burning" (also known as "dowry-death") staged as an accident by the husband's family. The National Crime Bureau of India reported 7,026 such killings in 2005.

BRIDE PRICE

Bride price refers to a payment made by the groom's family to the bride's family to compensate for loss of labor and fertility by the bride's kin group. There are various versions of the practice. For example, bride price might be simply a token exchange but is more likely to involve large sums of money, expensive goods, or perhaps livestock. Bride prices are mandatory in Islamic culture and are also paid in such countries as Thailand and China. In Thailand for instance, a bride price must include cash, very high-quality gold, and an adopted Western practice—a diamond ring.

The practice of bride price is another form of gender inequality because it is based on perceived monetary value of women. It is an economic transaction that sees women as commodities to be sold for profit. Some parents sell their daughter to the highest bidder. Men who cannot pay frequently abduct and rape the girl in order to force her family to give her to him for free. Parents usually comply because their daughter is now a family stigma and is no longer a virgin. Consequently, no other man will marry her.

WOMEN IN SOUTHEAST ASIA

In much of Southeast Asia, it is the custom for newlyweds to move in with the wife's family, which is headed by her father. Next in line, in terms of power, is not her son but the husband of her sister-in-law. Since she lives with her parents, she is not controlled by a jealous mother-in-law as is the case in South Asia. She mediates tensions between her father and her husband by passing messages and even money between the two antagonists. Consequently, she has access to a wealth of information regarding household affairs and becomes influential in decision making, which gives her a sense of empowerment.

Urbanization is changing the nature of the traditional family structure. Young couples become nuclear families as they migrate away from the countryside where extended families are the norm. In urban settings, women often hold jobs outside the home.

THE GENDER DEVELOPMENT INDEX (GDI)

The **Gender Development Index** is a statistic that indicates the well-being of women in a particular society. Ratings are expressed on a scale from 0 to 1; the closer to 1 the better. GDIs are based on various factors such as life expectancy, education, and earned income. As you can see in Table 3-3, GDIs range widely in Asia, with Japan having the highest (.992) and Nepal having the lowest (.520). Overall, the lowest ratings are found in South Asia. Six Asian countries fall below 100 in the GDI rankings of 177 countries around the world. You might wonder which one ranks first. It is Iceland, with Australia in second position. Notice that the United States ranks 16th.

Maternal mortality refers to the number of women who die in childbirth for every 100,000 live births. Numerous experts claim that the available data are unreliable and tend to be underestimated by governments. Nevertheless, in some countries, such as Nepal and Laos, the numbers are alarmingly high even if they understate the facts. In part, high maternal mortality rates reflect access to health care. However, they also reflect gender-based ill treatment. Particularly in Muslim cultures, women often are not allowed to be seen by a male physician. Too-early marriage and pregnancy, dietary deficiency, overwork, too-closely-spaced births, and late-in-life pregnancies take an enormous toll. Births are frequently unassisted or aided only by a midwife, as in Bangladesh where only 20 percent of women are assisted in childbirth by skilled personnel. In absence of proper care, complications typically mean death, and pre- and postnatal care are scant.

India is conducting an interesting experiment. A pilot program in some areas is paying US$15.00 to poor women to deliver in health centers. In addition, rural health-care workers receive US$5.00 for each woman

Table 3-3 Facts of Life for Women in Asia

Country or Region	Gender Development Index (GDI) 0-1 scale	GDI Ranking on List of 177 Countries	Maternal Mortality per 100,000 Live Births	Births Attended by Skilled Personnel in %	Literacy Rates Females/Males in %
United States	.937	16	7	100	100/100
Canada	.956	4	11	100	100/100
South Asia					
Bangladesh	.536	120	570	20	48/59
Bhutan	n.d	n.d	440	51	42/67
India	.600	112	450	47	54/77
Maldives	n.d.	n.d.	120		97/97
Nepal	.520	127	830	19	44/70
Pakistan	.525	124	320	39	40/69
Sri Lanka	.735	88	58	99	90/93
East Asia					
China	.776	72	45	98	90/96
Japan	.942	13		100	100/100
Mongolia	.695	99	46	99	98/97
North Korea	n.d.	n.d.	370	97	
South Korea	.910	26	14	100	100/100
Taiwan	.888	n.d.	n.d.	100	100/100
Southeast Asia					
Brunei	.886	31	n.d.	100	83/96
Cambodia	.594	113	540	44	68/86
Indonesia	.721	93	420	73	88/95
Laos	.593	114	660	20	67/80
Malaysia	.802	57	62	100	90/94
Myanmar (Burma)	n.d.	n.d.	n.d.	n.d.	90/94
Philippines	.768	76	230	60	86/94
Singapore	n.d.	n.d	n.d	100	94/93
Thailand	.779	70	110	97	93/96
Timor-Leste	n.d.	n.d.	380	19	
Vietnam	.732	90	150	88	87/94

n.d. No recent data available.
Sources: *United Nations Human Development Report 2007/2008; United Nations Development Program Report 2009*; UNICEF Educational Statistics; World Health Organization.

they bring in for delivery. Vouchers are provided for transportation to clinics. The initial results have been quite impressive. The proportion of women in the program that are having their babies under proper supervision has risen from 15 percent to 60 percent. Moreover, after the delivery, these women are more likely to return to the centers for birth control and postnatal services.

We can also see gender discrimination by looking at **literacy rates** and the differences between men and women. While measures of the ability to read and write vary widely from place to place, the figures do give us an idea as to educational levels. It is true that education and literacy rates in most of Asia have improved greatly over time. Whatever the case, women in general are not as literate as men.(Figure 3-6)

Education is vital for both girls and boys. It is essential for the empowerment of women. South Asia, with the exception of Sri Lanka, has a poor record of children staying

Figure 3-6

This is a boys' religious school in Pakistan's Northwest Frontier. Here the boys spend a few hours each day learning to read the Koran along with some other basic education. Note the visiting girls standing to the right. What are they doing? How will the lives of these boys and girls differ?
Photograph courtesy of B. A. Weightman.

in school. In many places, girls attend school for only one or two years and are the most likely to leave to help with domestic chores and to look after younger siblings. In Pakistan, the advancement rate of girls from primary to middle schools, which is poor to begin with, has declined in recent years. Lack of schooling, along with historical disparities between males and females, explains the male/female differential in literacy rates (Table 3-3).

Sri Lanka's Success in Education and Health

Sri Lanka invests carefully in both health and education, with particular attention being given to gender equality. Ninety percent of Sri Lankan women are literate compared to around 60 percent across South Asia. The government purports that investment in girls' schooling will result in their having higher economic value and, consequently, more influence in society.

An excellent civil registration system has recorded maternal deaths since 1900. Therefore, the country has reliable data compared to vague estimates from other places. With a nationwide public health infrastructure ranging from rudimentary clinics to provincial hospitals, women are much more likely to survive childbirth. The government has also established a wide network of trained midwives who provide prenatal care and refer risky cases to doctors. Since 1935, Sri Lanka has managed to halve its maternal mortality rate every 6 to 12 years. Today, 99 percent of births are attended by a skilled physician, and the country has one of the lowest maternal mortality rates in Asia.

Girls make up close to half the children attending school in China, but statistics do not show that they make up three-quarters of those enrolled but never attending. Some believe that education is wasted on girls, who will soon marry into another family's household. It is more practical that she help with chores at home as long as she lives there. It is well understood that, in more affluent circles, increasing educational achievement by women coincides with reduction in family size. An educated woman is more likely to understand the methods and reasons for family planning, and she is in a stronger position to make choices concerning her own life.

SEX RATIOS

Another demographic measure related to patriarchy and gender is **sex ratio**: the numerical ratio between men and women in a population. Normally, populations have a fairly balanced sex ratio. Sex ratios at birth are biologically stable (assuming no intervention). The norm is 105 male births to every 100 females. However, sex ratios in countries with long life expectancies reflect the fact that women live longer than men. Southeast Asia, where women are more highly valued, is the only Asian region to be female dominated.

In 2007, India had 108 boys for every 100 females. Pakistan had 106, Bangladesh 105, and China 107 males per 100 females. In China, among children age two or younger, there were 106 males for every 100 females. Demographers have discovered that millions of girls in these countries are "missing." In light of the importance of males in the extended and generational family, there is a strong preference for boys in many Asian cultures. In China, for instance, a boy is called a "big happiness" while a girl is a "little happiness." Some of the causes for imbalanced sex ratios are:

- Sex-selective abortions
- Excessive female child mortality mainly from deliberate neglect
- Female babies are breast-fed less than male babies
- Women often eat last and get the least
- Girls are less often taken to clinics if they are sick than are boys
- Female infanticide in some remote, tradition-bound regions

Sex-selective abortions are illegal in both India and China. If a woman has an ultrasound to determine the condition of her baby, the doctor is forbidden to inform her of its sex. However, doctors commonly use a hand signal to let her know. This is common practice in India.

The Chinese government banned sex-selective abortions in 1987 and reinforced the policy in 1994. Repressive measures are now taken against perpetrators. The 1990s saw the launching of a "Care for Girls" media campaign promoting the value of girls. Cash incentives are given to families that have girls. There are also special educational scholarships for them.

Female infanticide, the deliberate killing or neglect of girls, has a long history in Asia (and elsewhere, including the United States and Europe), particularly in times of strife. As we saw in the case of Pakistan and some areas of India, a girl might be perceived as a liability. She will offer little, if any, support to her own family because she will move into her husband's household. A boy will remain, support, and care for his own multigenerational household. In 2006, the Indian government reported that at least 10 million girls were killed by their parents in the past 20 years either by abortion or immediately after their birth. We used to hear grim reports about the killing of girls in China, but as of 2009 the Chinese government reports that the practice is very rare.

Some researchers argue that such seemingly appalling cultural practices are not simple cruelty but that they are strategies for survival in stressful life-world contexts. While this may be true, we do know that education and other improvements in quality of life witness dramatic reductions in female infanticide and infant mortality in general.

SEXUAL EXPLOITATION OF WOMEN

Poverty, lack of education for both sexes, and what might be considered medieval attitudes contribute to barbaric treatment of women worldwide. Rape of women and young girls is an all too common event, especially in South and Southeast Asia. In Pakistan women can be arrested for even minor infraction of male sensibilities. A woman who comes out of prison without having been raped is rare. Generally, if a woman is raped, it is thought to be her fault and she is cast out of society or even killed. This is particularly the case in Muslim and many Hindu regions.

In Pakistan for example, a rape victim is considered to have brought "dishonor" to the family. This notion has led to many so-called **honor killings** whereby the girl is killed by a male relative, typically a brother, an uncle, or even her father. The United Nations Population Fund has estimated that there are at least 5,000 honor killings a year, almost all in the Muslim world. Pakistan's government uncovered 1,261 such murders in 2003 alone. That estimate is likely too low because so many executions of women are disguised as accidents or suicides.

Inferior status is embedded into women's psyches in traditional, patriarchal societies from birth. Acceptance of a life of pain and drudgery is a part of life in these male-dominated worlds. Abused and angry wives and mothers-in-law turn their innate fury on their daughters. There are many cases where mothers hold their daughters down while they are raped and subsequently killed for some perceived infraction against the "family honor." Significant numbers of parents sell their daughters into virtual slavery to be severely mistreated servants or sex workers.

Violence against women is constantly mutating into new forms. One of the latest and increasingly common atrocities is for men to throw acid into the faces of women who spurn them. These are referred to as *acid attacks* in recent literature on the subject.

Women are attacked for a variety of reasons such as being "immodest" (not being covered properly) or rejecting a man's advances or marriage proposal. The first recorded acid attack took place in East Pakistan (now Bangladesh) in 1967. In 2008, a woman was blinded and disfigured by acid in Bengaluru (Bangalore), India's latest and greatest high-tech and call center.

Hydrochloric or sulphuric acid is the weapon of choice because it is commonly used as a household cleaner; it is available and cheap. Acid is usually aimed at the face although it often is poured over the whole body. It eats away skin and even dissolves bones. Victims suffer terribly and are horribly scarred for life.

It must be pointed out that not all victims are women. Class conflicts, property disputes, and other disagreements may drive men to attack other men. However, 80 percent of known attacks are against women and 40 percent of these are against girls under the age of 18.

As of 2009, there are many reliable reports of acid attacks increasing dramatically in India, Pakistan, Bangladesh, Nepal, and Cambodia. If you go online and look up "acid attacks in Asia," you will find stories about survivors who now devote their lives to helping other victims.

Millions of women and children under 18 slave as prostitutes in Asia. Roughly 60,000 girls offer sex services in the Philippines. Thailand has as many as 800,000 children working in the "sex industry," and it is estimated that 10,000 girls are bought from Myanmar each year to replace those overworked, diseased, or dead. In Phnom Penh, Cambodia, a six-year-old can be acquired for as little as three dollars. As many as 50,000 Nepalese girls work as prostitutes in India. Prostitutes are deemed essential for indigenous males. Mores in several countries restrict sexual behavior for females but not for males. Millions of men work away from their homes for weeks or months at a time, and prostitutes are seen as contributing to their well-being. Since local women are not readily available, imports are required.

Human trafficking of girls within and between countries for purposes of sex is a worldwide phenomenon. Countless millions of girls as young as six or eight years old are lured, kidnapped, sold, coerced, or otherwise forced into prostitution. In Asia, Myanmar (Burma), Laos, Cambodia, Nepal, and the Philippines are notorious source regions.

"Labor agents" recruit girls from poor regions by paying their parents a lump sum that then becomes a debt. Parents are assured that their daughters will be taken to work for a rich family, send money home, and attend school. In the face of abject poverty, this is seen as a blessing. In fact, the girl will be detained in debt bondage—forced to work in a brothel to pay her "debt" plus the cost of food and clothing.

In India, labor agents scour the region for children of the impoverished. Nepal is a virtual gold mine in this regard. There is also a tradition of declaring one's daughter a *devadasi*, or Hindu temple servant. Dedicated to a particular goddess, a *devadasi* participated in ritual singing, dancing, and other religious practices. However, tradition has been corrupted, and today many of these girls are forced into prostitution by the unscrupulous. The practice has been outlawed, yet it still accounts for at least 20 percent of prostitution in India. For example, there are an estimated 50,000 Nepali girls, mostly between the ages of 10 and 14, working in Indian brothels.

Girls are usually raped and beaten into submission by brothel owners. Very often they are forcibly turned into heroin or methamphetamine addicts. Girls "serve" customers for 12 hours a day, 7 days a week. They are rarely allowed out and are often chained to their cots. Police seldom do anything since they are bribed and often use the brothels themselves. Moreover, authorities frequently look upon the girls as less than human because they are usually from the poorest and lowest classes of society. Further, many believe that if men use prostitutes, "their own" women will be protected and remain pure/virginal for their inevitable arranged marriages. (Figure 3-7)

The story of Srey Rath of Cambodia illustrates what happens to so many young girls in Asia. This account is drawn from a must-read book recently penned by two Pulitzer Prize–winning authors: Nicholas Kristof and Sheryl Wudunn (*Half the Sky: Turning* Oppression *into Opportunity for Women Worldwide*. New York: Knopf, 2009).

The Saga of Srey Rath

When Rath was 15 her family ran out of money so she decided to get a job in Thailand as a dishwasher. A "job agent" offered her a job in a restaurant but took her deep into the Thai interior and handed her over to gangsters in Malaysia, where she was taken to a karaoke lounge that doubled as a brothel in Kuala Lumpur. "The boss" told Rath that she owed him money for her journey. He put her into a room with a man who tried to force her to have sex. She fought him off but when the enraged man complained to the boss, he and the other gangsters raped her, beat her, and threatened her with death. She was forced to take pills and soon became addicted. Constant batterings eventually forced her to smile for her customers. She was never allowed

Figure 3-7

I discovered this sign in Laos. It is informing people, especially "bridge populations," (men who work away from home) about the risk of AIDS from unprotected sex. Photograph courtesy of B. A. Weightman.

out and was kept naked to make it almost impossible for her to get away.

By a miracle of ingenuity, Rath finally escaped and went to the police where she was arrested for illegal immigration. After serving a year in prison, a Malaysian policeman drove her to the border with Thailand, where he sold her to a Thai brothel. Fortunately, she was not beaten there nor was she well guarded. After two months, she was able to escape and make her way back to Cambodia.

Once in Cambodia, she was fortunate enough to gain the attention of a social worker from the aid group American Assistance in Cambodia. She was given US$400 to open a souvenir stall in Poipet at the Cambodia–Thailand crossing—a town famous for its brothels.

When I got there on a jam-packed bus, I and the mob got off and dragged our bags along a dirt strip about the length of a football field. Part way along, little boys came running to carry the bags of foreigners and others they thought could afford the miniscule fee of a few coins. If you insisted on carrying your own luggage, you were swarmed, grabbed, and clawed by tough, desperate limbs and your stuff quickly thrown into rickety wheelbarrows or stacked on their head. There were plenty of young girls, their faces plastered with makeup, hanging around outside an array of shacks. I stayed at a medium-low end hotel that had a large teak-floored veranda with a bar. Several older, Caucasian and East Asian men sat around with their girls. I didn't dare interfere because I probably would have been arrested and thrown into some hovel never to be seen again. In 2008, the Thai government raided Poipet and half the brothels closed. As far as I know, Rath is still there selling her baseball hats and T-shirts.

What is the United States doing about human trafficking? As of this writing, only about one percent of the American foreign aid budget is directed toward women's issues around the world. However, the U.S. Department of State now is strongly involved in the issue of human trafficking. The agency is trying to coerce countries to take strict and legitimate actions against traffickers. Its 2009 *Trafficking in Persons Report* assesses a country's laws and programs and places it in one of four tiers of progress:

Tier 1: Countries whose government fully complies with the Trafficking Protection Act (TVPA)'s minimum standards. (South Korea)
Tier 2: Countries that do not comply but are making efforts to comply. (Brunei, Indonesia, Japan, Laos, Maldives, Nepal, Singapore, Taiwan, Thailand, Timor-Leste, and Vietnam)
Tier 2 WATCH LIST: Numbers of victims are rising and governments fail to provide valid evidence of efforts. (Cambodia, Bangladesh, China, India, Pakistan, and Sri Lanka)
Tier 3: Countries do not comply and show no significant efforts to do so. (Malaysia, Myanmar (Burma), and North Korea)

The American government is finally exerting meaningful pressure on these countries to get rid of corruption, make arrests, and dole out tough punishments to perpetrators. Bangladesh has assigned the death penalty to traffickers, but corruption has thwarted many of their efforts to stop the process.

Another insidious form of trafficking is becoming increasingly common in China. Because of a stringent population policy that has limited most people to having only one child, plus the fact that boys are preferred, there is a shortage of women to marry. Consequently, trafficking in potential brides has become a profitable business. Usually in collusion with local officials, traffickers kidnap young women from their villages and deliver them to another village far away where there is a man willing to pay for a wife.

Another lucrative racket in China involves the selling and trafficking of babies, usually female. A number of these children have been unknowingly adopted by American couples who pay more than US$25,000.

Globalization and the Sex Industry

Globalization is a contemporary and growing trend in the sex industry that has long been sustained by Europeans, North Americans, Australians, Middle Easterners, Japanese, and other groups. This flourishing enterprise has spawned a counterpart: "sex-worker/-entertainer" migration streams. The Philippines, Thailand, China, and other Asian sources provide thousands of sex workers to such key destinations as Japan, Taiwan, South Korea, Saudi Arabia, and other Middle Eastern countries. Some 35,000 women leave the Philippines yearly to work as "female entertainers" in Japan. Terms such as *entertainer* or *artist* mask the reality of "body as commodity." Although some nations such as Thailand and the Philippines have enacted restrictions and regulations for employment abroad, they are largely ignored.

SEX TOURISM

Globalization has spurred on **sex tourism**, which is big business in Asia. Brothels are ubiquitous, especially in cities. Some, including Bangkok, Ho Chi Minh City (Saigon), Taipei, and Manila, are known as **sex cities** because of the thousands of mostly heterosexual men who patronize bars, sex shows, massage parlors, private clubs, and brothels that cater to men's desires. Very young virgins are in high demand. No wish goes unmet. More than 60,000 girls offer sex services in the Philippines, and the number in Thailand is over 800,000. Some Japanese companies organize sex trips to these centers for their employees.

Songachi, which means "golden tree," is a sprawling red-light district in Kolkhata (Calcutta), India. Hundreds of multistory brothels lining dirty, narrow alleyways hold more than 6,000 prostitutes. Health workers started a sex-workers' union in 1992 with the assistance of WHO in an effort to increase condom use and reduce the incidence of AIDS. This is known as the Songachi Project, which promotes the "3 Rs"—Respect, Reliance, and Recognition—and trains older sex workers and madams to promote the use of condoms. Condom use did increase and HIV infections went down to only 9.6 percent in 2005. (This stands in sharp contrast to the 50 percent infection rate in Mumbai (Bombay) brothels.) However, all is not well in Songachi. New studies show that HIV prevalence is particularly high among new arrivals—27.7 percent among girls age 20 or younger. It turns out that girls who say they always use condoms actually don't. Since men don't want to use them for paid sex, girls are forced to lie to investigators by their pimps and brothel owners. (Figure 3-8)

Boys are not immune to the sex industry. They often see prostitution as a way out of poverty, and many pursue the urban sex trade. Both boys and girls participate in the exceedingly lucrative, worldwide, pornographic movie and video business.

PROSTITUTION IN CHINA

China has more prostitutes than India. Some estimates are as high as 10 million. However, there is a difference from other Asian countries. Few women are forced into brothels against their will. Most prostitutes are freelancers who have choices out of several options.

A well-educated woman might become a paid companion of a rich politico or businessman on a business trip. She might become the long-term mistress of a wealthy man. She might hold a job in bar, club, or massage parlor working on commission. Or, she might prefer to work as a *ding-dong xiaojie* calling hotel rooms at a late hour to offer a "massage."

A significant number of women manage to escape North Korea into the northeastern region of China formerly known as Manchuria. In fact 60 to 70 percent of defectors are female. Many turn to prostitution for lack of other opportunities to make a living. In that region of China, there are many large industrial cities with hundreds of thousands of potential customers. However, the Chinese government does not want North Koreans in their country. They arrest the women and deport them back across the border. There they are ar-

Figure 3-8

This arranged marriage took place in a Dayak longhouse in Indonesian Borneo (Kamantan). Thirty families and a three-legged dog lived in the longhouse. The longhouse, with its multiple rooms, had no modern facilities except for a very small TV that ran on a car battery. Bride price negotiations, which lasted five hours, were conducted between the respective fathers as the groom's mother shouted instructions to her husband from the background. The bride (age 15), who appeared to be terrified, and the groom (age 17) who seemed to be pleased, sat rigidly throughout the whole affair. The actual ceremony, conducted by a shaman, finally took place at 3:00 a.m. Photograph courtesy of B. A. Weightman.

rested again by the North Korean authorities and sentenced to years of hard labor in notoriously harsh penal colonies. If they have a Chinese-fathered baby, it is executed. If the girl is pregnant, she is forced to abort and the baby is killed.

AIDS AND PROSTITUTION

According to the World Health Organization (WHO), there were 30–36 million people with HIV-AIDS worldwide as of 2007. In that same year, an additional 2.5 individuals became infected. The major causes are unprotected sex and drug use, both of which are on the rise.

In Asia, there are an estimated 5 million victims including the 380,000 that were added in 2007. The highest rates of infection are in South and Southeast Asia where, according to UNAIDS (United Nations), sex workers make up an estimated 50 percent of the cases.

According to WHO and United Nations AIDS (UNAID), the rate of infection has reached epidemic proportions in India, which has at least 2.5 million cases of AIDS. Women comprise nearly 40 percent of the victims. A recent study of prostitutes in Mysore revealed that one-quarter of the women were infected. Other investigations reveal that 42 percent of sex workers believe that they can tell if a man had AIDS by his appearance.

The Indian government has been building AIDS testing centers and there are now about 4,000 of these across the nation. However, when you consider that there are more than 600,000 villages, let alone towns and cities in the country, 4,000 centers are merely a drop in the bucket. And, in Indian society, having AIDS is a stigma. Consequently, many people avoid testing.

There are 70,000 AIDS sufferers in Nepal. Seasonal migrant workers make up 41 percent and partners and wives of HIV infected men make up 21 percent. As is the case in all of Asia, sizeable numbers of men work away from home, many of them as truck drivers. These men are called "bridge populations" because while away from home, they contract AIDS from prostitutes and upon returning home, they infect their wives.

AIDS infection remains a serious problem in Southeast Asia. In fact, infection rates have risen in Vietnam and Indonesia. However, some progress has been made in battling the disease in Thailand, Myanmar, and Cambodia. Thailand, for example, began combating AIDS in the early 1990s. One of its successes is its 100% Condom Program, which mandates condom use in all brothels.

The Philippines has low rates of AIDS occurrence. The government began to screen and treat infected prostitutes in the early 1990s. However, there is evidence of complacency regarding the disease among Filipino youth.

Mother to child transmission is a very serious problem in Asia. Most mothers get AIDS from their husbands, who have contracted it from prostitutes or from homosexual encounters. Since homosexuality is scorned in most Asian cultures, many gay men get married and have children and their sexual encounters are clandestine. It was estimated in 2007 that 140,000 children in South and Southeast Asia and 7800 in East Asia were infected with AIDS. The vast majority contracted the disease from their mothers.

Until recently, AIDS in China was a taboo subject. Any evidence of the disease was denied. Then, the government said that only foreigners got AIDS. It was referred to as *alzibing*—"loving capitalism disease."

The first case to be recorded was in Beijing in 1995. At the same time, authorities finally admitted that there were significant numbers of AIDS cases in Yunnan Province in southwestern China, where drug use is widespread. Then the fact that thousands of people had contracted AIDS via contaminated blood transfusions hit the international news. The Chinese government had no choice but to admit that there was an AIDS problem. Official figures state that the country has 700,000 AIDS victims of which 200,000 are women. There is a media blitz against unprotected sex and drug use, and 75,000 individuals are in treatment programs.

Although the availability of antiretroviral drugs has increased in Asia, only about one-third of those infected are receiving treatment across the region. In India, in 2004, only 2 percent of HIV positive, pregnant women received antiviral prophylaxis, typically only one dose. Although appropriate medication is readily available over-the-counter at pharmacies, both doctors and pharmacists usually recommend a regimen of one pill three times a week for 21 days—totally ineffective treatment.

In 2007, the Indian government launched a new phase of its National AIDS Control Program (NACP) that will broadly communicate information on prevention and improve and expand treatment centers. Another part of NACP is directed at reducing stigma and discrimination against AIDS victims. It has been common practice for both government and private hospitals to throw patients out if they are found to be HIV positive. There is also a campaign to overturn laws that make homosexuality illegal and punishable by imprisonment.

Interestingly, Indian pharmaceutical companies are major suppliers of low-cost, generic antiretroviral drugs in Africa and elsewhere while HIV treatment needs in India are not being met. For example, Cipla, based in Mumbai, exports 18 times as many antiretroviral drugs as it sells in India.

Drugs required to abate the progression of AIDS are too expensive for most people in Asia. If more had access to cheaper, generic medications, treatment rates would most likely go up. However, government funding shortages, individual poverty, lack of roads and public transportation, and inadequate health facilities all combine to make it very difficult to organize lifelong treatment programs for millions.

The increase in HIV infection and AIDS cases has upped the demand for even younger "clean" sex workers. Men pay premium prices for virgins, believed to be disease-free. Once prostitutes are "too sick to work," they are left to the streets or "repatriated" to their home country to die. Some girls do manage to survive and even make money. Many of these return home to their villages as heroic "dutiful daughters" who have sacrificed themselves for their family. Their behavior is a model for the recruitment of younger sisters. However, the majority end up dead at a young age from brutality, drug overdose, or AIDS.

FEMINIST PERSPECTIVES

Most feminists agree that prostitution is not a conscious and calculated choice. Poverty, lack of opportunity, and domestic violence, among other factors, drive multitudes of women into prostitution. Force and coercion are employed by unscrupulous men (rarely women) to get girls into the business. This form of male domination over women results from a patriarchal social order that subordinates females to males.

In patriarchal societies, inequality between genders is embedded in all aspects of life. Prostitution is seen as vital for men who "cannot control themselves." It is acceptable for a relatively small number of women to be sacrificed—that is, used and abused to protect larger numbers of "chaste" ones from rape and harassment.

Prostitution is not only a woman's problem. Geographer Joni Seager (1997) states: "Prostitution is not a women's institution—it is controlled by men and sustained by violence." While local-level prostitutes might have a measure of control over their own destiny, "the global sex trade is almost entirely coercive, sustained by high levels of violence and predicated on the thorough subjugation of women."

Female Heads of State in Asia?

You probably know that several Asian countries have had female leaders. How is this possible in such patriarchal cultures? Look at Table 3-4, which lists Asian, female

Table 3-4 Female Heads of State in Asia

Country	Name of Leader	Term(s) as President or Prime Minister	Political or Family Connections	Married	Children
Bangladesh	Begum Khaleda Rahman Zia	1991–1996 2001–2009	**Widow** of Ziaur Rahman, President from 1977 to 1981 who was assassinated in 1981	Yes	Yes
	Sheik Hassina Wajed	Current	**Daughter** of Sheik Mujibur Rahman, founding father of Bangladesh and first President, who was assassinated in 1975	Yes	Yes
India	Indira Gandhi	1966–1977 1980–1984 Assassinated in 1984	**Daughter** of Jahawarlal Nehru, India's first Prime Minister who died in 1964	Yes	Yes
Indonesia	Megawati Sukarnoputri	2001–2004	**Daughter** of Sukarno, who was the first President of the world's most populous Muslim nation from 1949 to 1965	Yes	Yes
Pakistan	Benazir Bhutto (First woman President of an Islamic country and at age 35, the youngest chief executive in the world.)	1988–1990 1993–1996 Assassinated in 2007	**Daughter** of Zulfikar Ali Bhutto, Prime Minister from 1973 to 1977 who was executed in 1979	Yes	Yes
Philippines	Gloria Macapagal Arroyo	2001–2009	**Daughter** of Diosdado Macapagal, President from 1961 to 1965	Yes	Yes
	Corazon Aquino	1986–1992 Died 2009	**Wife** of Senator Benigno Aquino, who was assassinated in 1983	Yes	Yes
Sri Lanka	Madame Sirinovo Bandaranaike (World's first female Prime Minister)	1960–1965 1970–1977 1994–2000	**Wife** of former Prime Minister Solomon Bandarogo	Yes	Yes
	Chandrika Bandaranaike Kumaratunga	1994–2000	**Daughter** of Madame Bandaranaike	Yes	Yes

Figure 3-9

Corazon "Cory" Aquino was the Philippines' first female leader serving as President from 1986 to 1992. She oversaw the restoration of democracy and promulgated a new constitution after the harsh and corrupt rule of strongman Ferdinand Marcos. (You might have heard of his wife Imelda, who is reputed to have owned 2,000 pairs of shoes.) Promoting herself as "just a plain housewife," Cory was proclaimed the "Saint of Democracy" by TIME *Magazine. Sadly, she died of cancer in 2009.* ©Sandro Tucci/Time Life Pictures/Getty Images, Inc.

heads of state in the twentieth and twenty-first centuries. Let's see what they have in common. (Figure 3-9)

Several points are apparent. All these women are daughters, wives, or widows of politically influential and wealthy families. Their rise to power is linked to the notion of "dynasty," which generates a succession of hereditary rulers from prominent and powerful families through several generations. Furthermore, these women have "fulfilled their proper role" as wives and mothers. They also symbolize nonpartisan alternatives to corrupt male leadership even though they might become corrupt themselves.

It should be pointed out that substantial numbers of women have become active in programs to promote gender equality at the grass-roots level in most Asian countries, especially in India, Bangladesh, and various countries in Southeast Asia such as Indonesia and the Philippines. For example, the UN-supported International Fund for Agricultural Development (IFAD) assists women's groups that make efforts to increase female self-esteem and help them cope with family injustices including domestic violence.

In Pakistan, those women's organizations that do exist battle culturally embedded attitudes of male superiority throughout society. In 1996, the country's government ratified a UN convention to end gender discrimination. According to Amnesty International, women's rights continue to be routinely ignored and violated in Pakistan.

Given their sheltered and largely uneducated circumstances, most women in patriarchal societies are unaware of their inferior status. They accept 12 or more hours of hard labor in and out of the home, eating last and least, constant pregnancy, forced sex, beatings, and even death as their lot in life. These women do not know that they have even the potential to make a decision or control anything. Even if they are aware of their lesser position, most often there is little they can do about it.

Recommended Web Sites

www.amnestyusa.org
Amnesty International follows human rites violations around the world.

www.avert.org/aroundworld.htm
International AIDS charity that has an excellent and informative site with articles and global statistics.

www.brac.net
Bangladesh Research Advancement Committee for alleviation of poverty and advancement of the poor.

www.census.gov/ipc/www/idb/index/.php
U.S. Census international data base statistics and population pyramids.

www.ecpat.net
A network of groups fighting child prostitution, especially in Southeast Asia.

http//:genderindex.org/content/team
Social Institutions and Gender Index plus information and statistics by country.

www.path.org
Nonprofit international organization. Information on health, disease, family planning.
www.populationaction.org
Policies and programs to slow population growth and enhance quality of life for all.
www.prb.org
THE site for population statistics.
www.state.gov/g/tip/rls/tiprpt/2009/index.htm#
U.S. government HumanTrafficking Report.
www.unaids.org
United Nations data on AIDS by region and country.
www.unicef.org
United Nations Children's Fund home page.
www.un.org/en/index.shtml
United Nations home page.
www.who.int./hiv/data/en/
World Health Organization information on AIDS.
www.world.gazeteer.com
Population clock–watch population growth as it happens. Maps and data.

Bibliography Chapter 3: Population, Gender, and Disparity

Asia Watch Women's Rights Project. 1993. *A Modern Form of Slavery: Trafficking of Burmese Women and Girls into Brothels in Thailand.* New York: Human Rights Watch.

Barkat, Abdul et al. 2009. *Child Poverty in Bangladesh.* Dhaka: Human Development Research Centre/UNICEF.

Bhatnagar, Rashmi D., and Reena Dube. 2005. *Female Infanticide in India; A Feminist Cultural History.* Albany: State University of New York.

Chandiramani, Misra and Radhika. 2005. *Sexuality, Gender and Rights: Theory and Practice in South and Southeast Asia.* New Delhi: Sage.

Gore, Rick. 1997. "The Dawn of Humans." *National Geographic* 191: 84–109.

Guilmoto, Christophene, L. 2005. *Sex-ratio Imbalance in Asia: Trends, Consequences and Policy Responses.* Paris: Institute for Research and Development.

Hausmann, Ricardo et al. 2008. *The Global Gender Gap Report.* Geneva: World Economic Forum.

IFAD. 2000. *Gender Perspective: Focus on the Rural Poor.* Rome: International Fund for Agricultural Development.

Kristof, Nicholas D., and Sheryl WuDunn. 2009. *Half the Sky: Turning Oppression into Opportunity for Women Worldwide.* New York: Alfred A. Knopf.

Levine, Ruth. 2004. *Millions Saved: Proven Successes in Global Health.* Washington, D.C.: Center for Global Development.

Mortenson, Greg, and David O. Relin. 2006. *Three Cups of Tea.* New York: Penguin.

Population Reference Bureau. 2009. *Population Data Sheet.* Washington, D.C.

Riley, Nancy E. 1997. "Gender, Power, and Population Change." *Population Bulletin* 52/1. Washington, D.C.: Population Reference Bureau.

Seager, Joni. 2009. *The Penguin Atlas of Women in the World,* 4th edition. New York: Penguin.

Sen, Mala. 2001. *Death by Fire: Sati, Dowry and Female Infanticide in Modern India.* New Brunswick, N.J.: Rutgers University Press.

Steinbrook, Robert. 2007. "HIV in India: The Challenges Ahead." *The New England Journal of Medicine* 356/12: 1197–1198.

UNICEF. 2009. *A Report Card on Child Protection.* No. 8, September.

USDS. 2009. *Trafficking in Persons Report.* Washington D.C.: U.S. Department of State.

Chapter 4

Development, Urbanization, Migration, and Quality of Life

"If there were no contradictions and no struggle, there would be no world, no process, no life, and there would be nothing at all."

MAO ZEDONG (1893–1976)

Development

There are many ways of defining and assessing development. No two countries are exactly alike in measurable aspects of development such as per person income, calorie intake, or number of doctors. More importantly, different countries have different ideas about what development is. For example, the strength of family ties may be more important than the availability of a medical doctor. Access to sufficient food may be more important than monetary income. Whatever the case, development is a concept widely used by governments and aid organizations in determining needs, investments, and project applications. In this chapter you will learn about various perceptions and impacts of development.

MEASURING DEVELOPMENT

All nations have been assigned descriptors based on criteria such as **per capita income (PCI)** based on gross national or domestic product (GNP or GDP). PCI is quite misleading, in part because it does not provide information on the buying power of a currency. A more contemporary measure of income is PCI based on **purchasing power parity (PPP)**. PPP is based on GDP, accounts for price differentials among countries, and translates local currencies into international dollars. However, PCI measures do not recognize non-monetary transactions, such as bartering goods for services that are very important in many Asian societies. Also, they assume an unrealistic equal (per person) distribution of money Table 4-1 shows a variety of economic development measures.

More informative measures of development include the **Human Development Index (HDI)** and the **Gini Index (GI)**. The HDI is a composite index from 0 to 1 with 0 at the low end of the scale. HDI measures a country's average achievements in three basic aspects of human development: health, knowledge, and a decent standard of living. These categories look at various factors such as availability of doctors, level of education, and access to clean water.

HDIs suggest that people are better off in Japan or South Korea where the standard of living is very high. Thais are better off than Cambodians but not as well off as Singaporeans. Overall, people in East Asia have better standards of living than people in South or Southeast Asia.

The Gini Index measures relative income inequality within a country. A country rates a GI of 0 if it has absolute equality or a GI of 1 if it has absolute inequality. For example, Denmark, with a GI of 0.24, has the world's most equal society in terms of income distribution. There, the lowest 20 percent of society (in terms of income) controls 10.6 percent of PCI and the highest

Table 4-1 Selected Development Measures 2007/2008

Region Country	GDI per capita PPP/US$	HDI 0–1 (higher score is better)	HDI Rank out of 177	GI 0–1 (lower score is better)	CPI 0–10 (higher score is better)
Australia	34,923	.970	2	.352	8.7
Canada	35,812	.966	4	.326	8.7
United States	45,592	.956	13	.408	7.3
South Asia					
Bangladesh	1,241	.543	146	.334	2.1
Bhutan	4,837	.619	132	n.d.	5.2
India	2,753	.612	134	.368	3.4
Maldives	5,196	.771	95	n.d.	2.8
Nepal	1,049	.553	144	.472	2.3
Pakistan	2,496	.572	141	.306	2.4
Sri Lanka	4,243	.759	102	.402	3.1
East Asia					
China	5,383	.772	92	.469	3.6
Japan	33,632	.960	10	.249	7.7
Mongolia	3,236	.727	115	.328	3.0
North Korea	n.d.	n.d.	n.d.	n.d.	n.d.
South Korea	24,801	.937	26	.316	5.5
Taiwan	n.d.	n.d.	n.d.	.560	5.6
Southeast Asia					
Brunei	50,200	.920	30	n.d.	5.5
Cambodia	1,802	.593	137	.417	2.0
Indonesia	3,712	.734	111	.343	2.8
Laos	2,165	.619	133	.346	2.0
Malaysia	13,518	.829	66	49.2	4.5
Myanmar (Burma)	0,904	.586	138	n.d.	1.0
Philippines	3,406	.751	105	44.5	2.4
Singapore	49,704	.944	23	42.5	9.2
Thailand	8,135	.783	87	42.0	3.4
Timor-Leste	0,717	.489	162	n.d.	2.2
Vietnam	2,600	.725	116	34.4	2.7

Sources: UN Human Development Index Reports 2007–2009. Transparency International Corruption Index 2009.

20 percent controls 37.7 percent. Malaysia's GI is 4.92. This index reflects the fact that the poorest 20 percent of the population controls only 4.4 percent of PCI. More than half (54.3%) is controlled by the wealthiest 20 percent.

Income inequality is often referred to as the **rich-poor gap**. Income inequities are represented by GI scores in Table 4-1. GI scores indicate degrees of income inequity within a country but tell us nothing about spatial patterns of inequality. Marked regional disparities are apparent in every country. For example, northern Bangladesh is much poorer than the central and eastern regions.

Figure 4-1 is a cartogram that shows countries sized to reflect wealth. The smaller the country the poorer it is. You can see, for example, that Singapore, Japan, and Taiwan are much larger—that is, richer—than India or China. Look at the relative size of tiny Brunei. Why do you think that such a small country can be so wealthy?

The percentages of poor people are decreasing in Asia and around the globe. However, absolute numbers of poor are rising because of population growth. Rural people tend to be poorer than urban people, although there are many poverty-stricken people of

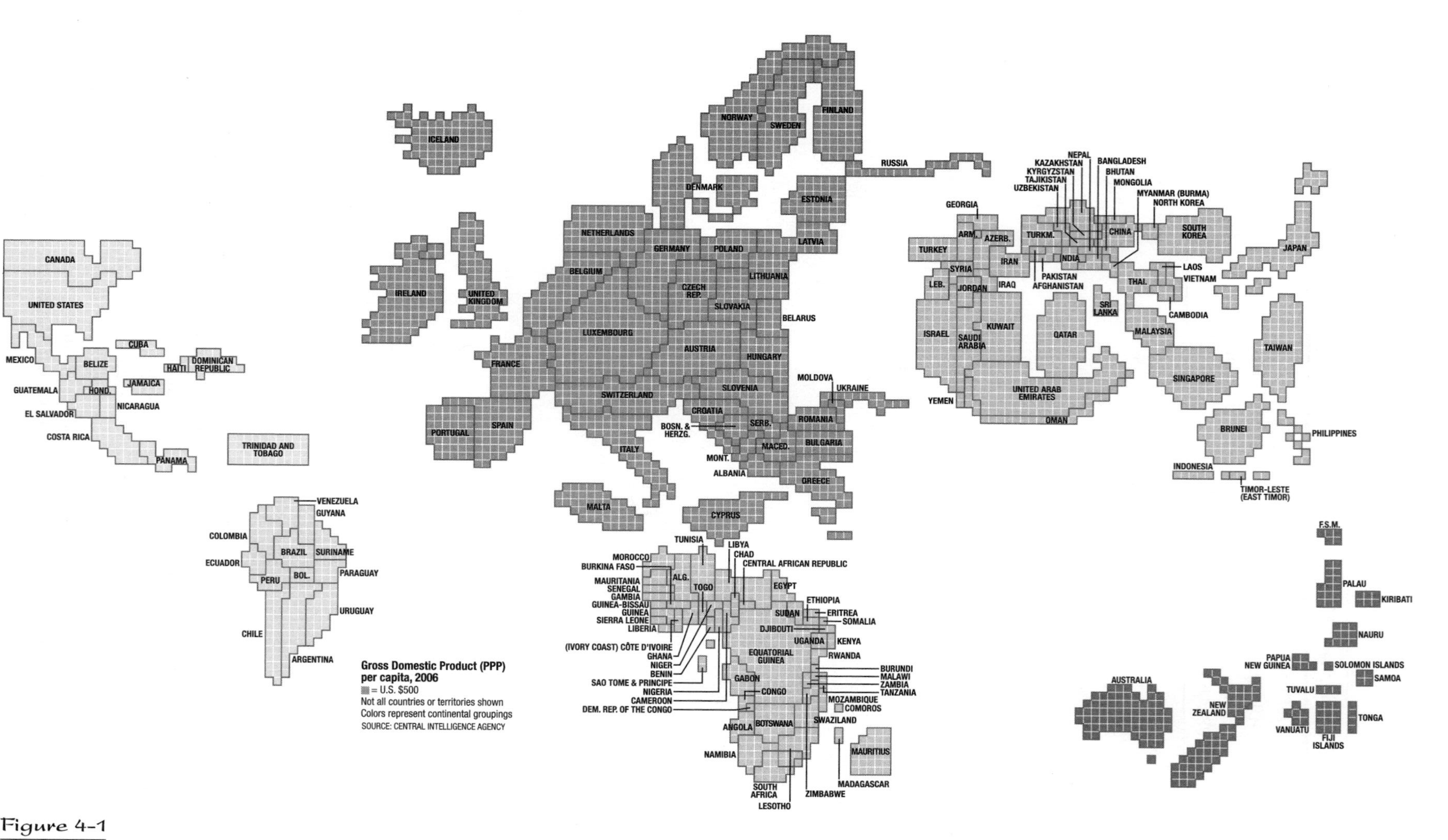

Figure 4-1

In this cartogram, countries are sized according to their relative wealth. Which countries are the richest and poorest in Asia?

From *National Geographic Earth Pulse*. Reprinted courtesy of John Wiley & Sons Ltd. ©NG Maps.

rural origins living in cities (see the section on urbanization below).

Corruption in poor countries can mean the difference between life and death because it means that money that is supposed to be used to improve people's lives is going into the pockets of business tycoons and government officials. High levels of corruption and poverty plague countries and regions around the world.

Transparency International assesses levels of corruption in business and government. It employs the Corruption Perception Index (CPI) to rank 180 countries. The CPI rates a country on a 0–10 scale with 10 being the least corrupt. As of 2008, Denmark ranked number 1 with a CPI of 9.3. Japan and the United States tie for number 18 with CPIs of 7.3. Note that 13 countries in Asia fall into the lower half of the rankings. Myanmar (Burma), for instance, with its CPI of 1.3 is almost at the bottom of the list at number 178.

SCHOOLS OF THOUGHT ON DEVELOPMENT

Assessment of a combination of development factors can used to position a country on a perceived continuum of socioeconomic development. This development spectrum is based on misleading pairs of opposites, such as less developed or more developed, overdeveloped or underdeveloped, and rich or poor. Geographers commonly use the relative terms *developed* and *developing* to indicate relatively rich or relatively poor countries.

After the rise of communism in Russia and China, development experts assigned countries ideologically based designations. Technologically developed, democratic nations formed the *First World,* the communist-bloc nations were called the *Second World,* and the remainder became the *Third World.* When the Soviet Union collapsed and China began adopting aspects of capitalism, and poorer countries such as Malaysia emerged as Economic Tigers, these assignations became obsolete. Nevertheless, you will still hear the term "Third World" being used by some people.

The **flying geese model**, developed by a Japanese economist in the 1930s, was used in the 1960s to describe industrial life cycles in recently industrialized nations. Industry- and setting-specific at first, the model was advanced to examine changing industrial structures and international shifts in location.

According to the model, an industry starts with a product import or technology transfer from an industrialized nation. An imitation is domestically produced and substitutes for the original product import. As domestic production exceeds domestic consumption, imports decline and an export surplus is created. Subsequently, exports and production begin to decline. As domestic costs increase, industry looks elsewhere and becomes a net importer again. Industries are thereby cycled in an orderly fashion, akin to the structured pattern of flying geese. Over time, industries rise and fall, with the old being replaced by the new. A country that produced textiles in the 1920s might have focused on steel in the 1940s, chemicals in the 1960s, and automobiles in the 1980s. Each industry is increasingly sophisticated and technologically complex, illustrating the development process.

The "flying geese" analogy can also be applied to a sequence of countries as each flows along its flight path of industrial development. For example, in the 1950s, a labor-intensive manufacturing focus (textiles) characterized Japan as a NIC. This has been replaced by capital-intensive machine industries to the extent that Japan was classified as an industrialized country in the 1980s. From an international perspective, the "geese" of modernization are seen flying from more developed countries to less developed ones. At the same time, a sophisticated division of labor emerges among Asia-Pacific nations.

The 1990s witnessed criticism of the flying geese model of development. Perhaps applicable in the past with its vision of Japan leading the Asian flock down the road of progress, the model overemphasizes Japan's role and underplays that of the United States as an export market for Asian goods. Further, it does not account for the fact that imports might be from one's own international subsidiaries or joint ventures, or exports might be dominated by primary (non-manufactured) products.

Through the latter part of the twentieth century, several schools of thought prevailed. For example, development trajectory proponents purported that all nations could proceed along an evolutionary path of economic advancement according to progressive stages already experienced in the West. Marxists stressed the ills of capitalism and class exploitation and the need to alter unequal power relationships within and between nations.

Another criticism rests on the fact that industries do not pass *in toto* from one country to another in a linear pattern. Rather, change is manifested through subsector specialization. Each time a new technology is commercialized, a new pattern of clustered economies and division of labor will appear. Furthermore, postmodern-era production is interdependent, at once concentrated and dispersed across hierarchies transcending multiple borders. According to economists Mitchell Bernard and John Ravenhill (1995), firms no longer simply seek comparative advantage in location relative to raw materials or labor force for an entire production process. They assign location in terms of "the comparative and technological advantage of *particular* territorial sites for *particular* parts of the production process." Parts of the same product may be produced within the same industry at the same time in different countries.

As you will see in the upcoming regional chapters, the meaning, practice, patterns, and consequences of development are in constant flux. What does *development* mean to you? How has it changed your community or city? Think about these things, because according to the Chinese philosopher Hsun-tzu (300–230 BC), "Through what is near, one understands what is far away."

A more recent theory argues that in an interdependent world, it is essential that balance be achieved to ensure mutual benefits through the control of trade between the many rich, industrialized countries that are in the Northern Hemisphere and the many poorer, former Third World countries that are in the Southern Hemisphere. This **North-South dichotomy** partitions the world into essentially superior and inferior realms and serves the notion of socioeconomic divergence between the haves and have-nots. It also connotes the notion of underdevelopment whereby the South's lack of development is at least partly a function of the North's development. Note that highly developed Australia and New Zealand are in the Southern Hemisphere.

You will remember from Chapter 1 that Immanuel Wallerstein proposed a *world systems theory* whereby all countries are inexorably connected through a global network of capitalism. Of course, as we have already seen, not all countries are equal. Neither are all regions within a country. Although there is an endless accumulation of capital within the system, it is distributed unevenly among many competing agents. Despite indigenous development models, inequity between and within countries remains apparent and is often fueled by foreign investment.

The *core-periphery* concept explained in Chapter 1 expresses the idea that the world comprises powerful centers and less-powerful edges. This concept applies in any context and at any scale. For instance, the world could be generally portrayed as having a Northern core and a Southern periphery. Or, in terms of Japanese investment in Asia, Japan is a core while its collective recipient countries form a periphery.

Although China and India have fast-growing economies and have increasing influence around the world, they are not yet global cores. However, they are not peripheries either. Both these countries are considered to be semi-peripheries within the larger world system. China, for example, imports raw materials from peripheral African countries into its own coastal core region to manufacture steel in order to construct a railroad into Tibet—one of its peripheral regions. It also exports vast amounts of products to the United States—the world's primary core.

An urbanized, capital region is usually the core of its own state, with rural areas comprising the periphery. For example, Bangkok is Thailand's core and Kuala Lumpur is Malaysia's. Some countries have several regional cores. For example, China's Beijing, Shanghai, Guangzhou, and Hong Kong are all cores with surrounding peripheries. India's Delhi, Mumbai, Kolkhata, and Chennai are similar cores. Cities also house core elite groups who hold relatively more power than the peripheral rural population. However, it is important to recognize that although effective power lies with the urban elite, potential power lies with the rural masses.

Geographer Debra Straussfogel (1997) explains that core and periphery bracket a continuum that "represents one horizontal level within the larger global structure." She describes development as a process "whereby one structural configuration is transformed into another." As development alters socioeconomic spaces, restructuring takes place at multiple levels, thereby repositioning the country on the scale of "coreness" or "peripheryness."

POST-DEVELOPMENT THEORY

Post-development theory holds that pigeonholing countries into Western-contrived categories is pejorative and associated with old colonialist paternalism and racism. Founded in "we-they" thinking, ideas of evolutionary progression along a path toward technological achievement and democratic institutions became moral justification for the application of Western and Northern social and economic principles to international relationships.

These principles have one thing in common: they begin with the notion of deprivation (judged by Western standards) and culminate with the concept of development incorporating measures of resource exploitation, urbanization, industrialization, and material consumption. Fraught with conflict and contradiction, such applications have long been contested by "recipient" nations.

Post-development adherents, such as Arturo Escobar and Jeffrey Sachs, believe that modern development theory is academically and politically construed. What once was a dream has progressively turned into a nightmare. Money and labor are directed at "problems" defined and evaluated in line with pre-existing social theory by academic "experts" in collusion with often Western-educated politicians. Thus, development is *socially constructed.* To quote Escobar (1995):

> For instead of the kingdom of abundance promised by theorists and politicians in the 1950s,

> the discourse and strategy of development produced its opposite: massive underdevelopment and impoverishment, untold exploitation and oppression.

And Sachs (1992):

> The idea of development stands like a ruin in the intellectual landscape. Delusion and disappointment, failures and crimes have been the steady companions of development and they tell a common story: it did not work.

Critics of post-development theory note that it ignores the very real improvements that have been realized in the lives of millions. On the other hand, deprivation is not an invention of the mind; it is an actuality and it is corrosive. A major portion of Asia's poor can be found:

- Living in remote forest areas (the upland poor, often indigenous people)
- Among the fishing communities (the coastal poor)
- On marginal land areas (the dry-land poor)
- Among those affected by regular floods (the wetland poor)
- In congested cities and towns with bad shelter conditions (the slum poor)

Whatever the situation, we cannot color the entire globe monochrome and explain it under a single rubric. Diversity of experience and the fact that there might be "good development" and "bad development" cannot be ignored. While post-development ideas might bring about deeper reflections about development and its ramifications, according to the critics, it is historically flawed.

The Meaning of "Poverty"

What does it really mean to be poor? Is being poor related to money? To food? To shelter? To material possessions? Who decides what poverty is? Poverty is usually defined by "experts" from the WB, UN, and other such organizations. The ability to buy a "bundle of goods," including commodities such as electricity, clothing, and bus fare, is being increasingly employed to measure poverty. As material development ensues in an area, inhabitants become drawn into the larger socioeconomic system with its Western concepts of consumption. However, why should poor people not be allowed to consume? Why is it acceptable for "us" to indulge in conspicuous consumption and not "them?"

In my travels over the years in developing countries from Afghanistan to Zimbabwe, I have observed changes as to what the "in" thing is to have as one's first "expensive" purchase. About 30 years ago, the rage was for a pair of sunglasses, next it was a wristwatch, then a transistor radio to be followed by a television, perhaps a motor scooter, and ultimately a cell phone, I-Pod, and so on (Figure 4-2). As people are increasingly exposed to Western

Figure 4-2

This TV was purchased at an open-air market in the northern hills of Vietnam. The man is toting it back to his village. Much of this tribal region has electricity, but the TV could also be powered by a car battery. The most popular items in the market were packaged food products and raw meat, tobacco and cigarettes, and individual packets of laundry soap and birth-control pills. Battery-run clocks were the most popular luxury items. The clocks were large and either square or rectangular. Most had pictures of natural phenomena such as snow-covered mountains, verdant forests, and stunning waterfalls overarched with rainbows. Photograph courtesy of B. A. Weightman.

ideals of what possessions one should have, their "needs" become commoditized.

Recent thinking regarding poverty is that it is best defined by the people themselves—and they should be the ones to decide what they need. Even the WB now considers the poor to be "experts on poverty."

The most important fact regarding poverty is that it is *multidimensional*. Participatory studies show that poor people are less concerned with income *per se* and more concerned with assets (Table 4-2). In addition, security from risk, crime, and violence; human rights and dignity; education, health, and well-being are fundamental issues. Participatory methods enable the poor to reveal their own conceptions of deprivation in light of their own values and priorities. For example, while literacy is viewed as important, *schooling* receives mixed reviews. Occasionally highly valued, frequently it is irrelevant in the lives of the poor.

Definitions and categorizations based on what people do *not* have and prescriptions as to how they *should* live, based on Western standards, have shaped perceptions and altered landscapes around the world. Moreover, students from developing countries who study abroad learn non-indigenous concepts of planning and development and may return to apply these within their homelands.

Crowded, winding "unplanned" streets, which were once functioning, multifaceted communities, have been seen (through newly trained eyes) as inappropriate spatial geometries that must be rearranged for greater efficiency and visual appeal. Squatter settlements, home to millions, perceived as unhealthy warrens of poverty and criminality, have been razed. These are rapidly rebuilt; most governments now realize that it is more expedient to implement self-help programs within existing squatter settlements. The social costs of such upheaval cannot be measured.

The vision of nations speeding along a trajectory of technological advancement and material consumption has been shattered in many cases. Things that were supposed to happen under this rubric did not. China, for example, has initiated broad economic reform and has dramatically reduced its population growth, thereby ending the specter of hordes of communist Chinese overrunning the planet. Burgeoning middle classes in both China and India are building mansions, driving luxury cars, and patronizing glittering shopping malls (Figure 4-3). Countries such as Singapore and Thailand

Table 4-2 Poor People's Perceptions of Poverty

Relatively Well-Off Households	• Solid and stable houses (usually renovated every 15 years) • Have either a bicycle or motorbike (usually both). • Own a TV and radio (usually both). • Can send children to school. • Never lack money and are able to save money. • Have a garden with useful plants and trees.
Average Households	• Have a stable house that does not need renovating for at least 10 years. • Own a TV or radio. • Have enough food all year. • Can send children to school. • Have a well, or easy access to water.
Poor Households	• Live in unstable homes, usually made of mud. • Have no TV or radio. • Are not able to save money. • Children cannot attend school or have to leave prematurely. • Have enough food only until the next harvest; go without for one or more months a year. • Are unable to use surrounding natural resources to their benefit.
Very Poor Households	• Live in very unstable houses that need to be rebuilt every 2–3 years. • Have no wells or easy access to fresh water.

Source: Deepa Narayan, *Voices of the Poor*, World Bank, 2000.

Figure 4-3

Shoppers in the Inorbit Mall in Mumbai, India.
©Prashanth Vishwanathan/Bloomberg/Getty Images, Inc.

have become regional economic powerhouses, and Malaysia and Indonesia are on the rise. However, social and economic protest movements have also been on the rise, especially in China. Common issues such as global resource depletion and environmental crises are drawing international attention. It should also be noted that, while it is true that development has helped millions of people to move out of poverty, large numbers of newly poor are appearing, especially in urban slums where the rural dispossessed have migrated. Moreover, income disparities are worsening.

The Middle-Class

This winter season (2009–2010), 3 million Chinese will swarm the country's 300 ski runs, some of which are indoors in the subtropical south. Thousands of young, English-speaking professionals will prowl the fancy shopping malls of Mumbai, New Delhi, Bangalore, and Hyderabad in India. These are the emerging markets' new middle classes at play.

The WB assesses that the middle class of emerging markets numbered about 250 million in 2000 and 400 million in 2005. That number is predicted to rise to 1.2 million in 2030. Despite these impressive numbers, the middle class made up only 6 percent of the global population in 2005 and is expected to comprise only 15 percent in 2030.

Two dollars a day is a commonly accepted definition of the poverty line in developing countries; people above this line are considered middle class in the respect that they have moved out of poverty. It is important to be aware that this is according to developing country standards, not American ones. (Thirteen dollars a day is the poverty line in the United States.) Based on this definition, the middle class in China leapt from 173.7 million in 1990 to 806 million in 2005. In India, it rose from 146.8 million to nearly 264 million.

Of course, in practical terms, what is middle class in one region of a country may not be the same as in a different region of that country. In China, for instance, US$3,000 may be enough to qualify in Chengdu, Chongqing, or other big cities in the west but not in Beijing or Shanghai, where living costs are much higher. So, defining the middle class in absolute terms is difficult.

In practice, emerging markets have two middle classes. One consists of those who are middle class by any standard-the *global middle class*. The second group incorporates those people who are middle class by standards of the developing world but not the rich one. Somewhere in the past couple of years, this *developing middle class* became the majority of the developing world's population.

The term "middle class" covers a multitude of differences. A recent study of China's middle class by Kellee Tsai, a professor at Johns Hopkins University, reveals that two-fifths of private-sector entrepreneurs come from farming families, while one-fifth were born to families of ordinary workers. Another 15 percent are children of government officials or enterprise managers. Moreover, middle-class commercial landscape expressions are quite varied.

In Sanlitun Village, a shopping mall in central Beijing, young Chinese couples dressed to the nines ogle the latest computers and BMWs. The Apple store, housed in a glass box and the Adidas one, encased in a jagged shard of ochre and orange, are only two of numerous cyber-age shops lining the requisite skating rink. Magnificent though it be, Sanlitun Village is not the norm for the average middle-class shopping experience.

A five-hour drive outside Hyderabad in central India is a more representative example—a shop in a dimly lit corner of a family house. Plastic jars hold 22 products such as chickpeas, teabags, and individual packets of washing powder. The family also runs a scrap-metal business and earns between US$2.00 and US$4.00 a day. The new bourgeoisie has created an enormous market for material goods. Moreover, they feed investment in new modes of production. For example, India's Tata Motors has produced the Nano, a car that costs only US$2,500, appealing directly to this new class of spenders.

Members of the middle class are keen on self-improvement. They try to lead healthy lives. They value education and are more likely than the poor to keep their children in school. They are more apt than the rich to invest in *new* businesses and are more willing to learn fresh ways of doing things. Many Asian economists argue that it is the middle class that will provide the bulk of hard-working business people along with a few exceptional entrepreneurs such as Nandan Nilekani—a founder of India's tech giant Infosys and *Forbes* magazine's Business Man of the Year in Asia in 2007.

CORES AND PERIPHERIES

Established core-periphery relationships have global impacts. Multinational corporations and capital investment by the United States, European Union countries, and Japan place recipient states in a position of dependency. These countries are now joined by South Korea and, most notably, China in terms of global investment. As part of its "going-out policy," China has spent an estimated US$115 billion on foreign acquisitions in the past decade. (Learn more about China's involvement in Africa and elsewhere in Chapter 12.)

Movement of labor-intensive manufacturing activities to peripheral locations to take advantage of cheap labor has been a hallmark of economic restructuring in the late twentieth century, often with unfavorable consequences. An assembly line might provide hundreds of low-wage jobs, but a downsizing or relocation decision in the core country can create instant joblessness. Many foreign firms in Asia are criticized for their exploitation of labor and resources and their cavalier attitudes concerning environmental quality.

The debatable concept of dependency has given rise to **dependency theory**. Dependency theory claims that cores can only grow through exploitation of their dependent peripheries. In order for cores to prosper, spatial imbalances must be maintained. And, to understand a region's low level of development, it must be viewed in context of both local and global core-periphery relationships. From this perspective, trade is not mutually beneficial.

People in the periphery compete to work for very low wages, which drives down the price of goods. However, price reductions are not passed on to the consumer because producers in the core have a common desire to keep prices high. As cores and peripheries become increasingly integrated on international scales, regional economies are structured into situations of dependency on the demands of the cores and are unable to develop their own economies. However, geographies are constantly changing.

In global cores, financial, investment, and assistance rationales and applications are in question. In peripheries, economic independence and local development progress have become paramount concerns. Non-governmental organizations (NGOs) and private concerns have taken on stronger roles in development. Such changes mean that dependency relationships are constantly shifting and increasingly complex.

Acquring Development Funds

FOREIGN DIRECT INVESTMENT (FDI)

Foreign Direct Investment (FDI) in Asia has increased with quantum leaps. Investment strategies are founded both in the needs of the investor and in the economic and political climate of the recipient country. Change in one area propels change in another. For example, India saw its foreign investment soar after its economic reforms in 1991. Huge investments in China by Japan, the Association of Southeast Asian Nations (ASEAN) countries, the United States, and the European Union, have been stimulated by the revision of China's tax code and its reassessment of incentives for foreign investment.

Developed countries are attracted by China's low wages and enormous market potential. In 2007, East Asia received one-third of all FDI inflows to developing regions. However, 80 percent of this went to China alone. While FDI decreased with the economic crisis of 2007 (see the discussion below), investments in India and China rose again in 2008.

Any sign of instability can induce investors to retreat. War and political chaos have kept many away from Myanmar, Laos, and Cambodia. Vietnam's mercurial investment policies and disregard for venture profitability have turned off many potential investors. Indonesia's 1994 deregulation improved the investment environment for small and medium-sized enterprises, and FDI rose significantly. However, political crises in 1997 and 1998 saw dramatic pullbacks. Nuclear testing by India and Pakistan in 1998 reverberated across the investment world. Current political and religious upheavals in Pakistan are causing investors to question their current and future financial involvement there. On the other hand, India now looks promising in terms of foreign investment.

Financial crises in Japan, Thailand, and elsewhere in Asia at the end of the twentieth century had far-reaching ramifications. Reductions in FDI were countered by mergers and acquisitions, with a gain in assets owned by foreigners. Still, the presence of a cheap labor force near Singapore keeps Malaysia attractive for investment in the electronics industry. And, with its skilled work force, an improved infrastructure around Manila, and a new government committed to economic liberalization, the Philippines has also become a positive FDI environment.

FDI can bring benefits and spatial transformation. Governments can use incentives to generate investment in less-developed, peripheral regions away from the core. Thailand, for example, encourages investment in areas remote from Bangkok. Moreover, Asia is no longer simply a cheap production realm; it is the fastest-growing consumer market in the world. Thus, there is a changing emphasis from fleeting use of a low-cost labor force to long-term commitment, including transfer of knowledge, skills, and technology.

Multinationals are powerful entities of global proportions, yet they generate paradoxical impacts. While providing employment and raising levels of disposable income, localization of production, and management, creation of subsidiaries, and linkages with established firms simultaneously strengthen their complex webs of influence.

Smoking Asia

The World Health Organization estimates that the twenty-first century will initiate 10 million tobacco-related deaths a year with more than two-thirds in developing countries. Developing countries already accounted for three-quarters of global tobacco production by 1980, and will consume 70 percent of the world's cigarettes by 2020. Tobacco use and production has declined in the industrialized world. However, exports have risen by 250 percent, and Asia has become a prime marketing target. Around 700 million, mostly men, light up every day. The habit is thought to kill 2.3 million Asians a year.

There are 120 million smokers in India. There you can buy 10 small cigarettes known as *bidis* for US$.05. One in five men is expected to die from smoking.

China now houses 25 percent of the world's smokers. Although male smoking has gone down a bit, more affluent young women are taking up "fashionable" smoking. Cigarettes are employed to ease social interaction. One is not supposed to refuse the offer of a smoke. A favor asked might include the offer of a cigarette. The person who accepts the cigarette (under pressure) is expected to carry out the favor.

Government actions are contradictory. Thailand banned smoking in all public buildings in 2003 and is considering banning it in all places of entertainment. At the same time, it is building a giant cigarette factory for its tobacco firm. China's government says that smoking is bad and that people should quit. Concurrently, it has plans for its state monopoly on tobacco to become the largest exporter of tobacco products in the region. China will manufacture Marlboro cigarettes and export Chinese tobacco to other cigarette-making companies in the region. India banned outdoor advertising for tobacco products in 2003, but tobacco firms are spending heavily on in-store promotional displays, which are still legal. There are some non-smoking restaurants in Japan such as Starbucks, but many people, including police, ignore the no-smoking rule.

Japan Tobacco Company is the third largest in the world. It manufactures Camel, Winston, Mild Seven, and Salem cigarettes. This company has invested in American and British firms for exclusive rights to market future-developed lung cancer vaccines! It claims that it is diversifying.

There is a Global Treaty on Tobacco Control that all Asian nations have signed with the exception of Indonesia. Its state tobacco firms generate huge revenues from sales and taxes. The government claims that it can't afford to sign the treaty, as tobacco taxes contribute 10 percent of its capital and the industry employs 7 million people. Nevertheless, it has promised to cap its cigarette production by 2010. You will be interested to know that, as of 2009, the United States has not ratified the treaty.

Most Indonesians smoke *kreteks* that are a complex blend of tobacco and cloves, among other ingredients. These are much cheaper than "white" cigarettes but have much higher amounts of tar. So-called "clove cigarettes," which, like all cigarettes, are linked to lung cancer and other respiratory diseases, were outlawed in the United States in 2009. Now, variations (with cloves) are being sold under the guise of "cigars."

Studies by the WB suggest that countries should raise cigarette taxes, which are among the lowest in the world. This would cut down on the number of smokers but would continue to raise revenue. They encourage governments to get farmers to grow oil seeds or some other in-demand crop instead of tobacco.

Although China National Tobacco Company, employing more than 5,000,000 people, is the world's largest tobacco product producer, global sales are dominated by just five companies. International markets are dominated by Philip Morris International (represented by Reynolds Tobacco in the U.S.), British-American Tobacco (selling such brands as Kent, Pall Mall, and Kool, Gold Leaf in Pakistan, Craven A in Vietnam, and Ardath in Indonesia), Altria (which makes Marlboro, the most popular brand in the world), Imperial Tobacco (a British company that owns an array of labels including Davidoff, the most popular brand in Taiwan), and Japan Tobacco.

Convincing U.S. officials to claim that foreign restrictions on tobacco product advertising were, in effect, trade barriers, tobacco companies were exempted from national controls in Taiwan, Japan, and South Korea. U.S. market shares jumped accordingly. Advertising bans elsewhere are circumvented through sponsorship of sporting and other crowd-attracting events. In Malaysia, tobacco companies held free rock concerts and distributed free cigarettes to attendees. Promotion is furthered through such popular items as Marlboro hats, T-shirts, lighters, and other paraphernalia. Such strategies are aimed at both selling cigarettes to existing smokers and expanding the smoking pool.

Particular target groups are individuals with rising incomes, women seeking greater autonomy, teens, and even children and the poor. Billboards appeal to status, power, and independent thinking. Cigarettes are often sold individually. This makes Marlboro (high-status), Salem, and other brands affordable and ensures that the buyer will not see any warning label. This strategy also allows children to buy cigarettes, and there are increasing numbers of child smokers in the five- to nine-year-old age group.

Cigarette smuggling is rampant, with one-third of all internationally traded tobacco landing in the black market. For instance, two out of every three packets sold in Vietnam have been illegally brought from Cambodia, Laos, Thailand, or China. Selling for "introductory" prices makes these more accessible and feeds the new smoker pool. Because there is money to be made, corruption is endemic in regard to the tobacco industry. For example, millions of cartons confiscated by the Vietnamese authorities are subsequently exported.

Some countries now have anti-smoking programs. Both Indonesia and South Korea target a portion of their tobacco revenues for anti-smoking campaigns. Under Singapore's very strict laws, smoking is prohibited in public buildings.

Although they know that cigarette smoking is harmful, tobacco companies produce more than 1,000 cigarettes every year for each man, woman, and child on Earth. Currently, 3 million people die each year from tobacco-related diseases. Costs of medical care and lost productivity are in the billions of dollars. Will WHO's prediction of 10 million deaths annually materialize? Should there be curbs on the activities of American or other multinational tobacco companies overseas?

PRIVATE INVESTMENT

This postmodern era has witnessed a dramatic increase in the amount of private capital flowing into Asia. Over 60 percent of international bank financing to emerging economies goes to ten countries, four of which are in Asia (South Korea, China, Thailand, and Indonesia). Deregulation of capital markets in Japan, the United States, and Europe, along with telecommunication networks permitting instant cash transfers, have facilitated this trend. International bank lending rose to its highest levels in 1996, with more loans to Asia than any other region. In that same year, EU banks surpassed Japanese banks as the paramount lenders in Asia. The two largest borrowers were South Korea and Thailand, followed by Indonesia, Malaysia, China, and the Philippines.

In recent decades, macro-financial flows have enabled some developing nations to reverse capital flows in their own favor and fostered the rearing of Economic Tigers. In 1997 and 1998, however, several countries such as Japan, South Korea, and Thailand experienced an economic downturn. Numerous banks had floated excessive loans for too many poorly evaluated development schemes. Many projects still stand incomplete or unoccupied, their loans outstanding.

FINANCIAL CRISIS 2007

Recoveries have been cut short by the current financial crisis that began in 2007 with the bursting of the American "housing bubble." Up until that time, significant amounts of foreign capital from fast-growing Asian economies came into the United States. The Federal Reserve kept interest rates too low, and millions of Americans were granted unrealistic housing loans. Interest rates rose and mortgage payments rose so that ultimately buyers couldn't pay their debt.

Loan defaults led to millions of foreclosures. As demand lessened, housing prices declined and global investors lost money. With no financial cushion to absorb losses, lenders cut back their investments. This resulted in a major decline in global economic activity—bank and business failures, unemployment and shrunken consumer wealth. Dramatically reduced spending generated even more business failures and more unemployment. India lost 500,000 exports in the last four months of 2008. Malaysia has revoked work visas for 55,000 Bangladeshis in order to boost job availability for locals.

The global economic downturn has hit the most vulnerable half of humanity—women—with exceptional force. Women, who make up 70 percent of the world's poor, are usually employed in low-paying and part-time jobs. In times of crisis, they are the first to be let go. Developing countries rarely have safety nets to help them survive, and increasing numbers of women are descending into poverty worse than ever before.

According to the Brookings Institution (2009), U.S. consumption accounted for more than a third of global consumption income between 2000 and 2007. In other words, world economies have been dependent upon American consumers to stay afloat. So when the U.S. economy crashed, there were reverberations throughout the entire world system. Japan's GDP fell 15.2 percent in the first quarter of 2009. With an economy that depends heavily on exports of machinery and electronics, Taiwan's exports slumped 34 percent and its GDP dropped in the first six months of that year. An estimated 9 million of 210 million migrant workers who work in China's cities have lost their jobs and been forced to return to their villages. Leading economists have called this economic downturn the worst financial crisis since the Great Depression (1929–1941).

Governments have responded to the crisis by providing stimulus packages to businesses to generate employment. They have given bailouts to financial institutions and funded social programs and infrastructure projects. China, for example, has poured money into rural development projects such as new irrigation facilities and the provision of safe water. It has also invested in energy-saving environmental engineering, as well as family planning programs.

The Asian Development Bank (ADB) predicts that the region's economic growth rate, which has dropped from its peak of 9.5 percent in 2007 to 3.9 percent in 2009, will only rise to 6.4 percent in 2010. This is already wreaking major consequences on the poor and vulnerable. The ADB estimates that the projected GDP decline in the region will plummet another 60 million people into poverty (living on less than US$1.25 a day) and an additional 80 million into a state of vulnerability (living on less than US$2.00 a day.

Nothing is really predictable, at least in Asia. While small economies like Malaysia, South Korea, and Thailand were crushed by the crisis, larger ones such as China, Indonesia, and India continued to expand. However, some of the smaller economies are on the rebound. South Korea's economy grew by 10 percent in the latter part of 2009, and Singapore's by 21 percent. According to many financial experts, Asia's emerging economies are clearly leading the global recovery. However, according to the naysayers, this new bubble will eventually burst.

THE ROLE OF OVERSEAS CHINESE

There are around 55 million **overseas Chinese**, most of whom live in Southeast Asia. Most are business entrepreneurs who are the driving force behind economic development in Asia. For instance, they contributed more than 65 percent of foreign investment in China in 2003. Currently, they own nearly half of the 28,000 foreign-invested firms in Shanghai. They own hundreds of conglomerates in Indonesia, Thailand, Malaysia, and elsewhere in Asia.

Although they live outside China, most overseas Chinese have long-standing connections with southern China. At least 2,000 years ago, some of their ancestors left the southern provinces to colonize Southeast Asia or to serve colonial interests demanding cheap labor, and tens of thousands have followed since.

Focusing on business and trade, they maintained, expanded, and intensified their connections with various power structures to the point of becoming gatekeepers for any form of economic interchange. After each of a series of upheavals in China, more would flee or at least channel their capital elsewhere. With the communist victory in 1949, billions of dollars left the mainland with the outflux of nationalist Chinese and their supporters. By prudent saving and investment, timely information, and fast action via closely guarded networks, the overseas Chinese have come to dominate the economies of East and Southeast Asia, with the exceptions of Japan and North and South Korea.

Overseas Chinese expand their businesses through acquisition of ever-increasing numbers of separate companies rather than enlarging existing ones. Consequently, they control an array of large, highly diversified conglomerates such as Salim Group in Indonesia, which comprises hundreds of companies spread across a wide range of markets in multiple countries.

Most overseas Chinese businesses are "family-owned"—controlled by members of a single family. Typically, sons and daughters are given a division of the conglomerate to manage. Financing is derived from their own private banks or through connections with foreign banks and joint ventures with foreign enterprises.

FOREIGN AID

The spatial duality of core-periphery development is frequently further entrenched with **bilateral** and **multilateral aid** programs. Most aid is bilateral—government-to-government transfers of cash, low-interest loans, material goods, food, or medical supplies. Technology and training are also transferred. Technical training, while improving skill levels, lays the groundwork for further capital input and even greater reliance on outside assistance. Aid might also take the form of preferential tariffs, whereby the donor nation agrees to buy goods produced in the recipient nation. Multilateral aid derives from international organizations such as the Organization for Economic Cooperation and Development (OECD), the World Bank (WB), or agencies of the United Nations.

If aid is concentrated in a core area, it strengthens the core and distances it further from peripheral realities. For instance, a full-service hospital is most likely to be located in the core, while poorly supplied clinics are placed in peripheral regions. Transportation development provides another example. Modern road construction typically radiates outward from the capital city. Starting as a paved highway, it eventually degenerates into an unpaved road, and ultimately into a dirt track, often impassable in the wet season. Transportation and other infrastructural conditions decline as distance from a core increases.

Tied aid, binding the source and destination in a not always mutually beneficial agreement, is a common practice. Much official development assistance (ODA) is tied to structural adjustment programs (SAPs). SAPs require privatization, increases in exports, and adherence to free market principles. ODA is frequently siphoned off to pet projects for political reasons or to mega-projects tied to purchases from donor nations. A country might be required to import capital-intensive equipment—road-grading machinery, for instance. Such equipment gets the job done faster but also replaces labor in circumstances of already high unemployment.

SAPs incorporate the notion that women's labor is elastic in the face of household survival needs. Poor urban women under SAPs had to work harder inside and outside the home to compensate for the rising cost of social services and diminished male incomes. Increased or new fees limited their access to education and health care.

Finally, because of administrative costs, less than a quarter of ODA ever reaches the poor people it is intended to help. Most experts, including the OECD and WB, have recently denounced SAPs and similar modes of "development" as failures.

NON-GOVERNMENTAL ORGANIZATIONS (NGOs)

Non-governmental organizations (NGOs) reflect contemporary culture-based perceptions of the development process. All have an agenda—some political, some humanitarian, others religious. Nevertheless, NGOs tend to focus in peripheral settings such as urban slums and disadvantaged rural areas. Many NGOs are involved in health and human resource assistance programs. Ideally, culturally derived development models and programs can be articulated with larger, international frameworks of change. Unfortunately, disparity between the ideal and the real seems unavoidable.

Because they address people's grievances and could become rallying points for political protest, NGOs are seen as potentially threatening by some governments. Indonesia, Malaysia, and other countries attempt to rein in NGO activities with tight controls and bureaucratic minutiae. Other countries, such as South Korea, have more comfortable relationships. Here, more than 20,000 NGOs maintain projects in areas ranging from environmental protection to political freedom and women's rights.

Some scholars argue that NGOs are really "soft imperialism" because major ones are captive to the agenda of international donors. Also, grassroots are dependent on the international NGOs. For all the glowing rhetoric about self-help, democracy, and strengthening of civil society, actual power relationships, at least in the large NGO realm, resemble traditional patronage.

PRIVATIZATION

As development proceeds, demand for public services, infrastructure, and shelter rises. Not only do sizable debts and shortage of revenues constrain governments, but also corruption, patronage, and mismanagement frequently render state-owned enterprises money losers. Efficiency is precluded by labyrinthine regulations and bureaucracies. For example, it can take as long as 10 years to get a private phone installed in Pakistan. Community phones are the rule in most Asian countries, and call completion rates are generally low. But with the advent of cheap cell phones, networks of interpersonal and business communication are becoming ubiquitous.

Growing dissatisfaction with central planning and government management of services has promoted *privatization* in various forms. Public assets or enterprises might be sold to private investors, as in Bangladesh. Or as in China, public agencies are *corporatized* and required to cover costs and manage operations more efficiently. More commonly, governments pursue policies encouraging private-sector participation in public service provision. Countries such as Thailand, Malaysia, and South Korea are promoting public service delivery by community, religious, or professional NGOs, cooperatives, and small operators. The Roman Catholic Church operates schools in the Philippines, and physicians' groups have opened clinics in crowded urban areas of Vietnam.

According to the WB in 2007, privatization has increased in Asia. For example, Myanmar (Burma) has privatized 288 businesses in an effort to attract foreign investment. Even communist Vietnam is allowing some degree of privatization, although the government retains over 50 percent of the company. Pakistan also has drawn in foreign money via privatization.

Air Space

Privatization has affected the passenger airline industry. State-owned Singapore Airlines and Malaysia Airlines were privatized in 1985, and Thai Airways International and Philippine Airlines were privatized in 1992.

Philippine Air Lines (PAL) offers an excellent example of privatization, globalization, and the role of overseas Chinese. PAL has had its ups and downs, and the government has had to take over in times of economic crisis. However, in order to cut costs, in 2000, it sold its maintenance sector to a German-led joint venture—Lufthansa Technik Philippines—the world's largest provider of aircraft maintenance services. PAL, again privatized, is buying more new planes and expanding its flight coverage around the world. It even has a flight to Las Vegas! The company is now controlled by PAL Holdings, which is part of a conglomerate owned by a Filipino business magnate who was born in China.

Deregulation encouraged the development of new airlines such as EVA Airways and Asiana. Geographers John Bowen and Thomas Leinbach (1995) note the importance of airlines in the development trajectories of emerging economies. By increasing access to places, they open up new domestic and international markets.

Asian airlines play significant roles in global interchange. Even so, the retention of vested interests by many governments indicates their reluctance to relinquish full control over what they perceive to be a strategic element of infrastructure.

Housing is another sector in which governments have sought private help. Sites and services programs have gained significance since the 1970s. Government agencies install infrastructure and provide essential services on cleared land that is sold with low-interest loans. Poor families contract with private builders or construct their own dwelling with subsidized materials. In India, the government registers and assists housing cooperatives to purchase land and procure financing for low-cost housing. In several countries, private corporations are responsible for the development of complete communities inclusive of residential, commercial, educational, and recreational facilities. Despite these ambitious programs, most shelter for the poor still derives from the informal sector as squatter or temporary housing.

Privatization is not readily accepted everywhere. Individuals and constituencies with vested interests in state-controlled enterprises strongly resist privatization efforts. In Indonesia and Malaysia, for instance, opposition is voiced by economic nationalists and Muslim fundamentalists, who detest Western consumerism and capitalism. Some people in these countries also fear that the already economically dominant Chinese will be the main beneficiaries. In Thailand, opposition comes from the military, which exercises significant control over public corporations to ensure employment for high-ranking retirees. Labor unions and other groups that depend on patronage, paybacks, and similar benefits also oppose privatization. In light of increasing urbanization, the provision of shelter and services will continue to challenge both private and public institutions.

MICROCREDIT

After decades of failure, global aid organizations believe they have found a solution to poverty in the developing world. It is called **microcredit**.

The modern version of microcredit was instituted in Bangladesh by banker and economist Muhammad Yunus, who founded the **Grameen Bank** in 1983. In 1976, Dr. Yunus visited a small village where women made bamboo furniture. In order to purchase the bamboo, the women had taken out a loan from a moneylender at usurious rates. Consequently, most of their profits were used to pay the loan, which kept on gathering interest. Dr. Yunus decided to lend the 42 women US$27.00 at a very low interest rate to buy their bamboo. As a result, each woman was able to net a profit of US$.02. This may sound like nothing, but two cents goes a long way in a Bangladeshi village.

The vast majority (94%) of Grameen's loans go to women who are more likely than men to spend their earnings on their family. Several women are united into a "solidarity group," and the group borrows money. Each woman acts as a co-guarantor to repay the loan and supports the other members in their efforts toward economic self-sufficiency.

Grameen has recently expanded into other nonprofit ventures such as fisheries and irrigation projects. Grameenphone is the biggest private-sector phone company in Bangladesh. With its village phone project, it has facilitated cell-phone ownership to nearly 300,000 rural poor in more than 50,000 villages. Aside from personal communication, cell phones are used by farmers to relay agricultural information and by women to run businesses.

The microcredit concept has spread around the world. The United Nations declared 2005 the "International Year of Microcredit," and Secretary General Kofi Annan said that the poor "are the solution, not the problem." Muhammad Yunus has been awarded many honors, including the Nobel Prize in 2006 and America's Presidential Medal of Freedom in 2009. In Houston, Texas, January 14 is "Muhammad Yunus Day."

REMITTANCES

Another type of income comes from what are called foreign **remittances**. This refers to the transfer of funds by a foreign worker to the worker's home country. Many families rely on remittances for day-to-day living expenses or emergencies. Some support small businesses or make investments. According to the WB, the system involves some 190 million migrants. During 2007 and 2008, US$60 billion was transferred to India from overseas workers. Twenty-six billion dollars came from the United States. The second and third largest recipients are China (US$40.5 billion) and the Philippines (30.8 billion). Now workers can make money transfers online through companies such as "Remittances to India."

Dual and Shadow Economics

Economies have differing but intertwining segments that are expressed on the landscape in various ways. Much is seen but much is not.

THE DUAL ECONOMY

One spatial duality, notably expressed in urban landscapes, is the **dual economy.** Developing nations are characterized by two levels of economic activity: a *formal sector,* functioning at a large scale in the realm of national and international trade, and an *informal sector,* functioning at a smaller scale at local levels. The formal economy exhibits permanent employment, sets hours of work and pay, and regulates other benefits.

The informal sector is characterized by family labor, small enterprise, and self-employment (Figure 4-4). It includes provision of services such as barbering, hawking of goods, and begging (Figure 4-5). These two sectors are not necessarily mutually exclusive and should not imply relative superiority and inferiority.

So-called "informal employment" is by its very nature exploitive because it lacks formal contracts, rights, regulations, and bargaining power. Foundations of the informal sector rest on crude technology, low capital investment, and tough manual labor. Informality ensures extreme abuse of women and children. Competition ensures that jobs are generated by fragmenting existing work, which furthers the division of income. Shortage of paltry resources creates communal friction and breaks down long-standing social networks. Even self-help groups devolve under the burden of grinding poverty.

In the informal economy, criminality is pervasive. Godfathers and landlords cleverly use coercion and even violence to regulate competition and protect their investments. In the absence of enforced labor rights, this is a realm of kickbacks, bribes, and hostility among ethnic, religious, and tribal groups. Moreover, urban space is never free. A place on the pavement, the use of a rickshaw, or a few hours work on a construction site—all of these require patronage or membership in some closed network, often an ethnic militia or street gang.

THE SHADOW ECONOMY

Increasing globalization masks a growing **shadow economy** or **black market** for illegal or at least shady activities. Millions of people are involved in the underground—unregulated, untaxed, often unseen business. Smuggling

Figure 4-4

Paper and wood umbrellas are made at this family enterprise in Myanmar (Burma). Three generations work in this small factory. Even the waterproofed paper is made on the premises.

Photograph courtesy of B. A. Weightman.

Figure 4-5

Street barber in Varanasi, India. Street barbers are usually Dalits *(untouchables) because they deal with dead hair.* Photograph courtesy of B. A. Weightman.

of "protected" wildlife, contraband, arms, drugs, and people is widespread, especially in Southeast Asia. In Thailand, the shadow economy contributed 52 percent of the country's GDP and involved 40 percent of its official labor force in 2001. In 2008, Cambodia and Myanmar (Burma) earned US$598 million and US$477 million, respectively, via the shadow economy.

Asian black markets generated US$252.22 billion with US$83.3 billion coming from China in 2008. In China, the "Counterfeit Goods and Piracy Market" generates an estimated US$60 billion a year. This involves the sales of illegally pirated items such as movie CDs and DVDs, video games, and computer software. Movie, music, and video game sales resulted in profits of nearly US$2 billion in 2008.

In many countries, rural poverty produces streams of migration to cities. Every day, tens of thousands of individuals and families leave the hopelessness of their rural origins seeking opportunities perceived to exist in urban destinations. Lacking requisite skills, they swell the populations of squatter settlements and slums and bolster the ranks of the unemployed and the shadow economy.

Crossing the border from Thailand to Laos, I was accompanied by a crowd of women, each with a few bars of soap, cigarettes, and other Thai and imported products, which they hid on their person or beneath other bundles under the vehicle's benches. Unavailable in Laos, these smuggled items were to be sold at roadside stands or local market stalls. Even apparently minor instances as these are part of the shadow economy.

Permeability of borders, official collusion, and lack of opportunity encourage these and other survival strategies. According to Ed Ayres, editor of *World Watch* (1996), "The growth of the shadow economy can be seen as pervasive evidence of human needs not being met by traditional institutions."

Growth of unregulated and illegal activities is also related to accountability. Not permanently based at a particular place, they are difficult, if not virtually impossible, to trace. Unregistered settlements, occupations, anarchic groups, and migratory populations transcend and escape the capabilities of extant legal strictures.

Urbanization

Asian urbanization has skyrocketed into the twenty-first century Table 4-3. Currently, Asia's urbanization rate is second highest in the world after Africa. Even so, most people in Asia still live in rural areas.

URBAN GROWTH

In recent decades, Asia has surpassed the rest of the developing world in terms of its integration into the global economy. This creates opportunities for urban development, and studies conclude that people are better off economically in cities. However, urban development has proceeded unevenly. Some cities, such as Seoul, Singapore, Taipei, and Shanghai, are becoming enmeshed into the global economic system. Others, such as Dhaka, Phnom Penh, and Vientiane, have more domestically oriented economies and the effects of globalization are less

Table 4-3 Percent Urban and Urban Annual Growth Rates (Selected Years)

Region Country	% Urban (2009)	% Urban Annual Growth Rate 1980–85	1990–95	2000–05	2005–10
South Asia					
Bangladesh	25	5.75	4.03	3.58	3.45
Bhutan	31	7.46	2.98	6.60	4.88
India	29	3.32	2.88	2.35	2.39
Maldives	35	5.71	2.63	5.62	5.29
Nepal	17	6.15	6.66	5.27	4.87
Pakistan	35	4.52	3.27	2.82	3.04
Sri Lanka	15	0.52	0.02	−0.31	0.46
East Asia					
China	51	4.47	3.79	3.10	2.70
Japan	86	1.01	0.79	0.36	0.24
North Korea	60	1.91	1.76	1.04	0.90
South Korea	82	4.05	2.13	0.75	0.62
Mongolia	60	3.86	1.43	0.93	1.22
Taiwan	78	n.d.	n.d.	n.d.	n.d.
Southeast Asia					
Brunei	72	3.66	3.59	2.94	2.63
Cambodia	15	10.46	5.56	4.84	4.64
Indonesia	43	5.36	4.54	4.04	3.34
Laos	27	4.66	5.18	6.02	5.56
Malaysia	68	4.36	4.82	3.69	3.00
Myanmar (Burma)	31	2.14	2.42	2.68	2.88
Philippines	63	5.21	4.31	3.45	3.04
Singapore	100	2.30	2.85	1.49	1.19
Thailand	36	2.60	1.73	1.49	1.66
Timor-Leste	22	4.88	4.33	6.75	5.00
Vietnam	28	2.51	3.89	3.13	3.08

evident. Even a relatively small but very poor country such as Laos will add 3.2 million people to its cities while it moves to a level of 43 percent urban in 2030.

Percentages of increase are falling in most cases. Still, because of large existing populations even smaller growth rates add huge numbers of people over time. Urban growth rates are expected to rise in Bhutan, Pakistan, Sri Lanka, and Thailand after 2010.

Population and the spatial extent of cities gain by rural-urban migration (see the discussion below), natural increase, and accretion of adjacent settlements. Some urban populations have risen dramatically only because of administrative boundary expansion. For example, Tokyo grew by 9 million people in the 1990s. However, 6 million of these were added by a boundary expansion that incorporated 87 surrounding towns and cities. Moreover, forecasts of urban growth concentrating only in existing giant cities have not always materialized. In the case of China and India, it is medium-sized cities (500,000 people) that are experiencing higher growth rates. Whatever the case, variation in city definition makes comparative urban data indicative rather than predictive.

Although cities grow significantly with incoming migration streams, growth from natural increase (NI) is even greater. These two factors are related as rural migrants have more traditional ideas about preferred family

Figure 4-6

This is Kali Besar, one of Jakarta's squatter settlements. Situated on a river near the sea, it is built on stilts to withstand the tides. Photograph courtesy of B. A. Weightman.

size. Consequently, NI rates in slums and squatter settlements are significantly higher than those in middle-class and elite districts.

SLUM DWELLERS

Across the region on average, the proportion of city dwellers living in slums is 33 percent (Figure 4-6). In India 40 percent of city residents live in slums. Slums can be found in both city interiors and peripheries. For example, in Karachi 34 percent of slum dwellers live in the inner city while 66 percent live in the urban periphery.

The five great metropolises of South Asia (Karachi, Mumbai, Delhi, Kolkata, and Dhaka) alone contain about 15,000 distinct slum communities whose total population exceeds 20 million. In Mongolia, Nepal, Laos, and Cambodia at least 50 percent of urbanites live in slums.

A slum dwelling might be a room in a dilapidated "apartment" building where 10 people share one room with few, if any, sanitary facilities or other essential services. Or it might be a primitive shelter of sticks and plastic sheeting. It might even be a large drainpipe. In India, even so-called middle-class dwellings often have running water for no more than two hours a day, usually after 8:00 PM. Maids (if a family has one) rush home in the evening to do the dishes or laundry.

Extreme health differentials are no longer between city and countryside, but between the urban middle classes and the urban poor. In Mumbai, slum death rates are 50 percent higher than in surrounding rural areas. There, forty percent of deaths are attributed to infections and parasitic diseases derived from wretched sanitation and water contamination. In Manila, more than 80 percent of people are not connected to a sewer system. In Manila, vendors sell bottled water with a markup as high as 3,200 percent, which puts clean water out of economic reach for the poor.

A 2002 UN study found that only 17 of more than 3,500 cities and large towns in India had any kind of primary sewage treatment. Jakarta still relies on open sewer ditches to disperse human and other waste to the sea. Thousands of individuals in slums have no access to a latrine. Others might share a deplorable facility with as many as 5,000 other people. Delhi has instituted mobile "toilet vans."

Being forced to tend to one's bodily functions in public while being expected to abide by rules of modesty is an important feminist issue. In many regions, women have to wait until very late at night or early morning to avoid harassment, molestation, or rape.

Pay toilets are a growth industry in developing countries. Obviously, these are rarely affordable for the poor. Men who cannot pay simply do their business in the street. Women's alternatives are unsanitary and even dangerous.

Accurate statistics on slum populations are difficult to come by since they are deliberately undercounted by officials. Indonesian and Malaysian governments are notorious for disguising urban poverty. At least 25 percent of Jakarta's population occupies slum dwellings, but the official figure is a mere 5 percent. An estimated

Figure 4-7

This family lives under a pedestrian overpass over a railroad line in Mumbai. This is an excellent location as it is sheltered and near a public "standpipe"—a water tap. Also, the family can scavenge the tracks for food such as banana skins that are thrown out of the train windows by middle-class commuters. Photograph courtesy of B. A. Weightman.

8 percent of Manila's slum residents remain uncounted. Moreover, slum populations do not include millions of pavement-dwellers who have no permanent home at all. Entire families, mostly in South Asia, seek shelter in railroad stations, under overpasses, in doorways, in parks, in alleys, or they simply lie down on the sidewalk (Figure 4-7).

Slum space has become commodified. For instance, the majority of slum dwellers in Bangkok actually rent the land where they build their shacks. In India, people who prop their shanties against high-rise apartment blocks pay rent to building or apartment owners. Landlordism is a fundamental and divisive social relation in the slum world. Already-poor slum dwellers rent to poorer slum dwellers and have no qualms about ousting those who can't pay. Even pavement-dwellers must pay someone for their space.

Slum communities continue to grow in scope and extent although most governments have tried to get rid of them. Consequently, there is a never-ending social war against the urban poor in the name of "beautification," and even "social justice for the poor." Persistent slum-clearance operations in Delhi, designed to create space for middle-class subdivisions, have produced endless cycles of settlement, eviction, demolition, and resettlement.

The most intense battle occurs in downtowns and major urban nodes. Globalized property values collide with the desperate need of the poor to be near central sources of income. Officials may redraw spatial boundaries to the advantage of landowners, foreign investors, elite homeowners, and middle-class consumers. The urban poor are a type of nomad—transient in a perpetual state of relocation.

Some cities such as Mumbai and Delhi have built satellite cities to encourage the poor to move from more valuable land nearer to the city centers. These new, perceptibly advantageous locations have emerged as magnets for rural–urban migrants and middle-class commuters. Squatters, renters, and even low-level landlords are summarily evicted without compensation or right of appeal.

It is poor city dwellers that many elite and middle-class groups consider to be the population "them," "those people," "the problem." Or they deny their existence altogether. (This attitude is confirmed by several Indian authors.)

I once had an Indian student in my Geography of Asia class who insisted that there was no poverty in India. When Geeta saw my slides she was appalled and couldn't believe what she was witnessing. She told me that she only left her (rich) family compound in Delhi in a chauffeured limousine with its windows blackened to keep "them" from looking in. The bulk of her outings were to private school, fancy malls, or social functions in luxury hotels. After a semester break back in Delhi, Geeta came to see me. She told me that she had never really looked at the landscape around her and that once she realized some of its realities, she tried to talk about them with her parents. Her mother told her that, "We don't discuss such things."

WHO considers road traffic one of the worst health hazards facing the urban poor and predicts that road accidents will be the third leading cause of death by

2020. A 2004 report in the *Hindustan Times* abhorred the fact that middle-class commuters seldom bother to stop if they injure or kill a homeless person or street child.

In urban China, where cars—owned by middle and upper classes—are filling streets formerly dominated by bicycles, hundreds of thousands of automobile accidents occur each year. For example, almost a quarter million Chinese were killed or seriously injured in the first five months of 2003.

URBAN PRIMACY

Today, most of the world's largest cities are in developing countries with increasing numbers of these in Asia. More than half of Asia's largest cities are in India and China. With some 170 cities with at least a million inhabitants, China has more large cities than any country on earth.

Some countries show a marked degree of **urban primacy**, whereby the population of the largest city is greater than the combined populations of the next three largest cities (Figure 4-8). The largest city is then referred to as a **primate city**. Countries exhibit varying degrees of primacy depending on the population differential between the largest and next largest cities. Primate cities house relatively large proportions of national populations. Greater Seoul, for instance, houses 35 percent of South Korea's urban population, while more than a quarter of city-dwelling Japanese

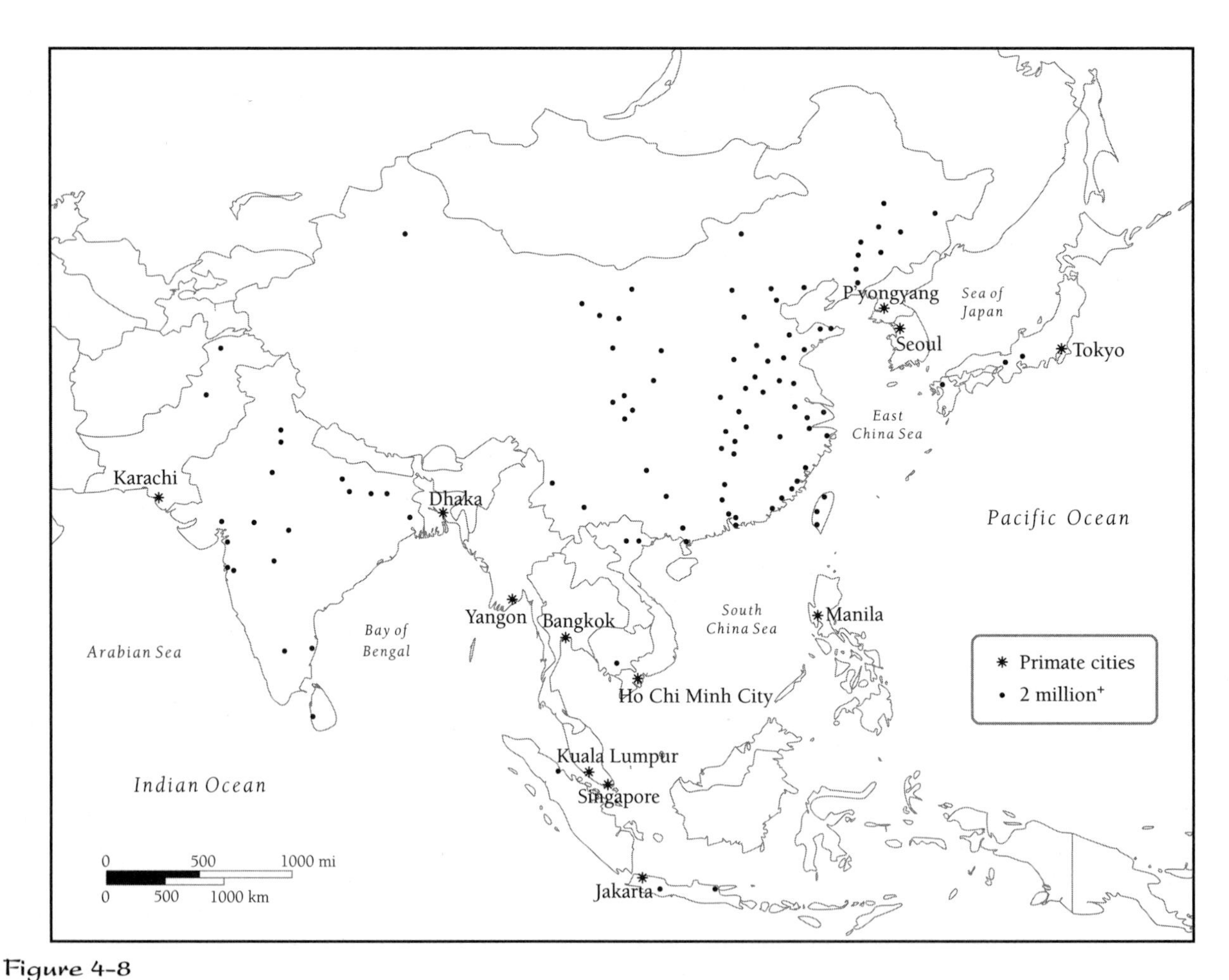

Figure 4-8

Urban primacy: cities of 2 million plus. Both India and China have many cities of a million or more inhabitants, but most of their people remain farmers.

Table 4-4 Changing Population of Asia's Mega-cities in the World's Top 30.

Asia's Mega-cities in the World's Top 30	Pop. In Millions 1950	Pop. In Millions 1980	Pop. In Millions 1990	Pop. In Millions 2000	2010 In Millions 2010	2015 In Millions 2015	Ranking In the Top 30 2000/2015
Tokyo	11,275	28,549	32,530	34,250	36,094	35,494	1/1
Mumbai	2,857	8,658	12,308	16,086	20,072	21,869	5/2
Delhi	1,369	5,558	8,206	12,441	17,015	18,604	8/6
Shanghai	6,066	7,608	8,205	13,243	15,789	17,225	6/7
Kolkata	4,513	9,030	10,890	13,058	15,577	16,980	7/8
Dhaka	336	3,226	6,621	10,159	14,769	16,842	15/9
Jakarta	1,452	5,984	8,175	8,390	9,703	16,822	12/10
Karachi	1,055	5,048	7,147	10,020	13,052	15,155	17/12
Manila	1,544	5,955	7,973	9,950	11,662	12,917	18/16
Beijing*	4,331	6,448	7,362	9,782	11,741	12,850	20/17
Osaka-Kobe	4,147	9,990	11,035	11,165	11,337	11,309	11/19
Guangzhou	1,491	3,005	3,918	7,388	9,447	10,420	26/22
Seoul*	1,021	8,258	10,544	9,917	9,762	9,545	19/26
Shenzhen	174	337	875	6,069	8,114	8,958	n.a./27

live in the Tokyo metropolitan region. Expansion and accretion absorb increasing shares of national populations into the urban world.

Primacy is not tied to a country's degree of urbanization or to its number of cities. Both South Korea and Japan are highly urbanized. Thailand is not highly urbanized, yet Bangkok is a primate city dominating the Thai economy. Shanghai, China's largest city, is not a primate city at the national level because the country has several similarly sized cities. However, Shanghai might be seen as a primate city in a regional context, as most of China's economic development is concentrated in the eastern provinces. Primacy indicates spatial dominance at varying scales of observation.

MEGA-URBAN REGIONS

In the fifth century BC, the Chinese philosopher and sage Confucius postulated 3,000 persons as an ideal city size. In the twenty-first century, the Beijing-Tianjin metropolitan region is expected to house 37 million persons. In the early twentieth century, a mere 26 million lived in India's cities. At the turn of the twenty-first century, India's urban dwellers totaled more than 222 million. Such dense accumulations of humanity have reached unparalleled proportions with entanglements of human and environmental problems beyond the scope of simple solutions.

Polarization of development has created **mega-cities**, giant cities of over 10 million people, often incorporating several cities in a cluster (Table 4-4). Today, 11 of the world's 19 megacities are in Asia, including 6 in the top 10. Although megacities are the most obvious manifestation of urbanization on the landscape, we must take note of the fact that more than half of the region's people live in cities with fewer than 5 million people. Figure 4-9 shows what the distribution of the world's largest cities will be like in 2050. Notice how these are concentrated in Asia.

Mega-cities in developing countries are predicted to be even larger in 2015. For example, Dhaka, Bangladesh, will continue to experience rapid growth although at decreasing rates of increase. Part of this increase derives from the fact that city boundaries were expanded to incorporate another million people. However, the majority of growth will still come from rural-urban migration. While the country has improved economically, most Bangladeshis remain farmers, and other types of job opportunities in the countryside are scarce. Since most migrants are poorly educated and relatively high-wage jobs in Dhaka are limited, they end up in low-end occupations. The fastest growing employment sector in the city includes rickshaw drivers and petty retailing.

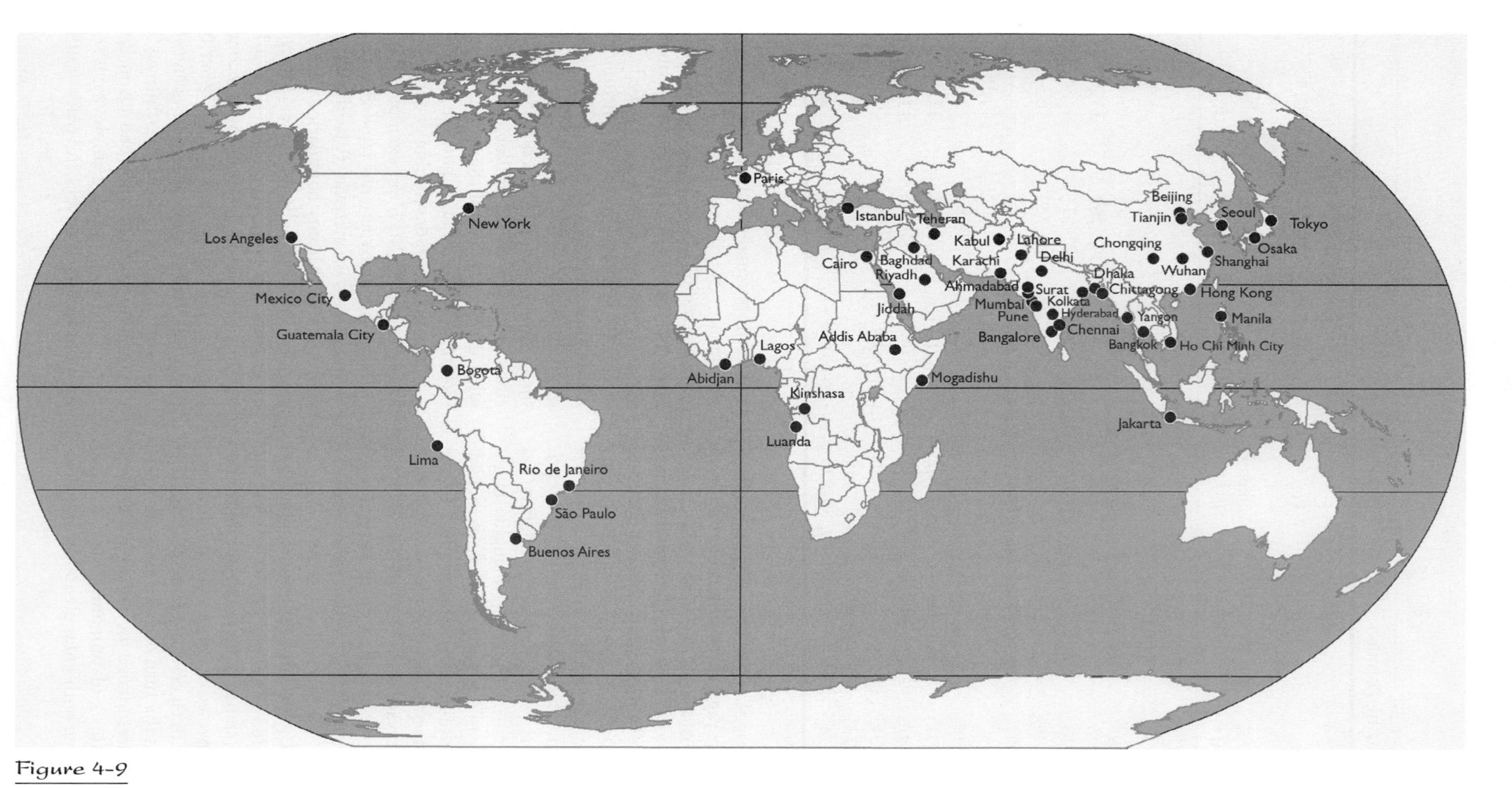

Figure 4-9

Global mega-cities in 2050. From Stanley Brunn et. al. eds., *Cities of the World,* 4th edition, 2008, p. 572. Reprinted courtesy of Rowman and Littlefield.

However, in developed countries such as Korea and Japan, populations in the biggest cities are expected to decline. In both cases governments have initiated decentralization programs by improving infrastructure and creating employment opportunities in other regions of the country. You can see by Table 4-4 that Seoul, Tokyo, and Osaka-Kobe's populations are shrinking. These cases provide an example of how circumstances can change over time.

Clusters of cities, or **urban agglomerations**, have evolved into **mega-urban or extended metropolitan regions** that extend far beyond metropolitan boundaries, incorporating towns, villages, and transportation corridors. These are complex fields of urban and rural interaction reaching 62 miles (100 km) or further beyond urban nodes. In other words, the distinction between urban and rural has become blurred as cities expand along communication corridors and either bypass or incorporate towns and villages in the process.

KOTADESASI LANDSCAPES

Kotadesasi zones are characterized by an increase of diverse nonagricultural activities in predominantly agricultural areas. Members of the same household participate in sundry occupations in both realms, making economic linkages within the *kotadesasi* zone as important as those with the dominant city. Larger numbers of women work in supply chain factories. Proliferation of cheap transport such as bicycles, motorbikes, and scooters, along with public buses and trucks, allows for great mobility of people and goods within and between these and other zones.

Mega-urban, *kotadesasi* landscapes are kaleidoscopes of activity, including rice paddies, water buffalo, bullock and horse-carts, factories, golf courses, and traffic-jammed multilane highways. Industries locate in and subsequently pollute surrounding agricultural land. Agricultural production shifts from mono-crop cultivation (principally rice) to diverse production of vegetables, livestock, and fruit for regional and even international consumption.

In this ambiguously rural and urban environment, governed by overlapping jurisdictions, management of infrastructure can only be a nightmare. However, since the definitions of "urban" and "rural" are highly variable and frequently change, *kotadesasi* zones are sorts of gray territories where urban and industrial regulations may not apply. Small-scale and informal sector operators along with squatters, who find it difficult if not impossible to conform to city regulations, see these as spaces of opportunity. The result, according to McGee (1989), is that "there is no clear-cut division between rural and urban relations; rather, activities in the two sectors are fused and complementary."

Globalization expands and intensifies *kotadesasi* landscapes. As we will discover in Chapter 12 on Southeast Asia, the ramifications of new socioeconomic complexities transform or even compromise long-standing familial and community relationships.

LIVABILITY OF CITIES

All city governments have to manage city growth and create comfortable and healthy environments for their inhabitants. These things are more likely to be addressed in richer countries such as Korea and Japan. But even their cities have significant levels of air pollution and large numbers of poor people that require assistance. In some East Asian cities such as Beijing or Taipei, people wear masks to fend off the horrendous air pollution.

Pictures of China's fantastic urban architecture dazzle the eyes and foster an image of modernity and well-being. Although this is the experiential world for some people, the reality is that millions live in less than ideal conditions. For instance, only three-fifths of residents in Shanghai live in buildings connected to a sewer. The United Nations says that dirty air is responsible for the premature deaths of at least 400,000 Chinese a year.

The worst case scenarios are found in poorer countries. For example, only a quarter of Dhaka's residents are connected to a proper sewage system. The majority use open latrines with the result that Dhaka has the highest rate of infectious disease of any city in Asia. Untold millions of Asia's city-dwellers have no access to potable water. Housing conditions are abominable (by Western standards) for most. New apartment structures and single-family housing tracts are surrounded by slums, especially in South Asia.

Fortunately, some city environments have improved. In Bangkok, a group of city officials (without the support of the national government) managed to reduce the city's air pollution levels by 20 to 50 percent despite a 40 percent increase in the number of vehicles in the past decade. They accomplished this feat by enacting stringent vehicle emission controls, raising taxes on two-stroke motorbikes, and making all taxis run on liquefied (subsidized) natural gas.

Expanding urban populations will continue to present an array of daunting challenges to local, national, and international governments. Geographer Terry McGee recommends the following steps to ameliorate ongoing and future urban problems.

- Creation of urban databases that enable monitoring and assessment of city progress.
- Reevaluation of relationships between national and city governments as urbanization proceeds. This might lead to bottom-up rather than top-down fiscal deployment.
- Provision of suitable housing, transportation, clean water, sanitation, and social services.
- Encouragement of civil participation in urban governance not only to increase private and other investment but also to garner the involvement of citizen's groups in planning.
- Development and enforcement of environmental standards.

Bureaucratic complexity, corruption, patronage, and lack of funding will render some of these steps ineffective or even impossible. Ad hoc approaches will not work either. As Terry McGee states, "The first steps to cope with the challenges of Asian urbanization are to recognize that urbanization is an integral part of development and give strategic priority to policies for the urban sector."

The Virtual Receptionist

TRG, based in Lahore, Pakistan, offers small American companies "virtual receptionists." Mussarat, also known as Margaret, is a receptionist for a six-person office in Washington, D.C. With her TV screen in Lahore, she can see visitors entering the D.C. office. She enters instructions on her computer and, miraculously, the American company's door opens for the visitors. There, her voice comes out of a speaker and instructs the visitors on how to get to the coffee room, etc. Meanwhile, she has notified company employees that their appointments are waiting.

Margaret performs all the tasks of a typical receptionist. She answers incoming calls from overseas, orders supplies for overseas offices, and completes small tasks for her overseas customers. For instance, when someone in the Washington office wants a pizza delivered or a cab, she calls Margaret, who makes the arrangements. Margaret has a hard copy of the D.C. Yellow Pages in her Lahore office.

Paid US$300 a month (a lot in Lahore), Margaret can function as a receptionist for several companies at the same time. With the cost of equipment and bandwidth at less than US$5,000 a month for six offices, she could replace six receptionists at US$3,000 a month. This translates into savings for the Washington, D.C., company and profits for the Lahore agency.

Urbanization and Employment

Cities and meg-acities are part of an all-encompassing international urban hierarchy topped by **world cities** such as London, New York, and Tokyo. The entire structure—from world cities, to powerful international economic centers such as Singapore, to regional market-oriented capitals such as Kuala Lumpur, through smaller regional and local urban centers—is interconnected by political and economic transactions conducted across Earth-space and cyberspace. Information and its technology disseminate from **technopoles** such as The Netherlands' "Randstad" region, England's "M-4 Corridor," Taiwan's "silicon valley," centered on Hsinchu, or India's "silicon valley" centered on Bangaluru. From such world cities and technopoles, global corporations manage their global networks and subsets of urban and rural production and processing sites.

Urban job creation is a gravitational pull on the unemployed in the rural hinterland. Rural to urban migration draws less educated and unskilled people into the urban employment realm. More people require additional infrastructure, which generates more jobs, which stimulates more in-migration. This ongoing process generally results in a permanent excess of workers over jobs and expands the informal and shadow economies.

Underemployment is also boosted with urban growth in developing countries. Underemployment is a Western notion referring to the underutilization of labor. It might take the form of jobs available only on a cyclical basis such as agriculture, construction, and tourism that operate on daily, weekly, or seasonal schedules. Or it might mean that a person takes ten hours to do a job that normally could be

Figure 4-10

I took this picture at about 14,000 feet (4,200 m) in the Indian Himalayas. No machines were available for the repair of this main road. Instead, three women operated a shovel. In this system, one digs the shovel into the dirt, then the other two pull the rope that pulls the shovel to move the dirt. Some shovel-rope teams comprise only two people. While this system may not seem very efficient, it provides more employment than a machine. Moreover, the shovel-rope system is appropriate to the terrain. Photograph courtesy of B. A. Weightman.

done in four hours. Another type of underemployment is the hiring of several individuals to do the job ordinarily held by one person. Also, there is *hidden underemployment*, based on group or family solidarity, where everyone participates in the work whether they are needed or not.

While unemployment denies means of livelihood, underemployment provides both income and sense of self worth (Figure 4-10). Homeless women in Mumbai chop rocks to make gravel, carry bricks, mix cement, and do other arduous tasks. Meanwhile, they can live within the project they are building. When it is finished, they will move on to another construction site. Children work alongside their parents. Economically, at the national level, this is not an ideal situation because the cost of services for surplus labor takes away from potentially broader expenditures. But, individually, at the local level, underemployment is preferable to no employment.

With disproportionate investment and capital concentration, cities exhibit dramatically disparate landscapes. Landscapes of power and privilege contrast with those of powerlessness and denial. Yet, many poor are employed by their rich counterparts as watchmen/guards, house servants, gardeners, drivers, and the like (Figure 4-11).

ECONOMIC DEVELOPMENT ZONES

Another vital urban phenomenon is the **trans-border development region.** Trans-border development regions

Figure 4-11

This beautiful house in India shows how the upper classes live. Note the gardener watering the plants. I took this picture on Christmas Day, and the star above the doorway indicates that this is a Christian household. The star is lit at night. Photograph courtesy of B. A. Weightman.

include **economic development zones (EDZs)** or export processing zones in adjacent political units. A region may include several countries, as in the case of the "development triangle" involving Singapore plus parts of Malaysia and Indonesia. Or it might involve several provinces, as in the case of southern China. Development zones are integrated across borders by transportation corridors and telecommunication links into the global system of production, commerce, finance, distribution, and consumption.

Globalization is not only about the hyper-mobility of capital and the rise of information economies. It is also about specific types of places and work processes. New labor dynamics have evolved in the context of global cities and their interconnectivities. Increasing numbers of women and migrant laborers are becoming enmeshed in the global economy.

CHILD LABOR

The International Labor Organization estimates that there are 250 million economically-active children in the world today. Sixty-four percent are in Asia, where at least 15 percent of all 10- to 14-year-olds labor in fields and factories.

Child labor is especially common in South Asia, where 30–46 million are exploited in the grimmest conditions (Figure 4-12). In Nepal, 40 percent of children from ages 10 to 14 work in mines, quarries, and carpet factories. In the case of carpet weaving, children are trained at a very young age (Figure 4-13). Bangladesh has as many as 30 million child laborers. Dhaka has the largest number of child laborers in Asia, with some 750,000 children between the ages of 10 and 14 providing

Figure 4-12

This is one of thousands of children working in South Asia.
©Deshakalyan Chowdhury/AFP/Getty Images, Inc.

Figure 4-13

These rug weavers are Tibetan refugees living in Nepal. Children quickly learn their trade by watching and helping their mother.
Photograph courtesy of B. A. Weightman.

half the income in poor, female-headed households and nearly a third in male-headed families.

Millions of children work 10 or 12 hour days in dangerous, medieval conditions in glass, match, textile dyeing, and carpet factories, gravel pits and brick kilns. They tote bricks for construction, carry baskets of raw sewage out of drains, and pick through garbage for anything salable. For the "boss" or factory owner, paying child labor a pittance is better than buying machinery for a fortune. Long hours, with lack of nutrition, combined with virtually no safety standards, means that accidents are commonplace. Dust, fibers, and chemicals cause respiratory damage and blindness. A villager in northern India proudly showed me his 7-year-old son at work weaving a magnificent carpet in a small, mud brick enclosure with one tiny window. The air was thick with fiber from the wool.

Another form of injustice is **bonded labor**. People who need money and have no assets to back a loan pledge their children's labor as security to the moneylender. Interest rates are usurious and borrowers rarely get out of debt. Therefore, their children are relegated to a life of toil. In the Hindu, sacred city of Varanasi, carpets and saris are woven by the bonded labor of more than 200,000 children under the age of 14, some as young as 5 according to Human Rights Watch.

Glass bangles, popular with Indian women, are often made by children in horrendous factory environments. They carry molten globs of glass on the ends of iron rods that they have drawn from furnaces with temperatures of 2,732°F (1,500°C). Factory floors are commonly strewn with broken glass, and many children have no shoes to protect their feet. If a child is hurt and can't work, she is fired and sent back to the streets.

Worldwide, the largest sector of child labor is domestic service. Surveys of middle-income society in Colombo, Sri Lanka and Jakarta, Indonesia revealed that one in three households had a domestic worker younger than 14. In Kuala Lumpur and other Malaysian cities, most domestic workers are children and teenagers from Indonesia who are forced to work 16 hour days, 7 days a week with no scheduled rest periods—ever!

Most people condemn these practices and decry the cruelty of child labor no matter what the circumstances. Sanctions and boycotts might be successful in removing significant numbers of children from deplorable work environments, but the story is more complex.

Realities often make child labor essential for family survival. Many children of poverty have been sold into servitude by desperate parents. Many have crippled or diseased parents, or none at all. Without income they will literally starve to death. Working children are facts of life in many parts of Asia.

Enforcing child labor laws is problematic given Mafia-like control over disguised factories, hidden children, corrupt officials, and dire poverty in certain regions. Furthermore, not all work environments are terrible. Significant numbers of children work as contributors to the family farm or business in relatively decent conditions. Many go to school as well.

Education works against child labor. In places where school attendance is enforced, child labor rates decline. Still, families in poverty must make a decision as to whether long-term gain, via education, is worth the wait in terms of their immediate needs. Education requires expenditures for books, uniforms, and food. This money and time spent must be balanced against lost wages.

Official and other organizations, while making efforts to end enslavement, brutality, and unsafe and unhealthy work environments, try not to further impoverish families or children by removing them from the workforce altogether. Instead, they attempt to get the children and keep them in school and improve working conditions. Half-day classes are common in such countries as Myanmar, and children work before or after school. In Bangladesh, garment workers' unions actually run schools.

Child labor may be unacceptable in societies flush with palliative programs for the less fortunate, but in societies characterized by large numbers of desperate poor and limited if any safety nets, a single working child might be the only solution to survival.

WOMEN AND SUPPLY CHAINS

In economic development zones, 80 to 90 percent of workers are women, resulting in the coinage of the term **feminization of labor**. Companies hire women, especially in assembly-line industries such as electronics, because they are considered to have "nimble fingers" and are more obedient than men. Moreover, in virtually all cases they are paid less than male workers. No benefits are provided. Twelve-hour days and seven-day work weeks are common. While "cheaper" women are employed in factories all over Asia, we will focus on this phenomenon in China and the Philippines in the regional chapters.

We have all had visions of hundreds of women slaving away on assembly lines—one person puts a sleeve on a shirt, the next one sews on a button, and so on. In the twenty-first century, everything has changed. Employees now work on what might be termed a "disassembly line." Companies now try to break up their products into specialized disassemblies to drive down costs, improve quality, and reduce the time it takes to get the product to market. This arrangement is referred to as a **supply chain**.

Here's how a supply chain works. J.C. Penney orders 100,000 copies of a shirt. It buys the yarn from a Korean producer that sends it to Taiwan to be woven and dyed into cloth. Buttons are ordered from a Japanese manufacturer located in China. Both buttons and cloth are shipped to Thailand to be cut and sewn into shirts. J.C. Penney wants the shirts on the shelves right away because fashions change so quickly. So the materials are sent to five different Thai factories that rush to make 20,000 shirts apiece. A few weeks later, you are trying on a supply chain shirt in your local mall.

While poorer countries like Bangladesh or Vietnam might not have the capability to produce a whole computer or car suitable for export, they can make relatively uncomplicated components. Such countries are often the origins of supply chains—doing the first basic steps of the production process.

The pace of urbanization and landscape change in general is exacerbated with global economic interchange that, in essence, ignores political boundaries, human costs, and environmental concerns. As we noted in Chapter 2, the majority of the world's most polluted cities are in Asia. Fortunately, there is growing awareness that we are running a race on a deteriorating environmental track. Development, whether indigenous or international in scope, is consuming the social and natural capital upon which it depends.

Migration

Population migration—the movement of people from place to place—is generated by *push forces* and *pull forces*. Push forces are conditions that drive people away from their home and pull forces are conditions elsewhere that look better. People are lured by real or perceived opportunities in other places. Globalization has led to one of the largest migrations in human history as increasing numbers of people move to cities and countries around the world. The numbers of people who change locations within their own country can only be guessed at. While we can track major streams of international migration, it is impossible to determine exact numbers of migrants because many move without documentation. Furthermore, untold numbers of refugees who are escaping war, persecution, or environmental catastrophes are not counted either.

Large numbers of people are **environmental refugees**. Some of these people are compelled to leave their homes due to environmental problems related to global warming and land degradation. Climatic factors, such as increased aridity or high winds, combined with overuse of soils to feed large populations, lead to land degradation. Entire families have little choice but to leave their land because it is so eroded or exhausted that it no longer can support them. Other driving forces are rising sea levels, weather-related flooding, desertification, and dried-up aquifers. Scholars estimate that there are at least 50 million environmental refugees worldwide.

There are two broad categories of migration:

- *Internal*, such as movement from rural to urban environments or changing residences within a city.
- *External*, such as working in or moving to another country or continent.

INTERNAL MIGRATION PATTERNS

As we have already seen, rural-to-urban migration is a significant cause of city growth in Asia. Important factors in the generation of these migration patterns are the improvements that have been made in agriculture (see Chapter 5). Increased productivity releases farm workers for alternative, better-paying jobs in or near cities. Other push factors include landlessness, hunger, and loss of social cohesion that is induced by growing competition for declining resources.

Such movement is of paramount importance in the evolution of cities, especially those in China and Southeast Asia. For example, 200,000 rural people moved to Jakarta, Indonesia, in 2008. These were relatives of people who already worked in the city and had returned to their villages for a festival. The government is making an all-out effort to create employment in the countryside in an attempt to curtail such movement. Family ties are important factors in migration everywhere in the world.

People also will move to cities in order to take advantage of services such as education and health care. In 2004, it was estimated that 60 percent of Cambodian peasants who sold their land and moved to a city were forced to do so because of medical indebtedness. Interestingly, prior to 1991 the official attitude of India's government was that expansion of the telecommunication infrastructure encouraged rural-urban migration. Therefore, telecom services should be curbed in cities to stem the population influx. This policy has changed.

One of the most important types of migration is **chain migration**, whereby the mover is part of an established migrant flow from a common origin to a prepared destination.

This flow might involve several generations of migrants who have established communities ready to absorb new ones. This creates ethnic enclaves or foreign-born communities in both cities and rural areas. Chain migrations typically involve relatives and friends and can be temporary or permanent, internal or international in nature.

Sometimes chain migration is specific to an occupational group. For instance, most construction workers in New Delhi in the north of India come from the state of Orissa in the east of India or from Rajasthan in the northwest. Newspaper sellers come from the state of Tamil Nadu in the south.

Coastal regions are growing faster than interior regions. In 1998, 60 percent of Asians lived within 250 miles (400 km) of the coast despite the continent's vast land mass. More than 75 percent of Japanese live along or near a coast. In fact, the government has declared 47 percent of the country as "depopulated" and has devised programs to encourage migration to the interior. Sixty percent of China's population lives near the coast. The Philippines has Asia's second longest coastline after Indonesia, and more than half of all Filipinos live within a day's walk of the coast.

Coastal development is produced by a variety of processes such as early urban patterns created by precolonial and colonial powers. Increased global interaction fostered the role of international trade in the development of ports and infrastructural linkages to hinterlands. Modern globalization has seen the intensification of integration of coastal cities into the world system of interchange.

India is the single exception to this phenomenon because a succession of rulers throughout history constructed cities in various regions of the interior. "Old" Delhi dates back to about 300 BC, while "New" Delhi was planned by the British in the twentieth century. Coastal cities such as Mumbai and Chennai, which remain magnets for migration, were actually created by the British during the colonial period. Environmental and other problems in coastal cities have led Indian planners to focus development in numerous interior cities such as Hyderabad and Bangalore. Moreover, there are serious efforts to clean up Delhi.

INTERNATIONAL MIGRATION PATTERNS

Asians have long comprised significant migration streams to other parts of the world. Today, it is the primary source of international **labor migration**. Most of this migration is voluntary—people seeking jobs outside their home country where unemployment is rampant. Many are recruited by foreign governments or employment agencies. For example, thousands of Chinese came to work in American and Canadian mining enterprises or railroad construction in the nineteenth century. "Chinatowns" are ubiquitous around the globe. Even Paris has two Chinatowns. One is populated by ethnic Chinese refugees who escaped from Vietnam, Laos, and Cambodia in the 1970s.

You can find Indian communities in almost every country. Great Britain is home to more than a million Indians. Durban, South Africa, has the largest Indian population in sub-Saharan Africa. One of the largest foreign-born populations in North America, Europe, and Australia is Indian.

Indian immigration to Canada and the United States began in the nineteenth century. Changes in immigration policies after World War II generated subsequent waves of Indian immigrants. The most recent and largest migration wave has been in the 1990s and 2000s fueled by the IT revolution. While early migrants took jobs such as taxi drivers, laborers, or small business owners, the recent wave has lodged itself in high-tech industries, academic and medical professions. There are more than 700,000 East Indians living in Canada and 2.7 million in the United States.

While the majority of international migrants are young males, significant numbers of women also migrate to other countries to work in global cities (Figure 4-14). Each year, some 100,000 women leave Asia's newly developing countries to work in domestic service, hotels, or factories in the continent's older industrial economies. Also, about 50,000 Thai women and Filipinas leave their homes (voluntarily or forcibly) to work in the sex and entertainment industries in Japan and South Korea.

Many women go to Middle Eastern countries. In 2003, the United Arab Emirates (UAE) granted an average of 300 visas a day to Southeast Asian women to meet their demand for domestics. The average UAE household hires three. Middle Eastern countries are also important destinations for an estimated 30,000 to 50,000 teachers from India, Sri Lanka, South Korea, and the Philippines.

The international migration of nurses, nannies, maids, and sex workers, especially from Sri Lanka and the Philippines, is outstanding in scope. Sri Lankans and Filipinas are in great demand, with thousands working as maids and nannies in places such as Hong Kong, Singapore, and cities in the Middle East. Unfortunately, many are mistreated and some are held in virtual slavery by their employers.

Qualified nurses are in great demand in the United States and elsewhere in the developed world. In 2009, there were 20,000 nursing jobs available for Filipinas in Saudi Arabia. Recruitment of nurses from developing

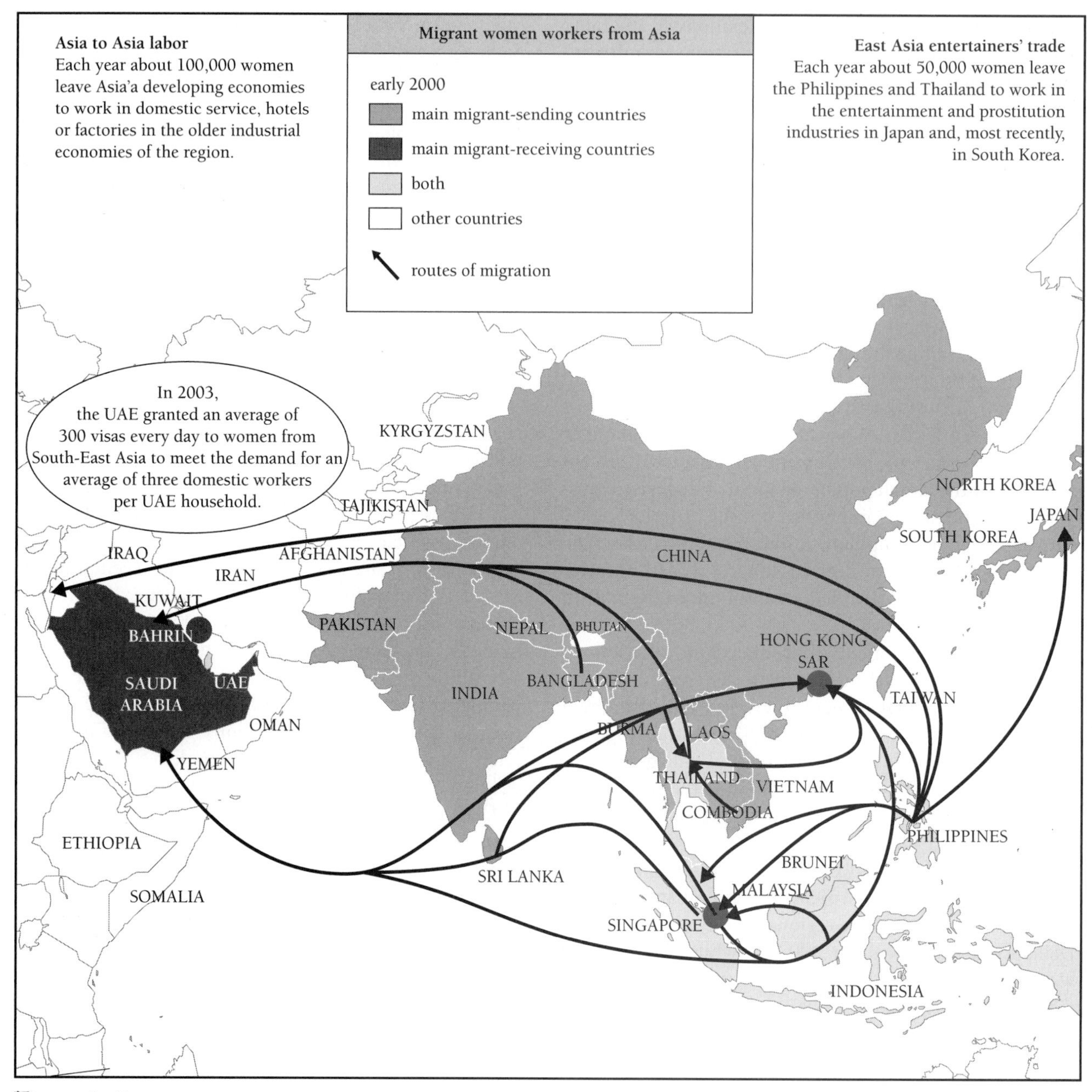

Figure 4-14

International female labor migration from and within Asia in 2008. From Joni Seager, *Penguin Atlas of Women in the World,* 4th edition, 2009, p. 72-3. Reprinted courtesy of Penguin Books.

countries is producing an acute shortage of trained health-care providers in those countries.

Many international migrants are students who are pursuing degrees in higher education. Some remain in their host country but most return home. The number of Indian students in American graduate schools is declining as India's institutions of higher education have been significantly ungraded. **Brain drain** refers to the process whereby developing countries lose their most educated people to jobs overseas.

Foreign-born populations often draw undue attention and prejudice, especially when they are perceived to be "taking jobs" from the majority population. Canada, Australia, and some European countries are in the process of enacting laws to protect "their own" people from job loss.

In Singapore, one in three residents is a non-citizen. Foreigners compete on every level, and citizens complain that they steal jobs and depress wages. High immigration has coincided with a widening of the income gap. Singapore's Gini Index rose from 0.444 in 2000 to 0.481 in 2008. The city's glitzy city center stands in sharp contrast to peripheral "heartland" areas. The government plans to slow its intake of immigrants and accentuate the benefits and responsibilities of citizenship.

Rapid urban growth intensifies machinations within battleground environments that are increasingly mystifying in their complex dynamics. While new techno-economic networks are emerging in the wake of development, new disconnects among competing political, economic, and social groups are being produced in the context of a torrent of globalizing forces. Decision makers at all levels mediate the interaction of humans and urban environments. Such capitalistic-political processes both benefit and disrupt different communities unevenly and unequally.

Recommended Web Sites

www.childlabor.in
Information from the International Labor Organization, which documents child labor across the globe.

www.cities.com
The latest news from cities all over the world.

http://city.jsc.nasa.gov/city/slash
NASA photographs of cities from space.

www.citypopulation.de
Population maps and statistics for cities around the world.

www.climate.org/index.html
Excellent site for information on environmental refugees, climate change and impacts on people. Asian country reports, climate games and negotiation simulations.

www.demographia.com/
Reports on cities and urban processes.

www.enterweb.org
An annotated meta-index of articles about the impact of business, finance, and international trade on both developed and developing countries in this era of globalization and cyberspace.

http://esa.un.prg/unup/p2k0data.asp
World Urbanization Prospects Report (2007 Revision).

www.focusintl.com/widnet.htm
WIDNET: Women in Development NETwork. A wealth of information about women in development.

www.grameen-info.org/
Information on the Grameen Bank.

www.havoscope.com/asia-black-markets/
Information and statistics about shadow economies/black markets in Asia and elsewhere.

http://hdr.undp.org.en/
United Nations Human Development Reports with interactive maps, graphs, statistics, etc.

www.imf.org
The International Monetary Fund (IMF) is intended to promote international monetary cooperation and facilitate the growth and expansion of international trade.

www.library.nwuedu/search
International government organizations.

www.nationsonline.org/oneworld/bigcities.htm
Virtual travel. Skyline photos of major cities with links to Google Earth satellite photos and information sites.

www.undp.org
United Nations Human Development Reports.

www.unsystem.org
Web site for all organizations in the United Nations. Information on problems, projects, and prospects in the developing world.

www.worldbank.org
The International Bank for Reconstruction and Development promotes sustainable development and investment in people. It provides loans, technical assistance, and policy guidance to developing countries.

Bibliography Chapter 4: Development, Urbanization, Migration, and Quality of Life

ADB. 2009. "Environmental Sustainability: Reducing the Vulnerability of the Poor to Degraded and Hazardous Conditions." *Poverty Reduction*. Manila: Asian Development Bank.

Agnew, John. 1987. *Bringing Culture Back*. In "Overcoming the Economic-Cultural Split in Development Studies." *Journal of Geography* 86: 276–281.

Ayers, Ed. 1996. "The Expanding Shadow Economy." *World Watch* 9: 10–23.

Bernard, Mitchell, and John Ravenhill. 1995. "Beyond Product Cycles and Flying Geese: Rationalization, Hierarchy, and the Industrialization of East Asia." *World Politics* 47: 171–209.

Boudreaux, Karol, and Tyler Cowan. 2008. "The Micromagic of Microcredit." *The Wilson Quarterly*. Winter Issue.

Bowen, John T. Jr., and Thomas R. Leinbach. 1995. "The State and Liberalization: The Airline Industry in the East Asian NICs." *Annals of the Association of American Geographers* 85: 468–493.

Chambers, Robert. 1995. "Poverty and Livelihoods: Whose Reality Counts?" *Environment and Urbanization*. 7/1: 173–204.

Cho, George. 1995. *Trade, Aid and Global Interdependence*. New York/London: Routledge.

Collier, Paul. 2007. *The Bottom Billion: Why the Poorest Countries Are Failing and What Can Be Done About It*. New York: Oxford University Press.

Davis, Mike. 2007. *Planet of the Slums*. New York: Verso.

Dossani, Rafiq. 2008. *India Arriving: How This Economic Powerhouse Is Redefining Global Business*. New York: American Management Association.

Douglass, Mike. 1995. "Global Interdependence and Urbanization: Planning for the Bangkok Mega-Urban Region." In *The Mega-Urban Regions of Southeast Asia,* ed. T. G. McGee and I. M. Robinson, pp. 45–77. Vancouver: University of British Columbia.

Drakakis-Smith, David. 2000. *Third World Cities*, 2nd ed.. New York: Routledge.

Easterly, William. 2007. "The Ideology of Development." *Foreign Policy*. July/August.

Easterly, William. 2006. *The White Man's Burden: Why the West's Efforts to Aid the Rest Have Done So Much Ill and So Little Good.* New York: Penguin.

Escobar, Arturo. 1995. *Encountering Development: The Making and Unmaking of the Third World.* Princeton, N.J.: Princeton University Press.

Farrell, Diana. 2006. *Driving Growth: Breaking Down Barriers to Global Prosperity*. McKinsey Global Institute. Boston: Harvard Business School Press.

French, Hilary. 1997. "When Foreign Investors Pay for Development." *World Watch* 10: 8–17.

Grant, Richard, and Jan Nijman. 1997. "Historical Changes in U.S. and Japanese Foreign Aid to the Asia-Pacific Region." *Annals of the Association of American Geographers* 87: 32–51.

Krishna, Anirudh. 2006. "Reversal of Fortune: Why Preventing Poverty Beats Curing It." *Foreign Policy* May/June: 62–63.

McGee, Terry. G. 1989. "Urbanisasi or Kotadesasi? Evolving Patterns of Urbanization in Asia." In *Urbanization in Asia: Spatial Dimensions and Policy Issues,* ed. Frank J. Costa, Ashok K. Dutt, Laurence J. Ma, and Allen G. Noble, pp. 93–108. Honolulu: University of Hawaii.

McGee. Terry. 2001. "Urbanization Takes on New Dimensions in Asia's Population Giants." Washington, D.C.: Population Reference Bureau.

McGinn, Anne Platt. 1987. "Cigarette Traffickers Go Global: The Nicotine Cartel." *World Watch* 10: 18–27.

Narayan, Deepa et al. 2000. *Voices of the Poor: Can Anyone Hear Us?* New York: Oxford University Press (for the World Bank).

Quarto, Alfredo. 1993. "The Vanishing Mangroves." *Third World Resurgence* 39: 4–6.

Ransom, David. 1996. "The Poverty of Aid." *New Internationalist* 285: 7–10.

Roberts, Brian, and Trevor Kanaley, eds. 2006. *Urbanization and Sustainability in Asia*. Manila, Philippines: Asian Development Bank.

Robison, Richard, and David Goodman. 1996. "The New Rich in Asia: Economic Development, Social Status and Political Consciousness." In *The New Rich in Asia: Mobile Phones, McDonalds and Middle-Class Revolution,* eds. R. Robison and D. Goodman, pp. 1–16. New York: Routledge.

Schneider, Friedrich, and Dominik Enste. 2002. "Hiding in the Shadows." International Monetary Fund Economic Issues Report No. 30.

Seagrave, Sterling. 1995. *Lords of the Rim*. London: Bantam.

Schmitz, Cathryne et al., eds. 2004. *Child Labor: A Global View*. Santa Barbara, Calif.: Greenwood Press.

Straussfogel, Debra. 1997. "Redefining Development as Humane and Sustainable." *Annals of the Association of American Geographers* 87: 280–305.

Stiglitz, Joseph. 2006. *Making Globalization Work*. New York: Norton.

Stiglitz, Joseph. 2006. "Social Justice and Global Trade." *Far Eastern Economic Review.* March: 18–21.

The Economist. 2009. "A PR Problem: Singapore." *The Economist,* November 14: 52.

The Economist. 2009. "Burgeoning Bourgeoisie." *The Economist,* February 14: 3–22.

The Economist. 2008. "How to Save a Billion Lives." *The Economist,* February 9: 65–66.

The Economist. 2009. "The Toxins Trickle Downwards." *The Economist*, March 14: 62–63.

Tinker, Irene. 1997. *Street Foods: Urban Food and Employment in Developing Countries*. New York: Oxford.

Tsai, Kellee. 2007. *Capitalism without Democracy: The Private Sector in Contemporary China*. Ithaca, N.Y.: Cornell University Press, Sage House.

UN-Habitat. 2003. *The Challenge of Slums: Global Report on Human Settlements*. London.

Weissman, Robert. 1997. "Move to Make U.S. Tobacco TNCs Accountable Internationally." *Third World Resurgence* 92: 7–10.

Chapter 5

Agriculture, Food, and Food Security

"If you are planning for a year, sow rice; if you are planning for a decade, plant trees; if you are planning for a lifetime, educate people."

CHINESE PROVERB

Agricultural Foundations

Agriculture is understood as the cultivation of crops and rearing of livestock. Early humans were hunters and gatherers but at some point learned to grow their own food crops and raise their own meat supply. **Vegeculture** most likely preceded seed cultivation; cultivating tuber and tree crops such as taro and bananas occurred prior to the **domestication** of rice and other grains. Land does not require clearing for root and tree crops, and a variety can be planted. These can be harvested at will; therefore, there is little need for storage facilities. Seed cultivation requires land clearance. Furthermore, grains have a specific growing season and must be harvested all at once, making storage necessary. Inadequate facilities invite losses to mold, insects, and rodents.

Primitive vegeculture developed in tropical regions around the world such as southern China, eastern India, and mainland and insular (island) Southeast Asia. Vegeculture was practiced as early as 9000 BC in Thailand and was definitely a multiregional phenomenon by 7000 BC. Evidence of vegeculture has turned up in many other areas in Asia, including Taiwan and Japan.

Asia possesses a wealth of indigenous plant species. In fact, many "familiar" foods originated there. However, crop assemblages were altered with increased interregional trade by land and sea. Plant diffusion was slow until the fifteenth century, but by the seventeenth century, botanical gardens had been established in Europe to foster plant exchanges around the world. As you can see in Figure 5-1, wheat, the second most important grain crop in Asian countries such as China and India, came from Southwest Asia. China is the world's largest growers of sweet potatoes, which originated in Mesoamerica. Mesoamerica also gave corn to Asia, where it is eaten by both animals and people and where you can even buy delicious corn ice cream. Soybeans diffused from China across the Pacific to America and Brazil. Now Brazil leads the world in soybean production. Improved transportation and increased trade have revolutionized diets around the world.

By the nineteenth century, the industrial world demanded non-food, **industrial crops** such as cotton and rubber. Radical change derived from the introduction of plantation crops such as rubber and oil palm, especially in Southeast Asia. Commercial variety rubber originated in South America and oil palms came from Africa. As we'll see later in this chapter, plantation agriculture has

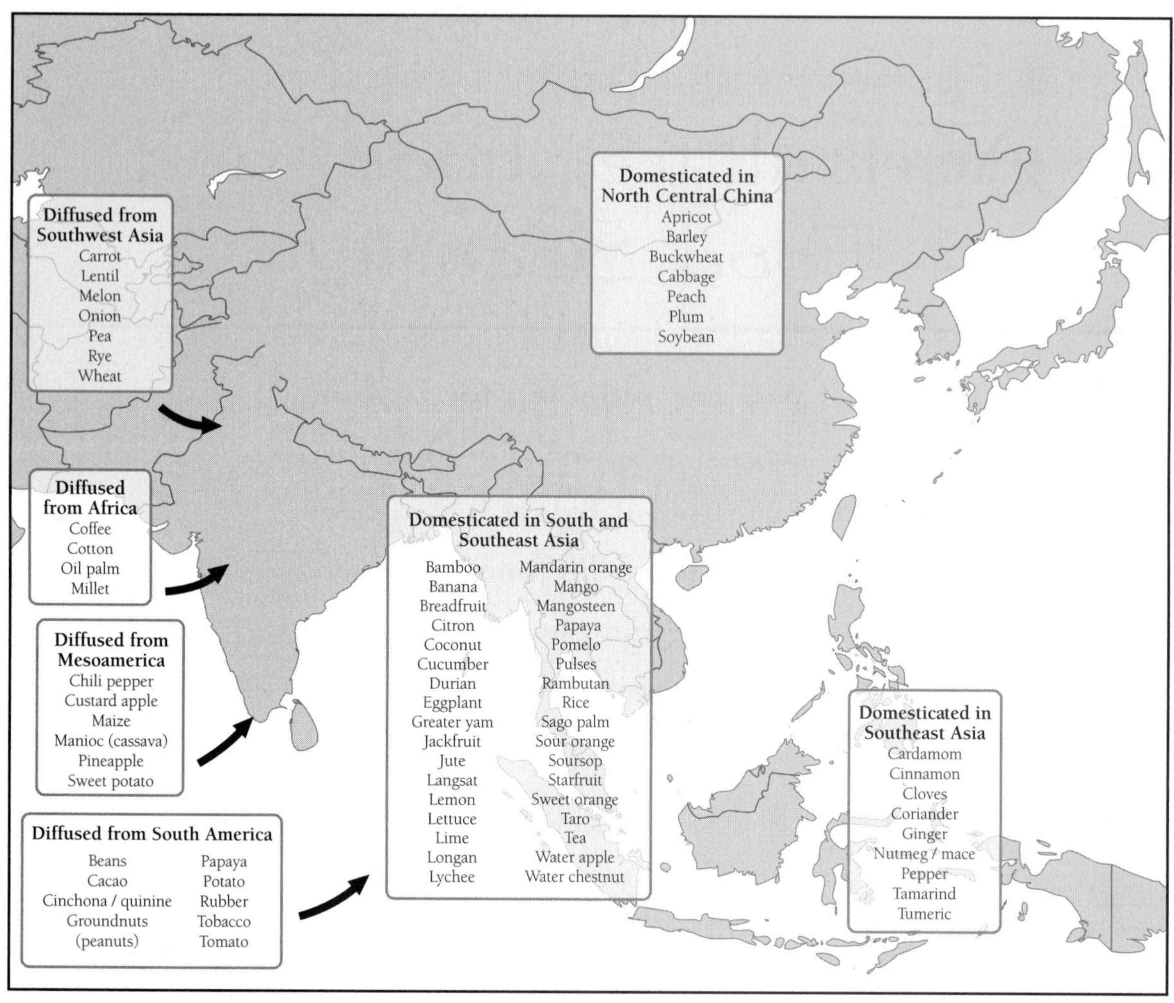

Figure 5-1

Crop assemblages, domestication, and introduction. Which of these grains, fruits, herbs, and spices are you familiar with?

become transnational agribusiness, dependent upon the machinations of the global economy.

SHIFTING CULTIVATION

Early farming methods included **shifting cultivation**, a migratory, field rotation system found mainly in forested regions and uplands with relatively sparse populations. Shifting cultivation is called **swidden** in most of Southeast Asia. It is referred to as *ladang* in Indonesia Figure 5-2.

An area is cleared by a **slash-and-burn method**—cutting down all but the largest trees and setting fire to the resultant debris. Burning brings nitrogen to the soil, which is then planted with the aid of a digging stick or hoe.

Shifting cultivation is usually dominated by women, who select the appropriate micro-environment, choose the essential seeds, and design the multilayered crop complex: below-ground tubers, surface plants, and above-ground plants and tree products. A field is used from one to three years depending on soil fertility. Then it is abandoned and allowed to remain fallow for 20 to 50 years or more, during which it regenerates its natural flora. People move to a new area and begin again.

Shifting cultivation can be an ecologically sound practice for maintaining soil fertility, reducing soil erosion, and encouraging forest rejuvenation. People participate in the ecosystem rather than superimposing their will upon it. However, this practice has drawn severe

Figure 5-2

In this ladang (swidden) plot in Kalimantan, Indonesia, the major crop is corn. A papaya tree stands in the foreground and a banana plant in the background. Eventually this plot will be abandoned for another. Photograph courtesy of B. A. Weightman.

criticism as being destructive of forests, as well as being responsible for damaging fires.

Problems reside not so much with the method as with the impacts of population and deforestation. Dramatic population increases, combined with the removal of forests for plantations and other uses, mean that there is insufficient land available for shifting cultivators. For example, every year in Southeast Asia, anywhere from 98,800 to 1,976,000 acres (39,520 to 790,400 ha) of land formerly used for shifting agriculture is transformed to permanent agriculture. Consequently, shifting cultivators are unable to fallow their plots long enough to regenerate soil fertility.

In shifting cultivation areas of Bangladesh's uplands, fallow periods have fallen to only three or four years. Consequently, the land cannot produce enough to maintain the family throughout the year. Instead, food reserves are exhausted in seven or eight months. Replanting too soon produces a downward spiral from diminished yields to total abandonment.

Until recently, shifting cultivation was considered a dying art—a disappearing phenomenon not worthy of formal development efforts. Critics now realize that in order to stem the flow of rural–urban migration, farmers need to stay on the land. However, farmers need cash crops as well as food crops, derived from cropping systems that maintain soil fertility.

The goal of finding ways to improve the productivity and sustainability of marginal lands, including uplands, was put forth at the 1992 United Nations Conference on Environment and Development. Agricultural scientists noted that shifting cultivation might be sustainable in some areas with external assistance. Ongoing plans include marginal lands being planted with perennial crops that can be sold. Perennial crops have permanent root systems that retain soil and stabilize slopes. They can even be integrated into shifting cultivation systems.

One successful program, based on a perennial crop, is the production of cardamom in Sikkim, an Indian state in the Himalayas. Cardamom is a spice that is indigenous to India and commonly used in Indian cuisine. A high-value, low-volume product crop that can be stored for some time, it is comparatively easy for hill-dwellers to market.

While swidden plots are being clear-cut and permanently planted, the issue of soil fertility remains. Sustained crop yields in most tropical soils require the application of large quantities of fertilizer, essentially ensuring reliance on external sources. For example, temperate-climate vegetables often grow well at higher elevations in tropical regions. However, they require heavy applications of chemical fertilizer and pesticides. Poorer farmers cannot afford chemical supplements, and many are forced to find subsistence elsewhere. Furthermore, chemicals ultimately contaminate water sources.

The United Nations and other experts recognize that most successful programs rely on outside support from governments or NGOs, especially during the initial years of a project. Without this support, farmers cannot generate enough funds for capital investment. Also, the needs of the environment must be balanced with the needs of upland farmers for increased production and a reasonable cash income. Further, conservation programs must be carried out on a large scale in order to be effective.

RICE CULTIVATION

Recent archaeological evidence from Zhejiang province in southern China indicates that rice was domesticated there around 5000 BC. Scientists think that its hearth area was in Southeast Asia, possibly Thailand, around 9,000 years ago. Gujarat in western India was another site of domestication. Swidden plots housed initial rice cultivation that was not irrigated. Wet-rice production did not become well established in China until the fourteenth century and in India even later. Rice cultivation diffused to Japan either via Korea or across the sea around 300 to 400 BC. It was introduced into Java by sea from India, but as elsewhere, irrigated cultivation was not well organized until much later. Eventually rice replaced millets as the main food grain in the warmer parts of Asia.

Today rice is the staple food of more than half the world's population, and 90 percent of it is grown in Monsoon Asia. The inclusion of Asian colonies into the international economy stimulated population movements and the establishment of commercial, wet-rice systems in major river deltas. The Irrawaddy (Ayeyarwady), Menam (Chao Praya), and Mekong deltas were occupied between 1850 and 1930 and exported rice to Asia and the world.

"Rice bowls," intensive rice-growing areas, were founded on environmental circumstances and human ingenuity regarding methodology and technology. Paddy preparation, water distribution systems, seedbed cultivation, sowing and/or transplanting techniques, along with terracing and irrigating slopes are only some of the requirements Figure 5-3. Farms are generally small, around 2 to 4 acres (1–2 ha). Plots are small as well, disallowing the use of large machines. Although the use of chemicals—fertilizers, pesticides, insecticides, and fungicides—are widespread, mechanization is not, except for some parts of Japan, South Korea, and Taiwan. Mechanization is increasing in parts of Southeast Asia such as Thailand and Indonesia.

In Asia, three types of rice are cultivated: *indica*, *japonica*, and *javanica*. *Indica* varieties have long grains that remain separated when boiled. *Japonicas* have shorter, stickier grains. *Javanicas* are somewhere in between. *Indicas* and *javanicas* dominate in tropical and subtropical areas while *japonicas* excel in cooler latitudes and at higher elevations. There are glutinous and nonglutinous varieties in all categories.

Farmers typically grow several varieties, depending on their own food preferences, market demand, and ceremonial requirements. They are also conscious of variation in maturation periods that can take from 90 to 260 days. Where temperatures permit, faster ripening varieties, like *indicas*, allow double or even triple cropping. Quick-ripening rice is critical where water supply is unstable.

Wet-rice cultivation relies on either rainfall or irrigation. In hilly areas, rice is grown on irrigated terraces. In deeply flooded areas like delta regions, floating rice is grown. It makes up a quarter of Bangladesh's rice output, for example. Elongated stems grow in concert with rising water in the wet season, keeping the grains above the water that may be several feet (meters) high. Floating rice is tended by wading or from boats.

Figure 5-3

These dramatic rice terraces in northern Luzon in the Philippines are testimony to the persistence of cultural traditions dating back 2000 years. The terraces follow the contours of the mountains and use mountain streams channeled into elaborate systems for irrigation. ©Tammy David/AFP/Getty Images, Inc.

Rice and Society

"Without rice there is nothing doing." (Malay Proverb)

The Malay proverb above signifies the importance of rice in much of Asian society. In Chinese and some Southeast Asian languages, "to eat" is the same as "to eat rice." Japan became known as *Mizumono Kuni*—the "Land of Luxurious Rice Crops." The planting of rice in Japan has an intimate bearing on Japan's indigenous religion—*Shintoism*—which makes a virtue of subordination of the individual to the group. Scholars believe that this may stem from the traditional labor-intensity of rice cultivation in which all village members were required to participate. Rice is the foundation of culture. It is requisite for survival and integral to daily, weekly, monthly, and yearly routines and permeates both secular and sacred worlds.

Rice is highly digestible and nutritious when eaten with its bran intact and/or with vegetable and fish supplements. Claims of dietary deficiencies due to rice are simplistic. It is polished rice that has reduced protein. As part of modernization, rice mills are increasingly common, and rice is polished to a glaze. Thoroughly devoid of bran, the nutritional value of the grains is reduced.

Rice wine, beer, and liquor are widely consumed. Japanese *saki* is distilled from rice wine. In some places, a common container emphasizes the role of rice as a social locus. On several occasions, I have been invited to share the rice wine pot. Sipping out of one of several reed straws protruding from large clay containers, I experienced a sense of unity with my Lao and White Tai hosts.

Rice husks are fed to animals. In the Mekong delta, rice bran is fed to fish on fish farms. Rice bran oil is a basic ingredient of soap and many insecticides. Noodles and cosmetics are made from rice flour. The straw provides bedding for livestock and can serve as a low-grade feed. It also is the medium for cultivation of straw mushrooms.

Aside from its practical uses, rice permeates ceremonial and ritual activities. For instance, there is an annual event in Bangkok, attended by the Thai king, called *The Royal Ploughing Ceremony* that calls for bounteous rice harvests. When Japanese Emperor Akihito was married in 1959, he carried a rice scoop to symbolize fertility. Folk dances still replicate stages in rice production, and rice motifs are frequent in weaving and basketry. Temples, goddesses, and gods are devoted to rice. It is integral to marriage, birth, and groundbreaking ceremonies. The importance of rice in Asian society can be summed up by a Balinese religious belief: "rice is the soul of man."

FARMING IN THE CITY

According to the United Nations, one-seventh of the world's food is produced in urban environments. Millions of city residents, most commonly in Asia, cultivate small garden plots on the ground, in garbage dumps, and even on rooftops. They also breed chickens, pigs, and other animals in backyards, and raise fish in small ponds. Not only are these entrepreneurs feeding their own families but they also are able to sell some of their products in local markets without the costly expenses of storage, spoilage, and long-distance transportation.

At least 60 percent of Bangkok's metropolitan area is cultivated. There you can find garden plots along *klongs* (canals) and in *sois* (small, crowded residential areas). Large metropolitan regions in China incorporate suburban agricultural areas that supply fresh produce to urban residents. Many city farmers have no other source of income.

There are both positive and negative environmental consequences of urban agriculture. City waste can be used as fertilizer or to feed animals. This recycling reduces the cost of wastewater treatment and solid waste disposal. Planting sloping banks of rivers, streams, and canals prevents soil erosion. In Kolkata (Calcutta), India, city sewage is dumped into some 7,400 acres (3,000 ha) of lagoons that produce around 6,000 tons of fish a year. More than 20,000 Kolkata residents farm the city's garbage dumps converting waste and rotting food to nutrition.

The downside of urban agriculture is that it exposes people to untreated waste such as human and animal feces, along with fly-covered and mosquito-infested rotting refuse. Consequently, handlers are exposed to often deadly diseases such as cholera and hepatitis. Moreover, when chemicals, such as pesticides, are applied indiscriminately by untrained users, water supplies can become contaminated. Diversion of water for irrigation can reduce the water supply for other city residents. For example, when groundwater intended for apartment dwellers is diverted elsewhere, pressure is reduced so

that running water does not reach the upper floors of buildings. Complaints regarding water supply and the stench of waste have caused some governments to place limits on the use of urban land for crops and livestock.

THE GREEN REVOLUTION AND BT CROPS

Since the 1960s, farmers have been planting *high-yielding varieties* of rice, wheat, corn, and other crops. These were developed at research institutions around the world in an effort to sustain burgeoning populations. High-yielding rice was developed at the International Rice Research Institute in the Philippines. These stronger, higher-yielding, pest-resistant seeds, along with appropriate management techniques, generated the **Green Revolution.** Tougher seeds enabled the use of hitherto vacant land. From the mid-1960s to the mid-1980s, land under rice production increased by 15 percent in Asia but crop yields jumped by 74 percent! Currently, every 2.47 acres (1 hectare) of paddy fields in Asia provides enough rice to feed 27 people. However, 50 years from now, according to some projections, each hectare will have to provide for 43 people.

In an effort to increase the food supply, scientists have invented what are called **biotech or BT crops**. BT crops are also referred to as GE (genetically engineered) crops. BT seeds are bred to resist disease and pests. Now BT rice, developed in China, is being sown in 50 percent of the country's rice paddies. BT rice requires only half the amount of pesticide as applied to regular rice. Today, China produces a third of the world's rice. While BT crops are promising in terms of increasing food supply, they have their downside.

The Cotton Fiasco

Cotton is commonly thought of as a fiber crop; we often wear it. However, cotton is also a food crop. New BT seeds can be fed to pigs, chickens, and fish. More importantly, cotton seed is 22 percent protein compared to only 7 to 10 percent for rice, wheat, and other grains. Kernels, when roasted and salted, have a nutty flavor. Further, they can be ground and combined with wheat or corn flour to make bread. Cottonseed oil can also be used in food processing.

BT cotton was introduced to India in the late twentieth century by the U.S.-based transnational Monsanto. It was promoted as being immune to the ruinous pest, bollworm. Farmers were convinced to buy the new seeds even though they cost 4 1/2 times the cost of regular seeds. Within two years, cotton blight destroyed the crop. Indebted farmers lost their land and homes and could no longer feed their families. As of 2009, more than 100,000 BT farmers have committed suicide, ironically by drinking the pesticides that were supposed to save their crops.

CROP LOSSES

All crops can quickly decimate by floods and droughts. Also, they are prone to physical damage as well as disease when exposed to pests such as rats, insects, and worms residing in the crevices of poorly or loosely woven, or old and torn grain baskets Figure 5-4. For example, Sri Lanka loses up to 40 percent of its food due to poor storage facilities. In 2009, Bangladesh lost 18 percent of its rice and vegetables because of inadequate storage facilities. Insecticides and pesticides cannot combat 175 types of insects

Figure 5-4

This rice storage bin in Nepal is made of straw plastered with mud. Note that it is raised above ground to be out of the rain and to deter snakes and rodents. However, many predators can climb and bore through the mud. The straw roof also does not keep out bugs or other creatures that will eat the grain. Photo courtesy of B. A. Weightman.

such as grasshoppers, crickets, and caterpillars, and at least 10 different diseases that exist in Bangladesh. As we have just seen, BT crops are not the answer either.

In India, 30 percent of fruits and vegetables (40 million tons) are wasted each year due to gaps in the "cold chain." Not only is there insufficient cold-storage capacity but also there are not enough refrigerated trucks or suitable roads that are even in close proximity to farmers. Consequently, there are millions of farmers who are not part of the beneficial "cold chain." Moreover, cold storage is very expensive—the cost is twice that of Western countries.

Avian Flu

Avian flu is a question of animal health. Avian influenza has been recognized as a highly lethal generalized viral disease of poultry since 1901. By 1955, it was called the "fowl plague." Since then, it has been found to cause a wide range of disease syndromes from mild to severe in domestic poultry. Disease outbreaks occur mostly in chickens, turkeys, and ducks. However, more virulent strains can emerge by genetic mutation.

In 2003 and 2004, avian flu infection spread widely to three continents, initially through East and Southeast Asia, and then into southern Russia, the Middle East, Europe, and Africa. By 2006 it had struck South Asia. There is considerable circumstantial evidence from Russia, Europe, and Mongolia that the virus was caught from wild birds. However, the epidemiological aspects of the disease are still not fully understood.

Once avian influenza is established in domestic poultry, it is highly contagious. Infected birds excrete the virus in high concentrations in their feces and also in nasal and ocular discharges. Once introduced into the flock, the virus spreads through the movement of infected birds, contaminated equipment, egg flats, feed trucks, and service crews. The disease spreads very quickly when birds are kept in close contact. The virus can also be airborne.

Ducks are seen as important vectors of avian flu. The presence of high quantities of virus in domestic ducks is found in a system where live poultry are moved long distances and sold in live-bird markets. Untested birds can easily spread disease across borders. The first recorded cases of infection with two types of viruses were found in 1997 in domestic ducks from live-bird markets in Hong Kong. It is thought that the ducks did not initiate the virus but rather contracted it from other types of birds such as chickens that had much higher rates of infection. Again in 2000, Chinese ducks imported into Hong Kong were found to be infected. However, fatal disease in ducks was not reported until 2002 for fear of ruining the export market. By 2005, scientists learned that infected ducks can spread the lethal disease for a period of at least 17 days via respiration and that their feathers are also contaminated.

The number of ducks reared in China increased more than three-fold from 223 million in 1985 to 725 million in 2005. Many of these birds are reared in ponds, potentially allowing contact with wild birds. Given the prevalence for live-bird marketing, lack of segregation of species, and unsanitary holding conditions, avian flu is more likely to spread to more birds and humans.

China and Vietnam account for 75 percent of the global duck population Figure 5-5. Ducks are a relatively high-value product and are often transported long distances to markets. Duck production in Thailand is no longer a local industry; Thai ducks are also shipped over long distances.

Several studies correlate the rearing of ducks with rice paddy cultivation. The relationship is not yet clearly understood, although contact with wild birds is still thought to be a factor in the spread of the disease. A similar association exists in the Mekong and Red River regions of Vietnam. Vaccination and regulated breeding programs have not eradicated the disease, and outbreaks occurred in 2007 and 2008. In Hong Kong, it is now illegal to ship ducks with other types of birds. Nevertheless, avian flu remains a problem, especially in southern China.

When severe avian flu outbreaks occur, hundreds, even thousands of millions of chickens and other poultry face death through disease or culling (killing infected birds). This results in an immediate loss of livelihood for hundreds of thousands of poultry smallholders. In addition, a significant number of people have died from the disease, and fear of a human pandemic is leading many consumers to reject poultry.

Significant reduction in purchases hits both large and small farmers heavily by creating monetary losses in regional and international trade. Even countries

Figure 5-5

Here in southern China, a farmer is preparing to take his ducks to a live bird market. Before being put into the basket, the ducks were force-fed rice paddy water via a plastic, liter-size, Coke bottle. The water will sustain them on their journey and make them heavier for sale. Photograph courtesy of B. A. Weightman.

that have not experienced avian flu outbreaks are feeling rebound shocks in their poultry markets. Until researchers discover the complexities of avian flu infection and devise means to prevent or at least cure it, the poultry industry will be negatively affected.

THE COST OF INNOVATION

Innovations are generally adopted by the most innovative but not necessarily the biggest landholders. However, larger operators are likely to have more money and connections and, therefore, easier access to loans, costlier BT seeds, farming cooperatives, irrigation systems, and so forth. Female-headed households and social minorities such as low class/caste or tribal groups seldom have equal access. Other impediments relate to environmental conditions or loss of cropland to urbanization. Nevertheless, the Green Revolution has generated quantum leaps in crop yields, lowered consumer prices, and made jobs available to landless unemployed. However, in some areas labor costs have increased to the point that farmers are forced to reduce labor inputs and use machinery and other cultivation methods. For instance, in Thailand, many farmers have turned to broadcast seeding (using machines called seeders to scatter seeds over large areas) to counter the costs of hand-planting seedlings.

In Vietnam, men and women cooperate in rice production but women still do most of the work. Men prepare the paddies, women plant and weed. Both maintain irrigation and both do harvesting. Women hand-pound the grain before winnowing and drying it (Figure 5-6). However,

Figure 5-6

This Vietnamese woman is winnowing rice. By tossing it into the air, the chaff will blow away and the heavier rice will fall into the basket below. Winnowing is regarded as women's work. Photograph courtesy of B. A. Weightman.

mechanized rice mills have recently displaced millions of women (and men) across Asia, hitting particularly hard in Vietnam, Indonesia, and Bangladesh. As rice mills eliminate perceived drudgery, they simultaneously eliminate what may be the singular means of family income.

Plantation Economies

Plantations are commercial, monocrop enterprises that focus on a single crop such as coconuts, bananas, coffee, tea, rubber, or palm oil. Plantations are often foreign owned and employ local or foreign labor. Plantations were introduced by colonial powers. For example, the British introduced tea plantations in southern India and Sri Lanka (Ceylon), and rubber plantations into Malaysia (Malaya). Indian laborers were brought into both Sri Lanka and Malaysia and remain minority communities in those countries. Plantations are designed for export. Typically, they are located near coasts or at least have access to ports. Most plantations are in tropical regions.

PALM OIL

The most popular plantation crop of the twenty-first century is the oil palm Figure 5-7. The two largest producers of palm oil are Malaysia and Indonesia, although the variety they grow (*Elaeis guineenensis*) is native to tropical Africa. The tree was introduced into Malaysia at the beginning of the twentieth century and was first planted commercially there in 1917. Today, almost half of Malaysia's cultivated land is planted with oil palms. Indonesia is currently the world's number one producer. Both countries are exporting large amounts of palm oil to China and India, the largest consumers. Palm oil is the most consumed edible oil in India. Thailand has also entered the oil palm business.

- Why is palm oil so popular? It is because it has unparalleled productivity. In fact, the oil palm seed is the most productive oil seed in the world.
- A single hectare of palm oil may yield 5,000 kilograms of crude oil—6,000 liters of crude. For comparison, soybeans and corn generate only 446 and 172 liters per hectare, respectively.
- Palm oil is a significant ingredient for the production of biofuel. The European Union imports about 3.5 million tons that could supply up to 20 percent of its biofuel needs in 2012.
- In 2007 alone, Malaysia exported palm oil worth nearly US$10 billion. It has opened dozens of plants to process the raw oil into fuel for export to China and Europe. The Finnish firm Nestle uses imported palm oil to make a biofuel that runs some of Helsinki's buses.
- Indonesia recently announced that it plans to devote 40 percent of its oil crop to biofuel production.
- Palm oil is widely used for cooking and in a variety of food products such as margarine, as a base for cosmetics, and to make engine lubricants.

Palm oil plantations are often controlled by multinational corporations. For example, KS Natural Resources (KSNR) is an Indian company headquartered in Singapore. KSNR, which is scheduled to produce palm oil, is a

Figure 5-7

This is the fruit of the oil palm. The fruit is crushed and the oil is processed for cooking and a variety of other products such as soap, cosmetics, margarine, and biofuel. Malaysia and Indonesia are the world's primary producers of palm oil.
Photograph courtesy of B. A. Weightman.

Figure 5-8

This Kalimantan fire has been burning since 1983. More than 9 million acres (80,000 ha) of forest and its resident wildlife have been destroyed. Photograph courtesy of B. A. Weightman.

subsidiary of KS Oils Limited of Madhya Pradesh, which specializes in mustard oil. KSNR acquired 50,000 acres (20,000 ha) of Indonesian forest in 2008 and has recently added another 35,000 acres (14,000 ha). All this acreage is to be cleared for palm oil production.

However, all is not well with palm oil production. *Science*, a leading research journal, has designated the crop as "filthy." In order to develop plantations, forests are cleared by burning and CO_2 and other greenhouse gases are spewed into the atmosphere, thereby increasing global warming Figure 5-8. Now Indonesia is the world's third largest carbon emitter after the United States and China.

More than 26 percent of both Indonesian and Malaysian oil palm plantations are on peat lands. Estimates are that for each ton of palm oil produced, an average of 20 tons of CO_2 will be emitted from peat decomposition alone. A United Nations study demonstrates that of the estimated 2 gigatons of CO_2 released by Indonesia, 600 million tons are generated by the decomposition of peat and 1,400 million tons are lost through the seasonal fires.

Aside from contributing to global warming, palm oil monoculture in tropical ecosystems has an array of other harmful environmental consequences. Water-catchment areas are ruined; animal habitats are destroyed; and unhealthy haze from fires spreads across large swaths of Southeast Asia. Without peat wetlands that store rain water, the water runs off, leading to erosion of precious topsoils and reduction of fresh groundwater for drinking.

Beyond the loss of forest ecosystems, palm oil production leaves piles of trash. In 2001, Malaysia's production of 7 million tons of crude palm oil generated 9.9 million tons of solid oil wastes, palm fiber and shells, and 10 million tons of palm oil mill effluent. The effluent is a polluted mix of crushed shells, water, and fat residues that are shown to have negative impacts on aquatic ecosystems.

Between 1985 and 2005, roughly a third of Borneo's rainforest has disappeared, felled by indiscriminate logging or cleared for palm oil plantations. One alarming 2001 study, from the University of Singapore, predicts that a third of Asian forest plant and animal species will become extinct in the twenty-first century!

IGNORING THE DOWNSIDE OF PALM OIL PRODUCTION

India, China, and Europe see biofuels as making up 10 percent of motor fuels by 2020. Indonesia argues that the palm oil industry will create as many as 3.5 million jobs, thereby alleviating poverty. An assistant to Indonesia's Minister of Energy notes that "all the forests have not been used yet." She observed that just in Kalimantan (the Indonesian part of Borneo), there are more than 13.6 million acres (5.5 million hectares) "available"—an area larger than Denmark. Also, there are another 22.2 million acres (9 million ha) available elsewhere in the country. Indonesia's environmental watchdog, *Walhi*, believes that coveting "available" land has nothing to do with oil palms but rather is an attempt to make money from harvesting valuable wood. *Walhi* notes that millions of acres have been cleared of trees but have not been planted with palm oil. Government efforts to curb illegal logging have been spotty at best. Corruption and collusion with multinational logging companies on the part of forestry officials render the program ineffective.

Local populations are reaping the consequences of legal and illegal forest clearance and plantation development. More than 500 communities in Malaysia have become embroiled in clashes with some 100 companies as they appropriate land that people have cultivated for centuries. The government describes these lands as "empty." The fact that the indigenous people grew rice, coffee, and cocoa doesn't seem to count—after all, it isn't palm oil! Meager compensation may be doled out, but most of it goes into the pockets of government officials.

The Indonesian industry says it is cleaning up its act. It claims that it is not burning down more forests or peat lands. Indonesian companies have joined the Roundtable on Sustainable Palm Oil, a World Wildlife Fund-led initiative to engage palm oil producers. However, now Malaysian companies are coming into Indonesia because their own suitable land is limited. They claim that they will use the areas that are already deforested and therefore will not harm the environment even though palm oil production (aside from deforestation) is inherently harmful.

Indonesia is planning to plant another 4.6 million acres (1.8 million ha) in northern Kalimantan near the Malaysian border. The designated region contains three National Parks that incorporate most of Indonesia's remaining, intact tropical forests. China is involved in the plan with an investment of US$7.5 billion for energy and infrastructure projects. The Chinese will control about 1.5 million acres (600,000 ha) of the oil palm plantations. Indonesians will run 3 million acres (1.2 million ha). The project eventually employ some 400,000 people and generate an annual inflow of US$45 million in taxes to the state.

Plantations are economically profitable enterprises, at least for a few years. Palm oil grows in clusters of fruit that weigh 88 to 110 pounds (40 to 50 kg). A hundred kilograms of seeds can yield 20 kilograms of oil. Fruit clusters are harvested by hand, difficult work in the heat and humidity. In Malaysia, much of the work is done by Indonesians. Wild oil palms can live 150 years and attain heights of 80 feet (24 m). The cultivated variety are clear-cut or poisoned after 25 years when they stand about 25 feet (7.5 m) and harvesting the fruit becomes too difficult. The land is abandoned and more land is cleared for new plantings.

Malaysia continues to claim that palm oil plantations are not harmful. However, liberal use of petroleum-based pesticides, herbicides, and fertilizers ensures that cultivation not only is polluting at the local level but also is contributing to greenhouse emissions. Malaysia believes that its production methods are not as harmful as Indonesia's. Indonesian plantations are so damaging that after a 25-year harvest, oil palm lands are often abandoned for scrubland. Soils are leached of their nutrients and devoid of anything other than weedy grasses that are prone to wildfires. World scientists predict the elimination of Indonesia's biodiversity as whole ecosystems are destroyed. Furthermore, indigenous groups such as the Dayak of Sarawak (a Malaysian State in Northern Borneo) who have lived in the forest for centuries will have to leave and give up their hunting and gathering way of life.

Thailand and the Philippines are also getting on track with palm oil. High prices for palm oil, driven by Bangkok's search for alternative fuels, have driven increasing numbers of farmers to convert rubber and fruit plantations to palm oil. Meanwhile, the Philippines has put about 63,500 acres (25,000 ha) under cultivation. Moreover, the government plans to add some 1.2 million acres (454,000 ha) of "disposable land"—now being used by locals for pasture and to collect edible berries and plants. If land is declared as "disposable" it, by definition, has no value. Consequently, it must be "developed" to give it "value." The fact that a few people might have to transform their lives or become worse off is a very small price to pay according to development strategists. What do you think?

Gender and Agriculture

Agriculture remains a significant part of most Asian economies in terms of employment. As Table 5-1 indicates, eight countries have more than half of their labor force employed in agriculture and an additional eight have at least a third working in agriculture. While agriculture does not account for large amounts of GDP in some countries, it counts for at least 25 percent of GDP in at least eight.

Table 5-1 also reveals the **feminization of agriculture** in that women represent a larger proportion of laborers than men in the agricultural sector in Asia. The number of paid women workers in agriculture is also increasing because of globalization, high-value agricultural production, and agro-processing for export. There has also been a rapid expansion in the use of contract laborers who are provided on a third-party basis to producers. Many of these jobs are temporary where women are subject to low

Table 5-1 Percentage of Female and Male Labor Force Employed in Agriculture (All data is 2008 unless otherwise indicated.)

REGION Country	% of Labor Force Employed in Agriculture	% of Employed Females in Agriculture	% of Employed Males in Agriculture	Agriculture as % of GDP
SOUTH ASIA				
Bangladesh	65	64.5	40	32
Bhutan	85	98 (02)	92 (02)	19
India	60 (05)	75 (01)	53 (01)	25
Maldives	.04 (04)	5 (02)	18 (02)	.07
Nepal	81	90	75	20
Pakistan	45	79	61	21
Sri Lanka	33	42 (02)	38 (02)	12
EAST ASIA				
China	40	38.9	35	13
Japan	4.8	2.6	3.9	1.3
North Korea	n.d.	n.d.	n.d.	21
South Korea	.07	12 (02)	9 (02)	3
Mongolia	36	42 (02)	38 (02)	19
Taiwan	.05	n.d.	n.d.	2
SOUTHEAST ASIA				
Brunei	.01	n.d.	n.d.	0.7
Cambodia	85	65 (02)	71?	31
Indonesia	40	43 (02)	43 (02)	14
Laos	78	81 (90)	70 (90)	45
Malaysia	14	14 (02)	21 (02)	10
Myanmar (Burma)	70	n.d.	n.d.	49
Philippines	36	25 (02)	45 (02)	35
Singapore	n.d.	n.d.	n.d.	0.01
Thailand	39 (07)	41 (07)	44 (07)	8.4
Timor-Leste	33 (07)	93 (02)	70 (02)	25
Vietnam	70	60 (06)	56 (06)	49

Sources: Asian Development Bank; Food and Agriculture Organization of the United Nations; International Labor Organization's Global Employment Trends for Women 2009; World Resources Institute/Earth Trends Organization.

levels of protection in terms of wage levels, employment security, and health, safety, and environmental standards.

Women's work in agriculture has become more visible over the past few decades as women farmers are becoming more involved in farming activities, increasingly assuming the responsibility for household survival and responding to new opportunities in agricultural production. This feminization of agriculture has made rural development in Asia considerably dependent upon women's capacity being enhanced as they increasingly provide vital contributions to the development of rural communities.

WOMEN IN TRADITIONAL, RURAL SOCIETIES

By tradition, commercial agriculture is primarily a male realm. Men prepare the land, irrigate crops, and transport produce to market. They own and trade large animals such as cattle and water buffaloes. They are the ones to cut, haul, and sell timber from forests.

Tradition holds that the primary responsibility of rural women is maintaining the household. They raise children, grow and prepare food, manage family poultry, and collect water and fuel such as wood or dung. In addition, women and girls play an important, largely unpaid, role in generating family income by providing labor for planting, weeding, harvesting and threshing crops, and processing produce for sale. Women might also earn a small income for themselves by selling vegetables from home gardens or forest products. They spend this income on meeting family food needs and child education.

FLEXING GENDER ROLES

As agriculture becomes increasingly commercialized, the dominant position of men is changing gender roles—often in men's favor. For example, as urban demand for vegetables increases, men are expropriating women's gardens and turning them into larger commercial enterprises. This diminishes the role of women in household security and their social standing in the community.

Another growing trend is outmigration of poor, rural men in search of employment, although a significant number have been returning because urban job opportunities have declined with the recent economic crisis. If men do leave, however, women are left with sole responsibility for food production and child-rearing. Consumed with planting and animal care, gathering water and firewood (often from faraway sources) along with the demands of children and households, women have little time to attend educational programs and they often have to withdraw their children from school to help them.

Women's experiences of economic growth and reform in agriculture and other economic sectors are mediated through their gendered position within and outside the household. As we discussed in Chapter 3, perceptions of gender are deeply rooted in society. They vary widely between and within cultures, and change over time. But in all cultures, gender determines power and access to resources for both females and males. Female power and resources are lowest in rural areas of the developing world and rural women comprise the majority of the global poor. They have the least education and the highest illiteracy rates. The poorest of all are female-headed households.

Rural women suffer systematic discrimination in access to resources essential for socioeconomic development. Credit, seed, fertilizer, and other supply facilities and services generally are not geared to dealing with women. Women are often excluded from training programs such as irrigation management. Rural women are rarely consulted in development projects that ultimately assist men and add to women's workloads. When work burdens increase, girls are more likely than boys to be removed from school to assist with farming and household tasks.

Since the bulk of women's work remains unpaid and unrecognized, it is usually not counted as "income." Since men are the "earners," they are perceived to work harder. In the latter part of the twentieth century, the United Nations, World Bank, and a variety of other influential organizations came to recognize the fundamental role of women in agriculture and food security. To illustrate this point, one study in the Indian Himalayas revealed that on a 2.47 acre (1 ha) plot over the course of a year on average, a pair of bullocks worked 1,064 hours; a man worked 1,212 hours; and a woman worked 3,485 hours.

Not only do women usually work longer hours than men, but also they earn less as paid laborers. Female farm workers' wages are lower than men's, while low-paid tasks in agro-processing are routinely "feminized." In other words, since it is understood that "women's work isn't as good as men's," it isn't worthy of equal pay. Therefore, low-paying jobs are "women's work." In Pakistan, women earn a mere 33 percent of men's wages. In Indonesia and Thailand, the figures are 51 and 61 percent, respectively.

Although women make substantial contributions to household well-being and agricultural production, men largely control the income. (Thai and matrilineal societies in Malaysia are exceptions.) Some 70 percent of Indonesian women are not permitted to keep their income or decide how it should be used. Failure to value women's work reduces them to virtual non-entities in economic transactions, allocation of household resources, and wider community decision making.

Money can bring respect to women. In rural Bangladesh, where Grameen Bank and other NGOs have helped some women, women use the phrase *Garam Taka*—"weighty money." Money has weight because it gives women control over their earnings and assets, which gives them weight within the household. One woman reports: "I gave 20,000 Taka to my husband for his fuel wood business, now my voice is louder than his." And another woman says: "If you have no money, there is no value for your choice. You are sitting in the corner like a little thief. If you have assets, everyone loves you."

Women have few property or land rights. With the death of a husband, land is often inherited by his family, who can leave the wife destitute. Women's exclusion from control and ownership of property or land can lead them to the following, in the words of young Nepali women, "If we are not taken care of at home, we go to our parental home. If our parents reject us, we go to India" (referring to the sex trade in Mumbai).

Today, numerous countries are making efforts to harness the knowledge and energy of women. For example, in the southern Indian state of Andhra Pradesh, government agencies have purchased land and transferred it, free of charge, to the landless Chenchu tribe. Transfers are made only to women. After much deliberation, Chenchu men conceded that this would be better for their households as there is little chance that women, unlike men, would lose the land because of drinking or gambling debts.

Social inequities are deep-seated in society. Culturally imposed household division of labor, gender norms on speaking out in public, and spatial constraints on mobility combine to prevent women from having an effective voice in decision making at home or in the larger community. Already debased and perceived by their husbands as not contributing enough, women often become victims of gender-based violence. As an Indian woman states: "If our husbands want us to eat, then we eat. It totally depends on them. If the land were in my name, he would no longer beat me, and he would take care of the children as well."

Violence, especially against rural women, is widespread in many Asian countries such as India and Thailand. If women are badly beaten or crippled, they are unable to work. This can push economically strained households into economic crisis. Moreover, beaten women are seen to deserve their punishment by their family members and others in the community. Men gain status by "controlling" their wives, who are "obviously" disobedient or lazy. Women's already poor self-esteem is thereby reduced even further. Tellingly, very few Asian countries have effective laws against violence including rape perpetrated on women. If laws do exist, arrests are seldom made and prosecutions are rare. In addition, women are usually considered to be at fault in the first place. Further, women are reluctant to report abuse because it is common for them to be abused again by the authorities.

Representation of women in traditional labor institutions is weak. Globalization, deregulation, and competition have contributed to the erosion of trade unionism and traditional modes of collective action. For instance, the Indian government did not formally recognize the abysmal working conditions of unorganized wage laborers, particularly women, until 2007. However, new forms of national and transnational movements have emerged. Women in Informal Employment: Globalizing and Organizing (WIEGO) was established in 1997 with India's Self-Employed Woman's Association (SEWA) as a founding member. WIEGO now has at least 35 member countries.

The fact that increasing feminization of agriculture has a deep and wide-ranging impact on agricultural productivity is now recognized by many governments and organizations. Women's roles in food and other crop production are becoming a leading focus in agricultural policies. Empowerment of women in rural communities must be supported by asset ownership, enhancement of agricultural management skills and knowledge, and widespread gender sensitization. A new discourse regarding the essentiality of women as producers, farmers, and economic contributors must be embedded within society in order to construct an enabling environment for strengthening the capabilities of rural women.

Fish Farming

The practice of raising fish, shellfish, and other marine products in various water bodies is generally called **aquaculture** although **mariculture** is a more recent term that refers to raising seafood in the sea or ocean. People have been practicing aquaculture for thousands of years. Chinese scholar Fan Li wrote a treatise on raising fish in rice paddies or ponds 2,500 years ago. By 600 AD, several species of carp were being raised in enclosures in rivers and lakes. Then, large fish ponds were constructed and subdivided by earthen dykes. Mulberry bushes (to feed silkworms) were planted on the dykes so that their roots were fed by the pond water. Leaves that fell off the mulberry were eaten by the fish.

Farmed seafood now provides 42 percent of the world's seafood supply. This number is expected to increase to 50 percent in the next decade. Seafood provides 30 percent of annual protein in the typical Asian diet. In Indonesia, it comprises 58 percent and in Cambodia, it makes up 75 percent. As dams and other human-made projects are decimating river fisheries and as marine fish stocks are being depleted by overfishing, fish farming is growing in importance in Asia and the world.

Originally, aquaculture was a relatively small-scale enterprise. However, small-scale endeavors have morphed into large-scale, scientific mariculture operations with innovations in feed technology, breeding strategies, and cage design. Properly done, aquaculture/mariculture is remarkably efficient in its use of feed and water. Farming smaller seafood products such as tilapia, shrimp, or seaweed is less resource-intensive than employing sophisticated boats to trawl the seas for large, predatory fish. According to a spate of experts such as the Food and Agriculture Organization, the Asian Development Bank, and Worldwatch, fish farming can be a critical way to improve the global diet.

China offers an example of the global boom in aquaculture. After the Communists took over the country in 1949, they pushed the idea of food self-sufficiency. Using mass mobilization of labor, more than 80,000 artificial water bodies were constructed for hydropower, flood control, and irrigation. Nearly 5 million acres (2 million hectares) of inland water areas led to a 50 percent expanse of aquaculture.

Today, China produces 70 percent of global, farmed seafood. More than three-quarters of China's seafood is derived from fish farming. China is the world's largest consumer of fish feed and the largest importer of fishmeal and fish oil. It is also the largest producer of predatory seafood such as black carp, eels, and marine shrimp—30 percent of the global product.

Reef Fisheries

Indo-Pacific coral reef fisheries harvest a wide range of seafood such as bottom-dwelling fish like snappers and groupers, small midwater fish such as anchovies and sardines, and large, midwater and oceanic fish such as tuna and barracudas. Tropical lobsters, crabs, octopus, sea cucumbers, and turtles are also taken. Reef sharks, harvested for their fins that are dried and sold on the East Asian market, are increasingly popular.

Many of these and other species are not associated with coral reefs for their entire life cycle. However, many deposit their larvae there and are vulnerable for capture during that time. Marine turtles, sharks, and tuna often spend time around the reefs in between their extensive migrations into deeper parts of the ocean.

Traditionally, coral reefs were fished by nearby coastal communities who often "owned" them under customary laws and practices. For instance, "ownership" might have meant "right to use." Typically, there were strict controls on who could use a particular area. Women were prohibited from using nets or boats. However, they often made the nets that were employed along with traps, fish fences, and enclosures. Some groups used spears and arrows in addition to traditional poisons.

Today, coastal fishing communities are largely the same as they were in the past and include men, women, and children harvesting reef products as a source of food and income. However, close to market outlets and urban centers, customary tenure of marine resources is being increasingly ignored in the face of commercial pressures and opportunities.

Large-scale, commercial reef mariculture is invading the Indo-Pacific region. Greater numbers of "outsiders"—people with no connection to traditional understandings of the resources in question—are increasingly involved in harvesting them. In many cases, this leads to conflicts with the traditional resource users. As indigenous groups have no written, legal rights to their land and water areas, it is relatively simple for "aquabusiness" to move in.

AQUACULTURE CRISES LOOM

As large-scale commercialization takes hold, traditional methods are displaced. Synthetic nets deprive women of their traditional role of net-making. Human-powered canoes and boats are replaced with mechanized fishing vessels, trawls, and dredges. Plant-based poisons, which didn't overfish an area, are giving way to more powerful chemical poisons and explosives. Aquabusinesses now are "reducing fish" to feed fish. This term refers to the practice of feeding whole or crushed up small species such as herring or anchovies to larger, predatory fish such as tuna and striped bass. ("Fish reduction" is also employed in salmon production in the Atlantic Ocean.)

As we have seen above, industrial aquaculture has social fallout. However, it also has environmental consequences. As increasing numbers of fish are bred within enclosures, ever greater amounts of manure pollute surrounding waters and harm other fish in the area. To ensure large harvests, fingerlings (baby fish) are infused with antibiotics and de-licing chemicals. Moreover, they are bred to be genetically uniform so that their parts, such as fillets, are all the same size for marketing. However, as we saw with BT crops, protecting fish from one or two diseases often results in their being prone to new diseases. For example, in 2008, Chinese shrimp farmers lost more than US$600 million to an unforseen disease. Furthermore, diseased fish that escape from human-made enclosures can devastate free-swimming ocean stocks.

Problems aside, fish farming remains an important source of food and livelihood for many coastal and other communities in Asia and around the world. A program to

Figure 5-9

Here, a man is dumping remnants from a fruit and vegetable market into the Hooghly River in Calcutta. This is excellent fish food and will boost the local fish catch. Photograph courtesy of B. A. Weightman.

increase small-scale aquaculture in Bangladesh by raising fish in rice paddies, for instance, increased incomes by 20 percent.

Eight thousand fish-farm workers manage traditional ponds called *bheris* in the Kolkhata (Calcutta) wetlands that produce 13,000 tons of fish yearly for urban consumption (Figure 5-9). Fish feed on 600 million liters of raw sewage every day. The World Bank calls this system "the city's sewage treatment plant." Water hyacinth ponds and algal blooms help to clean the water.

A tilapia-raising program, recently introduced to exceedingly poor and undernourished tribal and other rice cultivators in the Philippine highlands, has succeeded in doubling their annual incomes. This upland project is complemented with another one offshore. Another poverty-stricken tribe has been given a spacious, steel ocean cage along with 15,000 milkfish fingerlings. San Miguel Corporation, Southeast Asia's largest food and drinks conglomerate, has donated feed. Several NGOs have volunteered to help the tribe manage their new fishery and provide training in drying or smoking the fish for commercial markets. Part of the proceeds will go toward the purchase of more feed and fingerlings. These "pro-poor" upland and ocean ventures are the first of their kind in the entire country.

Fortunately, most aquaculture is still focused on raising seaweeds, shellfish, carp, and tilapia—species that are low on the food chain. Well-designed operations can be an effective and efficient means of adding to the global food supply. Moreover, they can help rebuild wetlands, soak up coastal pollution, and be used to restock wild fisheries.

Evidently, the problem does not lie with small-scale, self-perpetuating ecosystems such as rice–mulberry or rice–tilapia. Neither does it result from properly run, commercial enterprises. The problem lies with large-scale, carelessly run, entirely profit-oriented operations that rely on external inputs far removed from local ecological systems. To quote Brian Halwell of the Worldwatch Institute (2008): "As seafood shifts from being the last wild ingredient in our diet to being a heavily farmed commodity, ocean conservationists, commercial fishers, and public health advocates are sounding the alarm."

GENDER AND AQUACULTURE

Fisheries and aquaculture are diverse and often complex dynamic systems with men and women undertaking varying roles depending on local norms regarding resource access and control, type of technology involved, and extent of commercialization of a particular product. Many small-scale fisheries operate with men investing in boats, nets, and other essential gear to do the fishing and with women investing in processing equipment and being responsible for fish purchasing, processing, and sales. While this is the case in many regions of Asia, there are exceptions. For instance, in Cambodia and Thailand, women are involved in boat-fishing and in India and Bangladesh, women collect shellfish, including crabs, and produce shellfish seed.

Regardless of gender-role differences, wealthier individuals play dominant roles in the **value chain** within which they operate (Table 5-2). A value chain is a chain of activities in a production system in which each activity adds value to the end product. Poor members of the chain, who are unable to accumulate assets, have little bargaining or purchasing power, or control over various aspects of the chain or even their participation in it. For instance, in capture fisheries, most men do not own boats; they serve as crews on other men's boats. The largest and highest quality fish go to the boat owner.

The shrimp value chain is dominated by China, Indonesia, Thailand, Bangladesh, and Ecuador (in South

Table 5-2 Examples of Gender Roles in the Capture Fisheries Value Chain in Asia

Scale	Investment	Catch	Processing	Sales
Small	Women's savings. China: both women and men invest.	Boat owners are wealthy and older men. Crews comprise adult and young men. Women and men mend nets. Women collect shell-fish.	Women and men dry fish. Women might work in processing factory, where they are paid less than men and are the first to be let go.	Women and men sell in local markets and to contractors for national and international markets. Sales are more likely to be controlled by men in "conservative" areas. Middlemen and individuals higher on the value chain make more profit. Men might control family income.

Source: *Gender in Agriculture Sourcebook*. World Bank, 2009.

America). A considerable segment of this market is in the hands of large producers, supported by external capital, and destined for the international market. However, Bangladesh is an exception. Here, most shrimp production is in the hands of small producers although processing is completed in factories. As many as 1.2 million individuals are involved in the shrimp value chain in Bangladesh and another 4.8 million households are indirectly dependent on it for their livelihoods.

Profits from Bangladeshi shrimp exports are not shared equally throughout the chain. Middlemen and exporters realize more money than shrimp farmers and fry (baby shrimp) catchers, who are often locked in a cycle of debt with those further along the chain. The chain is also a highly sex-segmented market, with women and men receiving different wages for the work they do. For example, women fry catchers and sorters earn about 64 percent of men's earnings. Women are also concentrated in the least secure nodes of the shrimp chain—undertaking various low-paid tasks in and out of the processing plants.

Increased mechanization can radically alter the value-chain system. Motorization and mechanization of fishing vessels in the Indian states of Maharashtra and Orissa led to a concentration of fish landings at fewer harbors and landing sites and, in some cases, resulted in the takeover of the fish trade by merchants who were all men. Many women were displaced from fish retailing. Moreover, mechanization of any part of the system typically leads to job losses, and women are the first to be let go. This is particularly the case in processing factories.

Changing Asian Diets

Figure 5-10 shows the traditional Asian diet, which consists of much less protein than Western diets and places a greater emphasis on grains, vegetables, and fish. However, rapid economic and income growth, urbanization, and globalization are leading to a dramatic shift in Asian food consumption toward more meat and dairy products—significant components of Western diets. Expansion of global interconnectedness of urban middle classes is the driving force behind the convergence of diets. Also, the dissemination of supermarket and fast-food chains is reinforcing this trend.

Changes in Asian food demands are characterized by the following:

- Reduced consumption of rice
- Increased consumption of wheat and wheat-based products such as bread
- Increased diversity of food groups consumed

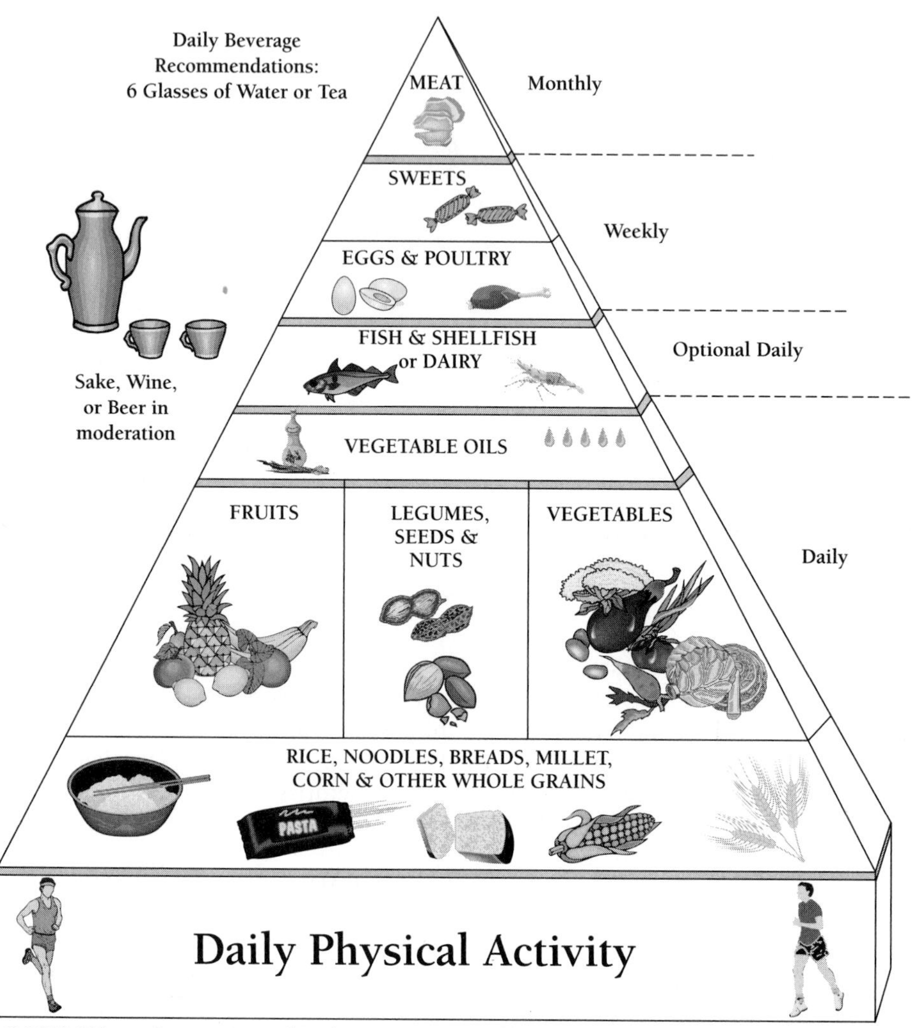

Figure 5-10

This is the traditional Asian diet pyramid. Notice how little meat is eaten relative to fish, vegetables, and grains. Meat consumption is increasing among more affluent populations.
Source: http://elderlynursing.com/foodpyramid.htm

- Rising consumption of high protein foods such as meat and poultry
- Increased consumption of temperate zone food products such as tomatoes
- Rising popularity of convenience food and beverages

Asian countries in economic and demographic transition are showing dramatic shifts in food consumption patterns. As has occurred elsewhere, rice consumption per capita has declined with income growth and urbanization. Affluent Chinese consumers now differentiate rice at different prices according to quality and attributes including stickiness, fragrance, and gluten and protein content. Traditional rice-eating societies are consuming increasing quantities of wheat in the form of bread, cakes, pastry, and so forth. Women who join the urban workforce have less time to cook or bake, and are prone to buy pre-sliced, packaged bread.

Those who can afford it are eating more meat. In other words, people are shifting away from carbohydrates to fat and protein. In 1997, Asia's meat imports were predicted to rise 389 percent by 2030. In 1900, China had 72 million head of cattle and produced 5,000 tons of beef. In 2005 it had 140 million head and produced 7 million tons of beef. Moreover, with continually rising demand for beef, the country is no longer self-sufficient and must import beef from elsewhere.

Increasing intensity of urban lifestyles has driven up demand for ready-made, packaged, and fast foods. Generational differences have evolved in dietary choices. For example, the younger generation of Japanese consume much more beef and beer than older people, who continue to eat more rice and vegetables. While dietary structures are becoming increasingly similar across developing countries, adoption of Western food consumption habits by younger people is generating a divergence of food habits across generations.

As countries become more globalized and urbanized, supermarkets and fast-food outlets tend to replace central food markets, neighborhood stores, and street-food sellers (Figure 5-11). By 2002, the share of supermarkets in the processed/packaged food retail market was 33 percent in Southeast Asia and 63 percent in East Asia. In 1999, the supermarket share of Chinese urban food markets was 30 percent. By 2001, it was 48 percent. Supermarkets and fast-food outlets are most visible in big cities, but they are spreading rapidly to medium- and small-size towns, especially in East and Southeast Asia.

Louisville, Kentucky, is home to Yum!—the world's largest restaurant corporation with more than 36,000 restaurants in 110 countries. Pizza Hut and Kentucky Fried Chicken are only two members in its diverse group of fast-food outlets. Pizza Hut was introduced into China in 1990 and as of 2009 had opened 440 restaurants in 100 cities. The fastest growing fast-food chain in China is Kentucky Fried Chicken, with some 2700 outlets in more than 550 cities. When Yum! China opened its 1,000th KFC in 2004, it gathered 1,000 managers on the Great Wall for "the world's largest Yum! Cheer!"

Fast-food restaurants are adapting their menus to Asian tastes. For example, in Japan, McDonalds has inspired "Mos Burger," which offers teriyaki chicken burgers and rice balls. It also sells a green tea dessert with sweet red beans, and a chestnut and green tea ice-cream shake. Another burger business is "Mamido," which makes its burgers from fruit and confectionary ingredients. For instance, you can buy a "fish" burger that consists of a banana shaped as a fish, topped with "tartar sauce" that is actually dairy cream. The French fries are deep-fried, elongated, custard cream.

Figure 5-11
The cook at this Thai food stall is adding soy sauce to her specialty dish. Photograph courtesy of B. A. Weightman.

Street Foods

Street foods are sold in almost every country in the world (Figure 5-11). Every kind of food imaginable is sold by men and women from permanent or semi-permanent stalls, mobile carts, or simply from containers set on the ground. Multiple vendors often hawk their dazzling arrays from under a giant tent where people can eat at tables. Some people use their homes as outlets, selling their wares out of a kitchen window to buyers who can then sit on benches outside the house to eat. Because they purchase their ingredients locally, these micro-enterprises support local market gardens and small farmers. In other words, they are integrated into local economies.

Street foods are important elements of the dietary regimen for millions of individuals in Asia. They supply lunches for workers who have no time to go home and eat. Both women and men pick up food to take home for dinner. Many school children rely on street food for their first and sometimes only meal of the day.

Street food vending reflects local culture, especially the appropriate roles of men and women. For instance in Bogor, Indonesia, the hawking of *sate* is a man's job. *Sate* comprises little pieces of chicken lamb or beef roasted on a bamboo skewer and coated with a spicy peanut sauce. The *sate* vendors carry their supplies on two trays, one at each end of a pole (*pikulan*) slung over one shoulder. One tray holds the food and a small stool for buyers to sit on and the other tray holds a charcoal brazier. Women sell fruits, cooked snacks, and sweets. Women vendors tend to own their own equipment, often bringing pots and pans from their home. Men frequently work at franchised outlets.

Over 200 types of foods are available in Bogor food stalls—from cassava (manioc) chips and egg rolls, to complete meals of rice with meat, fish, or chicken along with a side dish served on a banana leaf. The foods reflect the ethnic diversity of the country, and vendors often dress in regional garb.

Street foods also reflect the dietary preferences of a society. For example, to Indonesians, only rice is a meal; everything else is a snack. A filling chicken porridge might have the nutrition of a meal, but it is not considered a meal. It is merely a snack. Distinctly Indonesian is the spicy hot peanut sauce called *gado gado*, which is sold exclusively by women to be poured over fresh fruit or blanched vegetable salad sold by men. Indonesians love spicy foods to the extent that they have separate words for "hot" (*panas*) and "spicy hot" (*pedes*).

One of the most popular drinks is *jamu*, a traditional, medicinal, herbal drink that comes in 30 varieties, each concocted to treat anything from infertility to colds. Typically, women from central Java produce *jamu*, frequently returning home to their villages to gather ingredients. One *jamu* vendor that was studied walked more than 6 miles (10 km) on her daily route carrying a basket with bottles of 8 different drinks weighing 29 pounds (13 kg).

A year-round favorite in Bogor is noodle soup that consists of rice or wheat noodles, and vegetables with meat or chicken in chicken stock. This soup is in particular demand in the hot dry season because it replaces liquid and salt lost by sweating and provides energy from carbohydrates. This food is sold only by men, mostly migrants from West Java.

In the Philippines, university students patronize and work for street food vendors. After graduation, many start up their own businesses, which can be more lucrative than jobs in the low-paying government sector. Many women sell street food to pay for their children's school fees. Most of the street food enterprises are owned or operated by women, although many are run by couples. Whoever the owner, women control the business income and dominate decision making.

Most street vendors in the Philippines sell a variety of foods from self-produced items to already-packaged snacks and beverages, thereby acting as producers, traders, and servers. The custom of *merienda* or snacking is well entrenched throughout the Philippines. Consequently, food vendors are open all day long offering Western-style eggs, sandwiches, popcorn, fried peanuts, and the like. Local specialties include deep-fried local plantains coated with sugar called "bananacues" or when stuffed with ground meat, a "banana burger." Customers also like to chew on cooked pig's ears. Often, food is purchased from the street stall and taken home to eat with rice cooked in the family's electric rice cooker.

Women also dominate the street food trade in Thailand. Many vendors have contracts with schools and hotels and even with shopping malls. Children can only eat within the confines of the schoolyard. Because women's wages are considered merely supplementary and they earn only 60 to 85 percent as much as men in any regular job, food vending is an attractive alternative, especially since Thai women control their own income. Women support education for their daughters and sons and frequently give money to their husbands for leisure activities such as drinking and gambling, which are considered culturally appropriate for Thai men.

Thai women have more freedom of movement than women in many other Asian countries. They are often seen eating on the street, a practice that is not as common in Muslim Indonesia, for example. One of the busiest times of day for street food is in the late afternoon when most businesses have closed. Then, working women stop on their way home to buy ingredients or salty snacks to add to the evening meal. Vendors jokingly call these women *mae ban tung plastic* or "plastic bag housewives."

In Bangladesh, street foods reflect the poverty of the local population as well as the cultural aspects of Islam, most notably in the absence of women vendors. Moreover, because of the practice of purdah, few women are seen in the markets. However, it is women who produce the food for sale by men. Street food enterprises come and go as vendors find alternative work opportunities or, as is often the case, fall into bankruptcy. However, there are increasing numbers of micro-credit NGOs that try to help sellers stay in business.

Another feature of street food selling in Bangladesh is the prevalence of *periodic markets* that are held in different towns on different days of the week. Both buyers and sellers follow the markets in search of local products and home-made snack foods. Local governments have established all-weather structures with tin roofs where stalls can be rented. Those with simple carts set up outside the main market. Some male vendors do hire women assistants, especially little girls. Some pile up bricks for customers to sit on, and these are referred to by English-speaking Bengalis as "Italian restaurants" because the Bengali word for brick is "*it*."

Foods are designated as "wet" or "dry." Wet foods, including rice or sweets with syrup, are served in "hotels" (not a place to stay). Dry foods are served in a "restaurant," which means that egg dishes, bread, and fried spiced vegetables are available. Tea is served in a "tea shop."

Bangladeshis believe that certain foods should be eaten only at specific times. Foods that are considered hot "that warm the body" are sold in winter. Eggs and meat are considered to be hot foods. Fish and papaya are thought to be cold foods and therefore are commonly sold in the cool season.

In contrast to other Asian countries, few people patronize street food vendors or at least not on a regular basis. Snacks may be brought home but full meals are cooked at home. However, studies have shown that as affluence increases, more money is spent on street foods.

The most significant impact on the street food business in Bangladesh is Ramadan, the requisite sacred period of fasting for all Muslims. About half of the vendors close their shops for the month under social pressure not to eat from dawn until dusk. Those that stay open put up curtains so that nonbelievers such as tourists or Hindus or others who ignore the ban can eat in anonymity.

Numerous Asian governments have cited street foods as being unsanitary, their stands as eyesores, and their presence on streets as contributing to traffic congestion. In some cities, such as Jakarta, they have been banned from downtown areas. However, governments were soon persuaded that to ban street vendors from practicing their trade was to render multiple thousands unemployed and dismantle complex and essential social and economic systems that had evolved over decades.

In Indonesia, programs were set up to help vendors sanitize operations. Piped water was made available so that operators would discontinue washing their dishes in sewer water. Vendors were given lessons on disease and the importance of cleanliness and proper storage of foods to avoid contamination by flies and rodents. Special areas were set up for stalls and new structures were built and offered at low rents.

Most street food enterprises are family businesses with each member contributing at least for sometime in her life. It can be a means of independence for women who increasingly are able to control their income. An individual street food vendor can earn twice the minimum wage for work in Indonesia and three times the pay for agricultural labor in Bangladesh. Further, street food offers sustenance to many who cannot afford more expensive food, especially children. Well-fed people have more energy and perform better whether at school or on a job. Fortunately, governments are encouraging NGOs and other institutions to support mechanisms for the urban poor and explore the different roles and expectations of those men and women involved with street food.

Changing Food-Supply Systems

Asia is experiencing a dramatic transformation in its food-supply systems in response to rapid urbanization, diet diversification, and the liberalization of direct investment in the food sector. Feeding the ever-growing urban masses is one of the most important food policy challenges in the region today and in the foreseeable future. There are three specific dimensions to this challenge:

- Growing urban populations require increases in food supply in addition to the establishment of large suppliers in order to manage the increased level of activity in the market.
- Asia's most populous cities and towns tend to be coastally located. Importing food to these sites from external sources may be more practical and less costly than transporting it from rural hinterlands.
- Since preference for high quality is growing in importance, non-domestic foods might be more desirable. The probability is that both internal and external supplies will be in demand.

The rapid diversification (and Westernization) of the urban diet cannot be met by the traditional food supply chain. It requires modernization of the food retail sector and **vertical integration** of the food supply chain. Vertical integration exists when one company or conglomerate controls all stages of a process from the land on which to grow the crop; the seeds and required chemicals; the necessary equipment such as machinery or irrigation systems; the processing and packaging plant along with its equipment and supplies; the marketing and advertising media and perhaps even the commercial outlets that sell the product. Such organizational changes in the urban food supply drive the process of commercialization and diversification of domestic production systems.

Trade liberalization greatly facilitates the widespread establishment of global supermarket chains that speed up the diffusion of homogeneous foods and of a global diet in Asian markets. In Thailand, international retailers such as 7-Eleven, Royal Ahold, Tesco, Makro, and Sainsbury have been establishing supermarkets to serve the growing domestic market for fresh fruit and vegetables. Small farmers are being integrated into the fresh food supply chains via networks of contract farmers and buyers who are preferred suppliers, and via informal farmers' associations.

Similar cases of agricultural diversification and the emergence of contracts between farmers and large food outlets can also be found in India. For example, companies such as McCain (major supplier to McDonalds) negotiate with small farmers directly for the provision of potatoes. In these types of agreements, the large food outlet undertakes the required investment necessary to produce the specific product.

Vertically integrated supply chains are also focusing on the export market. For example, a vertically integrated vegetable export supply system has been formed in China in response to Japanese demand. However, since Chinese production standards are low, the Japanese trading companies typically provide the seeds, spores, and techniques of production and packing. Also, they monitor the harvest for the Japanese retailers. The increasing demand for better quality and safe products is leading to the creation of influential vegetable supply companies that stress quality and safety. As a result, there is a growing convergence between export standards and domestic retail product standards.

Small-holder production systems in Asia are under pressure to commercialize and diversify out of their traditional niche in cereal crop production. Rice, and rice-wheat (rice in summer/wheat in winter) systems account for about 80 percent of the farming population and 50 percent of the total agricultural area in Asia. These cropping systems witnessed rapid productivity growth during the Green Revolution and continue with high yields in the post–Green Revolution period. Yet, the pressure to diversify is the greatest in these areas because of low returns in grains relative to higher value alternatives such as vegetables or flowers. In regions that support two rice crops a year, such as southern China, many farmers are switching to one of rice (wet season) and the other of vegetables (dry season).

Commercialization and diversification demands are leading to a dramatic transformation of the rice monoculture systems of Asia. Some of the resulting changes include larger operational holdings, reduced reliance on non-traded inputs, and increased specialization of farming systems. Larger-scale production cannot function effectively on insufficient non-traded inputs such as locally gleaned seeds or animal fertilizer. Acquisition of commercial seeds and chemicals has become essential. Farm decisions are becoming increasingly responsive to market signals both domestic and international.

Diversified cropping actually increases work opportunities. Relative to rice production, labor requirements for vegetables and other high-value crops are substantially higher. Construction of irrigation systems, land preparation, planting, weeding, harvesting, and processing are all labor-intensive activities.

Diversification out of rice or wheat is constrained by market availability and size, land suitability, irrigation and other infrastructure, labor supply, and property rights. The opening of markets leads farmers away from protected domestic markets and government price supports and exposes them to increased risk due to the greater volatility of global prices.

Risk aversion is a significant impediment to diversification. Behavior, in the face of risk, is affected by attitudes of the farmer and the nature of technology. However, the fundamental problem is the failure of local credit and other institutions to provide alternative means for farmers to transfer their risk to other parties. If risk remains only an internal household strategy, it is less likely to be taken. Households will concentrate on growing a familiar crop with known technology and yields and guaranteed prices rather than risking their livelihood on unpredictable new crops with untested price patterns.

Food Insecurity

According to the UNFAO, the percentage of global hunger has decreased since 1990. However, the absolute numbers of hungry have increased significantly. Today, 1.2 billion people are without adequate nutrition. More than 24,000 persons die of hunger every day around the world, and 75 percent of these are children under five years of age. About 6.5 million of hungry people live in Asia and the Pacific. At least 40 percent of children in India, Bangladesh, and Timor-Leste are undernourished.

What does **food insecurity** mean?

- Not enough food for energy needed to lead active lives
- Food availability is only seasonal
- Food lacking in vitamins and minerals to maintain health
- Not being able to work or study as required
- Children not at the appropriate height for their age
- Mental deficiencies and weakened immune systems
- More prone to disease and infection
- Constant fear of not having enough food

The Food Policy Research Institute has developed a measure of hunger based on three criteria: percentage of children undernourished; percentage of children under five years of age undernourished; and mortality rate of children under five. These combined indicators are called the **Global Hunger Index (GHI).** Scores are based on 0 being the best and 100 being the worst. In other words, the closer you are to 100, the worse off you are (Figure 5-12). While the numbers appear small, they nevertheless indicate various serious situations for several countries.

- < to 4.9 = low hunger
- 5 to 9.9 = moderate hunger
- 10 to 19.9 = serious hunger
- 20 to 29.9 = alarming hunger
- 30+ = extremely alarming hunger

Some of the more developed countries such as Japan, South Korea, Malaysia, and Singapore have a "low hunger" rate. Seven out of ten Southeast Asian nations have a "serious hunger" problem. Six countries, mostly in South Asia, fall into the "alarming" category. While China exhibits only "moderate hunger," it is important to consider the fact that its index of 5 includes more than 6 million children that are underweight for their age!

The GHIs on the map imply that hunger levels are equal across the country. But, of course, this is rarely the case. Figure 5-13 shows incidence of extreme poverty and food related insecurity for various regions of Bangladesh. Higher incidence of poverty and hunger is found in areas that are subject to cyclonic storms and flooding, and tribal regions, hill regions, and other areas with poor soils, and places with poor infrastructure that are remote from markets and sources of assistance.

Bangladesh is divided into small areas called *upazilas*. Because the *upazilas* are very small, we have grouped them into larger areas by level of food insecurity to make the map more readable.

According to the World Bank, more than 400 million people in South Asia are chronically hungry. The worst affected areas are Nepal, Bangladesh, and Pakistan. This recent sharp rise in hunger and poverty derives from several factors. Given the current economic crisis,

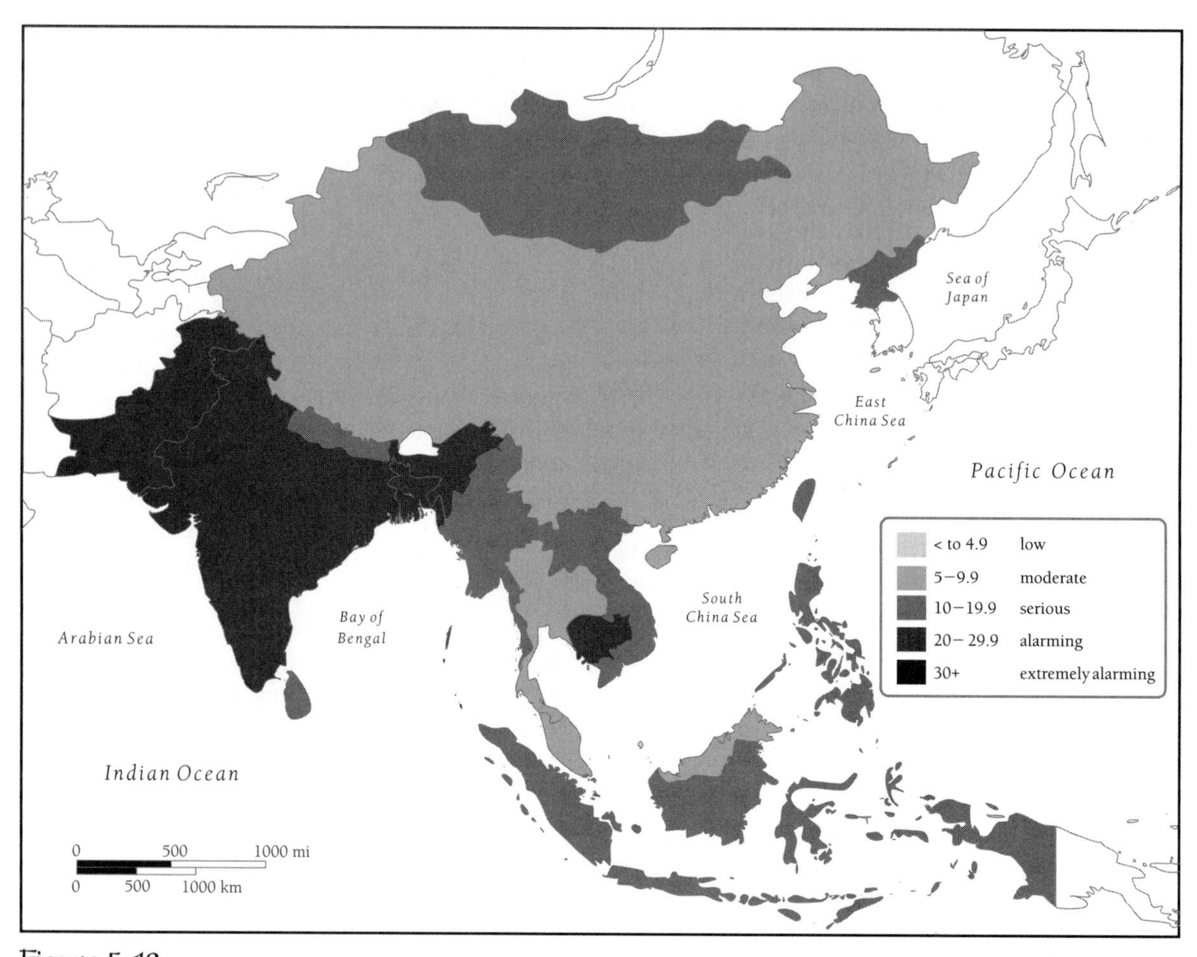

Figure 5-12

On this map you can see regional variation in the Global Hunger Index. Source: International Food Policy Research Institute.

unemployment is up and wages are down. People working abroad have lost their jobs and consequently, foreign remittances have fallen. Higher food and fuel prices have forced people to borrow money at higher interest rates, which deepens their debt. Many of those debtors lose their land to unscrupulous money lenders. Poor women go without food in order to feed their families.

The world food supply is dwindling and prices are soaring to historic levels according to the UNFAO. On the supply side, the early effects of global warming have decreased crop yields in some crucial places. So has a shift away from farming for human consumption to crops for biofuels and cattle feed. Demand for grain is increasing as the world's population grows and more is diverted to feed cattle as the numbers of upwardly mobile meat-eaters increase.

The World Food Program reports that its food procurement costs had gone up 50 percent from 2002 to 2007 and that poor people were being priced out of the food market. To make matters worse, high oil prices have more than doubled shipping costs, putting stress on poor nations that need to import food and the humanitarian agencies that provide it.

WHO ARE THE HUNGRY?

We know that by and large hunger is more of a rural than an urban phenomenon. For example, UNICEF estimates that in East and Southeast Asia, children living in rural areas are more than twice as likely to be underweight as compared with children in urban areas. This contrast is strikingly high in Nepal, India, Sri Lanka, and Vietnam.

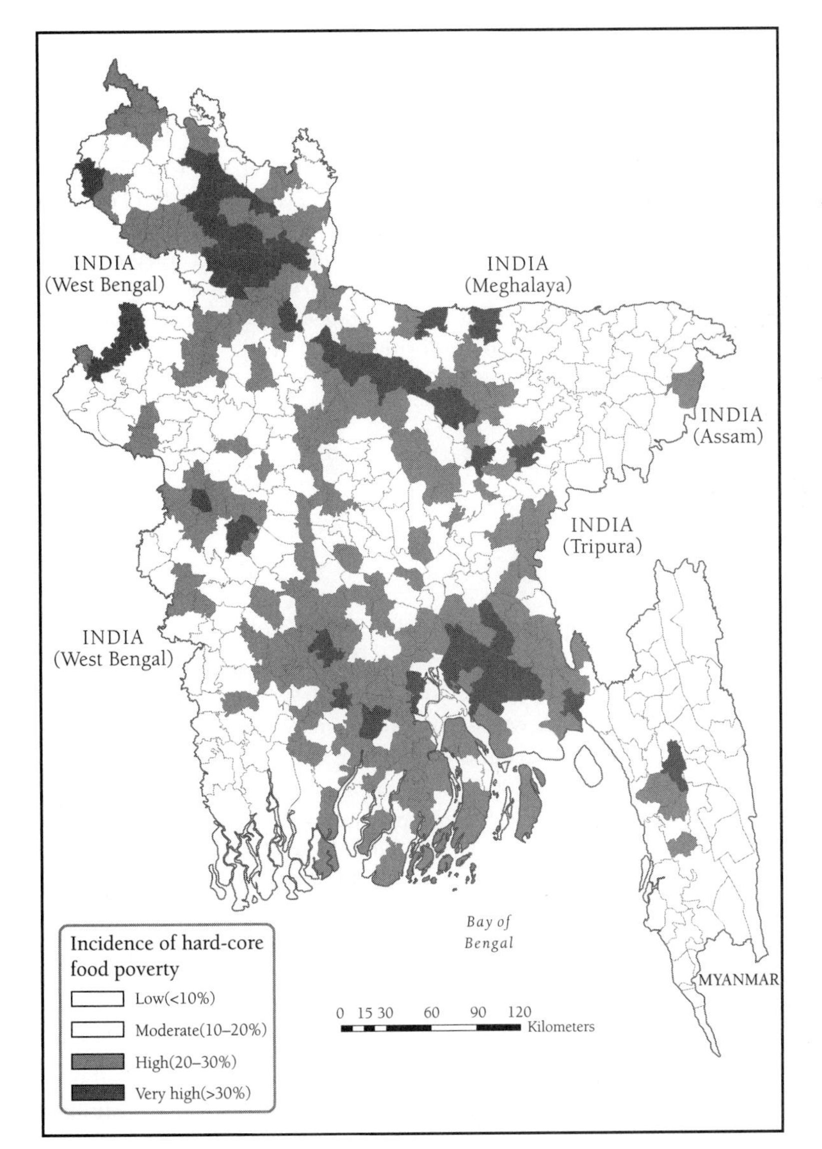

Figure 5-13

The previous map applies the GHI evenly within countries. This map of Bangladesh shows that hunger (like any other phenomenon) is not evenly distributed. Higher incidence of hunger is found in areas that are subject to cyclonic storms and flooding, tribal regions, hill regions, and other areas with poor soil, and places with poor infrastructure that are remote from markets and sources of assistance. Source: World Food Program.

The situation is bleakest for rural women who contribute about 65 percent of food production (Figure 5-14). Bangladeshi women's wage rates are so low that a day's earnings cannot maintain a family of three. Poorly paid migrant workers cannot afford rising food prices. Moreover, in the wake of the ongoing financial crisis, millions have lost their jobs. Also vulnerable are tribal groups, many of whom have lost access to their traditional lands and forests and other common property resources on which they depended for food and livelihood. Low caste (social rank) groups in India and Nepal account for more than 170 million individuals who are regularly discriminated against. The list of groups suffering food insecurity is a long one indeed.

IFAD and the WPF have found that the majority of poor and undernourished people are:

- Landless households
- Marginal and tenant farmers
- Indigenous people (tribes that are discriminated against)
- Scheduled castes (low social level groups in India, also discriminated against)
- Refugees and displaced persons from areas of conflict or environmental disaster
- Herders
- Coastal fishers

Figure 5-14

This Nepali woman and her baby live on this small sidewalk space in Kathmandu. Her husband has died and her family cannot afford to support her, so she must beg for food. This is the side of a small hotel where I stayed. The hotel will not allow her to sit out front where she might "upset" guests. In poor countries, many beggars try to locate near hotels or even banks where they are more likely to get donations. Photograph courtesy of B. A. Weightman.

- Forest dwellers (hunters and gatherers and swidden farmers)
- Highland farmers
- Women-headed households
- Females in general
- Migrant workers

ADAPTING TO HUNGER

The most general impact of food insecurity, and particularly rising prices, is an increase in poverty. In Indonesia, for example, each 10 percent increase in the price of rice has been estimated to reduce the spending power of the poorest tenth of the population by 2 percent. As anywhere that this is the case, people mortgage or sell their assets, migrate elsewhere in search for work, or remove children from school so that they can work to contribute to household income. Parents may even give up their children for adoption, or marry off their daughters early to reduce the number of mouths to feed.

When households come under immediate pressure as a result of rising prices or falling food supplies, the first response is usually to change how they eat, consuming less or different food. For instance, poor farmers in Fujian Province on China's southeastern coast reduced their consumption of pork by 15 percent and of eggs by 20 percent in 2008.

The poorest people generally buy the cheapest available carbohydrates. However, as we have already discussed, rising affluence for some increases demand for vegetables, fruit, meat, and dairy products. Farmers, therefore, feed more corn and other grain to livestock. This reduces grain supplies and consequently drives up prices of carbohydrates. Therefore, the poorest people do without.

Communities define food security and poverty in their own ways—and along many dimensions. Furthermore, the criteria will change markedly from one place to another. A 2003 participatory assessment in the Federally Administered Tribal Areas in northwestern Pakistan indicated four general categories of well-being (Table 5-3).

Clearly, there is a vast difference between those who are well-off and better-off and those who are poor or very poor. The first two groups have assets in land and livestock, a somewhat varied diet, and good health. The second two groups have little if any assets, are usually hungry, and lead unhealthy lives of deprivation.

Adapting to food insufficiency—eating more carbohydrates at the expense of protein, eating less, or not eating at all results in dietary deficiencies, especially in *micronutrients*. Micronutrients are vitamins and trace minerals that are essential for chemical processes that ensure the survival, growth, and functioning of vital human systems. People in poor regions of Asia frequently lack three key micronutrients: iron, vitamin A, and zinc. As a result, they are at greater risk of illness or death from infectious diseases and may not develop their full physical or mental potential.

Vitamin A and Blindness

The largest number of blind people in the world lives in Asia. India has 15 million blind individuals with another 52 million visually impaired. Cambodia has

Table 5-3 Categories of Well-Being in Pakistan's Northwestern Tribal Areas, 2003

WELL-OFF	BETTER-OFF	POOR	VERY POOR
Good physique	Some land	Drink black tea	Hungry
Land	50 sheep	Often hungry	Physically weak
Crops	Good health	Many dependents	Landless
100–150 sheep	Enough food	Bad health	No livestock
Surplus food (meat, butter, & milk)	Grain & bread	Very little land	No food
	Eat 2 meals a day	1 or 2 livestock	Low-quality food
		Insufficient food	Eats dry bread
			Depends on donations
			Begs

160,000 blind people and 20,000 are being added every year. In China, there are 9 million people that are blind and an additional 450,000 become blind every year. Approximately two-thirds of the blind are female. Most of this blindness is a product of vitamin A deficiency and is preventable with a proper diet. Vitamin A is found in liver, eggs, and green, yellow, and orange vegetables. Blindness can also be caused by infectious diseases. Women with certain types of infections give birth to babies that are already blind or who become blind by the age of six.

A large proportion of blindness is a result of cataracts. While this can be a childhood disorder, cataracts often form in the eyes of older people. Cataracts can be surgically removed, but there is an array of factors working against this. For instance, poor vision is often perceived as normal so nothing is done about it. There are very few qualified eye surgeons—one for every 100,000 Indians—and most people can't afford their services anyway. Some 25 percent of counties in China have no eye care facilities at all. In the country at large, there is only one cataract surgeon for every 150,000 individuals.

In many areas, poorly trained doctors tell parents to wait until the child is older. By then, it is often too late. There is also a general fear of surgery and anesthetics. Operations require antibiotics and other medications for up to two months. Many people can't afford this follow-up treatment. Moreover, they do not understand or pay attention to directions.

Other forms of blindness, such as glaucoma, or cytomegalovirus associated with AIDS, require medication for life. Medicines for these diseases are limited and costly. For example, in Indonesia, which has the highest rates of blindness in Southeast Asia, a type of herpes virus leads to blindness in people with compromised immune systems. A four-month treatment costs US$10,000.

Fortunately, there are a significant number of NGOs working to prevent blindness and other eye diseases. ORBIS, for example, is a nonprofit group that treats patients in Bangladesh, China, India, Pakistan, and Vietnam. It removes cataracts, unblocks tear ducts, repairs broken eye sockets, and helps children get glasses, among its many services.

HUNGER AND AIDS

The Asia Pacific region is home to some 6 million people living with HIV and AIDS. India has 2.4 million HIV victims. HIV infection is spreading rapidly in Indonesia, especially Papua, and in Vietnam and Pakistan. The AIDS epidemic and food and nutrition insecurity form a vicious cycle. Malnutrition heightens susceptibility to HIV infection, while HIV in turn undermines food security. People living with the disease have higher than normal nutritional requirements, needing up to 50 percent more protein and at least 15 percent more caloric intake. In addition, they require a variety of vitamins and minerals. Those suffering from HIV typically have a loss of appetite and exhibit anorexia, both of which reduce their dietary intake when their nutritional requirements are greatest. Resultant malnutrition expedites the onset of AIDS and ultimately death. In Asia in 2007, 380,000 people succumbed to AIDS-related illnesses.

HIV mostly affects sexually active young adults who are among those who are likely to be economically productive. Without proper treatment, these individuals will develop AIDS, resulting in repercussions for the food supply of other family members and even the community at large.

Women are biologically, socioeconomically, and socioculturally more at risk of HIV infection than men. They also tend to be economically dependent on men and have unequal access to resources, whether these are food or medical care. Further, the burden of looking after a relative with AIDS is considered a woman's duty.

Agricultural and Food Sustainability

Agriculture uses more land in the world than any other economic activity and employs more than 40 percent of the working population. It has been the principal enterprise of survival for all of human history. Fundamentally, it is an interactive system between environment and humanity. However, agricultural lands are disappearing with rapid population growth, urbanization, and industrialization. Overuse of land, forest removal, and technology have degraded physical environments, destroyed natural habitats, exacerbated floods and droughts—even to the point of famine—and increased social inequality. In addition, in the wake of globalization, integrated local systems of supply and demand have been fractured as they are impinged on by larger, more complex, and distant transnational enterprises intricately woven into the global network of production.

Sustainable development in agriculture for everyone, even at a global scale, means that there is enough nutritious food for everyone to meet their full potential in life. The United Nations calls for an effort to improve present-day living standards in ways that will not jeopardize the well-being of future generations.

DEVELOPMENT FOR WHOM?

Many geographers who consider themselves **political ecologists** ask the critical question: "Development for whom?" These scientists investigate the inner workings of "development" processes. They want to know just whose needs are being addressed and how "success" is being measured. Political ecologists have already exposed the travesties of oil palm plantations. As we have noted above, palm oil enriches growers along with government with taxes on profits. However, loss of forest resources, soil fertility, and substituting a single plant species in a multispecies forest not only devastates the physical environment but also eliminates ways of life that have been sustained for generations. People are forced to migrate to already overcrowded cities where their agricultural skills are of little use.

So-called "free market forces" encouraged by the WTO, means that wealthy countries' agricultural surpluses are exported to poor countries at prices that undercut local prices. Local farmers, who cannot compete in the market, fall into debt and many end up losing their land. Peasant farmers are displaced by large-scale, agribusiness ventures that are vertically integrated, owning all elements of production from seeds to final packaging. While some will argue that such shifts in production provide jobs, political ecologists will counter with the arguments that many of these jobs are demeaning of human worth, especially for women, and that they shatter traditional ties to physical and cultural environments.

PROMISING PROJECTS ON THE HORIZON

Outside Manila, in the Philippines, farmers are returning to traditional methods. There, farmers have had a long tradition of raising livestock, rice, and fish together. Manure from hogs and chickens fertilize algae in both rice and tilapia ponds. Very little waste is produced by this integrated system, which gives the farmer a sense of security if other food prices fluctuate. Moreover, animals that are raised outdoors seldom suffer from respiratory ailments common in factory farming.

NGO-operated projects in Bangladesh have demonstrated that integrated rice-fish culture reduces production costs by 10 percent. One 2.54 acre (1 ha) rice field produces 550–3,300 pounds (250–1,500 kg) of fish, which is enough to eat and extra to sell. The average farm income from this system has risen 16 percent.

In China, Taiwan, and elsewhere in Asia, the demand for shark fin soup has risen dramatically with growing elites. This soup costs around US$250 a bowl. Historically, this delicacy was consumed only by royalty and very-high-status individuals. Now there is a roaring market in shark fins, which can fetch US$700 for 2.2 pounds (1 kg). More than 100 million sharks are slaughtered each year. They are captured, their dorsal fin is chopped off, and they are thrown back into the sea to drown. Several major shark species are now extinct.

People are now beginning to speak out against this selfish act of cruelty. Asian celebrities such as director

Ang Lee and Taiwan's former president Chen Shui-bian have made public service announcements against the practice. Thai Airways and Singapore Airlines pulled shark fin soup from their first class menus in 2000. Hong Kong Disneyland and Hong Kong University no longer serve the soup after protests by animal rights and marine conservation groups.

SEWA (Self-employed Women's Association), mentioned earlier, is a 30-year-old organization dedicated to self-empowering Indian women. Fighting for women against being cheated by employers, charged usurious loans, being forced to bribe police among other indignities, SEWA now has 76,000 members in seven states. The organization was instrumental in the creation of India's National Policy on Street Vendors. More than 76 cooperatives have been formed for market gardening, tree growing, handicraft production, milk production, and salt mining in coastal areas. SEWA has also established networks of child care, health care, insurance, and housing provision. More than 300,000 women have taken advantage of the new health-care facilities and 110,000 are now covered by insurance.

All people need food, shelter, and clothing. The particular ways we fulfill those needs reflect our diverse cultures, and the climates and ecosystems we call home. But no factor matters more than our affluence. No matter where we live, the more money we have, the more we spend. The more we consume and the more we waste, the more we harm the Earth, including both the haves and the have-nots residing in it. Greater economic stability and food security for those living in emerging economies must be the goal for all of us. The most significant challenge of all is that we must achieve this state of well-being for all humankind without accelerating ecosystem collapse and total environmental ruination.

Recommended Web Sites

www.actahort.org
International Society for Horticulture and Science. Find out about scientific developments in agriculture.

www.adb.org
Asian Development Bank's information on fighting poverty in Asia and the Pacific. Gender, agriculture, economics, and plans for improvement.

www.afp.com
International news on a wide range of topics. Photos and graphics. Interactive global map.

www.agnet.org
Food, Fertilizer and Technology Center for Asia and the Pacific. Information on various topics and programs to improve agriculture in Asia.

www.alertnet.org/index.htm
Thomas Reuters Foundation. Timely articles, maps, and photos by region that cover social and environmental crises.

www.atimes.com
The Asia Times. Current articles on a wide range of topics including food and agriculture. Click on "China Business" for an interactive map of China.

www.cropscience.org
Crop Science Society of America. Scientific articles on crops, pests, soils, irrigation, and a wide variety of other aspects of agriculture. Access full text of articles in *Crop Science* and related journals.

http://earthtrends.wri.org
World Resources Institute. Information and statistics on employment, gender, agriculture, and environmental issues by region and country.

http://e360.yale.edu/
Publication of the Yale School of Forestry and Environmental Studies. Numerous environmental and human impact topics by region. NASA and other photo galleries.

www.fao.org
Food and Agriculture Organization of the United Nations. Information and statistics on crops, hunger, and food security around the world.

www.fao.org/gender/gender-home
Information on gender relations and agriculture.

www.fao.org/sd/
Information on sustainable development.

www.foodfirst.org
Nonprofit "people's" think tank focusing on root causes and value-based solutions to hunger and poverty around the world.

www.guardian.co.uk
Link to Guardian Environmental Network and other sites focusing on food, farming, health, biodiversity, and related issues.

www.ifad.org/
International Fund for Agricultural Development. Information on food and farming systems, crops, and related pests and diseases. Rural poverty and community-based projects.

www.ifpri.org/
International Food Policy Research Institute. Global Hunger Index and gender inequality.

www.ilo.org
International Labor Organization. Employment statistics for men and women in various economic sectors.

www.indiaenvironmentportal.org.in
Crops, environment, water, sanitation, irrigation, and dams in India.
www.ips.org
Inter-Press Service global news agency. Focuses on "the South" with articles about human rights, gender issues, development, and globalization.
www.ksoils.com/ and www.ksoils.com/subsidiary.htm
Read about a multinational edible oil producer.
www.mongabay.com
One of the world's most popular conservation and news sites. Focuses on preservation of wild lands and wildlife especially in tropical regions. Information and photos on impacts of climate change, economics, and technology on food production and security.
www.reefbase.org
Global information system for coral reefs and reef fisheries. GIS maps.
www.rff.org
U.S. think tank on environment and resources and the environmental aspects of global food and agriculture.
www.theoildrum.com/
Information on the growth and expansion of palm oil plantations in Southeast Asia. Human and environmental consequences.
www.unescap.org
Economic and Social Commission for Asia and the Pacific. Information and statistics for economic and social development. Downloadable publications.
http://unifem.org
United Nations Development Fund for Women. Information and reports on women, poverty and economics, and peace, security, and human rights.
www.wfp.org
World Food Program. Country profiles, hunger statistics, and food security.

Bibliography Chapter 5: Agriculture, Food and Food Security

Adriano, Joel D. 2009. "Philippines: Fish Farming to Reduce Protein Deficiency in Uplands." *Inter Press Service and International Federation of Environmental Journalists,* October 22.

AFP. 2008. "Indonesia Looks to Papua to Expand Palm Oil Plantations: Official." *AFP,* May 21.

AFP. 2007. "Southeast Asia Gears Up for Palm Forests." *AFP,* September 11.

Asia Times. 2009. "Illegal Logging Costing Indonesia Dearly." *Asia Times,* July 11.

Bourne, Jr., Joel. 2007. "Biofuels: Boon or Boondoggle." *National Geographic* 212/4: 38–59.

Bray, Francesca. 1986. *The Rice Economies: Technology and Development in Asian Societies*. Berkeley: University of California Press.

Butler, Rhett, A. 2006. "Why is Oil Palm Replacing Tropical Rainforests? Why are Biofuels Fueling Deforestation?" *Mongabay,* April 25.

Cohen, Dave. 2007. 'Palm Oil—The Southeast Report." *The Oil Drum: Discussions About Energy and our Future,* January 30.

De Haen, H., et.al. 2003. "The World Food Economy in the Twenty-First Century: Challenges for International Cooperation." *Development Policy Review* 21: 683–696.

Dixon J., et al., eds. 2001. *Global Farming Systems Study: Challenges and Priorities to 2030.* Rome: Food and Agriculture Organization of the United Nations.

ESCAP. 2009. *Sustainable Agriculture and Food Security in Asia and the Pacific*. New York: United Nations.

FAO. 2008. *Understanding Avian Influenza*. Rome: Food and Agriculture Organization of the United Nations.

FAO. 2009. *The State of Food Insecurity in the World.* Rome: Food and Agriculture Organization of the United Nations.

FATA. 2003. *Between Hope and Despair, Pakistan Participatory Poverty Assessment*. Islamabad: Federally Administered Tribal Areas Report.

FFTC. 1998. *Sustainable Agriculture in Upland Areas of Asia.* Taipei: Food and Fertilizer Technology Center for Asia and the Pacific.

Greger, Michael. 2006. *Bird Flu: A Virus of Our Own Hatching*. New York: Lantern Books.

Halweil, Brian. 2008. *Farming Fish for the Future*. Worldwatch Report 176. Washington D.C.: Worldwatch Institute.

Halweil, Brian, and D. Nierenberg. 2008. "Meat and Seafood: The Global Diet's Most Costly Ingredients." In *2008 State of the World: Innovations for a Sustainable Economy*, pp. 61–74. New York: Norton/Worldwatch.

Huang, J., and C. David. 1993. Demand for Cereal Grains in Asia: The Effect of Urbanization. *American Journal of Agricultural Economics* 8: 107–124.

ICSC. 2004. "Diet and Nutrition Change in Asia." *New Directions for a Diverse Planet*. Brisbane, Australia: Proceedings of the 4th International Crop Science Congress.

IFAD. 2009. "Gender in Fisheries and Agriculture." In *Gender in Agriculture Sourcebook*, pp. 561–567. Washington, D.C.: The International Bank for Reconstruction and Development/World Bank.

IFAD. 2009. "Gender and Food Security." In *Gender in Agriculture Sourcebook*, pp. 11–22. Washington, D.C.: The International Bank for Reconstruction and Development/World Bank.

ILO. 2009. *Global Employment Trends for Women*. Geneva: International Labor Office.

Kelkar, Govind. 2009. *The Feminization of Agriculture in Asia: Implications for Women's Agency and Productivity.* New Delhi: United Nations Development Fund for Women.

Knudson, Tom. 2009. "The Cost of the Biofuel Boom: Destroying Indonesia's Forests." *Yale Environment 360,* January 19.

McCarten, Brian. 2009. "Myanmar's Generals Plow a Rich Furrow." *Asia Times,* December 19.

Manderson, Lenore, and L. R. Bennett, eds. 2003. *Violence Against Women in Asian Societies*. Richmond, UK: Curzon.

Millstone, Erik, and T. Lang. 2003. *The Penguin Atlas of Food.* New York: Penguin Books.

n.a. 2009. "Engineered Edible Cottonseed Could Feed Millions." *Chicago Sun Times,* November 30.

Narayan, Deepa, and E. Glinskaya, eds. 2007. *Ending Poverty in South Asia: Ideas That Work*. Washington D.C.: World Bank.

Narayan, Deepa, et al., eds. 2000. *Voices of the Poor: Crying Out For Change*. New York: Oxford University Press.

Nathan, Dev, and Kelkar, G. 2004. "Chenchu Women of Andhra Pradesh, India: Empowerment in the Market and Household." United Nations Development Fund for Women (unpublished report).

Mikkelsen, Lene. 2004. *Gender Equality and Empowerment: A Statistical Profile of the ESCAP Region*. United Nations Economic and Social Commission for Asia and the Pacific.

Pingali, P., and M. Rosegrant. 1995. Agricultural Commercialization and Diversification: Processes and Policies. *Food Policy* 20: 171–185.

PIngali, P., et. al. 1997. "Supplying Wheat for Asia's Increasingly Westernized Diets." *American Journal of Agricultural Economics* 80: 954–959.

Piper, Jacqueline. 1993. *Rice in South-East Asia: Cultures and Landscapes*. New York: Oxford.

Preston, Gary. 2009. *Reef Fisheries: Indo Pacific Overview*. Penang, Malaysia: The World Fish Center.

Raja, Marthy. 2009. "Cotton Heads for the Dinner Table." *Asia Times,* December 8.

Reardon, T., et al. 2003. "The Rise of Supermarkets in Africa, Asia, and Latin America." *American Journal of Agricultural Economics* 23: 195–205.

Ritter, Peter. 2009. "Borneo's Green Gold." *Far Eastern Economic Review* July/August: 77–79.

Rosenthal, Elizabeth. 2007. "World Food Supply is Shrinking, U.N. Agency Warns." *The New York Times,* December 18.

Sachs, Carolyn. 1996. *Gendered Fields: Rural Women, Agriculture, and Environment*. Boulder, Colo.: Westview.

Seager, Joni. 2009. *The Penguin Atlas of Women in the World*. New York: Penguin.

Sen, Amartya. 2000. *Development as Freedom*. New York: Knopf.

Smil, Vaclav. 2000. *Feeding the World: A Challenge for the Twenty-First Century*. Cambridge, Mass.: MIT Press.

Starke, Linda, ed. 2007. *Vital Signs 2007–2008: The Trends that Are Shaping Our Future*. New York: World Watch.

The Economist. 2006. "Filling Tomorrow's Rice Bowl." *The Economist,* December 9: 84–85.

Tinker. 1997. *Street Foods: Urban Food and Employment in Developing Countries*. New York: Oxford University Press.

World Bank. 2007. *Changing the Face of the Waters: The Promise and Challenge of Sustainable Aquaculture*. Washington, D.C.: World Bank.

Chapter 6

South Asia: Creating Dilemmas of Diversity

"What is here is nowhere else;

what is not here, is nowhere."

THE MAHABHARATA (CA 1000 BC)

This chapter initiates the regional analysis that makes up the rest of this text. In it we will set the historical and geographical framework for subsequent discussions of specific countries in South Asia. For an understanding and appreciation of conditions, events, and swirling controversies in South Asia, historical knowledge is indispensable. Therefore, after a brief review of the physical setting, we will focus on South Asia's history up to and including the creation of India and Pakistan in 1947.

The Physical Setting

South Asia is a realm of spectacular highlands, vast deserts, rugged plateaus, fertile lowlands, and magnificent coastlines. Here mountains and coastlines combine to create one of the world's most sharply defined physiographic regions (Figure 6-1). Splaying southward from the Himalayan wall to the north, the Indian subcontinent is differentiated in the west by the Punjab, a region through which flow the Indus River and its four major tributaries: the Jhelum, Chenab, Ravi, and Sutlej. Punjab means "five waters." South of the Punjab is the Indus River plain. This entire region is guarded by the Hindu Kush and other mountain ranges to the north and west and the Great Indian or Thar Desert to the east. Further east, the North Indian Plain extends to include the sacred Ganges River and the amazing Ganges and Brahmaputra delta. Southward, the Central Indian Plateau is separated from the Deccan plateau by mountains and waterways such as the Vindhya range and the Narmada River. These physical features have presented barriers to north–south movements throughout Indian history. Further south is the highly dissected Deccan bordered by the Western and Eastern Ghats. The Deccan is cut by several rivers, most notably the Godavari, Krishna, and Cauvery. The island of Sri Lanka is located at the tip of peninsular India. It too is a perplexing and troubled region. You will learn more about these regions in this and the following three chapters.

Monsoon and Life Cycles

As we noted in Chapter 2, a large part of life, including harvesting and planting, corresponds with the dry and wet monsoons. This is especially the case in South Asia, where the landscape evolves in a never-ending cycle of brown to green, sparse to lush, and lethargic to energetic. For millions, these cyclical events imitate the greater, overarching cycle of birth, decay, and rebirth. For countless others, they reveal the divine will of Allah or hand of God. Forgotten or improper behaviors and rituals might trigger drought or flood with disastrous consequences.

In northeastern Hindu India, the great festival of Rathajatra at Puri celebrates the transition from the season of drought to the season of rain. A two-story-high vehicle, known as a juggernaut,

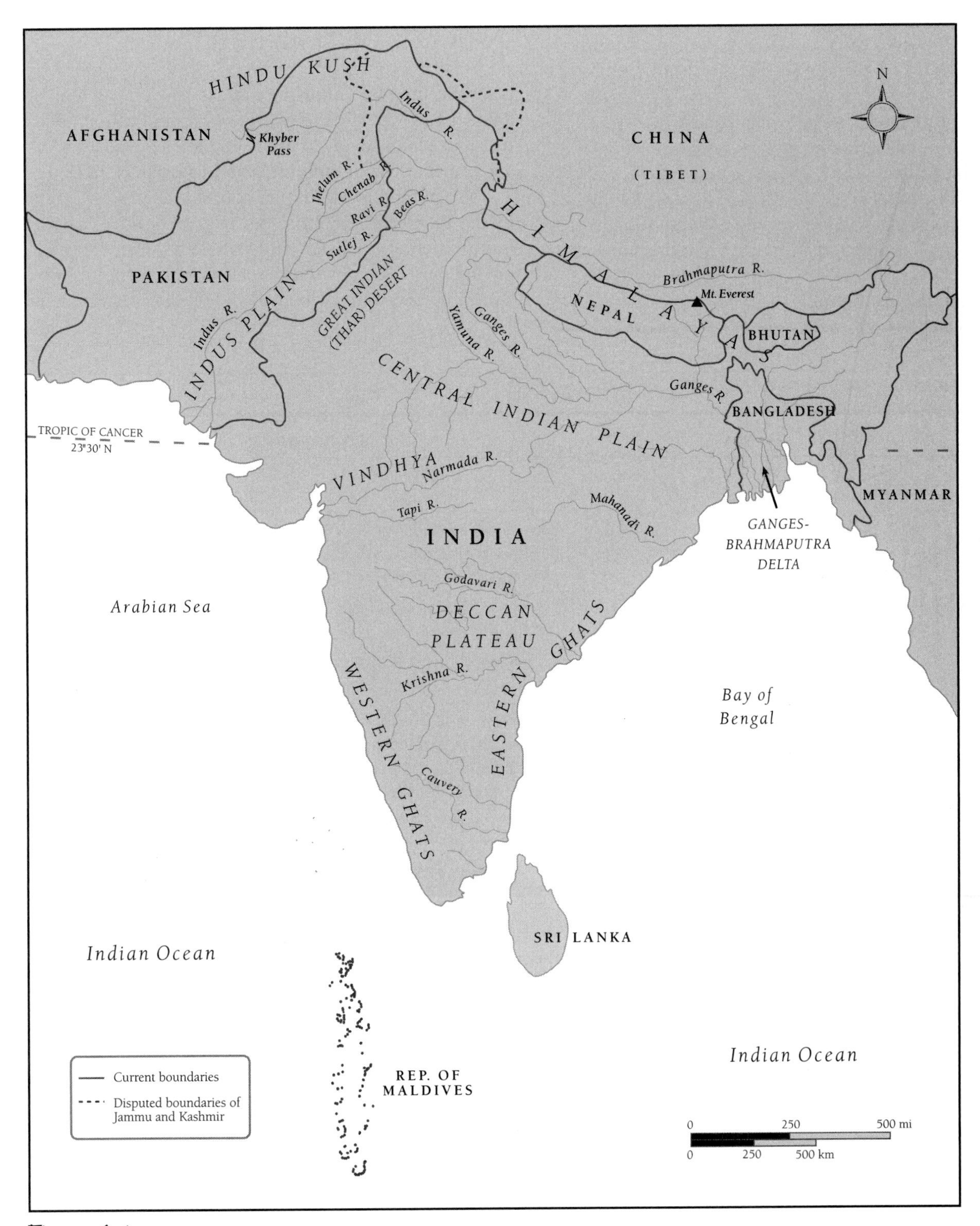

Figure 6-1

Landforms and waterways of South Asia. Compare this map with Figure 8-2 and note how topography influences settlement patterns.

is constructed to carry a likeness of Jagannath "Lord of the World," a form of Vishnu. To the clashing of cymbals and with a covey of chanting priests, the behemoth is pulled toward the excited throng. As Jagannath nears, people joyously leap and sing and chant "*Hari! Hari!*" (God! God!). Each creaking and turning of the great wheels speaks of impending change and assures the coming of the wet monsoon. Participation in this and other seasonal events reaffirms rebirth and reasserts one's union with nature.

Archives of Time and Place

Before reading the following discussion, you can look at Table 6-1, which gives an overview of significant individuals and events involved in the evolution of India and its subsequent division into India and Pakistan in 1947.

When the British Empire reached its greatest extent in the nineteenth century, the map of South Asia was quite different. Pakistan and Bangladesh, for instance, did not exist. Actually, when ancient Indian empires were stunning the known world with their spectacular

Table 6-1 Time Line of Selected Historic Events in India

Century	Year	Dynasty	Ruler	Important Events
4th–3rd C. BC	3000–2000 BC	Indus Valley Civilization	Harappans	Cities of Mohenjo-Daro and Harappa in Indus Valley
17th C. BC	1600 BC	Aryan invasions		Drove away Dravidians. Brought spoken Sanskrit, cows and horses, use of iron for tools. Expanded settlement on Ganges plain. Seeds of caste division.
11th C. BC	1000–500 BC	Vedic culture		Evolution of traditional Indian culture. One of the holiest scriptures, the *Rig Veda,* composed.
8th C. BC	700 BC			Beginnings of the caste system with Brahmins at the top.
7th C. BC	600 BC			*Upanishads* composed in Sanskrit.
6th C. BC				States form across Ganges plain. Caste divisions multiply to include *jatis* and untouchables. Economic specialization. Emergence of Indo-Aryan culture. Dravidians driven southward and develop their own languages and civilization.
6th C. BC	527 BC			Prince Siddartha Gautama attains enlightenment and becomes the Buddha.
5th C. BC	500 BC			The ascetic prince Mahavira founds Jainism in northern India.
4th C. BC	300 BC			*Ramayana*, a famous epic, is composed.
3rd C. BC	259 BC	Maurya dynasty	Asoka	Mauryan ruler Asoka converts to Buddhism. Extends empire from Hindu Kush to Bengal. Controls the Deccan, major trade routes, and coastal regions. First expression of political notions of a state. Wealth accumulates from trade.
3rd C. BC	268–233 BC	Mauryans	Asoka	Adopts Buddhism as state religion. Builds hospitals, wells, and water storage tanks. Inscribes moral principles on stone pillars. Most famous pillar at Sarnath is 40' tall with 4 lions back-to-back on top. These become the symbol of India.
3rd C. BC	200 BC			*Mahabharata*, another famous epic, is composed.
2nd C. BC	100 BC			*Bhagavata Gita* composed.
1st C. AD	60 AD			Thomas, an Apostle of Jesus, visits India. Christian conversions mainly among low castes and untouchables.
3rd C. AD	200			The *Code of Manu* puts down the rules of everyday life and divides Hindus into 4 major castes: priests, warriors, farmers/traders, and non-Aryans.

(*continued*)

Table 6-1 (Continued)

Century	Year	Dynasty	Ruler	Important Events
4th C. AD	320	Guptas	Chandra Gupta	North India unites under Guptas."India's Golden Age" of art and Sanscrit literature. Advances in science and mathematics. Concept that Earth was round and that a year had 365 days. Invention of 0 and numbers 1–9.
4th–5th C. AD	319–412	Guptas		Chinese Buddhist pilgrim Fa Hsien visits India and notes that Gupta kings were Hindu but that Buddhism flourished in the countryside. Chandra expands Gupta rule to Gujarat.
5th–6th C. AD	499–528	Guptas		Hindu mathematician writes the *Aryabhattiyam,* the first algebra book. Downfall of Guptas under continuous invasions.
6th–13th C. AD	500s–1200s			Multiple kingdoms. Most notable are the Pallavas and the Cholas in southern India.
7th C. AD	630–643		King Harsha	Harsha establishes a kingdom in the north India and Nepal. Chinese Buddhist pilgrim Huan Tsang visits domain of King Harsha that extends from Punjab to Orissa and south to Narmada River. Notes military importance of Rajputs. Notes strict caste and gender divisions, practice of *sati,* and oppression of untouchables. Returns to China with 657 Buddhist texts and relics. Translation of 74 of these furthers spread of Buddhism in China.
7th C. AD	628–651			Persian Zoroaster brings Parsi religion to region that became Bombay (Mumbai).
8th C. AD	712			Arabs establish themselves in Sind (now SE Pakistan), and then Punjab. Introduce paper and gunpowder.
10th–11th C. AD	997–998			Mahmud of Ghazni, "The Sword of Islam," from Afghanistan plunders India and conquers Punjab.
11th C. AD	1023			Mahmud of Ghazni sacks Sarnath and kills 50,000 Hindus.
12th C. AD	1192	Delhi sultanate		Beginning of Delhi sultanates. Succession of 35 sultans over 300 years. Rebellions see 19 sultans assassinated.
14th C. AD	1309	Delhi sultanate	Alauddin	Alauddin plundered southern India. Society tightly divided along religious lines. Cities and commerce dominated by Muslims. Hindus not permitted to own land. Arab men marry Hindu women and enforce their conversion and *purdah*. Persian culture favored by court. Fearing persecution, even Hindu women adopt veiling. Persian becomes the official court language. Urdu, with Islamic overtones, is written in a Persianized, Arabic script. Many Muslims speak Urdu.
15th C. AD	1490			Guru Nanak Dev Ji establishes Sikhism in Amritsar and promotes peace between Hindus and Muslims.
15th C. AD	1497	Mughul dynasty	Babur	Babur of Afghanistan establishes Mughul dynasty in India.
15th C. AD	1498	Mughals		Portuguese traders arrive at Calicut on southeastern coast.
15th–16th C. AD	1483–1530	Mughals	Babur	Babur "The Tiger" defeated Lodi, the Sultan of Delhi. Goa falls under Portuguese control in 1520.
16th C. AD	1526	Mughals	Akbar	Establishment of Mughal Empire, first modern territorial state.

(continued)

Table 6-1 (*Continued*)

Century	Year	Dynasty	Ruler	Important Events
16th–17th C. AD	1542–1605	Mughals	Akbar	Defeat of Rajputs and conquest of Gujarat and Bengal. Infusion of Persian culture. Persian merges with languages of Delhi and Agra to become Hindi.
17th C. AD	1600	Mughals	Akbar	Establishment of British East India Company.
17th C. AD	1616	Mughals	Jahangir	British get permission to develop trading post in Gujarat.
17th C. AD	1627–1657	Mughals	Shah Jahan	Shah Jahan "Ruler of the World" erects grand architecture such as Taj Mahal in Agra, Red Fort and Jama Masjid (largest mosque in India) in Delhi. Rich *zamindars* control vast swathes of land. Many peasants become indebted to *zamindar* moneylenders and lose their land. Series of famines; millions die. Mughals weakened by too rapid expansion of empire.
17th C. AD	1661	Mughals		Portugal cedes Bombay (Mumbai) to Britain.
17th C. AD	1639	Mughals		British buy land on southwest coast to build trading post. Later becomes Madras (Chennai).
17th C. AD	1627–1680	Mughals	Aurangzeb	Zealous Hindu, Shivaji "The Mountain Rat," and the Marathas attack Mughals.
17th–18th C. AD	1658–1707	Mughals	Aurangzeb	Shah Jahan's son Aurangzeb seizes power. Indian state at its greatest extent. Cruel anti-Hindu policies and brutalization of Sikhs, Jains, and Buddhists. Collapse of Mughal Empire with death of Aurangzeb in 1707.
17th C. AD	1690	Mughals		British found Calcutta (Kolkata) on the Hooghly river.
18th C. AD	1702–1703	Mughals		Famine and plague on Deccan cause deaths of 2 million. Rise of separate states ruled by Maharajahs, Nizams, etc. Decline of Delhi.
18th C. AD	1751			British make deals with various state rulers—protection for loyalty. Britain becomes leading colonial power in India.
18th C. AD	1761	Maratha Confederacy		Marathas rule most of northern India. Constant factionalism and civil war. Severe loss to Afghan invaders at Battle of Panipat northwest of Delhi. With the defeat of the only real power in India, the British take advantage to extend their territory.
18th C. AD	1769			Famine kills 10 million in Bengal, and East India Company does nothing to help.
19th C. AD	1853		Lord Dalhousie as Governor General of India	Railways, postal service, and telegraph system introduced into India by British. Development of ports.
19th C. AD	1857			First Indian revolt against the British known as the "Sepoy Mutiny."
	1858	The Raj		Formalization of India as a British colony. Indians call this period "The Raj" meaning "The Rule."
19th C. AD	1869	The Raj		Opening of Suez Canal shortens route from Asia to Europe.
19th C. AD		The Raj		Rise of colonial ports and building of railway network in order to bring raw materials such as cotton from the interior destined for textile and other factories in England. Manufactured goods to be imported and sold in India.

(*continued*)

Table 6-1 *(Continued)*

Century	Year	Dynasty	Ruler	Important Events
19th C. AD	1877	The Raj		England's Queen Victoria declared "Empress of India." India regarded as the "Jewel in the Crown."
19th C. AD	1885	The Raj		Rising nationalism. First meeting of the Indian National Congress in Bombay led by high-caste, English-educated Hindus.
19th–20th C. AD	1899–1905	The Raj	Lord Curzon Viceroy of India	As more British come to India, they bring their families. Separate areas, called civil lines, are set aside for them. British arrogance and racism increases. British clubs and other institutions do not accept Indians. Families go to hill stations such as Simla or Darjeeling in the heat of summer. Indians are servants who live in miserable conditions apart from British estates. Curzon enrages Indians by trying to divide the state of Bengal.
20th C. AD	1906–1911	The Raj	John Morley Viceroy of India	All India Muslim League is formed. Morley is more liberal and understands plight of Indians. Division of Bengal rescinded but southern parts become the provinces of Orissa and Bihar. This leaves Bengalis united and without linguistic or cultural minorities.
20th C. AD	1912	The Raj		The Imperial capital is shifted from Calcutta to New Delhi.
20th C. AD	1914	The Raj		Mohandas Gandhi returns from South Africa where he developed his concepts of passive resistance and nonviolent protest. Encourages boycotts of British imports. Appeals to Indians to spin their own cloth and become self-sufficient. Spinning wheel becomes a nationalist symbol. Eventually is incorporated into Indian flag.
20th C. AD	1914–1918	The Raj		More than a million Indians fight valiantly with the British in World War I.
20th C. AD	1917	The Raj		Russian Revolution spurs fear of communism influencing masses of unhappy and resentful Indians. British curb the press and pass new laws to control Indians.
20th C. AD	1919	The Raj		Indian protests and violence against British. British fire on, kill and wound unarmed Indians protesting at Amritsar. This event often regarded as the final blow of British arrogance and cruelty.
20th C. AD	1920–1942	The Raj		Gandhi launches first of several major nonviolent resistance movements. Various British commissions come to India to decide what to do.
20th C. AD	June 3, 1947	The Raj		Lord Mountbatten presents plan for partition of India.
20th C. AD	Aug. 15, 1947	Independent India and Pakistan	Congress Party India Muslim League Pakistan	Independence from British rule and partition of India into India, West Pakistan, and East Pakistan. India to be ruled by the Congress Party under the leadership of Jahawarlal Nehru. Pakistan to be ruled by the Muslim League under Muhammed Ali Jinnah.
20th C. AD	1948–1956	Congress Party	Jaharwarlal Nehru	Gandhi assassinated. Muhammed Ali Jinnah dies. Nehru proceeds to preside over the creation of 16 new language-based states. Hindi declared the national language.
20th C. AD	1964	Congress Party	Indira Gandhi	Nehru dies and his daughter Indira Gandhi takes over.
20th C. AD	1971	Congress Party	Indira Gandhi	Creation of Bangladesh. led by Sheikh Mujibur Rahman.

Figure 6-2

This is the Red Fort in Agra, India. Built by the Mughal Emperor, Akbar, between 1565 and 1573, its red sandstone exhibits Indo-Islamic architecture. It stands guard over the Yamuna River. Photograph courtesy of B. A. Weightman.

achievements, the British Empire was unheard of because it did not exist. A venture into South Asia's intriguing past is surely worth pursuing.

Geography plays a vital role in history; space and time are twisted together like a two-colored rope. Mountains, such as the Himalaya and the Hindu Kush, and deserts such as the Thar (Great Indian), marked defensive frontiers. In the days when territory was traversed on foot or by animal, terrain conditions could serve as barriers or at least make distances impractical to cross. Control over strategic sites such as river fords and mountain passes could make or break empires. For instance, a dozen or so rulers chose Delhi as their capital because of its strategic situation. Delhi guards the western gateway into the ***Doab*** (two rivers), the level and fertile land between the Ganges and its main tributary, the Yamuna (Figure 6-2). Much of India's history concerns mountain passes, especially those in the Hindu Kush. The Khyber is one of the most fought-over passes in the world (Figure 6-3).

With slow and difficult travel, overextension could easily lead to territorial losses. The Deccan, for example, was a difficult region to conquer and keep, and it typically was subdivided among several rulers. Many an

Figure 6-3

Here is the Pakistan side of the Khyber Pass. The road was built by the American Corps of Engineers. Note the hairpin turns, switchbacks, and tunnels. I remember traveling this route in a Russian car with a Pakistani driver who careened the road, usually on the wrong side, with his hand constantly on the horn. We had several near misses with huge trucks headed for Afghanistan. Photograph courtesy of B. A. Weightman.

empire or state bordered the Vindhya Mountains, which form the northern flank of the Narmada River, which runs westward into the Arabian Sea. Accessibility, or lack of it, played a key role in the relatively separate development of southern India.

Lack of consideration of the surrounding Indian Ocean blinded Indian rulers to the seriousness of maritime encroachment by European powers. Mere trading posts along the shores were not read as clues of impending conquest, as no one had ever conquered India by sea. The sea was seen as the black water, and its monsoons would certainly deter even the most daring intruders. Geographic perceptions along with physical realities were clearly integral to historic developments in South Asia.

Many historians divide the subcontinent's history into Hindu, Islamic, and British periods. However, as landscape expressions of these transcend both time and space, I will use ancient, medieval, and modern to designate critical shifts in historic evolution. History is a progression through time, which makes necessary the inclusion of dates. Think of these as markers of time, not as something to be memorized. Reference to text and atlas maps will help you envision the upcoming story.

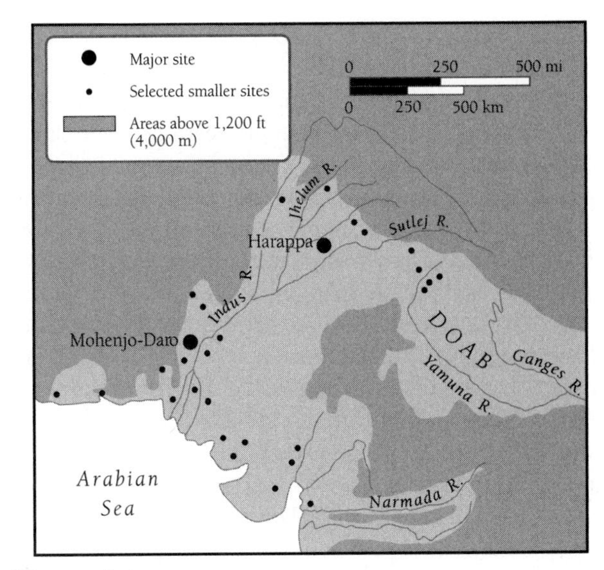

Figure 6-4

Indus civilization, ca 5000 BC.

ANCIENT SOUTH ASIA

The discovery of crude stone tools in the Punjab shows that human life in India can be dated to the period following the Second Ice Age (400,000–200,000 BC). Cultivation of grain commenced around 7000 BC in South Asia, but major settlement did not occur until around 3000 BC, when a notable increase in precipitation and water-control technology allowed large-scale, settled agriculture.

The Harappan or Indus civilization emerged as the world's third major civilization after those in Mesopotamia and Egypt. It is thought that the people were dark-skinned, proto-Dravidians who spoke a precursor of Tamil, now an important language in southern India. With more than 100 settlements spread over a 1,000-mile (1,609 km) stretch of the Indus plain, the region also supported cities such as Harappa and Mohenjo-Daro (Figure 6-4).

When unearthed by archaeologists in the 1920s, the ruins of Harappa and Mohenjo-Daro revealed a phenomenal level of sophistication. Both cities conformed to a general model of ancient city planning found elsewhere. Constructed of baked brick, each walled city had a fortified citadel and a large granary. Oriented to the cardinal directions, streets were grid-patterned. Courtyards shielded two-story, elite dwellings from the smaller quarters of common laborers.

Amazingly, these cities had garbage collection, an elaborate system of running water, and sewage disposal. Elite homes had baths and "toilets," water channels to carry waste to underground sewers and septic tanks. Water was of the essence to the prosperity of these cities and apparently took on ritual/religious significance. The "Great Bath" at Mohenjo-Daro might have been used for ritual ablutions, suggesting the existence of a powerful priesthood. To this day, Hindu temples have rectangular water tanks, and ritual washing is important for purification purposes.

Religious cults abounded, but the embryos of current beliefs were present. The Bo tree (a type of banyan), under which the Buddha attained enlightenment centuries later, was sacred then as now. There were solar symbols, such as the wheel of light rays called *suasti*, "well-being" in Sanskrit. In the twentieth century, this was corrupted and misused by the German Nazis as a Swastika, their emblem of Aryan purity. The *suasti* remains a popular symbol in both Buddhism and Hinduism.

Achievements in urban development imply a sizable agricultural surplus and a well-organized, authoritarian control structure. Harappa alone supported around 40,000 people. Wheat, barley, and rice, along with domesticated sheep, goats, and fowl, plus game provided food surpluses that allowed people to indulge in other pursuits. Some settlements were specialized in certain activities such as mining or pottery. Also, there was trade by land and sea with Mesopotamia. Indian crews navigated the Persian Gulf with the assistance of a "compass-bird"—a crow that would fly directly to the nearest point of land. There is evidence that Indic cotton was a valued trade item. To this day, cotton is India's most important export.

But the Harappans could not control all of nature's forces. Mohenjo-Daro was rebuilt at least ten times and apparently was abandoned in panic during a severe flood. Recent research shows that despite their mastery of water control, it was climatically-changed water balances that were instrumental in the fluctuating fortunes of the Harappan civilization. Steep decline took place from 1800 to 1700 BC, corresponding with a decrease in rainfall. Water problems and food shortages combined with other crises to drive the Dravidians southward.

INVADERS FROM THE NORTHWEST

The Aryans

Around 1500 BC, lighter-skinned, nomadic herders called **Aryans** appeared out of Central Asia and Afghanistan. The Sanskrit-speaking Aryans thrust through the Khyber and other passes of the Hindu Kush onto the plains of the northwest, where another wetter period nurtured pasturage suitable for their animals. It is important to note that while Aryans would have far-reaching influences in South Asian history, from the start they were largely incorporated into existing cultures. In other words, India was not "Aryanized." Rather, it became even more diverse and complex.

The Aryans brought cows and horses to India. In fact, cows were an early form of wealth and currency. Eventually these would become sacred in Hinduism. Apparently, horses were employed to pull chariots—the basis of cavalry warfare, which spread quickly throughout the continent. Horses made possible the expansion of empires, fostered the militarization of states, and enabled more expedient collection of taxes in kind. As horse-breeding in India was difficult, they were brought from Arabia and Persia at great expense.

While copper and bronze were already in use, the discovery of iron near Patna, combined with Aryan knowledge of iron-working, was instrumental in territorial conquests. Used for weapons, tools, and plows, iron facilitated both forest removal and agricultural settlement.

Settlement along the Ganges had been limited to the Doab and the northern foothills and tributary valley, possibly because rivers were more easily crossed there and there were dense jungles in the lowlands. Eventually another dry period narrowed the rivers and the jungle receded. Iron implements furthered settlement on the Ganges plain, and forest dwellers were either conquered or absorbed. In the *Mahabarata,* these non-Aryan forest dwellers were called *nagas* or "serpent people."

Unlike the Harappans, the Aryans did not have a script for their spoken word. For centuries, the *Ramayana* and other stories were memorized and passed on verbally. A script was not developed until the fourth century BC, but religious taboos prevented the writing of religious tomes. Nevertheless, Sanskrit is rich in pejorative terms, and these were willingly used to denigrate non-Aryan peoples. Many were described as physical freaks, and these images became part of medieval European perceptions of the East.

Seeds of Caste Division

The Aryans, whose own name means "noble" or "first born," used the term *dasa* for people with darker complexions. Initially, the term meant "enemy" and later, "subject." *Dasa* always implied dark-skinned people, and the preference for lighter skin persists. Even today, marriage advertisements often list "wheaten complexion" as a desirable characteristic.

Social norms including social hierarchies were clarified in the *Shastra* legal texts. These were the "Laws of Manu," Manu being the first man. Vedic societies became organized around four classes of people, or **varnas**, which means "color." The four classes were ***brahmins*** (priests); ***kshatriyas*** (warriors); ***vaishyas*** (peasant farmers); and ***sudras*** (serfs). Subjugated people, including the darker Dravidians, could only enter the system at the lowest level. Those who had closest contact with earth, fire, or water, such as artisans, smiths, and millers, were despised. The most skilled artisans were Dravidian. *Sudras* were excluded from ritual affairs because they were seen as impure, and ritual purity was of the essence. This was the precursor of the **caste system**, which still endures. The English word caste comes from the Portuguese word *casta*, used to designate people of different colors in Brazil.

By the sixth century BC, regional rivalries had produced a string of states across the northern plains. A sophisticated commercial sector evolved, with all the necessary trappings such as coinage, taxes, and bureaucrats. Prosperity fostered economic specialization and varna subdivisions, known as ***jatis***, multiplied.

A *jati* is typically defined and named by occupation. It is also an extended kinship group that practices group

endogamy. This practice of marriage within the *jati* ensures continuity of its principles and practices. Caste and *jati* are at once divisive and unifying. While fragmenting society at large, they provide structured mutual exchange and support systems at communal levels.

Within a growing mercantile class, the *varnas* had to be further modified. Commercial dealers were taken into the *vaishya* class and farmers were demoted to the *sudra* class. Serfs and other low-order *sudras* became casteless (i.e., outcasts whose designation ultimately hardened into **untouchables**). Much later in the twentieth century, the great Indian leader Mohandas (Mahatma) Gandhi decried the inequities of the caste system and said that the so-called "untouchables" should be called ***harijans*** meaning "Children of God." Today, many untouchables prefer the term ***dalits***, meaning the "oppressed."

After the establishment of Aryan supremacy in the Indus region and across the northern plains, an Indo-Aryan culture emerged. Large numbers of Dravidians were driven southward across the Deccan. There, they developed a very distinct civilization with their own languages. In fact, Dravidian languages are fundamentally different from Indo-Aryan languages. This difference has induced linguistic strife even into the twentieth century.

BUDDHA AND MAHAVIRA

Two major religious events occurred in the fourth and fifth centuries BC. First, **Buddha**, born Siddartha Gautama in the Himalayan foothills of what is now Nepal, preached the paths and precepts of Buddhism across northern India. Buddha's "Middle Way" was appealing as it lay between the extremes of worldliness and asceticism. He founded *sanghas* (communities for monks) and *viharas* (wet-season refuges). Buddha himself was not worshipped for another hundred years, and the first sculptures of him did not appear until the Christian era.

At first, his followers carved simple cells and halls out of mountain faces, and Indian stone carvers ultimately excelled in their craft. Many carvings and wall paintings still survive in the wake of Buddhism's diffusion throughout Asia. Later, lands were provided for monasteries, and a monastic code was devised. This code and the Buddhist scriptures were recorded in a script called *Pali*. Even today, ancient *Pali* prayer scrolls can be seen in Buddhist monasteries. After his death in 483 BC, Buddha's relics were distributed and encased in structures called *stupas* (Figure 6-5).

The second event was the founding of **Jainism** by **Mahavira**. Both Buddha and Mahavira were *kshatriyas*, both disdained social class distinctions, and both addressed the fears of never-ending rebirths. Mahavira was from the region of Bihar. He became a wandering ascetic at the age of 30. He preached in the Ganges valley until he died around 612 BC. In Jainism, the Buddhist injunction against killing became all inclusive: people, animals, insects, and even certain plants. Jainism was nonmaterialistic, but since at first it was only property holding that was forbidden, the religion became popular with the mercantile class. Cultivators and soldiers obviously failed to meet the nonviolence standards. Concentrated mainly in the state of Gujarat, Jains have managed to retain an unbroken tradition in India.

Figure 6-5

This impressive stupa is Swayambunath in Kathmandu, Nepal. Sacred relics are confined inside the stupa, from which the eyes of Buddha see all things. Devotees move around the stupa in a clockwise direction, chanting and twirling their prayer wheels on the way. Photograph courtesy of B. A. Weightman.

THE MAURYAN EMPIRE

Out of a period of uncertainty that followed an invasion and subsequent withdrawal by the Greeks arose a new leader named Chandragupta Maurya. By the end of the fourth century BC, he had acquired an empire stretching from the Hindu Kush to the Bay of Bengal. This was the first time the political gap between northern India and southern India had been bridged. Although not controlling the entire subcontinent, the Mauryas did control regions of the Deccan, important trade routes, and coastal regions. Great mercantile wealth was accumulated (Figure 6-6).

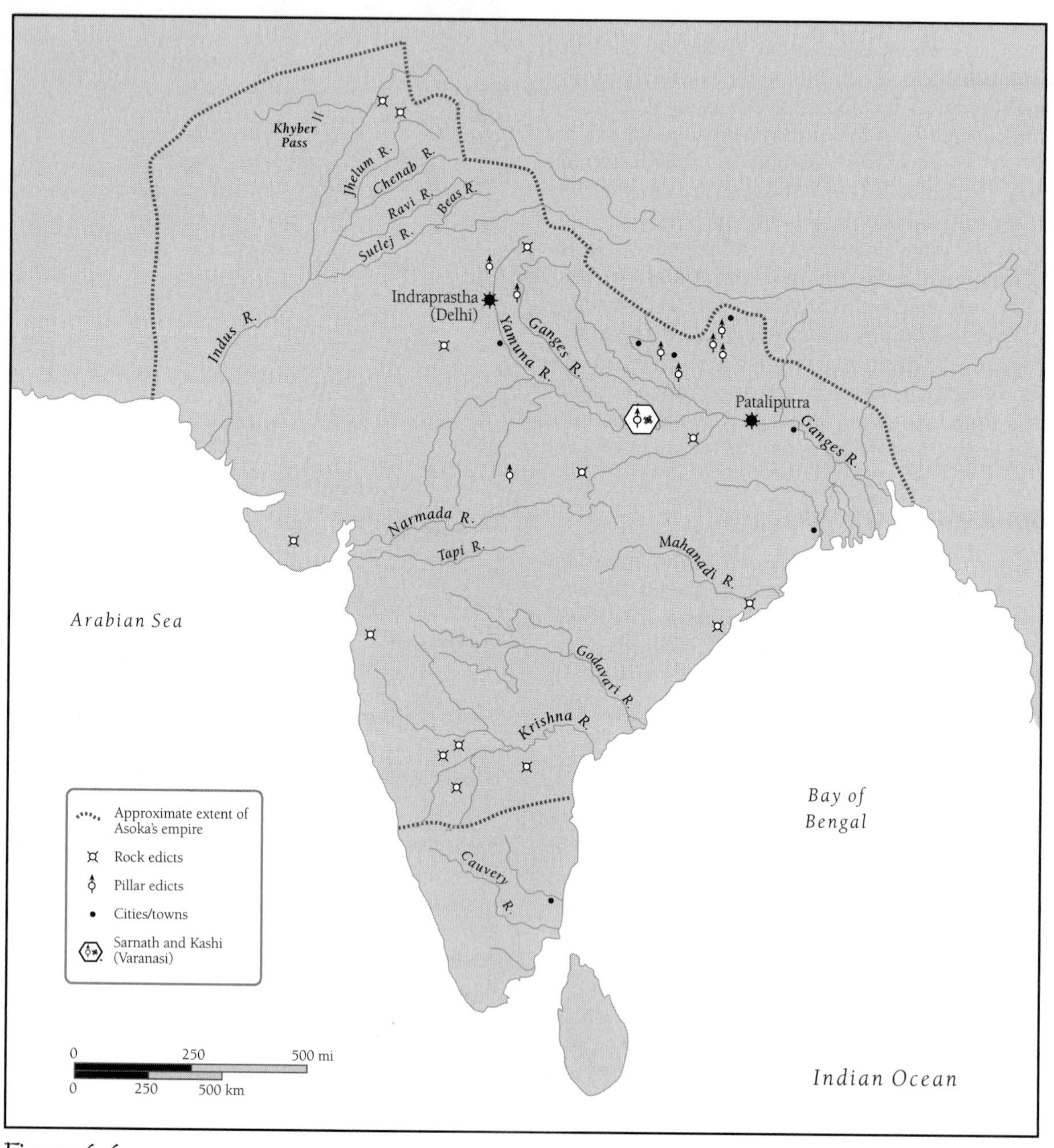

Figure 6-6

South Asia in the third century BC. The significance of the Indo-Gangetic Plain in Asoka's empire is unquestionable.

The Mauryan Empire is regarded as the first expression of the political notion of a state on the subcontinent. It was highly centralized, and many economic activities such as mining and land clearance were state monopolies. The fact that death was imposed for defrauding revenues or killing an elephant (a capital asset) suggests that justice was harsh.

The most memorable Mauryan ruler was **Asoka**, who ruled from about 268 to 233 BC, the first fairly well-documented period in Indian history. Asoka is remembered not only because of the size of his empire, but also because he adopted Buddhism as the state religion. He sent Buddhist missionaries, including his own son and daughter, to Sri Lanka. Asoka became a vegetarian. However, meat, including beef, was still available in the markets. The cow was not yet considered sacred. Asoka established hospitals, animal shelters, wells, and water tanks. His ruling principles and moral doctrines were inscribed on rocks and sandstone pillars around the country. The capital of the 40-foot (12 m) pillar of Samath, Gujarat, is a magnificent carving of four back-to-back lions. This official seal of Asoka was later adopted as India's national emblem.

THE GUPTAS AND INDIA'S GOLDEN AGE

When Asoka died, political conditions became chaotic. New Aryan migrants such as the Kushans arrived from the northwest. The Kushans put their capital at Peshawar, gateway to the Khyber Pass. Around 320 AD, northern India became united under the **Guptas**, under whom art and Sanskrit literature flowered. Derived from ancient sources, the Puranas were shaped and polished by the Guptas. These valued historic documents include Hindu myths, philosophical dialogues, ritual prescriptions, and genealogies. Art, science, and mathematics flourished, as well. The Gupta period marked **India's Golden Age** (Figure 6-7).

Under the Guptas, great strides were taken in mathematics. Numerals had been carved on some of Asoka's pillars, and these were further developed with the concepts of zero and nine numbers. The Arabs, who introduced them to the West, called them *Indisa* (Indian numbers). Europeans called them Arabic numerals. Aryabhata was the first to solve some basic problems of astronomy when he calculated pi to be 3.1416 and the length of the solar year as 365.358605 days. He knew that the Earth was round and that it rotated on its axis. He knew that an eclipse was caused by the moon's shadow blacking out the Earth. And all this was determined in India at least 1,500 years ago!

Fa-hsien, the first of three great Chinese Buddhist pilgrims to visit India, journeyed from China to India and Sri Lanka (399–412). He recorded that although the Gupta kings were Hindus, Buddhism flourished in the countryside. He commented on the well-being of the people and the fairness of the justice system. "Even those who plot treason only have their right hands cut off." He also noted that when certain people appeared, pieces of wood were struck to ensure avoidance of their evil presence. Fa-hsien had witnessed the caste system in action.

India's Middle Ages

North India was the focal point of Indian history until the beginning of the sixth century AD. Then, regional kingdoms became dominant, only to be reabsorbed after being toppled by more powerful ones. During this period, central and southern India rose into prominence, and events occurred that would alter the geography of the entire Asian region. You will learn about Indian exploits in Southeast Asia in later chapters.

The Chinese pilgrim **Hsuan-tsang** spent 13 years in India (630–643) and reported to the domain of **King Harsha**, whose empire spread from the Punjab to Orissa and as far south as the Narmada River. Harsha was the last north Indian ruler to encroach south to the Narmada for 600 years. When Hsuan-tsang returned to China, he brought 20 horses carrying 657 Buddhist texts and 150 relics. He personally translated 74 of these texts, furthering the spread of Buddhism in China.

Harsha brought attention to influential western tribes such as the Rajputs of Rajasthan, who had settled around Jodpur. As *kshatriyas*, the Rajputs were highly militarized and exalted chivalry (Figure 6-8). They were also noted for their patronage of the arts, including the construction of the famous temples of Khajuraho.

The Rajputs honored and protected women and even educated some. At the same time they promoted child marriage and female infanticide. Being married to a military chief was a dangerous affair because wives were expected to die with their husbands on or off the battlefield. *Sati,* being burned alive on a husband's funeral pyre, was enforced. A low-class wife had the option of having her head shaved by a male untouchable, eating one meal a day, and enduring the degradation of widowhood. These practices diffused throughout India.

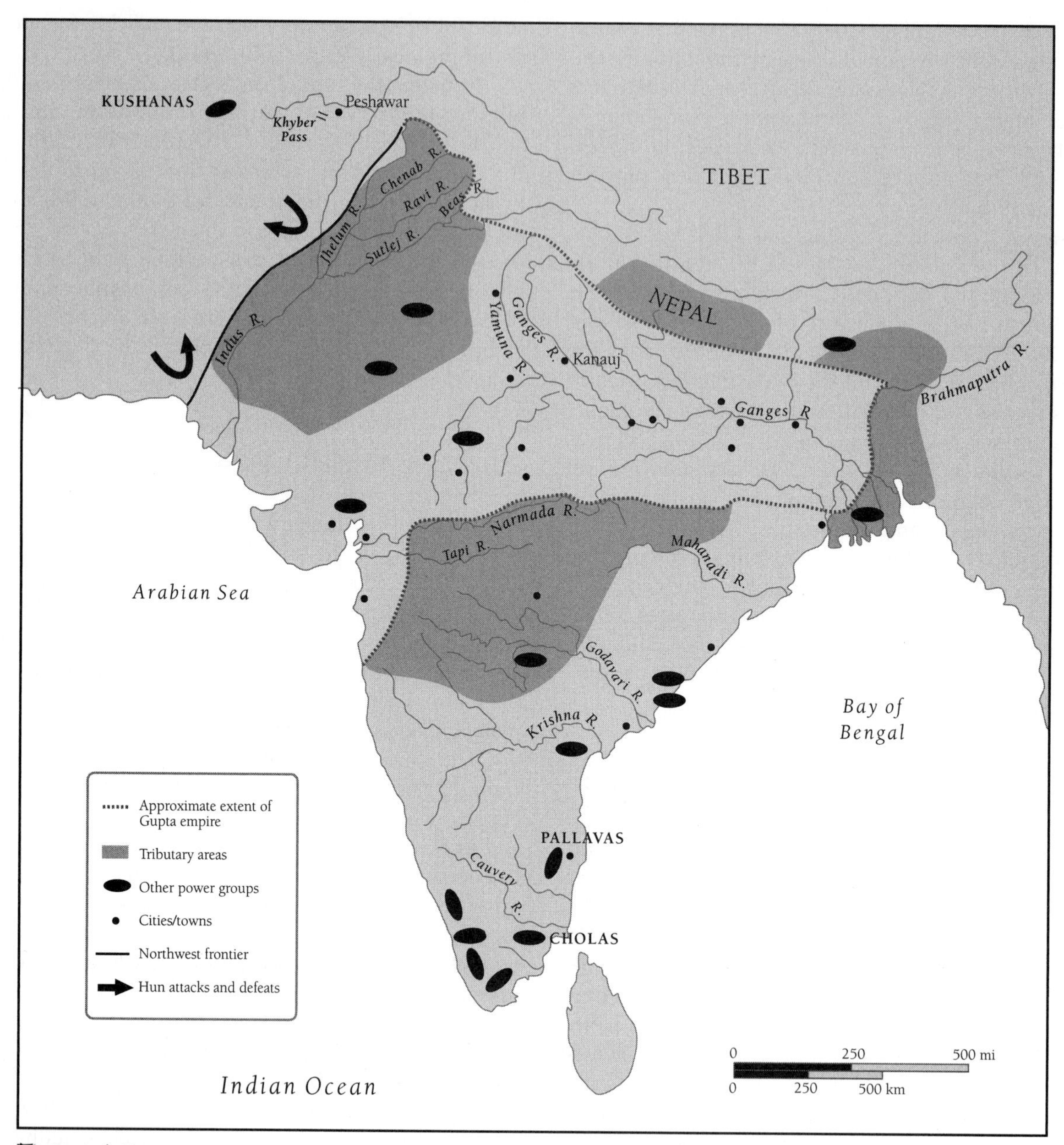

Figure 6-7

South Asia, ca 550 AD. While the Indo-Gangetic Plain remained important in the north, river and coastal regions fostered several powerful communities in the south.

Funeral flames dispel suspicion, honour lives when woman dies! The Ramayana

Women: Subjugation and Sati

Born into a male-dominated world, girls, if they survived the common practice of female infanticide, were worthy only as breeders of sons. No matter their caste, women were linked philosophically with *Sudras*, the lowest caste group. Morally and intellectually, they were seen as weak. Chastity, loyalty, submission, and subservience were the watchwords of a woman's life. The essence of a woman's being was her relationship with her

Figure 6-8

In Jaipur, India, these Rajputs are sword-makers. This occupation has been passed down through the family for generations. Photograph courtesy of B. A. Weightman

husband and his family. Without a husband, there was no point to her existence.

Child marriages were common. In the last census of the nineteenth century, there were 10,000 widows under the age of four, and over 50,000 between the ages of five and nine in and around Calcutta alone. In marriage, a girl "took away" wealth from her own family and was "gifted away" to the groom's household, where she was expected to be selfless and sacrificial (literally). A properly dutiful wife was expected to climb willingly onto the funeral pyre of her husband and burn alive to purge the sins of the entire family. A pregnant woman was expected to burn on a separate pyre after giving birth. With this act of courage and devotion, a wife would earn the title of *sati* and be venerated and honored forever. "Suttee" is the English version of *sati*.

Sati comes from the word *sat*, meaning "truth." A *sati* was a woman "true to her ideals." Since a woman's only value was to serve her husband, his death eliminated justification for her living. Those who got away with widowhood suffered countless indignities and lived the remainder of their lives in depredation and disgrace. Remarriage was forbidden to higher castes.

The origins of *sati* are unclear, although references to such an act are made in some of the post-Vedic scriptures such as the *Mahabarata*. Increasingly prevalent after 400 AD, actual dictates appeared around 700. By that time the life of a woman apart from her husband was declared sinful.

Sati soon became a "religious duty" for *kshatriyas* like princes and warriors, but soon spread to other castes. Although a wife was supposed to self-immolate voluntarily, most had to be forced onto the funeral pyre. In situations where there was a harem, all the wives would burn, commit mass suicide, or be killed. Some were buried alive.

Initially, *brahmin* women were forbidden from seeking this sort of salvation. A *brahmin sati* would be guilty of suicide and would therefore disgrace herself and her husband. However, eventually they too adopted the practice.

In the eighth century, Muslim rulers attempted to check *sati* but were unsuccessful. Islam had already introduced the customs of veiling and ***purdah***—the complete seclusion of women. By the 1200s, Muslim male and female social lives were completely separate. While the Hindu Laws of Manu instructed women to worship their husbands as gods, the Koran sanctioned similar behavior, and added the practice of polygamy. Hindu and Muslim customs regarding women became intertwined, especially in the north.

The British, who came to India in 1608, were able to do nothing about *sati* for 200 years. From 1815 to 1828, there were close to 8,000 (reported) cases in Bengal alone. The British finally banned the practice in 1829. However, the practice continued in most regions, including its stronghold in Rajputana (Rajasthan). Wife-suicide or killing was not unique to India. It was practiced in China, Egypt, Russia, Africa, and elsewhere.

The caste system was well entrenched in this era. Caste taboos meant that leather-workers, butchers, and scavengers were confined to mud quarters beyond the city wall. Moreover, they were compelled to walk on only the left side of the road.

Southern India was not entirely isolated from activities in the north. Cultural influences diffused from the northwest southward via the western highlands. Many Sanskrit words were incorporated into Dravidian languages. Contact with the Mauryas brought new ideas of statehood. Brahmin families were instrumental in the spread of Hinduism. Translation of sacred texts and orthodox practices strengthened the hierarchical structuring of society. Oppression of low castes and untouchables was frequently intense.

INVASIONS AND EMPIRES

Eighty years after the death of Muhammad in 632, an Arab empire extended from the Pyrenees in Spain to the Indus valley. The Arabs initiated their territorial gains in India when they established themselves in Sind at the mouth of the Indus in 712. Gujarati rulers tried to fend off conquest, in part by welcoming Zoroastrians who fled Persia and formed India's Parsi community, now concentrated in Mumbai (Bombay). Parsis, for whom fire is the essence of purity, would later make significant contributions in finance, trade, industry, and the arts. It is thought that the Persian water wheel was introduced at this time (Figure 6-9). The Arabs also introduced paper and gunpowder. By 871, they had established independent dynasties in the Sind and the Punjab and in 1000, a holocaust began.

Figure 6-9

This animal-powered Persian water wheel in Rajasthan, India, is evidence of Persian influences. Photograph courtesy of B. A. Weightman.

Mahmud (971–1030) of Ghazni (in Afghanistan), son of a Turkish slave, plundered India with 17 invasions over the next quarter century. Called "The Sword of Islam," Mahmud and his armies swarmed through the Khyber Pass during the dry monsoon. After devastating everything in their path, they sped back to Afghanistan before the wet monsoon could make the Punjab rivers impossible to ford. In 1023, they sacked the temple at Sarnath and slaughtered 50,000 Hindus. Interestingly, Mahmud was never interested in acquiring territory. For him, India was merely a treasure trove.

Why were these invasions so successful? First, fragmentation in terms of social and political divisions precluded any united front against them. The Cholas were far from the fray. The Rajputs with their independent group structure were never able to coordinate their efforts, despite their strong fighting abilities. The caste system was another hindrance to correlative action. In contrast, anyone could join Muslim armies. With ability and loyalty, even slaves could rise to top positions. Second, Indian armies fought with relatively cumbersome elephants. Muslim horse-archers could easily outrun them. With a rapid charge, they would suddenly swerve their horses sideways, release a hail of arrows, and race away.

Did you know that the game of chess was probably derived from elephant warfare?

Elephants and Chess

Around 500 BC, the elephant arrived on the Indian battlefield. Elephant warfare strategy is faithfully represented in the game of chess, probably invented in India in the sixth and seventh century AD. The king directs the battle from atop his elephant. His moves are restricted to protect him because if he is killed, the army is vanquished. Generals, cavalry, and runners conduct the battle. Infantrymen (pawns), primitively armed, slow, and untrained, are valued only for their numbers. Flanks are guarded by elephants, who take front-line positions in the last stages of the battle.

The game diffused via Persia to Europe. Chess is thought to be derived from shah, the Persian word for king. Shah mat approximates "checkmate," meaning "the king is dead."

Elephant warfare remained in place for 2,000 years, reaching its pinnacle under the Mughuls in the sixteenth century. By that time, firearms and artillery, brought from Europe by the Turks, had been introduced into India. The combination of raining arrows from mobile horse-archers, bullets from musket-men sharpshooters, and explosive barrages from ranks of artillery brought the era of elephant warfare to a close.

THE DELHI SULTANATES

The Arab conquerors inaugurated Delhi as their capital in 1229 and for nearly 90 years struggled to maintain control of northern India. The second sultan, Raziyyat, lasted only for three years. A chronicler of the time described her as being a wise ruler and competent military leader. "But of what use were these qualities to her as fate had denied her the favour of being born as a man?" Raziyyat was ultimately deposed and executed.

After defeating the Mongols in 1279, the notoriously cruel Balban had thousands of Mongol prisoners trampled to death by elephants as entertainment for the Delhi court. Afterward, the victims' heads were piled as a pyramid outside the city gate. Balban had copied the standard Mongol style of revenge.

In 1309, Balban's son, Alauddin, invaded southern India. The loot, supposedly including the famous Koh-i-noor diamond, was hauled back by 1,000 camels. Another profitable campaign required 612 elephants to carry plundered treasure.

During Alauddin's rule, society became even more partitioned along religious lines. Cities and commerce were predominantly Muslim. Entire castes of artisans converted to Islam to escape their low-caste status. By and large, farmers were Hindus, represented by their village headmen. They could not accumulate property as a consequence of Alauddin's belief that "the Hindus will never become submissive and obedient till they are reduced to poverty."

During the tenure of the Delhi sultanates and thereafter, many Arab Muslims married Indian women. The ruling class accumulated harems of inordinate size. Furthermore, they promoted the practice of *purdah*. In many countryside villages in northern India, even Hindu women adopted this practice of veiling and remaining secluded from nonfamily males and strangers in general.

Other changes occurred as well. The Delhi court favored the sophistications and fineries of Persian culture. Persian was the official language, influencing Sanskrit and maintaining its presence even in some parts of the south into the nineteenth century.

Persianization

For an 800-year period, Persian culture dominated elite culture in north and northwestern India as well as in other Muslim-controlled areas. The Persian connection was an ancient one. However, the process of Persianization was initiated with the Turkish invasions of the eleventh century and furthered by subsequent immigration of large numbers of Persian soldiers, administrators, scholars, and others into India. Urdu, cultivated with Islamic orientations and written in a Persianized, Arabic script, evolved as the national language of Indian Muslims.

Even the Mughals, the last Indian court of empire, retained Persian dress, literary, artistic, and architectural styles, and the Persian language. Without the later introduction of English by the British, Persian could well have remained the language of political and cultural elites in South Asia.

Two significant events took place in the sixteenth century. First, the Portuguese appeared in the south, where they destroyed Hindu temples at Goa and Madras (Chennai). Goa fell to Portuguese control in 1510. Second, the **Mughals** rose to power and ushered in India's modern era.

India's medieval period ended with the arrival of the Mughals, who introduced the modern territorial state. Until this time, India was more a collection of structures than of territories (Figure 6-10). Cities, towns, villages, and temples; rulers, soldiers, and priests; merchants and farmers; all were ensconced in not always well-coordinated networks. There was a large degree of local autonomy. All this was to change in the era of Mughal authority and British hegemony.

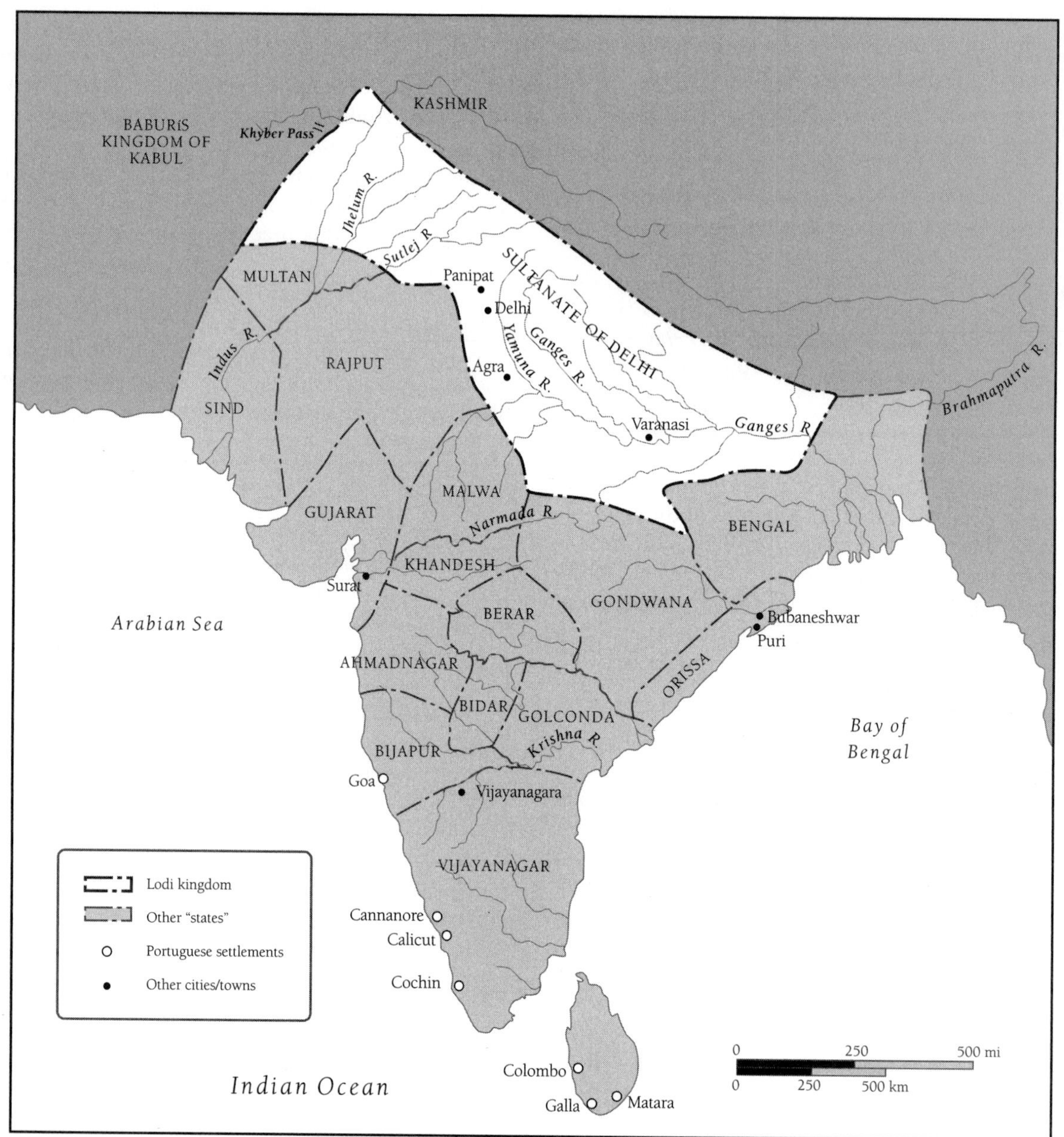

Figure 6-10

South Asia, ca 1525. While the north was unified under the Delhi sultanate, the west and south were fragmented among various rulers. The Portuguese were the first Europeans to set up trading posts, most of which were on the Arabian Sea.

India's Modern Era

In 1526, on the traditional battlefield of Panipat, 50 miles (80 km) from Delhi, thousands of elephants, horsemen, and footmen of the Sultan of Delhi, Ibrahim Lodi, amassed to face the impending terror, Babur (1483–1530). In only a few hours Lodi was dead, his massive army crushed by an onslaught of artillery, a cavalry attack from the rear, and elephants run amok. Babur's mobile artillery overwhelmed army after army, including the Rajputs. After besieging Delhi and other fortified sites, he swept victoriously into Bengal. Babur "The Tiger," direct descendent of Timur and Chinghis (Genghis) Khan, thus became the first Great Mughal. "*Mughal*" is derived from the Persian word for Mongol.

THE MUGHAL EMPIRE

Babur, while establishing order, treated the Indian people as subjects rather than prey and forbade his soldiers to maraud them. He and his son and successor, Humayun, initiated functional divisions in administration. Humayun died prematurely in a fall, leaving his young son Akbar to rule. Akbar (1542–1605) quickly consolidated his empire in further military campaigns, defeating the Rajputs and conquering Gujarat and Bengal (Figure 6-11).

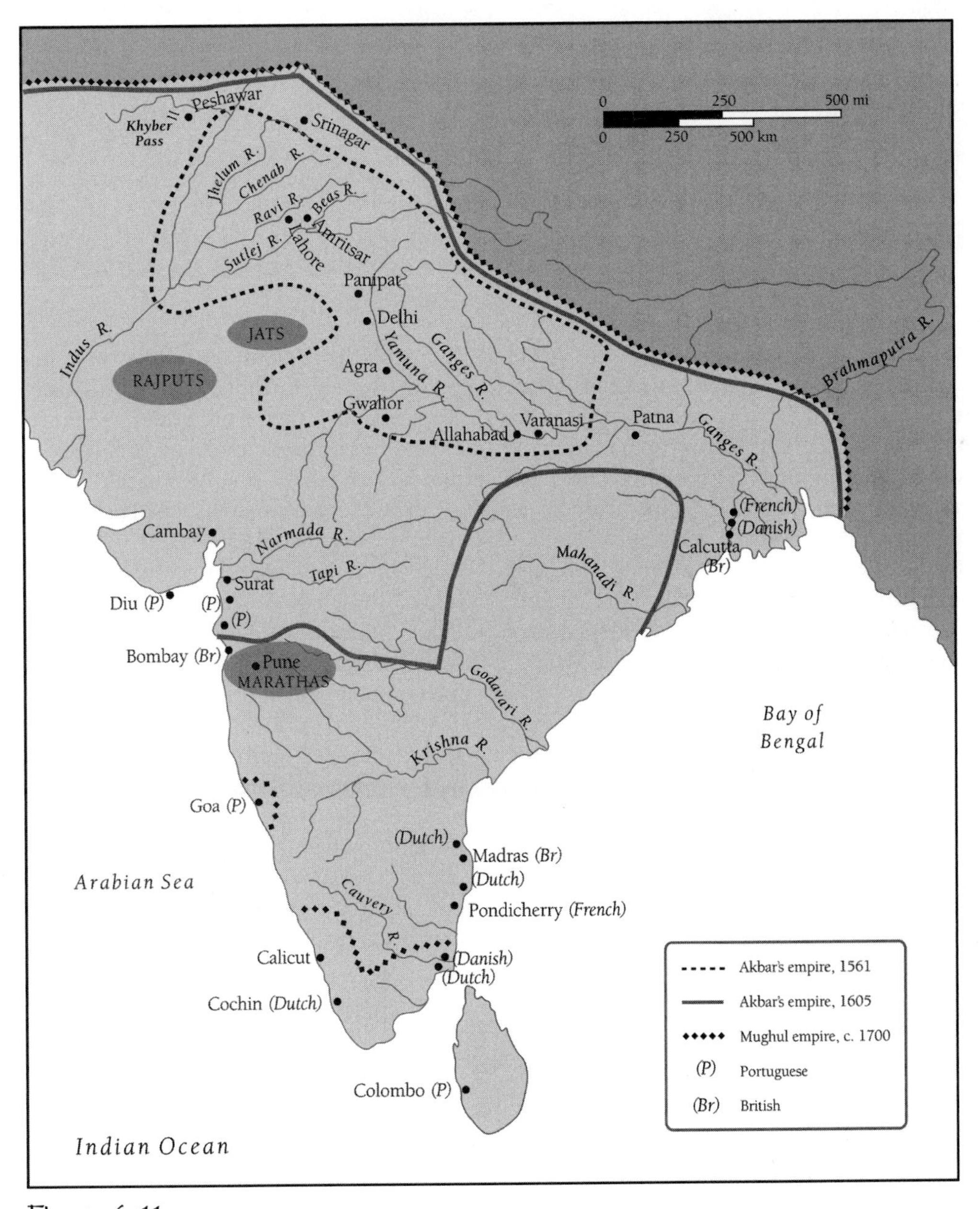

Figure 6-11

South Asia, ca 1700. Again the focus is the Indo-Gangetic Plain, from which the Mughal Empire expanded southward. More Europeans including the British, Dutch, French, and Danes are on the scene. Note the concentration of activity near Bombay (Mumbai).

Figure 6-12

This is the magnificent Taj Mahal in Agra, India. It was built from 1632 to 1653 by the Mughal Emperor Shah Jahan as a tomb for his wife Mumtaz, who died bearing her fourteenth child. Thousands of workers and designers were brought from Europe and Asia to erect this monument of marble and semi-precious stone. As the costly edifice progressed, one of India's worst famines ravished the countryside. Photograph courtesy of B. A. Weightman.

Akbar was poisoned by his son, Jahangir the "World Seizer." Jahangir's beautiful Persian wife, Nur Jahan (Light of the World), became very powerful and was instrumental in furthering Persian culture and building in northern India. Politically astute, she arranged the marriage of her niece, Mumtaz, to the next truly great Mughal: Shah Jahan (Ruler of the World).

Shah Jahan (ruled 1627–1657) extended Mughal sway into the south. Setting up his capital on the Yamuna at Shahjahanabad (Shah Jahan's city), he developed sites of architectural magnificence including Jama Masjid, India's largest mosque, and the Red Fort (in today's Old Delhi). Living in sumptuous splendor, Shah Jahan had 5,000 women in his harem. Nevertheless, when his beloved Mumtaz died, he had what has become India's most famous landmark—the Taj Mahal—erected in her honor (Figure 6-12).

For a while, the empire enjoyed relative peace and prosperity. Vassals of earlier regimes, tribal leaders, petty rulers, and the like were absorbed into the larger regime, making it increasingly cumbersome. Recognition as *zamindars* (landlords) bought their loyalty but encouraged peasant exploitation. Trade fostered urban growth supported by agrarian masses. Mughal culture was essentially urban and increasingly differentiated from the majority rural Hindu population. The Urdu language expanded, with contributions from Persian, favored by the elite; from Arabic, employed by religious scholars; and from Hindi, spoken by the general populace. Linguistic differences elsewhere exacerbated other divisions.

After the reign of Akbar, interest in science and technology waned in India. Court brilliance, architectural display, and military prowess took precedence. Lavish expenditure, constant warfare, and court intrigue peaked under Shah Jahan's son, Aurangzeb.

The Mughal empire reached its greatest extent under Aurangzeb (ruled 1658–1707). But Aurangzeb's cruel, anti-Hindu policies and uncontrolled expansionism promoted his empire's downfall. Non-Muslims—Hindus, Jains, Buddhists, Sikhs—were brutalized mercilessly. Non-Islamic structures were razed and conversions forced upon pain of torture and death.

Tax burdens in general became increasingly oppressive to finance court excesses and military ventures. Newly conquered areas of the Deccan yielded proportionately less revenue than the fertile northern plains. Revenue collection was stepped up and special taxes placed on Hindus. Discontent was rife, but when protesters gathered outside the Red Fort, Aurangzeb ordered them crushed by the imperial elephants.

THE SIKHS

Sikhism emerged in the fifteenth century. Its core area was the Punjab, a region of incessant Hindu-Muslim conflict. Its founder, **Guru Nanak** (1469–1538), taught that "Man will be saved by his works alone . . . There are no Hindus and no Muslims. All are children of God." He opposed all forms of discrimination, including caste. His teachings, compiled by his disciple and successor Guru Angad, became the Sikh holy book, the ***Adi Granth***.

Guru Angad met with Akbar, who gave him a temple site at Amritsar near Lahore.

In the sixteenth century, some Sikhs became involved in political intrigue, and their leader was executed by the Mughul ruler Jahangir. Subsequently, the Sikhs began to take on a militaristic character and by 1650, claimed the Punjab as Sikh territory. Persecutions by Aurangzeb intensified their militarism. Under Guru Govind Singh, they vowed not to cut their hair, which is still worn knotted under a turban. They swore to avoid tobacco and alcohol and renounced *purdah* and *sati*. Sikh women were accorded almost equal status with men. Sikhs also adopted the surname of Singh (Lion). There is a saying that "All Sikhs are Singhs but not all Singhs are Sikhs."

Sikhism grew as a force to be reckoned with, eventually becoming the last bastion of resistance to the British takeover in India. Only in 1849, after two British campaigns, were the Sikh armies defeated. Eventually Sikh regiments were incorporated into the British-Indian army. See the next chapter for more on the Sikhs and Punjab.

In order to alleviate the friction of distance that made central control difficult, Aurangzeb moved his capital to Aurangabad. Even so, encumbered with an army of hundreds of thousands of elephants, guns, artillery, cavalry, and at least half a million camp followers (a moving camp 30 miles (48 km) in length), Aurangzeb was no match for the guerrilla-style tactics of Shivaji, "The Mountain Rat."

Shivaji (1627–1680), the zealous Hindu leader of the Marathas, had erected a complex of fortifications at defensible sites in the fractured western Deccan. Headquartered at Poona (Pune), behind Bombay (Mumbai), Shivaji was able to lash out, then quickly retreat to his mountain strongholds. The Marathas turned out to be the Mughal Empire's most formidable opponent. In 1644, Shivaji sacked the main Mughal port at Surat. In ensuing confrontations, Aurangzeb the Muslim and Shivaji the Hindu stood as protagonists of their respective faiths, deepening the division between believers.

As disenchantment with the Mughals grew, common gripes and shared concerns bonded individuals in competing groups. The Sikhs—bound by faith, the Jats—farmers bound by kinship, and the Marathas—bound by nationalistic aspirations, challenged the Mughal hegemony. Many revolts were led by the *Zamindars*. Mughal retaliation was unforgiving. The situation worsened in 1702–1703 when famine and plague eradicated more than two million people from the Deccan.

The empire disintegrated in chaotic circumstances, with the strongest and cleverest manipulators and power-grabbers staking out their territories. And so arose the powerful **maharajas** (great rulers) of states such as Baroda, Gwalior, and Hyderabad. During this fray, the Persians invaded and sacked Delhi in 1739, carting off the jewel-encrusted Peacock Throne and other treasures. As Delhi declined, cities such as Lucknow and Hyderabad became centers of Mughal culture.

EUROPEAN SEA POWER

When Babur raged into India, the Portuguese already controlled the Indian Ocean with a monopoly on the pepper trade. Their fortified outposts at Goa and elsewhere served as customs stations where Asian merchants had to obtain letters of protection in order to avoid Portuguese attacks on the high sea. The land-based Mughals were interested in international trade insofar as it was a source of silver and gold. They remained unconcerned about the intricacies of its functioning as long as the precious metals kept coming.

With the Portuguese monopoly, European competitors were welcomed by the Indians, who could play them off against each other. In London, the British East India Company was founded in 1600. Two years later, the Dutch East India Company was founded in Amsterdam. Indian textiles were desired products, and factories (storehouses) were erected in the foreign concessions. However, as the textile trade boomed, factories became involved in ordering, directing, and financing the industry. Soon they became instruments of foreign control.

By 1740, the British had eclipsed the Dutch, who then focused their attention on Indonesia. Starting with a collection of trading posts, the British fought the Mughals, forged alliances with local rulers, and eventually ousted their European competitors. By exploiting local rivalries, jealousies, Hindu-Muslim and other tensions, they were able to expand their influence. Three key commercial centers emerged.

On the Coromandel coast, Fort St. George developed as Madras (Chennai). Madras soon flourished as an entrepot, an intermediary center of trade and transshipment.

With the dissolution of the Mughal empire, many merchant ships moved to Bombay. Its natural harbor was safely protected by the British fleet. As the empire further dissolved, Indian shipping networks floundered and interregional trade with Bengal and the Malabar Coast dissipated. The British in Bombay had their sights on trade with Europe.

At a swampy, unhealthy, deep-water anchorage 80 miles (129 km) up the Hooghly River, Britain's Fort William expanded to become Calcutta, the empire's capital until 1931. British Bengal trade boomed in the eighteenth century, and further inroads were made into the interior. There, factory agents established contact with the spinners and weavers, and British artisans came to train them in the ways of British fashion. However, little investment was made in the means of production, and the workers remained in poverty. Farmers were forced to grow industrial, cash crops in order to meet their tax obligations. This caused food shortages, increased indebtedness, added millions to the ranks of landless and unemployed, and enriched the *zamindari* land owners, tax collectors, and money lenders. Indian industry declined, and India became an importer of European manufactures and an exporter of raw materials such as raw silk, cotton, indigo, sugar, and opium. The British shipped the opium to China to satisfy their imperialist ambitions there.

By 1750, distrust of the British was widespread, and Calcutta had grown to 100,000 people. Even then, British racism marred the landscape. The city was divided into "Blacktown" and "Whitetown." Attempts to get rid of the British were immediately thwarted. With Colonel Robert Clive's connived victory over a vastly larger force at the Battle of Plassey in 1757, they achieved control over Bengal. British sea power was transformed into British land power.

For the next hundred years, the British struggled to expand their territorial control and suppress opposition on the subcontinent. Repeated wars were fought with various groups such as the Marathas and the Sikhs. In 1819, the British gained control of Rajasthan and became the largest land power on the subcontinent. By 1846, they had taken the Sikh capital at Lahore, the fertile lands of the Punjab, and the exquisite mountain vale of Kashmir. Then, in what seemed an insignificant move at the time, the British sold Kashmir to the highest bidder, a Hindu chieftain, Gulab Singh. By 1850, the British controlled an empire extending from the Indus to Bengal and from Kashmir to Ceylon (Figure 6-13).

Colonial transformation was unstoppable. As the landed aristocracy saw their lands appropriated, Western influences were instilling new ideas about development and missionaries were challenging Hindu beliefs and practices, especially the caste system. English replaced Persian as the official language of government and education. According to historian Lord Macaulay (1800–1859), "A single shelf of a good European library is worth the whole native literature of India and Arabia." Macaulay's ideas led the way in developing India. The British were to bring up a "class of persons Indian in blood and colour, but English in taste, in opinions, in morals, and in intellect." Schools and colleges were built to do the job.

As governor-general of India, Lord Dalhousie (1848–1856) launched an ambitious program of modernization. Expanding on the previous work of British engineers, he constructed the first 450 miles (724 km) of the Ganges Canal, which today provides hydroelectric power for a large region and irrigates two million acres of farmland. Under the Dalhousie administration, electric and telegraph services were installed, an efficient postal system developed, and the first 300 miles (483 km) of railroad track laid. While it is true that India ended up with Asia's best rail system, there were negative consequences.

On the one hand, railways permitted the relief of famine with grain shipments, while on the other hand they created famine in areas from which the grain was sent. Grain shortages drove prices beyond the reach of the impoverished masses, and millions starved. For example, 10 million in Bengal and Bihar starved in 1769–1770. At least 23 famines took their toll between the eighteenth and twentieth centuries. The last severe food crisis was in 1943, when 1.5 to 3 million died as a result of grain shortages, usurious prices, profiteering, and government negligence and incompetence.

Railroads also opened up India as a vast consumer market. However, instead of stimulating interior development, they favored coastal cities through new cotton milling and shipping facilities. Local textile mills closed. Towns and cities bypassed by the rail line declined. By 1879, three-quarters of India's textile mills were in and around Bombay.

In the minds of Indians, infrastructural change did not have the impact of social change. The British had outlawed the practice of *sati* in 1829. In 1856, another law permitted the remarriage of widows, anathema to Hindus. By British law, Hindu states could not be inherited by any other than a blood heir. Any "heirless" properties reverted to the Crown. Some states, even those loyal to the British, were annexed under the guise of mismanagement. Such attacks on Indian culture and Hinduism itself were cause for alarm, and in 1857 there was a mutiny.

The army of the East India Company comprised mainly British-trained Indian troops called **sepoys**. Intrigue sparked rumors that cartridges, which had to be bitten off to load the new Lee-Enfield rifles, were greased with pig and cow fat. With the pig being unclean to Muslims and the cow

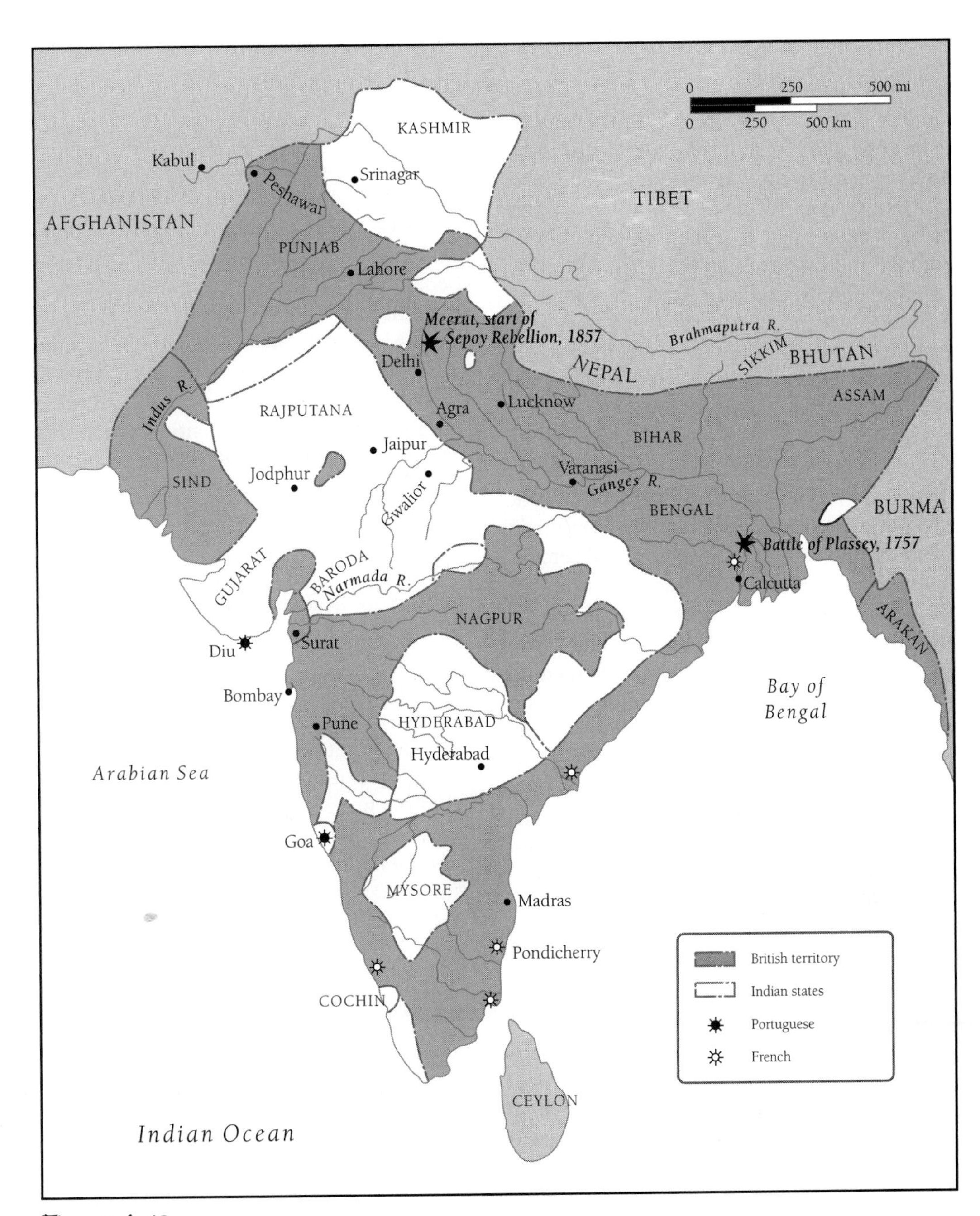

Figure 6-13

South Asia, 1856. By this time Britain had control over most of the region. Other colonial possessions were reduced to small coastal colonies.

sacred to Hindus, a mutiny erupted. The Sepoy Rebellion of 1857 triggered nationwide violence. Whole communities of Europeans or Indians were slaughtered, and cities were reduced to rubble. Some states such as Hyderabad, Baroda, Rajasthan, and Punjab remained loyal to the British and the rebellion was ultimately crushed.

In 1857, the British government, horrified at the extent of atrocities, abolished the East India Company and formally took over India as a colony, a status it would retain for nearly a century. The era known as the **Raj**, meaning the "rule," had begun.

THE RAJ

Under British government control, social reforms were abandoned and attention was given to maintaining the status quo. The army was reorganized under British officers. The British legal system was introduced, and the Indian Civil Service was formed. This initiated the cumbersome bureaucracy that, like the road and railway system, would tie the country together. In all of this, Indians were deliberately excluded from positions of power. The country was measured, cataloged, mapped, organized, and centralized. The Grand Trunk Road stretched more than a thousand miles between Calcutta in Bengal and Peshawar near the Afghan border.

In 1877, Queen Victoria was declared Empress of India, and her representative in India became the viceroy. New treaties offering varying degrees of autonomy were made with the various princely states. There were more than 500 of these native states, occupying nearly 40 percent of the landmass. Larger states such Hyderabad, Punjab, Bengal, and Assam had governors or commissioners who reported directly to the viceroy. Smaller states had advisors. But these political alliances with their incumbent privileges contributed to even greater division of this already deeply divided region.

The British avoided contact with their subjects. New towns, suburbs, and military stations were built across India. In cities, military cantonments and civilian districts known as **civil lines** were spatially and socially separate entities, complete with their own accommodations and services. These British mini-worlds in the lowlands had their counterparts in the highlands: **hill stations** (Figure 6-14).

Hill Stations

Hill stations are mountain-amenity landscapes offering opportunities for rest and recreation at relatively cool elevations of 5,000 to 7,000 feet (1,500–2,200 m). They existed throughout the colonial world and remain important centers of commerce and tourism in many parts of Asia.

In India, hill stations were of great significance, and not just as social, economic, and political centers for Europeans. They symbolized the paramountcy of the Raj and revealed prevailing attitudes of nineteenth-century Europe: racism based on beliefs grounded in environmental determinism.

At that time, it was believed that white people could not withstand the rigors of torrid, foreign lowlands for extended periods. The combined onslaught of heat, dust, and "natives" would surely damage European sensibilities and possibly cause offspring to be less than normal. It was essential that Europeans be restored in more familiar, temperate climes. As one Englishman remarked, "like meat, we keep better there."

Hill stations have been described as "island(s) of British atmosphere hung above the Indian plains," and "comforting little piece(s) of England." At the highest location, a protective nest of European architecture, trees, plants, and parks implanted British ideals of restorative havens,

Figure 6-14

Clinging to the mountainside is the hill station of Simla, founded by the British in 1819. Simla served as the summer capital from 1865 to 1939. It was much cooler than either Calcutta or Delhi. All government materials were hauled by train and caravan from Calcutta (Delhi after 1912) each year. Photograph courtesy of B. A. Weightman.

salubrious and picturesque. In contrast, the "eyesores and nuisance" of the Indian quarters and bazaars on the slopes below were discreetly hidden behind "attractive" buildings and hedges.

Hill stations thrived as social centers. They glittered with banquets, pageants, and balls and catered to European tastes with afternoon teas, activities at the Club, and the frivolities of games and sport. Indian servants graced only the background, ears cocked for their next command. In later years, educated Indians were permitted to join the fun but remained fundamentally disliked by the British.

The British perceived India's climate to have three seasons: the cold, the hot, and "the rains." They adjusted their schedules to suit regional variations. In the north, for example, the cold lasted from October to March. During that time woolens and warming fires were needed, and British ladies remained on the plains with their husbands. By April or May, the plains were roasting and **amenity-migration** was made to the cooler hill stations, where thick-walled bungalows reduced temperatures even further. These small, one-story houses and the name **bungalow** diffused from India throughout the Western world.

Some hill stations became so popular that they were used as summer capitals by government functionaries. For instance, the governments of Bombay repaired to Poona (Pune), Madras to Ootacamund known by the British as "Ooty," and Bengal to Darjeeling. In 1864, the Viceroy of India proclaimed Simla in the Himalaya as the empire's summer capital, and the entire government and all its files and documents were moved there from Calcutta each year from 1865 to 1939.

Hill stations did not boom until the opening of the Suez Canal in 1869 and railroad construction in the colonies. A narrow-gauge rail line carried its first passengers to Simla in 1903. Prior to that time, people and all their trappings were literally carried or carted up the mountain track by their Indian servants. As late as 1932, nearly a third of Simla's summer residents were porters or rickshaw pullers, most working to pay debts at home.

Increased population stimulated commerce. In 1830, Simla had only 30 houses. By 1881, there were over a thousand. The opening of the railway saw the population spiral to more than 30,000 by 1898. Market gardening contributed to increased commerce between hills and plains, and the town's commercial sector grew rapidly.

Indian areas frequently were targeted for improvements, which meant replacing their "unhealthy" warrens with environments suitable for European pursuits. Indians were forced by displacement to live in even more crowded conditions.

Simla, as the Raj capital for at least eight months a year, hosted salient political events. Here in 1913–1914, the British, Chinese, and Tibetans met to fix their common boundaries. It was in Simla that the Indian National Congress was conceived and shaped and that the Muslim League had its beginnings. Here both organizations participated in discussions leading to India's independence and partition in 1947. And here, too, Prime Minister Indira Gandhi granted statehood to Himachal Pradesh in 1971. Simla, with its viceroy's mansion and British architecture crowning the teeming bazaars below, is now a popular destination for Indian upper- and middle-class tourists and is still a popular site for national and international negotiations.

India had been Britain's major trading partner since 1840. With the opening of the Suez Canal in 1869 and the expansion of railways, the route to Britain was significantly shorter and India was drawn increasingly into the international economy. British capital was invested in rail works and coffee and tea estates (plantations). India exported these products, along with raw cotton, wheat, and jute (fiber). They imported machinery and textile manufactures (made in England from Indian cotton). Eventually an industrial base was established.

India's industry rose and fell with British policies. Her economy was used to make up trading deficits with other industrialized nations. Once-renowned manufactures such as cotton, brocades, and silks were eliminated. However, raw material–based industries such as textile mills and jute, cement, and sugar factories were eventually built. Old trading or religious centers evolved as industrial centers, attracting landless peasants as a cheap labor supply. By 1900, Calcutta was awash with slum dwellers and squatters. British India's most famous author, Rudyard Kipling, described it as the "city of Dreadful Night."

Indians were left to their own resources to develop other types of industry. J. N. Tata (1839–1904) was one of several Indian industrialists. Tata, for instance,

Figure 6-15

This Victorian-Mughal hotel was built by the great Indian industrialist Tata. Opening in 1903, it had granite staircases, electric fans, and 400 beds. It was the only luxury hotel in Bombay where Indians were permitted to stay. Photograph courtesy of B. A. Weightman.

founded the country's steel industry, with the Tata Iron and Steel plant opening in Jamshedpur, Bihar, in 1907. The Tata family continues to reign as India's largest industrial dynasty.

In order to show their superiority, power, and dominance, the British constructed imposing edifices. Grand architecture of what became known as the Victorian-Mughal style housed train stations, post offices, and other important buildings. Bombay's Taj Mahal hotel was one of the grandest of all (Figure 6-15).

In 1911, Queen Victoria's grandson, King (Emperor) George V, and Queen (Empress) Mary arrived in Bombay, the first British royal visit in history. At a key moment in the spectacular durbar (assembly) in Delhi, the monarchs announced from Shah Jahan's Red Fort that the imperial capital was to be transferred from Calcutta to Delhi, where a new city would be constructed.

Designed by British architects and finally inaugurated in 1931, New Delhi was designed with broad, straight avenues, well-suited for parades; prominent Victorian-Mughal government buildings; stately palaces for the viceroy and other notables; and spacious bungalows for the elite. European in plan, yet Indian in style, New Delhi was a contradictory counterpart to the traditional maze of Old Delhi (Figure 6-16).

THE ROAD TO INDEPENDENCE

The notion of self-rule had its roots in the nineteenth century, when there was a renewed interest in Indian history, culture, and religion. European scholars pursued translations of ancient works, and these were avidly read by growing numbers of educated Indians. At the same time, Indian religious scholars had reinvigorated Hinduism. Some even suggested modifications more appropriate for a modern society such as the abolition of caste.

The Indian National Congress, formed in 1865, provided a forum for nationalist sentiments, and became the main instrument for the independence movement. The **Congress Party** continues to be one of the most influential in India. Expressions of desire for greater autonomy were disdained or ignored by the British.

When the state of Bengal was halved into Hindu-dominated and Muslim-dominated parts in 1905, without consultation with any Indian leaders, Bengal erupted with huge demonstrations. Picketers were beaten and leaders were whisked away without trial. Although four years later the division of Bengal was revoked, nationalistic mind-sets had hardened and moved underground, from where a campaign of terror against the British was directed.

During the same period, Muslims were actively looking out for their own interests. The Muslim League was created in 1906. Although Muslims made up more than 20 percent of the population, they were scattered and not well represented in positions of power. Certainly the Congress Party was run by Brahmin Hindus. In 1916 it was agreed that Muslims and Hindus should have separate elections and representation.

World War I (1914–1918) was a significant event in that 1.2 million Indian troops served the Empire by fighting on the side of the Allied Forces. In the aftermath, British colonies such as Australia and Canada were granted significant control over their own affairs.

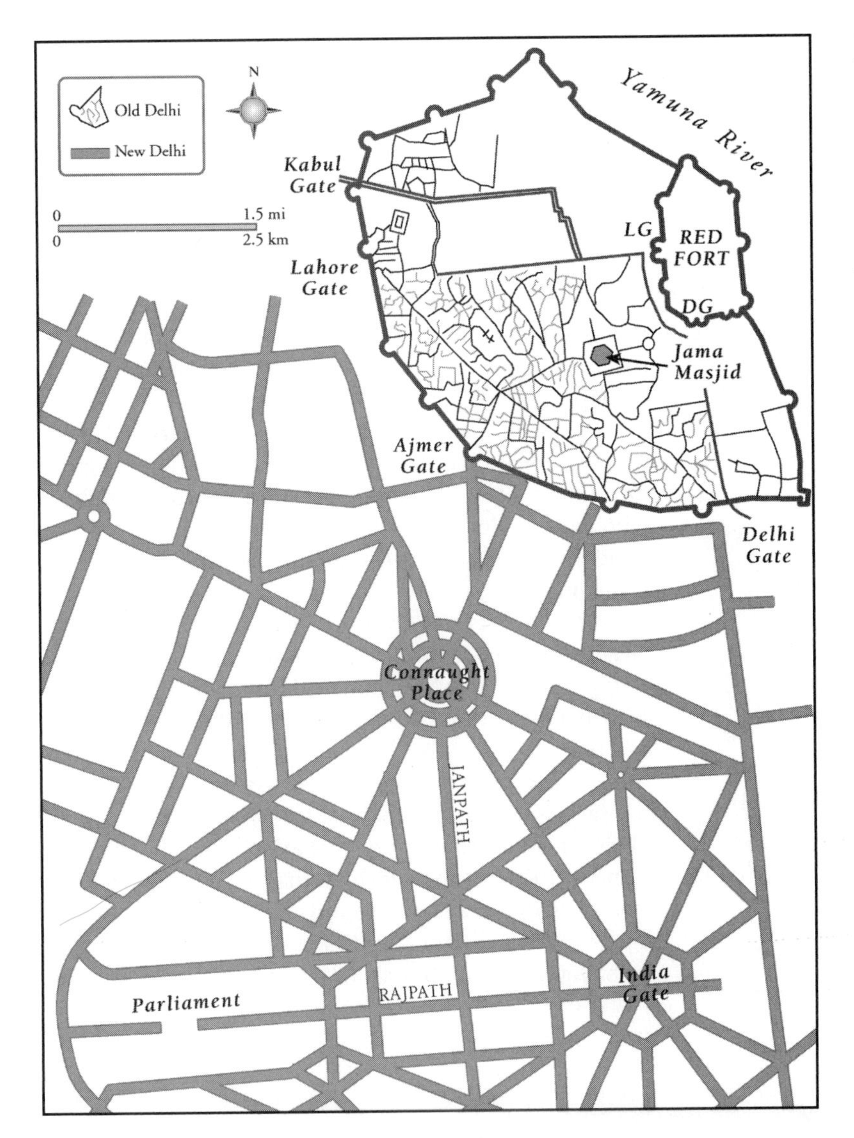

Figure 6-16

Old and New Delhi. Fascination with geometry and perspective, concern for order and accessibility, and desire for grand vistas and impressive layouts led to New Delhi's design. This plan stands in stark contrast to the narrow streets and winding passageways of the fortified old city. Whereas New Delhi was planned, Old Delhi evolved naturally.

There were those who thought that India would be rewarded in a similar manner. It was not—but imperialist ways would not prevail. Anything short of independence was unacceptable. After all, the Chinese Revolution of 1911 and the Russian Revolution of 1918 demonstrated that popular uprisings could overthrow despotic regimes.

British opposition reached its lowest point in 1919 with a massacre of unarmed Indian demonstrators at Amritsar in the Punjab. The toll was 400 dead and 1,000 injured. The British Raj lost the loyalty and support of millions. Mohandas Gandhi (1869–1948), considered a "saintly fanatic" by the viceroy, was profoundly affected by this event.

Gandhi: Great Soul

Did you know that Martin Luther King, who led the civil rights movement for African Americans in the United States, was strongly influenced by the philosophy of Gandhi?

Mohandas Gandhi (1869–1948), born in Gujarat, experienced discrimination as an Indian practicing law in South Africa. There, on behalf of the Indian community, he developed the strategy of nonviolent protest against injustice. Upon his return to India in

1914, he abandoned his European dress and adopted the lifestyle of a *saddhu*. Traveling far and wide, he used traditional Hindu values such as ahimsa to spread the concept of peaceful, mass protest against British oppression (Figure 6-17). He employed traditional symbols such as the spinning wheel to rebuild an Indian sense of nationalism.

Jawaharlal Nehru, in his autobiography (1942), said that Gandhi was "like a beam of light that pierced the darkness." Moreover, rather than direct from the top down, "He seemed to emerge from the millions in India . . . and the essence of his teaching was fearlessness and truth." Gandhi's notion of nonviolence was not simply the absence of violence; compassion and responsibility for the welfare of others were vital components of the concept.

Gandhi spoke out against Hindu-Muslim animosities as well as violence against women. He pointed to male chauvinism as being responsible for the genesis of *sati*. He abhorred the inequities of the caste system and as we noted earlier, he said that untouchables should be called *harijans*: children of God. Beloved by the masses and many political leaders, he was given the title Mahatma, meaning Great Soul. Mahatma Gandhi was instrumental in bringing about India's independence from Britain. He is often referred to as the "Father of Independent India."

Gandhi, who came to lead the Congress movement, launched three mass campaigns of civil disobedience in 1920, 1930, and 1942. These marches, sit-downs, boycotts, and blockades resulted in retaliatory violence and imprisonment for thousands. But the British did make concessions. In 1932, they began recruiting Indian military officers. In 1935, the Government of India Act and a new constitution put the power of 11 states in the hands of Indian representatives. Indians moved into higher ranks of the civil service. In sports, science, and literature, Indians began receiving world recognition—Olympic gold medals and Nobel Prizes.

In 1929, Gandhi nominated **Jawaharlal Nehru** (1889–1964) to succeed him as leader of the Congress movement. The dynamic, Western-educated Nehru, who could eloquently express his empathy for the Hindu masses, would eventually become India's first prime minister.

The only serious rival to Congress was the Muslim League, founded in the Bengali city of Dhaka. It was led by European-educated **Muhammed Ali Jinnah** (1876–1949), who believed that Muslims and Hindus had evolved as "two nations." As Nehru's Hindus gained power, thousands of Muslims joined the League. With Nehru's refusal to recognize Jinnah as the spokesperson for all Muslims, an attempt at power sharing failed. Jinnah then turned to convincing Muslims that a Congress election victory would be a Hindu victory.

World War II (1939–1945) was the final catalyst in bringing about India's independence from Britain.

Figure 6-17

This Delhi sign shows Gandhi promoting vegetarianism. Note that the sign mentions ahimsa*—the sanctity of all forms of life. Note also the use of both Hindi and English. Photograph courtesy of B. A. Weightman.*

As Britain had declared war on Germany without consultation, the Indian provincial representatives resigned. Being called upon to "Quit India," in 1942, the British arrested the entire Congress leadership and 60,000 activists. Then they invited Jinnah into the government. With Congress power diminished, Jinnah was able to put the Muslim League on an equal footing. The British had agreed to Indian independence at the end of the war, but now it was too late. Muhammed Ali Jinnah demanded a separate Muslim state: Pakistan "Land of the Pure." Unable to speak Urdu, he remarked in English, "If not a divided India, then a destroyed India."

Hindu-Muslim violence was widespread, and on August 16, 1946, the "Great Calcutta Killing" event cost the lives of 5,000 Muslims and Hindus. Bloodshed swept across northern India, with violence and atrocities sparing no community. Faced with the prospect of complete anarchy, Nehru and other Hindu leaders became convinced that a Muslim state was indeed necessary.

A new viceroy, Lord Louis Mountbatten, was sent to India in 1946 to negotiate an independence agreement. He quickly saw that the division of India into Muslim and Hindu entities was inevitable. Pakistan was to encompass heavy Muslim concentrations in the northwest and northeast, meaning that the new nation would consist of two separate entities over 860 miles (1,400 km) apart. Moreover, to give Muhammed Ali Jinnah his Pakistan, both Punjab and Bengal would have to be divided. The consequences were terrible.

THE PARTITION OF INDIA

As both Punjab and Bengal had mixed populations, no boundary line would be satisfactory to all. Punjab, in particular, was communally mixed with Hindus, Muslims, and Sikhs interspersed. The decision was made to award 62 percent of the land and 55 percent of the population to Pakistan. As of midnight January 14, 1947, millions found themselves trapped in a country hostile to their faith.

The partition of India in 1947 spurred the largest migration in human history. Around 12 million people moved between West and East Pakistan and India seeking refuge in their own religious communities. The ensuing savagery was unparalleled. Entire villages were massacred, and trainloads of people were butchered and torched. At least a million people were killed in the wake of India's partition.

These events created Hindu-Muslim animosities deeper than ever. A horrified Gandhi took up residence in the Muslim quarter of Old Delhi and vowed to fast to death if the communal violence did not cease. Although calm returned several days later, many Hindus were enraged by his seeming support for the Muslims. On January 30, 1948, Gandhi was assassinated by a Hindu fanatic. A distraught Prime Minister Nehru spoke to his nation: "The light has gone out of our lives and there is darkness everywhere."

In that same year, Muhammed Ali Jinnah, Pakistan's first political leader, died as well. Spurning Muslim prohibitions against pork, alcohol, and smoking, he died of lung cancer.

The nearly 600 princely states were diverse geographically, socially, and politically, precluding any uniform policy regarding their status. With the creation of India and Pakistan, each state was given the option of joining either country. Most acquiesced to India but some, such as Hyderabad, had to be coerced by the Indian army. Kashmir presented a special case, its complexities defying easy solution.

The Maharaja of Kashmir was a Hindu, but the majority of his subjects were Muslim. While he wavered on his decision, hoping to remain independent, Pakistan tribesmen rebelled. The Maharaja then acceded to India and asked for military help. Indian troops were unable to oust the Pakistanis. With stalemate in 1949, a "line of control" became the de facto boundary. This resolution was not a solution, and Kashmir remains a region of conflict to this day.

Nationalism and Regionalism

Complex, diverse, and paradoxical—this aptly describes the cultural mosaic of South Asia today. With ancient roots and an amalgam of influences from within and without, empires have risen and fallen, leaving indelible legacies on the landscape. Despite continual religious, linguistic, and political conflict and restructuring, India and the newer nations of South Asia remain intact. Nevertheless, the interminable dynamics of nationalism and factionalism continue.

The idea of the territorial state was introduced by the Mughals, who developed an ordered bureaucracy. This merged into the British civil service, an all-encompassing, multileveled network of control and regional division. Yet, persistent national agitation did reap change. Constitutional reform ended with a British, parliamentary-style, democratic system. And the federalism that the British employed in their devolution of power was retained after independence. Highly centralized power of the Mughals and British fed nationalistic tendencies and was furthered by the partition of India. Yet federalism, which allows for regional differences and varying degrees of local autonomy, has been fundamental to the stamina and resilience of the Indian and other governments in the region. At the same time, it has facilitated regionalism, localism, and communalism. National and regional ambitions and inequities underlay countless examples of instability and strife marring the subcontinent into the twenty-first century.

Recommended Web Sites

www.bing.com/videos/search?q=mahatmagandhi&qpvt=mahatmatgandhi&FORM=Z7FD1
Excellent site with videos and photos of Gandhi, Nehru, Jinnah, and other leaders.

www.csuchico.edu/~cheinz/syllabi/asst001/spring98/india.htm
Explanation of *varnas* and *jatis* of the caste system. Photographs of different caste groups.

http://edwebproject.org/india/mughals.html
The Mughal legacy with links to biographies of various leaders such as Babur and Akbar.

www.iloveindia.com/history/index.html
Cultural and economic history. Information on various rulers and dynasties.

www.iloveindia.com/timeline-of-india.html
Historic time line for India.

www.indhistory.com/
Ancient, medieval, and modern history.

www.indianchild.com/caste_system_in_india.htm
Cultural aspects of intercaste relations such as food rules.

www.infoaboutsikhs.com/sikh_symbols.htm
History and culture of Sikhs. Explanations of special Sikh symbols and the Punjabi language.

www.mapsofindia.com/history/
History by topic; emperors; maps; information by state.

www.mohenjodaro.net/
Information and 103 indexed images of Mohenjo-daro over 30 years of research.

www.sacred-texts.com/hin/
Internet sacred text archive. Text of books and epic poems such as the Vedas, Ramayana, etc.

www.stockton.edu/-gilmorew/consorti/1aindia.htm
Excellent site on the history of South Asia. Maps, photos, videos, and short articles on different eras in addition to information on languages, music, etc.

http://storyofpakistan.com/
India and Pakistan prehistory to present. Articles on leaders. Videos.

www.unigroup.com/PTIC/body_html
Information on the partition to the Islamic Republic of Pakistan.

Bibliography Chapter 6: South Asia: Creating Dilemmas of Diversity

Allchin, F. R., ed. 1995. *The Archaeology of Early Historic South Asia: The Emergence of Cities and States*. Cambridge: Cambridge University Press.

Basham, A. L. 1968. *The Wonder That Was India*, 3rd ed. London: Macmillan.

Brown, J. 1977. *Gandhi and Civil Disobedience*. Cambridge: Cambridge University Press.

Choi, Dong Sull. 1997. Origins and Realities of Suttee in Hinduism. *Comparative Civilizations Review* No. 36: 38–53.

Dalrymple, William. 2007. *The Last Mughal: The Fall of a Dynasty, Delhi 1857*. Chicago: University of Chicago Press.

Dalton, Dennis. 1995. *Mahatma Gandhi, Nonviolent Power in Action*. New York: Columbia University Press.

Davies, C. Collin. 1963. *An Historical Atlas of the Indian Peninsula*. London: Oxford.

Dickason, David G. 1975. The Indian Hill Station. Geographical Record. *The Geographical Review* 65: 115–116.

Dirks, Nicholas. 2001. *Castes of Mind: Colonialism and the Making of Modern India*. Princeton, N.J.: Princeton University Press.

Dutt, Ashok K., and M. Margaret Geib. 1987. *Atlas of South Asia*. Boulder: Westview.

Farmer, B. H. 1993. *An Introduction to South Asia*. New York: Routledge.

Kenny, Judith T. 1995. Climate, Race, and Imperial Authority: The Symbolic Landscape of the British Hill Station in India. *Annals of the Association of American Geographers* 85: 694–714.

Kulke, Hermann, and Kietman Rothermund. 1990. *A History of India* (revised edition). New York: Routledge.

Lapierre, Dominique, and Larry Collins. 1975. *Freedom at Midnight*. New York: Simon and Schuster.

MacMillan, Margaret. 1988. *Women of the Raj*. New York: Thames and Hudson.

Murphey, Rhoads. 1996. *A History of Asia*. New York: Harper Collins.

Narasimhan, Sakuntala. 1990. *Sati: A Study of Widow Burning in India*. New York: Penguin.

Nehru, Jaharwalal. 1942. *An Autobiography*. London: Chatto and Windus.

Pandian, Jacob. 1995. *The Making of India and Indian Traditions*. Englewood Cliffs, N.J.: Prentice Hall.

Pubby, Vipin. 1996. *Shimla Then and Now* (2nd edition). New Delhi: Indus.

Schmidt, Car. 1995. *An Atlas and Survey of South Asian History*. Armonk, N.Y.: M.E. Sharpe.

Sen, Amartya. 2005. *The Argumentative Indian: Writings on Indian History, Culture, and Identity*. New York: Allen Lane, Farrar, Straus and Giroux.

Smith, B. 1995. *Classifying the Universe: The Ancient Indian Varna System and the Origin of Caste*. New York: Oxford.

Sopher, David. 1964. Man and Nature in India: Landscapes and Seasons. *Landscape* 13: 14–19.

Spear, Percival. 1963. *A Study of the Social Life of the English in Eighteenth Century India*. London: Thames and Hudson.

Watson, Francis. 1979. *A Concise History of India*. London: Thames and Hudson.

Wolpert, Stanley. 1991. *India*. Berkeley: University of California.

Wolpert, Stanley. 1984. *Jinnah of Pakistan*. London: Oxford.

Chapter 7

South Asia: Pakistan and the Himalayan States

"From your past emerges the present,

and from the present is born the future."

MUHAMMAD IQBAL

(1873–1938)

In this chapter, we will focus on the Indian subcontinent's northern regions: Pakistan, the disputed region of Kashmir, Nepal, and Bhutan (Figure 7-1). These mountain zones are both physically and culturally fractured. High altitudes and deserts have limited accessibility and foster cultural differentiation, communalism, and regionalism.

Within the formidable Himalaya and Karakoram are the tenuous and oft-disputed boundaries between Pakistan and Afghanistan, Pakistan and India, and Pakistan, India, and China. Political instability has marked the evolution of virtually every state in the region. We will start with Pakistan at the time of India's partition in 1947 and examine the factors leading to the transformation of East Pakistan into Bangladesh in 1971. Then, we will explore current conditions and problems across the region, beginning with modern Pakistan.

Nationalism and Regionalism: The Devolution of Pakistan

In 1948, British geographer O. H. K. Spate wrote that he saw the creation of East and West Pakistan as "the expression of a new economic nationalism that has inevitably taken into its hands the immensely powerful weapon of immemorial religious and social differentiation." In 1947, Pakistan had two "wings," the eastern wing being only one-sixth the size of the western wing yet holding eight million more people. This was only one of a litany of differences between the two parts of this new nation—centrifugal forces that would ultimately result in its political demise.

For the split nation of Pakistan, unity could not be maintained with diversity. Contrasts were too numerous, and conflicts were too intense. West Pakistan is dry; access to water and drought is of critical concern. East Pakistan is humid; flooding is a yearly event. Aside from differences in size and population, the west possessed 80 percent of the cultivable land, and its share per farmer was more than double that of the east. West Pakistan had the country's capital and only port at Karachi. East Pakistan had far less infrastructure and was much poorer economically.

Cultural differences are exemplified in linguistic differences (Figure 7-2). There are several languages spoken in the west. These are from different language families and possess numerous dialects. Bengali, a branch of Sanskrit, is spoken in the former eastern wing of the country. In 1948, the Karachi government declared Urdu the national language of Pakistan, enraging Bengalis and deepening the chasm between east and west. Linguistic division would prove problematic in the years ahead.

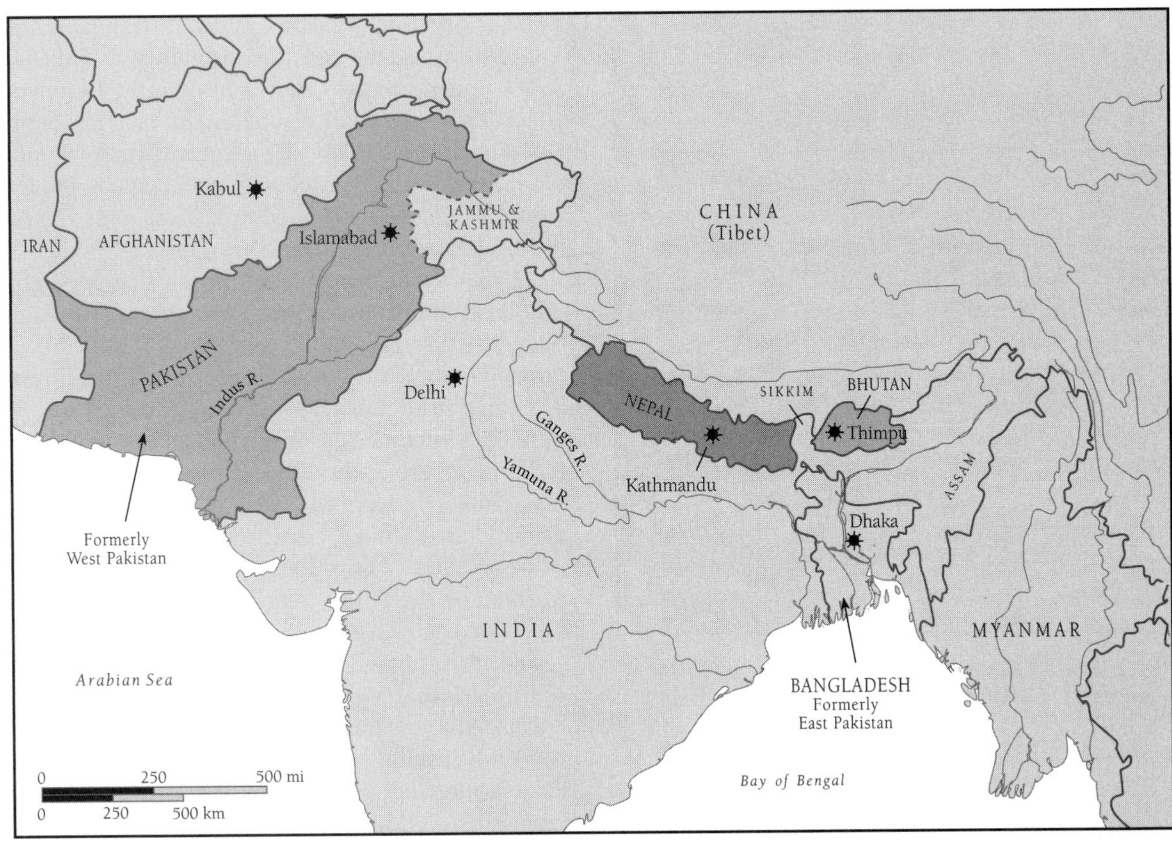

Figure 7-1

Northern South Asia. The political complexity of this region is compounded by its rugged physical geography.

Language and Nationalism

Pakistan (formerly West Pakistan) has reaped the consequences of lacking linguistic cohesion. About two-thirds of the population speaks Punjabi. Punjab is the most economically productive province and contains the nation's capital, Islamabad. Sindhi is spoken by about 12 percent, mostly in Pakistan's second industrial province of Sind. Urdu is spoken by only 8 percent, yet it was proclaimed the national language. Along with English, it is the language of the elite and government. Urdu-speakers are concentrated in cities, especially Karachi. Language riots broke out in Sind in the 1950s when university students were forbidden to write their essay questions in Sindhi. More riots occurred in 1965 when Urdu signs and radio programs replaced Sindhi ones in Hyderabad. In the 1970s, some of Karachi's Urdu-speaking Muhajirs, post-partition Muslim migrants from India, demanded equal representation in government even though they were a minority population. Others agitated for a separate state.

These incidents epitomize the sense of deprivation felt by Sindhi-speaking leaders and the emerging middle class. A Sindhi Language Authority was established in 1990. Language still fuels violence in Sind and elsewhere.

Several other languages are spoken, with the greatest diversity in mountainous regions. Pashto and Baluchi belong to the Iranian branch of the Indo-European language family. Sindhi, Urdu, and Punjabi belong to the Indic branch. Brahui is a Dravidian language. Balti is a Sino-Tibetan language.

Figure 7-2

Major languages of Pakistan.

> Since language is the context for thought and action, and conceptualizations in one language do not necessarily exist in another, differences can act as centrifugal or divisive forces in any centripetal drive for national cohesion.

Divisions sharpened with the economic and political policies pursued by the government in West Pakistan. There were arguments over exploitation of natural resources, industrial development, acquisition and allocation of monetary resources, communal and regional representation, and the role of Islam in government.

Although both parts of Pakistan were essentially Muslim, there were differences in practice. A history of communal strife marred the relationship between Muslims, Hindus, and Sikhs in the west. Moreover, some ethnic groups (e.g., the Pathans) were extremely fervent in their application of the mores and laws of Islam in their daily lives. For Bengalis, religious interaction with Hindus had become routine. They attended each other's festivals and reciprocated in temple offerings.

The question of Islam's role in government remains a critical aspect of political instability in Pakistan. In 1956, under pressure from conservative Islamists, the government declared Pakistan an Islamic Republic. Government was to operate according to the Quran and *sunna*, the words of Muhammed. Religious moderates, who had pushed for modernization and democracy, were sidelined.

Other problems centered on the rich-poor gap and other real and perceived inequities. After partition, many business-class Punjabis moved to Karachi's Sind Province, where they took over former Hindu businesses and came to dominate the economy. Added to the fray were the **Muhajirs**, Muslims who had previously lived in India. The urban-dwelling, Urdu-speaking Muhajirs, many of whom were educated and experienced in business and civil service, came to dominate administration. However, these Punjabi and Muhajir urban elites were resented by the Sindis. As the government struggled to establish some degree of equity under the weight of vested interests and expected patronage, the rich-poor and urban-rural gaps widened. Urban riots and peasant revolts led to the imposition of martial law in 1958.

With ensuing economic and political instability, East Pakistan looked to a new political party—the Awami League led by **Mujibur Rahman (Mujib)**. Mujib would fight for greater political autonomy for the east. The following years of hectic east-west relations served to strengthen the resolve of the Awami League.

In 1965, Pakistan went to war with India over border demarcation in the Rann of Kutch as well as in Kashmir. The year-long war was costly. Since both border disputes were with West Pakistan, East Pakistan felt unduly burdened. More disturbances again drew military rule, which lasted until 1970 when elections were to be held.

In Karachi, the Pakistan's People's Party (PPP), under the leadership of Zulfikar Ali Bhutto, ran on the slogan "Islam our Faith, Democracy our Polity, Socialism our Economy." Bhutto also declared a "thousand years" war with India.

Elections were delayed because of a crippling cyclone in East Pakistan. Then, entirely dissatisfied with West Pakistan's relief efforts, the Bengalis swept the Awami League to victory, a pivotal event because the proposed system of "one person one vote" would give the more populous east a majority in the National Assembly. As it became clear that Mujib rather than Bhutto might rule Pakistan, the National Assembly was "postponed" by the military.

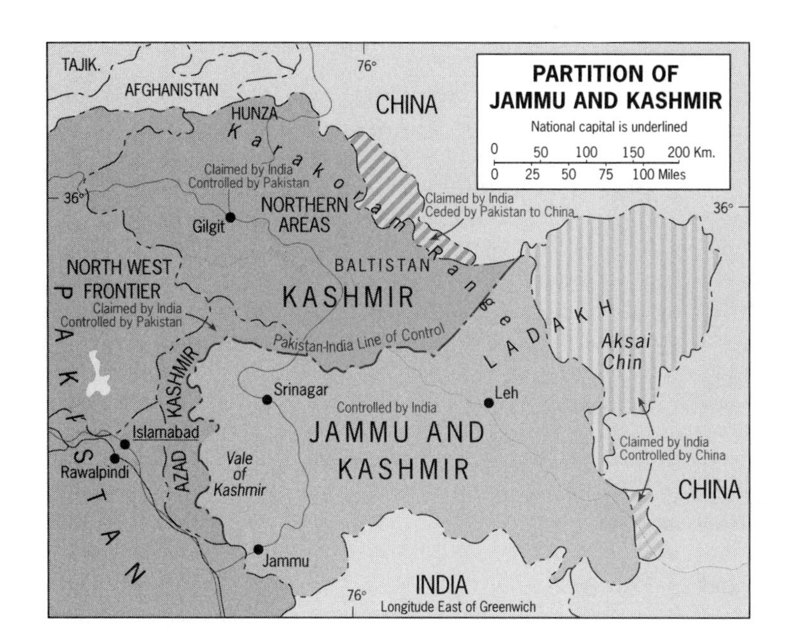

Figure 7-3

Geography of Pakistan and the Northern Areas including Kashmir. Note the relationship between settlement, croplands and rivers, especially the Indus.
From H. J. de Blij and P. O. Muller. *Geography: Realms, Regions and Concepts*, 14th Edition, 2010, p. 422.
Originally rendered in color H. J. de Blij and P. O. Muller.
Reprinted with permission of John Wiley & Sons, Inc.

Withholding power from Mujibur Rahman and his Awami League was a fatal error. East Pakistan erupted with strikes, demonstrations, and insurgencies. The Bengalis refused to pay taxes. West Pakistan countered with brutal military action—complete with rampant murder, pillage, and rape. Immediately, 250,000 refugees poured into India. Ultimately 10 million would seek refuge across the border.

Faced with hordes of penniless refugees in the already poor state of West Bengal, India seized the opportunity to retaliate against its arch enemy in Karachi. Indian intervention brought the east-west war to an end, and in 1971, East Pakistan became the independent nation of Bangladesh, with Mujib as its prime minister. Bhutto became prime minister of West Pakistan, which was renamed Pakistan.

Pakistan: Land of the Pure

DIVERSE LANDSCAPES

Pakistan of 1971 was politically divided into four highly differentiated and unequal parts: Northwest Frontier Province (NWFP), Punjab, Sind, and Baluchistan. Waziristan is not a province, but is part of the Federally Administered Tribal Areas (FATA) (Figure 7-3). To understand contemporary trends and problems we must look at the different characteristics of each region.

Northwest Frontier Province (NWFP)

The rugged hills of the NWFP are capped by a dramatic knot where the Karakorum, Pamir, and Hindu Kush have buckled together. Afghanistan's Wakhan Corridor, a 16,000- to 24,000-foot (4,880—7,315 m) buffer zone between Tajikistan (part of the former Soviet Union) and the former Indian Raj, lies in this region, extending to China in the east. Below glaciers and barren, rocky slopes are forests of pine, deodar, and evergreen oak. At lower and dryer elevations, these give way to acacia, dwarf palm, and coarse grasses. Wood and stone houses cling to steeply walled ravines, and finely constructed rock terraces support irrigated agriculture (Figures 7-4 and 7-5).

This dramatic and broken terrain is, in fact, a **shatter zone**, inhabited by a kaleidoscope of independent and feuding tribes. Many cross the border with Afghanistan (Figure 7-6). Most are Pathan, a general term for tribes such as the Afridis or Waziris. Most speak Pashto, an Iranian dialect. While some Pathans had accepted British rule and laws, others followed *pukhtunwali* (local Pathan law). Other groups such as the Chitralis are not Pathan.

Figure 7-4

This scene shows Mount Trich Mir, at 25,260 feet (7,578 m). I took the picture overlooking the Chitral River in northern Pakistan. The alluvial terraces are irrigated by gravity flow of water. Rice, fruit such as apricots, and vegetables are grown here. Photograph courtesy of B. A. Weightman.

Cutting through the mountains are the Kabul and Swat Rivers. These converge in the Vale of Peshawar, a lowland separated from the Indus River by a sandstone ridge. Under irrigation, the deep alluvial soils support wheat, corn, sugar beet, and tobacco (Figure 7-6). In Peshawar and other towns of the NWFP, Punjabis became prominent in business. Peshawar was also a budding industrial center.

Figure 7-5

Here in the northern town of Bahrain, Pakistani houses cling to the sides of the rocky slopes. The structures are made of wood and stone with no mortar. The building in the foreground is a grain grinding mill where water power turns the huge stone grinding wheels. Photograph courtesy of B. A. Weightman.

Waziristan

Waziristan, meaning "Land of the Wazirs," is a rugged, mountainous region bordering on Afghanistan. The region was independent until 1883 and was incorporated into Pakistan in 1947. Although it looks small on the map, Waziristan comprises 4,473 square miles (11,585 km^2). The area is divided into two agencies: North and South Waziristan. There are some 800,000 related tribes living here. Perceiving themselves as warriors, the Wazirs are noted for their love of fighting. Like all Pushtuns, they live by the code of *Pustunwali*. *Badal*—revenging blood feuds to protect *zan*—family, *zar*—treasure, and *zameen*—land are central pillars of this code.

The Wazirs, who have practiced *purdah* since at least 600 BC, also want to keep themselves pure—apart from the influences of outsiders. Consequently, they live in fortified, walled villages in the mountains. Since the 2001 invasion of Afghanistan, these villages have become havens for the Taliban. At the urging of the United States, the government of Pakistan has been shipping arms, reconnaissance planes, and troops into the region to root out militants—with limited success.

Figure 7-6

Here in Pakistan's Northwest Frontier, water buffalo pull carts of tobacco to the market at Mardan. The water buffalo are also suppliers of milk with a high fat content. Photograph courtesy of B. A. Weightman.

Punjab

In 1947, Pakistan's portion of the Punjab region (formerly India) held 60 percent of the population. It also dominated the armed forces and produced more than half of the GNP. Punjab long had been the breadbasket of the subcontinent. Here flow the Indus, Jhelum, Chenab, Sutlej, Beas, and other vital rivers. Out of the mountains and foothills, streams and their *doabs* (lands between the rivers) slope downward with masses of alluvium fanning onto the Indus plain. Wheat, cotton, rice, oil seeds, and sugar cane thrived under irrigation. This region has great potential for development.

Sind

Sind, more than any other region, is wedded to the Indus. To the north and west of Sukkur lie the Kirthar Mountains. Alluvial fans permit cropping, but this western valley is underlain with clay and therefore subject to water-logging and salinization—accumulation of salts. Northward, the land turns to clay desert.

Centuries of seasonal flooding below Sukkur have produced a *doab* between the Indus and the Nava Rivers. Each part of this eastern valley at some time has been traversed by an Indus stream. This fertile and productive region has invited water control works through the centuries. The *doab*, actually the ancient Indus delta, fades eastward into the Thar Desert (Figure 7-7).

Further south, small but significant limestone outcrops support the Sukkur water barrage and the city of Hyderabad. The delta of the Indus merges with the mud and salt flats of the Rann of Kutch. *Rann* means "salt marsh" in Hindi. This 10,000 square mile (30,000 km^2) region between the Gulf of Kutch and the mouth of the Indus, fills with water in the wet monsoon, when its marshes attract thousands of flamingoes and other exotic birds. In the dry season, it turns into a saline, clay desert. Since the river distributaries have long been in flux, the coast is dotted with dead ports. However, when Indus flooding converges with high, ocean tides, the coast floods inland for up to 20 miles (32 km). Mangrove and tamarisk thickets fringe old and new beach-ridges and bands of clay-like silts.

Sind epitomizes heat and aridity. Summer temperatures can be well over 100^0F (37.8^0C) at midnight! It rarely rains but when it does, it pours. Karachi, with an annual average rainfall of about 8 inches, has recorded 12 inches (305 mm) of rain in 24 hours.

Second to Punjab in size and wealth, Sind has great agricultural potential and in 1947 was in the process of being colonized mainly by immigrants from Punjab. However, much of rural Sind remains controlled by *zamindari*.. Originally, *zamindars* collected taxes for the Mughuls and later the Raj. Usually members of the elite, they were or became large landowners, especially in Sind and Punjab. They continue to control both land and peasant farmers under a feudal system. *Zamindari*, because they are relatively rich and come from prominent families, are often voted in as members of Pakistan's Parliament. Many voters are coerced. Also, several *zamindars* have their own prisons and armies. Polling stations

Figure 7-7

This aerial photo shows irrigated lands along the edge of the Thar Desert in India. You can see one of the many irrigation canals that are linked to the Indus. Photograph courtesy of B. A. Weightman.

may be overwhelmed or closed and the ballot boxes stuffed with votes for the "boss."

Typically, a parliamentary candidate visits villages and makes promises, giving money to village elders who distribute it among their clans who will then vote for the candidate. According to conventional wisdom in Pakistan, the only loyalty more powerful than that to one's clan is that given to one's *zamindar*. In recent years, as education and communication improve, some middle-class candidates, including women, have been elected. This certainly goes against the norm for many people. The rural peasantry is virtually excluded from higher levels of government. According to the writer Ahmed Rashid, "If the feudals put up their dog as a candidate, that dog would get elected with 99 percent of the vote." This has even greater meaning because generally, Muslims do not like dogs, regarding them as unclean.

Baluchistan

Although twice the size of Punjab, rugged and arid Baluchistan possesses the smallest population and is the poorest province. Sedentary farming, pastoral nomadism, and related crafts and trade continue to be ways of life among the various tribes of the craggy outcrops, plateaus, and mesas of the Suleiman and other ranges sweeping southward and westward to the Iranian border. The southern Makran coast has been described as "bizarre," comprising gigantic pillars and pedestals, layered and etched like old teeth. Northward to Afghanistan lie volcanoes, expanses of black pebbles called *dasht*, shifting red dunes, and salt-encrusted *playas* (dry lakes) called *hamun*. A local proverb refers to Baluchistan as "the dump where Allah shot the rubbish of creation."

Baluchistan's only notable urban center is Quetta, a hill station strategically situated between two passes into Afghanistan. In the hills above Quetta at 5,490 feet (1,650 m), juniper trees are cultivated. Oil of juniper berries gives gin (e.g., Bombay Gin) its distinct flavor. Coal reserves offer some economic potential, but these have since proved to be of poor quality. However, the province has rich oil and gas reserves and provides 75 percent of the marble for Pakistan's marble industries. This region has been long-known for its *karez*, underground irrigation systems using hand-dug, gravity-flow tunnels to divert water from aquifers to surface streams and ponds. *Karez* systems still exist in arid lands from Iran to western China and Oman, even further west.

Baluchistan has been a center of Islamist and sectarian violence since 2007. Nationalists claim that the region was forced to join Pakistan in 1947. They and others want a greater share of the profits reaped from mining and other industries. There have also been violent outbreaks between Sunnis and Shites. General Musharraff, prime minister at the time, sent in troops to put down the insurgency. The army has been accused of human rights violations including torture and murder. Thousands of people have lost their lives or have "disappeared." Now, there are at least 60,000 internally displaced persons, mostly women and children. The United Nations have set up refugee camps, but they face roadblocks every step of the way as the army doesn't want them to see what they have done. NATO volunteers have been shot and bombings are regular occurrences.

Roller Coaster Development

Pakistan was saddled with an infrastructure purposefully laid to connect cantonments with strategic frontiers and tuned to economic and military needs of the former Raj. As a result, the bulk of the road and rail network remained in India. As goods historically moved overseas via Bombay, Karachi was a rather pathetic port. Pakistan gained only 5 percent of the Raj's industry. Although the British had constructed canal projects, new water power and irrigation development schemes were essential. In essence, Pakistan received only the leftovers of the vast British system.

Long distances, rugged terrain, and cultural complexity, along with an inadequately developed infrastructure, did not foster the economic or political integration of Pakistan. In fact, its economic history has been turbulent. Wracked with political instability, policy shifts, and changing focus, Pakistan muddled its way from one crisis to the next into the twenty-first century.

Although Pakistan was an agricultural country, early stage planners, many of whom were urbanites from India, turned to industrialization. Cotton and jute mills were built for fiber processing previously done in Bombay (Mumbai) and Calcutta (Kolkata). Agriculture received little attention. Farmers were poorly paid for their crops, and lack of incentive hindered their productivity. By 1950, this decline, combined with rapid population growth, made Pakistan a net food importer for the first time in its history.

The sluggish economy was revived in the 1950s when landlords returned to the political arena. With a decade of support from the *zamindari*, and benefits of the Green Revolution, Pakistan's agricultural yields rose to be among the world's highest.

Under another military government, a major geographic shift took place. Karachi, with its immigrant population, squatter settlements, and other unseemly aspects, was perceived as unsuitable as the capital of a new Islamic state. For this and more practical reasons, a new capital, Islamabad, was to be built adjacent to Rawalpindi (Pindi) in Punjab, the true heart of Pakistan. Aside from its governmental capacity, it would function as a **growth pole**. With the evolution of Islamabad, industry and commerce would be decentralized from Karachi, Punjab would become developed, and a presence would be maintained close to the unstable regions of NWFP and Kashmir. Islamabad, meaning "City of Islam," also exemplifies the concept of a **forward-thrust capital**.

Islamabad took on its function as capital in 1963. As there are restrictive laws as to who can live in the new capital, most workers actually live in Pindi and are bussed back and forth to their government jobs.

International relations also affected Pakistan's roller coaster development. In 1949, the long-established economic interchange between Pakistan and India ground to a halt in a disagreement over monetary policy. But the Korean War (1950–1953) generated a market for leather, wool, cotton, and jute products, perfect timing to support farmers and the newly constructed processing mills. In 1954, Pakistan signed a mutual defense agreement with the United States, and this brought substantial amounts of American investment. However, the economic boom ended in 1965 with the outbreak of war with India over Kashmir.

The United States stopped military assistance to Pakistan and reduced its level of monetary assistance. With the breakaway of East Pakistan in 1971 and the perception of India as a "friend of the United States and enemy of Pakistan," Pakistan turned away from South Asia and toward the Islamic countries of the Middle East.

Beginning in 1974, hundreds of thousands of workers migrated to the oil-exporting countries of the Middle East to work in construction, oil, and other industries, and in services. Large numbers of these came from poor, *barani* (dry/not irrigated) regions. Pakistan continues to receive remittances from the Persian Gulf states but these have declined in light of the global recession and cutbacks in employment. Recently, it incorporated Iran and adjacent Islamic states into its sphere of interest.

Relationships with China were also a factor. Border disputes were settled in the early 1960s and both countries pursued closer ties, taking a stand against India. In 1965, a joint-venture project was begun to link Pakistan and China by a highway through the Karakoram. Islamabad would be linked to Beijing via the NWFP and Kashgar, gateway to China's ancient Silk Route. Built largely with human labor (more than 500 died), the 500-mile (805 km) road through the 14,928 feet (4478 m) Khunjerab

Pass and Hunza was officially opened in 1986. Environmental problems continue to plague this cold and windswept region. Not only is the pass closed in the winter months, but also landslides frequently render the highway impassable. The highway is currently under repair.

Another important event occurred in 1979 when the Soviet Union invaded Afghanistan. Suddenly, for the United States, Pakistan became a frontline state, a strategic barricade against the spread of communism and Soviet access to the Arabian Sea and Persian Gulf. Aid was restored, and Pakistan became the third-largest recipient of American aid after Israel and Egypt.

The war in Afghanistan brought millions of refugees into Pakistan, where they joined with their ethnic kin, not only in refugee camps but also in economic pursuits. In cities, Afghans replaced workers who had migrated to the Middle East. Afghans who migrated to the United States sent remittances to their relatives in Pakistan. In addition, the war stimulated commerce in guns and drugs that supported new economic investments. However, vital structural problems were still not addressed.

Zulfikar Ali Bhutto was ousted in a military coup in 1977 and was later executed for conspiring to murder a political opponent. In 1988, when Benazir Bhutto, his daughter, became the youngest and first female leader of a Muslim state, Pakistan was essentially bankrupt. When Iraq invaded Kuwait in 1990, triggering the Gulf War, a sudden leap in oil prices, loss of remittances, and reduction of exports put Pakistan in a critical position. Assistance was provided by Saudi Arabia in return for the service of 6,000 Pakistani soldiers deployed as part of Operation Desert Shield, sponsored by the United States. Even with additional support from the World Bank (WB), Pakistan has nowhere near the resources needed to overcome its economic problems. Moreover, these are confounded and exacerbated by an increasing rich-poor gap, rapid population growth (2.3 percent in 2009), increased religious conservatism, and political instability.

Years of military dictatorship, interspersed with short periods of unstable civilian government, have been characterized by mismanagement, patronage, nepotism, corruption, and violence. Leaders have been deposed, executed, or have died under mysterious circumstances. Benazir Bhutto held office twice (1988–1990 and 1993–1997), only to fall each time under corruption charges. General Musharraf became President in 1999 but in a fierce climate of insurgencies, political dissension, and periods of declared national emergencies, was forced to step down in August 2008. In 2007, Benazir Bhutto returned to run for president again but was assassinated by Muslim militants on December 27. This was a devastating blow to many Pakistanis who believed that she would save the country. Her husband, Azif Zardari, is now running the government. In recent years economic growth has faltered while communal strife intensifies.

The War on Terror

Following the terrorist attacks on 9/11, the United States began military operations in Afghanistan to destroy Osama bin Laden and his Al-Qaeda bases. President Musharraf withdrew support for the Muslim, fundamentalist Taliban who controlled most of Afghanistan. By November 2001, Al-Qaeda had escaped to Pakistan's Federally Administered Tribal Areas (FATA) such as Waziristan. Eager to solidify his military rule, Musharraf lent support to the Americans. However, in 2006 heavy troop losses drove him to make agreements with tribal leaders in Waziristan that his soldiers would stay away if the Waziris kept the Taliban out of FATA. This was virtually impossible, and Taliban youth strengthened their militancy in FATA. This indigenous terrorist group amassed a force of some 30,000 fighters under the command of Baitullah Mehsud in South Waziristan alone. Mehsud was killed in February, 2010.

THE TALIBAN

The Russians invaded Afghanistan in 1979 only to be fought by the fervent Taliban. The United States supported the Taliban in their defeat of the Russians in 1989 but then withdrew. This allowed the Taliban to take over the country. Schooled in fundamentalist Islamic ***madrassas*** and trained in the rugged mountains, the Taliban quickly killed most of the *maliks* (feudal tribal leaders) and enforced draconian rules of behavior, especially for women:

- *Sharia* law was imposed.
- Women could not work outside of the home. Even female doctors were forbidden from working.
- Women could not go to school.
- If women went outside, they had to be fully covered with a *chador* and be accompanied by a male relative. Exposure of the minutest amount of flesh brought a public whipping by the moral police.
- A woman who was raped or committed adultery would be stoned to death.

- Men could not cut their beards. Barber shops were closed.
- Music, movies, sports, and kite-flying (popular in Afghanistan) were forbidden.
- Anyone speaking out against the regime would be tortured and publicly executed in the former soccer stadium.

The Taliban are very active in Pakistan and have strongholds in many regions of the country. In 2009, they took control of Swat but the Pakistani army drove them out for awhile. Although they are in hiding, they still control the region (Figures 7-8 and 7-9).

JIHADIS

In 2001–02, Pakistan was home to 58 religious political parties and two dozen religious, armed militias popularly known as *jihadi* groups. *Jihad*, in its pure sense, means "striving" or "struggle." It does not equate to any Western notions of "Holy War." Moreover, there are many types of *jihad*, and most of them have nothing to do with war. Basically, it is a struggle within oneself to avoid sinful inclinations. It also means resistance to oppression. According to the prophet Muhammed, "The best *jihad* is speaking a word of justice to a tyrannical ruler."

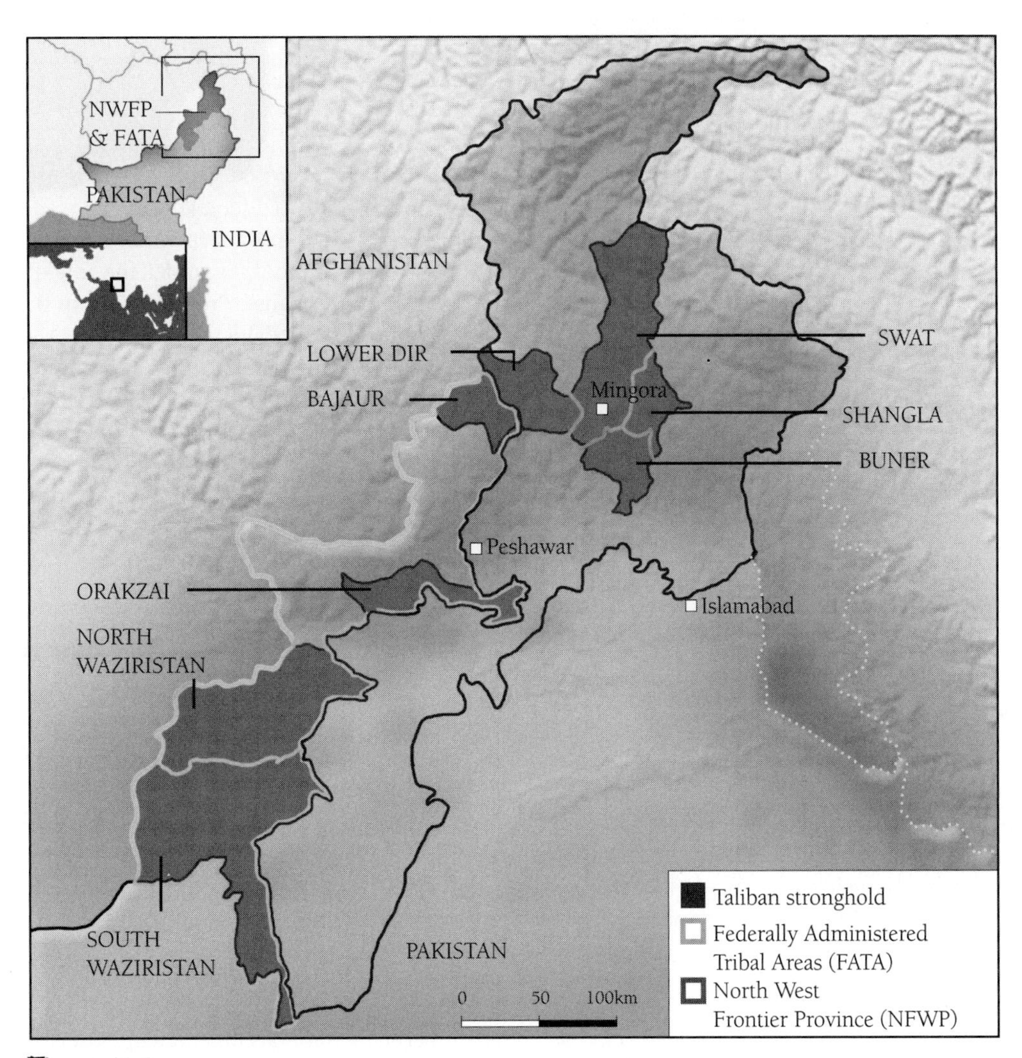

Figure 7-8

Taliban strongholds. Redrawn from a map produced by the BBC at http://news.bbc.co.uk/2/hi/south_asia/8046577.stm

Figure 7-9

Girls' school in Swat, NWFP. The Wali (king/ruler) of Swat in the 1970s was a very enlightened man in that he believed in family planning and education for girls. (A family planning clinic was just down the road from this school.) In 2009, the Taliban took over Swat and destroyed all girls' schools. Photograph courtesy of B. A. Weightman.

Today's *jihadists*, to quote the author Hassan Abbas (2005), "interpret *jihad* primarily in terms of the use of force to impose their version of Islam on others and to fight "infidels" [nonbelievers] to conquer the world." Even Muslims with more liberal views are to be exterminated. Although we might think that these people must be uneducated in the ways of the wider world, in fact, they derive from all social classes and come from many different countries such as Morocco, Somalia, Yemen, and even Great Britain and the United States. Furthermore, their numbers are increasing.

Aside from the attractiveness of their religious fervor to many Pakistanis, *jihadi* groups increase their popularity by financially supporting their members' families. They provide housing, food, religious education, and other services to otherwise poverty-stricken people.

Madrassas

In 1947, there were 189 *madrassas,* also known as "seminaries" in Pakistan. By 2006, there were 11,000. Today, there are more than 30,000. Most of these religious schools are for boys; only 448 are open to girls. In some girls' schools, not only do they have to be fully covered in black—looking out of a narrow slit-but also they have to wear gloves. More than 2 million Pakistani students attend *madrassas*.

Some Pakistani parents want their girls to have an education (Figure 7-9). In fact, research shows that girls have more interest in learning and perform better than boys. However, there are very few religiously educated female teachers regarded to be suitable as seminary instructors.

Madrassas bloomed under the rule of the religious fundamentalist, President Zia, in the 1980s. The failure of the Pakistani government to provide alternative forms of education and to require religious instruction for the masses has only encouraged an increase in religious schools.

The majority of religious schools have **Deobandi** roots. This rigid form of Sunni Islam emerged from a *madrassa* in Deoband, a town in India's Uttar Pradesh, in 1866. The Deobandis are traditionally apolitical but this stance has changed, particularly in Pakistan. Currently, Indian Deobandis are regarded as orthodox Muslims while Pakistani Deobandis are seen as radical extremists. They are particularly hostile to the Pakistani Sufis, who worship saints and dance as they practice their mystical form of Sunni Islam. Most Taliban are Deobandi-trained.

It is accepted that southern Punjab is home to the most aggressive and "poisoned" of all *madrassas*. These are funded generously by private citizens from Saudi Arabia, the United Arab Emirates, and Kuwait in an effort to further the spread of Islam. Many religious schools reject government funding because they think the money comes from America. However, numerous seminaries ingratiate themselves with poor families by providing funds, food, and even accommodations for both students and families.

Together with Peshawar, Islamabad is the most vulnerable city in reference to a takeover by the Taliban. In June 2009, the government discovered that there were 260 *madrassas* in Islamabad—many unregistered. Political scientist Christine Fair (2009) espouses that many of the madrassas are busy, "spreading hatred against the armed forces of Pakistan. . . and that in some ways they are centers of a civil war of ideas." Fair also points out that many religious scholars are marginalized by modernization and globalization and seek their own relevance via orthodoxy.

In the context of Islamic history, *madrassas* were the primary sources of Islamic and scientific learning. Contrary to Islam's call for contemplation and reflection, a majority of *madrassa* students are taught only to memorize Koranic verses in Arabic. They are also subject to harangues against non-Muslim societies, especially the United States and Israel. Students are exhorted to, "kill infidels and the enemies of Islam."

A recent survey (2009) indicates that some religious schools are changing their approach. As international organizations focus on women's education, empowerment, and economic independence along with criticism of Pakistan's low female literacy rate (42%), increasing numbers of families are demanding a more practical education for their daughters. Even boys' schools are stressing language and writing skills (Arabic, Urdu, and English) as well as computer literacy.

TERRORISM

Pakistani authorities have long had ties to militant groups in and out of their country. They have supported fighters in Kashmir (see below), the anti-Russian Afghan forces, and the rise of the Taliban in that country. However, once Pakistan threw in its lot with the United States after 9/11, it has earned the resentment and anger from militant groups. Beginning in 2002, deployment of troops in the tribal regions has further fueled anger and even generated hatred. There are five categories of terrorists:

- **Sectarian:** Sunni and Shi'ite groups who are hostile to each other.
- **Anti-Indian**: Groups that operate with the alleged support of the Pakistani military amd intelligence agency who fight in Kashmir and who may be responsible for recent bombings in India (see Chapter 9). The most well-known is *Lashkar-e-Taiba*.
- **Afghan Taliban**: The original Afghan Taliban under Kandahari (a southern Afghan province) leadership led by Mullah Mohammad Omar, believed to be living in Quetta.
- **Al-Qaeda and Affiliates**: Organization led by Osama bin Laden and others believed to be hiding in the FATA.
- **The Pakistani Taliban**: Groups including extremist outfits in the FATA and the NWFP. As mentioned above, one leader, Baitullah Mehsud, is believed to have been killed and Mullah Abdul Ghani Baradar, the Taliban's deputy leader, was captured in 2010.

Sectarian Violence

Neither Sunnis nor Shi'ites believe that the other is adhering to the "correct" form of Islam. Centuries of violence, mutual discrimination, and hatred erupted in 2003 when attackers, armed with machine guns and bombs, attacked a Shi'ite mosque in Quetta, Baluchistan, and killed 55 worshippers as well as themselves. Violence spread to Karachi and parts of Punjab, leaving more than 350 dead by 2004.

In another terrible incident in Quetta that year, machine-gun fire and grenades cut into a Shi'ite procession commemorating the death of Muhammed's grandson. Two suicide bombers blew themselves to smithereens in the middle of the procession. Shreds of their bodies hung from balconies and electric wires. This ghastly carnage left 44 dead and scores wounded.

The rising tide of sectarian violence is beyond the capability of the police and intelligence agencies to preempt and deal with. Many of their officers have been murdered. Also, terrorists have infiltrated both groups. For example, a 2004 suicide bomber in Karachi turned out to be a police officer and a member of a state security agency.

Al-Qaeda and the Taliban

To describe the "war on terror" as an American enterprise is to simplify the facts. The United States is involved in many such wars, in Pakistan, Afghanistan, Iraq, Yemen, and the southern Philippines, for example. The Pakistani government is fighting terrorism, not only in the frontier regions, but also in numerous areas throughout the country. While former President Musharraf and the current President Zardari are trying to quell discontent and terrorist atrocities, their roles have not been supported by the majority of Pakistanis who show their disapproval by voting for fundamentalist politicians, voicing support for remnant Taliban, and giving sanctuary to Al-Qaeda operatives.

In 2010, *The Economist* noted that "al-Qaeda is nebulous." In other words, their members are everywhere. "It is at once a secret organization, a network of militant groups and a diffuse revolt." They use mosques, *madrassas,* and the Internet to spread their doctrine of terror. Another tactic is to form "franchises" that include convicts who have converted to Islam. It is estimated that more than 100,000 Pakistani militants received training by Al-Qaeda in Afghanistan when the Taliban were in control.

Al-Qaeda is an international organization with cells and franchises in numerous countries. It has a network of allies that includes Afghan, Pakistani, and Kashmiri extremists, followers in Somalia, Chechen rebels, Indonesian militants, and more. A still broader social movement includes self-radicalized groups of young Muslims living in Great Britain, the United States, and other Western countries. Suicide bombings are popular with young men and some women who are indoctrinated with the belief that they will be going to Paradise in the name of Allah. Table 7-1 shows the impact of terrorism in Pakistan since 2006. Clearly, the future of Pakistan hangs in the balance.

Table 7-1 Terrorist Acts in Pakistan 2006–2010*

Year	Killed **	Injured ***
2006	907	1,543
2007	3,448	5,353
2008	2,267	4,558
2009	3,021	7,334
2010 (to August)	1,000	1,392
Totals	10,643	20,180

* Terrorist acts include: landmines, suicide bombings, sectarian violence, shootings, assassinations, government–militant battles, riots, and bombings.

** As of June 2009, American and Pakistani aerial bombings were responsible for 577 of the total deaths.

*** The numbers of injured do not include hundreds or perhaps thousands that were never found and so are not included in the counts.

Sources: Numerous sources such as the *The Daily Telegraph*; BBC News; CNN; and the *International Herald Tribune*, as reported at http://en.wikipedia.org/wiki/Chronology_of_terrorist_incidents_in_Pakistan

CONFLICTS IN THE NORTH

Azad Kashmir, and Jammu and Kashmir

Muslims first arrived in Kashmir in 711, and by the sixteenth century it was known as an "ornament of the Mughal Empire." After a period of Afghan and Sikh rule, the British, having defeated the Sikhs, sold the regions of Gilgit, Ladakh (now in India), and Kashmir to Gulab Singh, a Hindu, Dogra chieftain.

Muslim agitation began in the 1930s. With partition in 1947, Gilgit, Hunza, Baltistan, and other predominantly Muslim mountain states acceded to Pakistan. The British "encouraged" Kashmir's Hindu ruler to join India. That this mainly Muslim region would become part of an independent, mainly Hindu India was anathema to Pakistan. Religion was only one factor. Jammu and Kashmir contain the five headwaters of the Indus River (refer to Figure 7-3). Control of these rivers means control of Pakistan's fate.

With Muslims agitating to join Pakistan, NWFP tribesmen entered the scene to "liberate" Jammu and Kashmir' from Hindu India's grip. A revolutionary government was formed in northern Kashmir, which became known as Azad Kashmir. This stimulated a significant migration of Hindu Kashmiris to India.

Fearing for the lives of his fellow Hindus, the Maharajah fled from Srinagar to Jammu and signed an agreement to join India. The Indian army took control and promised a plebiscite. In 1949, a United Nations' Cease Fire Line dividing Azad Kashmir from Jammu and Kashmir was accepted by both India and Pakistan. In 1956, Jammu and Kashmir formally entered the Indian Union. The promised plebiscite was never held.

In 1965, with accusations of infiltration, spying, and sabotage, another war broke out and lasted six months. The following year, in a meeting at Tashkent in the Soviet Union, Pakistan and India agreed to settle their disputes peacefully.

Ideals of peace were shattered in 1971 with the East-West Pakistan war and Indian intervention on East Pakistan's behalf. Fighting in Kashmir continued until 1972, when a new "Line of Control" was drawn. A Kashmiri Accord was signed by India and Pakistan in 1974.

Since that time, military fighting has been sporadic. Skirmishes take place mostly on the Siachen Glacier. This ice mass is strategically important because it commands a China–India–Pakistan triangle formed by China's 1962 annexation of India's Aksai Chin region.

Civilian conflict increased in the 1980s and 1990s. Muslim uprisings, election boycotting, and anti-Hindu terrorism intensified. In 1993, the Jammu and Kashmir Liberation Front and Hizbul Mujahideen united for the liberation of Kashmir. The Indian military has been ruthless in its suppression of insurgents. One result is that tourism, once a mainstay of Kashmir's economy, has picked up to some degree. Overall, however, the industry is decimated.

Srinagar and the Vale of Kashmir are open for business even as inhabitants' vehicles, shops, and facilities remain damaged or closed for repairs. Unlike in the past, tourists are coming mainly from India and not from America and Europe. Many travelers see the potential for more violence. After all, this is still disputed territory.

A beacon of light appeared in 2003 when India and Pakistan started to mend their relations. A cease-fire has been called, and both countries are supposed to come to some agreement about Kashmir. India withdrew 30,000 of its troops in 2008 and 30,000 more in 2009. Nevertheless, the conflict has claimed at least 42,000 lives since 1990. Furthermore, some 350,000 to 450,000 people have been internally displaced.

To compound the situation, a massive earthquake struck near the Line of Control on October 8, 2005, affecting both Pakistani and Indian territories. More people perished on that day than in two decades of conflict. The estimated 7.6 quake killed at least 73,000 in Pakistan and 1,300 in India. More than 3 million Pakistanis and 150,000 Indians lost their homes.

Gilgit-Baltistan

Gilgit and Baltistan comprise the northernmost territory governed by Islamabad and are an important element in the struggle for Kashmir. New Delhi claims that the region belongs to India because it is part of Jammu and Kashmir. Pakistan considers the territory to be separate from Kashmir.

The two-part territory became a single administrative unit in 1970 under the name of "Northern Areas." This 28,174 square mile (72,971 km^2) region has an estimated population of 1.5 million. It is an important strategic area as it borders the contentious zones of Afghanistan's Wakhan Corridor, China's Uygur Autonomous Region (see Chapter 10), Pakistan's NWFP, as well as contested parts of Kashmir.

Gilgit-Baltistan was once a major destination for foreign tourists, especially trekkers and mountaineers who hoped to scale the slopes of the Karakorum and the western Himalayas, where there are more than 50 mountains above 23,000 feet (7,000 m). These include the challenging peaks of K2 (Mount Godwin-Austen) at 28,250 feet (8,611 m) and Nanga Parbat at 26,660 feet (8,125 m). Also, three of the world's longest glaciers outside of the polar regions can be found here.

Prior to 1978, Gilgit-Baltistan was cut off from Pakistan due to its harsh terrain and lack of roads. Existing roads went into Kashmir. However, people could walk through mountain passes to get to Rawalpindi in the summer. The Karakorum Highway (noted above) connects Islamabad to the main towns of Gilgit and Skardu (Figure 7-10).

This rugged and largely inaccessible region is populated by a patchwork of ethnic, linguistic, and religious groups. Identity is of utmost importance to the many and diverse ethnic groups that are often referred to as Kashmiris by Pakistan's government. In fact, they are not at all related to the Kashmiris nor do they speak their language.

While Urdu is the lingua franca (understood by men), 40 percent of the population, mostly in Gilgit, speaks Shina (with several dialects). The people of Baltistan speak Balti, a Tibetan language. There are at least six additional languages (with numerous dialects) spoken in the region. One interesting language is Domaki, which is spoken by traveling music clans.

The majority of the population is Shia, unlike the rest of Pakistan where the majority is Sunni. There are also significant numbers of Ismaili Muslims—a moderate sect that follows Ali Khan (son of the Aga Khan) of Egypt.

In a further effort to secure control of the north, Pakistan's President Zardari signed the Gilgit-Baltistan Empowerment and Self-Governance Order on August 29, 2009. This grants self-rule (but not provincial status) to the people of the Northern Areas and officially renames the region "Gilgit-Baltistan." A governor will be appointed and an Assembly partly elected by the people. The government has also promised a development package to improve the very poor economic and social conditions. In September 2009, Pakistan signed an agreement with China for a mega-energy project that includes the construction of a 7,000-megawatt dam at Bunji.

The offer of "province-like" status does not appease the Gilgit-Baltistan people at all. They insist on true provincial status equal to that of the other provinces. They see that a centrally appointed governor will function to maintain Pakistan's control of the region. Moreover, they claim that the "package" has been imposed on them without consultation.

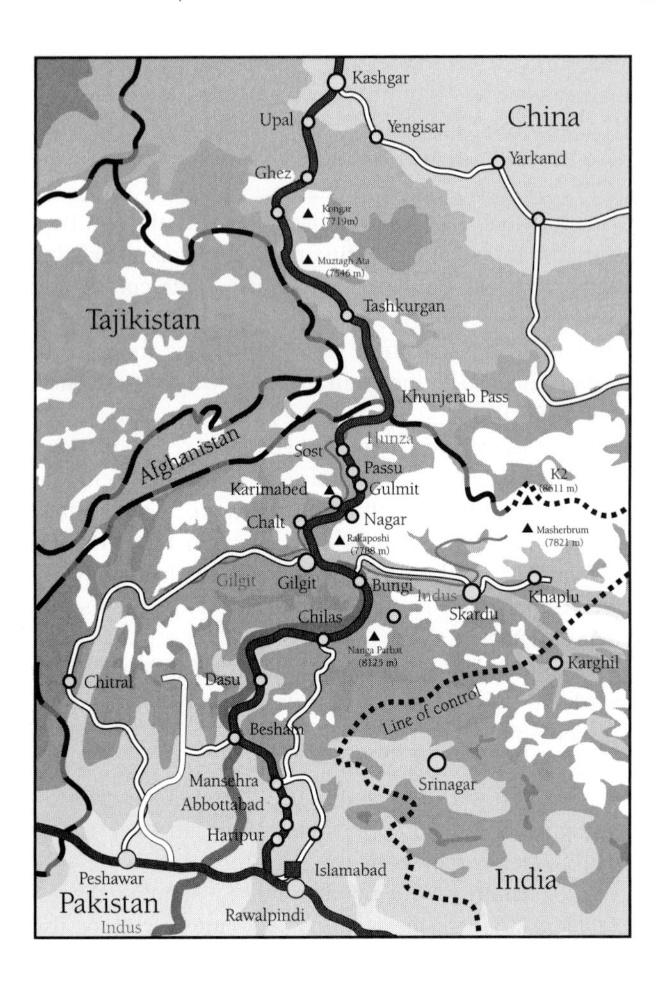

Figure 7-10

The Karakorum Highway that links Pakistan with China.
Redrawn from a map at http://en.wikipedia.org/wiki/Gilgit-Baltistan

Opposition has also come from Kashmiris who see the plan as an insidious move by the central government to dilute their cause by giving a region, internationally considered a part of the Kashmir issue, greater status within Pakistan. India is also protesting the moves as it sees Pakistan as solidifying its position in Kashmir. However, it has rankled the people of Gilgit-Baltistan even more since it implies that the Kashmir controversy is "solved."

There are other arguments against the plan as well. Some say that it is intended to assuage the people to allow dam construction without protest. Others espouse the notion that China is the driving force because they want stability along their western border. Appeasement of Western donors who are concerned about human rights issues in the region is also irksome. Over all, the Gilgit-Baltis see the plan as a "package of gimmicks." Clearly, trouble is brewing in this region.

People Working the Land

Agriculture comprises the largest sector of Pakistan's economy, contributing 24 percent of its GDP. It accounts for nearly half of labor force employment and is the largest source of foreign exchange earnings. Only about 40 percent of the country's land is suitable for cropping and 90 percent of this is on the Indus River plain in Punjab and Sind. The most important food crops are wheat and rice, and the most important commercial crops are cotton and sugar cane. Unfortunately, agriculture in Pakistan is fraught with problems that have resulted in increasing food insecurity.

MAJOR CROPS AND SEASONALITY

Traditionally, crops fall into two seasonal categories: ***kharif*** and ***rabi*** (Figure 7-11). *Kharif* crops such as rice,

KHARIF and RABI CROPS

KHARIF CROPS
- sown after the onset of the wet monsoon (June–July)
- harvested in autumn
- dominate southern Pakistan in Sind and Baluchistan
- floodplains and tank irrigation

RABI CROPS
- sown at the end of the wet monsoon
- harvested in spring
- dominate northern Pakistan in Punjab and NWFP
- doabs and uplands

	TYPE OF CROP	MAIN PRODUCTION REGION (B = Baluch; P = Punjab; S = Sind; NWFP = Northwest Frontier Province; IRV = throughout Indus River Valley)
KHARIF	RICE	P & S
	JOWAR (MILLET VARIETY)	P & IRV
	BAJRA (MILLET VARIETY)	P & IRV
	SESAMUM (SESAME)	S & P
	MAIZE (PULSE: LEGUMINOUS GRAIN)	NWFP
	GROUNDNUTS (PEANUTS)	P
	COTTON	IRV
	VEGETABLES	EVERYWHERE
	PEARS	B & NWFP
	PLUMS	B & NWFP
	GRAPES	B
	DATES	S

	TYPE OF CROP	MAIN PRODUCTION REGION
RABI	WHEAT	P
	JOWAR	P & IRV
	BARLEY	P & NWFP
	GRAM (PULSES—GRAM = CHICKPEAS	P & S
	—MASOOR = LENTILS	P
	—MATTAR = PEAS	S
	—MUNG = KIDNEY BEANS)	P
	UNSEED (OIL SEED)	P
	RAPE (OIL SEED)	IRV
	MUSTARD (OIL SEED)	IRV
	TOBACCO	NWFP
	VEGETABLES	EVERYWHERE
	CHILIES	S
	APPLES	B & NWFP
	APRICOTS	NWFP
	BANANAS	S
	CITRUS	P
	GUAVA	P
	MANGO	S
	PEACHES	NWFP

NOTE: SUGARCANE IS GROWN IN RABI AND HARVESTED IN KHARIF IN PUNJAB.

Figure 7-11

Kharif and rabi crops.

corn, and cotton are sown after the onset of the wet monsoon (June–July) and are harvested in autumn. *Rabi* crops such as wheat, barley, and oil seed are sown at the beginning of the dry monsoon and harvested in spring.

Pakistan is noted for its **Basmati rice**. The name "basmati" originated from a Sanscrit word—*bash,* meaning "smell." Pearly-white, long-grained Basmati has a nutty aroma and flavor. Legend says that it was meant to be fed only to maharajahs. Grown in Punjab during summer, it is carefully harvested by hand.

Wheat is the main food crop in Pakistan and covers more acreage (37 percent) than any other crop. It is used to make flour for such things as leavened, flat, and steamed breads. *Chapatis*, similar to *tortillas,* are a meal staple (Figure 7-12). A limited amount of wheat is planted as a forage crop for livestock and the straw can be used as fodder or as roofing material.

Cotton is Pakistan's chief industrial crop. In fact, it is the world's fifth largest producer of "white gold." More than a million farmers are involved in cotton production. Moreover, the textile industry, centered in Karachi, accounts for 65 percent of the country's exports. According to the government's "Textile Vision 2005" plan, the industry is designated to be further mechanized and modernized to improve quality and enhance product appeal to Western economies.

Another significant commercial crop is sugarcane. Not only does it produce sugar, but also it is an important contributor to industry. *Bagasse*—cane fiber—is used to make chipboard and paper. Cane components are also used in the manufacture of chemicals, paints, insecticides, and detergents.

Productivity and Poverty

Agricultural production in Pakistan is stagnant. Sixty-five percent of the population, directly or indirectly involved in agriculture, is unable to feed themselves plus the 35 percent of nonfarmers. Crop yields have fallen in almost every sector for reasons discussed below. Yields are significantly below potential output. For instance, the gap between actual and potential rice yield is 50 percent, wheat 40 percent, maize (corn) 28 percent, and sugarcane 35 percent.

Consequently, poverty has increased, especially in the countryside where most Pakistanis live. According to an official 2005 survey, overall poverty in Pakistan is 24 percent. While 15 percent of city dwellers are considered to be poor, 28 percent of rural people cannot make ends meet. In a more recent inflation-adjusted study (2010), rural poverty has risen to 31 percent.

The government estimates that the country now needs US$2–4 million to import enough wheat to feed its growing population. Other rising imports include milk and milk products, meat, vegetables, dried fruit, edible oil, tea, spices, and sugar. In 1999, Pakistan produced enough grain for a per capita annual availability of 570 pounds (259 kg). By 2007, per capita availability fell to 418 pounds (190 kg), a drop of 81 pounds (37 kg) per person. In contrast, per capita yearly food consumption in the United States was 1,950 pounds (886 kg) in 2003.

Food shortages and rising prices have produced an illegal commodity trade that has exacerbated the situation. For example, in 2008, in context of a 40 percent increase in global wheat prices and price controls at home,

Figure 7-12:

This street-food vendor in Peshawar is preparing to put lamb and vegetables into a chapati. The end-product looks like a taco and is very spicy. Photograph courtesy of B. A. Weightman.

many Pakistani farmers sold their bulk wheat illegally across the borders in India and Afghanistan. Selling wheat outside of Pakistan has created massive shortages within the domestic market and near-riot conditions in many cities, especially in NWFP and central Punjab. Paramilitary forces have been employed to oversee wheat and flour distribution.

ONGOING PROBLEMS IN AGRICULTURE

There are numerous underlying causes for Pakistan's declining agricultural production and food insecurity. Some of these are:

- Inequities in land ownership.
- Water shortages and salinization.
- Insufficient and misuse of fertilizers.
- Use and misuse of BT seeds.
- Failure of government policies.

Who Owns the Land?

Pakistan continues to have a feudal-landlord system. In fact, a mere 2 percent of the richest landlords control 25 percent of the agricultural land. About one-third of Pakistani farmers are tenant farmers, required to give 50 percent of their crop to the landowner. Also, despite 1992 laws opposing the practice, **indentured servitude** is widespread, especially in Sind and parts of Punjab. Poor farmers, forced by circumstance to borrow from the *zamindaris*, become victims of usurious interest charges and can remain eternally indebted to a land-owning family willing to exploit this situation for generations.

Pakistan has struggled with land reform in efforts to reduce absentee landlordism, shrink large farms and enlarge small ones, consolidate fragmented holdings, and increase peasant ownership. However, these efforts have met with strong opposition from the landed rich, who also hold political power. Improvements are virtually negligible.

In 1947, a mere 1 percent of farmers owned a quarter of the agricultural land, and 65 percent of farmers held only 15 percent of the land. More recent studies (2001–2002) reveal that in Sind, 80 percent of the people own no land and in Punjab, 75 percent are landless. Conditions are even worse in the remoter provinces.

While there has been some farm-size reduction and an increase in the number of medium-sized farms, 34 percent of farms are still under 5 acres (2 ha). A mere .08 percent of households, representing the richest 20 percent of the population, own farms larger than 5 acres (2 ha).

Where's the Water?

Since independence in 1947, Pakistan has increased its cultivated land by one-third, largely through the expansion of irrigation. Acquisition of water has always been a challenge–waterworks date back 4,000 years. Both the Mughals and the British constructed irrigation canals.

In 1960, the **Indus Waters Treaty** gave Chenab, Jhelum, and Indus water rights to Pakistan and gave Ravi, Beas, and Sutlej water rights to India. Then, the World Bank–sponsored Indus Basin Development Fund was established to build new waterworks and link western rivers with the fertile soils of eastern Punjab. The giant Mangla Dam on the Jhelum and the Tarbela Dam on the Indus provided flood control, hydropower, and irrigation. The 2 mile (3.2 km) wide Tarbela, completed in the 1970s, is the world's largest earth-filled dam. Today, Pakistan has the world's most extensive, continuous irrigation system. Unfortunately, most of it is in a state of disrepair. Funds allocated to fix it either have not been allocated yet or never reach the designated recipients.

Shortages of water have generated tribal feuds over its use. Reduction in irrigation water results in reduction in crop yields. This means that farmers cannot pay water usage fees for the next season and yields are lowered again. Flooding in 1992, 2007, and 2010, brought too much water to some areas. Lack of drainage and containment facilities meant that excess water was not stored for the future. According to UNICEF, the 2007 flood killed at least 300 people and nearly 4 million were displaced.

Early July 2010, Pakistan experienced the worst floods in its history. Fuelled by global warming, the monsoon rains became torrential, bringing unprecedented volumes of water to the northern mountains that contain the headwaters of the Indus River. The subsequent southward deluge inundated the river basin all the way to the Arabian Sea. In the wake of the flood, at least 1,600 people were killed, 900,000 injured, and countless millions rendered homeless. Many of these people, including 3.5 million children, were subjected to water and insect-borne disease such as cholera and dengue fever. More than 8 million acres (3.2 million ha) of cropland were destroyed, devastating Pakistan's already struggling economy.

Increasing water supply to croplands has a downside. Added water raises the water table—the upper level of groundwater—and root systems can be inundated. In Punjab and Sind for instance, the Indus water table used to be 53 feet (16 m) below the surface. With increased irrigation, it has risen to within 10 feet (3 m) of the surface.

Salts accumulate with rapid evaporation of moisture in the country's arid and semi-arid climate. To counter the loss of cropland from salinization, thousands of pumps, tube wells, and conduits had to be installed (mostly private). Nevertheless, about 17 million acres (7 million hectares) remain saline, especially in Sind, Punjab, NWFP, and Baluchistan. The magnitude of the problem can be gauged by the fact that salinization continues at the rate of 988 million acres (40,000 ha) annually. Measures to ameliorate this condition such as planting green manure like clover, or adding chemicals such as gypsum are not affordable for most farmers. Furthermore, many farmers have little if any knowledge of these methods.

What about Fertilizer?

Another factor responsible for poor production is nutrient deficiency. Even though government support of the industry has made more chemical fertilizer available, consumption remains very low in the countryside. Many farmers cannot afford enough and some do not apply it properly. Animal manure can be used if farmers raise livestock. Farm animals such as cattle, sheep, and goats can graze on wheat, maize, and other stubble after harvest, but if yields are poor, the animals have less to eat and consequently produce less manure.

What Is Happening with BT Seeds?

As discussed in Chapter 5, the use of BT seeds does not always live up to its promise. In a desperate attempt to increase cotton yields, 70 percent of farmers are sowing these genetically engineered seeds. However, most of the 33 varieties being sold in the markets are not approved by government scientists. Many are fake but they are bought because they are relatively cheap. The more expensive approved types are sown on only 20 percent of cotton acreage. Pakistan is now looking to India, where cotton output had doubled, for legitimate BT seeds.

Unfortunately, the BT gene normally disappears after three years, thereby exposing the crop to every conceivable sort of disease. Cotton-leaf curl disease, which appeared in NWFP in the 1980s, continues to be a problem and has spread to Punjab, the major cotton-growing region.

What Is the Government Doing?

Pakistan is importing more food than ever before at a cost of more than US$5 billion in 2007. In 2001, the cost of food imports was significantly less at US$1 billion. Critics say that domestic issues are being allowed to worsen, both at the policy and management levels. A procurement price mechanism is used to keep prices depressed. Allocated funds are not disbursed, or simply disappear. Water shortages are worsening. Agricultural research is in disarray. No disease-free cotton has been developed, and the country regularly misses production targets in this and other crops. Government policies are allowing imports of less expensive foods such as vegetables from India, which hurts the domestic industry and puts numerous farmers out of business. Government functionaries argue that they are simply expanding the economy.

Agricultural credit programs, while beneficial for many farmers, are actually increasing regional disparities. In Sind, for example, farmers must have "passbooks" to receive credit but apparently they are difficult to get. Local officials often are loath to verify these documents and often charge a fee to do so. At the same time, counterfeit books are available for a price. Allocations are now based on the size of farms as well as levels of production. From a 21 percent share in national agricultural credit disbursements, Sind's funding is down to 11 percent. The position of Baluchistan's and NWFP's ability to attract credit is dismal. Drought has decimated crop production and few funds are available. At the same time, Punjab receives 82 percent of disbursements.

Much agriculture is controlled by large landowners and *sardars* (tribal chiefs) who manipulate land-title records in connivance with the Revenue Department. Big landlords fail to pay even a tenth of their taxes. Repeated complaints to the Revenue Minister in Sind in 2010 met with no response. In central Punjab, farms are owned by *waderas* (the elite) and members of the civil service and businessmen who are well positioned to access credit from a network of 4,000 banks. Many of these landlords are absentees and only appear at their farms during elections. Clearly, corruption and patronage are systemic in the agricultural sector.

The Village Agricultural Center (VAC)

The Government of Pakistan has an idea: village agricultural centers to be placed in all villages with at least 1,000 people. It is recommending that as of 2010, all agricultural revenue should be reinvested in improving agricultural production. VACs, managed by graduates in agriculture, should be extended to

20 percent of villages each year. Each center is supposed to have all inputs such as fertilizer, pesticides, and certified seeds. Repair facilities for machinery and training for tractor driving are only two of the services to be provided. Centers should have exhibition plots, training videos, and information about such activities as animal husbandry, poultry raising, bee-keeping, and floristry. Interest-free loans should also be available.

Experts claim that farm productivity could be raised by 80 to 100 percent if VCAs are established and run properly. While VCAs sound promising, their efficacy will depend on action instead of promises as well as equitability in services and disbursements instead of patronage and corruption.

FEMINIZATION OF AGRICULTURE

Increasing numbers of women are working in agricultural production, livestock raising, and related cottage industries. Women are especially important in rice production, and cotton-picking is considered to be exclusively women's work. Female farm labor is essential in the wet season when many men leave to find work elsewhere. Whatever the case, women do more farm tasks and spend more time doing them than do men. Pakistan's Labor Force Survey of 2004 indicates that male and female participation in agriculture is 38 percent and 67 percent, respectively.

Women participate in all operations relating to crop production such as sowing, transplanting, weeding, and fertilizing; harvesting, threshing, grinding, winnowing, husking, and storage; and building storage bins and feeding animals. Rural women carry out these duties in addition to their regular domestic chores of cooking, cleaning, caring for their husbands, children, and elderly family members, as well as collecting water and firewood.

In some areas of the country, especially in the south, men may marry several women in order to get more workers. After all, female labor is essentially free! Whatever the situation, women, even with their 7 to 15 hour daily workloads, eat alone, eat last, and eat the least.

Pakistan's Industries

With government backing, industrialization has progressed at a rapid pace. Even so, heavy industry is limited. Pakistan has only one integrated steel mill, a Soviet Union–aided venture at Port Qasim 22 miles (35 km) east of Karachi. Pakistan Steel Mill, which depends on imported iron and coal, was privatized in 2006. The country manufactures and exports such items as farm and industrial machinery, buses, motorcycles, and refrigerators.

Port Quasim is Pakistan's second largest port after Karachi and is the center of the country's auto industry. Together the two ports carry more than 60 percent of Pakistan's imports and exports. Port Qasim is situated in a coastal mangrove environment that once had eight species. As a result of general pollution and a massive oil spill in 2003, only four species survive.

In 2002, construction began on a new port at Gwadar, 46 miles (75 km) east of the Iranian border. Funded by China, the facility opened in 2008 and is now supervised by Singapore's Port Authority. As Singapore has one of the world's greatest ports, it is assumed that their experts will ensure that Gwadar will be run effectively. The first ship to enter the new port carried wheat from Canada.

Much of Pakistan's industry is tied to agriculture. Manufactured goods account for about two-thirds of exports. These are mostly cotton products and processed foods.

The government has a strong "political will" to upgrade and modernize the textile industry. However, declines in cotton production have driven down exports. Moreover, the industry is facing new, stiff competition from China, India, and Bangladesh. In fact, Bangladesh, which is not a cotton-growing country, is earning more than Pakistan in garment and other textile exports.

Pakistan's major trading partners are the United States, Japan, and Persian Gulf nations. In recent years Pakistan has pursued trade relationships through supranational organizations. In the mid-1980s, the South Asian Association for Regional Cooperation (SAARC) was formed to disassociate political conflicts from economic cooperation. SAARC includes Bangladesh, Bhutan, India, Maldives, Nepal, Pakistan, and Sri Lanka. The Economic Cooperation Organization (ECO), established by Pakistan, Iran, and Turkey in 1985, now includes Afghanistan, Azerbaijan, Kyrghizistan, Tajikistan, Turkmenistan, and Uzbekistan.

Pakistan is furthering its ground linkages with neighboring countries. In 1994, construction started on a highway connecting Pakistan with Turkey via Iran. However, progress has become mired in litigation on the part of Turkish investors. In 2009, the first container train arrived in Istanbul, Turkey, via Tehran, Iran, from Islamabad. Pakistan has no rail link to China, but contracts for feasibility studies were awarded in 2007. The proposed line would connect NWFP with China's western rail head at Kashgar.

Energy has been of paramount concern both for industrial production and improvement of quality of life. Water-generated power is of great significance but alternative sources are being sought. Unfortunately, Pakistan's coal is relatively poor quality, but natural gas sources have been developed in Baluchistan and oil is produced in Punjab. Natural gas pipelines link Islamabad, Faisalabad, Peshawar, and Quetta. Additional oil fields have been discovered in the south. At least 20 international companies are involved in their exploration. However, development has been limited by lack of security in both Sind and Baluchistan.

Geopolitics is a vital factor in international trade. For instance, up until recently, Pakistan has had very little to do with India because of hostilities from the time of partition and the clash over Kashmir. After six decades of turmoil, a bus route via Kashmir was reopened between the two countries in 2005. Imports and exports of food and other products have also increased.

A natural gas pipeline has been proposed to run from the Caspian Sea in Tajikistan via Iran to Pakistan and India. Since the United States wants to isolate Iran, it has pressured India to quit the deal, which it did in 2009. In 2010, the United States told Pakistan that if it did not quit the deal, it would not be given energy assistance. The proposed alternative offers Pakistan electricity from Tajikistan via Afghanistan's Wakhan Corridor, thereby avoiding Iran. This plan would strengthen both America's hold on Afghanistan and its geopolitical linkages with Pakistan. Pakistan has yet to reply to the offer.

Nuclear power is one alternative. Pakistan's first heavy-water nuclear reactor was built near Karachi with Canadian aid, in 1965. In 1974, when India exploded its first atomic device, President Ali Bhutto pledged an equal response: Pakistan would build an "Islamic Bomb" if it had to "eat grass" to do it. In the 1990s, another plant opened at Chashma on the Indus about 149 miles (240 km) south of Islamabad. Fuel is provided from a uranium enrichment plant at Kahuta, also near Islamabad.

In 1998, Pakistan exploded an atomic device. India responded with its own nuclear event. Non-nuclear foreign aid has been repeatedly diverted to the country's nuclear program. Moreover, China continues to support the program. This is all the more reason for India and Pakistan to mend their relations and solve the problem of Kashmir.

Hyperurbanization

The spectacular growth of cities in Pakistan can be summed up with the example of Karachi, which had a million people in 1950. This rose to 8 million by 1991 and 12 million by 2007. Lahore, the second largest city, had 850,000 people in 1950, and now has nearly 7 million. City growth in Pakistan is fast enough to be called **hyperurbanization**.

Under colonialism, cities were created and supported to serve colonial purposes. Along the Grand Trunk Road, towns were developed at river crossing points to function as centers of control, administration, collection, and distribution. With the introduction of railroads, urbanization gravitated to rail stations and lines. Bypassed, small communities were isolated, and many became deserted. Expanded irrigation projects increased agricultural production, but the Green Revolution and land reform favored medium-sized farms. Very small farms were absorbed. Large-scale, rural–urban migration ensued, and towns became specialized in various agriculture-related processing and manufacturing activities.

Three-quarters of Pakistan's urban places are situated in the Indus Basin (refer to Figure 7-3). Most of the remainder, in the uplands of Baluchistan and NWFP, are more like overgrown towns. In fact, it is the largest cities that are experiencing the greatest growth. For example, Karachi has grown by over 90 percent in the last 60 years. Most of this growth was and continues to be from in-migration. Afghan refugees have been major contributors to rapid growth in Quetta and Peshawar. Peshawar, capital of NWFP and flush with Afghan refugees, is growing at an annual rate of 3.3 percent and now has a population of at least 4 million.

Karachi, the country's primate city, continues to grow by accretion of ***katchi abadis***—squatter settlements on the periphery. Efforts to decentralize people into new "metro-villes" outside the city have been largely unsuccessful. Land intended for the poor is purchased by speculators, who leave it vacant. Even some of the low-income participants, after moving into metro-ville housing, sell their properties to upper-income buyers for immediate monetary gain. Almost half of Karachi's slum dwellers reside in these pirated subdivisions.

The city government now concentrates on improving services and supporting self-help programs in the *katchi abadis* and other poor areas of the city. Meanwhile, the larger infrastructure is crumbling, resulting in contaminated water, accumulation of refuse, and power shortages for everyone. Acquiring potable water is a major problem for the poor, who usually have their water delivered from a cart or truck carrying water-filled oil barrels. They pay 600 percent more than richer people. Frustrations arising from these problems are exacerbated by rivalry between the richer, Urdu-speaking Muhajirs and relatively poorer Sindhis, Pathans, and other ethnic groups.

Disillusionment breeds discontent and draws people to activism. All too often, this means turning to fundamentalist Islam and terrorist organizations to solve problems. Karachi, like so many other places in Pakistan, has become a city of violence.

Violence in Karachi

From the *Washington Post* in 2002:

"One block down and just around the corner from the U.S. Consulate, which was shaken by a car bomb, stands the restaurant where *Wall Street Journal* reporter Daniel Pearl disappeared. (Daniel Pearl was tortured and beheaded by al-Qaeda terrorists in 2002.) Three blocks farther along is the Sheraton Hotel, where another car bomb killed 11 French engineers. From there it's just a hundred yards to the bridge where a man with an AK-47 shot dead four Houston oil company auditors on their way to work. And this would be Karachi's very best neighborhood."

A significant characteristic of urban populations in Pakistan is that they are largely people from rural backgrounds. Consequently, urbanization is couched in rural tradition. Most work is in the manufacturing and service sectors and most is labor-intensive. The informal sector of the dual economy is critical to the survival of millions.

The government is convinced that improving rural conditions will stem the ongoing rural-to-urban migration streams that have turned Pakistan's cities into unmanageable, crowded behemoths. Electrification of villages has been a high priority. Power in the countryside can light homes, pump water, and grind grain. The number of villages electrified increased to 126 thousand by 2007, an increase of 11 percent from 2005.

Urbanization is obviously tied to population growth. Although Pakistan has had family planning since 1952, programs have not been very effective. The rural fertility rate is 6.6 children per woman (average TFR 4.0 in 2009; the average age at marriage is 17). With only 11 percent of women using any form of contraception, most are exposed to childbearing for 33 years! Needless to say, infant and maternal mortality rates are high. The already-low sex ratio of 106 males per 100 females is decreasing, indicating better care and increased life expectancy for males versus female infanticide, neglect, and high infant mortality rates for girls.

Part of the problem is the repression of women and lack of education. Enforced seclusion, (except for farm work) and traditional attitudes prevent most females from attending school or even learning to read. Pakistan's literacy rate is 69 percent for men but only 40 percent for women. According to *The Economist*, in 2009, Pakistan ranked as the 14th least literate country in the world.

Although schools have been built for girls in several areas, less than 10 percent of girls ever attend. More girls go to school and attend university in urban centers, but life roles of "suitable positions," arranged marriage, and general subservience to men are still expected for most. The status of men among men is vital in South Asian, Muslim society and must be maintained, in part, through modest and deferential behavior of women, especially those in a man's immediate family.

Muhammed Ali Jinnah had been a strong supporter of rights for women. Gains were made such as women getting the right to hold property. The election of Benazir Bhutto in 1988 suggested that progressive strides would be made. But modernization has impacted primarily elite women. Very few women work in formal or informal sectors full-time. There are scores of women's groups such as the Women's Action Forum (started in Karachi in 1981), but opposition to women's rights has only grown stronger. With Islamization being carried forward in a relatively conservative fashion, women's movements (largely urban and elite) face tough obstacles. Increasing violence against women in recent years has put women's groups into a position of defending the few rights they have rather than acquiring new ones.

Pakistan's rich-poor gap has not narrowed significantly. In 1999, the poorest 20 percent had 8.8 percent of the wealth. The richest 20 percent possessed 42 percent. With nearly 80 percent of Pakistan's population over 25 having little or no schooling, this situation is unlikely to change.

Endemic poverty has spurred the informal economy. In recent years the drug and arms trade, strengthened by upper-level collusion and mafia-like operations, has

Figure 7-13

In Saidu Sharif, these boys are rolling balls of opium to be sold in local shops for processing elsewhere. Drugs are an important source of income for militants such as the Taliban.
Photograph courtesy of B. A. Weightman.

become a critical issue within the international community (Figure 7-13). Not only does this trade generate income, but also it allows Pakistan a measure of political pull in this and other regions of turmoil.

Today, Pakistan is beset with problems. Regional and structural economic imbalances; rapid population growth and urbanization; and sociopolitical conflicts riddle this turbulent land. When considering Pakistan's volatility, keep in mind its rocky beginnings and increasing complexity.

Pakistan is a country at the crossroads, with major political, economic, and social problems, along with immense inequalities. While the elite seem to want real democracy, the disillusioned masses are turning to religious fundamentalism and even supporting the Taliban and other terrorists. Civil unrest, suicide bombings, and insurgency are plaguing the nation now more than ever.

Himalayas: Abode of Ice and Snow

Both Nepal and Bhutan are landlocked, isolated, and very mountainous domains (refer to Figure 7-1). The Hindu Kingdom of Nepal is no longer a kingdom and has been beset with political violence in recent years. Bhutan is trying to maintain its traditional culture in the face of globalization and modernization.

The Himalayas extend across eight nations: Afghanistan, Pakistan, India, China, Nepal, Bhutan, Bangladesh, and Myanmar. The relative isolation promoted by the world's highest peaks and deepest gorges has sustained both natural and human diversity among the 120 million who live here. Many people speak Sino-Tibetan languages and practice Tantric or Vajrayana Buddhism Figure 6-11).

The critical positioning of these various peoples between the political giants of India and China has deeply affected their ways of life and threatened or even eliminated their autonomy. Only Nepal and Bhutan remain independent states in the Himalaya. Ladakh and Sikkim, once independent, are now part of India. Border conflicts and unstable political boundaries are facts of shatter-zone existence.

NEPAL: A KINGDOM NO MORE

Landlocked Nepal is pinched between Tibet (China) and India. Nepal's history has encompassed both survival in context of environmental constraints and maintenance of autonomy and cultural identity within the clutches of competing world powers.

Ten Himalayan peaks mark Nepal's boundary with Tibet (Figure 7-14). Here glaciers and treeless tundra give way to **cloud forests** that coat the slopes from 8,000 to 13,000 feet (2,400–39,000 m). These mountain rain forests are blanketed year-round in fog and mist. Up to 140 inches (356 cm) of rain fall from June to September, but even in the dry season, the forest drips with moisture. Oaks, laurels, maples, and magnolias mix with ancient spruce, fir, and cedar trees. Orchids, morning glories, azaleas, giant ferns, and rhododendrons color the gray-green forest mass. Flowering stands of rhododendron, standing 60 feet tall, turn acres of land crimson. This once-pristine forest houses rare and unusual animals such as the red panda, sun bear, clouded leopard, fishing cat, and goat-antelope. Human encroachment has put this habitat in danger.

Southward from the high Himalaya, the land drops precipitously into lower mountains and foothills. This region contains the 12-by-15-mile (19 by 24 km) alluvial Kathmandu Valley. Here are Nepal's main urban centers including the capital of Kathmandu, with more than 1.5 million people, having grown 62 percent in a decade

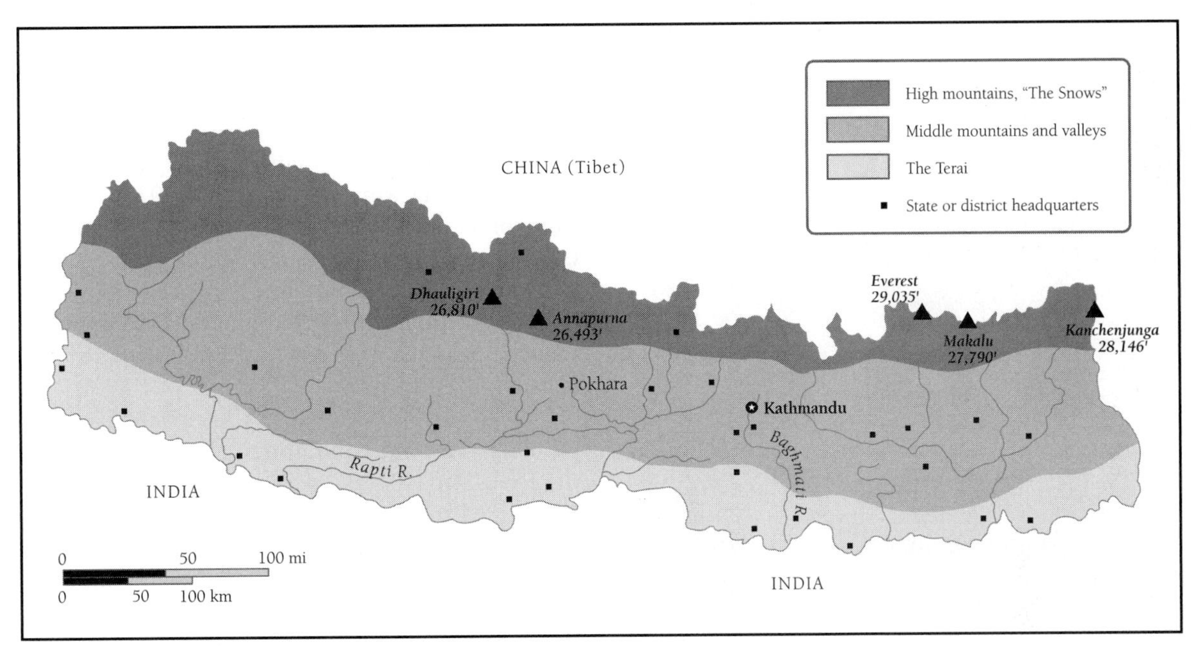

Figure 7-14

Physical geography of Nepal. Nepal serves as a buffer state between India and China.

(Figure 7-15). About two-thirds of Nepal's people live in this hill region, terracing the slopes with rice and wheat. Yields are generally poor. Below lies the marshy plain known as the Terai at 900 feet (270 m). In recent years large numbers of migrants from the hills have descended into Terai to find jobs (Figure 7-16).

The Terai is an extension of the Indo-Gangetic plain. Approximately one-third of Nepal's population lives here. Many of these are high-caste Hindus of Rajput extraction from India. The Terai constitutes only 17 percent of the land but generates 60 percent of Nepal's GNP from commercial agriculture and industrial concerns. Competing for space is the Royal Chitwan National Park, with its tropical monsoon forests, giant *sal* trees, 12-foot (3.6 m) high elephant grass, and diminishing populations of tigers, one-horned rhinos, sloth bears, and an array of other animals and birds. Here too flows the sacred Rapti River, a tributary of the Ganges.

One of my most memorable journeys was in a dugout canoe moving silently along a narrow channel of the Rapti. Resting on the narrow banks, in the shadow of the towering elephant grass, were mugger crocodiles. A one-horned rhino stood belly deep in the water munching on river plants. As the sun rose higher, the muggers slithered into the water, under my canoe! I was further startled by a ruckus in the upper branches of the tall *sal* trees that suddenly came alive with bird cries. Hundreds of parakeets soared overhead, streaking the blue sky with their brilliant green plumage.

Nepal, with its population of 27 million, is culturally and socially diverse Figure 7-15. Nepali is an Indo-European language related to Hindi but infused with Tibeto-Burman. Another indicator of the crossroads situation is religion. Although most people are Hindu, Buddhism and Hinduism intermingle in practice. Life is further complicated by the influence of the Hindu caste system. Here people are differentiated not by occupation but by group customs. The caste system affirms group cohesion and its flexibility adds to the confusion.

Historically, the region was fragmented in competing states. In the eighteenth century the western Gorkhas defeated the ruling Newars of Kathmandu, and in 1790, the country was unified. Border disputes with British India led to the Anglo-Nepalese war of 1814–1816, a war that was fought with the rhythm of the monsoon. In the end, Nepal had to accept a British resident.

From 1816, Britain ruled Nepal as a buffer state. Nepalese troops—the **Gurkhas**—fought for the British against the Indian Mutiny of 1856 and made their way into the annals of military history for their fighting ability

Figure 7-15

This photo of Kathmandu shows a Hindu temple backed by a modern office building. Notice the cloth market held on the temple steps. Shawls, scarves, and rugs are woven in factories and small, back-alley shops and homes. Photograph courtesy of B. A. Weightman.

and heroism. Gurkha regiments would fight for Britain in many battles, which contributed to the granting of Nepal's independence in 1924.

Gurkha regiments stationed overseas have contributed significant remittances to Nepal. However, the return of Hong Kong to China in 1997 and pullbacks from other areas of the world have reduced British Gurkha troops by thousands. While they are still stationed in Singapore and elsewhere, most Gurkhas now fight for the Indian army. Remittances have been severely reduced, and many older Gurkhas, dismissed without pension, live in poverty.

In recent years Nepal has been mired in political problems. Several contentious elections were held in the 1990s. In 2001, Crown Prince Dipendra grabbed an AK-47 and murdered his parents and five other members of his family because he was not given permission to marry the woman of his choice. The king's brother was hastily installed but brought nothing except additional political strife.

Meanwhile, a dissident group of communist Maoists, who were opposed to the monarchy and the corruption and oppression of the poverty-stricken masses, began to take over the country. Inspired by Mao's revolution in China (see Chapter 11), the Naxalite movement in India (see Chapter 8) and the extremist Shining Path movement in Peru, they started their guerilla activities in 1996.

They began by exacting "fees" from the rural population and trekkers in order to "protect" them. A reign of terror ensued with gruesome torture, maiming, and execution of thousands. The Royal Nepali Army didn't do

Figure 7-16

This is a Tharu house in the Terai region of Nepal. The Tharu migrated here from India many years ago. The house is of wattle and daub (reeds and mud) with few openings, to repel insects. The inside of the house is very dark, but this is not so important because most work is done out of doors. The roof is a convenient place to air bedding and dry clothes. Photograph courtesy of B. A. Weightman.

much better. While the Maoists had gained control of almost 70 percent of the countryside, political disarray still characterized Kathmandu. Eventually, in 2006, a cease-fire agreement was made with the auspices of the United Nations. Successful elections were held in 2008 with the Maoists winning a plurality of parliamentary seats. The new government abolished the monarchy and declared Nepal to be a democratic republic. The Maoist insurgency displaced more than 100,000 people and was responsible for the death of at least 12,000.

Poverty and misery remains the reality for most Nepalis, with a third living in abject poverty (see Figure 5-14). According to the United Nations, 90 percent of the labor force is in agriculture and 47 percent are unemployed. Incidence of malnutrition, with consequential mental and growth retardation, and blindness remains critical. Five thousand girls are trafficked to the brothels of India every year. Limited resources and isolation stifle the formal economy.

India has invested and continues to invest in Nepal's industrial and commercial realm. There are two major reasons for this: Nepal provides a buffer between India and China, and India needs the electricity derived from dam development that will be incorporated into its expanding power grid. It is also involved in Nepal's production of items such as garments, carpets, bricks, cement, sugar, soap, and matches. Apparel and carpets are exported primarily to the United States and Germany. Imported manufactures come mainly from India.

The most recent significant entry into Nepal's economy is China. Chinese officials and tourists are increasingly frequent visitors to Kathmandu. This is very disturbing to India, which considers Nepal to be a part of *its* sphere of influence. Like India, China is providing funding for several hydroelectric projects.

Researchers are beginning to recognize that the people of Nepal (and elsewhere) managed to survive in their difficult environment for centuries prior to Western impacts and that modernization and commercialization have severely damaged both fragile environments and cultural integrity. Many "solutions" have backfired, primarily because they are couched in the industrialized world's "scientific expertise" and "development" (read economic growth and consumerism) mentality. As we pointed out in Chapter 3, top-down programs are no longer in vogue. The new government is courting NGOs to return to the country now that peace has been reestablished.

Tourism is Nepal's number one industry, accounting for 60 percent of foreign exchange. However, it benefits primarily the hill and mountain regions because most tourists are trekkers. Internecine violence has severely affected the industry. In 2002, 275,000 tourists visited Nepal but this figure fell 27 percent in 2003.

Numbers have been rising (even with the global recession) as the new government is actively promoting tourism appealing to non-trekkers and trekkers alike. Of the approximately 400,000 people who visited Nepal in 2009, trekkers accounted for 130,000. Numbers of annual foreign visitors are expected to reach well over half a million by 2016. In 2010, the government declared 2011 as Nepal's "National Tourism Year."

Until they are certain of their safety, few North Americans are taking their chances in Nepal. These days, most of the country's visitors are from Asia, especially Indians, Chinese, Japanese, and South Koreans.

Environmental problems abound in Nepal, especially in the Kathmandu Valley. Kathmandu continues to grow at an annual rate of 6 percent. By 2020, the population will reach 2.5 million. About one-third of the valley remains forested. Most forests have "protected status" but are disappearing anyway with city expansion. In the latter half of the twentieth century, valley forests declined by 40 percent.

Kathmandu is literally a pit of pollution. It is situated in a topographical bowl and temperature inversions are common. The city and the surrounding mountains are bathed by ever-thickening layers of dust and smog. In winter and spring, tons of dust are blown in by monsoon winds from India. Air pollution from cars and buses is compounded by the dirt spewed out of brick and carpet factories. Effluents from cooking oil, leather, and other factories pour into open sewers and streams. Only 15 percent of houses are connected to a municipal sewer system. Most people throw their waste into the rivers, resulting in bacterial levels far above even basic public health standards.

Because of the needs of the ever-burgeoning population in the valley, the two sources of potable water—rivers (tapped at their source) and groundwater—are being greatly reduced. In the dry season, rivers provide less than half of the daily water demand. Groundwater is being withdrawn at double its sustainable rate.

The Kathmandu Valley is the main source of farm products. However, as urban phenomena such as shopping centers, bus parks, and squatter settlements sprawl on the city periphery, agricultural land is disappearing. Farmlands decreased by one-third over the past two decades and if this trend continues, there will be no more by 2050. As geographers David Zurick and Autumn Rose (2009) note: ". . . the prospects for a vibrant farming life are bleak."

In 1979, UNESCO classified seven valley locations as World Heritage sites. By 2003, UNESCO had reclassified them as "World Heritage in Danger" sites. The new government has dedicated itself to cleaning up popular tourist attractions such as the great Buddhist stupa at Bouddhanath (Bodnath) on the outskirts of Kathmandu (Figure 7-17). Vehicular traffic has been banned and teams have been organized to clean up the streets. Thanks to these efforts, Nepal's historic sites have been removed from the Danger list.

Figure 7-17

This Tibetan refugee center of Bouddhenath (Bodnath) is a Buddhist stupa containing relics of the Kashyapa Buddha who is believed by Tibetans to have preceded (the original) Gautama Buddha. Here you can see people spinning prayer wheels and moving clockwise around the stupa. The eyes of the Buddha are on worshippers at all times.

Photograph courtesy of B. A. Weightman.

Bhutan: Land of the Thunder Dragon

Bhutan (also called **Drukyul**), is a Lamaistic Buddhist kingdom struggling to keep its culture and identity in the face of outside influences and modernization. Landlocked and isolated, it was virtually closed to the outside world until the 1960s. Like Nepal, Bhutan is one of the world's least developed countries.

This Switzerland-size kingdom once comprised monastic power bases centered on *dzongs*—fortified monasteries. Its deep valleys were linked by cultural and trade contacts facilitated by iron-chain suspension bridges constructed in the fourteenth century. Given the highly differentiated terrain and natural communities ranging from glaciers and tundra in the north to tropical rainforests and malarial swamps in the south, cultural variety is to be expected (Figure 7-18).

The Bhote tribes of northern Bhutan are originally from Tibet and adhere to the Drukpa Kagyudpa sect of Tibetan Buddhism. Commonly called **Drukpas**, they revere the king as their spiritual leader (Figure 7-19). The Nepalese of the south are primarily Hindu. Most people speak Dzongkha, a Tibetan dialect. Others speak Sharchop or Nepali or one of at least 18 other different languages or dialects.

In 1731, Tibet imposed suzerainty over Bhutan. Forty years later, in light of constant regional struggles and raids into Assam and Bengal, the British became involved. After a lengthy civil war one family triumphed and in 1907, the first hereditary king was enthroned with British support. Four years later, Bhutan signed a treaty with the British to counter Chinese designs on the region and later agreed to rely on the advice of India (Great Britain) for action in foreign affairs. This agreement was formalized by India in 1949.

While Bhutan's current foreign policy leans toward isolationism, China's 1950 claims to Bhutan and its 1959 occupation of Tibet cemented Bhutan's stronger ties with India. The first road from India to the capital of Thimpu was opened in 1962. Now, several roads from India are furthering cultural intrusion and the trappings of development. For India, Bhutan is a convenient buffer state between it and China. It is a strategic training and staging ground for troops that could be quickly employed against China or separatist groups in Assam.

Subsistence farmers cultivate grains and vegetables, and herders raise yaks and cattle and practice transhumance (moving livestock between the highlands in summer and the lowlands in winter). A mere 6 percent of the

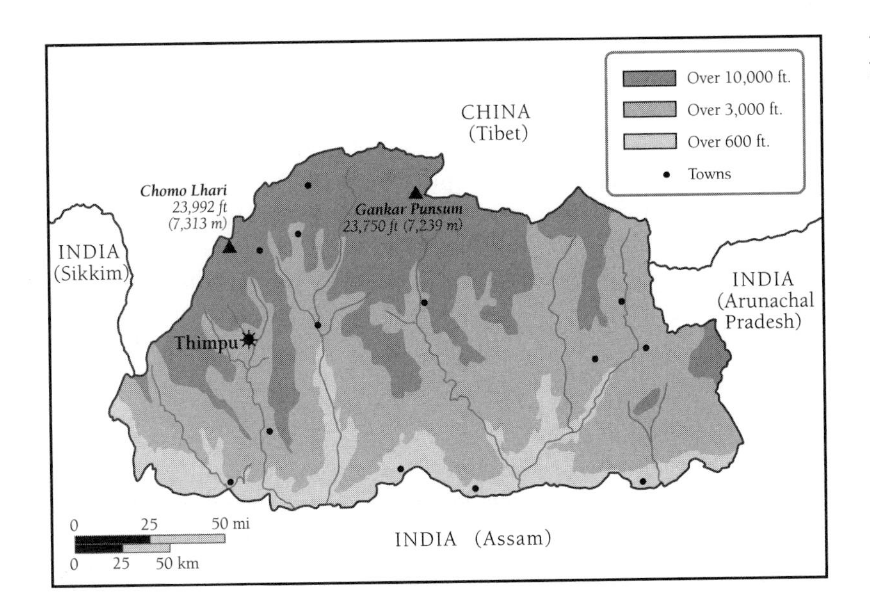

Figure 7-18

The Kingdom of Bhutan.

land is cultivated. Most cropping is confined to the flatter alluvial valley floors and narrow terraces. However, Bhutan is no longer self-sufficient in cereals. Efforts are being made to diversify agriculture from traditional maize and rice to fruit and vegetables, especially for export to India. One long-standing export crop is **cardamom**, a spice commonly used in South Asian cuisine. There is evidence that some commercial crops are culturally and environmentally inappropriate.

Bhutan is also a source of hydroelectric power, convenient to India's industrial region west of Kolkhata. India is active in dam construction and has negotiated power-sharing agreements. In fact, most of Bhutan's generated power is used by India. As most foreign aid and development assistance is also from India, Bhutan finds itself in an ever-tightening dependency relationship.

Modernization began in the 1950s. However, the country is too small to develop any economies of scale.

Figure 7-19

These Drukpar are herders in the Bumthang Valley of east-central Bhutan. The Drukpar are nomadic yak herders who move seasonally with their animals between the high pastures and the lowland forests. Photograph courtesy of David Zurick, Eastern Kentucky University.

For example, Drukair (Royal Bhutan Airlines), is possibly the smallest national airline in the world with its two planes. However, rural-to-urban migration and natural increase in the capital city of Thimpu has resulted in a 100 percent growth in its population since 2001. Today, Thimpu has close to a million residents.

Bhutan's King Jigme Singye Wangchuck has pointed out that for his people, "gross national happiness" supersedes "gross national product." Although there is some small-scale industry in the south, including chemicals, cement, and wood products, the majority of Bhutan's inhabitants work directly with the land.

"Gross National Happiness"

The philosophical basis of life in Bhutan is upheld by a Buddhist precept that stresses both spiritual and emotional fulfillment. It also emphasizes that people have necessary material necessities and a respect for natural order. Consequently, the kingdom's development policies focus on enrichment of people's lives. Meeting basic needs, increasing social and economic choices, ensuring cultural preservation and environmental protection are fundamental. Bhutan's ruler and his Drukpa subjects are very concerned about the precarious state of their Buddhist culture and the endangered state of their physical environment in light of modernization. Consequently, tourists are limited to 3,000 a year and must be on guided tours that are very expensive for most people. Eighty percent of the population participated in Bhutan's first election in 2008, which swept the "Peace and Prosperity Party" into power.

Sixty percent of the land is designated as forested, and sustainable forestry is national policy. Nevertheless, reforestation programs are not keeping up with deforestation. Fuelwood consumption is high in Bhutan—among the world's highest. A lumber industry introduced in the 1960s has been curtailed and farmers are given free tree seedlings to plant on private lands. Unfortunately, the logging industry, legal and illegal, continues.

Large tracts have been set aside as floral and faunal preserves. With its array of biomes, Bhutan houses the world's only population of golden langur and some of the few remaining populations of one-horned rhino, tiger, and Asian elephant. A greenbelt and an elephant corridor have been created along the Indian border.

Nepalese have resided in Bhutan since the nineteenth century. In the mid-twentieth century, many of Nepal's soaring population have illegally migrated across the border, settling in the southern lowlands of Bhutan alongside the existing Nepalese community. When Bhutan conducted a census in 1988, it discovered that ethnic Nepalese made up an unspecified but significant proportion of the total population. The former kingdom of Sikkim had already been overrun by Nepalese and, for protection, elected to join India. Bhutan's king did not want a repeat performance in his realm.

In an effort to salvage Bhutanese culture, the king launched ***Drignam Namzha***, a "traditional values and etiquette" policy that promoted Dzongkha in the schools and required all people to wear traditional Bhutanese dress on certain occasions. In 1985, it instituted the Citizen Act, which requires that all residents prove their historical roots in the country.

From 1990 to 1992, 100,000 illegal immigrants were expelled: 70,000 to Nepal, and 30,000 to India to join the 4 million Nepalis already there. Again, the problem was not solved, and numerous incidences of violence and human rights abuses occurred thereby drawing the attention of the United Nations. In 2009, the first batch of Nepalis left for UN-supervised refugee camps in eastern Nepal. The remaining 86,000 signed up for resettlement in the West—mostly in the United States.

Even with sustainable development policies, Bhutan has numerous social and environmental problems. Poverty, disease, illiteracy, high infant mortality, and low life expectancy are universal. However, Bhutanese women have more freedom and power than in most other South Asian societies. Environmental problems are increasingly widespread and are similar, if not as intense, as in other Himalayan communities.

Mountain Development and Eco-Destruction

Magnificent, awesome, challenging—such descriptors belie the reality of the Himalayas' geological instability and eco-vulnerability. In this oft-perceived "barren wasteland," millions struggle daily to eke out a living. Long subject to earthquakes, landslides, and flash floods, Himalayan populations more and more are confronted by problems related to environmental fragility and degradation.

There are three geo-ecological zones. The Outer Himalayas have an average elevation of between 3,280 and 6,560 feet (1,000–2,000 m). This lowest region extends almost unbroken from the Indus River to the Brahmaputra and includes numerous long, flat-bottomed valleys called duns. These are filled with gravelly alluvium (look again at Figure 2-1).

In between this zone and the Great Himalayas are the Middle Himalayas. About 129 miles (206 km) wide and between 9,840 and 16,400 feet (3,000–5,000 m) in altitude, this zone comprises the Himalayan foothills. Ranges run obliquely from the Great Range or stand disconnected. Most people live here, concentrated in such valleys as the Kathmandu Valley of Nepal and the Vale of Kashmir. This is the region of most critical environmental and human damage.

The Great Himalaya, the highest zone, is sparsely populated. However, in recent years people from the Middle zone have come into the region, placing additional stress on limited resources. This is also the environment affected most by tourism. The colonial era brought British explorers to the Himalaya. Thereafter, the region was open to commerce with the densely peopled plains of India. Then imported mass-produced goods were introduced to mountain communities, and the region was exploited by foreign power structures for resources such as minerals, timber, and cheap labor supply. Himalayan peoples soon became enmeshed in political and economic dependency relationships outside their control.

Tibetan-influenced Buddhism predominates in the region of the Himalayas, although Hinduism and Islam are important in places closer to India. Monasteries, *chortens* (structures containing relics of a holy person), *mani* walls (walls of stones with prayers carved on them), prayer flags, and pilgrimage tracks are Buddhism's impress upon the Himalayan landscape. Dwarfed by Chomolungma, "Mother Goddess of the World" (Everest), Kanchenjunga, Dhaulagiri, and other pillars of the gods, it is little wonder that millions believe in the sanctity of Himalayan peaks.

Mount Kailas

Not the highest, but most significant in human terms, is 22,028 feet (6,608 m) Mt. Kailas. Believed by both Hindus and Buddhists to be the sacred center of the universe, Mt. Kailas is central to religious conceptions of the world known as *mandalas*. From this sacred pinnacle flow the four great rivers: the Sutlej, Indus, Ganges, and Tsang Po (Brahmaputra). In reality, the sources of these important rivers lie within a 60-mile (96 km) radius of Mt. Kailas.

The mountain also figures prominently in pilgrim circulation. Thousands of pilgrims converge on this site each year to complete the 34-mile (55 km) circumambulation upward to 18,600 feet (5,580 m). In 1900, Kawaguchi Ekai, a young, Zen monk from Japan, walked across Nepal and traversed an east–west trail across the Tibetan Plateau. As recorded in Scott Berry's book *A Stranger in Tibet* (1989), the monk wrote of his first vision of Mt. Kailas and its sacred lakes. "Verily, verily it was a natural mandala. The hunger and thirst, the perils of dashing stream and freezing blizzard, the pain of writhing under heavy burdens, the anxiety of wandering over trackless wilds, the exhaustions and lacerations, all the troubles and sufferings I had just come through, seemed like dust, which was washed away. . . ."

Although the Himalayas were apparently remote and inaccessible, isolation was not complete. Traders plied precipitous routes, transferring goods such as salt and wool from Tibet and silk from China to eager merchants in India. Through trade and religious pilgrimage, connectivity was maintained inside and outside the region. Although contact with Tibet is now limited, yak caravans continue to trek remote areas, linking them with newly built truck routes. Aside from salt and other raw materials, the caravans and trucks now carry cans of cooking oil, tanks of propane, and other products. Pilgrimage continues to be a means of information and economic interchange. But the pilgrims now talk of how much their lives have changed, perhaps about the new satellite dish atop the local monastery.

ROADS AND DAMS

Development schemes such as roads and dams are often based on the interests of outsiders. India has been very active in the region, understandably in light of its strategic interests regarding Pakistan, China, and Bangladesh.

After a border war between India and China in 1962, India built numerous road networks as high as 17,000 feet (5,100 m). Road construction—especially dynamiting—destabilized the hills and soil, and the roads gave lumber companies access to virgin forests.

Forests and wildlife quickly vanished. The threat of tigers was replaced by the threat of landslides.

Road construction has also delivered sweeping cultural change. Truck routes funnel in new products and new ideas. They are also conductors for increased prostitution and the diffusion of AIDS. Commercialization, resource extraction, and mass tourism have effectively devastated natural and human landscapes to the point that the integrity of traditional human–environmental relationships has been destroyed.

The Himalayas function as the water tank of Asia. It has the highest runoff of any mountain range, providing water for one-tenth of the world's population. However, torrential rivers invite dam construction to generate hydropower for electricity and water for irrigation. Since the 1950s, dams have proliferated, particularly in the upper Ganges and Indus tributary areas. Dam construction destroys forests, degrades land, and displaces mountain and hill people, forcing them to move to other already stressed regions. Pakistan's Mangla Dam, completed in 1967, receives so much silt and debris from the Jhelum that its operational life has been cut from 100 to 50 years.

Soil erosion is accelerated by development projects and general overuse of natural resources. Rivers run brown with silt, precious topsoil needed to sustain crop yields. Some 300 cubic meters of topsoil annually wash from the Nepal Himalaya to the plains of the Ganges and Brahmaputra below. The stain of Himalayan silt can be seen as far as 400 miles (645 km) from shore in the Bay of Bengal.

LOGGING

Forestry has exacted the greatest toll from the ecosystem. Large-scale cutting commenced in the latter nineteenth century when the British hauled off 6.5 million railway ties from one of the world's finest deodar forests. Since 1950, logging, agriculture, and urbanization have destroyed nearly half of the forest cover in the Indian Himalayas alone.

Throughout the Himalayas, original village forests are so degraded that people (mostly women) must spend hours every few days to find and gather firewood. Imagine carrying at least 50 pounds of wood on your back, secured by a single wood-fiber strap around your forehead, trudging barefoot to get home to cook a meager breakfast for the family, do household chores, feed the animals, and do farm work. Every year, a Nepalese family spends at least two months' worth of time acquiring firewood.

In 1970, a huge storm raged over the great mountain Nanda Devi (25,650 feet, or 7,695 m). Tons of earth, boulders, and trees crashed down slopes. A tributary of the Ganges rose more than 60 feet (18 m). Six road bridges, 24 buses, and some 600 homes were swept away, and nearly 200 people were killed.

Mountain dwellers were stunned at the magnitude of this disaster. Realizing that vast forest areas had been cleared by commercial companies, they were determined to fight back and save their environment. In the north of India's Uttar Pradesh state, where leopards and tigers once prowled the forests, a group of women literally held on to trees as loggers wielded their axes. This grass roots movement became known as *Chipko Andolan*, meaning "the movement to embrace the trees." As women in other areas engaged in "tree-hugging" protests, the **Chipko Movement** drew international attention and became instrumental in the formulation of India's conservation policy. Although village preservation groups have become ubiquitous, population and other pressures limit their effectiveness.

TOURISM

Tourism and other forms of commercialization have had far-reaching impacts. The tourist count on the slopes of Mt. Everest alone was 18,000 before the 2001 political crisis in Nepal reduced that number. Tourism is a double-edged sword. It does generate revenue and employ significant numbers of locals. On the other hand, tourist facilities and activities add trash and pollute land and water, trample delicate vegetation, and place even greater demand on wood supplies than local populations. Villagers strip forests to provide hot water for "needy" tourists. Tourists deface rocks and religious sites, and some even go so far as to steal the inscribed stones of sacred *mani* walls. The high Himalayas have been termed "The loftiest garbage dump in the world." Tourism also compromises cultures. As noted above, this is why Bhutan has imposed strictures on its tourism industry.

One people to be drastically affected by tourism are the Sherpas of Nepal. When Sir Edmund Hillary became the first Westerner to ascend to the top of Mt. Everest from Nepal in 1953, he was lauded throughout the world. Standing in the shadow of his glory was a Sherpa named Tenzing Norgay. The Hillary expedition is now known as the Hillary–Norgay expedition.

Although Sherpas are famed for their mountain-climbing abilities, they are not traditionally mountain climbers. Seeking new pastures for their yak herds, their ancestors migrated into the Everest region from Tibet around 500 years ago. They began trading with people in lowland Nepal and India. When the Tibetan border was closed by China in the 1950s, Nepal was opened to foreign climbing expeditions.

Sherpas had already served as servants and porters for British climbers attempting to reach Everest's summit from Tibet. However, when Hillary and Norgay achieved the summit from Nepal, the word *Sherpa* was heard around the world.

Sherpas are now dependent on tourism. Although some lead expeditions to conquer peaks, others guide trekking groups at lower elevations. Some Western climbers admit that they pay Sherpas to take risks they are unwilling to take. On some occasions, injured Sherpas have been abandoned by climbers too anxious to reach a summit. There are other negative consequences. Having earned money to be there in the first place and then dirtying the land is offensive enough to the gods. A terrible death in the snows even furthers the detriment to one's karma and diminishes chances of a good rebirth.

Master–servant relationships are now being transformed into client-guide relationships. Today, many climbers go lightweight and even without oxygen. Sherpas are employed to maintain base camps. However, media attention is usually given to the large-scale expeditions, and those guides who do climb seldom receive recognition equivalent to that of their clients. Have you ever heard of Sundare Sherpa, who has ascended Mt. Everest five times without oxygen? Or Kaji Sherpa, who in 1998 climbed the peak in 20 hours and 24 minutes, two hours faster than the previous record?

About Yaks

Imagine a one-ton, hairy ox standing 6 feet high at its shoulder. Imagine a beast that is able to carry as much as 300 pounds for 20 or more days at an elevation of 18,000 feet! (I've trekked at 15,000 feet and could barely breathe. I had to have a little boy carry my camera equipment.) This is a yak, the most useful animal for residents of the high Himalayas (Figure 7-20).

Aside from their packing abilities, they provide meat, milk, butter, cheese, wool, and leather. Yak butter is offered in copious amounts in Tantric Buddhist temples and is even used to make sculptures of Buddha and various other gods. If you visit one of these temples, you will be offered butter-tea. This is tea with seemingly rancid yak butter floating on top. Yak products are also used in traditional medicine and their tails make fine dusters for the sacred scrolls in Buddhist temples.

First domesticated in Tibet, yaks cannot survive well below 10,500 feet (3,500 m). A blood hemoglobin is thought to be responsible for a yak's ability to exist in thin air at high altitudes. It begins to lose its condition around 7,000 feet (2,100 m). Females, called *dimos*, are smaller than males and give about one quart of milk a day. *Dimos* produce a calf once every two years; populations do not reproduce quickly. Yaks can live 25 years.

There are around 12 million domesticated, black, brown, and gray yaks. Wild yaks are much larger and are always black. Wild yaks, however, are very rare, their populations diminished by illegal hunting and loss of habitat.

Yaks can be crossed with cattle to produce animals better suited to lower elevations. This animal is called a *dzho* and looks like a cow with long hair.

Figure 7-20

This is your friendly "rent-a-yak." Tourism is increasing in the remote regions of the Himalayas. Photograph courtesy of B. A. Weightman.

The Impact of Population and Modernity

For thousands of years, Himalayan villagers have survived by subsistence farming and herding. For many, transhumance was a way of life. But all this has changed as the region becomes more populated and entangled with the forces of modernity.

Populations increased with immigration of Indians from the twelfth century and again with eighteenth- and nineteenth-century food surpluses. However, road construction in the 1950s allowed the advent of modern, although limited, health care. This resulted in a drastic decline in death rates, an increase in life expectancy, and, inevitably, a population explosion. The populations of Pakistan, Nepal, and Bhutan are predicted to mushroom by 85, 67, and 46 percent, respectively, by 2050 (refer to Table 3-1).

With the increasing population, land use intensified on already limited farming and grazing areas, resulting in reduced crop yields and ruined pastures. Greater wood requirements meant that woodlands were rapidly depleted. Productive land use systems that sustained smaller Himalayan populations for centuries were unable to support the added numbers.

The average population growth for the entire Himalayan region is 2 percent. This, combined with a decrease in the area of cultivated land and a mere 1 percent increase in farm yields, means that local food production can no longer keep pace with human requirements. Because living conditions are deteriorating, many poor women feel the need to have more children to help them.

With self-sufficiency destroyed, ecological migration began in earnest. People sought new land in higher and even more fragile environments, where too-steep slopes and thin soils cannot support intensive cropping or grazing. A second, mostly male ecological migration to lower elevations is an ongoing process. Cities and towns of the hills and lowlands swell with unemployed males, while women and children assume added burdens in their mountain villages.

Many outsider "experts," agencies, and national and local governments have held the view that the people were "destroying their own environment" with their primitive and harmful methods. But proposed solutions have been drawn from experiences elsewhere. These often do not appreciate the intricacies of varied niche environments of mountain settings. Consequently, modernization efforts and palliative programs have worsened the situation and increased dependency relationships with outside political and economic forces. Attitudes take a long time to change.

Commercialization has transformed a barter economy to a cash economy. This, along with public intervention, improved accessibility; media exposure and tourism have combined to raise expectations. Simultaneously, social and economic inequities have intensified, with little opportunity for improvement.

Until recently, efforts to improve conditions have been top-down and rarely implemented effectively if at all. Even with changing attitudes of large organizations such as the World Bank and the United Nations and the presence of more NGOs, funds continue to be funneled into the foreign bank accounts of the politically powerful. As we observed earlier in this chapter, corruption and vested interests reign supreme in this region.

Research in the region of the Himalayas has made clear at least three important facts. First, a society's inability to live within the usable limits of a biophysical resource base generates responses that inevitably cause environmental alteration and degradation. Second, market forces can be singled out as the most significant factors underlying those responses. Third, policy makers and farmers perceive neither the problems nor the solutions in the same way. This derives from differing degrees of closeness to the problem and differing stakes in outcomes of both problem and solution.

While the situation is critical in many areas, especially environmental, for many people general well-being has improved. Land reform, health care, income subsidies, increased education, self-help programs, and so forth have exerted some positive impacts. However, most of these have bonded mountain communities to outside power structures more tightly than ever.

Clearly this northern region of South Asia is rife with complex problems. Some of these arise from environmental constraints. Others emerge from transitional political settings. At the turn of this twenty-first century, however, two trends are apparent. On the one hand, international and global economic and strategic concerns are increasingly overriding the impacts of local geographies. On the other hand, global frameworks are counterbalanced by cultural and territorial identity consciousness. This is an era of rampant and proliferating nationalism. Unfortunately, in many regions environmental integrity and social well-being have become lost in the shuffle.

Recommended Web Sites

www.cfr.org
Council on Foreign Relations. Links to Asian regions with current articles on foreign affairs.

http://countrystudies.us/pakistan/49.htm
Agriculture and industry to early 1990s.

http://en.wikipedia.org/wiki/Chronology_of_terrorist_incidents_in_Pakistan
Compendium of terrorist attacks in Pakistan in 2001–2010.

www.fao.org/sd/WPdirect/WPre0111.htm
Sustainable Development Department of the U.S. Food and Agriculture Organization (SD Dimensions). Articles about agriculture, women's participation, and sustainable development in South and Southeast Asian countries.

www.finance.gov.pk/finance_economic_survey.aspx
Pakistan economic survey 2008.

www.kashmirstudygroup.net
Information on the Kashmir conflict to 2003. Excellent map of the region.

www.lib.berkeley.edu/SSEAL/South Asia/
University of Berkeley Library with links to online resources on South Asia.

www.nation.com.pakistan-news-newspaper-daily-english-online/
Pakistan's *The Daily*-a major newspaper in English.

http://news.bbc.co.uk/
British Broadcasting Corporation news with links to South Asia and Asia-Pacific.

www.pakistan.com/english/index.shtml
Information on Pakistan's agriculture: data on crops, productivity, related industries, food imports and exports.

www.pakistaneconomist.com
Pakistan's leading business magazine.

www.satp.org
South Asia terrorism portal. Keep up on the latest insurgencies, acts of terrorism, responses, and economic, social, and political consequences.

www.statpak.gov.pk/depts/fbs/publications/publications.html
Pakistan's Federal Bureau of Statistics. Demographic, labor force, gender, and various social indicators. Statistical Yearbooks 2007–2008.

www.textileasia.com.pk/
Information on the textile industry.

www.wttc.org/eng/Tourism_Research/
World Travel and Tourism Council. Useful site with Information and data on tourism by country.

Bibliography Chapter 7: South Asia: Pakistan and the Himalayan States

Abbas, Hassan. 2005. *Pakistan's Drift into Extremism: Allah, the Army, and America's War on Terror*. Armonk, New York: M. E. Sharpe.

Ahmed, Akbar S. 2002. *Discovering Islam: Making Sense of Muslim History and Society*. London: Routledge.

Akbar, Malik S. 2010. "Balochistan's Unattended IDP Crisis." *DailyTimes* (Pakistan), January 31.

Allan, Nigel J. R. 1990. "Household Food Supply in Hunza Valley, Pakistan." *The Geographical Review*, 80: 399–414.

Arif, Jamal. 2009. *Shadow War: The Untold Story of Jihad in Pakistan*. Pacific Palisades, Calif.: Melville House.

Ayesha, Jalal. 2008. *Partisans of Allah: Jihad in South Asia*. Cambridge, Mass.: Harvard University Press.

Bajoria, Jayshree. 2008. "Pakistan's New Generation of Terrorists." *Backgrounder*, Washington, D.C.: Council on Foreign Relations.

Bray, John. 1993. "Bhutan: The Dilemma of a Small State." *The World Today*, 49: 213–216.

Brook, Elaine. 1988. "Through Sherpa Eyes." *Geographical Magazine*, 60: 28–34.

Brower, Barbara. 1993. "Co-Management vs Co-Option: Reconciling Scientific Management with Local Needs, Values, and Expertise." *Himalayan Research Bulletin*, 13: 39–49.

Byers, Alton. 2005. "Contemporary Human Impacts on Alpine Ecosystems in the Sagarmatha (Everest) National Park, Khumbu, Nepal." *Annals of the American Association of Geographers*, 95/1: 112–140.

Casella, Alexander. 2009. "Nepal Finally Waves Away Refugees." *Asia Times*, December 15.

Chopra, Pran. 1986. "The View from Bhutan: The Temptations and Traps of Development." *Ceres*, 19: 24–27.

Cohen, Stephen. 2006. *The Idea of Pakistan*. Washington, D.C.: Brookings Institution Press.

Dalrymple, William. "Pakistan Reborn?" *New Statesman*, February 25: 26–28.

Davis, Mike. 2007. *Planet of the Slums*. London: Verso.

Denniston, Derek. 1993. "Saving the Himalaya." *World Watch*, 6: 10–21.

EDF. 2005. "A Drought of Accountability." *Environmental Defence Fund*.

Elahi, Asad. 2006. *Compendium on Gender Statistics 2004*. Islamabad: Government of Pakistan.

Fair, Christine. 2009. *The Madrassas Challenge: Militancy and Religious Education in Pakistan*. Lahore: Vanguard Books.

Ganguly, Sumit, and Devin T. Hagerty. 2006. *Fearful Symmetry: India Pakistan Crises in the Shadow of Nuclear Weapons*. London: Oxford University Press.

Gerrard, A. John. 1990. *Mountain Environments: An Examination of the Physical Geography of Mountains*. Cambridge, Mass.: The MIT Press.

Haggani, Husain. 2005. *Pakistan: Between Mosque and Military*. Carnegie Endowment for International Peace.

Halvorson, Sarah J. 2002. "Environmental Health Risks and Gender in the Karakorum-Himalaya, Northern Pakistan." *The Geographical Review*, 92/2: 282–306.

Hausler, Sabine. 1993. "Community Forestry: A Cultural Assessment: The Case of Nepal." *The Ecologist*, 23: 84–90.

Hussain, Akmal. 2003. *Pakistan National Human Development Report: Poverty, Growth and Governance*. Karachi: Government of Pakistan.

Hussain, Zahid. 2007. *Frontline Pakistan: The Struggle with Militant Islam*. New York: Columbia University Press.

Jodha, N. S. 1995. "The Nepal Middle Mountains." In *Regions At Risk: Comparisons of Threatened Environments*, eds. Jeanne X. Kasperson, Roger E. Kasperson, and B. L. Johnson, pp. 140–185. London: Heineman.

Jones, Owen B. 2002. *Pakistan: Eye of the Storm*. New Haven, Conn.: Yale University Press

Khan, Ashraf. 2003. "Gulf Gateway." *Far Eastern Economic Review*, October 8: 54.

Karan, P. P., and Cotton Mather. 1985. "Tourism and Environment in the Mount Everest Region." *The Geographical Review*, 75: 93–95.

Karan, P. P, and Shigeru Iijima. 1985. "Environmental Stress in the Himalaya." *The Geographical Review*, 75: 71–92.

Mandelbaum, David G. 1993. *Women's Seclusion and Men's Honor: Sex Roles in North India, Bangladesh, and Pakistan*. Tucson: University of Arizona.

McColl, Robert W. 1987. "House and Field in the Karakorums: The Interaction of Environment and Culture." *Focus*, 37: 15–19.

Margolis, Eric. S. 2000. *War at the Top of the World: The Struggle for Afghanistan, Kashmir and Tibet*. New York: Routledge.

Metz, John, J. 1990. "Forest Product Use in Upland Nepal." *The Geographical Review*, 80: 279–287.

Mortenson, Greg, and David O. Relin. 2006. *Three Cups of Tea*. New York: Penguin Group.

Norton, James H. 2009. *India and South Asia (Global Studies 9th edition)*. New York: McGraw Hill.

Owen, Nicholas. 2009. "Nepal's Democracy Honeymoon Ends." *Far Eastern Economic Review* June: 21–24.

Pomeroy, George, and Ashok Dutt. 2008. "Cities of South Asia." In *Cities of the World,* eds. Stanley D. Brunn, Maureen Hays-Mitchell, and Donald J. Ziegler, pp. 385–427.

Rahman, Mushtaqur. 1996. *Divided Kashmir*. Boulder, Colo.: Lynne Reinner.

Rahman, Tariq. 1995. "Language and Politics in a Pakistan Province: The Sindhi Language Movement." *Asian Survey*, 35: 1005–1016.

Rashid, Ahmed. 2008. *Descent into Chaos: The United States and Their Failure of Nation Building in Pakistan, Afghanistan and Central Asia*. London: Viking Press.

Rauniyar, Durga, S. 1997. "Spotlight: Nepal." *Population Today*, 25: 7.

Ring, Laura A. 2006. Zenawa: *Everyday Peace in a Karachi Apartment Building*. Bloomington, Ind.: Indiana University Press.

Scarborough, Vernon L. 1988. "Pakistani Water: 4,500 Years of Manipulation." *Focus*, 38: 12–17.

Schmidle, Nicholas. 2009. *To Live or to Perish Forever: Two Tumultuous Years in Pakistan*. New York: Henry Holt.

Schofield, Victoria. 2000. *Kashmir in Conflict: India, Pakistan, and the Unfinished War*. London: I. B. Tauris.

Shaikh, Farzana. 2009. *Making Sense of Pakistan*. New York: Columbia University Press.

Siddiqi, Akhtar Husain. 1989. Urban Development in Pakistan. In *Urbanization in Asia: Spatial Dimensions and Policy Issues*, eds. Ashok K. Dutt, Laurence J. C. Ma, and Allen G. Noble, pp. 47–74. Honolulu: University of Hawaii.

Spate, O. H. K., 1948. "The Partition of India and the Prospects of Pakistan." *The Geographical Review*, 38: 5–29.

Spate, O. H. K., and A. T. A. Learmonth. 1967. *India and Pakistan: A General Regional Geography*, 3rd edition. Bungay, Suffolk: Richard Clay (Chaucer).

Stephenson, Bryan. 1994. "Bhutan: The Development Dilemma." *Geographical Magazine*, 66: 51–55.

Taylor, George C. Jr. 1965. "Water, History, and the Indus Plain." *Natural History*, 74: 40–49.

Thapa, Gopal B., and Karl Weber. 1991. "Deforestation in the Upper Pokhara, Nepal." Singapore *Journal of Tropical Geography*, 12: 53–67.

The Economist. 2010. "The Resurgence of al-Qaeda: The Bombs that Stopped the Happy Talk." *The Economist*, January 30: 69–71.

United Nations. 1988. *Population Growth in Mega-cities. Karachi*. Population Policy Paper No. 13. New York: United Nations.

Urban Resource Center. 2002. "Urban Poverty and Transport: A Case Study from Karachi." *Environment and Urbanization* 13/1.

Vick, Karl. 2002. "In a Dangerous City of Dreams, Survival Rules: Karachi's Lawless Streets Attract Mix of Militants." *Washington Post*, June 24.

Walcott, Susan. 2009. "Urbanization in Bhutan." *The Geographical Review*, 99/1: 81–93.

Weaver, Mary Anne. 2002. *Pakistan: In the Shadows of Jihad and Afghanistan*. New York: Farrar, Straus & Giroux.

Weiss, Anita. M. 1998. "The Slow yet Steady Path to Women's Empowerment in Pakistan." In *Islam, Gender, and Social Change*, eds. Yvonne Haddad and John Esposito, pp. 124–143. New York: Oxford.

World Watch. 2007. "Kashmir: Physical Tremor, but no Political Earthquake." In *Beyond Disasters*, pp. 32–36. Washington, D.C.: World Watch.

Young, L. J. 1991. "Agriculture Changes in Bhutan: Some Environmental Questions." *Geographical Journal*, 157: 172–178.

Zurick, David, and Autumn Rose. 2009. "Landscape Change in Kathmandu Valley." *Focus on Geography* Spring: 7–16.

Chapter 8

South Asia: India, Giant of the Subcontinent

"We wear the dust of history on our foreheads and the mud of the future on our feet."

SHASHI THAROOR (1997)

Indian scholar and writer Shashi Tharoor (1997) believes that "The central challenge of India as we enter the twenty-first century is the challenge of accommodating the aspirations of different groups in the Indian dream." The reality of 1.12 billion individuals, belonging to countless ethnic, linguistic, religious, caste, and other defined groups, makes India's continued existence as a federation amazing. *Federalism* is a governmental system whereby central and regional authorities are voluntarily tied in a mutually interdependent relationship. In this light, India is a classic paradox; it is a unity of disunity. It is multi-textured, a cauldron of competing interests, a collection of centers and peripheries (Figure 8-1).

Independent India was conceived with diversity in mind. Its ability to change, redrawing boundaries along linguistic lines for instance, is a large part of its *raison d'etre*. Federalism thrives on the tension between cores and peripheries, as long as the links are not severed. With this in mind, India's Constitution grants extraordinary powers to federal authorities when unity and the integrity of the nation are threatened. Thus far, separatist movements have been quashed. However, after years of independence, the voices of dissension are more frequent, much louder, and more powerful. These arise from social inequity, regional disparity, and political instability. These are some of India's challenges.

Demographic Challenges

Figure 8-2 shows that India is already densely populated, especially in the Ganges River valley. At its partition, India's population was 344 million. According to the 1991 census, it was 844 million. By 2003, it had reached more than a billion. Although India's NIR has fallen from 2.1 in 1978 to 1.6 in 2009, overall population growth is phenomenal. India's highest growth rate was in the 1970s and 1980s at 28 percent. In the 2000s, it has dropped only to 22 percent. India adds at least 18 million people a year—2,000 babies are born every hour! Its area is just half that of China, but it will probably exceed it in population by 2020. Moreover, as we will see below, food production is not keeping pace with population growth.

FAMILY PLANNING AND FERTILITY

India has a less-than-successful history of family planning in terms of programs and their impacts. In 1952, it became the first country in Asia to establish an official family planning program with the aim of curtailing population growth. Various national and state programs were instituted, including promotion of IUDs and sterilization. Three million men underwent vasectomies in 1970–1971.

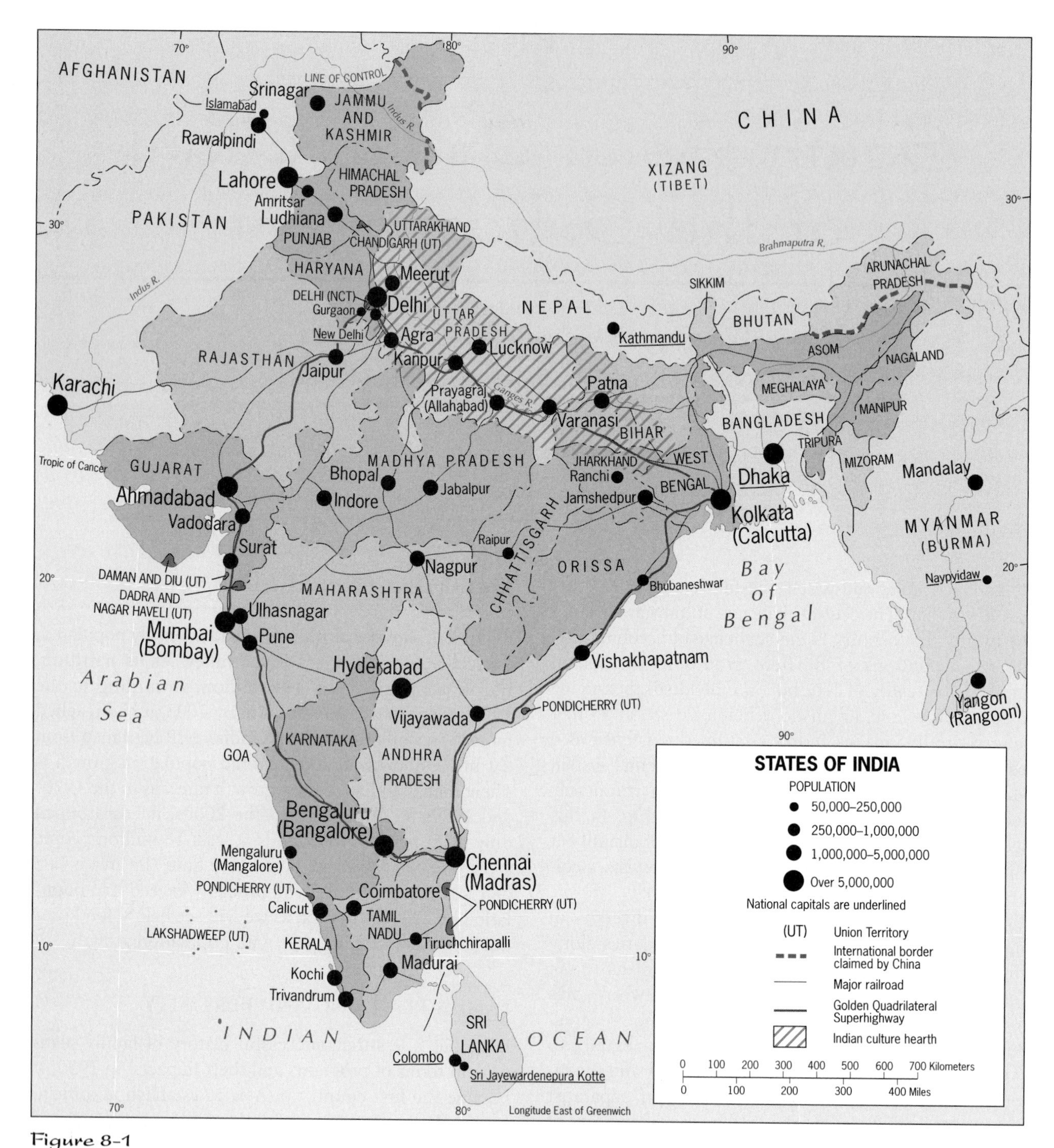

Figure 8-1

States and cities of India. From H. J. de Blij and P. O. Muller, *Geography: Realms, Regions and Concepts*, 14th edition, 2010, p. 428. Originally rendered in color. H. J. de Blij and P. O. Muller. Reprinted with permission of John Wiley & Sons, Inc.

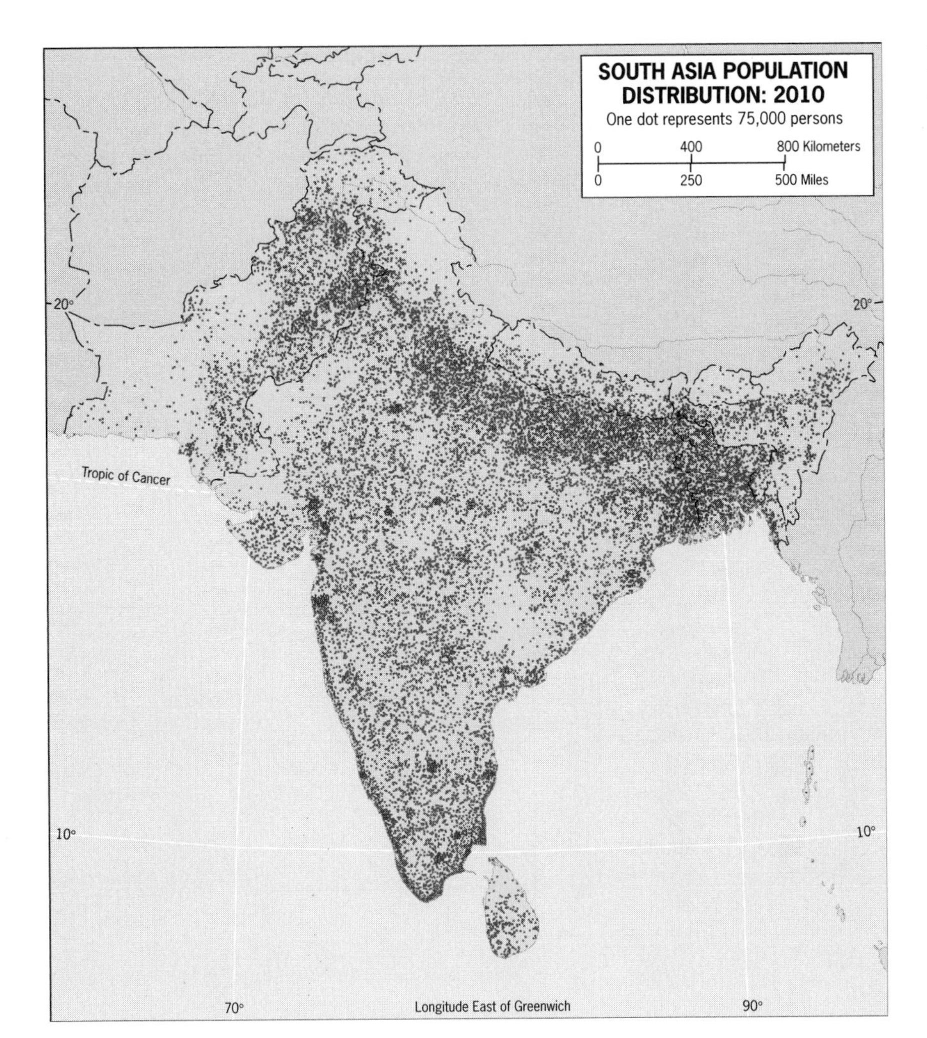

Figure 8-2

Population density. Note the intense concentration in the Ganges River valley. From H. J. de Blij and P. O. Muller, *Geography: Realms, Regions and Concepts*, 14th edition, 2010, p. 404. Originally rendered in color. Reprinted with permission of John Wiley & Sons, Inc.

Economic incentives were common; food rations were withheld from some families with more than three children; others lost a month's pay unless a worker was sterilized. Such programs are believed to have averted more than 85 million births from 1956 to 1986. Even so, progress was problematic.

Ethnic, religious, and caste rivalry fostered fears of persecution and genocide, and violent resistance became frequent. Capitalizing on these fears, politicians were quick to blame family planning for other social ills. Many organized programs have fallen by the wayside. I recall visiting a village where a "two-child" poster had been painted on the side of a school building. As the children crowded around me, I noticed that the poster was in a terrible state of disrepair. It was barely legible. One boy jumped up, pointed at it, and said in English: "Small family very funny." Everyone laughed, the teacher louder than anyone.

Fertility behavior is strongly influenced by prevailing social and economic conditions as well as by religious/cultural norms and expectations, such as early marriage. Of the 4.5 million marriages each year, an estimated 3 million brides are between 15 and 19 years old. India has a young population, with 32 percent under the age of 15.

Less than half of couples of reproductive age use contraception. The most popular form of birth control is surgical sterilization, but this is elected by couples who already have three or four children. It is virtually always the woman that gets sterilized. Less than 0.1 percent of males undergo the procedure. India's TFR has fallen from 6.0 children in the 1950s to around 3.1 in the early 2000s, but periods of quickly falling fertility have been

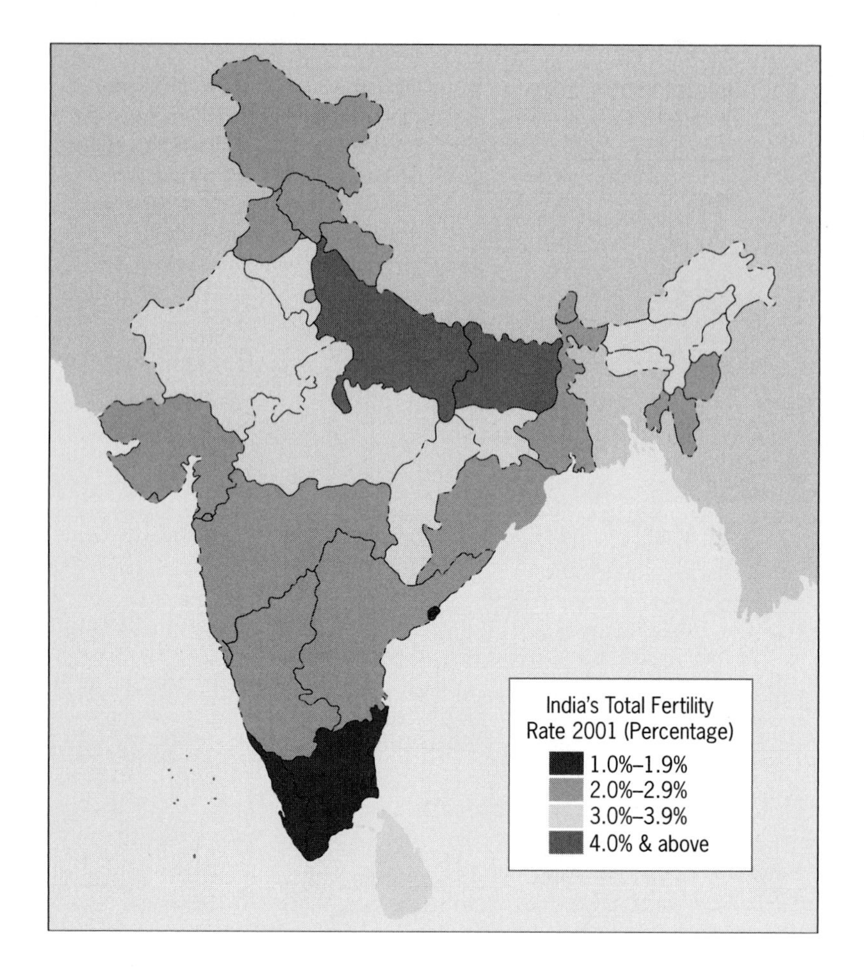

Figure 8-3

Total fertility rates by state. Note that lower fertility rates are in southern India, where women are literate and have more education.

followed by periods of stable levels. Whether India has reached another plateau is not clear.

Also, as we saw in Chapter 3, poorer families tend to have more children. Poor, uneducated women, who have less control of their bodies or have no access to birth control information or devices, also have higher TFRs You can see by Figure 8-3 that TFRs are much lower in India's southern states where families are less poor and women more educated (Figure 8-4). In fact, Kerala and Tamil Nadu have TFRs lower than that of the United States!

Not all populations are growing at the same rate; regional variations are apparent. Remember that growth rates account for both natural increase and migration. Most migration within India comprises males or entire families. A woman rarely moves anywhere on her own unless it is to the home of her husband. In fact, 75 percent of female migration in India is related to marriage.

In 1996, the Indian government eliminated national targets for acceptance of fertility control methods. The government promulgated its first National Population Policy in 2000. This policy contains a variety of population growth-reduction socioeconomic programs such as lowering maternal and child mortality, promoting later marriage, universal immunization of children, and AIDS prevention.

A new "Target-Free Approach" puts *voluntary* family planning at the community level, where grassroots workers set targets according to individual needs. Acceptance of all forms of birth control has risen since that time, but the effects are yet to be seen. Moreover, some states such as Himachal Pradesh and Maharashtra have objected to these government efforts to encourage couples to have

Figure 8-4

This young woman is approximately 15 years old. She is a Dalit and a member of the untouchable, sweeper caste. She lives in a temporary tent on the outskirts of Udaipur in Rajasthan. Photograph courtesy of B. A. Weightman.

only two children because they believe the policies to be too strict and invasive of internal affairs.

SEX RATIOS DECLINING!

One of the most striking aspects of India's population is the fact that far more boys are born than girls. While about 105 boys are born per 100 girls in most countries, the ratio is about 113 per 100 in India. For every 1,000 men in India, there are only 929 women—a figure on the decrease, which means that the situation is worsening.

This declining trend has existed unabated since the first census in 1861. The deficit of young girls among children under 7 years of age increased in every major state with the exception of Kerala in the 1990s, according to India's most recent census in 2001. The gender gap reflects widespread discrimination against women and is more apparent in the north, where sex ratios are considerably lower than in the south. The states of Rajasthan, Haryana, and Uttar Pradesh, for instance, have much lower ratios than Kerala or Tamil Nadu (Figure 8-5).

The major reasons for India's severely imbalanced sex ratio are cultural preferences and the availability of abortion. Sex-selective abortion has been illegal since 1994, but the ban is rarely enforced. The practice has increased in urban areas and among more affluent people who have easier access to prenatal clinics.

Caste differences play a role in these discrepancies. There are more higher-caste women in the northern states, especially of the *kshatriya* or former military groups. These castes are noted for their seclusion of women from outside contacts, including education. Female illiteracy rates are high, and women are expected to produce sons. Female children are not usually killed at birth but rather neglected in their early years. Fed last and less than their male siblings, their ailments ignored, many girls never reach the age of five.

HEALTH AND LIFE

India's mortality has declined at a sluggish rate. For example, maternal mortality rates have declined since the 1970s, although at 450 maternal deaths per 100,000 births in 2000, the rate remains higher than in many other developing countries, and nearly 10 times higher than in China.

Some 300,000 children (more girls than boys) die of diarrhea every year. This is attributable to lack of sanitation and neglect, especially of girls. According to the United Nations, only 33 percent of under-fives with diarrhea receive oral rehydration and continued feeding. A mere 18 percent of rural dwellers have access to improved sanitation facilities. This figure is higher in cities but still only 52 percent.

Chronic diseases such as diabetes, high blood pressure, and cholesterol and other "Western-type" disorders are on the rise. This is because socioeconomic improvements have increased life expectancy in many areas. One 2006 study in Andhra Pradesh found that villages that were more developed than average had progressed further down the path of "Westernization" of their disease patterns. Most deaths now occur from middle-age onward, and the majority of these deaths are due to chronic illnesses. Existing health-care infrastructures are unable to cope with this trend.

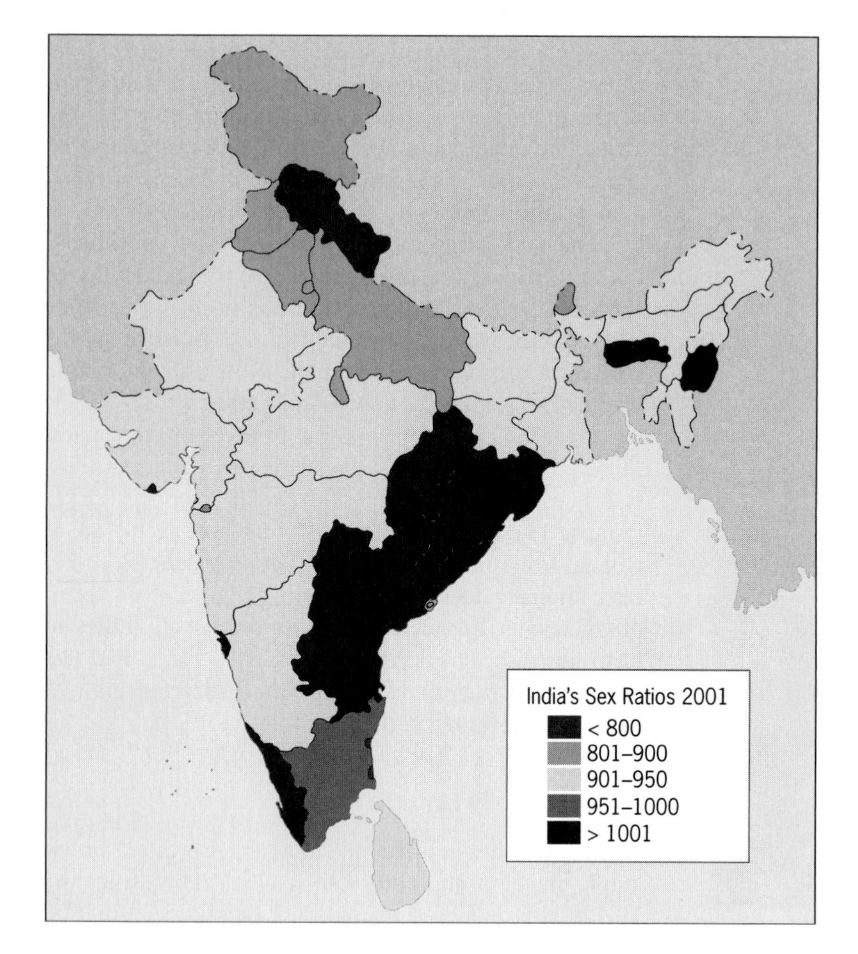

Figure 8-5

Sex ratios by state. Again, the lower (worse) numbers are found in the poorest states that remain steeped in patriarchal traditions. In these societies, women receive little if any education and have little say in their lives.

The earliest cases of HIV were reported in Mumbai and Chennai in 1986. HIV/AIDS is being combated by prevention and treatment programs as well as publicity campaigns that are expanding geographically. Many of these are aimed at truck drivers; the incidence of AIDS is significantly higher along major highways. Numerous programs focus on commercial sex workers, but sex workers are difficult to track down because many do not operate from fixed locations.

Although it has narrowed, the knowledge gap between urban and rural populations remains substantial. As of 2001, on average, 65 percent of rural women and 86 of urban women had heard of HIV/AIDS. But in Bihar, just 22 percent of rural women had heard of HIV/AIDS compared with 63 percent of urban women. In the important industrial state of Gujarat, only 25 percent of rural women and 62 percent of city women had heard of it. Given the size and diversity of India's population, tackling this devastating disease will continue to be a daunting task.

India has a long tradition of herbal medicine based on *Ayurvedic* and other ancient systems. The government is spending US$40 million on a "Golden Triangle Partnership (GTP)" to assess the country's herbs scientifically, conduct clinical trials, and introduce them into the mainstream. This is a difficult task because there are at least 80,000 *Ayurvedic* treatments involving the products of some 3,000 plants. More than 7,000 firms make herbal compounds for medical use. Most plants are cultivated but many in the wild are endangered. Herbs are used to treat countless ailments such as malaria and diabetes.

While health care remains woefully inadequate, it has improved overall. Indeed, death rates have been lowered all over the country. Killer diseases such as smallpox have been eradicated, and infant mortality rates have been lowered. More than 70 percent of one-year-olds have been vaccinated against prevalent diseases such as tuberculosis and measles. The life expectancy of Indians has risen from an average age 32 in 1947 to age 63 for men and 65 for women in 2009. In fact, many more Indians are living longer and are capable of having more children.

2001 CENSUS SHOCK

The 1991 and 2001 census results, which revealed that India's population had increased by 21 percent in a decade, shocked many into realizing that fertility rates must be further reduced at all costs and that women must be empowered to have more control over their lives. However, women remain last in line for education, job training, credit, medicine, and even food. Constantly changing governments cannot provide the stability needed to sustain long-term empowerment strategies. Yet experts agree that this is the most effective mode of reducing population growth rates.

India's population growth will continue its spectacular rise, given its current size and youthfulness. Despite the use of family planning, more than 40 percent live in areas where crude birth rates (CBRs) are in the mid and upper 30s. World Bank predictions indicate that even if India achieves a two-child family by 2015, it will still have a population of 1.9 billion!

Kerala—a Women's State?

Kerala, in southern India, is a small, not particularly prosperous state. However, it has the highest literacy rates in India, even higher than those for Delhi. Although women still lag behind men, more than 80 percent of both genders are able to read and write. Prior to India's partition, only 11 percent were literate. Clearly, this state has made great strides in educating its population (Figure 8-6). Why is this state so different?

Kerala's government spends 40 percent of its budget to maintain a vast network of schools and colleges. Education has long been valued in society, especially among higher castes. Since most of the higher castes were traditionally matrilineal, many women continue to enjoy freedoms unattainable elsewhere. During the Raj, most of Kerala was ruled by independent maharajas who promoted education in the Malayalam language. Literacy rates were much higher in Kerala than in states with English-based schooling. About 20 percent of Kerala's citizens are Christian. This group attaches more importance to equal education for both males and females. Access to education at Christian schools was an attractive lure for the conversion of low-caste Hindus. In addition, knowledge of record keeping was essential to maintain the status of Kerala's international traders.

In 1990, Ernakulum (Cochin) became the first administrative district to be officially declared fully literate. This was the result of a literacy drive entailing motivational strategies so appealing that the project's founder received commendation from the United Nations. Other communities followed suit with the support of various official agencies.

Figure 8-6

A middle-class family in Kerala. This family owns a lumber business. Note the difference in dress between the more traditional parents and the more modern children. Photograph courtesy of B. A. Weightman.

Volunteer teachers were mobilized, two-thirds of them women. Local religious leaders were recruited to back up the media blitz. Interestingly, this was accomplished by a communist state government. Most impressive has been the progress achieved by tradition-bound Muslim women.

Linguistic Conflicts

With 15 national languages and more than 100 tribal tongues in India, it is no wonder that language has been a divisive force. This is especially significant because language is even more important than ethnicity or nationality for individual and group identity. Unlike Europe, where Latin long served as a medium of cultural diffusion, in India, Sanskrit was replaced by Persian as the court language. Persian was later replaced by English. Educated elites, versed in one or more of these languages, became divorced from the masses, which remained steeped in their own linguistic worlds.

Four language stocks, each with its own set of language families, produce a complex language map (Figure 8-7). The Munda and Tibeto-Burmese stocks are relegated to relatively isolated areas, including the Eastern Ghats and the Himalayas. The main cleavage is between the Indo-European and the Dravidian language stocks. The most widely spoken language is Hindi, a member of the Indo-European language family. Hindi speakers also occupy India's central states, a core ringed by people speaking unrelated tongues. Dravidian includes 4 main languages and 18 lesser ones. These occupy the southern states, having been pushed out of the Indus valley with historic migrations. The most well known Dravidian language is Tamil.

Hindi is one of India's official languages, a source of aggravation to Muslim Urdu speakers. In 1994, riots broke

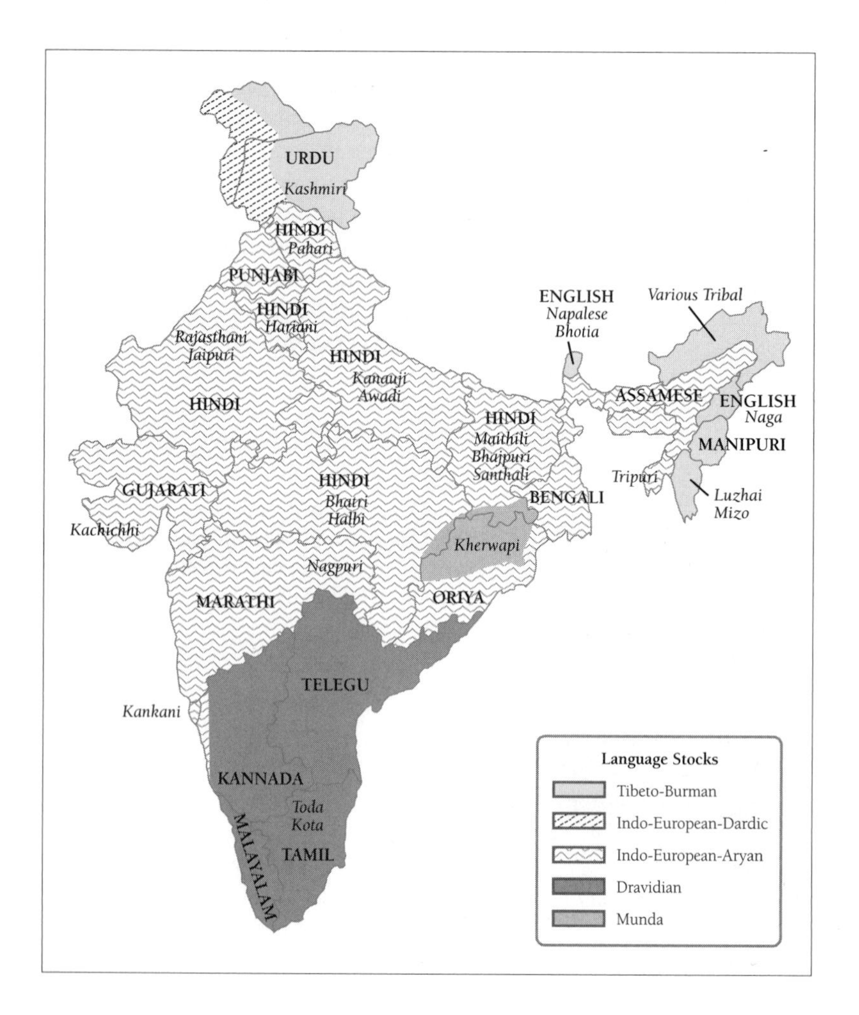

Figure 8-7

Language stocks. Note the division between Indo-European and Dravidian linguistic regions.

out in several states over the broadcast of a 10-minute news program in Urdu in Bengaluru (Bangalore). Hindi derived from Hindustani, a language used in Mughal bazaars and military camps. It incorporates numerous Persian words, as this was the Mughal court language. There are several dialects, but Indian schools teach "Standard Hindi" written in the *Devanagari* (Sanskrit) script.

Urdu is similar to Hindi in grammar, but from a political perspective it is different. It borrows from Arabic and Persian and is written in a Persian script. Muslims in northern India see Urdu as their "special language."

Transcending this linguistic web is English, which was made the language of high school education in the 1830s. Contrary to Indian advice, the Imperialistic British ignored linguistic boundaries when establishing state boundaries. In 1953, there was a Telugu movement for a linguistic state. In 1956, there were language riots in Bombay between Marathi and Gujarati speakers. Boundaries had to be changed.

Bombay became part of Maharashtra. In the south, Kerala and Tamil Nadu were formed to benefit Malayalam and Tamil speakers, respectively. Orissa was defined in the eastern tribal region, and Karnataka was created for Kannada speakers. These are only a few examples of boundary restructuring following partition.

The Nagas, who reside in the mountainous region of India's northeast along the Myanmar (Burma) border, were incorporated into Asom (Assam) at independence. They had since insisted upon having their own state. Nagaland was created in 1961.

In 2000, three additional states were created. A number of poverty-stricken districts in southern Bihar formed Jharkhand. To assuage tribal groups that had been agitating for independence since the 1930s, Chhattisgarh was formed. A highland region of Uttar Pradesh known as Uttaranachal became the state of Uttarakhand.

Since then, numerous other groups have demonstrated publicly and made demands for their own states. Uttar Pradesh is agitating to be cut into four parts. Several ethnic groups in the Asom (Assam) region want their own states. Most significantly, northern Andhra Pradesh wants to separate from the rest of the otherwise impoverished state. If Telangana were to be created, it would be centered on the city of Hyderabad—the "IT City." The remaining Andhra Pradesh would be left without this major source of state income.

Boundary alterations did not mask the fact that India's elites spoke and read in English—a "foreign tongue." Government efforts to force Hindi as India's only national language were met with strong resistance. It has spread, nevertheless, via media such as schools and television, and interregional migrations. Hindi is even a *lingua franca* (common language) for mixed migrant worker groups. English remains the language of private schools, colleges, and universities. Moreover, India's modern, urban social class is differentiated by English more than any other cultural force. Many intellectuals argue that an English education should be the means employed to "uplift" the poor masses.

Language issues remain significant. In Goa, recent attempts to force English speakers to learn Konkani caused mass demonstrations. Politicians representing Uttar Pradesh and Madhya Pradesh are calling for the eradication of English. In 2010, the government of Karnataka abolished English classes in more than 2,000 schools. Primary-school students are to be taught only in Kannada (the state language). Schools and English-speaking higher institutions that do not comply will be closed. The state already has a dearth of skilled labor, and Bengaluru (Bangalore)—focus of India's high-tech industry and call-centers—attracts educated, English-speaking workers from elsewhere in the country. Disagreements and conflicts over language clearly have far-reaching impacts on political and socioeconomic development.

Caste Complexities

Even though there is a growing ambivalence toward caste in India's upper social strata, caste rules and regulations apply among virtually all people. Brahmins, Rajputs, Marathas, and Jats are examples of regular castes. But more than two-thirds of Indians belong to so-called scheduled castes and tribes. Scheduled tribes are also called *adivasis*, meaning "original inhabitants." These include such groups as the Nagas of Nagaland and the Bondas of Orissa (Figure 8-8). Another term employed by the government is other backward classes (OBCs) that include the bulk of the population—the "lower middle classes."

The Indian government has published lists including 405 scheduled castes, 255 scheduled tribes, and 3,500 OBCs. These groups are entitled to special social and political benefits. Discrimination in favor of the "backward" classes did not begin with independence. It began in the state of Mysore (Karnataka) in 1921. In this state, 92 percent of the people belong to the backward classes for which university admissions and jobs are reserved. In contrast to predictions at the time, the state did not collapse, and the capital of Bengaluru (Bangalore) has become the center of India's computer industry.

Figure 8-8

Bonda woman and child. Numbering about 6,000, the Bonda live in geographic isolation in the Eastern Ghats. They are hill cultivators, growing mangoes and jackfruit, but still hunt with bows and arrows. Theirs is a female-dominated society. The very independent Bonda speak Koui, an Austro-Asian language. Photograph courtesy of B. A. Weightman.

Tribal Troubles

India has 255 scheduled tribes plus numerous unofficial ones. Many of these groups live in the Eastern Ghats and the hills of Asom (Assam). In 1944, a little-known but decisive battle was fought in the hills and jungles around Kohima in Nagaland. Here, the British broke the Japanese advance from Burma, ending their plans to capture the Raj. This remote region has remained fractious, with insurgencies against the Indian government mixing with intertribal conflict. Seven states cover the upper basin of the Brahmaputra River. Incursions by Christian missionaries and Bengali Muslims add to the fray.

The Nagas, who became Christian in the nineteenth century, have been up in arms since 1946. They were granted statehood in 1963. In 1993, 316 people were killed in ethnic massacres between the Nagas and the Kuki tribes in the Manipur hills. In Meghalaya, the Khasis are rebelling against outsiders, and Asom has witnessed rioting between ethnic Assamese and Bengali Muslim migrants from India and Bangladesh. If this seems very confusing, be assured that this is only part of the story of tribal troubles.

Smuggling is another part of the picture. Teak, aromatic agar wood, jade, gems, and heroin are among the smuggled goods. Eastern India is a key route for the movement of heroin out of Myanmar. Drug lords cum politicians rule their territories with personal armies. Border towns are hotbeds of crime and prostitution. Burmese girls, retired from the Thai brothels because they are HIV positive, now work here. Addicts number anywhere from 15,000 to 40,000 in Manipur alone. The government claims that the HIV infection rate is more than 50 percent.

Why does the Indian government spend so much money on maintaining control of this region? After all, it is attached by only a narrow, corridor of land at Siliguri. New Delhi's concern lies in the fact that the region is of strategic and economic importance. It borders on four nations: China, Bangladesh, Bhutan, and Myanmar. Further, it is rich in oil, timber, and tea. It also has great hydroelectric potential. Culturally, it is the meeting point between Hindu-based culture groups and the animist and Christian Tibeto-Burman people of the hills. Keep your eye on this region as one of South Asia's hot spots.

Caste as a form of kinship or marriage lineage circumscribes who can marry whom and delineates mutually supportive behaviors and responsibilities among members. In this context, hundreds of thousands of castes exist. As members and nonmembers are sharply defined, favoritism and nepotism to one group are expected by other groups.

Caste is not static. Castes' positions in relation to one another are constantly being renegotiated on the basis of fluctuating power, status, ritual behavior, political mobilization, education, and geographical location. Caste holds diverse peoples together via a hierarchical structure permitting flexibility within the caste. Upward mobility is potentially possible. Caste may be associated with occupation—*jati*. While one's *jati* is fixed at birth, its position within the hierarchy is not. Moreover, modernization has confounded the association of birth assignation with practical occupation.

Abolition of discrimination has applied only to untouchability, not to caste. Those in low-level or polluting *jatis*—the Dalits—seek recognition as regular members of Indian society. Whatever their origin, their subordination derives from perceived ritual impurity rooted in antiquity. Some scholars argue that untouchability was exacerbated by the British, who were overly concerned with caste inequities and thereby drew undue attention to them.

Today, Dalits demand their rights, and government programs have facilitated their move into many upper echelons, including politics. Education, in particular, has invested them with expectations for equal treatment. Indeed, "untouchables" now have many opportunities historically denied them. Nevertheless, stepping outside the bounds of untouchability is no easy task; deeply rooted social taboos are difficult to expunge.

Affirmative action policies have created privileged sections within underprivileged groups. An interesting point is that pro-action policies are part of the official Constitution, and they were written by a former untouchable. The situation that those classified as "backward" go forward by virtue of their "backwardness" is resented in many circles.

The progression of these formerly subordinated classes has political impacts. Many politicians avidly seek the loyalty of entire socially categorized populations. Appointments are made with an eye toward potential mobilization of caste followings. Just as significant is the fact that 85 of 545 seats in India's Parliament and 22.5 percent of government jobs are reserved for scheduled castes and tribes.

Caste then, in terms of establishing hierarchical relationships and interactions, is not disappearing, although it is exhibiting greater flexibility. The poor and lower-middle-class masses are seeking economic improvement, while the privileged are trying to retain the status quo. Brahmins find themselves taking orders from or eating food prepared by Dalits. Increasing numbers of marriage ads state "caste no bar." Although quotas and special privileges for untouchables have propelled many into upper echelons, these have sometimes triggered violence. Most untouchables continue to exist at the very bottom of Indian society.

India's Nomads

Most people do not realize that there are some 80 million nomads in India. The best known are the *Gadulia Lohar,* who once forged armor for Hindu kings. Their name derives from Hindi: *gaadi* meaning "cart" and *lohar* meaning "blacksmith." Other groups include camel herders, salt traders, fortune-tellers, snake charmers, acrobatic and musical groups, tattooists, basket-makers, and hunters and gatherers. Anthropologists have identified around 500 nomadic groups in India.

Once part of India's mainstream, they established mutually beneficial relationships with the villagers who lived along their migratory routes. However, when the British disparaged them as vagrants and criminals, the seeds of discrimination were sown. In this information age, young people have little interest in or use for snake charmers and the like. Hunters and plant gatherers and herders are being axed out of their traditional environments as these disappear in the wake of urban and industrial expansion. Also, politicians ignore this minority who are fragmented by caste, language, and religion.

Many groups have clustered in slums where they eke out an existence in an environment of exclusion and misery. The few who do try to help them make efforts to get them shelter and some kind of address so that they can get government assistance and enroll their children in school. Unfortunately, this is met with resistance from town-dwellers who regard them as social vermin.

Religious Conflicts

In December 1992, a rampaging mob demolished an unused mosque in Ayodhya, Uttar Pradesh, claiming that it had been built on the birthplace of Lord Rama. This inflamed the sensibilities of India's largest cultural minority: 130 million Muslims. Communal riots broke out nationwide, and hundreds were killed and injured. This incident was merely a spot on the tapestry of religious conflict in India's officially secular democracy.

The overwhelming majority of Indians are Hindu. In fact, in some languages such as French and Persian, the word for "Indian" is Hindu. It once meant "people beyond the Indus," but history has rendered it, correctly or incorrectly, a term designating a nation and a religion. However, like India itself, Hinduism is a polyglot of practices and lifestyles. Unfortunately, in the twentieth-first century, this kaleidoscope has become blurred under the forces of socioeconomic inequities and political ambitions.

The unlikely notion of Hindu chauvinism emerged in the political arena with the Bharatiya Janata Party (BJP) that was in power from 1998 to 2004. This Hindu nationalism is not simply religious. It incorporates the resentments and animosities of caste, belief, and regional sectarianisms. It cannot be simply religious, because Hinduism is not a singular faith divorced from all other faiths. Hinduism is not a monolith; it is inclusive in nature. However, it is convenient to appeal to differences in practice or detail of belief for those with a particular political or economic agenda.

In the early 1990s, the BJP called for a ban on English, the promotion of Sanskrit, and a ban on proselytizing by other religions. By 1999, the BJP had achieved national power and Hindus were killing Christians in Gujarat, Orissa, and elsewhere. Gujarat is also a hotbed of Hindu-Muslim violence. Religious insurgencies continue to plague this vast and complex country.

HINDUTVA: GENDER AND SOCIETY

Hindutva means "Hinduness"—a desire to make India a country where Hindu principles prevail. This is the guiding precept of the BJP and is expressed as Hindu nationalism, Hindu patriotism, and Hindu heritage. *Hindutva* believers want to foist a Hindu curriculum on schools, inhibit activities of non-Hindu religious proselytizers, change family law in ways that would upset Muslims, and render non-Hindus as outsiders.

This philosophy worries Muslims and other non-Hindus. It also concerns those who believe that India's secularism, its separation of religion and state, are indispensable to the survival of its democracy. Moderate Hindus and non-Hindus oppose these divisive ideas. However, they do acknowledge the appeal of the BJP, which swept into power on a Hindu nationalist platform. They also realize that it is only the constraints of a coalition government that keeps the excesses of *Hindutva* from being put into practice.

Whenever there is a patriarchal system, control of women's honor becomes a primary symbol for the strength and preservation of both family and state. The most reliable form of ensuring honor is embedding the concept into the female psyche from birth. A woman must be imbued with a creed of self-discipline. Entrenched into the notion of self-discipline is reverence for such precepts as chastity, fidelity, maternity, domesticity, humility, and self-sacrifice.

Because women have always been instilled with these essential fundamental values in the context of patriarchal structures, they become custodians of such value systems that must be passed on through generations, thereby ensuring piety of the household. *Sati,* as discussed in Chapter 6, was the epitome of a woman's subservience to and honor of her husband and the entire community.

Women who have internalized these values and shaped their lives around them find their very self-esteem threatened by the social and economic changes they see around them. Risqué television programs, showing a couple holding hands or mildly embracing, for instance, are perceived as outright scandalous. The "brazen" behavior of middle- and upper-class city girls in public and such radical ideas as dating or choosing one's own husband are anathema to conservative women as well as men.

Ironically, Hindu nationalism and conservatism have created new spaces for women's activism. Upper- and middle-class families who, in many areas, tend more toward to female seclusion, are permitting their daughters to take part in protests and even riots against perceived forces of societal evil.

Upper-caste Hindus see themselves as under attack on two fronts: from the outside by "immoral minorities" (read Muslims) and from the inside by "unclean castes"—having to work with *them,* for instance. Under this onslaught, it is clear that women must be protected from the rapacious intent of "others."

Prominent women in the *Hindutva* movement are very few in number and varied in style but common to all of them is an emphasis on personal strength. Moreover, all stress vigilance against Muslim "perpetrators" and even the potential evils of Christianity.

In 2007, a *Hindutva* political official announced that, "we Hindus believe in peaceful coexistence, Muslims do not." This proclamation seems ludicrous to India's some 150 million Muslims who well remember the riots of 2002, which began after a train in Gujarat, carrying Hindu activists on their way back from Ayodhya, caught fire in a Muslim neighborhood. Muslims were blamed for the many deaths and in the pogrom that followed, 2,000 people died. Also, to embed the atrocities in the minds of Muslims forever, some Hindus poured kerosene down the throats of Muslim women and children and set them afire.

A good part of anti-Christian animosity derives from the fact that the vast majority of Christians are either converted Dalits or tribal people trying to escape their low status under the caste system. Many Christians, including missionaries, have been burned alive, especially in remoter areas of the country.

Even under the Congress Party government, internecine violence is ongoing with riots, bombings, individual and mass killings, and rape on the part of both Muslims and Hindus. Many Muslims carry the burden of past and ongoing injustices compounded by religious fundamentalism.

Hindu fanatics do not have faith in the institutions of Indian democracy. They see the state as "soft" and pandering to minorities out of a misplaced and Western secularism. Hindu author Shashi Thakoor (2007), who is proud to belong to a religion that has taught the world both tolerance and universal acceptance, says that Hindu fanatics are not fundamentalists (like Muslims) because they do not root their Hinduism in the spiritual underpinnings of the faith. "They seek revenge in the name of Hinduism-as-badge, rather than Hinduism-as-doctrine."

The Naxalites

The term "Naxalite" comes from the village name of Naxalbari in West Bengal, where the movement originated around 1967. The **Naxalites** are a group of far-left, radical communists who support Maoist ideology. The Naxalites claim to fight poverty and injustice while providing poverty-stricken villages with facilities and services that the federal and state governments are unable or unwilling to provide. For example, they have installed wells and electric lines, and built houses and roads in some areas. However, they act with force and anyone who denies them is brutalized and murdered. Of the few police that are available in Naxalite-controlled areas, hundreds have been killed and the rest are terrified to do anything.

In Jharkhand, to fund their operations, the Naxalites are encouraging the cultivation of opium poppies in hundreds of villages. Many poor farmers find this very lucrative because they get high prices from local heroin traders who also train them in heroin refining and packaging.

The Naxalites also recruit child-soldiers. First, they bomb schools and threaten teachers, then they force the students into their army. In one region of Chhatisgarh, officials estimate that the Maoist militancy has denied at least 100,000 children their education.

An estimated 20,000 Naxalites are present in varying degrees of strength throughout a swathe of eastern India known as the "Red Corridor" (Figure 8-9). Due to the ineffectiveness of authorities to address severe regional and local inequities, the group has millions of sympathizers.

National and state policies have been inconsistent and spotty in their conception and application. Consequently, the spread of Naxalism is causing justifiable alarm. While their power base remains on the margins of Indian society, in the ever-deepening holes of poverty and destitution in the remote countryside, the movement is growing.

Naxalism attacks wherever the system is weakest—delivering on never-kept promises. They do not threaten the government in New Delhi but they do deter investment into some of the country's poorest regions, which also happen to be rich in vital resources such as coal and iron. Ironically, their presence, in effect, sharpens inequity.

In a speech in 2006, India's Prime Minister Manmohan Singh observed that Naxalism is India's prime security threat. Many Indians are appalled that he would grant the movement such a high-priority position. They see Naxalism as a primitive, peasant rebellion based on an outmoded ideology that has no room in an India of soaring economic growth, Bollywood dreams, and call centers. Even so, according to the BBC, more than 6,000 people had died up until 2010 in the ongoing strife.

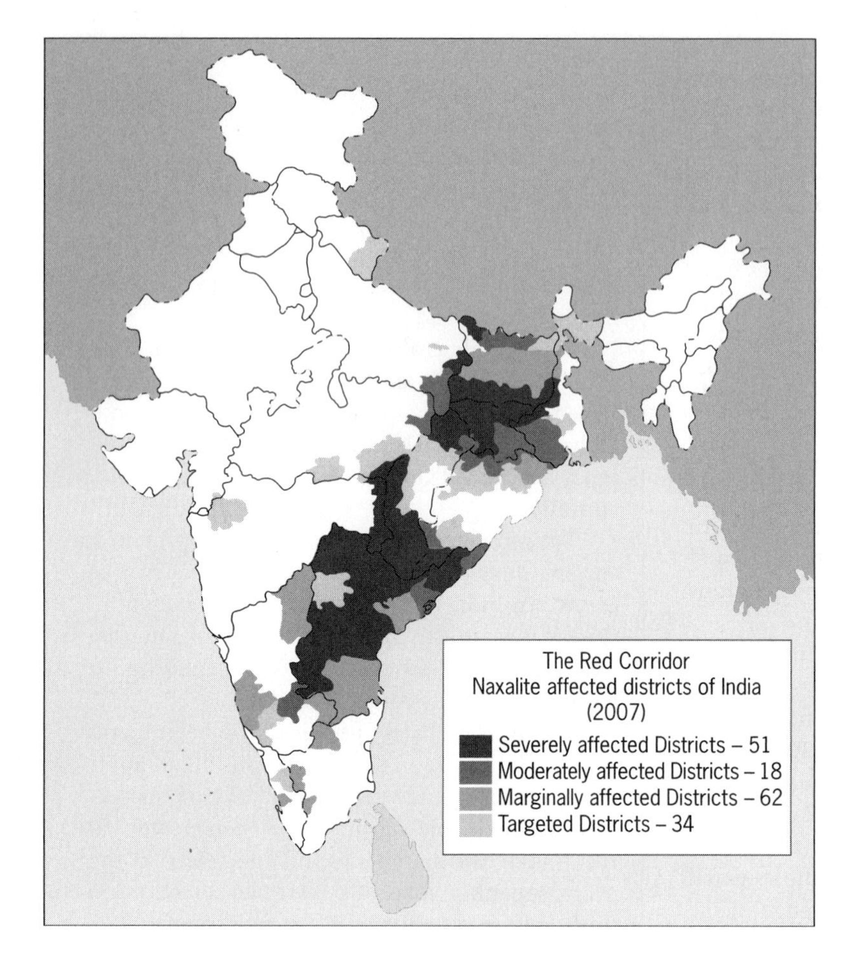

Figure 8-9

Areas of Naxalite activity known as the "Red Corridor." These exceedingly violent Maoist rebels have varying degrees of control in some of the poorest regions in northeast and eastern India.

Agricultural Landscapes

Figure 8-10 illustrates the wide variety of crops cultivated in India. While crop yields have risen significantly in the latter half of the twentieth century, they do not equal those achieved elsewhere nor, because of inadequate distribution facilities, are they sufficient to meet population requirements. Human and animals continue to dominate labor even as the use of tractors and other machines becomes more common (Figure 8-11).

India manufactures 30 percent of the world's tractors, more than any other country. However, India's farmers do not have many tractors. The United States has 27 tractors per 247 acres (1,000 ha) of arable land. India has 11. The global average is 19.

Most of India's farms are too small for large machinery. Most farmers would have to go into serious debt to buy a tractor. Government grants for mechanization frequently are diverted to purchase lifestyle enhancements such as a television. Technical improvements may not be appropriate in all settings. What about cases where different religious or caste groups are not willing to cooperate in the acquisition and sharing of such equipment?

A fundamental problem in India's agricultural sector is farm size. There are too many marginal and small farms. In fact, nearly 60 percent of farms are too small to be efficient and are often on poorly producing land. Small and medium-size farms make up 40 percent and large farms comprise a mere 2 percent. Most farmers do not own their own land but are at the mercy of *zamindars* (landlords similar to those that we discussed in the previous chapter).

Water availability has increased via tube wells, storage tanks, and canal irrigation. More pumps have been installed, and wells now are the major source of irrigation water. Farmers have been able to put more acreage under production and use high-yielding seeds. Some

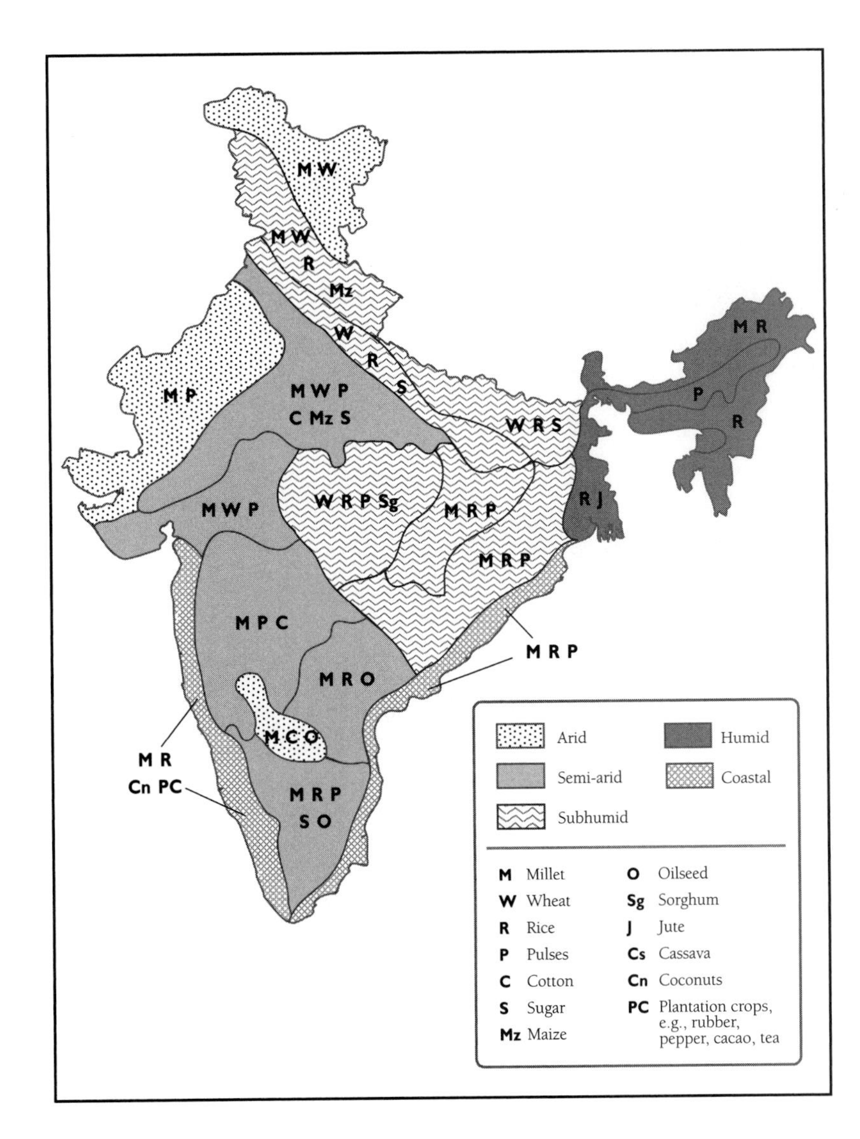

Figure 8-10

Patterns of agriculture. What is the relationship between crop distribution and water availability? Note that millet can be grown in many different environments. Millet is a common food in many parts of Asia. In North America, it is fed to livestock and sometimes used as an ingredient in multigrain bread or cereal.

well owners have started water markets, selling their water to others to open up even more farm land.

Timely application of water can increase yields markedly, but supplies are overexploited in many areas. Other areas have never had enough. Management agencies lack coordination, and the entire issue of water, like land, is very political. According to Robert Repetto, writing for the World Resources Institute (1994), "In India, the rights to an immensely valuable resource [water] were distributed gratis in a pattern even more unequal than that of land distribution, reinforcing rural inequalities in income and wealth."

Thirty percent of India's land is dry-cropped. With irrigation, these lands could become relatively affluent. In Maharashtra, experiments with water conservation, tree planting, and grazing control have led to agricultural improvements and the creation of a viable market system.

Demise of the Jajmani System

In Punjab the custom was for Harijan laborers to work for their Jat landlords in exchange for grain and fodder. This was called the jajmani system. Harijan laborers received one-twentieth of the wheat harvest. This was much more than an

Figure 8-11

Man and water buffalo harrowing a rice paddy in Orissa. This action helps to break up the lumps of clay-like soil. Rice cultivation relies heavily on human and animal labor. Photograph courtesy of B. A. Weightman.

economic system. It was a centuries-old social contract rooted in caste relationships and practiced among Sikhs and non-Sikhs alike. It required a certain decorum and specific exchange relationships including water, tea, and food.

With the advent of high-yielding varieties (HYV), the one-twentieth share became greater from the HYV fields than the traditional variety fields. Furthermore, some farmers could afford tractors and no longer needed Harijan labor on their large plots. With concomitant economic growth in towns and cities, and extended bus routes, many Harijans sought urban employment. Newfound wealth meant that Harijans no longer had to accede to Jat requirements or even requests. Jats purchased motor scooters; Harijans bought bicycles. Activity spaces were expanded, and Harijan-Jat social distance was extended. The intergroup ceremony and camaraderie surrounding the harvest disappeared.

More tube wells were installed, and underground pipes reduced evaporation. Rice was grown in addition to wheat. With quick-maturing seeds (e.g., 60-day lentil), triple-cropping became possible. When labor was needed, migrant workers from the Gangetic plain were hired until that source dried up. Now many Jats do their own labor and must work longer hours to support all their essential equipment. There is a rising level of debt—some Jats sell their land in fear of bankruptcy. Rich landowners buy these parcels, circumventing property-holding limitations by registering titles among family members. Small farmers cannot compete and ultimately sell out. Large landowners cannot function efficiently without tractors, and many become overburdened with debt. Technological change alters social circumstances. Clearly, the pros and cons of these changes are in the eyes of the beholders.

INDIA'S FOOD SECURITY

Outright famine has not plagued India for decades. However, food insufficiencies and malnutrition remain troublesome in many areas, especially when distribution networks fail or costs are prohibitive. A critical contributor to improved food supply in India was the 1966 Green Revolution, which began in Punjab with high-yielding varieties (HYV) of wheat and in Tamil Nadu with HYV rice.

But scientific improvements have not been equitably distributed, and they benefit different crops in different ways. In the case of wheat, for instance, Punjab, Haryana, and western Uttar Pradesh have benefited most because of suitable irrigation water supply. In the case of rice, Tamil Nadu, Kerala, Andra Pradesh, and Jammu and Kashmir have profited most. Other rice-growing states or regions with variable rainfall and inadequate water storage and distribution systems use a different approach. There, farmers plant a *rabi* (dry season, winter) HYV wheat crop.

Uttar Pradesh, where one in six Indians reside, has seen variable consequences of the Green Revolution. Uttar Pradesh is India's largest grain producer, yet yields are only half those in Punjab. One government and World Bank–supported countermeasure is the reclamation of sodic land—land where crusts of salt or other alkalis prevent nutrients and water from going below the surface. Application of gypsum and flushing with water can bring such soils into production. Farmers are given free gypsum, loans to buy water pumps, and counseling on crop and farm management.

Agricultural improvements have not erased food import requirements, which are now at 14.1 million tons (as of 2010). Food imports are expected to increase by at least 2 million tons in each subsequent year. These requirements reflect the needs of an ever-increasing population, deceleration of crop yield improvements, economic growth and increased consumption, and greater consumption of non-vegetarian foods. Such factors pose challenges to Indian agriculture that does not make the best use of its land in terms of potential productivity. Storage, packaging, and shipping to avoid losses and disease are further issues to be grappled with.

At the same time, India is an international exporter of food but must continue to increase crop yields to feed its own ever-growing population. However, there are obstacles. How can India sustain increased yields? Will there be an endless supply of ever-better HYV seeds? Policies regarding the ownership and use of HYV seeds have become a global issue. How far can biotechnology go?

Nearly 500 transgenic plants (human-made) were developed in 1994–1995. Will these be available to Indian farmers? Would Indians even want to eat these products? (Think of the beautiful tomato developed to be machine picked; it looks great but has no taste.) Who would buy India's exported food of a transgenic nature? Several countries in Asia have already banned human-engineered foods.

Bio-tech (BT) *brinjal* (eggplant or *aubergine*) seeds were introduced in 2010 by Mayco—an Indian hybrid-seed company, and Monsanto—an American biotech giant. BT *brinjal* is genetically modified with a soil bacterium (*Bacillus thuringiensis*). This is purported to cut insect damage in half and reduce fertilizer requirements by 80 percent. However, environmental groups claim that cross-pollination could wipe out thousands of indigenous *brinjal* varieties. India's Environment Minister has declared that BT *brinjal* will not be grown until independent studies show that it would have no impact on human health, the environment, or biodiversity. Half of India's states have now concurred with this decision.

WOMEN AND THE WHITE REVOLUTION

Millions of rural Indian women work in agriculture, and 93 percent of them are involved in the country's **White Revolution** as its dairy industry undergoes transformation. Western planners introduced new techniques to increase milk yields as they had improved crop yields. The development plan, introduced in 1970, was called Operation Flood (OF). India has more than 240 million bovines including cattle and water buffalo. Here, 18 percent of the world's cattle stock produces only 6 percent of the world's milk. OF determined that India had too many low-milk-yielding zebu cattle and that new breeds and dairy processing plants should be introduced. The results are mixed.

Urban dwellers now consume more milk, but rural people often cannot afford higher prices and therefore drink less. New cattle breeds require grain, and this means less grain is available for human consumption. Moreover, women who frequently tend and milk zebu were deprived of this central role. Women traditionally collect milk and make *ghee* (clarified butter), curds, buttermilk, and cottage cheese, which they can sell. Dairy processing plants deny women these activities. Farmers take their milk to the factory, where it is reduced to products for purchase by the urban elite. Thus, nutritional sources are diverted from rural areas.

New breeds and cross-breeds are not always suited to tropical or even subtropical environments and an investment can be easily lost to disease and death. Dairy cattle require high-quality feed and clean water—items not available to millions of people. They more easily succumb to parasites. Further, milk cows are not well suited as draft animals.

The sacred zebu is well suited to India's environmental conditions. It does give some milk, but more important, it provides dung for fertilizer and fuel, and it is India's main beast of burden. Further, it devours anything from straw to newspapers (Figure 8-12).

The goal of getting more milk from an animal fails to recognize the whole animal and its integral role in traditional cultural complexes. Many scholars attack this reductionist approach to development. While OF claims to have "emancipated" millions of women, in fact it has eliminated an important role and source of income for them. Very few women are directly involved

Figure 8-12

Cows are sacred in India but they are the primary work animal and are important sources of milk and dung. It is common to see typically skinny zebus wandering around city streets eating anything they can find, including cardboard and newspapers. These cows are hovering around a garbage bin filled with rotting vegetables from a nearby market. Women and girls collect their dung, which, if not used for fertilizer, is used as cooking fuel. Dung cooking fires are major sources of pollution. Photo courtesy of B. A. Weightman.

in the modern dairy industry. Modern dairy enterprises provide better care for dairy animals than many farmers can for their children. Kanti George (1985) notes that in certain parts of India, "it must be more comfortable to be a large landholder's crossbred cow than to be a small farmer's child."

Why Waste Waste?

Judicious use of animal and even human waste can improve the lives of millions. Cow dung, when fermented or digested in an enclosed, cement container, produces methane gas, carbon dioxide, and nutrient-rich slurry. From this biogas plant, methane can be piped into homes and used as cooking gas, or it can fire a diesel engine to generate electricity or pump water. The slurry is an excellent fertilizer. Generating biogas is culturally appropriate as well, because all products of the cow are considered sacred.

Biogas was produced as early as the sixteenth century in Persia and has been used in India for almost 100 years. Now, more than two million small "digesters" and a thousand community-size plants have been put in place in rural, agrarian regions. Biogas provides light to millions who are not linked to the larger electrical grid and is a boon to the 80 percent of rural Indians who lack access to sufficient cooking fuel. The Indian government hopes to construct digesters for the 12 million households that have enough cattle to maintain a sufficient supply of dung.

Human waste can also be put to good use. Sewage-fed lagoons produce 10 percent of the fish consumed by Kolkata (Calcutta). A natural wetland east of the city has been turned into a 29,652-acre (12,000 ha) aquaculture operation. While the sewage provides nutrients for the algae that carp and other fish eat, the process dilutes the high concentration of fecal coliform bacteria in the sewage. The resulting effluent can be safely used for irrigation in urban gardens. City gardens in the rich, composted soils of old garbage dumps produce 150 to 300 tons of vegetables a day and employ around 20,000 people in East Calcutta.

Deceleration in crop yields is now occurring in key areas such as Kerala and Punjab. This can be attributed to several factors. Even with the widespread installation of tube wells, water management is often ineffective. Frequently, water control projects have been installed in regions where crops do not require irrigation. Fertilizers and pesticides may be adulterated. Seed multiplication farms are not operating efficiently. Roads have fallen into disrepair, disrupting distribution of both inputs and outputs. In addition, crop yields have suffered from the unpredictability of the monsoon.

Having weathered the financial crisis, rural Indians have to face the weather. Fickle monsoons have foisted drought on at least half of the country's farm districts. Also, floods ensued as the monsoon dissipated. Agricultural output has shrunk and food prices have risen. Life will be very difficult for millions. However, the economy will not collapse as agriculture now accounts for only a 17 percent share of national output, down from 40 percent in the 1980s.

Rural electrification, expanded to all but 80,000 villages (out of 580,000) by 2004, plays a significant role in pumping and other aspects of water control. Unfortunately, mismanagement has resulted in intermittent power supply in some areas and weeks of blackouts in others.

Those without access to the power grid—about 55 percent of the population—must often walk long distances to buy a few liters of expensive kerosene. The United Nations Foundation states: "Kerosene used by the poor for lighting is often unaffordable, unavailable, unsafe, and unhealthy...."

A solar-voltaic pilot project in Karnataka has transformed the lives of some 100,000 people living in poverty-stricken rural regions. The systems supply a few hours of power in homes or shops to run small appliances and provide improved reading light. This lighting has been credited with improvement of grades for schoolchildren. Two leading Indian banks, the project's original partners, are supplying low-interest, small loans for the personal power systems that could be paid over five years at their 2,000 branches throughout the country.

THE VILLAGE

Two-thirds of Indians live in rural villages. Unfortunately, most of them are very poor. Nearly 300 million Indians suffer from malnutrition and do not know where their next meal is coming from. A major failing of the government is its inability to provide food subsidies for those living below the poverty line. People are supposed to acquire special ration cards but having to pay bribes to get them is a common occurrence.

Kerala and Tamil Nadu do well in terms of getting food to their poor. Other states such as Bihar do not. In fact, 40 percent of 75 million Biharis who hold ration cards are not even classified as poor. Furthermore, more than 80 percent of available food is stolen. The all-India average varies between one-quarter and a half of all food being stolen. Government warehouses hold billions of tons of food, much of which is rotting, in case of widespread famine. When there are food distributions, they all too frequently do not reach the right people because corruption, ethnic rivalry, and caste discrimination pervade and pervert the system.

Probably the most significant problem for rural dwellers is debt. Fueled by crushing debt, failing crops, and government indifference, thousands of desperate farmers have been committing suicide in the hope that their family will get a small government pension. An estimated 17 thousand farmers took their own lives in 2006, many by drinking pesticide, jumping down a well, or self-immolation. These men epitomize India's "agrarian crisis." In 2006, Oxfam, an international charity, published a study arguing that the farmers' plight was exacerbated by their, "indiscriminate and forced integration" into an "unfair global system."

Murari's Debt

Probably the most significant problem for rural dwellers is debt. Look at the case of Murari, a 30-year-old farmer who lives in a village with his family. He began his career as a contract laborer five years ago for a *Thakur*—a dominant-caste farmer. The *Thakur* is a moneylender for many of the area's villages. Murati took out a loan of 1,000 rupees and contracted to work for five years on the Thakur's land for 5,000 rupees a year. Murari's food, housing, and miscellaneous expenses were provided by the moneylender, who kept a record of these things.

After two years of labor, Murari owed the *Thakur* 250 percent more than he had originally borrowed due to accumulated interest and charges for his housing, etc. Despite this situation, Murari was not permitted to go elsewhere for a job that paid more money. The moneylender had "people" who would track him down with serious consequences for him or his family. After five years of work as a farm laborer and now a house-servant as well, Murari owes 8,000 rupees to the *Thakur*. Murari, and others like him, find that they are virtually powerless once they enter the vicious cycle of contractual labor where they are subject to the exploitation and tyranny of the landlords.

Bhimrao Ambedkar was the country's first Dalit to be educated abroad. A graduate of Columbia University, he was the principal author of India's 1950 Constitution. In fact, Ambedkar is more well known than even Gandhi within India. This is what he thought of the Indian village: "The love of the intellectual Indian for the village

Figure 8-13

This poor village in Madya Pradesh has just received a well for potable water. Getting water is a twice-daily chore for women and children. Going to the well gives women opportunity to catch up on the latest gossip. Photograph courtesy of B. A. Weightman.

community is of course infinite, if not pathetic . . . What is the village but a sink of localism, a den of ignorance, narrow mindedness and communalism." In India, "communalism" refers to allegiance to one's own culture group. He also described caste hierarchy as "an ascending scale of hatred and a descending scale of contempt."

While Ambedkar denigrates villages in the context of his boyhood experiences, the vast majority of Indians are very attached to them, even after they migrate elsewhere. The village is a *life-context* that shapes *life-ways* such as reciprocal social relations among kin and neighbors (Figure 8-13). However, exclusion and deprivation, embedded within unequal power structures, reduce social cohesion thereby disrupting the connectivity among individuals and groups that is essential to equitable resource distribution at the household, community, and state level.

When social solidarity crumbles, collective action is difficult, and social norms and sanctions no longer regulate behavior. Societal breakdown produces strife and even greater inequalities.

THE IMPACT OF GLOBALIZATION

In light of globalization, import duties on crops such as cotton have been reduced, leaving Indian growers at a disadvantage against cheaper American cotton, which, as critics point out, is heavily subsidized by the U.S. government. Further, Indian farmers have been encouraged to use costlier, genetically modified seeds. Although these are pest-resistant, they have turned out to be unsuitable for small, nonirrigated plots. Nevertheless, India remains the world's second largest cotton producer after China and is the United States' fiercest rival for exports.

The Digital Village

In 1999, the Karnataka state government launched a program to computerize the land records of 6.7 million farmers in 30,000 villages. The program is called *Bhoomi,* which means "land" in both Hindi and Kannada. Farmers can now go to government-owned, computer kiosks and retrieve information in Kannnada (the local language) about their land and its potential productivity. They can find out about input costs, market prices, and even get weather forecasts.

With equal access to information, upper castes have lost their advantage over lower castes. Many land deeds used to be fraudulent and cost poor farmers US$20 million a year. Now, the problem has essentially disappeared.

One program in Madhya Pradesh has practically eliminated corruption in the sale of soybeans. Traditionally, farmers sold their beans to intermediary traders at open auctions called *mandis*. The traders then sold the produce to food-processing companies. As the traders knew what these companies would pay and the farmers did not, they could hold farmers to unfair prices for their soybeans.

An Indian company (ITC Limited) introduced a network of internet kiosks called *e-choupal* into several villages in 2002. Two years later there were close to 2,000 kiosks, each serving its host village and four others within a 5-kilometer (3 mile) radius. *E-choupal* are allowing farmers access to all the various stages of agricultural pricing. Furthermore, they no longer have to sell to intermediary traders.

There are hundreds of such programs across India. Many of these are private initiatives, connected by a common theme of finding inexpensive, digital solutions to the critical problems of the poor. In rural India, where the majority is semi-literate and live in remote communities unconnected by road or phone, this is indeed a digital revolution!

Urban Landscapes

Indian cities have their roots in Harappan civilization. Later, they centered on the Indian princely states. Such indigenous cities were nodes in functional regions, offering protection in return for food. With British intrusion, powerful economic systems shifted the emphasis to coastal cities. Interior cities became peripheral in importance to the larger international economic system in which spatially peripheral (coastal) cities became central to economic interchange. Furthermore, two separate administrative systems operated: that of the Indian princes and that of the British.

Similarly when British institutions and infrastructure were superimposed on traditional landscapes, two urban structures emerged. English cantonments and civil lines operated differently from the rest of the city, and straight avenues brought order to the perceived chaos of the Indian areas. Calcutta, Madras, Bombay, and later New Delhi and others emerged as dual cities.

In 1947, there were 582 princely states and estates of which 115 were gun-salute states. The number of gun salutes—21, 15, 10—indicated the importance of the prince, not the economic system tied to the princely city. When states were reorganized in 1950 and 1956 along linguistic lines, princely states lost primacy, prestige, and power. Only five of these are still state capitals: Bhopal, Hyderabad, Jaipur, Srinagar, and Thiruvananthapuram (Trivandrum).

According to geographers Brian Berry and Howard Spodek (1982), in the residential geography of the traditional Indian city, socioeconomic status was the dominant theme. However, modernization is transforming communal caste status to communal class status. High-status areas are central by location; low-status areas are peripheral by location. With rapid growth and change after independence, by the 1960s, distinct familial areas and male migrant areas had emerged around factories and port facilities. And there was some suburbanization.

In the 1960s, planning direction shifted from controlling metropolitan growth to diffusing it. Greenbelts and satellite towns were installed but were largely ineffective in stemming urban growth. Many of the new towns were too close to the main city and were quickly engulfed by squatter settlements. Construction was shoddy and buildings quickly deteriorated (Figure 8-14).

More recent efforts include relocating squatters to new housing even further away. There, families find themselves without basic services, including transportation to the city where they can conduct their informal activities. Many men are forced by circumstances to find shelter in the main city. Many of these men start new families, thereby fracturing their original family, leaving women and children destitute.

Mumbai (Bombay) is located on a series of small, joined islands and reclaimed land. Traffic congestion is exacerbated with north–south commuting. The government decided to develop a New Mumbai (Navi Mumbai) on the mainland to decentralize the ever-choking stream of commuters. Success was deemed likely because similar plans had been effective in Japan, China, South Korea, and Singapore. With jobs and housing available in Navi Mumbai, congestion would be relieved and more industry would be attracted away from the central city. Similar schemes were planned for all of India's major cities. Unfortunately, success at decentralization has been limited, at best.

India is a place where conflicting interests and powerful lobbies can distort any plan. Industry and business, real estate and construction, middle class and poor: all have their different goals. Besides, the Indian government has generally stood as champion of the poor, so how could it legitimately support projects benefiting the other groups? Maharashtra state government showed little interest in the plan. Moreover, it permitted further land reclamation and building development at the southern end of Bombay Island, adding to the existing congestion.

Figure 8-14

This apartment is one of hundreds of buildings constructed in the 1960s in Bombay (Mumbai) when the government was desperate to house millions of poor urban migrants. The quality of the construction was poor so the apartments became dilapidated very quickly. People in these structures share cooking areas and toilets. Often, there is no water or electricity and the toilets are blocked. Photograph courtesy of B. A. Weightman.

In Navi Mumbai, 50 percent of the housing is for the poor. Tiny apartments house extended families in large apartment blocks. Hurried building has resulted in poor-quality structures, which now appear no better than inner city *chawls* (slums). However, middle-class housing shows sensitivity to both space and privacy needs. This reveals the class bias on the part of the designers. Today, half the residents of Navi Mumbai live in slums, and the city is ringed with squatter settlements. Nevertheless, in the 2000s it has become a center for computer-related industries. Some call it "Silicon City."

Indian cities are notoriously crowded, dilapidated, and polluted. Kolkata (Calcutta) is regarded by many as a lost cause. New Delhi is one of the world's most congested and polluted cities. With 16.5 million, Mumbai adds 300 migrant families every day. Many end up as rag-pickers in Dharavi, Asia's largest slum; others join the city's throng of 120,000 prostitutes (Figure 8-15).

Government decentralization with developments in interior cities and rural areas has not stemmed this tide of migrants to larger urban centers. Thousands sleep in the streets, yet millions—the wealthy, the middle class, and the poverty-ridden—do manage to survive in often-appalling conditions (Figure 8-16).

Figure 8-15

New middle and upper-class apartment buildings are quickly surrounded by squatter settlements in major cities. The apartment dwellers try to ignore these surroundings. © INDRANIL MUKHERJEE/AFP/Getty Images, Inc.

Figure 8-16

Homeless woman in Mumbai. Hundreds of thousands of individuals and families live in the streets of India's major cities and towns. In Mumbai alone there are at least 800,000. Photograph courtesy of B. A. Weightman.

Mrs. Hiyale—the Rag-Picker

In India, recycling provides a living for millions. Most urban households keep all their "trash" such as plastic bags and sell it to itinerant *kabiri-wallahs* who call at regular intervals. The rest of the garbage is usually collected by men and boys and sorted through by female "rag-pickers." Once the "good stuff" is removed the remainder goes to the dump where even poorer women and children sort through it again.

Mrs. Hiyale became a rag-picker 10 years ago when a drought forced her and her family to abandon their tiny plot of land in Mumbai's rural hinterland. They migrated to the city to find work. "Every day was a bad day," she said. After a period of misery, she came across a feminist charity that seeks to find jobs for female rag-pickers. Now, she retrieves rubbish from apartment blocks that have an arrangement with the charity and sorts it in a shed where the swarming flies and stench do not seem to bother her.

Everything is sorted and resorted by size, thickness, weight and color: copper wiring, plastics, paper, pieces of metal, cloth, leather, coconut shells, and the like. Sorting is important because, for example, envelopes fetch more money than a piece of paper. Mrs. Hiyale is very thankful to have her job, which brings in a regular income: about 150 rupees a day (US$3.00), a very good wage for an informal activity in India.

Turning the Elephant into a Tiger

Indian politicians claim that they are turning India from an elephant into a tiger. Old manufacturing regions are burgeoning with industrial production and are expanding to incorporate secondary cores (Figure 8-17). The primary cores associated with Mumbai, Kolkata, and Chennai (Madras) are products of historic inertia, but associated nodes such as Pune (a former hill station) and Indore in the western industrial region are products of the late twentieth and the twenty-first centuries.

MUMBAI

Mumbai is the largest port on the Indian subcontinent and handles about 25 percent of India's foreign trade. It also accounts for about 11 percent of India's industrial employment. Mumbai has recently become an important center for the IT industry that includes research and development as well as call centers. Its metropolitan area contributes a staggering 38 percent of India's taxes.

Mumbai has the world's largest movie industry, producing more than 700 films a year. Made in **Bollywood** (Bombay/Hollywood) films are in Hindi and plots are filled with music, dance, and clichés. Many are romance stories but whatever the theme, good always overpowers evil. Movies typically exhibit nine cultural elements—love, hate, sorrow, disgust, joy, compassion, pity, pride, and courage—as instructed by the ancient text *Naya Shastra* (Science of Theater).

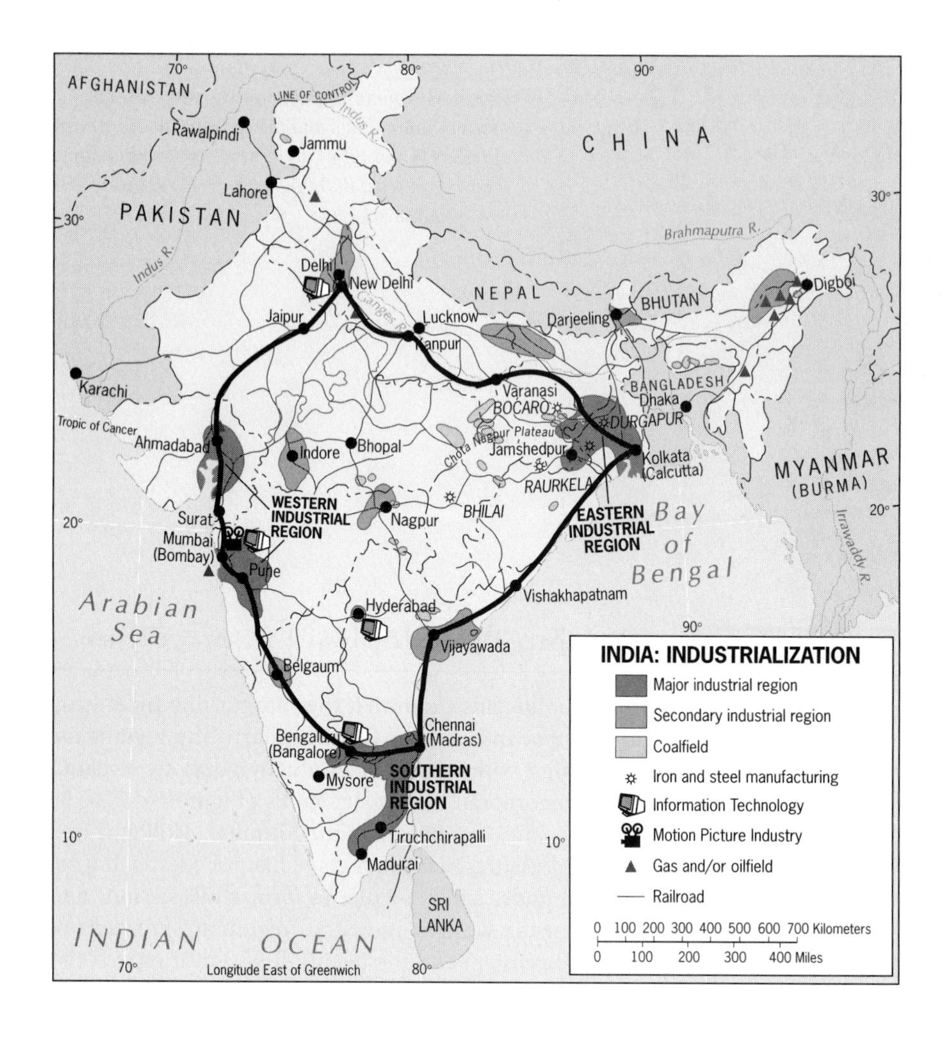

Figure 8-17

Raw material and manufacturing regions. Note the role of former colonial cities as centers of activity. The Golden Quadrilateral highway connects these with hundreds of other places. From H. J. de Blij and P. O. Muller, *Geography: Realms, Regions and Concepts*, 14th edition, 2010, p. 443. (Modified by B. A. Weightman). Originally rendered in color. Reprinted with permission of John Wiley & Sons, Inc.

After showing in India, Bollywood films are exported to other parts of Asia and to Africa and the Middle East where they are very popular. Interestingly, only one in five films turns a profit. Audiences are frequently poor and pay only 25 cents to see one. Obviously, large audiences are requisite for profitability. The entire industry generates US$1.3 billion a year.

Financing is difficult and comes with exorbitant interest rates—24 to 36 percent. Mafia-type organizations have stepped in to fill the financial void. These organizations extort protection money from producers, directors, and stars. Like Mumbai's construction industry, its film industry is riddled with organized crime. Nevertheless, Bollywood's studios and stars add to the city's magnetism and bolster its ranking among the world's great cities.

In 2004, Prime Minister Singh said that Mumbai should become another Shanghai by 2010. A report—*Vision Bombay*—provides a strategic framework for improvements in such areas as housing, education, and infrastructure. In 2006, the government demolished some 90,000 "makeshift" homes on prime economic land designated for modern development. The government stopped the demolition only when it realized that poor people vote! Three out of every five residents live in slum housing, yet thousands pour into the city every day.

The goal of becoming another Shanghai appeared quite ridiculous with the collapse of Mumbai's infrastructure in 2005 when 37 inches (94 cm) of rain fell in 24 hours. People walked home in chest-high water floating with waste. Some disappeared in bottomless potholes or sewers. At least 500 people died and countless homes were destroyed.

The civic administration fell apart again within a year when several coordinated bomb blasts destroyed seven commuter trains during rush hour in 2006 killing 200 and injuring 700 others. The fact that the trains were running again in six hours was amazing, but people complained about the state's utter failure to coordinate relief efforts. India blames Pakistan for these and subsequent attacks.

Kolkata now anchors an industrial district that focuses on jute and cotton manufactures, engineering, and chemical industries. Coal mining and related iron and steel manufacturing center on Nagpur. The region incorporating Maharashtra and Gujarat specializes in textiles, food processing, chemicals, and engineering.

INDIA'S SILICON VALLEY

Chennai, specializing in textiles and light engineering, has expanded into India's southern industrial region, which incorporates what has become known as **India's Silicon Valley**, centered on Bengaluru (Bangalore). Bengaluru's story illustrates some of the problems and trends in modern economic development in India.

Bangalore, an army cantonment during the Raj, later became known as India's Garden City. In the 1980s, a spate of colleges and skilled workers, along with a pleasant climate, drew the attention of foreign corporations. Multinationals like Lipton Tea and Unilever, and American firms such as Texas Instruments, IBM, and Dell opened plants in the area. Indian companies such as Infosys and Wipro have become highly respected multinational corporations.

The 1990s brought more companies, stores, high-end shopping malls, and restaurants. Then an array of woes ground the city to a halt. Population increased from 2.9 million in 1981 to more than 6 million in 2009. Now, there are shortages in supplies of water and electricity. Cars and motor scooters vie with pedestrians in a dangerous rat race. Bengaluru is now one of India's most polluted cities.

Suburbs are built and occupied before service facilities are installed. Commuting times have doubled. Slums that sprung up to house migrant construction workers have become permanent. Some say that the city is collapsing under the weight of its own success. Nevertheless, liberalized government policies and continued availability of highly trained workers, a relatively high literacy rate (56 percent), the use of English for business, and a new international airport combine to make Bengaluru's Karnataka State desirable for investment.

India remains the world's most attractive country, ahead of China, Malaysia, and Thailand, for back-office functions. This is according to A. J. Kearney's "Global Services Index," which evaluates 50 countries according to their financial attractiveness, skilled-labor supply, and the business environment. India tops China because of significantly lower wages, infrastructure, and regulatory costs.

THE NEW CONSUMERS

According to a National Research Council study in New Delhi in 1994, India's consumers can be divided into five classes:

- The "very rich"—6 million people or 1 million households
- The "consuming class"—150 million people (half the bloated, conventional estimate)
- The "climbers"—the lower-middle-class of 275 million
- The "aspirants"—275 million (who would be classified as poor in Europe or America)
- The "destitute"—210 million

Since then, another 100 million have joined these groups, but the relative balance among these five classes, despite some progress, has not changed dramatically. The first three groups are the major consumers of modern trappings such as appliances, electronics, and Western-style dwellings (Figure 8-18).

The middle-class is expected to expand to 583 million by 2025. Even though most Indians will remain rural, consumption in urban areas by the rich and the middle-classes is expected to rise from the current 43 percent to 62 percent of the total. If this does occur, India may overtake Germany as the world's fifth largest consumer market.

The Indian mobile phone market is adding more than 3 million subscribers a month. In 2005, there were 39 million subscribers. In 2009 there were 500 million! In 2006, Nokia built a new factory in India and before the plant was completed the company had commitments for a million phones. Motorola is making India its headquarters for sales in emerging markets. Acer, HP, and IBM are slashing prices to gain their market share. Dell is also building a factory there. Not to be outdone, China's Lenovo has named India its fifth market target. With computer sales expected to reach 20 million by 2010, this market is too big to ignore.

The expansion of mobile communications in India is also driving growth in personal computing. As prices come down, notebook sales are skyrocketing. While computer sales rose 30 percent in 2006, notebook sales rose 168 percent.

Figure 8-18

It is now very trendy for the elite and the upper-middle classes to live in Western-style houses. Many of these are gated communities in new suburban developments. © Ed Kashi/NG Image Collection.

Wireless, broadband PC networks are being expanded. Microsoft plans to construct 50,000 Internet cafés over the next four years.

Television is playing an equally powerful role in India's transformation. There was only one state broadcaster in 1991. The station's mission was to unify the country. Color TV arrived in the 1980s. In 1983, television signals were available to 28 percent of the population; now it reaches more than 90 percent. CNN and MTV arrived in 1991. As of 2008, Indians could choose from as many as 350 channels. Now "Tellywood" stars are challenging the fame of Bollywood icons.

Despite jarring economic inequities, television has generated a sense of national community within one of the world's most diverse populations. This is critical in an era of intense political, social, and economic transformation.

THE ROLE OF GOVERNMENT POLICY

For years, government policies founded in socialist ideals served to block economic development. Government-operated institutions, state-run enterprises, and entangled regulations made efficiency unachievable. As one Indian author noted, "If there is to be an Indian tiger, it must be freed from the shackles that keep it languishing in its small cage."

In 1991, the Congress Party began a program of economic liberalization, including industry deregulation, privatization of state monopolies, and easing of foreign investment rules. The transition to a more market-oriented economy accelerated with the BJP, which came into power in 1999. The notorious system of quotas and import licenses for machinery and consumer goods has been dismantled. Foreign ownership of Indian firms is now possible, and transnationals such as Pepsi, Coca-Cola, Sony, and Phillips have entered the Indian marketplace.

Foreign trade is growing, particularly in the service sector. India's software industry is spreading from its Bengaluru and Hyderabad base and finding customers abroad, especially in the United States, which outsources many services there. Software exports have been growing at an annual rate of 50 percent.

Fresh from its success in the global software and information technology markets, India is fast becoming a key outsourcing center for making everything from cars, to steel, to pharmaceuticals. India has a plethora of engineers and designers capable of creating low-cost, high-end products. China and Southeast Asia cannot compete in this arena.

Transnationals come to India to establish their presence in the huge domestic market. The South Korean car giant Hyundai provides an example. Hyundai built a car assembly plant in Chennai in 1996. In 2002, the company grabbed a 20 percent share of India's rapidly growing passenger car market. A confident Hyundai is investing millions to increase its production capacity to 250,000 cars a year. Hyundai is not alone in India. Ford and Suzuki are using their Indian operations to produce compact cars for markets in Asia, Latin America, and Africa.

Indian car manufacturers are selling vehicles designed for emerging markets around the world. Some 8 million Indian households are able to afford a car in the US$5,000–8,000 range. Maruti-Suzuki, India's largest car maker, is meeting their needs with its Maruti models. Pitching to an even lower income strata, Tata motors had released a bare-bones car called the "Nano" that is selling for US$2,200. (This is a fortune in India.) Tata Motors has also bought the British companies Jaguar and Land Rover. India's automobile industry is predicted to overtake South Korea's by 2015.

India still has problems in its manufacturing sector. Its infrastructure is rickety and high tariffs on many imported products saddle manufacturers with bloated costs. Nevertheless, as of 2003, India's economy has been growing at a rate of 7.5 percent a year. A slowdown has ensued in the current economic downturn and the rate of growth in 2009 was 6.1 percent. Even so, India now has the second fastest growing economy in the world after China.

India's economy is rapidly becoming more enmeshed in the larger Asian economy. Japan now invests more in India than anywhere else. It has also promised to finance a Mumbai-Delhi industrial corridor. South Korea has entered the Indian market with its LG electronic brands. LG advertises its products in a dozen Indian languages. Meanwhile, India is investing in Southeast Asia.

The government has also come up with a plan for Special Economic Zones (SEZs) like those in China. However, the idea of SEZs has come under stiff opposition from a variety of government, social activists, and environmental groups who envision loss of farmland and natural habitat. The plans are currently frozen.

Getting Around the Country

India's roads are notorious for being jagged and potholed. Road linkages between the country's 41 metropolitan areas are grossly inadequate. I took a bus from Delhi to Agra and remember the journey as a life-threatening experience. Along the way, I counted seven trucks and buses that had apparently fallen off the road where it had broken away.

Now, India is building a nationwide four-lane expressway that will link the major nodes in its urban system: Mumbai, Delhi, Kolkata, and Chennai. Fifteen additional cities will be woven into this national transport web called the **Golden Quadrilateral** (Figure 8-17). The project will expand city hinterlands, link once remote towns and villages to markets, speed up economic interchange, and accelerate rural–urban migration.

Millions of Indians travel by train. Boarding a train is a major accomplishment. I remember journeying from Delhi to Varanasi. At each stop police were literally beating the hopefuls with sticks to prevent them from trampling each other. Windows have bars to stop people from crawling in. With the exception of upper-class seating, the carriages are bulging with humanity. Male passengers hang onto the window bars and sit on the roofs. Ticket-takers roughly push through the masses.

Three million individuals commute from Mumbai's suburbs to work every day. At peak hours, 5,000 cling to the trains designed for 1,700. Hundreds fall off and are killed on the tracks every year. Clearly, huge improvements in urban infrastructures are essential.

Getting Energy

THE PETROLEUM BACKBONE

The petroleum industry forms the backbone of India's national economy. One of the oldest oil fields in the world was established in 1889 at Dibrugrh in Asom. Exploration and discovery expanded from the folded mountainous regions to the alluvium-covered shelf in the Brahmaputra Valley. After independence, exploration was expanded to nearly all the sedimentary basins of the country. Several discoveries of oil and gas were made in the Bombay Basin (Gujarat), Assam and Arakan Basin, and the Bombay Offshore Basin. In the 1990s, additional fields were discovered offshore Mumbai (Bombay). The "Bombay High" is located 161 miles off the coast of Mumbai and produces half of India's oil but only 15 percent of its requirements.

Another recent development has been to commercialize reserves of natural gas. Over the years there has been a significant shift in the domestic pattern of energy consumption. In 1947, more than two-thirds of India's total energy consumption was of noncommercial fuels such as wood, harvest waste, animal dung, and commercial coal. Now, natural gas provides more than half of India's commercial energy, and the private sector has diversified into infrastructure construction, marketing, import facilities, pipeline manufacturing, and the production of an array of byproducts such as naptha, toluene, and lube oil.

India must still import fuel. Multinationals as well as private firms are contracted to seek imports in addition to finding new sources. The Asian Development Bank has proposed a pipeline that will bring natural gas from the Caspian Sea to India via Turkmenistan, Afghanistan, and Pakistan. Instability in Afghanistan and Pakistan is hindering the plan.

DAMS AND THE DAMNED

Inadequate water control—timing, amount, and distribution—has been a bane of Indian farmers for centuries, and the rapid expansion of irrigation systems has been one of the country's highest priorities since independence. Dam construction aims to alleviate these stresses, thereby increasing food production. Numerous projects have been installed, mainly large-scale operations involving foreign aid and modeled on America's Tennessee Valley Authority's (TVA) principles of regional hydroelectric planning. Tremendous investments have been made in several areas, but the consequences have not always been positive.

The Krishna River, most important for irrigation on the Deccan, was developed with a series of dams, tanks, and canals dating back to the Raj. Unfortunately, planners did not take into account farmers' perceptions of their water requirements or desired crops. State boundaries divided stored water supplies from users' lands. Moreover, construction integrity was poor, resulting in silted dams and leaky canals. In fact, dams and canals are rarely able to provide water to even 50 percent of their designated service areas.

Another earlier dam project concerns the Damodar River, which drains Bihar and West Bengal, an old industrial core founded on iron and coal supplies. For years, excessive rainfall in the upper valley produced flooding in the lower valley, with devastating consequences. The Damodar Valley Corporation (DVC) was designed with the assistance of the TVA. Floods have been reduced, but the DVC's greatest success is in the area of power generation. Unfortunately, increasing industrialization of the Damodar region with chemical factories and mining operations has rendered the river one of the most polluted in India.

The Narmada River project, the largest ever undertaken, includes two super dams: the Sardar Sarovar in Gujarat and the Narmada Sagar in Madhya Pradesh. In addition, 30 large dams, 130 medium-sized dams, and 3,000 minor ones are planned. The scheme should help feed 20 million, foster an industrial boom, and protect 750,000 downstream residents from floods.

Critics claim that the Narmada scheme is a disaster in the making. Both forests and agricultural lands are being flooded and up to 1.5 million people will be displaced. The destruction of forests will be particularly devastating because they have housed a diverse community of tribal peoples such as the Bhils and the Gonds, as well as endangered wildlife, including the tiger. Tribal representatives say they have nowhere to go, and when the valley is flooded, they will join India's growing number of environmental refugees.

THE POWER OF WIND

India has committed itself to energy independence by 2020. In 1994, Tulsi Tanti set up two wind-power machines to support his family's textile mill in Gujarat state. This venture evolved into the founding of Suzlon Energy, which manufactures and installs wind turbines. It has developed the largest wind park in Asia, at India's southern tip where the trade winds blow consistently. The company is experiencing rapid growth with contracts from North America, Australia, and China. Suzlon already has a large wind-power park and a rotor-manufacturing plant in Minnesota.

India's Oustees

The twentieth century was a century of uprooted people: **oustees**. These are people who flee from war and violence, political and/or ideological oppression, natural and other disasters, economic stagnation, and human-induced environmental problems. Dams produce the largest segment of environmental oustees, followed by mines.

About half of oustees are landless or sharecroppers with no recorded or "legitimate" claims to the land they may have worked for generations. No more than half of oustees are ever resettled. Some are compensated monetarily, but many squander their new-found "wealth" on consumer items and end up landless, jobless, and moneyless.

Many oustees become seasonal laborers in agriculture or construction. In northern India, husbands tend to leave their wives and children with a male relative, perhaps for years at a time. In southern India, husbands tend to take their family with them. These workers join the vast sprawls of squatter settlements on urban fringes. Many become indebted to ruthless bosses and spend the rest of their lives virtually enslaved.

An important point to note is that as India's population increases, every project displaces greater numbers of people. The Narmada River project is only one of many. The Sardar Sarovar in Gujarat will flood out 193 villages in Madhya Pradesh. All the associated works will displace an estimated 200,000 people. More than half are scheduled tribes. Thousands of people have marched in protest, but the waters continue to rise as the project continues.

India is one of the world's largest dam-building nations, and estimates of those displaced by large dams in India in the last 50 years vary from 21 to 56 million people. Around 40 percent of these are *adivasis*—tribal people. The government claims that 30 percent of India's food production increases can be directly related to improved irrigation from dam projects.

Cheap electricity attracted a number of industries into this once-stagnating region. Urbanization and industrialization increased with clusters of cities focusing on power plants. The area around Chandrapura became a primary growth pole at the expense of some of the other industrial centers. Devised initially to control floods, the DVC has stimulated industrial development in one of India's most heavily populated regions.

South Bihar is one of the richest areas of India in terms of raw materials such as coal, iron, bauxite, and mica. These minerals have been systematically exploited, first by the British, then by Indian industrialists such as Tata, as well as the Indian government with money from the WB. The results have been lucrative for these outside developers but disastrous for the region's inhabitants and the environment.

Mining displaced thousands of poor people. For example, Tata Iron and Steel displaced nearly all the original inhabitants to make room for its massive complex at Jamshedpur. Other consequences are deforestation, soil erosion, and groundwater pollution. In addition, the bulk of government expenditure is plowed into industries producing consumer goods for the urban wealthy and upper-middle classes.

The industrial region around Jamshedpur straddles the borders of four states: Bihar, West Bengal, Orissa, and Madhya Pradesh. Development has bypassed or negatively affected millions in this otherwise poor tribal region (see Figure 8-1). In the 1980s people began to organize politically. Ultimately, a separatist movement emerged. Protesters advocated a new state to be called Jharkhand. Jharkhand was created in the year 2000.

Ganga Ma Is Sick

Ganga Ma—Mother Ganges, 1,560 miles (2,510 km) long—is the ancient symbol of creation, preservation, and destruction. Its sacred waters emanate the powers of Brahma, Vishnu, and Shiva (Figure 8-19). But it has become fouled with the profane evils of poison, death, and disease. More than 400 million people depend on its monsoonal waters to flood their fields with nutrient-rich silt. By 2030, there could be as many as a billion people living within its reach. However, an increasingly sick river could spell catastrophe.

Microorganisms that break down organic wastes require oxygen. Biological oxygen demand (BOD) is the measure of this need. Imagine how much BOD is needed to destroy the harmful components of the raw sewage of more than 30 cities, nearly a hundred towns, and thousands of villages. Although sewage comprises most of the waste material, industrial waste is far more dangerous (Figure 8-20).

Chromium and organic wastes from tanneries, bleaches and dyes from textile factories, thick goop from sugar processing, toxic farm chemicals (many of which are banned in the United States), sludge from oil refineries, human ashes, body parts, and whole bodies—all become part of the sacred river.

An estimated 35,000 bodies a year are brought to the sacred city of Varanasi alone for cremation. But the required sandalwood is expensive. Many bodies are only partly burned. Others are simply placed in the river by the devout.

Pollution worsens as the Ganges makes its way to the Bay of Bengal. The World Health Organization standards for drinking water call for no more than 10 coliform counts per 100 milliliters. In Varanasi, coliform counts are as high as 100,000. Typhoid, cholera, and viral hepatitis are common here. In this region, one person dies of diarrhea every minute. Amoebic dysentery, gastroenteritis, and tapeworm infections are part of the lives of millions. About 150 factories are lined up along the Hooghly River at Kolkata. These contribute 30 percent of the waste present in the mouths of the Ganges. An experiment showed that fish put in the water upstream from Kolkata survived only five hours.

Figure 8-19

Here on the ghats (steps) of Ganga Ma at Varanasi (Benares or Kashi—City of Light), throngs come at dawn to assuage their sins and suffering. Each supplicant has her own way of worshipping. Some totally immerse themselves; others drink the water. Most make offerings of marigolds and take some sacred water home with them. Sewerage and factory waste along with dead animals and partially-cremated human corpses moil about in this sacred water; Hindus believe that the water is pure.

© David Zimmerman/Masterfile.

Industrial discharge from the Ganges is growing at 8 percent a year. With increased population and industrialization, how can the river possibly possess enough BOD to clean itself and flush out the dangerous chemicals?

Consider these facts. Nearly 70 percent of India's available water is polluted, and waterborne diseases such as typhoid and cholera are responsible for 80 percent of all health problems and a third of all deaths. Only 7 percent of India's cities have any kind of sewage treatment services.

In 1985, there was a plan to clean up the river. However, corruption, mismanagement, and technological errors rendered the plan useless. "A fundamental reason for the failure was that most of those who have a stake in the river's health were never included in the planning" (Sampat, 1996).

Figure 8-20

Untreated effluent pours into Ganga Ma all along its banks. This scene is at Varanasi (Benares or Kashi—City of Light).

© John McConnico/AP/Wide World Photos.

Having the Bomb

Gandhi called the atomic bomb the "most diabolical use of science" ever. Prime Minister Nehru proposed a "standstill agreement" on the nuclear arms race in 1954. These Indian ideas were essentially ignored by the world.

Geopolitical considerations are crucial to understanding why India "needs" military nuclear capability. In 1962, China invaded the disputed territory east of Kashmir. Two years later, China conducted its first nuclear test. India-Pakistan wars ensued from?1965 to 1971. China provided assistance to Pakistan. By 1974, India had exploded its first "peaceful" nuclear device. By 1988 it had tested a missile capable of delivering a nuclear warhead into Pakistan. In 1996 both countries refused to sign the International Test Ban Treaty. Two years later in 1998, Pakistan tested its first nuclear missile. India responded by conducting five underground tests.

Pakistan's missile is named the *Ghauri* after an Afghan, Muslim king who defeated a twelfth-century Hindu ruler named Prithviraj Chauhari. India's missile is called the *Prithvi.*

Kashmir has been a flashpoint in stimulating the nuclear competition between India and Pakistan. The Bharatiya Janata Party (BJP) has long advocated a more militant India and is committed to making the nation a nuclear power. Pakistan is just as committed to proving its nuclear strength.

Military expenditures on the part of both nations are exorbitant in light of their socioeconomic needs. The Indian government spends 3 percent of its GDP on defense—the same as that spent on education and twice that spent on health. In Pakistan, defense spending gobbles up almost 30 percent of the national budget, far more than is spent on education, health, and other social programs. Clearly, military costs are devastating to social and economic needs.

China's role is pivotal in this ongoing arms race. Not only has China supported Pakistan in its anti-India hostilities, but also it has sent it missiles capable of carrying nuclear warheads. Moreover, it has stepped up its military assistance to Myanmar (Burma). China is displeased that New Delhi views Tibet as an "autonomous" region within China. To make things worse, Tibet's Dalai Lama and his anti-communist, expatriate followers are headquartered in India at Dharamsala. You can read more about the "Tibet question" in Chapter 11.

Pradox: Reality Belies India's "Developed Nation" Image

India, with its nuclear capability and its average annual economic growth rate of 6 to 7.5 percent, is the leading economic power of South Asia. Scores of millions comprise a dynamic middle class, and urban India, by itself, could be counted as the world's third largest country. Thirty million households are already in the market for various durable consumer products, and millions more are climbing into or aspiring to this category of economic behavior. Fired by the engine of its burgeoning middle class, India is on the move toward becoming a "developed" nation. This is the message delivered by an array of Indian government studies and politicos since the 1990s. These perspectives are at once correct and misleading. The truth is a paradox.

That the middle class exists is undeniable, but it is the middle class of an overall poor country. Only one in five Indians owns a wristwatch; only 48 percent of households own a bicycle; less than 4 percent of households have a telephone. In this country of over a billion, only 1.4 million are regarded as rich. The middle class is, in effect, a poor one unable to generate the kind of demand that can sustain rapid growth, even in elite consumption sectors.

Economic liberalization policies of the 1990s have granted the middle class a role it cannot assume. The credit card industry has mushroomed. Consumer giants such as Hindustan Lever are spending millions on advertising in the consumer sector. But as advertising expenditures have grown 70 percent, sales have risen only 30 percent. Much of the so-called middle class is not able or willing to march to the drum of consumerism.

Satellite TV has exploded, offering multiple channels, foreign programming, and images of a well-off, cosmopolitan, and urban-centric upper-middle class. This has reinforced the notion that a consumption-driven middle class will lead the masses down the road of development, progress, and prosperity. But those in the middle class watch television and confirm their lifestyles and ambitions. This does little for the masses. With respect to TV sets, India only has 76 sets per 1,000 people. Malaysia has 167. For those convinced that India is on the move, the fact that the country has the largest number of illiterates in the world is glossed over. Glitzy commercials mask harsh realities of urban slums, poor villages, and wasted countrysides.

Deprivation is sanitized by statistics and TV-screen "realities." Even the Indian government admits that

39 percent of the people live below its own definition of poverty. In effect, the small elite and the larger middle class live far removed from the worlds of the masses of destitute Indians. According to one critic, economic liberalization, "gave a flamboyant ideological justification for the creation of two Indias, one aspiring to be globalized, and the other hopelessly, despairingly marginalized" (Varma, 1998).

India's rich-poor gap is being perpetuated. We know that education is an important vehicle for change because through education comes opportunity, empowerment, and self-realization. In Malaysia, the government spends US$128 per person on basic education. India invests US$9 per capita. Furthermore, the bulk of India's huge workforce is far below the rest of the world in productivity and quality. For example, the value added per worker in India's manufacturing sector is a tenth that of Japan's.

Although multinationals and joint venture enterprises manufacture cars, electronics, machinery, chemicals, and war material, the sad fact is that poverty and ignorance prevent at least three-fifths of Indians from participating in any direct way in bettering their condition in any national context. Many of the middle class "have consigned the poor to being a fixture on a landscape they do not wish to see" (Varma, 1998).

Feel-good journalism has blasted India into the sky of global success. In 2006, magazine covers around the world trumpeted: "The Rise of India," "India Inc., and "The New India." But Indians talk freely about *two* Indias: "India" meaning the new India, and *Bharat* meaning the old India. India gets celebrated in the media, and Bharat is shrouded away behind the glitz, effectively rendering invisible the still-mean lives of millions. Simon Long, Asia editor for *The Economist*, observes that: "The broad-brush picture of India is glorious; the details can be sordid...and that the shining dreams evoked by the world's recent recognition of India as a great emerging power have always seemed at odds with the messy reality of the country itself."

Recommended Web Sites

www.censusindia/net
1901–2001 India censuses population statistics.

www.globalsecurity.org/military/world/war/naxalite.htm
Keep track of Naxalite activities in India.

www.iipsindia.org
Renowned demographic research institute in Mumbai. Links to other related Web sites.

www.indiastat.com
Variety of statistical data on India.

www.mapsofindia.com
Excellent site with hundreds of maps providing geographical and socioeconomic data. Click on states for detailed information.

www.nrda.in/
Official site of the Narmada Valley Development Authority. Government information on the project. Glosses over related problems.

www.popfound.org
New Delhi organization that conducts population research and conducts advocacy programs.

www.savefamily.org
Information about domestic and other forms of violence and crimes against women, men, and children in India. Coverage of suicides and laws on dowry killings.

http://timesofindia.com
The Times of India. One of India's major newspapers.

www.unicef.org/infobycountry/india_statistics.html
United Nations Children's Fund's statistics about children by country.

Bibliography Chapter 8: India: Giant of the Subcontinent

Adams, Paul C., and Emily Skop. 2008. "The Gendering of Asian Indian Transnationalism on the Internet." *Journal of Cultural Geography*, 25/2: 115–136.

Adlakha, Arjun. 1997. *Population Trends in India.* Washington, D.C.: U.S. Department of Commerce.

Beteille, Andre. 1996. "Caste in Contemporary India." In *Caste Today*, ed. C. J. Fuller, pp. 150–79. Delhi: Oxford.

Blij, Harm, and Peter Muller. 2010. *Geography: Realms, Regions and Concepts*. New York: Wiley.

Christophe, Jaffrelot. 2005. *Dr. Ambedkar and Untouchability: Analysing and Fighting Caste*. New Delhi: Permanent Black.

Chu, Henry. 2010. "Hope Has Withered for India's Farmers." *Los Angeles Times*, August 11: A5.

Dossani, Rafiq. 2008. *India Arriving: How This Economic Powerhouse Is Redefining Global Business*. New York: American Management Association.

Dreze, Jean, and A. Sen. 1996. *India: Economic Development and Social Opportunity*. Delhi: Oxford.

Dutt, Ashok K., and George M. Pomeroy. 2003. "Cities of South Asia." In *Cities of the World*, eds. S. Brunn, J. Williams, and D. Zeigler, pp. 331–371. New York: Rowman and Littlefield.

Dyson, R. Cassen, and L. Visaria, eds. 2004. *Twenty-First Century India: Population, Economy, Human Development, and the Environment*. New York: Oxford University Press.

Easterly, William. 2006. *The White Man's Burden: Why the West's Efforts to Aid the Rest Have Done So Much Ill and So Little Good*. London: Oxford University Press.

Engardio, Pete, ed. 2007. *Chindia: How China and India are Revolutionizing Global Business*. New York: Mc-Graw Hill.

Etienne, Gilbert. 1998. "Alarm Bells for South Asia." In *Feeding Asia in the Next Century*, eds. G. Etienne, C. Aubert, and J. L. Maurer, pp. 35–55. Delhi: MacMillan.

Farmer, B. J. 1981. 'The "Green Revolution" in South Asia.' *Geography*, 66: 202–207.

Gardner, James S. 2002. "Natural Hazards Risk in the Kullu District, Himachal Pradesh, India." *The Geographical Review*, 99/1: 81–93.

Haq, Khadija, ed. 2002. *Human Development in South Asia 2002: Agriculture and Rural Development*. (Mahbub ul Haq Human Development Centre, Islamabad) Karachi: Oxford University Press.

Haub, Carl. 1996. "Spotlight: India." *Population Today*, 24: 7.

Haub, Carl, and O. P. Sharma. 2007. *The Future of Population in India: A Long Range View*. New Delhi: Population Foundation of India (for the Population Reference Bureau).

Haub, Carl, and O. P. Sharma. 2006. *India's Population Reality: Reconciling Change and Tradition*. Washington, D.C.: Population Reference Bureau.

"India Bombs the Ban." 1998. *The Bulletin of the Atomic Scientists* (July/August).

Jacobson, Jodi. 1991. "India's Misconceived Family Plan." *World Watch*, 4: 18–25.

Jayaraman, N., et al. 1996. "GloomTown." *Far Eastern Economic Review*, 18 (January): 40–2.

Kalam, A. P. J., and Y. S. Rajan. 1998. *India 2020: A Vision for the New Millennium*. New Delhi: Viking.

Kamdar, Mira. 2007. *Planet India: How the Fastest-Growing Democracy Is Transforming America and the World*. New York: Scribner.

Karan, Pradyumna P. 2004. *The Non-Western World: Environment, Development, and Human Rights*. New York: Routledge.

Karp, Jonathan. 1995. "To Have and Have Not." *Far Eastern Economic Review*, 17 (August): 62–65.

Khan, Nizamuddin, and Almatar Ali. 1998. "Periodic Markets and Rural Transformation in Gonda District, Uttar Pradesh." *Focus*, 45: 34–73.

Lancaster, John. 2010. "Lost Nomads." *National Geographic*, February: 102–121.

Laquuian, Aprodicio A. 2005. *Beyond Metropolis: The Planning and Governance of Asia's Mega-Urban Regions*. Washington D.C.: Woodrow Wilson Center Press.

Long, Simon. 2007. "Back to Earth." *The World in 2007: Asia*. London: The Economist.

Luce, Edward. 2006. *In Spite of the Gods: The Rise of Modern India*. New York: Random House.

Mandelbaum, Paul. 1999. "Dowry Deaths in India." *Commonweal*, October 8: 18–20.

Marwah, Bhushan. 1995. "India Striving for Self Reliance in Oil." *India Perspectives*, 7: 24–6.

McDonald, Hamish. 1994. "Rebel Redoubt: New Delhi Fights Tribal Insurgencies in the Northeast." *Far Eastern Economic Review*, 9 (June): 31–35.

McKibben, Bill. 1996. "The Enigma of Kerala." *Utne Reader* (March/April): 102–112.

Mehta, Arun. 2010. *An Analysis of Census 2001 State-Specific Population Data*. New Delhi: National Institute of Educational Planning and Administration.

Mehta, Pratap, B. 1998. "India: The Nuclear Politics of Self-Esteem." *Current History*, 97: 403–406.

Mehta, Suketu. 2004. *Maximum City: Bombay Lost and Found*. New York: Random House.

Meredith, Robyn. 2008. *The Elephant and the Dragon: The Rise of India and China and What It Means for All of Us*. New York: W. W. Norton.

Mendelsohn, Oliver, and M. Vicziany. 1998. *The Untouchables: Subordination, Poverty and the State in Modern India*. London: Oxford.

Mohan, Rakesh. 1999. "What Ails the Indian Tiger?" *Far Eastern Economic Review*, 8 (April): 61.

Mukherjee, Ishani. 2007. "Solar Power Reaches 100,000 in Rural India." *World Watch*, September/October: 5.

Nair, Janaki. 2005. *The Promise of the Metropolis: Bangalore's Twentieth Century*. New York: Oxford University Press.

Narayan, Deepa. 2000. *Voices of the Poor: Can Anyone Hear Us?* New York: Oxford University Press.

Neal, Bruce. 2006. "No Easy Health Cure: Asia's Health Challenges." *Far Eastern Economic Review*, 169/10: 46–49.

Noble, Allen G., and Asok K. Dutt, eds. 1982. *India: Cultural Patterns and Processes*. Boulder, Colo.: Westview.

O'Rielly, Kathleen. 2004. "Developing Contradictions: Women's Participation as a Site Struggle within an Indian NGO." *The Professional Geographer*, 56/2: 174–184.

Raaj, Neelam. 2006. "Sex Selection Is a High-Volume, Low-Risk Biz." *The Times of India*, March 29.

Registrar General and Census Commissioner, India. 2005. *Series 1 India: Slum Populations India*. New Delhi: Government of India.

Repetto, Robert. 1994. *The "Second India" Revisited: Population, Poverty, and Environmental Stress Over Two Decades*. New York: World Resources Institute.

Roberts, Gregory David. 2005. *Shantaram*. New York: St. Martin's Press.

Saha, Suranjit K. 1979. "River-Basin Planning in the Damodar Valley of India." *The Geographical Review*, 69: 273–287.

Sally, Razeen. 2009. "Don't Believe the India Hype." *Far Eastern Economic Review*, May: 46–47.

Sampat, Payal. 1995. "India's Low-Tech Energy Success." *World Watch*, 8: 21–23.

Sampat, Payal. 1996. "The River Ganges' Long Decline." *World Watch*, 9: 24–32.

Sarangi, Satinath, and Carol Sherman. 1993. "Piparwar: White Industries' Black Hole." *The Ecologist*, 23: 52–56.

Shurmer-Smith, Pamela. 2000. *India: Globalization and Change*. London: Arnold.

Sharmer-Smith, Pamela. 1984. "The Sikh Identity." *The Geographical Magazine* (September): 442–443.

Singh, Prakesh. 2006. *The Naxalite Movement in India*. New Delhi: Rupa and Company.

Solomon, Jay. 2003. "India Catches Up with East Asia." *Far Eastern Economic Review*, 9 (October): 50–52.

Srinivasan, Padma, and Gary Lee. 2004. "The Dowry System in Northern India: Women's Attitudes and Social Change." *Journal of Marriage and Family* 66/5: 1108–1117.

Tharoor, Shashi. 1997. *India: From Midnight to the Millenium*. New York: Arcade.

Tharoor, Shashi. 2007. *The Elephant and the Tiger and the Cell Phone: Reflections on India in the Emerging 21st Century*. New York: Arcade Publishing.

Tripathi, Salit. 2007. "Bombay's Growth Gets Shanghaied." *Far Eastern Economic Review* 170/3: 45–49.

The Economist. 2010. "India and GM Food: Without Modification." *The Economist* February 13: 46.

The Economist. 2009. "India's Naxalites: A Ragtag Rebellion." *The Economist* June 27: 47–48.

The Economist. 1997. "A Survey of India: Time to Let Go." *The Economist* 22 (February): 1–26.

The Economist. 2010. "Ending the Red Terror." *The Economist*, February 27: 14–15.

The Economist. 2007. "Bridging the Divide." *The Economist* November 3: 14–16.

The Economist. 2009. "Indian States: Divide but not Rule?" *The Economist*, December 19: 74–75.

The Economist. 2007. "The Great Unraveling: Is Globalization Killing India's Cotton Farmers?" *The Economist*, January 20: 34.

Varma, Pavan, K. 1998. *The Great Indian Middle Class*. New Delhi: Viking.

Wallach, Bret. 1984. "Irrigation Development in the Krishna Basin Since 1947." *The Geographical Review*, 74: 127–144.

Wallach, Bret. 1996. *Losing Asia*. Baltimore: Johns Hopkins.

Chapter 9

South Asia: Bangladesh, Sri Lanka, and Islands of the Indian Ocean

"When unity is evolved out of diversity,
then there is a real and
abiding national progress."

MANHAR-UL-HAQUE (1866–1921)

Bangladesh: Nation on the Edge

Imagine 162 million people eking out an existence in a country the size of Wisconsin (Figure 9-1). Think about the facts that 59 percent of males but only 48 percent of females are literate, 27 percent of children under five are malnourished, and 39 percent are stunted (under-height for their age). Add the fact that storms and floods wipe out people and livelihoods by the hundreds or even thousands yearly. Then, consider that this impoverished, primarily agricultural country is one of the largest exporters of ready-made garments in the world. What is the story of this relatively new nation of Bangladesh, and what are its future prospects?

Established in 1971, Bangladesh occupies a drainage basin where the Brahmaputra, Padma (Ganges), and Meghna rivers converge (Figure 8-1). Most of the country is a deltaic, alluvial plain although there are hill districts in the northeast and southeast. Its marshy coastline extends 370 miles (600 km) into the Bay of Bengal.

BENGAL NATION

Bangladesh means "Bengal Nation." This region, once called Bengal, is renowned for its tigers and as South Asia's "rice bowl." Aside from tribal areas in the Chittagong hills, Bangladesh is linguistically homogeneous. Bengali, known locally as *Bangla,* serves as a unifying force within this politically fractious nation.

Although not officially a Muslim state, most of its people are Islamic. Bengal, especially in the east, had a tradition of Tantric Buddhism and Hinduism. This was conducive to the introduction of new faiths such as Sufi Islam brought by Arab sea traders in the thirteenth century. The Hindu caste system made Islam and its notions of equality an attractive option. By 1947, more than half of all Bengalis were Muslim. Today, around 85 percent are Muslim, 12 percent are Hindu, and the remainder are Buddhist, animist, or Christian.

Christianity was introduced by Portuguese traders in the sixteenth century. Again, many low-caste Hindus were drawn to Christian ideas of equality and salvation. The Portuguese also introduced cashew nuts, papaya, pineapple, and guava to Bengal.

The region has a history of political subjugation and fragmentation dating back to the sixth century. British divide-and-rule policies only added to this legacy. Calcutta had served as the capital of British Bengal since 1772, but this changed in 1905 when eastern Muslim regions were linked with Assam and were designated as East Bengal

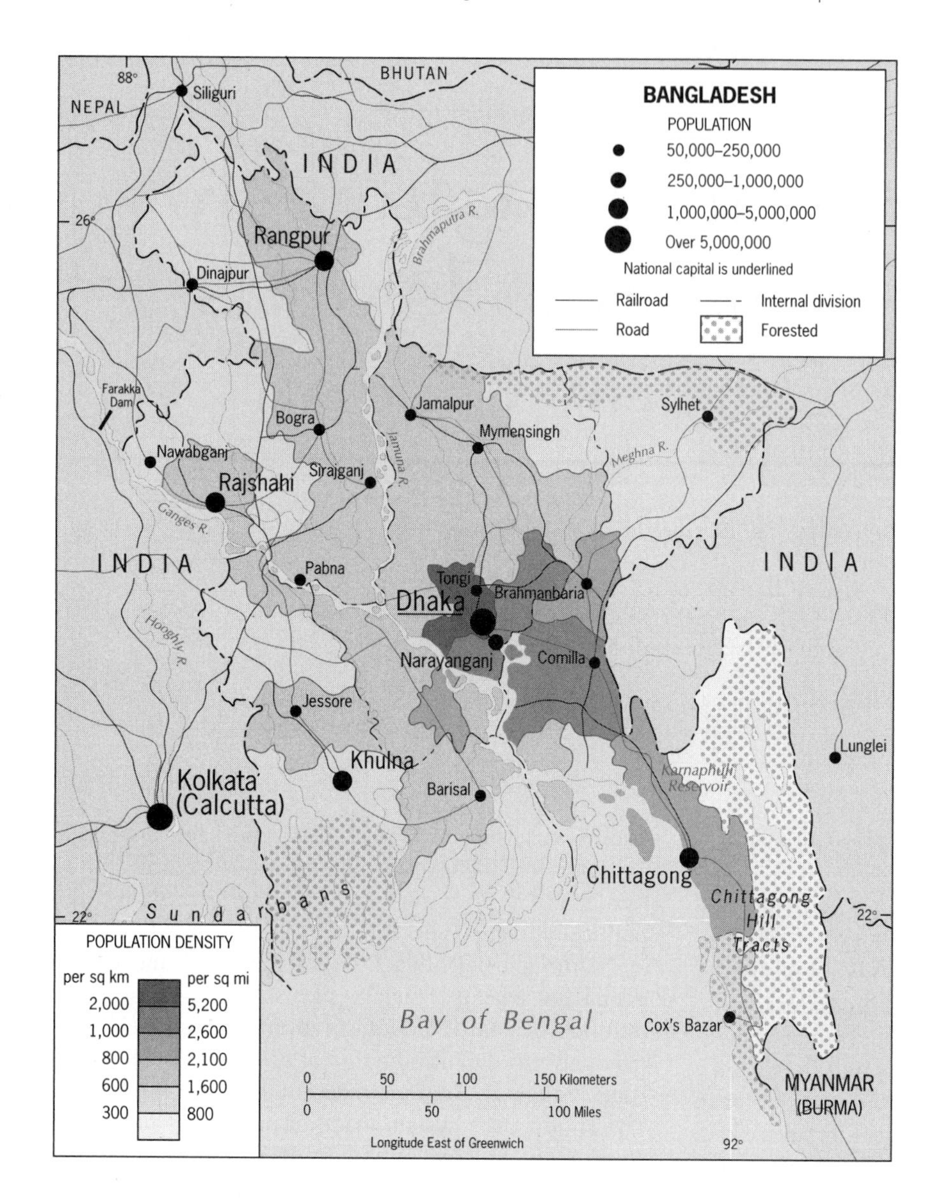

Figure 9-1

Bangladesh and city populations, 2001. From H. J. de Blij and P. O. Muller, *Geography: Realms, Regions and Concepts,* 14th edition, 2010, p. 446. Originally rendered in color. © H. J. de Blij and P. O. Muller. Reprinted with permission of John Wiley & Sons, Inc.

and Assam, with Dacca (Dhaka) as the capital. At the same time, majority Hindu districts of West Bengal, Bihar, and parts of Orissa were designated as Bengal.

In 1912, following years of mass resistance, East Bengal was divided into three provinces and West Bengal was reunited with Calcutta. Muslims insisted on having their own region and so boundaries were again redrawn with the creation of East Pakistan in 1947. Consequently, Bangladesh's borders are politically, not naturally, defined. Surrounded by India on three sides, Bangladesh also shares a 120-mile (193 km) frontier with Myanmar. Bangladesh has signed agreements with both countries, but territorial control remains controversial among some tribal groups. Marine boundaries are still undefined and problematic because of undersea oil discoveries in the Bay of Bengal.

This history of jurisdictional change has created regional dissension and political factionalism. For example, dissension continues to be problematic in the Chittagong Hill Tracts (CHT) situated in the southeastern corner of Bangladesh, bordering on both India and Myanmar. The indigenous Jumma, incorporating more than a dozen ethnic groups, are being overwhelmed by the Bangladeshis who have continued to move into this territory.

Requests for a degree of sovereignty were denied, and the nationalistic Jana Sanghati Samiti (JSS) was formed.

Rebellion ensued, and the Bangladeshi army went on a rampage of killing and destruction in the northwest CHT. Thousands of refugees fled to India's Tripura state. At least 8,000 rebels have died in this two-decade struggle.

A peace accord was signed in 1997, and about 50,000 refugees returned. The Jumma were given some autonomy in the border areas, but the JSS continues to press for further changes. The future of this region remains uncertain.

Regional instability also engendered several core areas focusing on such centers as Calcutta and Dhaka. When East Pakistan was created, the key industrial center and port of Calcutta was lost to India, leaving East Pakistan without a viable industrial base or seaport. Cotton and jute mills were built in Dhaka. Ports were developed at Chittagong in the far south and Mangla in the southwest. Dhaka and Chittagong remain the focus of the modern economy. This economy is only in its initial stages relative to other countries in South Asia.

The Impact of Floods and Cyclones

"Bangladesh" is often perceived as synonymous with "disaster." It is true that millions of lives have been lost to natural forces. In August 1998, 10 million people were rendered homeless by floods! Countless others were affected in 2002, 2003, and 2004. As geographers, we want to know why this happens and what can be done about it.

There are three types of floods that affect South Asia:

- Seasonal Monsoon Floods. These increase slowly and decrease slowly but cover large areas (Figure 9-2).
- Flash Floods. These might be caused by global warming. Glacial and snow melt in the Himalayas along with deforestation causes streams and rivers to rush down valleys, sweeping away everything in their path.
- Tidal Floods. These can occur with cyclones or tsunamis and are typically 10 to 20 feet (3–6 meters) high.

The erratic nature of Bangladesh's climate produces uncertainty. Normally, the monsoon generates 60 to 80 inches (150 to 200 cm) of precipitation from April to October, with rainfall typically peaking from June to September. Excessive rains cause rivers to overflow their banks and inundate surrounding regions. With deforestation and rapid runoff, rivers become deluges. In 1974, two-thirds of the country was under water; the subsequent famine killed 30,000 people. This degree of inundation occurred again in 1998, although with only a few thousand losing their lives.

The 2007 South Asian floods are said to have been the worst floods in memory. A deluge of rain and snow melt during an abnormal monsoon affected some 30 million people in India, Nepal, and Bangladesh (Figure 9-3).

In Bangladesh, rising waters from the Brahmaputra and Padma Rivers forced 7.5 million people to flee their homes and 500,000 were marooned on little bits of high ground. At least 100,000 suffered from waterborne diseases and 500 people died. Help was long in coming as

Figure 9-2

These homes outside of Dhaka are built on stills to avoid anual flooding. However, they have been washed our on several occasions, including the 2004 tsunami. Photograph courtesy of B. A. Weightman.

Figure 9-3

This poor woman is trying to escape the 2004 tsunami with a pot of potable water.

all roads from Dhaka were impassable. Bangladeshi's very existence is closely tied to the floods that inundate at least 18 percent of the country every year.

It is understandable to associate Bangladesh and flood disaster. But Bangladesh and drought? In fact, the country does suffer from drought, especially in the northwest. Since governments are geared to dealing with floods, little if any aid reaches drought-stricken regions. Victims are left to their own devices and must rely on friends and relatives for support.

Bangladesh is also beset by **cyclones**. These tropical storms swirl in a northeasterly direction from the Bay of Bengal and wreak havoc across the nation. In 1970, a cyclone accompanied by a 20-foot (6.10 m) **storm surge** took 300,000 lives. A storm surge is water pushed toward shore by the force of the winds swirling around the storm. When combined with a tide in an increasingly narrow channel across low, flat terrain, it can cause an increase in water level of 15 feet or more. In the case of the Bay of Bengal, as the landform funnel narrows, the wave heightens and wind velocities increase to hurricane force.

The 1970 storm surge wiped out fishing communities, killed fishers, and destroyed boats, to the extent that 65 percent of the nation's fishing capacity was eliminated. This spelled human disaster in light of the fact that two-thirds of the country's protein is derived from seafood.

While rain and run-off floods can occur anywhere, the impact of cyclones is most evident in the south. In April 1991, Bangladesh faced the worst cyclone of any since independence. The death toll was 138,000 people, most of whom lived in the south.

The Sunderbans and the Tiger

The delta comprises 3,720 square miles (6,000 km^2) of ever-shifting mangrove, mudflats, islands, and distributary channels. This region is called the **Sunderbans.** Settlement, although discouraged by the government, is virtually immediate after a flood. Land and water configurations may be different, but any land is space to be seized at the first opportune moment. Even islands barely above sea level are occupied, at least until the next flood. Tigers are also a threat to those living in the Sunderbans.

Those few Royal Bengal tigers that are left after poaching are seriously threatened by rising sea levels caused by global warming according to the World Wildlife Fund for Nature (WWF). The Projected rise of 11 inches (28 cm) above 2000 levels in the Sunderbans will likely outpace the tigers' ability to adapt. This is expected to reduce the tiger population by 96 percent. This means that there will be only 20 breeding tigers left!

THE PLIGHT OF CLIMATE REFUGEES

Concern over sea level rise (SLR) associated with global change has drawn the attentions of scientists, politicians, and concerned citizens around the world. According to the *World Development Report (2010)*, about 18 percent of Bangladesh's land will be submerged if the sea rises by

39 inches (1 m). Should this occur, almost 30 million people will be displaced.

Bangladesh's government has few resources to cope with this pending disaster. At the 2009 Climate Change Conference in Copenhagen, Denmark, the country's leaders urged richer countries to take in *climate refugees*. It tried to drive home the point that industrialized countries have a responsibility to help because they are responsible for global warming in the first place.

Few if any developed countries are prone to absorbing thousands of poor, uneducated, Muslim Bangladeshis. Some Bangladeshis think that the Middle Eastern, Muslim countries have an obligation to take them in.

Many experts hold the view that this will not be necessary because those who lose their land will simply move to other parts of the country, especially the cities that are already overburdened with people. Large numbers of people will likely cross the border into India. India already is displeased with the numbers of illegal migrants from Bangladesh and has built an 8.2-foot (2.5 m) wall along the 2,500-mile (4,100 km) border between the two countries.

The wall is not meant to simply deter Muslim migrants from Bangladesh. India argues that Bangladesh is becoming increasingly militant and that it is sending terrorist militias across the border to support insurgents in its northeast. Because of intense population pressure in poverty-stricken Bangladesh, the wall is unlikely to stop cross-border movement.

WATER, LAND USE, AND SETTLEMENT

India affects the water supply in Bangladesh. In 1974, India installed the Farakka water diversion channel on the Ganges (Padma in Bangladesh) near the Bangladeshi border, allowing less water to flow to Bangladesh. Bangladesh claimed that reduction in water levels increases salinity in the Padma River basin, thereby preventing some 35 million people from getting irrigation water. After years of deadlock, the two countries signed a treaty in 1996 to share the water from the Farakka Barrage.

Instability of land and water patterns affects settlement and land use patterns. As rivers sway between valley walls, they assume a braided pattern separated by small, sandy islands called *chars*. As river waters shift laterally, erosion is exacerbated on the right banks and the whole system exhibits a westerly shift. *Chars* disappear and new channels are formed.

Populations move with *chars*, taking advantage of naturally fertilized land. If a *char* is developed, it can act as a magnet for more people who inhabit the adjacent riverbanks in makeshift shelters. Over half of these displaced people are sharecroppers or rent from large landowners. Others survive in the informal economy. Whatever system is established can be quickly erased in a subsequent flood.

In 1972, it became policy for newly emerged *chars* to revert to the government. The idea was to redistribute these lands to the landless and poor, small farmers. However, the power of landlords, inequities in land ownership, intergroup conflict, and political intrigue have hampered the implementation of resettlement. In fact, over half of rural householders are landless. With little chance of participating in the larger economy, they remain as semi-permanent clusters in the countryside or become urban squatters.

What is being done to mitigate these yearly disasters? In 1967, tidal gauges were installed along the coast. A Flood Action Plan was developed in 1990. A year later, a Cyclone Warning Service was established. However, the system was overused and people became complacent. When warnings of "great danger" failed to materialize on several occasions, many people began to ignore the warnings.

Today, with help from the World Meteorological Organization, Bangladesh has its own Space Research and Remote Sensing Organization. With links to Indian systems, it can broadcast more accurate warnings ahead of time. In addition, concrete shelters, channels, and embankments have been built across the country. Unfortunately, many of these fail to hold up in the floods.

Embankments are also magnets for human habitation. Impromptu settlements cling to their sides. In fact, the right bank of the Brahmaputra has been described as "the world's longest linear housing development."

Agricultural Rhythms

As in all of South Asia, the rhythm of life in Bangladesh follows the seasons in which flooding is integral. Flooding also has a spatial pattern and temporal rhythm. For example, the Brahmaputra and Meghna rise in March and April, fed by Himalayan snowmelt. At this time, premonsoon rains occur in Assam and northeastern Bangladesh. The Ganges rises in May. By June or July, all rivers are swollen with monsoon rains. The Ganges reaches its peak flow anywhere from August to October.

In terms of planting crops, there are three main seasons (Figure 9-4). In addition, cropping practices are adapted to flood conditions. For example, virtually thousands of rice varieties have been planted over the centuries

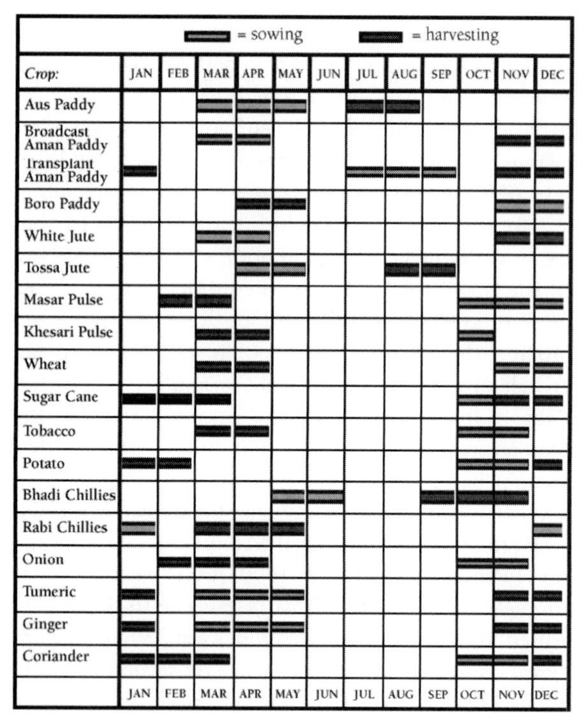

Figure 9-4

Agricultural calendar. Note that June and July are the least busy months. Why do you think this is?

in the context of variation in water levels. These include deep-water types that grow in 15 feet (5 m) of water. This rice is harvested from boats.

Bangladeshi farmers produce a wide array of crops, yet rice covers 80 percent of the cultivated area. Unfortunately, the high-yielding rice of the Green Revolution is not suited to flood conditions. Consequently, high-yielding rice is limited to *rabi* or dry season and other varieties are planted for the rain-fed and flooded *kharif*. Bangladesh must import food to meet the needs of its ever-growing population.

Given climatic factors and the limited availability of irrigation facilities and fertilizer, most of Bangladesh is limited to one crop a year. About 40 percent of the land can be double-cropped and 8 percent triple-cropped.

Recent efforts to diversify have seen regional specialization. Virginia tobacco is grown in the north, and the Sylhet region in the northeast produces 94 percent of the nation's tea. Harvested nine months a year, exported tea is an important earner of foreign exchange. Rubber and coffee are concentrated in the CHT.

With the installation of tube wells, high-yielding wheat has become prominent in the north, with yields exhibiting a ten-fold increase between 1970 and 1983 and subsequent significant increases ever since. However, in much of Bengal overuse of tube wells is drawing arsenic from pyrite bedrock into the water system. In Bangladesh, between 40 and 60 million people are developing symptoms of arsenic ingestion. This causes ever-worsening skin lesions, various cancers, and leads to death. The World Health Organization calls this the largest mass poisoning in human history.

Fiber crops are critical to the Bangladeshi economy (Figure 9-5). For instance, two-thirds of all the manufacturing jobs in the country are related to the cotton industry. Second to rice in terms of acreage is **jute**, a fiber crop used for sacking, twine, rope, carpets, car interiors, and lining for asphalt. Green jute is finding increasing use in

Figure 9-5

These mats are made of coconut fiber (coir). The picture was taken in Dhaka's jam-packed downtown. Photograph courtesy of B. A. Weightman.

fabrics. In fact, special jute blankets have been manufactured to protect poor, northern peoples from the winter cold. Moreover, jute stalks are used as fuel, the leaves can be eaten, and any remaining debris can be plowed back into the soil as natural fertilizer.

Sown from March to May and harvested in August, jute can be grown in uplands or lowlands. One variety can withstand 5 feet of flooding. In fact, Bangladesh produces two-thirds of the world's jute. Until recently, synthetic fibers threatened the jute industry and production declined. Fortunately, environmental concerns have made jute desirable once again.

Livestock are of increasing significance in the economy. In fact, livestock rearing—employing about 20 percent of the labor force—is the most important activity for the rural poor. Cattle and water buffalo are not only draft animals, but also provide dung for fertilizer, milk, and meat for food, and leather, horn, bone, and oil for export. Bovine animals work as much as 80 percent of the agricultural land and are indispensable for threshing grain and crushing oilseeds and sugarcane.

Despite the large numbers of cattle, buffalo, goats, and chickens available, Bangladesh is deficient in meat, milk, and eggs. This can be attributed to genetic deficiencies of livestock, ignorance of breeding, feeding, and maintenance procedures, and lack of veterinary care. As a result, Bangladesh is a net importer of live cattle and milk products.

While Green Revolution technologies have increased yields, beneficiaries are primarily middle-income and rich peasants who dominate cooperatives and local power structures. It is important to note that more than half the land is owned by 10 percent of households and that 75 percent of rural households are either landless or possess tiny plots around 2 acres (0.8 ha). Geographer Abu M. S. Ali (1998) found that small farm size is the leading cause of unemployment and underemployment in agriculture (i.e., frequency of unemployment rises with decreasing farm size). Land reform or any other improvements in agriculture must be viewed in the context of population requirements.

Population and Gender Issues

Bangladesh experienced a population explosion after 1947 with the introduction of health care and improved sanitation (Figure 9-6). The death rate fell dramatically. Increasing numbers moved into marginal lands, where their needs could not be adequately met. This situation was compounded by migration streams from India.

Bangladesh has a natural increase rate of 1.6 percent. In 2003 the TFR was 3.6; as of 2008 it was 2.5. Nevertheless, only 53 percent of rural Bangladeshi women (61 percent urban) are using some form of contraception and just short of half are illiterate (59 percent for men).

In this Muslim society, women are traditionally subjugated by men, marriages are arranged, and acid attacks, burnings, and dowry deaths still occur. The sex ratio remains imbalanced with 105 males per 100 females. Although there is a tradition of *purdah,* most women must work outside. Consequently, strict *purdah* is practiced primarily among the middle classes.

Figure 9-6

These Bangladeshis are trying to get onto an already crowded train. Trainmen crawl over the throngs to collect tickets. © RAFIQUR RAHMAN/ Reuters/Landov LLC.

What is life like for millions of Bangladeshi women? Here are some facts from a Bangladeshi government study in 2009:

- The incidence of low birth weight is significantly higher for girls. Nearly 40 percent of girls in the countryside are born underweight. This figure is 34 percent for boys. Figures are lower in cities.
- Up to 40 percent of females are married before the age of 15 and 70 percent are wed prior to age 17. However, child marriage rates are going down. Of women between the ages of 45 and 49 today, 57 percent were married before they were 15. Of those between the ages of 15 and 19 today, only 17 percent were married before the age of 15.
- In 2003, 58 percent of women had no schooling (48 percent of men). By 2006, the figures were 51 and 37 percent, respectively. Rural–urban differences are apparent. Forty percent of urban women versus 54 percent of rural women have no education, while only 25 percent of urban men and 41 percent of rural men are without any schooling.
- When it comes to decision making, some 48 percent of men decide on food expenditure and what is to be cooked, and do the food procurement. As to whether a woman can go to a clinic for health reasons, only 9 percent of women can decide on their own.
- Most women are not permitted to go out of their village or community by themselves. For 20–24, year olds, only 32 percent can go outside the village and only 27 percent can go to a clinic on their own. For women ages 30–34, both these trips are denied to more than half without a male or much older female escort.
- More than 50 percent of rural men believe that it is acceptable to beat their wife if she goes out by herself without his consent (39 percent urban). A quarter of men believe that they can beat her if she argues with him.

Changing Female Activity Space

Geographer Bimal Kanti Paul (1992) studied mobility and employment opportunities for rural women in Bangladesh and found that as poverty increased, more women were forced to seek work outside the home. This trend jeopardized traditional spatial contexts defined by village structure, *purdah*, and patriarchy.

The basic spatial unit is the *bari*, two to six houses built around a square or rectangular courtyard. Within the *bari* are patrilinearly-related, nuclear families. Clustered *bari* form a neighborhood, and several of these make up a village.

Tradition holds that women's work is confined within or close to the *bari*. However, various factors such as religion and economic status influence the spatial extent of **activity space**. Also, the status and position of women change at different times in their life cycle.

Rich women, young wives, and spouses of Islamic religious leaders have the smallest spaces and rarely venture from the home or *bari*. Other women are confined by more extensive neighborhood boundaries. Elderly, Hindu, and poor women have an even greater range of excursion; they can go farther away from the *bari*. Males and tribal women, with freedom of movement, have the largest activity spaces. Of course, this scheme is a model, and there is much variation in reality.

Patterns of women's activities and associated spatial realms are in flux. Certainly programs intended to empower women have encouraged and facilitated the expansion of female activity spaces (Figure 9-7). However, poverty is the basic cause of working outside the home. While paid labor within other households remains the major employer of women, increasing numbers are seeking wage employment beyond *bari* and village boundaries.

As landlessness and destitution increase, traditional familial support structures fail and the rates of divorce, abandonment, and wife-beating rise. Consequently, more women are working alongside men in the fields, in road and building construction, in brick-making, and numerous manufacturing enterprises. Usually, increased activity space outside the home is accompanied by improvement in living standards, but as Dr. Paul observes, "Unfortunately, this relationship has not emerged in Bangladesh."

Recent governments, headed by women (refer to Chapter 3), have taken great strides in improving women's lives through education and employment. Widespread media propaganda promotes the necessity of empowering women to fully participate in the country's development. Employment in labor-intensive industries

Figure 9-7

Bangladeshi women. Self-help programs are improving the lives of many women in Bangladesh. These women are middle class. Photo courtesy of B. A. Weightman.

and craft and other cooperatives have markedly advanced living standards for some Bangladeshi women.

Bangladesh's population dilemmas derive more from distribution than from sheer numbers. Rural-to-urban migration has overwhelmed cities, especially Dhaka and Chittagong. In 1961, a mere 5 percent of Bangladeshis lived in cities. In 1991 it was 18 percent. Currently, 25 percent live in urban places, and by 2015 that will increase to 37 percent. In other words, 80 million Bangladeshis will be urban residents by 2020. This is an amazing number in the context of infrastructural requirements.

Take power availability as an example. Only 16 percent of Bangladeshis have access to electricity. The country often faces riots over blackouts. At the opening of the largest generating plant, 30 miles (50 km) northwest of Dhaka, Prime Minister Sheikh Hasina urged people to save electricity at home and at work by shutting off lights. Public and private sector projects were expected to boost the national grid's power to 4,700 megawatts by 2002. But there is another problem: People tap into power lines illegally, and nearly 30 percent of power production is lost to theft.

Dilemmas of Growth

CONDITIONS IN THE CITIES

Rural push factors drive urban migration streams while growth in production is not expected to meet employment needs. As population experts Professor Abdul Barkat, U. R. Mati, and M. L. Bose (1997) point out, "urbanization in Bangladesh will remain poverty-driven." Poor migrants typically end up living in slums, squatter settlements, or as pavement dwellers. There are already about 25 million of these individuals, more males than females.

Dhaka, the country's capital and primate city, attracts the most migrants. In 1992, Dhaka had 7.4 million inhabitants. It now has 10 million. By 2015, it is expected to have 17 million residents, with at least 50 percent living in slums or worse. Dhaka's Old City can only be described as a hive of humanity (Figure 9-8).

Every crumbling building, room, cubicle, alley, doorway, and indentation reveals people living in every imaginable condition. The narrow streets are crammed with people, carts, bicycle rickshaws, trucks, and buses, all carrying stupendous loads of oil drums, cotton bales, vegetables, sacks of ice, sheets of metal, thousands of rubber thongs (flip-flops), and baskets of poultry. The city is a limitless kaleidoscope of colors, sounds, and smells.

Child labor is common. This is reflected in the fact that only 20 percent of slum children attend school. In addition, at least one-third of slum dwellers are sick at any one time. Slum communities have the highest infant and maternal mortality rates in the nation, in part because health-care programs have been geared to rural as well as more affluent urban areas.

Outside Dhaka's Old City are still relatively old areas that are becoming increasingly densified. Here, unchecked settlement outpaces government efforts to develop housing and other infrastructural needs. Incentives to decentralize and create communities outside the city have met with limited success.

Figure 9-8

Traffic crush in Dhaka. In all my travels, I have never seen anything to equal the traffic jams in Bangladesh. Photograph courtesy of B. A. Weightman.

In the 1960s, industries were given incentives to locate outside Dhaka. However, they located just beyond an established boundary to claim "less-developed area" status and still enjoy the advantages of urban location. These and other efforts at decentralization were disrupted by the 1971 war of independence.

In the 1980s, subdistricts called **upazilas** were created in Bangladesh. Each of these 500 subdistricts was to house an average of 186,000 people and have revenue-raising powers and measures of self-government. However, related employment development has not slowed out-migration.

Although foreign investment and new construction are changing Dhaka's skyline, labor strikes called *hartals* are frequent, and infrastructural problems remain overwhelming. Power outages are daily events. Inadequate garbage and sewage removal results in clogged canals and drainage backups and overflow. Raw sewage gushes into the Bay of Bengal. Vehicles belch diesel fumes, and industries pour raw effluents into canals and poisonous chemicals into the air. Dhaka is now one of the most polluted cities in the world. Lead poisoning is more than double unsafe levels. Thousands of premature deaths and millions of cases of illness result from the city's pollution.

There are more than 175,000 motor vehicles in Dhaka, including more than 40,000 auto rickshaws that use a deadly combination of engine oil and leaded gas. Beginning in 1999, the government took steps to reduce lead pollution by banning the import of leaded fuels and calling for catalytic converters on new cars. Auto rickshaws were to be phased out. However, these are mostly still plans. Population and urban growth, natural calamities, and political or other instabilities or inefficiencies can easily slow or halt implementation.

AQUA-BUSINESS AND THE TEXTILE INDUSTRY

Bangladesh's frozen food industry is founded on fish, prawns, and similar products. More than 11 million individuals, mostly women, are connected either directly or indirectly with the fisheries sector for their livelihood. Oceans, rivers, lakes, canals, and more than a million human-made ponds and tanks are fishery sources.

Shrimp exporting began in the early 1970s, earning US$3 million. By 1996–1997 this had increased to US$279 million. Increasing numbers of fish farms and hatcheries along with offshore cage fishing have made this into a US$500 million industry. Chittagong and Mangla are the most important processing centers, and aqua-business is now second to textiles in terms of foreign exchange earnings.

The textile industry is the most important sector of Bangladesh's economy, accounting for nearly 80 percent of the country's exports. Garment exports to rich countries accounted for US$8.9 billion in 2005. Competition from China has not hurt the industry because Bangladeh still has the lowest wages in the global textile sector.

Bangladesh is most competitive in knitwear, which makes up half of the garments manufactured. This is because three-quarters of inputs are made locally. This saves the firms storage and transport costs, import duties, and long lead-times that come with imported fabrics used to make shirts and trousers. It also provides them with duty-free access to the EU.

Well-known brands such as Nike, Ralph Lauren, Tommy Hilfiger, and others are made in Bangladesh. The United States and the EU are the largest purchasers of Bangladeshi-made garments. Clothing factories and related activities employ four million workers, 90 percent of whom are women.

Until recently, children were also employed in the textile industry. In 1993, the U.S. Congress passed a bill to ban Bangladeshi clothing made with child labor. Consequently, 50,000 children were dismissed from the industry. Bangladesh has since agreed with international organizations to ban children from the garment sector.

Women Working Outside the Home?

If women are not supposed to go out on their own, how come they can go to work away from home at a factory? Social-economist Naila Kabeer (2000) explains how women in Bangladesh have reconstituted their situation in the context of a patriarchal society. This includes the decision to "go out" to work—a renegotiation of *purdah*.

Public disfavor of women working for wages outside the home provides the context for women's decision making. Many believe that women are taking jobs away from men and that their exposure to the (male) outside world is a "threat to moral order." The view that "garment girls" are of "loose moral character" is widespread. In order to sustain the employment of women in this industry, both employers and women have had to create a scenario where "chaste" women are working in "chaste" environments.

Employers, who think that women work harder and are more "docile" than men, assure their female workers that they are safe. In fact, women are commonly guarded and locked in their workplaces—symbolic of their assured "protection." Several fires have erupted and women killed because they could not get out of the factory. The most recent in 2010 took the lives of 21 women.

Since female workers see themselves as "protected," they can conceive of themselves as not breaking *purdah*. There are males working in the factories but women have internalized the notion that their male and female coworkers are "like brothers and sisters." In other words, they have *domesticated* the workplace.

Since clothing manufacturing is regarded (globally) as women's work, women feel that they are in the "right" area of employment. This type of job redeems itself in that it not only brings extra income but also brings self-respect, dignity, and status to women of their class. Higher-class women would prefer more "suitable" teaching or government positions, but lack of such jobs mediates their decisions to join the ranks of garment workers.

Even so, many women still believe that they are being sinful as there are also men working in the factories. Moreover, they are exposing themselves to males on the street as they walk to work. As one interviewee proclaimed, "Allah does not want women to mix with men. He asks us to remain within four walls, wear a *dosh-hather*, [10-foot (3 m) long)] sari, and a *burkah* if we go outside." Another said, "What else can I do, I have to live somehow.... But we are being sinful...."

NON-GOVERNMENT ORGANIZATIONS (NGOs)

There are more NGOs in Bangladesh than in any other country. More than 80 percent of Bangladeshi villages have them. According to most reports, NGOs function far more effectively than government organizations such as the Bangladesh Rehabilitation Assistance Committee (BRAC). Yet, BRAC provides health services to more than 100 million. It educates 1.5 million children and creates jobs for landless peasants. With a mostly female staff of 108,000, it operates like a state within a state.

Proshika organizes environmentally sensitive human development programs such as organic, sustainable farming and tree planting. It also distributes special filters to remove arsenic from water.

The most well known NGO in Bangladesh is the **Grameen Bank**, which offers small loans to the poor—again mostly women—for shops, sewing machines, and other labor-saving devices. They also loan money to purchase livestock or anything else to empower women. Grameen has given credit to 7.3 million individuals—93 percent of whom are women. Women are regarded as better managers of household finances and more responsible in paying back loans. Loans are also made to small groups. Both peer pressure and group responsibility are critical in ensuring repayment.

World Bank (WB) studies indicate that 50 percent of borrowers get out of poverty in five years and that both divorce and birth rates decline. With Grameen as a model, thousands of other organizations have provided micro-credit to the poor worldwide.

Critics point out that the bank is heavily supported by foreign funds and argue that it would collapse otherwise. Some say that Grameen is really controlled by the government. Some say it is anti-Islamic. Others argue that peer pressure drives out the weakest members, leaving them destitute and compelled to a life of bonded labor. Case studies indicate that perhaps two-thirds of Grameen loans are actually controlled by male family members. Stringent payback requirements drive some women to usurious moneylenders. Grameen authorities say that such stories are rumors circulated by a jealous BRAC or larger organizations like the WB, which are promoting their own schemes. Whatever the case, microloans apparently have helped millions of women in Bangladesh.

The Rise of Islamism?

Bangladesh's image as a secular and tolerant Muslim nation was shattered on August 17, 2005, when more than 400 explosions went off simultaneously in 50 cities and towns across the country. Arrests were made and the main suspects were found to be Islamic extremists. Subsequently, threats were made to blow up the British and U.S. embassies and a suicide bombing occurred in Dhaka that killed 10 and seriously injured 21 people.

Several years of political chaos have helped to create an environment for extremism. Two women vie for power in Bangladesh: Begum Khaleda Zia and Sheikh Hasina. Begum Zia and her BNP political party are pro-Islamic and isolationist; they propose an Islamic state. Sheikh Hasina and her Awami League want to establish a modern, secular state and improve relations with India.

Because of violence among these and other political parties during Begum Zia's tenure as Prime Minister (2001–2007), the military declared a state of emergency and took over the country. Both Zia and Hasina were arrested on charges of graft. They were released the following year. Elections were held in 2008 and Shaikh Hasina won handily. She took office as Prime Minister in 2009.

Some people say that Bangladesh is destined to become the next Afghanistan. It is a fact that some 50,000 radicals, belonging to at least 40 groups, have received or are receiving training in 50 camps around the country. In recent years many fundamentalist Islamic organizations have appeared that are funded by Saudi Arabia and Kuwait. However, it must be noted that the majority of Bangladeshis remain moderate in their approach to religion and that militants are a small minority.

Prime Minister Hasina says that poverty is a threat to the nation and its development. Her government has instituted a national services program whereby hundreds of thousands of unemployed male and female youth will be given training in a trade and a job for at least two years. It is hoped that these apprenticeships will lead to permanent employment. She hopes to engender a "happy and prosperous Bangladesh free from hunger and poverty." Only time will tell.

Sri Lanka: Teardrop Weeping

Arabs called it **Serendip**, the Portuguese named it *Cilao*, and the Dutch spoke of *Zeylan*. All were referring to *Simhala Dvipa*—Island of the Lion Tribe. Known as Ceylon until 1972 when it was renamed Sri Lanka, this tear-shaped island is divided from the Indian subcontinent by the 22-mile (35 km) wide but shallow Palk Strait. While separation fosters cultural distinction, propinquity favors spatial interaction. In the case of Sri Lanka and India, interrelationships have been filled with inherent antagonisms.

THE PHYSICAL LANDSCAPE

The island of Sri Lanka includes both up-country and low country. About 30 percent is up-country, where mountains rise above 8,000 feet (2,500 m). From these heights, rivers flow in a radial pattern downward through lush forests to the coastal plains. The longest of these is the Mahaweli Ganga, running in a southeasterly direction for 206 miles (332 km). In the low country, hills are separated by abrupt escarpments and cut by deep valleys. They descend in a step-like fashion to lowlands, mostly under 1,000 feet (300 m). In the north, these lowlands extend as an undulating plain to the Jaffna Peninsula (Figure 9-9).

The island experiences four seasons in a yearly cycle. From May to August, the southwest monsoon brings orographic rainfall to the southwestern hills. This season culminates with tropical cyclones from the Bay of Bengal, which can affect all areas from October through November. Next comes the northeastern monsoon, which supplies 40 to 80 inches (102–203 cm) of rainfall in the north and

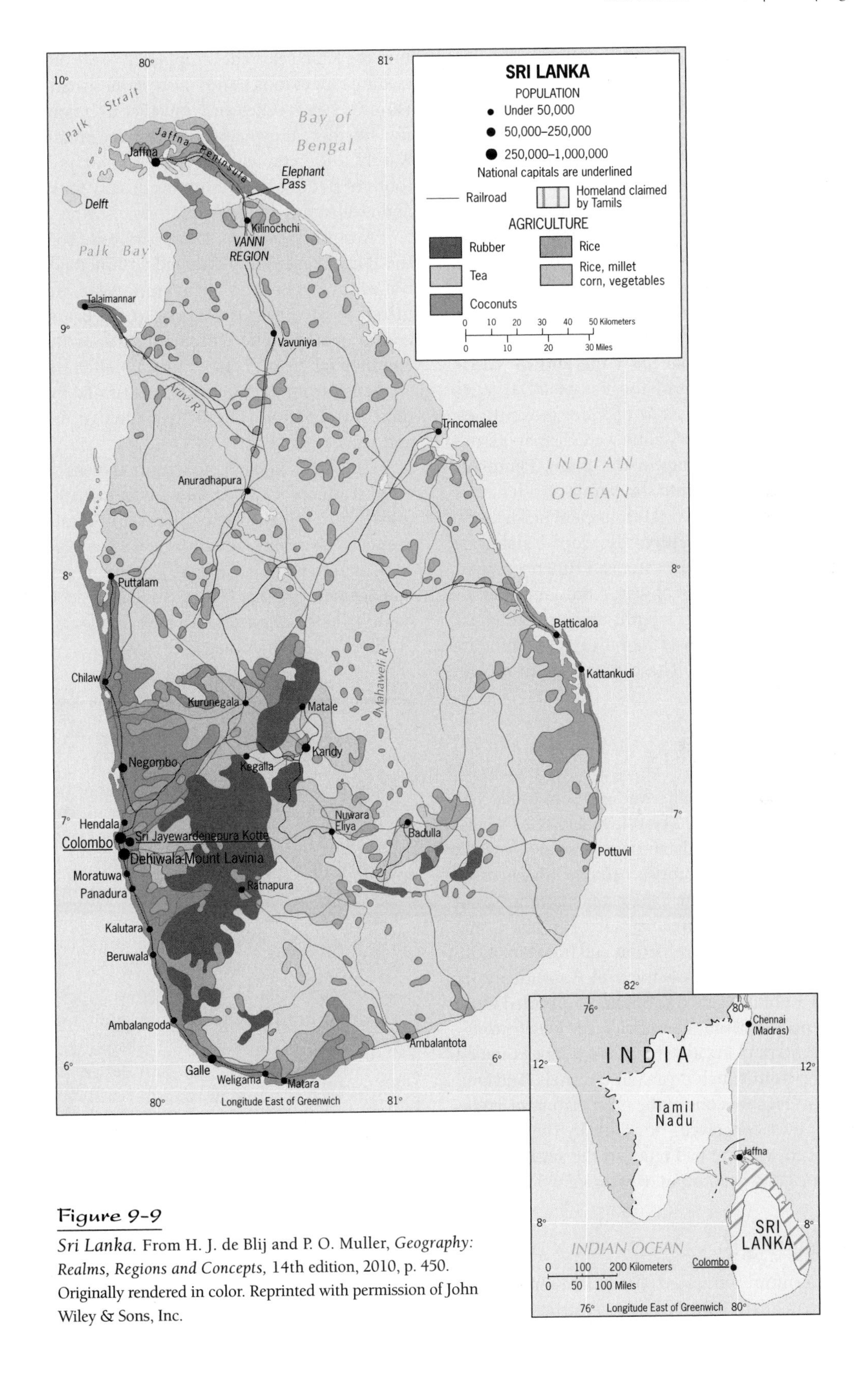

Figure 9-9

Sri Lanka. From H. J. de Blij and P. O. Muller, *Geography: Realms, Regions and Concepts,* 14th edition, 2010, p. 450. Originally rendered in color. Reprinted with permission of John Wiley & Sons, Inc.

northeast through February. From March to April, summer insolation and forested uplands generate **convectional rainfall.** Moisture-sodden warm air rises, cools, and condenses, resulting in late afternoon downpours.

Sri Lanka is tropical with little temperature variation. In fact, nocturnal–diurnal variations are greater than seasonal ones. Although temperatures rarely rise above the low 80°F (25°C), the sun's verticality induces high evaporation rates. This makes irrigation essential, especially in the east. Aridity is a traditional problem. Ancient civilizations fell not simply to invasion but to the destruction of their irrigation systems. Moreover, dry zones were malarial until the end of World War II and the introduction of the use of DDT. With lack of moisture and the presence of killer mosquitoes, colonization schemes of the dry zone were dismal failures before 1945. Karst topography in the Jaffna Peninsula allows the drilling of wells into limestone aquifers.

Water availability was pivotal to ancient settlement. Around 500 BC, Sinhalese settlers developed elaborate water storage and distribution systems. Property ownership was based on the control of water, not land. Canals reached 50 miles (31 km) in length, and the Sinhalese devised water-control valves at dams to accommodate variation in water pressure. These preceded European devices by 1,500 years!

THE INDIAN CONNECTION

Sri Lanka's history is replete with invasions from India. The Sinhalese, descended from the Aryan invaders of Northern India, migrated via the islands and sandbars of the Palk Strait from the fifth to third century BC. In time they became Buddhist. In the seventh century there were Hindu, Tamil invasions. In 1017, Ceylon became part of India's Chola Empire. Its history is marred by friction and fragmentation.

The Buddhist Sinhalese complex of Anuradhapura, with its 15-mile (24 km) walled perimeter, supported more than a million people based on an irrigated rice economy. In an elaborate interlocking system, rivers were channeled to storage tanks, from which diversion channels delivered water to villages. This was one of the most elaborate irrigation complexes ever conceived. However, by the time the Europeans arrived, warfare had reduced the settlement to ruins and the area had become an overgrown wilderness.

EUROPEAN IMPACTS

The sixteenth century witnessed dramatic alterations in Sri Lanka's state of affairs. The Portuguese, striving to capture the Arab sea trade, stopped at Galle and set up spice-trading operations. They also demolished Buddhist and Hindu religious sites and forced conversions to Christianity. Further, they captured the mountain kingdom of Kandy and removed its most sacred relic—an alleged tooth of Buddha. The tooth eventually found its way back and is currently housed in Kandy's Temple of the Tooth.

A century later the Dutch arrived (1658) and ousted the Portuguese. They pursued commercial development, but as a result of the Napoleonic Wars in Europe, were forced to turn over the island to the British in the nineteenth century. The Dutch introduced the island's first commercial crop: coffee. But the coffee trade was under challenge from Brazil, and it was ruined by blight during the British period. The British then turned to **cinchona** and tea.

Cinchona is extracted from tree bark and was the only treatment for malaria, widespread at the time. Eventually numerous other plantation/estate crops were grown (e.g., rubber, coconuts, cacao, and pepper). Estates were nationalized in the 1970s but reverted to private management in the 1990s. Sri Lanka is still one of the world's top tea exporters. Tea accounts for about 75 percent of the country's income (Figure 9-10).

Tea: The Global Beverage

More people drink tea than any other beverage (aside from water) in the world. Ancient Chinese texts describe tea as an elixir. Today tea is touted for its antioxidant properties. Tea is widely employed as a social medium. I have enjoyed tea on many occasions around the world: "China" cups of black tea in England; small glasses of mint tea in Morocco, cups of steaming milk-tea in Pakistan; glasses of spicy *chai* in India; and cups of aromatic green tea in China. Socioeconomic interchange is eased via the ceremony of tea drinking.

Tea diffused from China to Japan, where at first it was used medicinally and in Buddhist monasteries to enhance meditation. Subsequently it spread to Korea. The British introduced tea to India and Sri Lanka in the nineteenth century, and the tea trade contributed significantly to the fortunes of the British East India Company.

Tea flourishes best in the cooler air from 3,000 to 7,000 feet (900–2,150 m). Tea trees are kept small with hand pruning, which stimulates growth.

Figure 9-10

Picking tea in Sri Lanka. Typically, supervisors are male and the pickers are female. Photograph courtesy of B. A. Weightman.

Leaves are hand picked and graded according to size and position on a branch. The bud and top two leaves are the best and earn the label "Orange Pekoe." Tea bags contain small, broken leaves known as "fannings."

The flavor of tea is derived from local growing conditions and processing. For example, Darjeeling tea has a peach aroma and is called the champagne of tea. Earl Grey tea is flavored with oil of bergamot, extracted from an orange-like citrus. Jasmine tea is green tea mixed with jasmine flowers.

There are three basic categories of tea: "black" tea is fermented; "oolong" tea is partly fermented; and "green" tea is unfermented. Fermentation includes bruising, wilting, and drying, followed by rolling. As more surface is exposed to the air, tea juices oxidize. Hot-air drying stops the process. In the case of green tea, the freshly picked leaves are dried immediately.

China, India, and Sri Lanka are the world's largest tea producers. However, it is also grown in Bangladesh, Kenya, Indonesia, Japan, Georgia, and Azerbaijan. Are you a tea drinker? Investigate the variety and origin of the types you enjoy most.

The British brought Hindu, Tamil laborers from Tamil Nadu in India into predominantly Buddhist Sri Lanka. These nineteenth-century plantation workers were distinguished from the Sri Lankan Tamil settlers of earlier times and are known as **estate Tamils**. In fact, because of pressure from the powerful Buddhist *Sangha* (a non-government group representing all forms of Buddhism), they were disenfranchised in 1948. Since India refused to recognize them as citizens, these Tamils were virtually stateless.

Other minorities are the aboriginal **Veddas** and the **Burghers**, descendants of the Portuguese and Dutch. The Burghers are Christian and speak English. There are also Muslims, a religious minority in this Buddhist country. In 1931, Britain granted Ceylon limited self-government and in 1948, full independence.

A controversial development project was launched with British and other aid in the 1950s, an era when dam building was at the forefront of foreign aid and regional development. Incorporated in the Mahaweli Development Program, the Victoria Dam was the world's largest aid project at the time. Although the dam was to provide electric power and irrigation, it required the displacement of 1.5 million people. Typical of many grandiose schemes, construction costs tripled and the government was forced to raise taxes. Displacees were settled in isolated areas with poor infrastructure and adverse climatic conditions. They also suffered psychologically from being uprooted from their traditional homeland. Eventually they became known as *Victoria's victims*.

The Victoria Dam was part of a larger scheme for agricultural colonies in the eastern zone. Here, the climate was drier and the hilly land less fertile. More significantly, Sinhalese colonists clashed increasingly with Tamil inhabitants, presaging the problems to come.

CAUSES OF DISSENSION

Religious, linguistic, political, and economic differences fomented dissension between Sinhalese and Tamils. Political assassinations and violence continue to revolve around these fundamental differences, although the cleavage is not simply two-sided. Anthropologist S. J. Tambiah (1986) makes a salient point: "Sinhalese and Tamil labels are porous sieves through which diverse groups and categories of Indians intermixed with non-Indians . . . have passed through." Moreover, this already complex situation is exacerbated by international involvement.

The Sinhalese are of Aryan extraction with ancient roots on the island. They are Buddhist and speak **Sinhala**, an Indo-Aryan language with its own script. Sinhalese make up 74 percent of the population and are predominent in administration.

The Tamils, who make up 20 percent of the population, are a Dravidian people. They are Hindu and speak **Tamil**, a Dravidian language with a script different to the Sinhala one. It was not until the 1970s that Tamil was given recognition as a national language.

Since the Sinhalese dominate government, they have tried to inculcate Sinhalese history and values into schools, textbooks, and the like. There have also been efforts to divorce Sinhalese and Tamils from any common history, such as their relationship with India. These practices have encouraged social distancing and hostility.

English is regarded as a **link-language**; it is supposed to serve as a *lingua franca*, or common language. But English is spoken and understood by an elite minority, and years of anti-English and pro-Sinhala policies mean that it is not a common language. Students have rioted and burned books at Colombo University because 90 percent of the books are in English. For many rural Sinhalese, English is known as *Kaduwa*—the sword that separated the privileged from the masses.

Violence erupted in the 1950s with a succession of Tamil uprisings against the Sinhalese. After her husband was assassinated, Sirimayo Bandaranaike became the first woman head of state in the world. She nationalized oil fields, companies, and plantations. Communal strife worsened in the following years as Tamil nationalism hardened. In 1972, Ceylon was declared a republic and renamed Sri Lanka. Nevertheless, organized terrorism continued.

In the 1980s, transnational capital entered in the form of an International Monetary Fund (IMF) development program and U.S. aid "to stabilize the region." Subsequently, U.S. warships acquired refueling rights at Trincomalee on the east–west sea route via the Suez Canal. Sri Lanka was to be developed as an "export center" like Hong Kong, and this led to the creation of a free trade zone in Latunyabe. However, interethnic and other discriminatory practices channeled most benefits to the Sinhalese.

CIVIL WAR AND ITS AFTERMATH

The killing of 13 Sinhalese soldiers by Tamils in Jaffna in 1983 sparked a spiral of violence that has only tentatively ended. Clashes between Sinhalese and Tamils resulted in 40,000 Tamils seeking refuge in India's state of Tamil Nadu. A series of political assassinations followed, and foreign investment faded.

More fighting led to the intervention of Indian troops. In 1987, the Tamils were granted a degree of autonomy in the north and northeastern provinces, and their language rights were guaranteed. Meanwhile, Tamil militancy had intensified in the form of a strong guerrilla force with international ties. Known as the **Tamil Tigers**, the group called for the establishment of a Tamil state to be named **Tamil Eelam.** In 1991, India's Prime Minister Rajiv Gandhi was assassinated by a suicide bomber. The Tamil Tigers are blamed for this act.

December 1995 witnessed the largest military operation in the island's history. Government forces routed the Tamil Tigers and took control of Jaffna. Perhaps 2,000 died, and half a million civilians were displaced in this conflict. A cease fire was broken in 1996 when a bomb exploded in Colombo, killing 91 and injuring 1,400 people. The explosives were determined to be Ukrainian RDX smuggled via Dhaka. Kandy's holy Temple of the Tooth was bombed in 1998.

With international mediation, another cease fire agreement was signed in 2002. Nevertheless, hostilities broke out again in 2005 and the conflict escalated. The government swept the Tamil Tigers out of the eastern part of the island in 2006 and moved on to the northern region the following year. With the destruction of a number of arms-smuggling ships belonging to the Tigers and an international crackdown on illicit funding, the Sri Lankan army was able to defeat them in 2009. The conflict has left more than 100,000 people dead and the east and especially the north devastated.

Tsunami 2004

On December 26, 2004, a huge earthquake struck off the coast of Sumatra in Indonesia (see Chapters 2 and 16) thereby generating one of the biggest tsunamis the world had ever seen. On January 3 in

Figure 9-11

Residents sift through the trail of destruction along the coastal railway line in southern Sri Lanka after tsunami tidal waves lashed more than Sri Lanka's coastline in 2004. © SENA VIDANAGAMA/AFP/Getty Images, Inc.

Sri Lanka, wild animals scrambled up hill slopes to escape the impending disaster that humans were as yet unaware of. Later came the roar and then came the wave and all was lost for millions of Sri Lankans along the southern and eastern coasts (Figure 9-11). The death toll was at least 35,000, and 1.5 million people became environmental refugees.

Aside from destroying lives directly, the 14-foot (4.3 m) tsunami smashed the economy. Thousands of acres of rice paddy were inundated with saltwater rendering them useless for cultivation. Many hotels and roads were destroyed or damaged, which ruined the important tourist industry. Fortunately, international aid arrived quickly although distribution was hindered because of infrastructural damage.

Self-help efforts were notable. Thousands of Sri Lankans loaded their pick-ups with food, clothing, household goods, and medical supplies and tried to get them to those who had lost everything. Some, who thought that they had excess supplies from relief agencies, set up shop on sidewalks and roadsides to sell cheaply what they didn't need in order to buy what they did need.

Many organizations became involved in the relief effort. BRAC (see Bangladesh above) offered thousands of small loans to restart businesses mostly related to *coir* processing, lace-making, and other handicrafts along with retail sales of fruit and vegetables. The United Nations Development Program (UNDP) created women's programs under the acronym "Strong Places" that were intended to support community-based organizations to distribute supplies and help people

Figure 9-12

Here in Uva, a remote mountain village in Sri Lanka, a woman is editing a radio show at the community radio station. This is one of many activities supported by the Area Based Growth With Equity Program lauched in Sri Lanka by the United Nations. The program aims to promote sustainable human development by creating employment opportunities and sustainable livelihoods in context of poverty reduction and increased social integration. Photograph courtesy of the United Nations Development Program (UNDP).

recover their means of livelihood (Figure 9-12). This has been very successful and loan repayment has been 100 percent.

Housing has been rebuilt in most areas with the exception of the war-torn north, where progress is much slower. While 150,000 families lost their livelihoods after the tsunami, about 75 percent have regained their main source of income. Clearly, there is still much to be done.

WHAT DOES THE FUTURE HOLD?

New elections after the war's end brought the same government into power. The country is now called the Democratic Socialist Republic of Sri Lanka. The government is working on rehabilitating the north and east, removing land mines, and relocating and housing displaced people. It is doubtful that the Tamils will get their own state, but they may gain some degree of autonomy.

Since the United States and other Western countries were critical of the government's handling of the civil war, Sri Lanka is deepening ties with other Asian countries such as India, Myanmar, Iran, and China. China is now the country's largest investor, overtaking even the World Bank and the Asian Development Bank. Foreign funding will underpin rebuilding, road, rail, power, and port development.

As of 2010, the economy, crushed by a slump in garment exports and tourism because of the war, is picking up. The *New York Times* has named Sri Lanka as its top tourist destination for 2010. Annual remittances, mostly from Sinhalese and Tamil workers in Arab countries, have rebounded. The Sri Lankan stock market has risen to be one of the best performing in the world.

More Islands of the Indian Ocean

Four island groups punctuate the Indian Ocean near the subcontinent: the Maldives and Laccadives to the west, and the Andaman and Nicobar to the east. Only the Maldives is independent. The latter three island chains belong to India. Virtually all of these islands were devastated in the 2004 tsunami. Thousands were killed and the damage to infrastructures is inestimable.

MALDIVES

Twenty-seven atolls including 1,200 islands known as Maldives are located atop a submarine ridge rising from the depths of the Indian Ocean. Each atoll is a ring-shaped, coral reef supporting 20 to 60 islands with 5 to 10 inhabited. Dry during the northeast monsoon, the islands experience heavy precipitation with the southeast monsoon—from April to October. Although coral reefs serve as breakwaters, they could not protect the islands from death and destruction caused by tsunamis in 1991 and 2004. The force of the wave in 2004 flooded most of the islands, and two-thirds of Male (the capital) was submerged (Figure 9-13).

Hills and rivers are nonexistent in Maldives. Breadfruit trees and coconut palms shade expanses of dense scrub, shrubs, and flowers. Soils are poor, and only about 10 percent of the land is cultivated.

Agriculture, mainly taro, a starchy root crop, plus bananas and coconuts, is concentrated on the highest island of Fua Mulaku. At an elevation of 9 feet (about 3 m), its fresh groundwater is less prone to the intrusion of salt water. However, with global warming and rising sea levels, all the islands are experiencing a depletion of fresh groundwater. This situation threatens their existence.

Situated on historical maritime routes, Maldives became inhabited by diverse ethnic and linguistic groups including Dravidians from the subcontinent, Sinhalese speakers from Sri Lanka, Arabs, Australians, and Africans. Maldivians speak Dhivehi, derived from an archaic form of Sinhalese. The word *atoll* is Dhivehi.

Maldives was once the major source of *cowrie* shells, widely used as currency in Asia and East and West Africa. The renowned explorer Thor Heyerdahl discovered that Maldives was integral to the trade routes of the Egyptian, Mesopotamian, and Indus Valley civilizations as early as 2000 BC. He postulated that the first settlers were the Redin, sun-worshipping seafarers. Even today, many mosques in Maldives are oriented to the east rather than to Mecca.

Islam of the Sunni school was introduced by Arab traders in the twelfth century and quickly replaced Buddhism, established by Sri Lankans in the fourth century. Both civil law and Islam-inspired law or *sharia* are used. Under *sharia*, the testimony of one man equals that of two women. However, women do not veil, nor are they rigidly secluded.

In 1558, the Portuguese became the first Europeans to occupy Maldives, administering it from their colony at Goa. The Dutch followed in the seventeenth century but were subsequently forced out by the British. Maldives was a British protectorate from 1887 until it achieved independence in 1965. It became a republic in 1968. In 1988, a mercenary coup was squelched after military intervention by India. This move symbolized India's hegemony over Maldives.

About 30 percent of Maldives' 359,000 people are concentrated in Male, which is just seven-tenths of a square mile (2.59 km^2). High birthrates and lack of an

Figure 9-13

More than 100,000 people live in Male on this 7/10 square mile island. © mediacolor's/Alamy

official family planning policy means intense population pressure. As rice, the dietary staple, is not grown in the islands, food shortages can be problematic.

Maldives has an active merchant shipping fleet, but fishing employs almost 25 percent of the islanders. Coconut wood ships have been mechanized, leading to depletion of fish stocks. Additional refrigeration and canning facilities are being built with aid from the World Bank and Japan. The Japanese have long participated in Maldives' fishing industry.

In the industrial sector, small-scale cottage industries are dominated by women who weave rope and mats from reeds and coconut and palm fiber. Women also work in Hong Kong–owned apparel factories at the former British air base on Gan.

The most important source of foreign exchange is tourism. As most Maldivians are uneducated, Sri Lankans, Indians, and other groups fill all but the menial jobs. Tourist resorts, deliberately isolated from settlements, offer gambling, alcohol, and Thai "massage girls." It is government policy to keep Maldivians away from Western influences in order to protect local Islamic values. The majority of tourists are from Germany, Italy, Britain, and Japan. Unfortunately, the seasonal employment pattern for the tourism industry coincides with that of the fishing industry: from April to October during the northeast monsoon.

While economic circumstances have improved for some, Maldives remains one of the world's poorest and most isolated nations. Even so, extended restoration and education programs as well as an emergency medical rescue service among outlying islands have been established. An adult literacy rate of 97 percent has been achieved. Also, the government continues its efforts to improve water supplies through desalinization and purification. Maldives' government also has a 10-year carbon-neutral strategy. The idea is to eliminate the use of fossil fuels by turning away from the consumption of oil to renewable power production by 2020.

ANDAMAN AND NICOBAR UNION TERRITORY

The Andaman and Nicobar islands, in southeastern Bay of Bengal, are considered an Indian Union Territory. The Andamans, consisting of more than 300 islands, sit on an ancient trade route between Burma and India. The islands served as penal colonies in the nineteenth century. Only a few indigenous hunters and gatherers remain, with the majority population being Indian. In 2001, the population of the Andaman and Nicobar islands was 356,265.

The Andamans comprise a series of highly dissected longitudinal valleys and are densely forested with mangroves to the north. There are few rivers, and fresh water is difficult to get. Only the south Andamans have roads. The main activity is agriculture, focusing on coconut, betel, and other palm nuts, cassava, and spices such as tumeric.

The Andamese language is interesting in that it has no numeric component. There is a word for "one" and another word for "more than one." Andamese is unrelated to any other language.

LAKSHADWEEP UNION TERRITORY

The Laccadive islands, a collection of coral reef islands in the Arabian Sea off the southwest coast of India, were renamed Lakshadweep in 1973. There are only 13 islands of which 8 are inhabited. The inhabitants are of Indian and Arab descent and practice Islam. The islands are covered with a sandy coral soil. Coconut groves are everywhere, and the production of *coir* (coconut fiber) is the major industry. Coir production is conducted by women. Men build boats and indulge in fishing.

Recommended Web Sites

www.alertnet.org/thefacts/countryprofiles/asia.htm
Thomson Reuters Foundation. News on natural disasters, political turmoil, terrorism, disease outbreaks, and related humanitarianism.

www.bangladesh-web.com/
News from Bangladesh. Daily news monitoring service.

www.bbs.gov.bd/
Bangladesh Bureau of Statistics. Lots of information and data. Special report on gender issues. Yearbook of Facts, etc.

www.bssnews.net
Bangladesh *Sangbad Sangstha* (National News AGENCY OG Bangladesh). Information on culture, environmental issues. Search archives by place and date. Excellent site.

http://en.wikipedia.org/wiki/2004_Indian_Ocean_earthquake
Excellent source for geographic/scientific account of the 2004 tsunami. Hundreds of links to related sites. Maps, diagrams, and photos.

http://en.wikipedia.org/wiki/Sri-Lankan-Civil-War
Detailed account of the Civil War 1983–2009.

www.fastmr.com/about/
Market research data and short articles by country.

www.lankapage.com/index.php
"Lanka Page." Various articles, mostly political and economic.

www.priu.gov.lk/
Official Web site of the Sri Lankan government. Look up daily news.

www.thedailystar.net/newDesign/index.php
Bangladesh's main newspaper.

www.worldnewspapers.com
Links to all Bangladesh current and archival newspapers.

Bibliography Chapter 9: South Asia: Bangladesh, Sri Lanka, and Islands of the Indian Ocean

Ali, Abu Muhammad Shajaat Ali. 1998. "Toward an Ecological Explanation of Agricultural Unemployment in Bangladesh." *The Professional Geographer* 50: 176–91.

Amin, Sajeda, and Sara Hossein. 1995. "Women's Reproductive Rights and the Politics of Fundamentalism: A View from Bangladesh." *The American University Law Review* 44: 1319–1343.

Barkat, Abul, Mati Raham, and Mark Bose. 1997. "Family Planning Choice Behavior in Urban Slums of Bangladesh: An Econometric Approach." *Asia-Pacific Population Journal* 12: 17–32.

Bigelow, Elaine. 1995. *Bangladesh: The Guide*. Dhaka: ABP.

Daneels, Jenny. 1998. "Defying Mother Nature." *Asian Business* 34: 6–8.

Davis, Anthony. 1996. "Tiger International." *Far Eastern Economic Review* 22: 30–38.

Elahi, K. Maudood. 1989. "Population Displacement due to Riverbank Erosion of the Jamuna in Bangladesh." In *Population and Disaster*, eds. John I. Clarke et al., pp. 81–97. International Geographical Union. London: Blackwell.

Foster, Gerard. 1969. "The Concept of Regional Development in the Indigenous Irrigation Systems of Ceylon." *Association of Pacific Coast Geographer's Yearbook* 31: 91–100.

Fritsch, Peter. 2003. "Bangladesh Stares Into the Abyss." *Far Eastern Economic Review* 27 (November): 46–49.

Haque, C. E., and MD. Z. Hossain. 1988. "Riverbank Erosion in Bangladesh." *The Geographical Review* 78: 20–31.

Hye, H. A. 1992. *Case Studies on Rural Poverty Alleviation in the Commonwealth of Bangladesh*. London: Commonwealth Secretariat.

Islam, M. Aminul, and Nazrul Islam. 1997. "Urban Environmental Issues and Governance in Bangladesh." In *Urban Governance in Bangladesh and Pakistan,* eds. N. Islam, and M. M. Khan, pp. 69–80. Center for Urban Studies Dhaka: University of Dhaka.

Ives, Jack. 1991. "Floods in Bangladesh: Who Is to Blame?" *New Scientist* 13 (April): 1–4.

Joehnk, Tom Feliz. 2007. "The Great Wall of India: The New Iron Barrier with Bangladesh." *The World in 2007*. London: *The Economist*.

Kabeer, Naila. 2000. *The Power to Choose: Bangladeshi Women and Labour Market Decisions in London and Dhaka*. London: Verso.

Karlekar, Hiranmay. 2005. *Bangladesh: The Next Afghanistan?* London: Sage Publications.

McDonald, Hamish. 1991. "Learning From Disaster." *Far Eastern Economic Review* 30 (May): 23–4.

Ministry of Planning. 2009. *Gender Statistics Bangladesh 2008*. Dhaka: Bangladesh Bureau of Statistics.

Murphy, Colin. 2007. "Final Curtain Call for Dhaka's Divas? *Far Eastern Economic Review* July/August 2007: 21–25.

Murphey, Rhoads. 1957. "The Ruin of Ancient Ceylon." *The Journal of Asian Studies* 16: 181–200.

Oakley, Emily, and Janet Henshall Momsen. 2005. "Gender and Agrobiodiversity: A Case Study from Bangladesh." *The Geographical Journal* 171/3: 195–208.

Paul, Bima Kanti. 2010. "Climate Refugees: The Bangladesh Case." *The Daily Star* (Dhaka), March 6.

Paul, Bima Kanti. 1998. "Coping Mechanisms Practiced by Drought Victims (1994/95) in North Bengal, Bangladesh." *Applied Geography* 18: 355–373.

———. 1992. "Female Activity Space in Rural Bangladesh." *The Geographical Review* 82: 1–12.

Pearl, Mary C. 2006. "Ecologists in Sri Lanka Assess the Impact of the 2004 Tsunami." *Discover,* December: 29–30.

Potten, David. 1994. "The Impact of Flood Control in Bangladesh." *Asian Affairs* 25, Part II: 156–162.

Rain, Nick. 1994. "Living with the Landscape." *Geographical* 66: 24–27.

Rao, Radhakrishna. 1992. "On Fertile Ground." *Far Eastern Economic Review* 13 (August): 70.

Renner, Michael and Zoe Chafe. 2007. "Patching Up Paradise: Sri Lanka Struggles to Recover from Conflict and Disaster." *Worldwatch,* September/October: 22–28.

Saenger, P. 1993. "Land From the Sea: The Mangrove Afforestation Program of Bangladesh." *Ocean & Coastal Management* 20: 23–39.

Sally, Razeen. 2009. "Sri Lanka at the Cross Roads." *Far Eastern Economic Review* July/August: 30–35.

Tambiah, S. 1986. *Sri Lanka: Ethnic Fratricide and the Dismantling of Democracy*. Chicago: University Press.

Taylor, Alice, ed. 1968. "Ceylon." *Focus* 18: 1–10.

The Economist. 2010. "Victory for the Tiger-slayer." *The Economist* January 30: 27–29.

The Economist. 2007. "Garments in Bangladesh: Knitting Pretty." *The Economist* August 18: 54.

Worldwatch. 2008. "Sri Lanka: A "Double Blow" to Development." *Beyond Disasters*, Washington D.C.: Worldwatch.

Chapter 10

East Asia: Center of the World

"To understand is hard. Once one understands, action is easy."

SUN YAT-SEN (1866–1925)

For some 7,000 years, East Asia has been dominated by *Chung Kuo:* the Middle Kingdom. Not only is China the largest country in the region, but also its cultural, economic, and political impact has been profound. Japan, Korea, and Taiwan—each owes significant aspects of its heritage to the wonder that was China. In this chapter we will first discuss the physical characteristics of the region. Then we will then introduce you to the Chinese worldview relative to those within its sphere of influence. Subsequently, we will review China's history up to its inception as the People's Republic. Although Taiwan, Korea, and Japan will be mentioned where appropriate in this chapter, we will save the more salient details of their histories for Chapters 12 and 13. Be sure to examine the pronunciation guide before reading further (Figure 10-1). Pinyin, a system of romanization (phonetic notation and transliteration to Roman script), will be used in this chapter as the primary spelling. Pinyin pronunciation is based on the Mandarin dialect of Chinese. Following the Pinyin word, the Wade-Giles system of spelling is given in parentheses.

Since Chinese is written with characters, it is necessary to transliterate these phonetically into the Roman alphabet. The nineteenth century Wade-Giles system of romanization was commonly used before the 1970s. At that time a new system was introduced: Pinyin. Pinyin was made official in 1979.

While Pinyin is found in contemporary books and atlases, Wade-Giles remains in pre-1950 publications. Therefore it is important to be cognizant of both systems. If you learn to pronounce the names of all China's provinces and autonomous regions, you will have a head start on pronouncing other Chinese words.

Before reading the following discussion, take a look at Table 10-1, which gives an overview of significant events in China's history.

Topography

Mountains, hills, and plateaus dominate the East Asian landscape. For example, mainland China is 65 percent and Japan is 75 percent mountainous. These physiographic barriers, while permitting the evolution of unique cultures, were penetrable enough to allow outside influences to wend their way from one valley settlement to another. Mountain passes, river arteries, and ocean passages were instrumental in the diffusion of people and ideas across the East Asian landmass to the Korean peninsula and the islands of Japan and Taiwan.

PRONOUNCING AND SPELLING CHINESE PLACE NAMES

Since Chinese is written with characters, it is necessary to transliterate these phonetically into the Roman alphabet. The nineteenth century Wade-Giles system of romanization was commonly used before the 1970s. At that time a new system was introduced: Pinyin. Pinyin was made official in 1979.

While Pinyin is found in contemporary books and atlases, Wade-Giles remains in pre-1950 publications. Therefore it is important to be cognizant of both systems. If you learn to pronounce the names of all China's provinces and autonomous regions, you will have a head start on pronouncing other Chinese words.

PROVINCE IN PINYIN	PRONUNCIATION	WADE-GILES SPELLING
ANHUI	AHN-WAY	ANHWEI
ZHEJIANG	JUH-JEEONG	CHKIANG
FUJIAN	FOO-JEEEN	FUKIAN
HEILONGJIANG	HAY-LOONG-JEEONG	HEILUNKIANG
HENAN	HUH-NON`	HONAN
HEBEI	HUH-BAY	HUPEH
HUNAN	HOO-NAN	HUNAN
HUBEI	HOO-BAY	HUBEI
NEI MONGGOL	NAY-MUNG-GOO	INNER MONGOLIA
GANSU	GAHN-SOO	KANSU
JIANGXI	JEEONG SHE	KIANGSI
JIANGSU	JEEONG SU	KIANGSU
JILIN	JEE-LYNN	KIRIN
GUANGXI	GWANG-SHE	KWANGSI
GUANGDONG	GWANG-DOONG	KWANGTUNG
GUIZHOU	GWAY-JOE	KWEICHOW
LIAONING	LEEOW-NING	LIAONING
NINGXIA	NING-SHEAH	NINGSIA
SHANXI	SHAHN-SHE	SHANSI
SHANDONG	SHAHN-DOONG	SHANTUNG
SHAANXI	SHUN-SHE	SHENSI
XINJIANG	SHIN-JEEONG	SINKIANG
SICHUAN	SSU-CHWAN	SZECHWAN
XIZANG	SHE-DZONG	TIBET
QINGHAI	CHING-HI	TSINGHAI
YUNNAN	YUOON-NAN	YUNNAN
CITIES		
BEIJING	BAY-JING	PEKING
GUANGZHOU	GWONG-JOE	CANTON
CHONGQING	CHONG-CHING	CHUNGKING
SHANGHAI	SHONG-HI	SHANGHAI
TIANJIN	TEEN-JIN	TIENTSIN

Figure 10-1

Pronunciation and spelling guide for Chinese place names. Be sure to study this guide. Once you know how to pronounce and spell the province names, pronouncing and spelling city names will be easier.

Figure 10-2 shows the topography of East Asia. The mainland comprises rivers, plains, and basins separated by mountains. Geographers demarcate the rugged and arid environment of the west—China Frontier—from the gentler and more humid environment of the east—China Proper.

CHINA FRONTIER

The region demarcated as China Frontier is dominated by spectacular mountains and plateaus, as well as forbidding deserts and swampy basins. The Altai Mountains and the Tien Shan (Heavenly Mountains) encase the dry Junggar

Table 10-1 Time Line of Selected Historic Events in China.

Century	Year	Dynasty	Important Events
23rd c–18th c BC	ca 2000–ca 1500 BC	Xia (Hsia)	First dynastic state. Stone tools.
18th c–11th c BC	ca 1766–ca 1122 BC	Shang	Neolithic to Bronze Age. Beginning of Chinese writing.
11th c–3rd c BC	ca 1027–221	Zhou (Chou)	Irrigation systems, iron works, horses, chopsticks. *Daoism* and *Confucianism*. Writing system established. Major towns develop.
3rd c BC	221–206	Qin (Ch'in)	Much of Great Wall built. "Yellow Emperor's" tomb discovered at Xian in 1976 with 6,000 life-size terra-cotta warriors.
3rd c BC–3rd c AD	206 BC–220 AD	Han	Major territorial expansion. Flowering of culture. Xian becomes one of the greatest cities of the ancient world. Chinese call themselves "People of the Han."
3rd c –6th c AD	220–580 AD		Period of disunion and division. Arrival of *Buddhism* in East Asia. Diffusion of *Confucianism* into Korea and then Japan.
6th c–7th c AD	581–618 AD	Sui	North and south China reunited. Construction of key sections of Grand Canal linking Huang and Yangzi. Xian rebuilt and expanded.
7th c–10th c AD	618–907 AD	Tang	China's "Golden Age." Enormous bureaucracy. Xian is cultural capital and largest city in the world. Arabs and Persian seafarers visit China's ports. Silk Road very busy. *Islam* arrives in Central Asia and Tang armies are defeated by Arabs in 751 after which Silk Route breaks down. Chinese influences penetrate Korea, then Japan. Buddhism thrives.
10th c AD	907–960 AD		Period of fragmentation and political instability. Chinese culture perseveres.
10th–13th c AD	960–1279 AD	Northern and Southern Song	Capital moved from Kaifeng to Hangzhou. Mongol invasion. Kublai Khan establishes his capital at Beijing (1272). Preoccupation with administration. Creation of an efficient civil service. Achievements in mathematics, astronomy, map-making. Paper, movable type, and gunpowder invented. Foot-binding practiced. New rice varieties feed 100 million Chinese. Several cities of more than a million inhabitants.
13th–14th c AD	1264–1368 AD	Yuan (Mongol)	Controls all eastern Asia from Siberia to Vietnam border and as far west as Inner Mongolia and Yunnan. Mongols acculturate to Chinese way of life. Attacks Japan but fails. Marco Polo visits.
14th–17th c AD	1368–1644 AD	Ming	Chinese dynasty rules all of eastern China from Amur River in north to Red River in Vietnam; sometimes northern Korea and even Burma. Explores Pacific and Indian Oceans and reaches coast of Africa. Builds defensive walls around cities. Climate change brings about food shortages. Grand Canal expanded to redistribute food.

(*continued*)

Table 10-1 *(Continued)*

Century	Year	Dynasty	Important Events
17th–20th c AD	1644–1911 AD	Qing (Ch'ing) (Manchu)	Largest China-centered empire ever. Incorporates vast area including much of Turkestan, Mongolia, Xizang (Tibet), Burma, Indochina, Korea, and Taiwan. The Qing map of this expanded territory justifies China's claim to Tibet. Retains Ming administrative framework and officials. Population pressure, land concentrated with big landowners, floods, and famines. European powers force concessions. Opium War 1839–42. Britain acquires Hong Kong. Taiping Rebellion 1850–65. First railroad out of Shanghai in 1875. Boxer Rebellion 1890s–1901. Treaty ports number 100 by 1910.
20th c AD	1911–1949 AD		Collapse of Qing Dynasty. Period of Warlords. Sun Yat-sen forms Nationalist Party. Civil war between Chiang Kai-shek and the Nationalists and Mao Zedong and Communists. People's Republic declared.

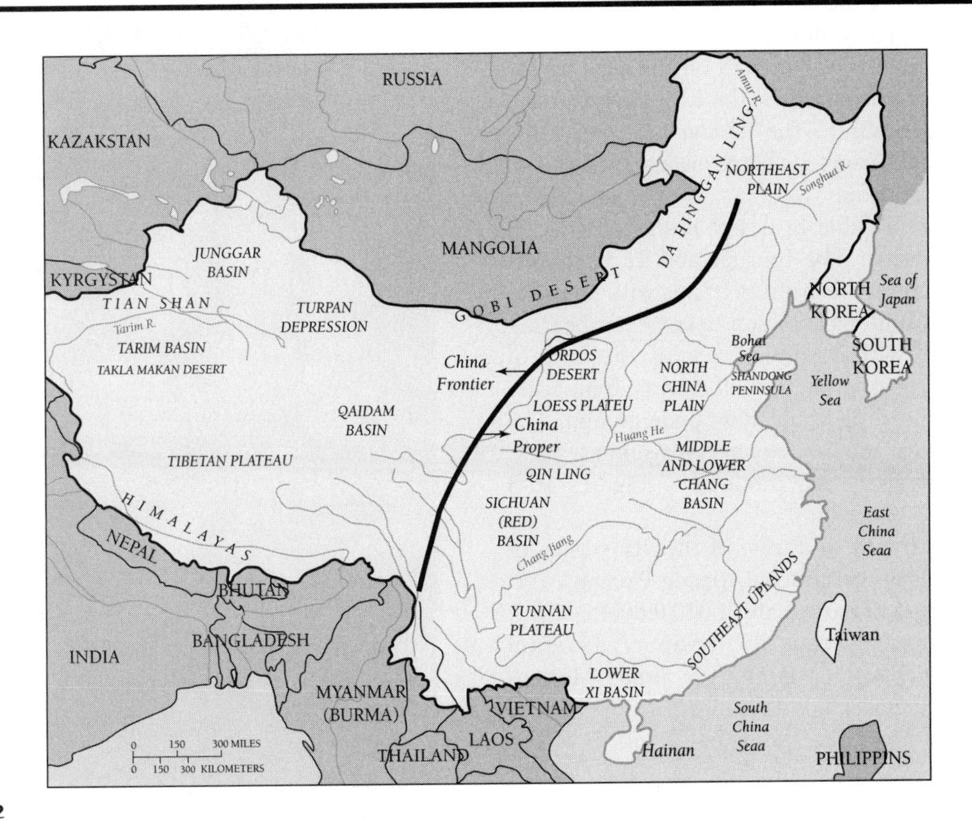

Figure 10-2

Physical geography of East Asia. Study this map carefully and use it in reference to the text. Your atlas map will give more detail. Notice how mountainous China is and how distinct lowlands are. Look at a population distribution map and check the relationship between population distribution and China Frontier and China Proper. From Douglas S. Johnson, Viola Haarmann, Merrill L. Johnson and David L. Clawson, *World Regional Geography,* 10th edition, 2010, p. 508. New York: Prentice-Hall.

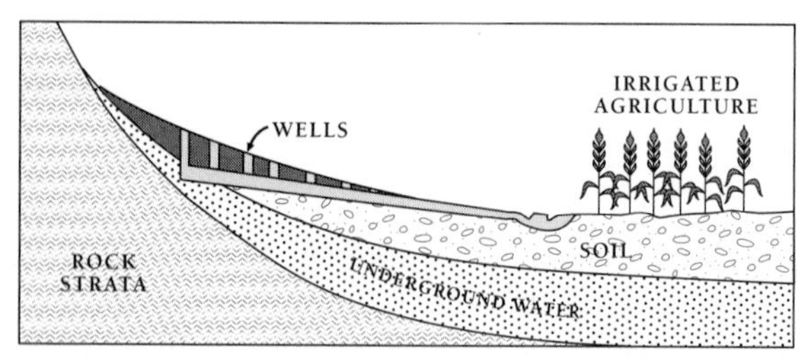

Figure 10-3

Karez wells are vertical shafts connected to an underground horizontal canal that penetrates an aquifer. The water flows by force of gravity to a surface location, where it is diverted to homes and fields. Both vertical and lateral canals must be kept clear of material that could interrupt the water flow. Both digging and clearing wells are dangerous ventures.

(Dzungarian) Basin. Just to the east of the Tien Shan is the Turpan (Turfan) depression. At 505 feet below sea level (155 m), this is the second lowest place in Asia, after the Dead Sea. The southern slopes of the Tien Shan fade into the Tarim Basin, bordered on the west by the Karakoram and on the south by the Kunlun and Altun Shan. Within this basin lies the Taklamakan, one of the world's largest sand deserts. Its name means: "You can go in but you can't come out."

Settlement is possible in desert basins because of the presence of water. The Junggar and Tarim basins and the Hexi Corridor are artesian basins with plentiful water trapped in underground aquifers (water-bearing rock), especially within alluvial fans at the base of mountains. With seasonal rainfall, rivers frequently disappear from the surface and flow underground. The use of *karez* see page 186 in Chapter 7) irrigation systems render mountain-base oases productive (Figures 16-3 and 16-4).

Directly south and southeast of the dry wastes of the Taklamakan lies the Qinghai-Xizang Plateau, with an average elevation of more than 12,000 feet (4,000 m). The Qilian Shan separate the swampy Tsaidam Basin from the steppes of Nei Menggu (Inner Mongolia). These short grasslands fade northward into the Gobi (desert) of Mongolia. Immediately north of the Qilian is a narrow strip of land known as the Hexi or Gansu Corridor, an integral part of China's famous trade route known as the Silk Route. To the northeast are the Da (Greater) and Xiao (Lesser) Hinggan (Khingan) mountains, sloping inland to the Manchurian Plain and the Liao River basin. Northward is the Amur River, forming the border between China's northeast and Russia. The Yalu and Tumen rivers separate China from North Korea.

CHINA PROPER

China Proper is dominated by rivers and plains. To the north is the Huang He, or Yellow River. Flowing 3,395 miles from the Qinghai Plateau, it sweeps northward, wrapping the Ordos Desert in a big loop, then cuts southward across the loess region of Shaanxi and Shanxi provinces. Loess, resulting from the last glacial period, is fine, windblown soil that is readily eroded. At the confluence of the Wei River, the Huang He turns northeast across the North China Plain and empties into the Bo Hai Gulf of the Yellow Sea.

Figure 10-4

Road and water channel in Dunhuang. Poplar trees line this dirt road. Straight poplar wood is important for construction. Note the irrigation canal alongside the road. The water comes from a karez and is fundamental to irrigation in this arid region. Photograph courtesy of B. A. Weightman.

The Huang He is literally yellow, due to the immense amounts of silt it carries in suspension through erosion of the loess plateau. At one time, the Shandong (Eastern Mountain) Peninsula was an island, but eons of alluvial deposition have joined it to the mainland. Its highest peak is the sacred Tai Shan at 5,056 feet (1,517 m).

Just south of the Wei River is the Qin Ling Range, an extension of the Kunlun. East of the Qin Ling is the Huai River. As you will discover, both the Qin Ling and the Huai are sentinels of change in both climatic and agricultural patterns in China.

The North China Plain is the largest alluvial landform region in eastern China. It covers 125,000 square miles (323,750 km^2). It blends into the Central Chang Jiang (Yangzi River) Plain and its associated lake basins. The Chang Jiang, the world's third longest river, flows an incredible 3,900 miles (6,279 km) from Tibet to the East China Sea. To the west, surrounded by mountains, is the fertile Sichuan Basin and the Yunnan Plateau. Mountains dominate southern China until breaking at the Xi River Basin by the South China Sea. Southwestern China is dominated by the longitudinal extensions of the Tibetan-Qinghai ranges.

Climatic Variation

East Asia is characterized by continentality modified by the monsoon. Seasons are accentuated in the interior due to its remoteness from the moderating effects of water bodies. Mountain patterns add to these contrasts. Winter and summer temperatures are extreme in the interior of China, while coastal regions have more moderate climes. Latitude is another consideration. While rice is being harvested in the Xi Basin, ice sculptures are being created in the Liao Basin.

The summer monsoon (detailed in Chapter 2) brings ever-decreasing amounts of precipitation from south to north (Figure 10-5). Precipitation ranges from around 120 inches (305 cm) on Hainan Island to less than 10 inches (35 cm) in the Tarim Basin.

The Qin Ling, with peaks reaching 12,000 feet (3,600 m), form a topographic barrier to monsoon rainfall from the south and to dust-laden winds from the north. Significantly, the Qin Ling and Huai River approximate both the 20-inch (51 cm) isohyet (a line on a map joining points of equal rainfall) and the northern limit of double-cropping.

In winter, winds sweep outward from the Siberian high pressure cell near Lake Baikal, bringing relatively dry, cold air to the continent as well as to Japan. Sometimes the chilling air is filled with loess, carrying yellow dust as far as Japan's west coast.

Typhoons (the same as hurricanes in the United States) are important sources of moisture during the summer and fall. These develop in the South and East China seas, striking China's southern coast around May and reaching Taiwan and Japan in July or later. Typhoons bring a succession of rain squalls and winds that can

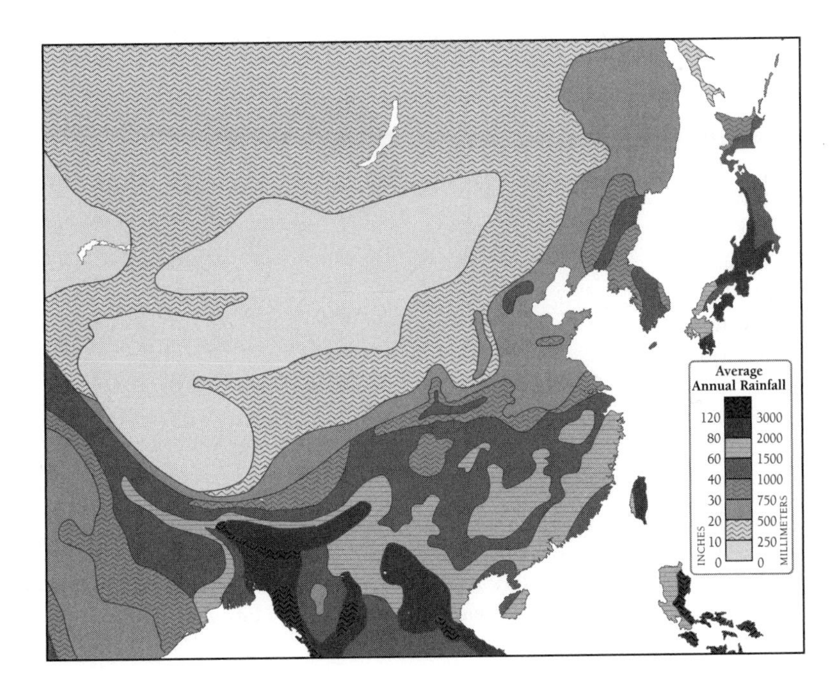

Figure 10-5

Yearly precipitation in East Asia. Note how the amount of precipitation decreases as you move inland. How does this pattern compare with population distribution? Compare this map with Figure 11-18 and see the relationship between precipitation and agricultural patterns.

reach velocities of 150 miles (242 km) per hour, causing severe damage.

Seasons are highly differentiated between both hot and cold and wet and dry. Temperatures contrast greatly with latitude, averaging 21.38°F (−5.9°C) in northern Manchuria in January and 68.8°F (20.4°C) in Hainan in the south. July temperatures average 70.8°F (21.6°C) and 84.8°F (29.3°C), respectively. Summers are oppressively hot and humid, in Taiwan, Korea, and Japan as well, except in the dry west.

Peoples of East Asia

It is a common but erroneous perception that all East Asians are of Mongoloid stock. Although this is true of Chinese, Koreans, Mongols, and Tibetans, it is not true of all groups. The Japanese, for example, are thought to be a Mongol-Malay mixture. Also, in China Frontier, there are several Turkic populations, such as the Uyghur. As we will discuss in the next chapter, China has many minorities, and Sinicizing (making Chinese) minority groups is a major policy issue for the Chinese government.

A brief look at language patterns reveals the dissimilarities of the peoples of East Asia. Mandarin and Cantonese are Chinese languages belonging to the Sino-Tibetan language family. Tibetan is included in this family but belongs to the Tibeto-Burmese group. Some of the minority people in southwestern China speak Mon-Khmer (of the Austro-Asiatic language family), which is related to languages in Southeast Asia. The Uyghur, Kazak, and other Turkic people speak Altaic languages of the Turkish group. Mongolian belongs to a different group of Altaic languages. Neither Korean nor Japanese are related to any of these!

The Chinese see themselves as Han Chinese, descendants of the glorious Han Dynasty (202 BC–220 AD). All others are non-Han people. China's historic perception of its own non-Han people mirrors its perception of other non-Chinese, Asian people. Furthermore, China's perception of itself relative to outsiders has been a major force in shaping its relationships with the world.

China's World View

The written characters for China translate as *Zhong Guo* (*Chung Kuo*), meaning central or middle kingdom. For much of China's history, China perceived itself as central to the known world. This notion has its roots in antiquity. To appreciate it we must venture back in time 4,000 years.

An ancient creation myth has the confusion of heaven and earth divided into light and bright, and heavy and dark segments, with a primeval being called **Pan-gu** (Pan-ku) growing between them. Pan-gu's body parts transmuted into various parts of the cosmos. Pan-gu's function prepares us for the important role of humans in the transformation of chaos into order and explains the Chinese view of apparently opposing forces as complementary in an organic harmony. These ideas are seeded into the philosophy of Daoism (Taoism) discussed later.

One seminal work regarding the origins of human society is the ***Yi jing (I-ching)***, a work in which hexagrams depict the subtle harmonies of complementary opposites (Figure 10-6). According to the Yi jing, humankind was ruled by a succession of sage kings such as Fu-xi, who invented nets and baskets for fishing. His wife invented feminine arts and patched a rip in the fabric of heaven. Shen-nong, the Divine Farmer, invented agriculture and herbal medicine, and taught people how to hold periodic markets.

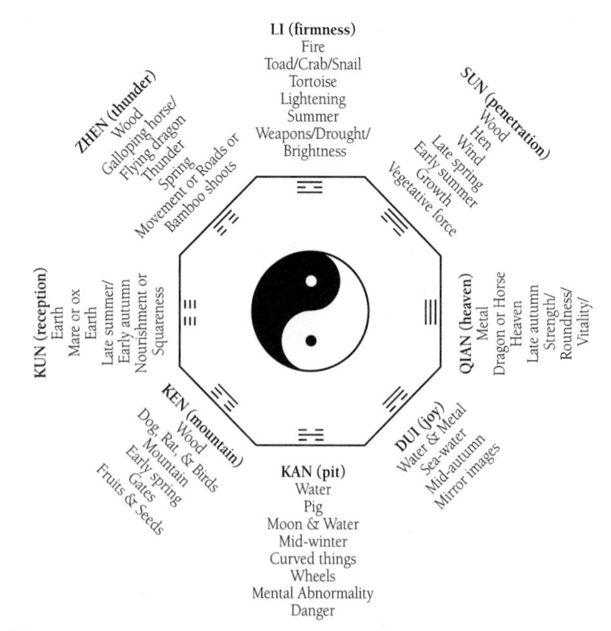

Figure 10-6

This diagram expresses harmony and interrelationships among all things. Understanding these relationships is important in interpreting East Asian landscapes. This diagram is part of the Korean flag attesting to its importance in East Asian culture.

Oracle Bones

The rulers of the ancient Shang Dynasty (originally thought to be 1766–1122 BC, now 1600–1046, practiced divination by heating certain bones such as animal scapulae (shoulder blades) and interpreting the resultant cracks. Questions and interpretations were often inscribed on the bone. These **oracle bones** were first discovered by scholars in 1899 near the last Shang capital at Anyang in the Wei River valley.

The character Huang Di was of particular importance. Huang Di, the Yellow Lord (emperor), constructed roads and set up a bureaucracy to run his domain. He distributed grain at appropriate times and clarified knowledge of heaven and nature. Educated Chinese believe that humanity reached a peak of order about 2350–2200 BC during the reigns of subsequent emperors. This belief in human order is a fundamental theme in Chinese thinking and behavior.

Another fundamental belief is that an ordinary person can achieve great things and that the best government is government by virtuous individuals. Shun is held as an example of a man who rose from humble origins. The Confucian thinker Mencius (ca 390–305 BC) wrote that when Shun "heard a single good word, witnessed a single good deed, it was like water causing a breach in the dikes of the Yangzi or Yellow River. Nothing could withstand it."

Water control has been central to China's development since the earliest of times. Shun appointed a man named Yu, who assembled a massive labor force and spent 13 years supervising water control and irrigation works. Many of China's unruly rivers, including the Huang He, were rechanneled. Subsequently, agriculture flourished. The concept of water management and agricultural productivity as responsibilities of the ruler is another salient theme in Chinese history.

Yu, the model of a competent minister, also surveyed the empire, compiling information on soils, local products, and land revenues. He reported these in a lengthy document called the Yu Gong (Tribute of Yu). The obligation of ministers to report to and advise the ruler is another important element of proper governance and social order. As Shun said to Yu, "The mind of a man is perilous; the mind of the True Way is hard to discern. Concentrate! Be single minded! Hold fast to its center." Years later, Yu's son came to power and his descendants ruled China by hereditary right for 400 years. Yu is regarded as the founder of China's first dynasty: the Xia (Hsia).

Control by diminishing degree, according to distance from the capital, has haunted Chinese history and, by association, the history of East Asia. An important component of the model of Chinese rule was the tributary zone. People of this zone acknowledged Chinese sovereignty, through trade and giving gifts (paying tribute) to the Son of Heaven, the official title of the Chinese emperor.

Tribute payments were not always a one-way process. China often paid tribute to barbarians to keep them at bay or to employ their assistance in thwarting other barbarian groups. The northern frontier was of constant concern.

The northern frontier formed a zone where opposing ways of life—farming and herding—met and commingled. Sometimes pasturage would give way to sedentary farming. Historian Jacques Gernet (1983) observes, "The problem of defense against incursions from the steppe must be seen in a context which is as much cultural, political and economic as military, thanks to the phenomena of assimilation, diplomatic combinations, and commercial exchanges." Here were allied tribes who collaborated in the defense of the empire, outposts, forts and garrisons, military colonies, lands developed by deportees, horse breeding, and nomad encampments.

Two types of troops were located in the frontier: farmer-soldiers, known as "soldiers of the irrigation canals" or "soldiers of the granaries," and soldiers on garrison duty in forward posts. Each post could contact the other through a system of signals: red and blue flags, smoke by day, and fires by night. Transmission of information was swift, and all was recorded in writing. Precise information about troop movements is given in texts as early as the second century BC.

The Ways of Writing Chinese

Chinese characters originated as pictograms, where a symbol stood for an object. Over time, pictograms evolved more complexly as ideograms, where symbols represent ideas. There are thousands of Chinese characters. To be reasonably literate, you need to know at least 10,000. Chinese is a tonal language.

Different dialects have different numbers of tones. Mandarin has four tones while Cantonese has nine, although most of the time only six are used. The same sound, pronounced with different inflections, can mean different things or ideas. For instance ma might imply "mother" or "horse" depending on what tone you give it. However, these two words are written with their own unique characters.

Writing Chinese characters in the Roman alphabet requires "transliteration," a way of transforming one to the other. One system, employed until well after the establishment of communism in China (1949), is the Wade-Giles system. More recently, the Communists have introduced a more simplified character system. How Chinese words are transliterated into English is called **Pinyin**.

Shan-tung under Wade-Giles becomes *Shandong* in Pinyin. Ch'ing Dynasty becomes Qing Dynasty, and *Mao Tse-tung* becomes *Mao Zedong*. Pinyin more closely approximates the sound of the word in the Mandarin dialect of Chinese. While most books and atlases now use Pinyin, English publications from Taiwan continue to use Wade-Giles.

Perhaps the most obvious evidence of China's relationships with outsiders is the Great Wall (Figure 10-7). Dates for initial construction and subsequent connections of the wall(s) are more suggestive than accurate. It is said to have been built during the Qin (Ch'in) Dynasty (221–206 BC) when the Yellow Emperor consolidated existing sections with watchtowers along the northern frontier. More or less permeable throughout history, it was consolidated again under the Ming Dynasty (1368–1644 AD). It is the Ming wall, located north of China's capital, Beijing (Peking), that most tourists visit today. The western sections have either vanished or are in a state of disrepair. The measured length of the Great Wall also varies, depending on which sections are measured. Estimates range from around 1,500 to 4,000 miles (2,415–6,440 km).

For most of Chinese history, the Great Wall was actually many walls—sections constructed as the need arose in time and space. Their positioning depended on the state of China's relations with the barbarian peoples of the time. China spent millions in terms of finances and manpower in warding off such nomadic hordes as the Khitan, Xiongnu (Hsuing-nu), Mongols, and Manchus, not always with success. The Mongols and Manchus, respectively, defeated Chinese forces and took control as the Yuan Dynasty (1279–1368) and the Qing (Ch'ing) Dynasty (1644–1912). China's influence in East Asia peaked under the Mongols, who sent expeditions against Japan, Burma, Vietnam, and Java using both Chinese and Korean troops and fleets.

The Terra-Cotta Army

In 1974 the discovery of an entire life-size army that had been buried for more than 2,000 years astonished the world. Thousands of terra-cotta soldiers, each with unique facial features, along with war-horses and chariots, are regarded by some as the "eighth wonder of the world." The army is believed to be guarding the tomb of Emperor Qin Shi Huang of the Qin Dynasty (221–206 BC). This site, near the city of Xian (Hsi-an), is one of China's top tourist attractions.

Figure 10-7

The Great Wall of China was begun in the Qin (Ch'in) dynasty (221–206 BC). "China" comes from Ch'in. The wall actually linked several older walls, and in its full length reached nearly 4,000 miles (6,000 km). It averages 25 feet (7.8 m) in height and 19 feet (5.8 m) in width. Only parts of the wall survive today. This section of the wall near Beijing was refurbished under the Ming dynasty in the fourteenth century. Photograph courtesy of B. A. Weightman.

CONFUCIUS

K'ung-fu-tsu (Kong Fuzi)—or **Confucius**—was born around 551 BC in the ancient state of Lu in Shandong Province. After holding several political posts, he became a teacher of ethics, ritual, and philosophy. His ideas were written down by his students as the **Analects**. In addition to transmitting Confucius' teachings over time, the Analects state his belief in hierarchical structures of responsibility and duty: of ruler to Heaven (the source of all power and authority); of subject to ruler; of son to father; of younger brother to older brother; and of wife to husband. Each person within society should fulfill the responsibilities implied in their name be it ruler, son, or subject. Rulers should behave like rulers and take care of the people, sons should behave like sons and respect their father, and subjects should act like subjects and obey the ruler.

Another important aspect of Confucian philosophy is filial piety or filial devotion. This is one of the "right relationships" that govern family life and, ultimately, the entire social order. By filial piety, Confucius meant the respect and care children owe their parents during their lives and the reverance they pay to their memories after death. Being an obedient and loving brother to one's older brother is also a kind of filial piety. All behaviors within the family and society must be accompanied by a loving consideration or kindness (called humaneness by Confucius) for the other person. If all individuals behaved according to their name, practiced filial piety, and treated others as they themselves would wish to be treated, the society would enjoy peace, harmony, and prosperity.

Self-cultivation and education as the way to achieve a virtuous character were also key elements in Confucian philosopy. The cultivated person enjoyed life in the form of music, painting, good food, the company of his family, and the attainment of a wise, old age. The greatest sadness was to have no descendants (i.e., male offspring) to perform funeral rites and rituals of remembrance. It was not necessary to be rich or noble to practice self-cultivation, according to Confucius. Each person could learn to be good. Also, through dedication to learning, one could become a sage. No priests were necessary to guide one on this path. Self-development was the key to total fulfillment.

Confucius also affirmed the right of the people to rebel against an unjust ruler. China's rulers held power through the Mandate of Heaven. If they failed to fulfill their role as ruler and lapsed from virtue, or if a disaster occurred, they forfeited the Mandate of Heaven. Loss of Heaven's favor could lead to the downfall of one ruling family and its replacement by another.

Confucius was also a fervent believer in the importance of culture and learning, as well as hard work. A good and just ruler would surround himself with ministers who achieved their goodness through moral education. Regarding government, Confucius said: "If you set an example by being correct, who would dare to remain incorrect?" Confucius also recognized the need to exert control over distant peoples: "Ensure that those who are near are pleased and those who are far away are attracted." In all these matters, appropriate rituals were critical to facilitating and maintaining reciprocal relationships.

Confucius created the role of the scholar-minister, a calling of the highest value in ancient China. He believed that poverty was not a barrier to advancement and that any young man could obtain an education and achieve an administrative or even a court position. Confucius died in 479 BC, but has always been known in China as the "First Teacher." His example underpins the extraordinary respect for education and teachers in China today. Under the Han Dynasty, Confucianism was adopted as the state philosophy and a system of rigorous examinations was established for admittance to the Imperial Civil Service.

In the twelfth century, a Confucian philosopher Zhuxi (Chu Hsi) (1130–1200) speculated about the existence of a Supreme Ultimate, a cosmic force. However, Confucius never discussed priests or religious rituals, leaving any religious aspect in question. Nevertheless, faithful Confucianists built temples for the purpose of worshipping the sage. It was this neo-Confucianism that diffused to Korea, Vietnam, and Japan, where it became the prevailing philosophy, especially for the educated.

In the twentieth century Confucian ideas were disparaged as contributing to China's backwardness relative to the industrialized world. Under communism, the sage was ridiculed and vilified as a symbol of all that was wrong with China's imperial past. In recent years the rigors of communism have relaxed a bit and Confucius has been extolled for his teachings of loyalty and obedience to government. "Confucian principles" are frequently referred to in context of the economic successes of Singapore, Taiwan, South Korea, and other Asian countries.

DAOISM

A contemporary of Confucius called **Laozi** (Lao-tze), meaning "The Old One," supposedly debated with Confucius on the right way of being. Daoist philosophy as recorded in the **Dao de Jing** (Tao-te-ching or Classic of the Way), is concerned with the Dao, or the way the universe works. The sayings contained in the Dao de Jing are quite cryptic and even mystical. For example: "The Way is void, yet inexhaustible when tapped" or "Men should be bland, like melting ice, pure and peaceful like a block of uncarved wood."

Daoism evolved as a religion as it mingled with folk beliefs centered on the worship of nature and spirits. It stressed the desirable unity of people and nature and the futility of worldly strivings. A person's place in nature's enormous, eternally changing expanse was insignificant, "a drop of water in a flowing stream." Individuals should practice a kind of effortless action as water does when it flows around obstacles. This is the ideal of the Way. Daoist priests and monastic orders developed and temples were built. What started as an esoteric school of thought became a mass religion, eventually incorporating a pantheon of gods and immortals.

Later Daoists practiced alchemy—the mixing of metals to produce gold. They also searched for the elixirs of immortality, roaming hills and mountains in search of appropriate roots and plants. Their search for medicinal herbs contributed significantly to the evolution of Chinese medicine.

Even Confucianists found Daoism attractive. Daoism was particularly suitable when things were not going right, or in old age. Confucian activism and social reformism were counterbalanced with Daoist passivity and "go-with-the-flow" attitude. This dualism was very attractive in the context of even older ideas concerning the integral nature and harmonious balance of all things and the fundamental principles of **yin** and **yang** (Figure 10-8).

Yin and yang are two opposing, yet complementary, primordial forces that govern the universe and symbolize harmony. For instance, without dark there is no light. Without life there is no death. All phenomena contain elements of yin and yang that interact and create cyclical change. Night rolls into day, winter into summer, and so forth. This is the process of the Dao—the universal situation.

Did you know that East Asia has its own "geographic tradition?"

Feng Shui: The Chinese Art of Placement

The Chinese have always stressed relative position in the landscape, the world, and the cosmos. The forces of **feng shui** (wind and water) must be dealt with appropriately. For example, Chung-kuo infers the Middle Kingdom, a nation at the hub of the universe. Cities were usually oriented according to the cardinal directions. Numerology was also employed, with the number three and multiples thereof considered auspicious. The Heavenly Altar in Beijing, where the emperor conducted ceremonies for rain and good harvests, is reached by nine steps. The top surface comprises nine rings of stones (Figure 10-9). Gates to the Forbidden City are covered with nine rows and nine columns of brass studs. Four, which sounds like the word "death" in Cantonese, is not a lucky number. So you would not buy a house numbered 444. This belief is practiced mostly in southern China where Cantonese is spoken.

Decisions regarding location are made in the context of yin and yang forces in addition to the force of **Ch' i**. *Ch' i* is the most important element of *feng shui*. It is the vital force, the cosmic breath that gives life to all things. Without *ch' i*, there would be nothing. *Ch' i* is a pervasive concept in Chinese traditional medicine and acupuncture, and martial arts such as Gung fu (Kung fu).

The land most influenced by *ch' i* is the most habitable. *Ch' i* spirals around the earth, sometimes "exhaling" mountains or volcanoes at the earth's crust. When *ch' i* falls short of the surface, there is no water or flowers, but bad luck. Mountains are viewed as dragons exhaling cosmic breath. Mountains and hills are important as defensive sites and are typically kept to the north in orientation of settlement. "Back to the north; front to the south." The worst location would be on a featureless plain with no protection from anything. Atmospheric *ch' i* molds human *ch' i*. Balance and harmony of yin, yang, and *ch' i* are essential to repel negative forces.

Individuals called **geomancers** are experts in *feng shui*. They are consulted to determine the proper siting of buildings and even interior design and placement of furniture. Once a house is carefully and correctly designed and oriented, a Daoist priest is called to perform a moving-in or consecration ceremony. The Hong Kong and Shanghai Bank stages dragon dances when opening a new branch, even in overseas locations.

Even colors are subject to *feng shui* principles. The south is associated with red, representing joy and festivity. A south-facing house is auspicious, and the placement of red door guardians will ward off evil spirits. Green is related to wood and foliage and evokes youthful energy and growth. It represents the dragon and the east. Yellow was a royal

FENG SHUI RELATIONSHIPS

The concepts of the five elements or forces with their yin and yang relationships were introduced as early as the fourth century BC. All of the elements are connected: Wood makes fire, which burns it to make earth. Earth has metal (ores). Metal has condensation—water—and water nourishes plants and trees—wood.

The relationships of the five elements are in accordance with the laws of nature and heaven. It is important not to interfere with the order of nature or disrupt the natural conditions designed by heaven.

The ten stems and twelve branches were introduced prior to 2000 BC. Each stem and branch is related to the five elements and yin-yang principles. All these forces are significant for proper *feng shui* location, and a geomancer must be consulted.

FIVE ELEMENTS	YIN-YANG	TEN STEMS	TWELVE BRANCHES	DIRECTION	ANIMAL
WOOD	YIN	SPROUTING	BUD; BEGINNING	N	RAT
	YIN	SPREAD OF GROWTH	GROWTH	N/NE	OX
FIRE	YANG	BLOOMING	SPREAD OF GROWTH	E/NE	TIGER
	YIN	MATURITY	ABUNDANCE	E	RABBIT
EARTH	YANG	ABUNDANCE	PROGRESS	E/SE	DRAGON
	YANG	ORDER	RENEWAL	S/SE	SNAKE
METAL	YANG	FULLNESS	MATURITY	S	HORSE
	YANG	RESTORATION	SMELL OF MATURITY	S/SW	SHEEP
WATER	YIN	HEIGHT OF FUNCTION	EXPANDED MATURITY	W/SW	MONKEY
	YANG	PREPARATION FOR SPRING	RIPENESS	W	COCK
	YIN		DEATH	W/NW	DOG
	YIN		NUCLEUS	N/NW	PIG

Figure 10-8

Here the concept of yin and yang are outlined. Yin and yang balance opposite forces to ensure harmony between humans and nature. Yin and yang represent the interrelationship of all phenomena and are critical to feng shui orientation.

color and is worn by Daoist priests for burying the dead and for geomantic blessings. White is the color of autumn. It stands for metal, represents purity, and denotes western orientation. Black is the north and winter. Its element is water. Black denotes the consequence of humans, death, mourning, and penance. It is the color of calamity, guilt, and evil.

Feng shui's goal is maximum harmony with natural forces to induce happiness, prosperity, and longevity. Widely practiced among Chinese communities worldwide, it was banned in Communist China as harmful superstition. However, in the new, modernizing China, rules have relaxed, and feng shui and the practice of geomancy have reappeared.

Figure 10-9

This is the Temple of Heaven in Beijing where the emperor would conduct ritualized prayer for good harvests. Photograph courtesy of B. A. Weightman.

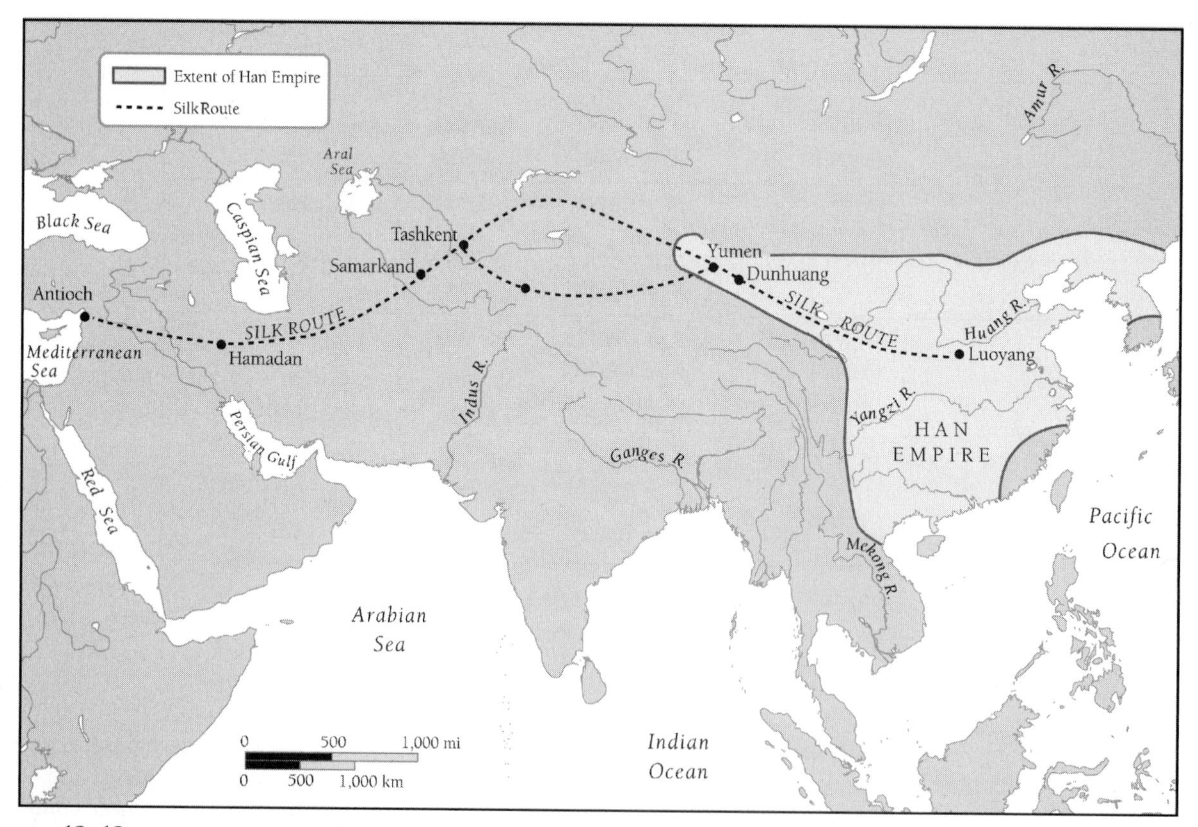

Figure 10-10

Note how the Han Empire focused on major river plains. Compare this map to Figure 10–2 and a physical geography map in your atlas. Trace the arduous journey along the Silk Route where thousands died attempting to complete the trip. Note the preponderance of mountains and deserts.

Figure 10-11

This is called "White Horse Pagoda." It was built by a Buddhist monk in honor of his faithful horse who carried him along the Silk Route. Photograph courtesy of B. A. Weightman.

When I visited rural and urban homes in several regions of China, I noticed that many new pieces of furniture were painted either red or green.

THE COMING OF BUDDHISM

Buddhism entered China from India in the first century AD. This new religion spread quickly along the Silk Route (Figures 10-10 and 10-11). By the time of the Sui Dynasty (589–618), it was the dominant faith among the merchants and soldiers travelling the trail. Buddhist monks constructed monasteries at oases, and these became inns and even banks for travellers. Dunhuang became a significant staging point: this is where travellers would choose between the northern or southern route as the safest passage westward.

Dunhuang housed a large and elaborate Buddhist art treasury. Around 366, monks began to cut caves into cliffs and then carving and painting the interiors. This labyrinth of caves contains nearly 500,000 square feet of murals and 2,000 statues of various sizes. In the Tang Dynasty (618–906), there were more than 1,000 caves. Today there are 486. These Mogao Caves are one of China's prized tourist attractions.

In the political chaos following the Han Dynasty (see the next section), Buddhism's nonsecular belief system appealed to many people and Buddhist monks were able to assert their authority. Because there were few texts and many questions, in 399 the pilgrim Fa Xian (Fa Hsien) journeyed to India for manuscripts and advice. His account of his trip, *A Record of the Buddhist Countries*, revealed developments in Mahayana Buddhism, as well as China's communications with the outside world.

Fa Xian traveled the Silk Route to Afghanistan and later went on to India. From the Ganges Delta, he visited Sri Lanka and Sumatra before returning to Shandong in 414. Then, with the support of various rulers who hosted thousands of visiting monks from India, Buddhism made a phenomenal rise in China.

Dynasties Rule the Middle Kingdom

For more than 3,000 years of its history, the vast region of China was held together with varying degrees of success by a series of **dynasties**. These were periods of rule by one family. Under the dynastic system an emperor ruled for life. Upon the emperor's death, his heir became the new emperor. If the dynasty kept the Mandate of Heaven, its future was safe. If it lost this all-important favor, it could cease to exist altogether.

Natural disasters and disturbances of order were seen as portents of Heaven's displeasure with the mistakes of rulers and as cause for rebellion. During the Han Dynasty (206 BC–220 AD) the emperor presided over special rituals at the imperial capital to intercede with Heaven for appropriate rain, bountiful harvests, and freedom from natural and human-made catastrophes. Even so, many dynasties, having become weak and corrupt, collapsed in coincidence with calamitous events. Floods were often precipitated by the emperor's failure to maintain water-control works. Eventually, a just leader would come to power and bring order to the land and the **dynastic cycle** would begin again.

THE HAN DYNASTY

The gathering of territory and coping with barbarians are epitomized by the conquests of the Han Dynasty (206 BC–220 AD). At first, the Han made agreements with the barbarian Xiongnu, mollifying their leader with gifts, and later became powerful enough to wage war against them. Real expansion was made under Emperor Han Wu Di (Han Wu-ti), who ruled from 140 to 87 BC. He established control over the regions through military colonies whose job was to clear and irrigate the land and defend the area.

Wu Di developed "forward policy" regarding the west, which was occupied by the Xiongnu. By 115 BC, the Chinese had forced the Xiongnu north of the desert and had occupied the Tarim Basin. The Great Wall was extended and fortified with outposts.

Wu Di was particularly interested in the relatively large horses of the "Western Regions" because they could be used to carry heavily armed soldiers against the Xiongnu and their smaller Mongolian ponies. In 101 BC, Wu Di's forces conquered the Ferghana Valley and the "Western Regions" were declared a Chinese Protectorate.

Wu Di also brought northeastern and southern regions under his control. In Korea, a large military colony was established near present-day Pyongyang. The southern kingdom of Nan Yueh around Canton was also captured by 111 BC. Kingdoms in the border areas of Sichuan and Tibet also accepted Chinese supremacy. Yunnan, the terminus of an important trade route to India, was subjugated as well. In the years 42–43 AD, southward expeditions captured Hainan Island, the Tonkin region (North Vietnam), and Annam (central Vietnam).

Assisted or forced population movement was another policy of the Han. Mass movement of populations to border regions was both a means of control and of evening out the distribution of people. More than 2 million settlers trekked to northern areas under Wu Di's reign alone. Even more migrated to the fertile lands of the Yangzi.

The original Han capital was at Ch'ang-an (near present-day Xian) in the Wei River Valley. The capital was later moved to Loyang. By 140 AD, Loyang had a population of more than a million.

Throughout the Han and later dynasties, borders waxed and waned with the relative strength of Chinese and barbarian. Furthermore, China itself was unified or fragmented depending on internal strengths or weaknesses. Overspent and overburdened, the Han dynasty collapsed in 220 AD. The Han dynasty was followed by several others. One of the most important of these dynasties was the Tang.

THE TANG (T'ANG) DYNASTY (618–906)

Under the Tang, China became the richest and most powerful empire in Asia. Its frontiers extended even further than those of the Han (Figure 9-12) and its population numbered more than 50 million. Two million people lived in and around their capital at Ch'ang-an, then the largest city in the world! This was a period unparalleled in poetry, painting, pottery, printing, and other art forms. The Son of Heaven—the title of the emperor—was perceived as the lord of "all under heaven" (i.e., the known world). This was acknowledged by all others who offered praise and paid tribute, and imitated everything Chinese.

Ch'ang-an was likely the largest planned city ever built. Laid out in a checkerboard pattern, it covered some 30 square miles (78 km^2). Its wide avenues ran north–south and east–west to great gates at the cardinal compass points. The walled city was further divided into quarters and lesser administrative units. The people lived in rectangular wards, each with gates that were closed at night. Markets and entertainment districts served residents and visitors alike. Ch'ang-an was notably cosmopolitan, housing an international group of merchants, traders, artists, pilgrims, students, scholars, adventurers, and others. Official tributary emissaries came from Korea and Japan and regions of the south on a regular basis, returning to their homelands with all they could carry.

Despite its accomplishments, the dynasty failed to endure. Tang forces were defeated by a coalition of Turks and Arabs at the Talas River near Samarkand in 751. The Silk Route was thereby closed and trade with the west cut off. This battle is significant because certain captured Chinese transmitted the recently developed art of wood-block printing and an older method of papermaking to the West. By 1030, the Chinese had developed movable-type printing, at least 400 years before it appeared in Europe. In fact, China hosted a wide spectrum of inventions long before the West (Figure 10-13).

A great drought in the north precipitated a series of rebellions against the Tang, exacerbating the dissatisfaction already apparent throughout the empire. Rival generals vied for control, but the Tang empire collapsed in disarray in 907.

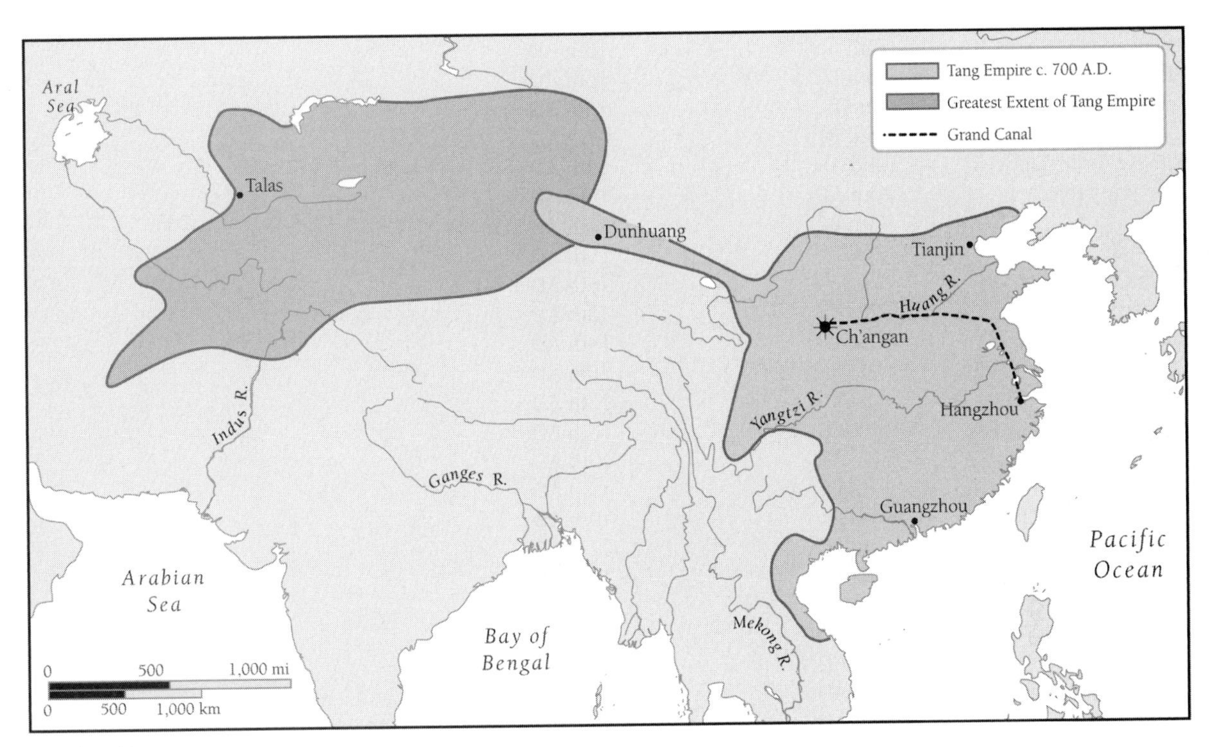

Figure 10-12

Note the expansion of the Tang Empire into Central Asia. The empire, although larger than the Han, was still basically confined to the lowlands and river valleys of the east. Note how the Grand Canal links the region of the Huang He and the Yangzi River basins.

The Golden Lotus

Golden Lotus is a Chinese euphemism for the **foot-binding** of women. The first reliable reference to this atrocity is from the Southern Song (1127–1278). Supposedly, it began in court circles and spread among the upper classes of society. The idea was to keep women's feet very tiny by folding the toes under the foot and tying the bent foot tightly with cloth. In this manner, the feet would be deformed, but small. This very painful process was started between the ages of five and ten. Needless to say, the victims were incapable of walking normally. Judging by poetic references, the bound foot was an erotic sight to some men. Moreover, foot-binding created a gait that would eventually restructure the pelvic bones to enhance men's pleasure during sexual intercourse.

The Mongols of the Yuan Dynasty (1279–1368) did not allow their own women to have bound feet but encouraged it among Chinese women. Foot-binding was at its height during the Ming Dynasty (1368–1644). The Manchu or Qing Dynasty (1644–1911) tried to ban the practice but it continued into the early twentieth century. Very few women with bound feet are alive today.

THE YUAN DYNASTY (1279–1368)

Genghis (Chinghis or Ghenghiz) **Khan** (1155–1227) was a military genius who amassed his nomadic hordes into the most deadly cavalry armies the world had ever seen. In 1211, he crossed the Gobi into northern China. Carrying sparse rations of grain and mare's milk, his horsemen could cover 100 miles a day. They could fire arrows at full gallop—all in all, a terrifying force. A Chinese army of at least 150,000, stationed behind high walls and moats and armed with catapults and flamethrowers, was effectively massacred by the mobile Mongols who swarmed into Beijing in 1215 to enjoy "a most glorious slaughter." Genghis Khan died in 1227.

50 AMAZING CHINESE INVENTIONS / DISCOVERIES

INVENTION	IN CHINA (century)	IN THE WEST (century)
Row Cultivation of Crops	6th BC	16th AD
Iron Plow	6th BC	16th AD
Multi-tube Seed Drill	2nd BC	16th AD
Horse Collar	3rd BC	7th AD
Quantitative Cartography	2nd AD	15th AD
Mercator Map Projection	10th AD	16th AD
Cast Iron	4th BC	13th AD
Crank Handle	2nd BC	9th AD
Steel from Cast Iron	2nd BC	18th AD
Water Power	1st AD	13th AD
Suspension Bridge	1st AD	19th AD
Deep Drilling for Gas	1st BC	20th AD
Belt Drive	1st BC	13th AD
Chain Drive	10th AD	18th AD
Underwater Salvage	11th AD	19th AD
Lacquer (Plastic)	13th BC	19th AD
Petroleum and Gas as Fuel	4th BC	19th AD
Paper	2nd BC	12th AD
Wheelbarrow	1st BC	12th AD
Fishing Reel	3rd AD	17th AD
Horse Stirrup	3rd AD	6th AD
Porcelain	3rd AD	20th AD
Biological Pest Control	3rd AD	19th AD
Mechanical Clock	8th AD	14th AD
Printing Movable Type	11th AD	15th AD
Playing Cards	9th AD	14th AD
Paper Money	9th AD	17th AD
Spinning Wheel	11th AD	13th AD
Circulation of Blood	6th BC	12th AD
Deficiency Diseases	3rd AD	19th AD
Thyroid Hormone	7th AD	19th AD
Immunology for Smallpox	10th AD	18th AD
Compass	4th BC	11th AD
First Law of Motion	4th BC	16th AD
Seismograph	2nd AD	16th AD
Modern Geology	2nd AD	17th AD
Kite	5th BC	15th AD
Relief Map	3rd BC	13th AD
Parachute	2nd BC	18th AD
Ship's Rudder	1st AD	12th AD
Multiple-Mast Ships	2nd AD	14th AD
Watertight Compartments	2nd AD	19th AD
Canal Pound-Lock	10th AD	14th AD
Rotor and Propeller	4th AD	19th AD
Paddle-wheel Boat	5th AD	15th AD
Chemical Warfare	4th BC	19th AD
Crossbow	4th BC	2nd BC
Gunpowder	9th AD	12th AD
Fireworks	10th AD	12th AD
Land Mines	13th AD	14th AD

Figure 10-13

How many of these inventions are you familiar with?

Genghis Khan's three grandsons ruled the North China Plain, Mongolia, and parts of the MIddle East. One of these grandsons, **Kublai Khan**, mastered the North China Plain and moved his headquarters to Beijing. He began his invasion of southern China in 1271. Sweeping into Hangzhou, he crushed the last Chinese resistance at Canton in 1279. Kublai Khan thus became the Mongol Emperor of China.

Korea, northern Vietnam, and the previously non-Chinese southwest were also conquered, driving forced migrations southward into Southeast Asia. Southern Vietnam, Siam (Thailand), Burma, and Tibet were then compelled to accept tributary status.

Kublai Khan did have one important defeat. In 1274, he launched a force to occupy Japan. His main army was prevented from landing on the southern island of Kyushu by an impending storm, which forced the Mongols to beat a hasty retreat. They returned to China, where they continued to subdue their enemies. In 1281, they launched another invasion of Japan. More than 100,000 Mongols landed on Kyushu, where they battled for 53 days. Then a typhoon struck the island and destroyed most of the Mongol transport ships. The Japanese commemorated the failed Mongol invasion by naming this typhoon ***kamikaze***: the divine wind.

The Mongols were now in control of the largest empire in the world (Figure 10-14). While they filled some top administrative posts with their own people and abolished the civil service exams, they left much unchanged in the traditional methods of Chinese government. They reconstructed defenses, extended the Grand Canal to link Beijing and Hangzhou, and turned battle-scarred Beijing into a beautiful city. The period 1280 to 1300 was marked by peace and prosperity: *Pax Mongolica.*

Merchants, traders, diplomats, and myriad others, including Marco Polo, journeyed the Silk Route to the cities of the vast Mongol Empire. Exchange of goods and ideas between East and West proliferated. But while commerce and the arts flourished once more, hatred of these foreign overlords did not wane. Kublai Khan was followed by increasingly inept leaders, and by 1300 revolts against Mongol rule were widespread. Famine, earthquakes, and floods took the lives of millions. Populations were further decimated by the Black Death that was ravaging Europe. One peasant-rebel leader welded the Chinese forces together and drove the Mongols back to the

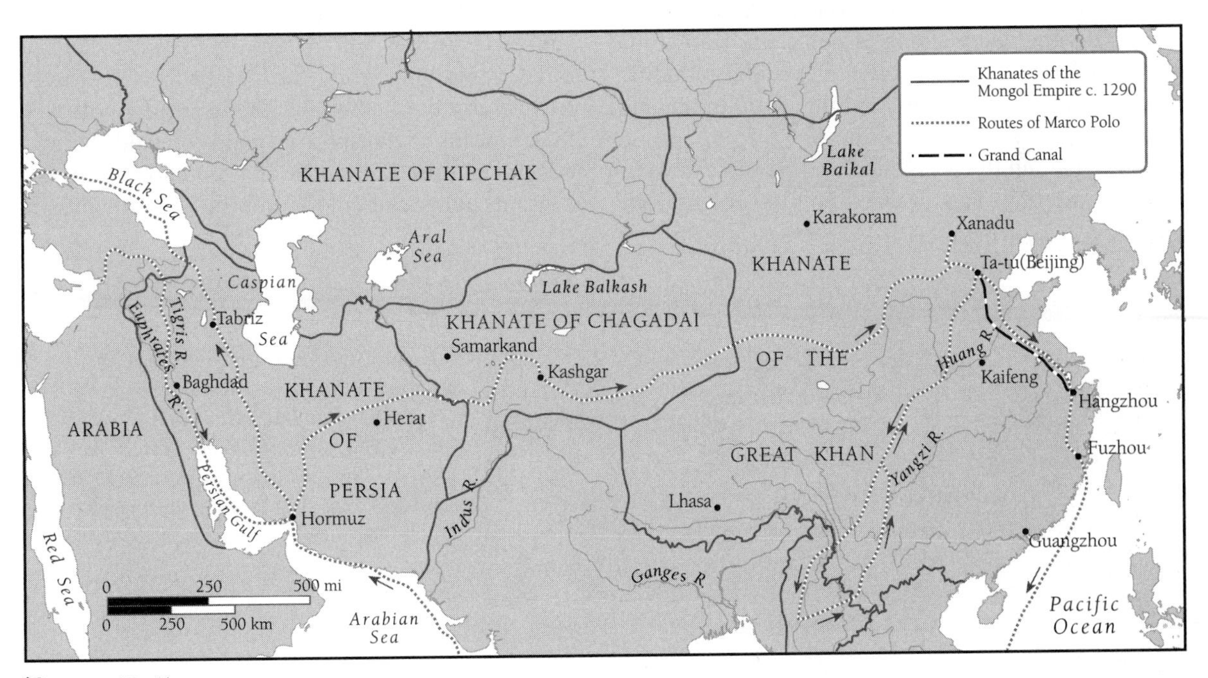

Figure 10-14

China under the Mongols. At this time China was linked with other parts of Asia ruled by various allied khans. Some scholars believe that Marco Polo never made his journeys. He failed to mention chopsticks in his diaries, which seems strange given the detail of his other observations. Read more about Marco Polo and see what you think.

steppes and, in 1368, announced the formation of the Ming Dynasty.

THE MING DYNASTY (1368–1644)

What was Chinese culture like at this time? With the Sinification begun under the Han and solidified under subsequent dynasties such as the Tang, Chinese culture had reached great heights. The scholarly bureaucracy was in place, as was a landed gentry. Government was administered through the multiple bureaus of the civil service and the military. The countryside was managed largely by the gentry and county magistrates. Strict Confucianism was enforced at all levels of society. Most important was the family that, in Confucian form, controlled members' actions and settled disputes.

The Yung-lo Emperor commissioned 3,000 scholars to produce an encyclopedia of all knowledge, a task that took five years. This was followed by a great medical encyclopedia that listed more than 10,000 drugs and prescriptions mostly unknown to the West. This text also recorded the use of inoculation to prevent smallpox, a medical advance not discovered in Europe until the eighteenth century. There was also a handbook of industrial technology describing techniques in canal lock construction, weaving, the manufacture of porcelain, shipbuilding, papermaking, and many other items and pursuits. China during the Ming Dynasty led the world in science and technology.

In Ming Beijing the walls surrounding the city were 40 feet (12 m) high and nearly 15 miles (24 km) in circumference. There were nine double gates with watchtowers. Inside the city was the Imperial City with its own walls, 5 miles (8 km) around. Inside the Imperial City was the moated Forbidden City—2 miles around—containing the emperor's palace, throne halls, audience halls, and other buildings situated along a north–south axis. Buildings with gold-colored tile roofs were set on terraces of white marble. The entire complex was designed to awe and impress (Figure 10-15).

Rice was the dominant element in the diet, supplemented by wheat noodles (brought back to Venice by Marco Polo as spaghetti). Food was eaten with chopsticks, a method quickly adopted in Korea and Japan. Bean curd (*dofu* or *tofu*) was a major source of protein, as were fish and chicken. Oxen and water buffalo were used as draft animals but eaten only if they died naturally. Whatever was put on the rice or noodles was sliced finely and cooked quickly in a wok. The shortage of fuel (as a result of deforestation) fostered the need for stir frying.

Humans altered the landscape with irrigated and terraced rice paddies and fish ponds. New towns and villages emerged as craft centers and periodic markets. From Tang times, the house became a common center for socializing and negotiating business deals and marriage contracts. It was not necessary to travel far—people rarely ventured beyond the nearest village. The wheelbarrow and the flexible bamboo carrying pole were the major means of transporting goods on land. These methods are still used in China and much of Southeast Asia. Peasants adhered to a folk proverb: "Work when the sun

Figure 10-15

Hall of Supreme Harmony viewed from the Gate of Supreme Harmony. This is the largest and most important building in the Forbidden City. Leading to the Hall is a three-tiered "dragon pavement" over which the Emperor was carried. Dragon motifs cover the pavement. Photograph courtesy of B. A. Weightman.

rises, rest when the sun sets. The emperor is far away." In essence, while dynasties rose and fell, the fundamental order of Chinese society persisted.

The first Ming emperor made valiant attempts to reopen the Silk Route, sending repeated expeditions against the Mongol rulers. Further, he repaired the northern frontier and made Korea a vassal state. More important, however, were Ming sea voyages.

Cheng-ho, a Muslim eunuch, sailed south to Malacca (Melaka) in 1403 in order to establish diplomatic relations and expand China's tributary system. Tributary states were required to trade with China and to pay homage to the Chinese emperor. Two years later, Cheng-ho's junks reached Ceylon, East Africa, and the Arabian Sea. But it was Malacca that would become the main regional entrepôt. Through it, thousands of Chinese merchants spread out to Malaysia, Sumatra, and Java. Then, in 1509, six Portuguese ships dropped anchor at Malacca. Soon they were seeking the land they called "Cathay." In 1514, the Portuguese were in Canton's harbor.

Between 1400 and 1600, China's population rose from 70 to 140 million. This growth spurt was accompanied by an agricultural revolution that made rice 70 percent of food production. The 100-day maturing Champa rice covered valleys and hillsides. Then, in the 1550s came the "American" crops: potatoes, maize, and peanuts. The Chinese planted formerly undeveloped hills with these crops and established new farming communities. This agricultural revolution was not one of mechanical devices, it was one of crops. Unfortunately, clearing hillsides caused so much erosion that the rivers silted up and flooded more than ever before.

Meanwhile, the Portuguese pressed their case for trade. In 1557, the Chinese allowed them to build a colony on the isthmus of Macao. Spanish colonists in the Philippines were also trying to break into the lucrative China trade. They even contemplated attacking China in 1586.

The Japanese had been harassing China's coastal trade for years. Things were so bad that all settlements 30 miles (48 km) inland were removed, and maritime trade was officially forbidden. This edict was largely ignored, however, and in 1592 an invasion force of 200,000 Japanese attacked Korea and occupied Pusan, Seoul, and Pyongyang. Since China held sway over Korea, the Chinese in 1593 crossed the frozen Yalu River and began years of fighting on the Korean peninsula. The extraordinarily talented Korean Admiral Yi Sun-sin, who had developed iron-covered ships called turtle ships, managed to sink 70 Japanese ships in a single engagement. As the Ming armies contrived to defeat the Japanese on land, they were suddenly faced with an invasion from the northeast. The Chinese defeated the Japanese in Korea in 1593 but the Manchu tribes stood on the horizon. They would ultimately take over all of China.

THE QING (CH'ING) DYNASTY (1644–1911)

Internal chaos, peasant rebellions, and Manchu invasions combined to destroy the Ming, China's last native dynasty. The Manchus ruled as the Qing (Ch'ing) Dynasty. Qing means "pure." In many ways the Manchus became more Chinese than the Chinese themselves. However, there were looming problems facing Manchu rulers: the population was growing at an unprecedented rate, and food was in short supply. Foreign powers with their gun boats were eroding Chinese sovereignty and missionaries were pursuing converts. Western merchants were bargaining for silk, tea, and art treasures. Meanwhile, peasant uprisings threatened the entire fabric of imperial government.

The Manchus were cousins of the Mongols, and their appearance, clothing, customs, and writing system were different from those of the Chinese. They insisted that all males adopt the Manchu queue (pigtail), otherwise they left Chinese culture alone. In fact, the Manchus tried to inject more honesty and enthusiasm into a government that had become corrupt and complacent.

The Russians had been penetrating China's northeast borders, and Russian frontiersmen clashed with Manchu border patrols from 1650 to 1689. Then the Manchu emperor negotiated the Treaty of Nerchinsk, the first of many attempts to stabilize the Chinese-Russian frontier. Emperor Ch'ien Lung settled colonists in Xinjiang and sent troops to Tibet to counter Gurkha incursions from Nepal. He also forced Annam and Burma to recognize Manchu authority.

Meanwhile, China's population was increasing, probably doubling between 1700 and 1800. Ch'ien Lung authorized huge public works to provide peasants employment in times of crisis. He encouraged farmers to grow more wheat and, wherever possible, to plant cash crops such as cotton, sugar, and indigo. Related industries were also established. His foresight provided a long period of peace. Unfortunately, this was not to last. European powers and overseas Chinese were pressing at the gates of commerce.

A New Breed of Invaders

THE CHINA TRADE

The Chinese intelligentsia did not regard trade as an honorable or worthy profession. Nevertheless, thousands of Chinese were involved in trade within and without China. Portuguese, British, Dutch, Spaniards, and Danes were all anxious to gain trading footholds on the mainland. By the time the first American ship sailed into Canton in 1784, the emperor had already decreed that all trade must be channeled through Canton and that factors or traders must confine themselves to factories (storehouses) leased from local landlords. Foreigners were thus crowded into small compounds the Chinese called hongs.

In 1793, a British trade delegation arrived at the court of Ch'ien Lung. The emperor accepted the gifts of clocks, guns, and telescopes but, in a letter to England's King George III, pointed out, "There is nothing we lack. . . . We have never set much store on strange or ingenious objects, nor do we need any more of your country's manufactures." Ordinary Chinese saw things differently, and thousands went abroad to participate in the business of trade. The favorite destination was the island of Luzon in the Philippines, which became the first major center of the China trade. Silks, tea, porcelain, and rhubarb were exchanged for high-grade silver mined in Spanish America.

The emperor would not deign to have contact with the "foreign devils." Instead, civil servants carried on negotiations and reported their activities via horse relay to the Son of Heaven. Misleading reports and mixed interpretations of information soon caused trouble between the traders and port officials. The accidental deaths of three Chinese led to the execution by strangulation of a British and an American sailor according to Chinese law.

Meanwhile, the British discovered that the most lucrative import of all was opium, which was grown in abundance in Bengal (then under British rule) and could be sold in China. When the emperor tried to stem its import because of rapidly increasing addiction among Chinese, demand increased and the illegal trade boomed. Several efforts were made to end this pernicious trade, but to no avail. At one point, the British were forced to seek refuge on their ships anchored in the Hong Kong harbor. There, they used the ships as trading headquarters for opium. When Chinese junks tried to intercept the British transport ships, British gunboats fired upon the junks, sinking four.

Britain responded to China's obstruction of the opium trade by attacking the Chinese coast and threatening Nanjing (Nanking). This first Opium War (1839–1842) culminated in the Treaty of Nanking in 1842, whereby Britain acquired the island of Hong Kong, a cash indemnity, and permission to trade at five treaty ports: Canton, Xiamen (Amoy), Fuzhou (Fuchow), Ningpo, and Shanghai. Shortly thereafter, Britain was given "most favored nation" status, which meant that it would enjoy the provisions of treaties signed with any other nations. In 1844, the United States signed a trade treaty. China's long period of isolation was over.

The Doctrine of Extraterritoriality

Extraterritoriality refers to consular jurisdiction. This means that a foreigner is subject to the laws of his own country. For example, a British sailor who committed a crime in China would be punished under British law. This practice was introduced to China by the Russians in the Treaty of Nerchinsk in 1689. Under the terms of that treaty the privilege was mutual. Extraterritoriality was one of the provisions of the Treaty of Nanking of 1842. In that case, however, it was not a reciprocal arrangement. Europeans were subject to European laws and so were the Chinese in their own country. This was one reason why trade treaties with the West were called the Unequal Treaties.

The Nationalist Chinese began to negotiate the removal of extraterritoriality in 1928. Unsuccessful, they abolished it by decree in 1932. New treaties, minus the Doctrine of Extraterritoriality, were signed with Britain and the United States in 1943.

THE MISSIONARIES

Another intrusion involved Christian missionaries. By the end of the eighteenth century, the Jesuits had amassed around 300,000 converts among the peasantry. Although they failed to impress scholars steeped in Confucianism, they did introduce the Chinese to new ways of looking at the world. Initially, British and American missionaries built their missions, schools, and hospitals within the confines of the treaty ports. They and other missionary groups later spread to many regions such as Sichuan and Manchuria. In Sichuan, those who accepted food from the missionaries were called "rice Christians."

The French Sisters of Mercy set up an orphanage in Tianjin (Tientsin) but unwisely offered to pay the Chinese for "orphans" brought under their care. In Chinese eyes, these foreign women were really kidnappers. When the Chinese gathered to protest, the French opened fire. This sparked the 1870 Tientsin Massacre of Christians. Western gunboats steamed into port ending any semblance of good relations between China and Europe.

Heavenly Kingdom of Great Peace

At the time of the Opium Wars, a Christian schoolmaster named Hung Chiu-chuan started a revolutionary movement in southern China. Hung taught that land should belong to the people, crop surpluses should go to areas of shortage, and women should be equal to men.

Resistance to the ineffective and corrupt Manchus began in 1850. After announcing the coming of the Taiping Tien Kuan—Heavenly Kingdom of the Great Peace—Hung and his followers attacked along the Yangzi valley and occupied Nanjing. By 1860, the "God Worshippers" had brought a third of China under their control. This was the **Taiping Rebellion.**

It seemed that Hung and his forces would soon rule all of China. However, they underestimated the reaction of the Western powers. The British decided to give the emperor just enough guns and ammunition to defeat Hung. Nanjing surrendered in 1864 and Hung eventually committed suicide. The Taiping Rebellion, which cost China 20 million lives, was over. However, its real significance, in terms of modern history, is the fact that it demonstrated the potential power of the peasantry. Properly equipped and led, China's peasant masses could achieve anything.

RESPONSE TO THE WEST

After the Opium Wars, many Chinese emigrated to Australia and the United States. Thousands of Chinese coolies (laborers) were hired to work in mines and build railroads. Australians resented their capacity for hard labor over long hours. Restrictive laws were passed, race riots took place, and widespread anti-Chinese feelings persuaded the government to adopt a "Keep Australia White" policy. In the United States, Chinatowns emerged and race riots were frequent. America's first zoning laws were designed to keep the Chinese out of "white" areas. People spoke of the "Yellow Peril," and in 1882, strict immigration laws were passed to stem the Chinese influx. Nevertheless, Chinese communities and business ventures prospered in the United States.

Within China a self-strengthening movement was taking place. Certain senior scholars recommended that China should learn from the West in order to get rid of the foreigners, including the corrupt Manchus. They pointed to Japan, a small nation that was busy industrializing like the West. Many young Chinese went to Japan, Britain, and the United States to study. However, there were few lasting results in terms of reform.

Work began on China's first railroad out of Shanghai in 1875. Peasants believed that the steel road was disturbing the *feng shui* so they smashed it to bits. Others, acting out of fear of modernization, destroyed textile machinery. The government did, however, manage to buy guns and warships for the beginnings of a navy.

But the self-strengthening movement failed to communicate any sense of urgency to the common people. Even the reclusive Empress Dowager Ci Xi (Tzu-Hsi) seemed unaware that China was disintegrating—losing territory to the "foreign devils."

DOWNFALL OF THE MANCHU

Figure 10-16 shows China's territorial losses in the nineteenth century. Vietnam was lost to France in 1885 and Burma fell into British hands the following year. Russia, desperate for an ice-free port in the Pacific, gained access to Manchuria. It also occupied the chaotic Xinjiang border regions. In fact, the Russians stayed in the Ili Valley for ten years. To get rid of them, the Chinese gave the Russians navigation rights on Manchurian river systems. The Russians then began road and railway construction in Manchuria.

Between 1875 and 1880, Japan forced its way into two more tributary states: Korea and the Ryukyu Islands. This and related events precipitated the Sino-Japanese war of 1894–1895. Chinese troops were decimated on land and at sea while the Empress Dowager spent money repairing the Summer Palace.

The war ended with the Treaty of Shimonoseki (1895), whereby China recognized Korea as an independent kingdom. China also ceded the island of Taiwan, the Pescadores, and the Liaodong (Liao-tung) peninsula to Japan. Later, the Russians pressed the Japanese to leave the peninsula because of their interest in the ice-free port of Dalian (Dairen or Port Arthur). The Japanese obliged, and with that move, the Chinese saw

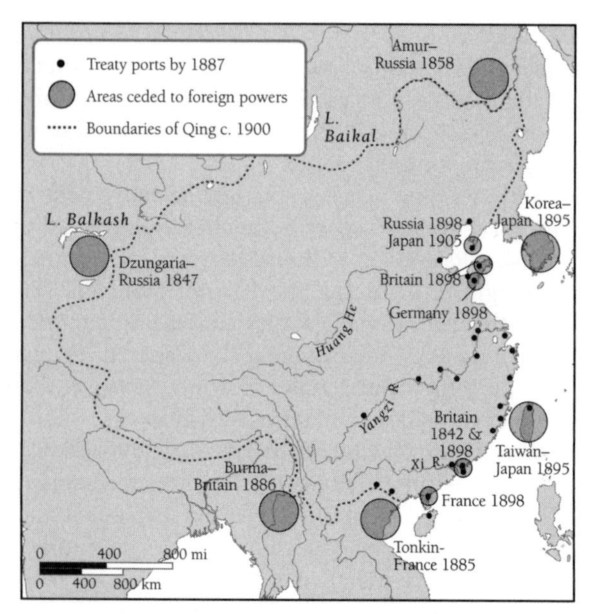

Figure 10-16

The Qing (Manchu) Empire in the nineteenth century. This map shows the extent of the Qing Empire including areas that paid tribute to China. It also shows treaty ports and other areas forcibly ceded to foreign powers in the era of Imperialism.

the Russians as their friends. A secret treaty was signed giving the Russians rights to build railroads across Manchuria as far as Vladivostok.

After the Treaty of Shimonoseki concessions, there was a torrent of requests for land, mining rights, and other trade deals. The Germans took territories in Shandong in 1897, Russia took Dalian the following year, and France claimed Guangxi as a French sphere of influence. Britain wanted control of the Yangzi and a region north of Hong Kong known as the New Territories.

Most Chinese wanted the foreigners to disappear and were prone to support a secret anti-foreigner society called The League of Harmonious Fists—the Boxers. The Boxers were also anti-Christian and believed themselves to be invincible to bullets. Efforts at reform by the Manchu government were dismissed by the Boxers.

The country was in a state of disruption and distress worsened by Huang He floods. In 1900, the Boxers entered the capital and stormed the foreign legations. After 55 days of siege an international relief force arrived. The Empress Dowager fled in a cart to Xian, leaving the Imperial Palace open to looters.

Once again, the foreign powers humiliated China with war reparations. Scores of new treaty ports fell firmly into the hands of foreign bankers, shipping and railway companies, merchants, and missionaries. The vast interior of China was now open to foreign commercialization and exploitation. Western gunboats patrolled China's rivers and Western soldiers guarded settlements. Then the Empress Dowager returned to Beijing, took up her role as empress, and began to negotiate with the foreigners. Many Chinese now began to seriously consider the overthrow of the Manchu Dynasty.

The Empress Dowager entertained Western diplomats, promised to outlaw foot-binding, permitted Manchus to marry Chinese, and abolished the state examinations that had supplied the Imperial Civil Service for 2,000 years. She sent military officers to train with the Japanese who had just won the Russo-Japanese War (1904–1905). But it was too late for the "Old Buddha," as she was known in the north. Ci Xi died in 1908 as did her nephew, her heir and the new emperor. His successor was the two-year-old Pu-yi, whose father became Regent. The Manchu Dynasty was near its end. A modern revolution was at hand.

Asia's First Nationalist: Sun Yat-sen

Sun Yat-sen was born of a peasant family in 1866. He attended a mission school in Hawaii and became a Christian. He then studied at Hong Kong Medical School and graduated as a doctor in 1892. In 1895, he joined a revolutionary group, but this ended with the arrest and execution of most of his compatriots. Sun spent the next 16 years in exile trying to gain support for the liberation of China. He founded the Nationalist Party (known as the Kuomintang) to promote democracy, nationalism, and equalization of land rights. His efforts made him known around the world.

Between 1908 and 1911, the Manchus encouraged European firms to build new railroads. This policy infuriated the people of Sichuan and a peasant revolt ensued on October 10, 1910, later referred to as the Double Tenth. Revolution spread to central and southern China. Sun Yat-sen returned from abroad and was quickly elected as provisional president. He proclaimed a Chinese Republic on January 1, 1912. But Sun Yat-sen stepped down after six weeks in favor of a powerful warlord whom the Manchu regent had named as successor.

In 1910, Japan annexed Korea and in 1914 seized Shandong and extended its influence in Manchuria. Rebellions broke out in practically all regions of China as warlords fought one another for territorial gain. Interminable military campaigns ruined the harvests, causing widespread starvation and suffering.

Europe was locked in the battles of World War I (1914–1918), and China hoped that by supporting Britain and France against Germany, it would gain their support against the Japanese. The terms of the 1919 Treaty of Versailles therefore came as a shock to China. Japan was allowed to keep its territorial gains in China and the Europeans kept all their treaty ports and privileges both along the coast and along the Yangzi.

Students led a nationalistic demonstration that sparked a nationwide strike. One of the strike leaders was a young teacher named **Mao Zedong** (Mao Tse-tung). Communism attracted many young students with its promises of equality, and the logic of Marx and Engels appealed to many Chinese Confucian intellectuals. A Chinese Communist Party (CCP) was created and held its First National Congress in Shanghai on July 1, 1921. Warlordism continued, however, as the Kuomintang (KMT) tried to revitalize itself with Russian assistance. Unfortunately, Sun Yat-sen, often referred to as the first great Asian nationalist, was unable to complete his mission. Suffering from cancer, he died on March 12, 1925. KMT leadership was assumed by **Chiang Kai-shek**.

The Road to Communism

Outwardly allies, Mao Zedong's CCP and Chiang Kai-shek's KMT proceeded to fight the warlords. Strikes, riots, and rebellions wracked the countryside, yet foreign powers continued to pour money into urban industrialization. Disruption in the rural areas drove millions of refugees to these cities, which became breeding grounds for secret societies, strike units, and rebel organizations. Meanwhile, the CCP was working toward a Communist revolution and had formed peasant organizations in the countryside to carry it forward. These would eventually rise up and annihilate the KMT.

Following a northward sweep by the combined KMT and CCP, which together conquered most of southern China by 1927, Chiang Kai-shek decided to rid China of the CCP. Launching a reign of terror in newly captured cities such as Shanghai, he routed out and executed as many Communists as possible. The Communists fled the cities and vowed to garner the loyalty of the peasants and regroup. Chiang set up his government in Nanjing and proclaimed it capital of the Nationalist Republic of China in 1928.

Chiang Kai-shek and his troops followed the Communist forces, determined to exterminate them. But in 1931, the Japanese invaded Manchuria, with profound consequences for all factions. The following year they made a savage attack on Shanghai. Because Chiang was unwilling to give up his campaign against the Communists to fight the Japanese, the Japanese were able to set up the puppet state of **Manchukuo** in Manchuria—China's most important industrial region.

The Nationalists continued their extermination campaign against Mao and his Red Army. In eastern China, the Communists, who were not faring well in battle, decided to move elsewhere to reorganize and reconsider their options. In October 1934, 85,000 made the now famous **Long March** (Figure 10–17). Pursued relentlessly by the Nationalists, they trudged 6,000 miles (9,660 km) through jungles, swamps, hostile tribal areas, across torrential rivers, and through 16,000-foot (4,800 m) mountains. About 7,000 people survived the march and set up their capital at Yenan in the loess hills of Shensi Province in October 1935.

The Long March became a legend, an epic story of the heroism of soldiers and loyal peasants who placed national interests before individual concerns. The Long March exemplifies struggle for an ideal.

In 1937 the Nationalists and Communists decided to form a temporary united front against Japan. As the Japanese advanced, Chiang moved his capital to Chongqing (Chungking), which he proceeded to industrialize with dismantled factories from the east.

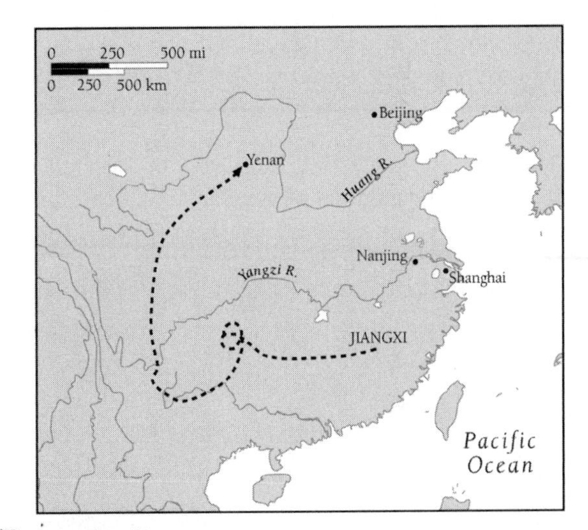

Figure 10-17

The Long March, from October 1934 to October 1935, was one of the most arduous journeys ever made by people. Mao and his fellow Communists travelled more than 6,000 miles (9,660 km) across snow-covered mountain ranges with passes at 16,000 feet (4,800 m), into lands inhabited by wild tribal people, and through deserts and swamps. Only 7,000 of Mao's forces completed the route to the relative safety of the loess caves of Yenan.

The Japanese pressed on with "Kill all, burn all" tactics. In the now infamous Rape of Nanjing, they butchered an estimated 300,000 people. Nevertheless, Chiang Kai-shek continued to expend much of his energies rooting out Communists. After a failed attack against the Japanese, Mao and his troops fell back to Yenan. In 1940 the war in China was a stalemate. The British and French were occupied fighting Germany, and the Japanese pressed southward to French Indochina.

Guerilla Behavior and Tactics

Mao Zedong advocated strict rules of behavior for his Red Army: 1. Speak politely to the people and help them when you can; 2. Return doors and straw matting to the owners; 3. Pay for any damage you cause; 4. Pay a fair price for any goods you buy; 5. Be sanitary—establish latrines well away from houses; 6. Do not take liberties with the women folk; 7. Do not ill-treat prisoners; and 8. Do not damage the crops. Unlike the relatively undisciplined Nationalists, Communist soldiers were much better able to win the support of the peasants.

Mao's guerrilla tactics can be summarized as follows: "The enemy attacks, we retreat; The enemy camps, we harass; The enemy tires, we attack; and, The enemy retreats, we pursue." The Nationalists were unable to cope with this type of warfare.

The Japanese attack on Pearl Harbor on December 7, 1941, changed the conduct of the war. China became allied with Britain and the United States. With U.S. aid, Chiang prosecuted the war against the Japanese with the aid of a motley group of warlords from Chongqing. Much U.S. support was wasted in battle losses and mass defections to the Communists ensued. However, Chiang's propaganda machine declared nothing but success to his Western supporters. Meanwhile, Mao was consolidating his position in Yenan. While masterminding guerrilla warfare against the Japanese, he spent most of his energy wooing the peasants of North China with Communist propaganda. With the Japanese war's end at hand in 1945, American efforts to bring Communists and Nationalists together failed. Their war continued.

Industrial Manchuria was regarded as an essential prize of war by both Mao and Chiang. In 1946, Chiang moved north. It was a fatal decision because at that point, the Americans—fed up with bogus stories of military successes and not wanting to take sides in what was now a civil war—decided to cut off aid to the Nationalists. Meanwhile, inflation and constant threat of conscription ruined morale among civilians. Their despair spread to the troops, who defected to Mao's Red Army by the millions. Chiang Kai-shek, conceding defeat on the mainland, nevertheless transferred his "government" to Taiwan, where he remained until his death in 1975.

By September 1949, Mao was ready to declare a new Republic of China (Figure 10-18). Speaking to the Chinese people from the Gate of Heavenly Peace in Beijing, he said, "Our work shall be written down in the history of mankind, and it will clearly demonstrate the fact that the Chinese, who comprise one quarter of mankind, have from now on stood up... we announce the establishment of the People's Republic of China."

Figure 10-18

This picture represents both old and new China. The tile art shows a Silk Route caravan, while the bicycles represent the major form of transportation in rural China today. I shot this photo in Dunhuang, a major gateway to the Silk Route. Photograph courtesy of B. A. Weightman.

Recommended Web Site

www-chaos.umd.edu/history/references.html
Universityof Maryland site with detailed account of Chinese history through 1988.

Bibliography Chapter 10: East Asia: Center of the World

Ballas, Donald. 1968. "An Introduction to the Historical Geography of Han China." *The Professional Geographer* 22: 155–162.

Blunden, Caroline, and Mark Elvin. 1983. *Cultural Atlas of China*. New York: Equinox.

Cotterell, Arthur. 1988. *China: A Cultural History*. New York: Penguin.

de Bary, William Theodore 1999. *Sources of Chinese Tradition*, Volume 1. New York: Columbia University Press.

Fung Yu-Lan. 1948. *A Short History of Chinese Philosophy*. Ed. Derk Bodde. New York: The Free Press.

Gernet, Jacques. 1983. *A History of Chinese Civilization*, 2nd edition. Trans. J. R. Foster and C. Hartman. New York: Cambridge.

Kitigawa, Joseph M. ed. 1989. *The Religious Traditions of Asia*. New York: Macmillan.

Levathes, Louise. 1994. *When China Ruled the Seas*. New York: Oxford.

Levy, Howard. 1966. *Chinese Footbinding: A History of a Curious Erotic Custom*. New York: W. Rawls.

Lip, Evelyn. 1989. *Feng Shui: A Layman's Guide to Chinese Geomancy*. Union City, Calif.: Heian.

Murphey, Rhoads, 1996. *A History of Asia*, 2nd edition. New York: HarperCollins.

Nakamura, Hajime. 1997. *Ways of Thinking of Eastern Peoples*. New York: Kegan Paul International.

Nemeth, David. 1995. "Feng Shui." *Asian American Encyclopedia* 2: 414–416.

Ronan, Colin, and Joseph Needham. 1981. *The Shorter Science and Civilization in China*: 2 (Abridged). New York: Cambridge.

Spence, Johnathan D. 1993. "Confucius." *Wilson Quarterly* 17: 30–38.

———. 1999. "Paradise Lost." *Far Eastern Economic Review* 15 (April): 40–45.

Temple, Robert. 1986. *The Genius of China: 3,000 Years of Science, Discovery, and Invention*. New York: Simon and Schuster.

Waldron, Arthur. 1991. *The Great Wall of China: From History to Myth*. Cambridge: University Press.

Wills, John E., Jr. 1994. *Mountain of Fame: Portraits in Chinese History*. Princeton: University Press.

Xinzhong, Yao. 2000. *An Introduction to Confucianism*. Cambridge, UK: Cambridge University Press.

Chapter 11

China: Great Dragon Rising

"The comrades must be helped to preserve the style of plain living and hard struggle."

MAO ZEDONG (1949)

"To get rich is glorious."

DENG XIAO-PING (1980)

These contradictory quotes illustrate the political and economic transition that China has made in the latter half of the twentieth century and into the twenty-first century. A grim situation faced the new government in 1949. The economy was severely disrupted; 40 percent of the arable land was flooded; and industrial and food production was well below pre-war levels. This chapter will tell you about China's rollercoaster ride from the strains of rural-based communism to the relative prosperity experienced under the current policy of market socialism.

Agrarian and Industrial Revolution

In 1950, the government abolished the "land ownership of feudal exploitation" and redistributed land to landless peasants. Two years later, 300 million peasants owned a sixth of an acre apiece and thousands of landlords had been eliminated. The following year witnessed **collectivization**, which put an end to individual land ownership.

Collectivization involved several stages. Farmers first were organized into mutual-aid teams, pooling their tools and helping one another during planting and harvest seasons. The second stage involved producers' cooperatives, which pooled tools, labor, and land, although theoretically retaining individual land ownership. The third stage was the fully socialized cooperative, modeled after the Soviet collective farm in which all members collectively owned the land. By 1957, there were close to 800,000 cooperative farms, each averaging 160 families, or 600 to 700 persons.

Large industry was seen as the foundation of socialist society. Plans were made for the building of hundreds of industrial projects, many with Soviet assistance. Spectacular advances were made in both industrial and agricultural production, and by 1962, the national income had increased by fifty percent.

GREAT LEAP FORWARD

The **Great Leap Forward** was announced in 1958. Cooperatives were reorganized into **people's communes**. There were 26,000 of these, embracing 98 percent of the farm population. Each comprised about 30 cooperatives, totaling approximately 5,000 households or 25,000 people. Communes took on the responsibilities of villages, collecting taxes and operating schools, banks, and health clinics. The commune was seen as "the morning sun above the broad horizon of East Asia." However, unrealistic goals, lack of expertise, ill-founded programs, and

Figure 11–1

This is Hohhot, the "Green City." Hohhot has been the capital of Inner Mongolia (Nei Menggu) since 1952. Part of Mao's decentralization and industrialization plans, the city now has more than 2 million residents. Hohhot is one of China's most polluted cities. These are workers' apartments. Chimneys attest to the presence of industry, mostly iron and steel and related products. Hohhot is dominated by ethnic Chinese. Why do you think this is the case? Photograph courtesy of B. A. Weightman.

general mismanagement caused agricultural production to actually decline.

Communization was expanded to include urban and industrial regions (Figure 11–1). Collective living, public eating halls, child-care centers, and the like, served workers who were organized into brigades, battalions, and platoons to produce every conceivable kind of output.

Everyone was to participate in industrialization. Thousands of backyard steel furnaces sprang up across the country. Production quotas were raised sky high, and outputs burgeoned. However, quality was sacrificed for quantity—in 1959, the government admitted that almost a third of the steel made was unfit for industrial use.

SINO-SOVIET SPLIT

The problematic circumstances of the Great Leap Forward were compounded by the Sino-Soviet split. A formal Moscow-Beijing axis had been established in 1950, and thousands of Soviet scientists, technicians, and advisors were sent to the PRC. In 1957, the Soviets agreed to help China in its nuclear development. It sent the Chinese a heavy water reactor the following year. But by the end of the decade, strains were evident in the relationship.

The Sino-Soviet pact, once described as "lasting, unbreakable, and invincible," degenerated into a bitter ideological and territorial dispute. While Mao counted on the rural masses as the basis for communist transformation, Soviet Marxists scorned their potential, stressing the importance of the urban proletariat in organizing the revolution. Disagreement over the path of international communism meant that Mao was challenging Soviet hegemony over the movement.

This conflict led to the withdrawal of Soviet technicians—along with tools, supplies, and blueprints—in 1959. China, determined to go it alone, successfully completed the major construction projects. In 1964, China exploded its first atomic device in Xinjiang. However, all was not well. Bad weather, poor harvests, and the departure of the Soviets created severe economic dislocations.

Overall, the Great Leap Forward was a disaster for China. Normal market mechanisms were disrupted, distribution systems became dysfunctional, and products were shoddy and useless. Most serious was the fact that famine and disease swept the countryside, resulting in anywhere from 14 million to 26 million deaths between 1959 and 1961.

GREAT PROLETARIAN CULTURAL REVOLUTION

The Great Leap Forward, communes, and the Sino-Soviet rift fomented dissension in the Chinese Communist Party (CCP). Some members questioned the wisdom of campaigns stressing "redness" over technological expertise. Others held that Soviet aid was essential for continued economic, military, and scientific development. In 1966, ideologies collided in the **Great Proletarian Cultural Revolution**—a giant social, political, and cultural upheaval.

Pure communization in agriculture was altered, with a policy of "Three Privates and One Guarantee." This allowed peasants to cultivate their own plots (about 5 percent of the arable land), operate private handicraft enterprises, and sell their products at rural free markets. The One Guarantee called for the fulfillment of agricultural quotas set by the government.

Worried about the lack of revolutionary zeal within the party apparatus, among young people, and within the educational system, Mao ordered a Socialist Education Movement in 1962. Suspected anti-Maoists were "sent downward" to the countryside to experience peasant living and working conditions. This was known as the *hsiafang* (*xiafang* in Pinyin) (downward transfer) movement. Thousands of cadres and intellectuals were "rehabilitated" in this manner. In addition, millions of urban-based middle-school students were dispersed to "strike roots firmly" in the countryside. By May 1975, nearly 10 million youths had been resettled in rural areas.

Cultural revolutionary groups were established at all levels to promote Mao's mass line and class struggle, which promoted the equality of all members of society. The youthful Red Guards saw themselves as true revolutionaries, dedicated to the obliteration of old thought, customs, and habits. Knowledge was deemed the source of reactionary and bourgeois thought and action. Eliminating their perceived enemies, the Red Guards rampaged cities, ransacked private property, and attacked anything remotely modern. Ultimately, Mao employed the army to restore order.

Although the Maoists proclaimed the Cultural Revolution a great victory, it actually ushered in a decade of turmoil and civil strife that drove the country to the edge of bankruptcy. The CCP was decimated through purges. Members were killed or imprisoned. Industrial and agricultural production suffered major setbacks, and disruptions in education reduced the availability of trained workers. It is estimated that nearly 100 million people were targeted during the Cultural Revolution. Tens of thousands were killed. Added to this was a series of natural disasters. The Huang He flooded seven times, and a major earthquake wiped out Tangshan and killed some 470,000 people.

Mao Zedong's death in 1976 set off a scramble for succession. Purges, including that of Mao's wife Jiang Qing and her associates, culminated in 1977 with the installment of Deng Xiaoping as party chairman. Deng, a colleague of Mao and a survivor of the Long March (Chapter 10), had been purged twice during the Cultural Revolution. His "rehabilitation" ushered in a new era for the PRC. But the China inherited by Deng exhibited widespread changes in light of agricultural and industrial policies pursued since 1949.

ALTERED SPATIAL PATTERNS

Spatial change was evident in both rural and urban landscapes. Communes, although reduced in size, had transformed the face of agriculture. Significantly, private plots were producing the bulk of vegetables and pork. Rice acreage had been dramatically increased, acknowledging rice's significance as China's most important crop of production and consumption (Figure 11-2). However, as peasants devoted more energy to private plots, rice had begun to decline as a preferred crop in favor of more profitable cash crops. The most important cash crops are cotton and tobacco (for both of which China is the world's leading producer), as well as silk, tea, ramie, jute, hemp, sugarcane, and sugar beets.

Industrial expansion resulted in changing spatial patterns in the distribution of both industry and infrastructure. From 1949, the government sought to decentralize

Figure 11-2

Threshing rice. Rice and other grain quotas are set by the government. Farmers work in groups to fulfill these quotas before spending time on their personal agricultural pursuits. While many farmers use gas-powered rice threshers, manually operated machines remain common, especially in poorer regions. This machine is operated by a foot pedal. The harvesting operation involves cutting the rice stalks; threshing to remove the grain; bundling the stalks for animal feed and other uses; sacking the rice kernels; and carrying the sacked rice to the road for truck pickup to a rice mill.
Photograph courtesy of B. A. Weightman.

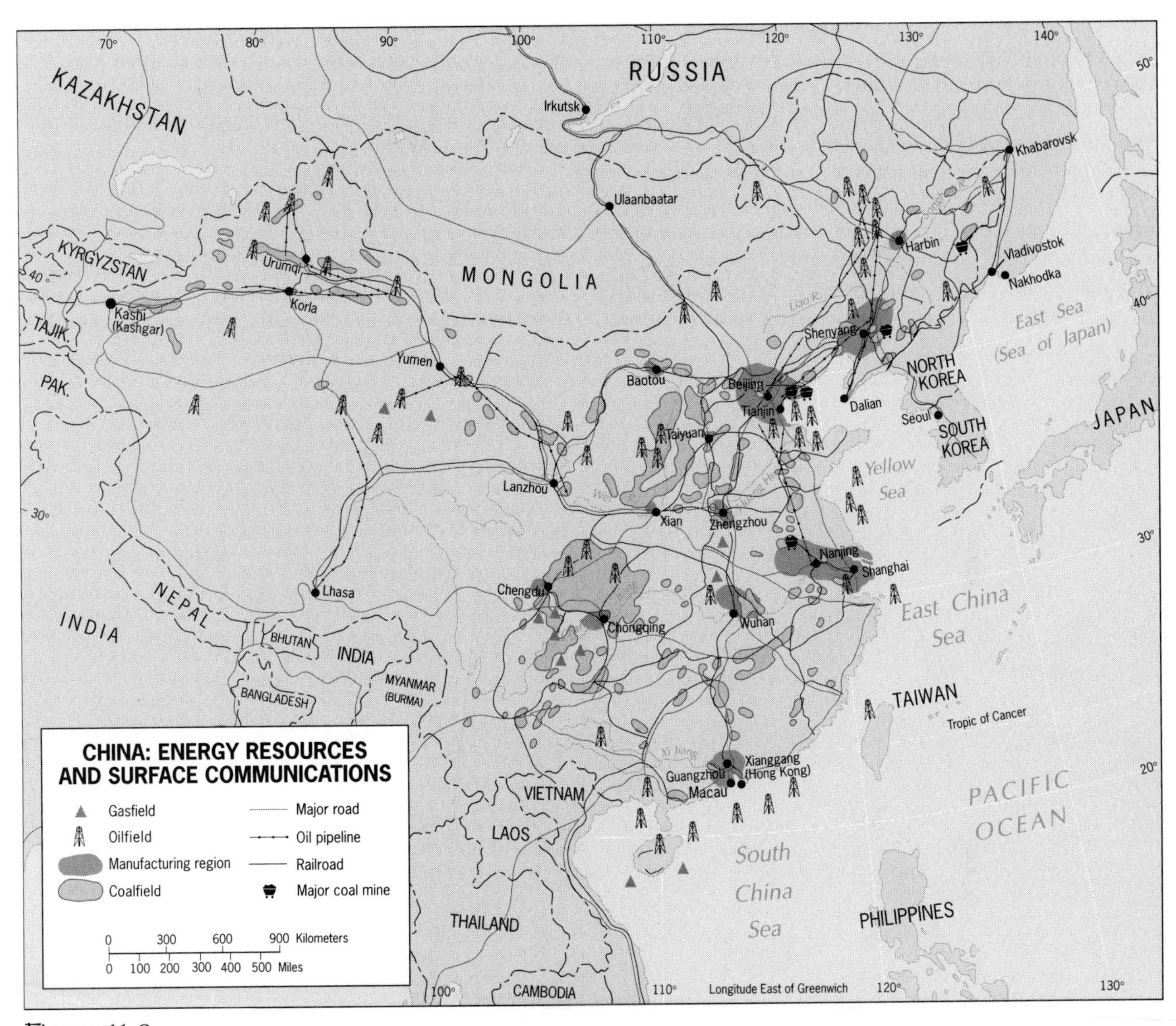

Figure 11-3

Note how surface communications are concentrated in China proper. Why is China Frontier so devoid of road and rail transportation? From H. J. de Blij and P. O. Muller, *Geography: Realms, Regions, and Concepts*, 14th Edition, 2010, p. 489. Originally rendered in color. Reprinted with permission of John Wiley & Sons, Inc.

industry to the interior. At that time, egalitarianism in regional development was deemed more important than efficiency. In 1949, most industrial centers were concentrated in the northeast from southern Manchuria through Shanghai. Subsequently, other centers were developed in line with discoveries of new resources such as coal and oil. As a result, the share of industrial output increased in the interior provinces. China's single oil well at Yumen became only one of hundreds across the nation. Xinjiang became a cornerstone of natural resource exploitation (Figure 11-3).

Railroads were extended into the interior. No longer was the network largely concentrated in northern and northeastern China. New links were established between north and south and extensions were built to the northwest and southwest. Railroads, although built for strategic reasons, were critical facilitators of population redistribution in concert with the *hsia-fang* program and industrialization strategies. They were also important channels of rural–urban migration streams deriving from the dislocations of the Great Leap Forward. Lanzhou, a city of 190,000 people in 1949, mushroomed to nearly 2 million by 1960. Urumqi, 80,000 in 1949, burgeoned to 200,000. Towns became cities, and new cities appeared on the landscape.

Under Deng Xiaoping's platform of reform and economic liberalization, China has become an economic and political force to be reckoned with on a global scale. Whereas Mao Zedong decried foreign interference and stressed "self-reliance," Deng Xiaoping pursued more pragmatic, flexible policies, opening the nation to the outside world. Deng said, "I don't care whether the cat is black or white as long as it catches mice." Foreign investment and even capitalism propelled China into the twenty-first century.

Administrative Framework

Figure 11-4 shows that the PRC has four categories of administrative structure. There are 22 provinces; 5 autonomous regions (ARs); 4 central government-controlled

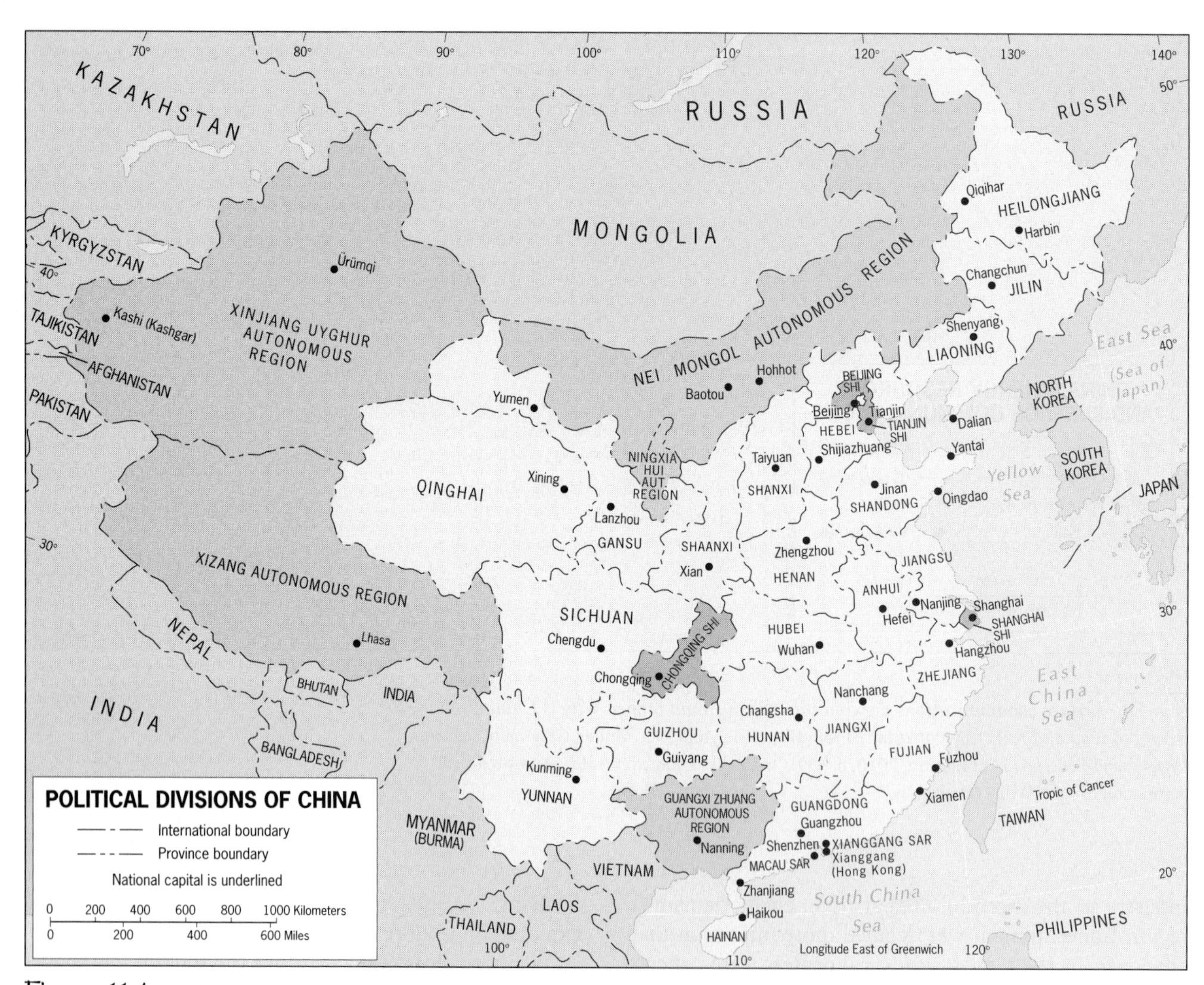

Figure 11-4

Note how China's autonomous regions are on the periphery. Why do you think this is the case? Which cities are designated as shi? From H. J. de Blij and P. O. Muller, *Geography: Realms, Regions, and Concepts,* 14th Edition, 2010, p. 477. Originally rendered in color. Reprinted with permission of John Wiley & Sons, Inc.

Figure 11-5

This Kazakh woman and daughter were spending the summer in the Tien Shan where they were tending animal herds. Their way of life is being altered as the Chinese government is promoting tourism in this region. I asked the daughter what she was going to do with her life. Her eyes filled with tears as she sadly proclaimed, "I don't know what will happen to me." Photograph courtesy of B. A. Weightman.

municipalities known as *zhixiashis* (shortened to *shis*); and 2 special administrative regions (SARs)—Xianggang (Hong Kong) and Macau.

Autonomous regions were created to recognize the existence of non-Han peoples. There are 55 officially recognized minority groups in China (Figure 11-5). Although minorities constitute only a small percentage of China's overall population, they occupy large regions such as Xinjiang and Nei Mongol (Menggu) ARs. China's population is 94 percent Han Chinese. Minorities make up a mere 6 percent, yet they are spread over 60 percent of China's territory, mostly concentrated in the international border regions.

Autonomous regions have a legacy of population transfers in the context of sinicization policies. Han Chinese actually now outnumber minorities in some places such as Nei Mongol AR (Inner Mongolia). Only the larger minority groups such as the Uyghurs and Chuang have their own ARs. Numerous other non-Mandarin speakers such as the Hakka, the Min, and the Yue (Cantonese) live in the provinces, especially in southern China.

The four *shis* are the capital, Beijing; its port, Tianjin; the largest city, Shanghai; and the Chang/Yangzi river port of Chongqing. When Chongqing was designated a *zhixiashi* in 1996, its municipal boundaries were extended to incorporate a huge hinterland totaling 30 million people! In reality, the central urban area has around 4 million inhabitants. However, because Chongqing *zhixiashi* is separate from the provincial administrative area, its creation actually reduced the population of Sichuan Province.

Xinjiang—New Frontier

Occupying one-sixth of China, the Xinjiang Uyghur Autonomous Region is the country's largest administrative area. Divided into a series of basins and ranges (Chapter 10), it is home to more than 40 different ethnic groups. The largest ethnic minority is Uyghur. Uyghurs are a Muslim, Turkic-speaking people who constitute 41 percent of Xinjiang's population and make up 45 percent of Xinjiang's populaton (Figure 11-6). Other groups include the Mongols, Hui (Chinese Muslims), Kazakhs, Tajiks, Tatars, Russians, and Xibo.

Due to sinicization efforts, Han Chinese now comprise one-third of Xinjiang's population. In 1949, the Han numbered 200,000. In 1993, there were 6 million Han in a population of 16 million. About 250,000 to 300,000 Han enter the region yearly. In 2000, Xinjiang had 18.25 million people, more than 41 percent of which are Han. Sinicization is important to keep any notions of separatism at bay.

Separatist movements on the part of the Uyghurs have been part of the history of Xinjiang. The broader issues include discrimination in religion, education, and employment. As part of a campaign against "the three evil forces"—terrorism, religious extremism, and separatism—the Chinese government has taken ruthless action against all

Figure 11-6

This man is clearly not Chinese. In fact, he is a Uyghur and speaks a Turkic language. Uyghurs are one of China's largest minority groups and dominate in Xinjiang Uyghur Autonomous Region. I took this photo at the Sunday market in Kashgar. Photograph courtesy of B. A. Weightman.

forms of dissent. The government insists on *enforcing* a "harmonious society.'"

The Chinese began the transformation of the Silk Route city of Kashgar in the 1990s. Kashgar was famous for its Sunday bazaar when people came from all over the countryside with their various products and services. When I was there in 1992, you could buy such things as giant, juicy yellow figs, twisted buns filled with spicy lamb, richly decorated golden-colored horses, fat-tailed sheep, and copies of the Koran. Women sold brightly painted baby cribs and wardrobes along with wildly striped cloth. Animal-drawn carts pressed through the crowds with drivers yelling "Posh! Posh!" Sitting in the dirt were barbers, shoe-repair men, traditional medicine herbalists, and a dentist who ran his drill with a foot-pedaled cable. There were loudspeakers playing American rock music and posters of Rambo, Muhammed Ali, and Madonna. Men and women mix as they drink chai (tea) and eat nan (flat bread) under plastic sheets or tarpaulins that define *chai hanas* (tea houses). Here, TVs scream Indian films for entertainment.

The Uygers of Kashgar live in mud brick (adobe) houses that are clustered along carefully swept narrow, stone-lined alleys. Houses have elaborately carved wooden doors. Women wear colorful sequined hats, and dresses of the striped material sold in the market. Others wear scarves to cover their hair in a show of Muslim modesty. Then there are those that cover their entire head with a blanket, leaving in question how they can see anything. Many men also wear embroidered hats and some wear fur ones and long, wooly, overcoats and boots, even in summer when temperatures reach 130°F (39°C).

The Uyghur way of life is disappearing. The old market has been replaced by Chinese shops and "Jewellery Street"—a boring round building that few show interest in. In a traditional residential neighborhood, a huge "Wenzhou Mall" is looking for vendors. Chinese authorities claim that the old buildings are dangerous and therefore must go. In their place are high-rise apartments designed for ethnic Chinese. Decent housing and services will draw even more Chinese into the city and the Uyghurs will continue to be sidelined. The fact that Kashgar is a popular tourist attraction apparently is of no account.

In 1992, the capital of Urumqi was decreed a port and given tax incentives like other port cities such as Shanghai. This is interesting because Urumqi is one of the most landlocked cities in the world. However, joint-venture economic development makes sense in light of the fact that Xinjiang gained three new (and nearby) trading partners with the collapse of the Soviet Union: Kazakhstan, Kyrgyzstan, and Tajikistan. Joint markets and factories have emerged along these borders.

Because of its dry climate, Xinjiang relies almost entirely on irrigation for cultivation. Both spring and summer wheat are grown, along with corn, rice, cotton, millet, and kaoliang (sorghum).

Xinjiang is also noted for its tomatoes, Hami melons, seedless Turfan grapes, and Ili apples. Sheep and horses are the principal livestock.

Unfortunately, the Chinese government has deemed nomadic ways of life politically impractical and economically inefficient. Consequently, it has made every effort to sedentarize (settle) groups such as the nomadic Kazakhs, many of whom have resisted. Provision of electricity is one of the tactics employed, as people are more likely to stay settled in an electrified, permanent dwelling. Only by settling down can the nomads "shake off poverty." However, since 1996, the government is focusing more on tourism to engage the Kazakhs in the regular economy. Increasing levels of cultivated land, encroachment by Han farmers, tourism, ånd other entrepreneurial activities all ensure the decrease in opportunity to pursue a nomadic lifestyle.

Similar to schemes of the eighteenth century, in the 1950s, soldiers organized the Production and Construction Corps. Continuing to the present day, this structure of state farms now known as the *Bingtuan* runs an empire of 2.2 million, mostly Han Chinese with direct loyalties to Beijing. State farms are complete entities with their own factories, oil refineries, processing plants, and marketing networks. As many as one million migrants come to Xinjiang each year for seasonal work and some decide to stay on permanently. Today, one in seven residents is a member of the *Bingtuan. Bingtuan* are heavily subsidized by Beijing; Subsidies are considered necessary to keep Han Chinese in this strategic region.

Xinjiang is rich in natural resources, especially in the Junggar and Tarim basins with their wealth of oil (Figure 11-7). The Junggar Basin is of particular significance because of its rail links to Kazakhstan and Central Asia's oil and gas deposits. Pipelines between the two regions are already under construction.

Xinjiang is a potentially strategic region especially in light of its new connections with the Central Asian Republics. Pan-Islamic movements threaten stability, and historic Russian interests in this region always weigh in the background. Xinjiang is also the heartland of China's space program and nuclear weapons development. This is one of many regions of China worth keeping an eye on.

Agricultural China

Overall crop patterns have remained essentially the same (Figure 11-8) but within this larger framework, landscapes are being notably transformed. Now, food needs are not only aligned with an increasing population but also with heightened consumption, including more meat, eggs, and poultry.

As affluence increases, especially among urban populations, the demand for meat and dairy products has skyrocketed. In 1980, China had a mere 70 million head of beef cattle. In 2005 it had 140 million. Dairy cattle have increased from only 5 million in 1995 to close to 25 million in 2007. It is now reckoned that China will become a beef importer in the coming decade.

Large-scale commodity production has become the essence of Chinese agriculture. Double-cropping (growing two crops a year) is practiced almost everywhere south of Beijing, where rice is the dominant crop. While two rice crops are garnered in the south, in the Yangzi valley rice is alternated with wheat. Wheat dominates the north. On the North China Plain, where soils do not retain water well, wheat is grown through the winter. It is replaced by cotton, maize, millet, tobacco, or sweet potatoes in the summer. In the north and northeast, it is too cold for winter wheat so maize and millet are single-cropped. Vegetables are largely concentrated in market gardens around settlements.

Crop yields are higher in the south than in the north but in both areas diminish westward. Nevertheless, some eastern provinces such as Shandong, Hebei, and Henan have made spectacular gains under the new policies.

Chinese Pigs

China has the world's largest agricultural economy and one of the most varied. One of the most significant elements is the pig. In 1955, China had an inventory of 88 million pigs. This number had increased to 508 million by 2005. More than 75 million pigs live in Sichuan Province alone, almost three times as many as in the United States hog belt of Illinois, Indiana, Iowa, and Missouri. Today, China has 126 of the 300 breeds of pigs in the world and over half of the world's pig population. Of the world's total meat increase in recent years, China's pork accounts for two-thirds.

Figure 11-7

Xinjiang Uyghur Autonomous Region. Note how most of the irrigated farming areas are at the base of mountain ranges. Ice melt feeds streams that flow down the slopes and typically disappear beneath the desert sands to form oases. This water is tapped to irrigate crops and support settlements. The silk route followed these oasis communities around the Tarim Basin from Gansu to Kashi (Kashgar). From H. J. de Blij and P. O. Muller, *Geography: Realms, Regions, and Concepts*, 14th Edition, 2010, p. 493. Originally rendered in color. Reprinted with permission of John Wiley & Sons, Inc.

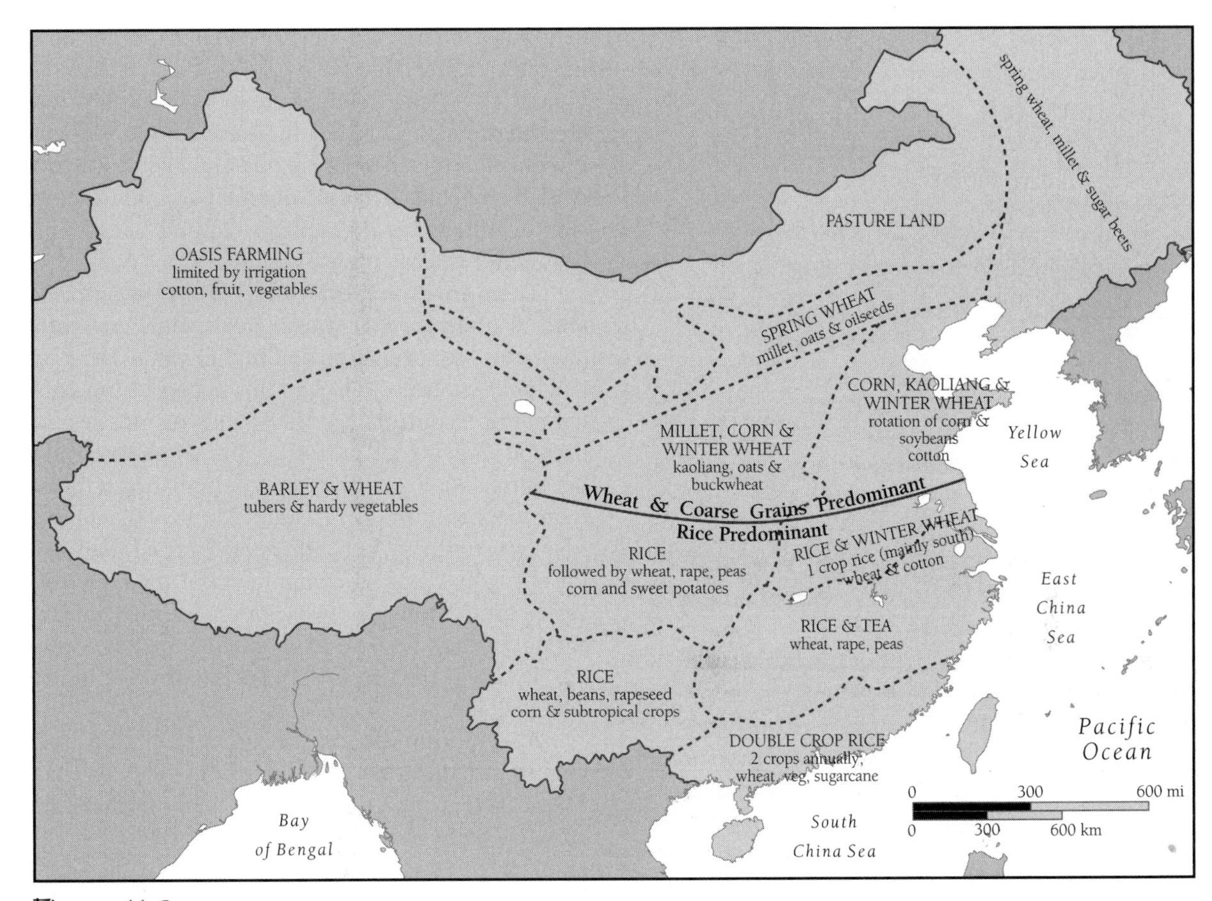

Figure 11-8

This agricultural map of China shows the division between north and south in terms of crops grown and the ability to double-crop. The north grows more wheat while the south grows more rice. The division coincides with the 20-inch isohyet, which follows the Qinling Shan and the Huai He.

China has had a long-standing relationship with the pig's ancestor, the wild boar. Farmers hated the boar for destroying their vegetable crops but admired it for its strength, speed, and the perceived therapeutic value of its body parts. Once the boar was domesticated as the pig, it provided meat, bristles, fertilizer, and countless other by-products. It required no large space and ate anything. Its fast growth and reproductive ability made it a symbol of wealth and prosperity. Even today, you can purchase gilded, golden pigs to bring prosperity to your home or office.

Before 200 BC, social stratification defined what one could eat. Pork could be eaten by the emperor and lower ministers. Others, such as feudal lords, state ministers, generals, and commoners, were required to eat other meats. Pigs also played a role in life and death rituals, and pig effigies and bones were often included in burials. Pigs were also thought to be efficient guardians of children.

Pigs are raised on large, state-run feedlots as well as on privately owned farms. Collectively, there are about 4,000 of these enterprises. As incomes rise with spiraling economic development, the demand for pork has increased dramatically.

Pork production can be a risky business. For example, bad weather in 2007 was responsible for the death of 3.7 million pigs. The 2008 earthquake in Sichuan killed another 4 million. Since then, government price supports have brought the industry back to its normal levels.

Figure 11-9

In the era of market socialism, many people have accumulated new wealth and put it on display. This is the home of a farmer-businessman in Guangdong Province. It shows a lot of traditional symbolism, things that would have been forbidden 30 years ago. The dragon stands for power and change, the phoenix for virtue and grace, the clouds for fortune and happiness, and the pearl for purity. Photograph courtesy of B. A. Weightman.

Beginning in 1978, a series of reforms and experiments have been taking place in China's agricultural sector. In a move away from collective farms, the government instituted the **household responsibility system,** which represented a return to family farming. Under the reforms, families are able to lease land for up to 30 years, and the land can be inherited. Farmers must fulfill state quotas in grain or industrial crops. Surpluses can be used or sold to the state or free markets for even higher prices. Such incentives have increased crop yields and rendered labor more efficient and productive. Many farmers have become relatively rich and are constructing new and larger homes (Figure 11-9). Countless farmers have gone into rural business enterprises such as pig farming, fisheries, or small factories.

By 1991, the production of major crops had increased dramatically. Most increases came about through the use of fertilizers on increased grain acreage. However, overall cropland decreased in area. Generally, the responsibility system has boosted agricultural production and brought prosperity to many rural areas.

There are several drawbacks to the system. Peasants tend to underproduce most cereal grains and cotton in favor of more specialized and higher-value crops such as fruit and vegetables. They do this in part because government price controls are less stringent on these crops, thereby allowing for greater profits. Moreover, the number of free markets has increased dramatically, providing more opportunities for peasants to sell their products. Also, more affluent consumers have raised their demand for specialty crops as components of their increasingly varied diet. Another drawback derives from the fact that agriculture has become more efficient calling for less labor. By 2001, the surplus of rural labor was estimated at 200 million!

A significant amount of agricultural land has been lost to urbanization and environmental degradation. These factors plus the fact that many farmers are not growing grain have resulted in a substantial decline in grain production. Current figures place the output at 1990 levels. Recently, the government began to subsidize grain farming in order to make it as valuable as vegetables and other high-end crops. Grain production is rising again, but China still has to import wheat.

In the mid-1980s, the government realized that not enough food was being grown to feed a growing population and increasing demand. Consequently it began to experiment with other farming systems that reflect regional diversity in agricultural conditions.

- The **two-land system**: "Food-land" is used for personal consumption and "contract-land" is used for commercial production.
- Farmland **share-holding system**: Land is distributed to peasants who turn their entitlements over to an administrative village, which founds an agricultural company. Groups of farmers pool their resources and bid for contracts. The resultant large-scale farms have been very successful but thus far are confined to the Pearl River Delta in Guangdong Province.
- **Collective system**: Small, inefficient plots called "noodle strips" are consolidated into large farms. Farmers work for wages. These farms, highly

mechanized and very efficient, are located near highly demanding urban centers. Most of them are in the more developed coastal provinces.

With these efforts to balance social equality and economic efficiency, some problems have arisen. For example, many farmers work part-time in local industries that manufacture everything from pencils to motor scooters. Others find full-time employment in cities. Many young people do not want to farm. This means that there is a shortage of farmers, which is all the more reason to consolidate land and mechanize.

The government is continually formulating policies to increase food production. Aside from increasing production at home, China is looking elsewhere for farmland. In recent years, it has leased agricultural land in other countries such as Kazakhstan, Laos, Brazil, and Cuba.

"COLLECTIVE INCIDENTS"

All is not well with China's 800 million farmers. Just as the urban-rural income gap has widened, so has the gap between rich and poor farmers. With monetary inequity comes power inequity. Both educational and health-care facilities in the countryside are entirely inadequate. Students must buy their own books and people pay a fee for health care. Poorer farmers cannot afford these things.

Accessible lowland is becoming scarce in the east because of demands of state and city occupants, peasant house construction, soil erosion, and consumption by local industry. Farmland in the east has been lost at the rate of 16,556 acres (6,700 hectares) annually since 1979. Another problem is the return to fragmentation of land with family farming. Pollution, from increased use of chemical fertilizers and pesticides, as well as the accumulation of refuse, especially plastics, has become a serious problem in small towns and villages. New economic enterprises are also polluters.

Free-flying development without building or maintenance of rural infrastructure has caused serious problems. For example, 40 percent of rural villages have no access to running water. Pollution of soil, groundwater, and air is worsening as factories and coal-power plants expand into the countryside. Essential irrigation systems are in disrepair because farmers think of them as public property and have no sense of individual responsibility to fix them. Increasing infringement on farm space is a critical issue with regard to the feeding of China's still burgeoning population.

Probably the most volatile issue is illegal land appropriation by corrupt local officials to sell to developers. Since the early 1980s, 30 to 40 million farmers have had their land appropriated with little if any compensation. Since the mid-1990s, some 17 million acres (42 million ha) have been lost. This is possible because in China, people have "user-rights" to land, not "ownership rights." Although such actions are illegal, a snarled bureaucracy and corruption prevents farmers from recourse.

"Collective incidents" is a euphemism for "protests." In 2005 alone, there were more than 5 million protesters in 87,000 collective incidents over land appropriations especially in the rapidly industrializing and urbanizing coastal provinces. The government wants to "build a new socialist countryside" and says that such land grabs are against the law. Nevertheless, protesters are arrested, many are beaten, and some have their homes burnt to the ground by local officials.

The Hukou System

Socioeconomic position and opportunities for mobility are controlled by a household registration system known as *hukou*. Every Chinese citizen is registered as nonagricultural (mostly urban) or agricultural (mostly rural). Nonagricultural persons are heavily subsidized by the state in grain rations, housing, medical care, and so forth. Registration includes one's particular geographic location, and it is very difficult to alter one's classification. *Hukou* denotes one's identity and place in society.

Until the 1980s, the system rooted people to their birthplace and allowed little movement elsewhere. A migrant was excluded from most desirable jobs and subsidized housing and benefits. It is extremely difficult to change one's *hukou*, especially from agricultural to nonagricultural and to thereby shift location from rural to urban. Nonagricultural *hukou* are considered superior to agricultural hukou.

Since market reforms have produced a sizable labor surplus, the state has allowed temporary migration to towns and cities, although permanent change of status remains strictly controlled. Millions of men have moved to cities such as Shanghai to work on new construction projects. Millions of others migrate from rural farms to work on state farms in the eastern provinces. Also, hundreds of thousands of women migrate to new manufacturing centers such as Shenzhen in Guangdong Province.

Geographers Cindy Fan and Youqin Huang (1998) have discovered a trend of marriage migration. Many women in poorer areas employ kinship and friend networks as well as marriage brokers to marry someone located in a more desirable area. Women gain access to better economic opportunities, and socially or otherwise disadvantaged men are able to get brides. Marriage migration streams have been demonstrated from poorer southwestern provinces such as Yunnan, Gweichou, Sichuan, and Guangxi to the more affluent coastal provinces of Guangdong, Hebei, Jiangsu, Zhejiang, Fujien, and Anhui. Brides then take on the *hukou* of their spouse and are able to participate in the local economy.

TVEs and "Privatization"

Rural industry continues to spring up around cities and towns, and it is estimated that 60 to 100 million people have left the land for better pay in small factories around towns and along transportation lines. Others work in industry on a part-time basis, as in Korea and Japan. Farm types, based on income sources, can be categorized as traditional agricultural, mixed activity, and industrially focused. The latter two categories include **Township and Village Enterprises (TVEs)**.

TVEs were designed to provide employment in the countryside to boost revenues and to hold people to specific areas under the rubric, "Leave the land but not the township or village." This has resulted in a pattern of partially urbanized rural people who engage in non-farm sector employment without migrating to cities. Rural industry contributes more than 40 percent of rural total output value. Moreover, the growth in non-farm activities is significantly greater than farm ones. These enterprises are important in absorbing surplus rural labor.

Originally TVEs were held collectively. However, economic reforms since the 1980s have led to increasing privatization. While actual ownership stays with "the people," the rights of the majority of TVEs have been "sold" or "rented" to share-holding managers. Industrial clusters compete with each other and those that fail are shut down. Even though private enterprise does not conform to socialism or communism the state government has acquiesced to it. TVEs in their original collectively-owned format are disappearing as they are transformed into "informal" private businesses.

Regional Development

Since 1978, China's development policy has stressed efficiency over equity and an open door instead of self-reliance. In 1992, Deng Xiaoping toured several southern Chinese cities. This well-publicized tour energized his reforms as he spread the word to dare to *xiahai*—"Jump into the sea of business."

In order to defray the impact of radical change on the political system, Deng Xiaoping focused his new economic policies on the country's Pacific Rim. The plan incorporated a system of special economic zones (SEZs) and economic development zones (EDZs) designed to attract technologies and foreign investment from abroad (Figure 11-10).

SPECIAL ECONOMIC ZONES (SEZs) AND ECONOMIC DEVELOPMENT ZONES (EDZs)

SEZs and EDZs are founded on a variety of incentives such as relaxed import and export regulations and low taxes. Land can be leased and labor can be contracted. Goods produced can be sold in China or abroad. Investors can reap profits at home.

There are many open economic zones and development zones in China, but formally there were only five SEZs: Shenzhen, Zhuhai, and Shantou in Guangdong Province, Xiamen in Fujian Province, and the entire province of Hainan Island. Initially, there were four SEZs along the coast of China. Each of these possesses particular locational advantages. Shenzhen is adjacent to Hong Kong (Xianggang) on the Pearl River estuary in southern Guangdong Province. In the same region is Zhuhai, adjacent to the former Portuguese colony of Macau. Shantou and Xiamen are former treaty ports across from Taiwan. Hainan Island, approaching Southeast Asia, was added in 1988.

Fourteen additional cities were opened to foreign enterprise in 1984. Most of these were once treaty ports with a history of foreign relations. They also had established industry and transport infrastructures as well as local labor pools. Moreover, all the open cities had connections to overseas Chinese communities.

In 1988, three open economic regions were designated as special zones for foreign investment: the Yangzi (Chang Jiang) Delta Economic Region around Shanghai; the Pearl River (Zhujiang) Delta Economic Region around Guangzhou; and the Minnan Delta Economic Region around Xiamen. These represent efforts to diffuse the benefits of an open policy from SEZs to other parts of the country. China now has hundreds of central-government-regulated economic zones like Shenzhen and more than a thousand run by provincial governments.

More freedom and opportunity for development and profit have generated unrest. University students

Figure 11-10

China's special economic zones (SEZs). From H. J. de Blij and P. O. Muller, *Geography: Realms, Regions, and Concepts*, 14th Edition, 2010, p. 497. Originally rendered in color. Reprinted with permission of John Wiley & Sons, Inc.

and other dissidents decried the greed and corruption they saw among CCP officials and affiliates taking advantage of the new economic situation. In 1989, a huge protest demonstration took place in Beijing's **Tiananmen Square** and in several other cities. Deng Xiaoping called in the military, and the demonstrations were ruthlessly crushed with a significant loss of life. Whether this incident is interpreted as a "cry for democracy" or a "counterrevolutionary rebellion," it is a black spot on China's modern history.

"REINFORCING BARS" TO REDUCE REGIONAL IMBALANCES

While overall incomes have risen across the country, regional imbalances have actually increased. Rich-poor gaps have widened within cities, between urban and rural areas, and among regions. Figure 11-11 shows regional differences and emphasizes the fact that the eastern provinces have fared significantly better under the new economic policies.

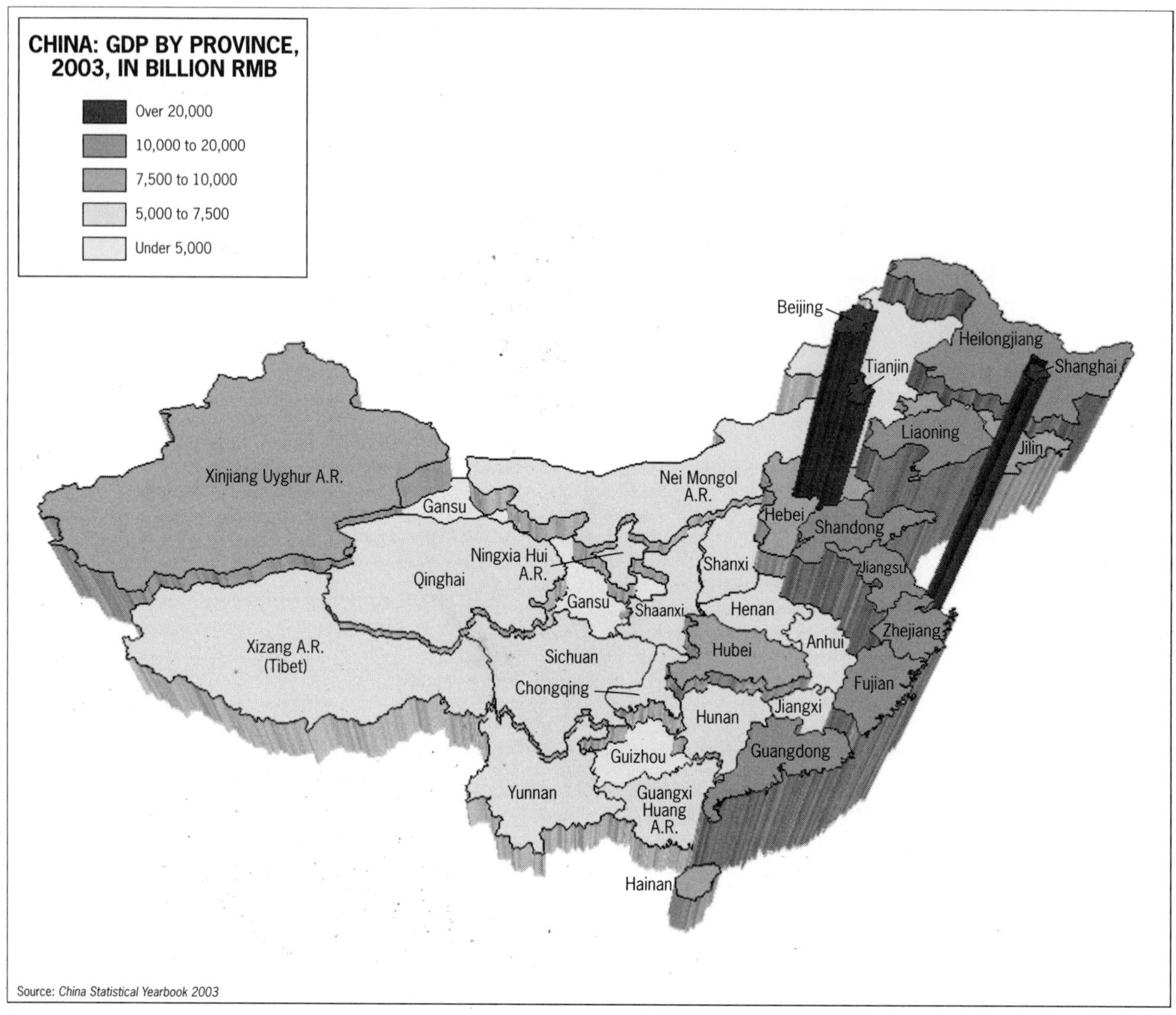

Figure 11-11

Which provinces have the highest GDPs? How do these correlate with the location of China's SEZs? Which cities have the highest GDPs and why is this the case? From H. J. de Blij and P. O. Muller, *Geography: Realms, Regions, and Concepts*, 14th Edition, 2010, p. 495. Originally rendered in color. Reprinted with permission of John Wiley & Sons, Inc.

One way to address regional inequities is to improve the communication and transportation network that will facilitate economic interchange and supply "reinforcing bars" to territorial cohesion. The world has not seen a national road-building effort such as China's since President Eisenhower supported the United States' National Interstate system's construction in the 1950s. China adopted a 30-year National Expressway Network Plan in 2001 and by 2004 it had became second in the world in total expressway mileage and third in total road mileage.

The vast majority of China's population moves by train—more than six times the world average. Moreover, China moves more freight by rail than any other country. The country has added some 10,540 miles (17,000 km) of new track and upgraded 8,060 miles (13,000 km) of existing track. Another 4,340 miles (7,000 km) of new lines are being laid to accommodate high-speed passenger trains that have a top speed of 198 miles (320 km) per hour. With high-speed connections between Shanghai and Beijing, Shanghai and Hangzhou, and Shenzhen

Figure 11-12

This rail line links Tibet's capital Lhasa with Beijing. At least 100,000 construction workers laid this engineering marvel across rugged and icy mountains at elevations of 16,640 feet (5,072 m). The train carriages, which hold 900 passengers, are pressurized like an airplane so that people can breathe properly. This train will ensure that Han Chinese have easy access to Tibet, thereby continuing the government's sinicization policies. Copyright AP/World Wide Photo/Color China Photo.

and Guangzhou, the volumn of passenger traffic is expected to grow exponentially.

In July 2006, considerable publicity accompanied the opening of the highest railway in the world connecting Beijing and Lhasa, Tibet (Figure 11-12). Reaching a height of more than 16,400 feet (5,000 m) above sea level, the passenger cars on the train must be pressurized. The new Qinghai railway into Tibet will enhance China's access to vast reserves of iron and copper ore. Tourism revenue from travel to Tibet is expected to double.

New and improved road and rail transport is expected to reduce regional inequities. It will bring services to remote towns and villages, facilitate trade, and improve access to natural resources, thereby increasing employment opportunities in mining, forestry, dam construction, and other activities. However, there are downsides to increasing accessibility. In many cases villages are being wiped out in the wake of construction, environments are being devastated, pollution is on the increase, and people are complaining about belching smoke and noisy trains. But protests are of no avail.

"Open Up the West"

Since the founding of the PRC, the Chinese government has been active in designing and implementing regional policy that has far-reaching impacts on the geography of development. In 1999, a new policy called "Open Up the West" (*xibu da kaifa*) was formally adopted. In general terms, "the West" refers to the western interior portion of China and comprises ten provincial-level units defined as the "western region" in the Seventh Five-Year Plan (1986–1990): Xinjiang, Tibet, Qinghai, Gansu, Ningxia, Shaanxi, Sichuan, Chongqing (which became a centrally-administered municipality and provincial-level unit in 1996), Yunnan, and Guizhou. More recently, two other provincial-level units—Inner Mongolia and Guangxi—were added to the grouping. Most of these provincial-level units share relatively low levels of development and adverse environments such as high relief and dry climate. Among them are the five autonomous regions and other areas that have high proportions of ethnic minorities.

Announcement of the campaign invited much attention from the public in part because of business potential in the West. In 2001, under the arrangement of the Hong Kong government, a business delegation of more than 200 toured the West and initiated a number of investment projects.

The campaign is also noteworthy because it signals an apparent departure from the reform era's regional policy that focused heavily on the eastern coastal region. Since the early 1990s, the Chinese government has been under severe criticism for allowing and even enabling regional inequalities to worsen. Observers warn of political and social instability if this trend continues. These criticisms echo those voiced against the persistent inequality between peasants and urbanites that is institutionalized by the *hukou* system and those highlighting the widening gap between the haves and have-nots. Moreover, unrest in the inland region, where ethnic minorities do not identify with the Communist regime and do not share the culture and history of Han Chinese, can seriously threaten the integrity of the PRC. It is believed that development of the West can not only alleviate political and social tension but also accomplish the goal of sinicization of minority regions.

One focal point of development is the geographically isolated city of Urumchi, capital of Xinjiang, Uyghur Autonomous Region, which has become a booming metropolis of 2 million people. With massive in-migration of Han Chinese, the city has taken on the profile and character of cities in the central and eastern parts of the country. Notable in new Chinese urban developments are white tile and blue glass facades in downtown areas. Although geographically isolated, Urumchi is a focal point of large-scale tourism, industrialization, and development of the region's oil and other resources. Power to control Uyghur separatist movements also emanates from this important center. In fact, Urumchi's geopolitical significance may exceed its economic importance.

Despite the fanfare and the government's large investment accompanying the campaign, its likelihood for success is questionable. In addition to environmental challenges, the West suffers from poor human resources, as its population has the lowest level of education in China. Moreover, migrants leaving the West have led to a loss of skilled labor. Though the resettlement of Han Chinese into minority regions can alleviate labor loss, it also induces conflict due to cultural and social difficulties of assimilation.

ENGINES OF GROWTH

In 1990, the Chinese government opened the Pudong New Area in Shanghai as an SEZ, transforming what had been farmland and countryside into a financial hub (Figure 11-13). Pudong, a district of Shanghai, was created as an SEZ following the opening of other coastal areas to further expansion as open economic zones. Pudong has succeeded in attracting foreign capital and developing into China's financial capital. The special preferential policies granted to Pudong include reducing or eliminating customs duties and income tax, the right of foreign business people to open financial institutions, and the right for Shanghai to set up a stock exchange. Pudong's Gross Domestic Product (GDP) in 2002 was US$25.13 billion. By 2009, it had reached US$54 billion!

China's latest "new engine" is the Binhai New Area that is centered on the country's fifth largest city—Tianjin. Larger than Pudong, this development will stretch along the Bohai Gulf for 90 miles (150 km). Inexpensive land and tax incentives are underpinning factory development.The area is intended to become a manufacturing powerhouse for products ranging from cars and microchips to chemicals and aircraft.

CHINA'S NEW CONSUMERS

Runaway development in China's coastal provinces has seen incomes soar for many people. This is the heartland for a growing middle class that currently numbers around 100 million. Although the economic boom in the coastal provinces has exacerbated regional inequalities, a new consumer class has arisen.

There are several types of consumers in China today:

- Blue-collar factory workers
- Pink-collar secretaries, flight attendants, and other service workers
- White-collar office workers
- Gold-collar executives in multi-national corporations

China has a new monetary elite that can also be classified:

- *Dan shen qui zu*—the "single aristocrats," usually women over 30 working as managers or bankers for foreign firms. These women buy high-end clothing and use cosmetics such as Lancome and Shiseido. They can afford to take beach vacations with their friends.
- *Bo pu zu*—the well-off intellectuals who indulge in books.
- *Yue guang zu*—those who earn middle-class salaries but spend money as fast as they get it. They are considered to be the "tapped-out class."
- *You pi*—the yuppies also known as the *xiao zi*—the "little capitalists," workers who own apartments and cars and spend their off-hours in cafes.

Figure 11-13

Pudong SEZ in Shanghai. Pudong's development began in 1990. Landmark buildings such as the Oriental Pearl TV Tower (the center tower continues for 1,535 ft (478 m) mark Pudong's new status as one of the most productive economic zones in China. Tibor Bognor/Corbis Images.

Chinese society is becoming increasingly stratified. However, the middle classes are expected to include more than 800 million in the next five years. The combined disposable income of these groups should reach US$1.6 trillion!

Globalization

Deng Xiaoping's economic policies opened China to the global economy and spurred sharp increases in tourism, imports and exports, and foreign investment. In 2002, there were 98 million tourists, 54 times the number in 1978. By 2007, the number of overseas tourists had reached 132 million. The leading countries of origin are Japan, Korea, Russia, and the United States. The World Trade Organization (WTO) predicts that by 2020, China will be not only the largest recipient of foreign tourists but also the world's leading sender of tourists abroad.

Because of these economic policies, China has become the world's largest exporter, trading extensively with the European Union, Japan, and the United States. China, now consuming more steel, coal, meat, and grain than any other nation, has outpaced Japan to become the second largest economy in the world after the United States and is a prime force for economic growth around the globe. For instance, China is one of Germany's most important trading partners. In 2004, China had 130 shipping companies in Hamburg, Germany's largest port.

Exports make up 80 percent of China's foreign currency earnings, and there has been a shift in the structure of these as industrial products have overtaken agricultural products. China's top exports are electrical machinery and equipment, power-generating equipment, apparel, toys, and processed foods. China is also increasing its market share in the furniture, appliances, and electronics sectors.

Japan is a big market for processed foods, especially instant noodles (*ramen*). China's relationship with Japan is interesting in that the Chinese despise the Japanese because of their atrocities in World War II. The Chinese word for "devil" now means "a Japanese person." The first contract given to a vehicle manufacturer was not given to the Japanese but rather to Volkswagen. Nevertheless, Japan is a huge investor in China, making an array of products such as automobiles and electric machinery. Sony, Toshiba, and Panasonic have opened plants that manufacture electronics. China promotes the idea of "Chinese-Japanese friendship" but it is well understood (at least by the Chinese) that this denotes Chinese superiority.

Cars, Cell Phones, and the Web

China now has the same number of cars per capita that the United States did in 1915. But production and sales are burgeoning. In fact, China surpassed the United States as the world's biggest auto market in 2009. Sales are expected to top 10 million in 2010.

The government began encouraging private car ownership in 1994. Now there are 11.5 million privately owned cars in China. China expects to have more cars than the United States by 2025. Many people who have cars didn't know how to drive three years ago and they remain poor drivers. Deaths from auto accidents are 4.5 times those in the United States. The hottest car in China today is the Rolls Royce.

Both foreign and Chinese brands are sold in the domestic market. Most global auto manufacturers have plants in China. These include such companies as Volkswagen, Honda, General Motors, and Ford. However, the Chinese are making their own cheaper cars. There is a spate of companies with such names as Chery, Great Wall Motors, Brilliance China, and Polarsun.

Many Chinese have adopted American car culture. For example, there are car clubs, self-drive vacations, and drive-through restaurants. The down side includes eye-stinging smog, traffic jams, and high accident rates.

Cell phones have grown from 87 million in 2000 to some 450 million today. Local brands such as Bird and Panda until recently have outsold Nokia, Motorola, and other global brands. However, by making phones more interesting with clever gadgetry and Chinese writing, foreign phone desirability has increased. For example, gem-studded phones are very popular among the elite.

Chinese text messages come in a variety of colors: yellow, gray, black, and now, with official endorsement, red. Yellow refers to the smutty type, and grey or black to spam messages, many of which offer products or services of dubious legality. To steer public thinking, the government is encouraging the sending of politically correct "red texts."

China has the world's largest number of Internet users—220 million. However, there is no such thing as free speech. For example, complaining about the authorities on a blog can result in your being arrested. China employs around 35,000 cyber police to monitor the Internet and help with censorship. Google admits that it censored itself in China where its corporate motto was, "Don't be evil." However, as of March 2010, Google opted not to censor its material. The Chinese government stepped in to prevent the free flow of information under the rubric: "If you leave the door open, flies will come in." Officials do not want the people to acquire any information that might reflect negatively on them. For instance, knowing about repression in Xinjiang or Tibet might be detrimental to China's cohesion, let alone its image. Google is in the process of moving its site to Hong Kong, where there is more freedom.

As of 2010, in order to set up a personal Web site, one has to apply in person with proof of identity and a photograph to regulators. Applications are then submitted to the Ministry of Industry and Information Technology for review. The Chinese government claims that this is to stop pornography. Users say it is to discourage anti-government commentary.

Initially, foreign investment was concentrated in the south but ultimately spread northward when the 14 cities were opened in 1984. A second spatial pattern emerged with the infusion of investment in interior cities, especially along the Yangzi and Huang He. Developments in Wuhan and Chongqing have given a new name to the Yangzi valley: "China's Soaring Dragon." Most of the foreign capital has come from Asian sources, directed into places with Chinese family ties. Most Western investment pours into large cities such as Shanghai, Beijing, and Tianjin.

A quarter century of open-door policy has paved the way for China to embrace globalization and become a powerful trading economy. China is now the largest trading nation in the world, with total trade (the sum of imports and exports) greater than US$ trillion. (The United States is third after Germany.) Moreover, China has surpassed Japan as the country with the largest trade surplus with the United States. More than 40 percent of China's exports go to the United States, which buys not only garments, electronics, toys, and Christmas ornaments, but also a wide range of other products from fine furniture to swimming pool supplies.

China's exports are competitive primarily because its large population keeps labor costs low. The large population, likewise, makes China an attractive market. After years of debates and negotiations and after China agreed to further open its economy, the country was admitted to the WTO in 2001. This symbolizes and formalizes China's prominent status in world trade.

Entering the WTO means that China's market is more accessible to the world. Multinational corporations searching for populous and unsaturated markets were among the first to advertise their products to the Chinese. At the same time, the Chinese economy has become increasingly dependent upon the world market, and its industries must improve in order to meet competition and survive.

CHINA IN SOUTHEAST ASIA

In a press for resources and interest in expanding its spheres of influence, China is investing in an array of countries around the world. One of these areas is Southeast Asia, where China is employing its new "smile diplomacy." China has resolved almost all of its border disputes and has committed to a specific code of conduct in the South China Sea. It work's closely with ASEAN and participates in several multilateral agreements. Also, it fosters cultural exchanges with each country in the region.

Part of China's appeal lies with the fact that it portrays itself as a potential ideal—a model. It emphasizes top-down development and poverty reduction in the context of sidelining political reform for economic reform. This model appeals to rulers in authoritarian states such as Myanmar (Burma) and Vietnam. With the Chinese model, regimes have time to figure out how to co-opt businesspeople and other elites that they need to remain in power.

China fosters communication between itself and other cultures in Southeast Asia. It has upgraded its newswire Xinhua to report in a variety of languages other than Chinese and English. It has expanded broadcasting from CCTV—Chinese state television. Confucius Institutes have been founded at various universities, and Mandarin lessons are being given in elementary schools in countries like Thailand and Vietnam.

To further cement China's presence, Chinese migration is transforming the demographic makeup of northern mainland Southeast Asia from northern Myanmar to Vietnam. Because of out-migration from Yunnan and other border provinces, ethnic Chinese now dominate entire towns in places like northern Myanmar and Laos. China is Cambodia's biggest aid-giver, and thousands of Chinese migrants are creating new villages in that country as well.

China has more hydroelectric power dams than any other country on Earth, but it needs more power to fuel its cities and industry. In fact, China is in the process of building 216 large dams in 49 different countries. In Southeast Asia, numerous dams, operated by Sinohydro, will emit power to link up with the China Southern Power Grid. Read more about dams and their impacts in the Southeast Asia chapters.

Mineral and timber resources also draw Chinese investment. For example, the Chinese are mining bauxite (aluminum ore) in Vietnam's Central Highlands. Myanmar is a source of timber and gemstones, and according to plan it will soon serve as a transit corridor for oil from the Indian Ocean to southern China. In return for various infrastructure developments Cambodia is allowing China to explore for natural gas in its territorial waters.

CHINA IN AFRICA

China is running short of certain resources such as oil to fuel its rapidly expanding industrial network. The country was self-sufficient in oil until the early 1990s. Now its oil imports have multiplied six-fold. At present, a large chunk of imports (44 percent) comes from the Middle East, but China is looking elsewhere for this and other mineral resources. The most significant ventures are in Africa, where there are more than 800 Chinese state-owned enterprises operating (Figure 11-14). For example, Angola is now China's top oil supplier, and the Gulf of Guinea will provide at least 20 percent of China's oil by 2020.

As we mentioned earlier, China takes a no-strings-attached perspective in its foreign investment activities. It does not criticize other governments for their human rights abuses. This attitude is despised by the EU, the United States, and a host of other countries that are concerned with human rights.

One very controversial issue is China's involvement in Sudan in light of the despicable situation in Darfur

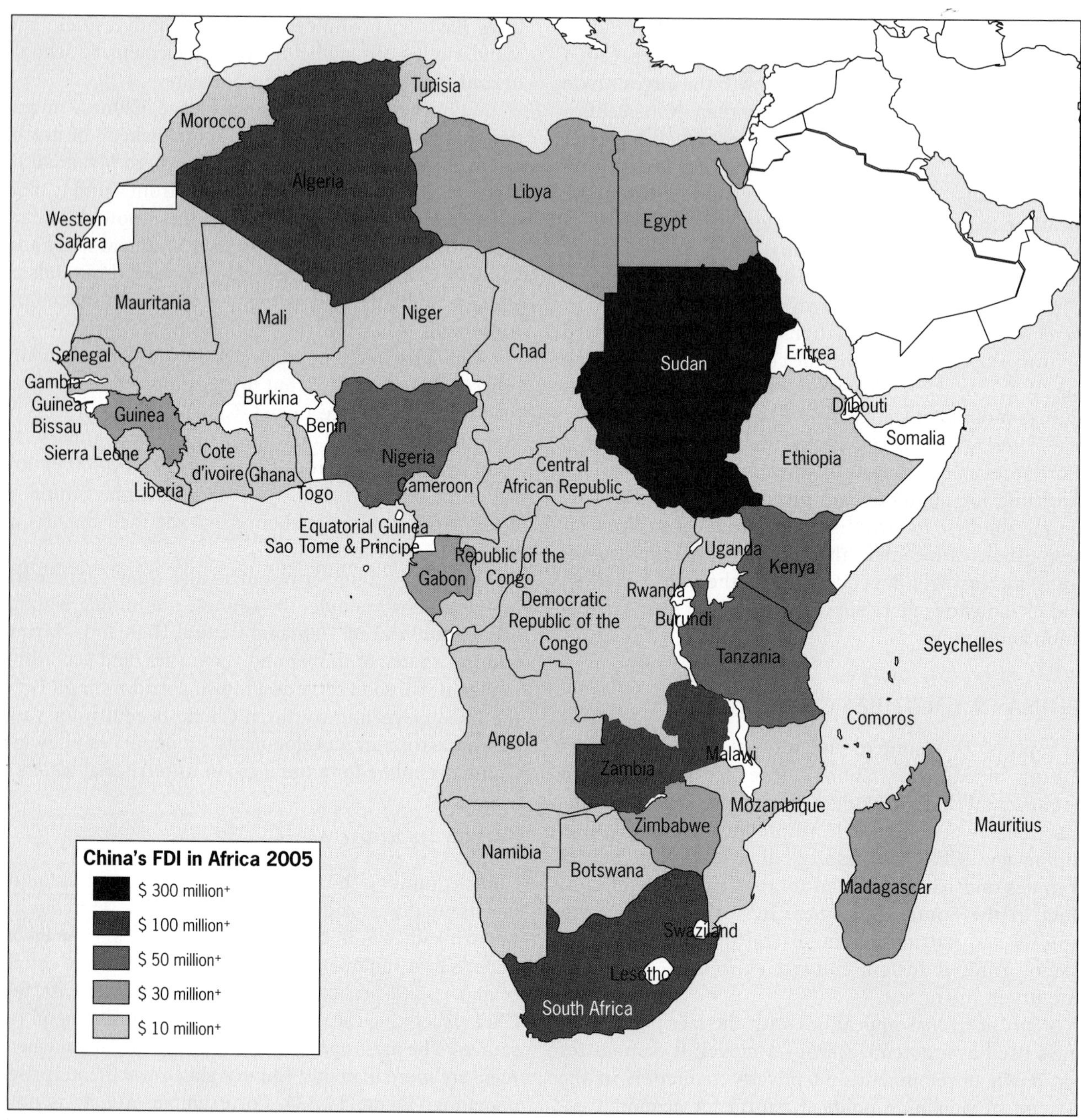

Figure 11-14

China's investments in Africa. Why is China investing in Africa?

where hundreds of thousands of black Africans have been slaughtered by Arab militias. China is heavily ensconced in this country in order to acquire oil. It has even installed arms factories for the corrupt and genocidal Sudanese government.

China typically combines aid projects, such as road and dam construction, with its extractive activities. For the most part, China hires locals providing much-needed jobs. In some cases however, projects are manned by Chinese workers who are alien to African culture. The resultant denial of jobs to locals and cultural clashes sow the seeds of resentment.

China is looking to African markets for its goods. Everything from chandeliers and appliances to blankets and plastic buckets proliferate across the continent. For many, this gives access to relatively cheap products that they never before could afford. For others, Chinese goods represent an economic malignancy—displacing African-made goods and destabilizing African industries. In Zimbabwe, Chinese goods are called "*izhing zhong*," a derogatory, epithet that mimics Chinese speech and highlights African dissatisfaction with Chinese goods entering their markets.

Both Chinese and Africans view each other with suspicion. Africans worry about Chinese intentions, an unease reinforced by their perceptions of their wealth, stinginess, and social distance. An online response to a BBC report about Chinese investment in Africa captures this anxiety: "Asian invasion. They are everywhere. Infiltrating like a colony of ants. . . . They hate other races and they are ruthless. Watch out for them." Similarly, Chinese characterizations of Africans tend to be negative. Interestingly, the Chinese characters for "Africa" translate literally as "negative continent."

In Guinea, where China extracts iron ore, bauxite, gold, and diamonds, people have looted and vandalized Chinese shops. While African governments do not like some of China's policies, most are corrupt and are happy to siphon off funds for themselves, typically into overseas bank accounts. Now China is backing off in some unstable countries like Guinea and Congo that have uncertain market climates.

Another way in which China is locking into networks of globalization is by sending troops overseas to conflict regions to help keep peace. China's participation in UN-sponsored peacekeeping missions has spread across Africa. For instance, China has sent a large contingent to Liberia and smaller ones to places such as the strife-ridden Democratic Republic of Congo and Sudan. It is a major financial contributor to UN peacekeeping operations as well as to the African Union contingents. China has also expanded its volunteer Communist Youth League program to place experts in areas such as agronomy, animal husbandry, health, and linguistics in African countries.

WORLD EVENTS

In 2006, billboards in Beijing publicized messages that welcomed the world to come to the city as it was selected to be the site of the 2008 Olympics. On December 3, 2002, the sky of Shanghai was sparkling with fireworks and the streets were filled with festive banners and Chinese national flags. The Shanghainese were celebrating their winning the bid to host the 2010 World Expo.

An unprecedented 86 world records were set at the 2008 games. Although the United States won 10 more medals than China (110 to 100), China's pride was boosted by the fact that it earned more gold medals than the United States (51 to 36).

Shanghai's World Expo's theme is "Better City—Better Life." This signifies the city's new status in the twenty-first century as a global economic and cultural center. More than 190 countries and 50 international organizations are participating. Almost 100 foreign leaders and 70 million visitors are expected. This will be the largest Expo in history.

Beijing and Shanghai are global cities attaining world-class status. Most importantly, their selections depict a resounding international recognition of China as a powerful nation, which is in no small part a product of its increasingly central role in the global economy. To many Chinese, their successful bids to host world events are evidence that once again "the Middle Kingdom" is occupying the center stage of the world. Against the backdrop of one century of failed resistance against Western imperialism and an even longer period of domestic turmoil, the perception and image of global-scale successes are hotly sought after by the Chinese and contribute further to a heightened sense of nationalism.

Urban China

One of the most spectacular achievements since the 1980s is the increase in the number of cities in China. Policy was relaxed to permit the formation of more cities, and city status became the mechanism to attract foreign investment.

NEW CITIES

The total number of cities rose from 194 in 1978 to 467 in 1990, to 660 in 2005 with a concomitant increase in city population. A relatively large number of small and medium-size cities have been added to the system.

In 2005, China had an urban population of 542 million. Of a total of 660 cities across the nation, 171 are mega-cities with more than 1 million people, 279 are big cities with half to 1 million, and 210 are small and medium cities with populations between 200 thousand and half a million (Figure 11-15).

Figure 11-15

Cities and inland waterways in China proper. Here you can see that China's major cities stretch from the northeastern rust belt to the North China Plain, and along the coast to Guangdong Province. Other major cities are situated on or near China's major rivers—the Huang He and the Yangzi. Note that the Yangzi is the major navigable river. You can also see the location of the famous Three Gorges Dam. From H. J. de Blij and P. O. Muller, *Geography: Realms, Regions and Concepts,* 14th edition, 2010, p. 484. Originally rendered in color. Reprinted with permission of John Wiley & Sons.

China's 53 metropolitan regions—anchored by a city of more than 1 million and incorporating adjacent counties—hold 29 percent of the population, yet produce more than half of the country's GDP. The largest metropolitan regions are Shanghai, Beijing, and Guangzhou.

China's largest cities are concentrated along the coast. However, numerous medium and small metropolitan regions are located inland. Here, in cities such as Lanzhou, Xian, and Chengdu, per capita GDP is significantly less than for coastal cities.

In the next five years, China expects to build over 300 new cities—bringing the total number of cities with populations of over 200,000 to 1,000. For each new city, China will have to spend at least US$35 billion in order to provide housing, potable water, power, transportation, and related services.

CHANGING CITY STRUCTURE

Upgrading, relocation, and new configurations are hallmarks of the modern Chinese city. Modernization means destruction. The demolition process follows precise rules. Behind carefully constructed palisades, tearing down begins in areas where land values are highest—the central city. Unfortunately, this is where architectural heritage is most valuable and best maintained.

People are given compensation packages and, with little choice, head out to new suburban developments where social and spatial segregation is based on affordability. Well-off families can move into middle-class privatized estates, townhouses, and gated, Western-style communities. Poorer people move into government-subsidized apartments. Migrant workers are typically confined to slums. Outer residential zones are often in "the middle of nowhere' with sparse infrastructure and a two-hour commute from the central city.

Beijing provides an example of current restructuring activities in Chinese cities. The city has been undergoing major reconstruction and renovation since the 1990s.

Situated only a couple of miles from the city center is the 9.7 square mile (25 km^2) Shixing *hutong* (traditional neighborhood). In 2006 the *hutong* was destined to be replaced by luxury shops and art galleries as a draw to tourists attending the Olympic Games in 2008. Protests over the loss of homes and the ruination of neighborhood character led to a hiatus in demolition. The district is now classified as a "protection area." This status does not prevent demolition, but requires the architects to design new buildings in a so-called "neo-traditional style" that must be based on Qing Dynasty style. The fine residences remaining will be kept and restored.

Of the 15,000 dwellings originally designated for destruction, only around 5,000 remain. In the last three years, two-thirds of families have been exiled to zones more than 9 miles (15 km) away from the city center. A few recalcitrant families hang on to their shattered homes. At nightfall, they sit in the dust and speculate about their uncertain future to the distant sounds of churning cement-mixers that will soon cover all physical vestiges of their lives. They know that this is inevitable and as they say regarding the power of officialdom: "An arm cannot twist a leg."

Shanghai: Head of the Dragon

The "Paris of the East." Asia's most sophisticated city. The world's most crime-ridden city. This was Shanghai before 1949. Today, Shanghai is China's largest city, its foremost economic center, and a rival to Beijing as the nation's cultural capital. In fact, Shanghai rivals Hong Kong in economic importance.

Shanghai thrived long before the arrival of Europeans. It had a population in excess of a million in the Ming Dynasty (1368–1644) and three million by 1816. From 1895 to 1945, the city's population increased ten-fold. Migrants accounted for a significant portion of this growth because the city was safely distant from most battlefronts and had economic opportunities associated with the multiple foreign concessions. While migrants came from the surrounding provinces, they also came from Guangdong to pursue business enterprises. This connection remains significant in China's new economic landscape.

In 1949, Shanghai accounted for one-quarter of China's industrial production. The city housed a highly educated population and continued to be a migration target in Mao's era. However, when large numbers of people were forcibly moved out of Shanghai to populate China Frontier as part of the *xiafang (hsiafang)* movement (from 1958 to 1976), the city experienced a net out-migration of more than 1.2 million. With its educated population, Shanghai had a tremendous impact on development in the northeast and western regions. With the end of the Cultural Revolution, the city once more experienced net in-migration.

Since 1992, when liberal policies were "given" to Shanghai, the city has undergone substantial reform and is in the midst of an economic boom. One advantage Shanghai has is that it is of sufficient size to become the "Dragon Head" of the Huangpu and Yangzi River commercial ribbon. It is already a key commercial center with proximity to both the north and south coasts. It has the largest pool of experts in science, technology, and management, as well as skilled workers.

These advantages led to the 1990 decisions to promote both Shanghai on the west bank of the Huangpu River and Pudong on the east side of the river. Pudong was to be developed and opened to foreign banks, finance, and insurance companies, department stores, and supermarkets. Since 1992, Shanghai's growth rate has exceeded that of all China!

While the Shanghai economy is mainly industrial, the service sector is becoming increasingly important. The tertiary sector is designed to serve the entire nation. Within the industrial sector, there are six pillars of industry: automobiles, iron and steel, petrochemicals, power equipment, telecommunications, and electrical equipment. The Jinqiao Export Processing Zone in Pudong has played an essential role in this development.

The high cost of land and labor in Shanghai has forced it into high-tech and high-value industry. Other kinds of industry are gravitating to the Nanjing and Hangzhou areas, where a new freeway is expected to enhance their competitiveness. A new Ring Road, six-lane elevated highways, and subway lines now facilitate spatial interaction in the web of industry and commerce. Pudong International Airport opened in 1999, and a maglev train connects the new airport with a downtown metro stop.

Housing and services are of great concern. Shanghai has five times the density of Beijing with 12,000 persons per square kilometer. Many residents lived in *lilongs,* densely packed neighborhoods inhabited by several generations of families and friends (Figure 11-16). Now the *lilongs* are almost gone—demolished for urban renewal. Many people are demoralized about what is happening to their neighborhoods. To be forced into isolated apartments far away from their social and cultural space is devastating, particularly for older people.

A new housing plan is based on the Singapore model, whereby the residents own their own housing. An owner must occupy the dwelling for five years before renting or selling it. New houses and apartments are rising everywhere, especially in the former market gardening suburbs.

Figure 11-16

This is a traditional housing area in Shanghai known as a lilong. *These apartments do not have modern facilities. Bathrooms are in a building down the street. Laundry is hung outside for lack of space inside. Most of Shanghai's* lilongs *have been demolished in the wake of modernization. As neighborhood and extended family groups are forced to move to the city outskirts, they are separated and their social networks are fractured.* Photo courtesy of B. A. Weightman.

Shanghai is one of the greatest urban renewal stories in history. Nearly 17 billion dollars in foreign investment has transformed the once drab skyline to one of glittering skyscrapers and neon. But all is not perfect. There is a vast influx of "floating" and "temporary" populations. These people live in slums and do not have equal access to education or services. There is a growing income disparity between rich and poor.

> Observers use Pudong as an analogy. Hardware in Pudong means infrastructure; software means services. Both must exist in tandem for Pudong to work. Change is in the offing and as Shanghai becomes more affordable for foreign investors, even greater progress is expected.

CHINA'S RUST BELT

Home of the Manchus, founders of China's last dynasty; a region of historic contention among China, Japan, and Russia; a northeastern extension of the agriculturally productive North China Plain. This is former Manchuria, resource-rich and responsible for 25 percent of China's industrial output in the 1970s. Today, the northeastern provinces of Heilungjiang, Jilin, and Liaoning are trying to overcome their rust belt status.

In the early years of communism, planners concentrated state-run industries around the coal and iron fields of Shenyang, Fushun, and Anshan. Shenyang became the nation's leading steel-making complex. Discovery of the Daqing oil fields west of Harbin rendered China self-sufficient in oil at that time. Directed on giant state farms, agriculture boomed and further illustrated what communist planning could accomplish.

With the new market reforms of the post-Mao era, state-run industries stood as monoliths of obsolescence, unable to cope with privatization and restructuring. Now the northeast contributes a mere 10 percent of China's industrial output.

Nevertheless, there is light on the horizon for former Manchuria. Its resource wealth and its proximity to Russian, Korean, and Japanese ports make it a potential core of development in East Asia. The launching of an economic development zone at Dalian, as part of the Liaoning Economic Development Area, has drawn the attention of foreign investors, especially the Japanese.

In 1994, one-third of Japan's investment in China was pumped into Dalian to manufacture everything from motors to chopsticks. Spatial association, economic incentives, a huge harbor, and a cheap labor pool have drawn more than 800 firms to Dalian.

Although much of the northeast still struggles with its rust belt heritage, some economic geographers see economic potential in light of future possibilities. A reunified Korea or a prosperous Russian Far East could alter northeastern fortunes.

GUANGDONG: REGIONAL STATE

Guangdong fits the definition of a regional state. It is a natural economic zone shaped by the global economy in which it participates. Guangdong Province—and especially the Pearl River Delta's agglomeration of Guangdong, Shenzen, and Zhuhai—has become a place apart from the rest of China. Guangdong's southern location has particular political-economic and sociocultural implications. Proximity to Hong Kong and the incentive of low production costs have transformed the Zhujiang Delta into an outward processing zone. Since many Hong Kong residents once resided in Guangdong, business ties are facilitated. This explains the flows of labor-intensive industries and investment into Guangdong from Hong Kong. Economic structural change has seen a decline in agricultural participation in favor of industrial participation.

Shenzhen has become Guangdong's wealthiest area and the most important channel for foreign investment. It has established trade relations with 120 countries and has attracted some 30,000 foreign investment projects. At least 250 of the world's top 500 corporations have invested here. Shenzhen was destined to align with Hong Kong in its transition from a British colony to a socialist state, a showcase for exhibiting the concept of "one country, two systems."

The state has continued to expand open zones in Guangdong. More rural and peripheral counties are being drawn into the picture as investors continue to seek lower wages and subcontracting opportunities in the context of established social relations and linkages. The province now has 75 economic function zones, many for the processing of high-tech equipment.

These development dynamics have generated new migration streams in Guangdong and elsewhere. Although migration was once strictly controlled by the government, recent economic reforms have created new push-and-pull forces fostering population movement. The household responsibility system with its boom in agricultural productivity has released a surplus labor supply. Foreign investment has instituted a plethora of labor-intensive industries looking to recruit workers. In 1984, a State Council directive allowed rural residents to move to market towns and even larger urban areas, provided they did not require the state-rationed grains to which the nonagricultural population was entitled. The eastern region, including Guangdong, is now the recipient of migration streams seeking economic opportunity in the industrial production system.

Changing employment opportunities have characterized the current economic crisis. Unemployment rose to an official level of 4 percent in 2007 and 2008, although experts say it was more likely 9 or 10 percent. With the economic downturn, thousands of workers returned to the countryside. But in 2009, China's economy picked up again, and suddenly there was a shortage of workers especially in the coastal provinces. One of the driving forces was the rush to finish construction for Shanghai's World Expo, where 400,000 laborers are needed.

Another region of labor shortages is Guangdong, where the economy has recovered and demand for exports has increased dramatically. Now employers are raising wages, storming job fairs, and staking out train stations hoping to snare migrants. The *China Daily*, the state-run English language newspaper, said that the city of Dongguan, where most of the world's toys are made, was short of its usual population of 5 million workers. The paper also reported that companies are offering higher wages, better housing, and more benefits to attract employees.

Continual reduction in China's population growth portends a demographic crisis. By 2050, a third of the population will be 60 or older (compared to 26 percent in the United States). Where will all the needed younger workers come from?

"FLOATING" POPULATIONS

Temporary residents obtain permission to stay in a city for a fixed amount of time to work in factories or on construction sites. These **floating populations** make up a quarter of some cities' populations. In Guangzhou they comprise 38 percent, and they outnumber permanent residents in Shenzhen. Most of the temporary people work in the building industry, followed by retailing, especially individual street-side enterprises. Temporary populations place great stress on social facilities, transport, and housing.

In 1989, there were 30 million migrant workers in China. By 2006, that number had reached 132 million. While migrants are poorly paid, they can earn more in cities than they can in rural areas. In 2005, these workers sent home US$65 million.

Migrants take jobs that are *zhan*—dirty; *lie*—physically demanding; *ku*—bitter; and *xian*—dangerous. They work 11 hour days and 26 days per month and their hourly wage is only a quarter of that paid to permanent urban laborers.

When migrant workers leave their small towns and villages, they give up their entitlements such as education and health care. Many migrants have started their own schools, which are of significantly lower quality than official schools.

Migrant housing for both men and women is typically abominable. Apartments and dormitories are rudimentary and overcrowded. Thirty workers may be crammed into a single room with no ventilation or showers. Shared sanitation facilities are limited. Women are often locked into their dormitories at night.

Many migrant workers suffer from poor health. Most live off vegetables and noodles and suffer from protein deficiency. The most prevalent disease is silicosis, which reduces lung capacity. Silicosis is caused by exposure to silica dust, which fills the air in stone and glass, ceramics, and fiberglass factories. In 2006, some 616,000 workers had the disease. Accidents, exposure to silica, textile dust, and toxic fumes from dyes and other chemicals kill as many as 130,000 women and men a year.

WOMEN'S WORK

"Women hold up half the sky." So said Mao Zedong in the 1950s. Women achieved many gains under communism, especially during the Cultural Revolution (1966–1976) when people tried to be gender neutral and participation of women in political and economic activities reached nearly 100 percent. Now, as communism gives way to capitalism, women are falling back into old situations of inequality in line with traditional Confucian and Daoist thinking, which defined women as culturally inferior to men.

In China, didactic thinking is common and is reflected in the fact that experience is classified in paired concepts that are frequently polar opposites. Furthermore, contrasting concepts often have one that is more highly valued. Gender differences reflect this system.

One fundamental dichotomy is that women are associated with "inside" domains of work and men are associated with "outside" domains of work. Women do "light" work while men do "heavy" work. "Skilled" work is done by men, and "unskilled" work is done by women. The introduction of new technology has exacerbated these divisions. Men dominate in work involving labor-saving machinery while women are relegated to performing more menial, labor-intensive tasks.

Today, women work in the fields while many men leave the village to work in a new factory enterprise. Agriculture has been reclassified as lesser "inside" work. Factory work is more important "outside" work, although millions of women do work in factories. Moreover, women are responsible for family members and interactions within the household, but external representations of family business to the outside world falls to males.

China expert Tamara Jacka (1997) found that although male-female work patterns have changed with recent reforms, "expectations embodied in the traditional saying 'Men rule outside, women rule inside' . . . continued to exert an important influence on work patterns." It is true that modernization has given women opportunities to work outside the home. Still, a woman is entirely responsible for household affairs and so carries a double burden of responsibility. Women are still expected to be mothers and keepers of social stability in the home. Although the content of male and female work is changing, the conceptual divisions remain.

Extra assembly or piece-work done in the household courtyard is also included in the inferior "inside" domain. Commodity production in the household realm has become important, yet women are poorly paid for this work. When "courtyard economy" expands into a larger enterprise, management is typically seized by men.

The government mandates equal opportunity, but things are different in practice. At present, large numbers of women work in township and village enterprises yet they are poorly paid, do the most menial tasks, and are the first fired. Very few are involved in management over male workers. As much as 70 percent of laid-off workers are women. Further, women are forced to retire at age 50, five years earlier than men.

The number of women working in state or government enterprises has fallen 24 percent since 1994. While job training centers teach men to become chefs' mechanics and carpenters, women are taught to be hairdressers and manicurists.

In any industrial city in China, women dominate low-end jobs. Some are sidewalk vendors who peddle cheap calculators or "Hello Kitty" notebooks. Others take up such tasks as cleaning tables in restaurants or sweeping streets. Many are reduced to rag-picking or hauling oil drums full of garbage.

In the SEZs, female workers are desirable because they will work longer hours for lower pay. Many women are working in deplorable conditions, locked in their unsafe factories at night. Hundreds have died in fires, unable to escape the barricaded structures. Managers find that women, who make up nearly half of the floating population streaming into the coastal regions, especially Guangdong, are subservient, docile, and easy to replace.

There is a new social body in modern China: the *dagongmei*. Transience is the dominant feature of the lives of these female workers. Their stay in factories is typically only four to five years. Women are constrained by both the *hukou* system and patriarchal attitudes toward their leaving home to work in a town or city. Moreover, short wage-earning terms are expected to occur only in the premarital stage of a young woman's life.

Many factories have codes of conduct for women. For example, going to the bathroom requires a written permit. Anyone who leaves the premises without permission is subject to firing or at least a hefty fine. Arriving five minutes late results in a deduction of two hours wages. No talking is allowed on the shop floor. Anyone receiving a phone call will be dismissed immediately. Permission must be obtained to miss work because of illness, and so on.

Newfound affluence has also revived the notion of women as property. This has led to an upsurge in sexual harassment, prostitution, concubinage, and the abduction of women into forced marriages. Garment, shoe, toy, and electronics factories, staffed primarily by women, are often referred to as "peach orchards" where women are "ripe to be picked."

The Global Sweater Factory

Shanxi Shuofang Flax Textile Company is a Chinese state-owned enterprise. The beginning of its supply is flax (linen fiber) from France—softer than Chinese-grown flax. The fiber is processed by men in multiple stages on secondhand, French-manufactured machines. Other machines turn the combed flax onto spools that are dipped 144 at a time into vats of bleach for six hours. The bleaching room equipment is from Russia. Next, a set of machines made in Germany transforms the thick strands of flax into thin, color-dyed thread and winds it onto six-inch spools—2,000 a day. Flax dust coats the machinery and fills the air, but few workers wear masks.

The spools of thread are delivered to the factory's Everbright plant, the knitting and sewing machine building where women piece together the sweaters, taking turns for three 8-hour shifts a day, seven days a week. Here too, dust and the smell of dye permeate the air. The sweaters are washed and dried and laid out for inspection. Every 15 seconds a woman attaches an Eileen Fisher label to each garment. Others attach U.S. price tags. Also, there are other assembly lines in the Everbright factory that produce Guess jeans and various other garments for the Gap, the Limited, and Target. The Eileen Fisher sweaters cost US$148—many times more than the monthly wages of Everbright workers.

Inside the factory compound are brick-walled dormitories. A smell of noodles mixed with the odor of urine fills the halls. Each room has six bunks with stained, thin mattresses. The few squat, hole-in-the-floor toilets are shared by many and are perpetually clogged. Nonetheless, dorm rooms are decorated with dreams—posters of pop singers, movie stars, sports cars, and the like. On the wall across the hall is an advertisement showing a shiny Phillips mobile phone.

Workers earn as little as US$37 a month—barely a dollar a day. Those in supervisory positions can earn up to US$8 a day, turning them into the major breadwinners for their entire family; often they can earn more in a few months than their family can earn in a year. When they return home during the week-long Chinese New Year holiday each spring, they come back from the countryside with friends or cousins eager to land jobs in the sweater factory.

Hard Times Since 2007

Many companies have fallen on hard times in the recent global economic crisis. A collection of factors such as rising energy and other input costs, workers clamoring for higher wages, higher tax rates, and tougher regulations are leading a swath of factories to close, especially in Guangdong Province. In 2007, nearly 1,000 Hong Kong-owned businesses left Guangdong. Some relocated to cheaper areas inland. Others are moved to less-costly Vietnam and Cambodia. Most of the closures involve, small, low economies-of-scale plants.

Businessmen are encountering resistance in several areas. Younger generations are more willing to stand up for their rights. No longer accepting of pittance pay, they are more interested in pursuing a career than simply having a job. The yawning income gap, rapid turnover, and disgruntled migrant workers have led to higher crime rates. Senior leaders in Beijing want to propel the economy upward from polluting, resource-draining, labor-intensive light industry toward innovative, high-tech and service businesses. Moreover, local governments are beginning to monitor environmental impacts and workers' rights.

HONG KONG (XIANGGANG)

Here on 400 square miles (1,000 km^2) of fragmented, rugged land live 7 million people, many of them refugees from the mainland (Figure 11-17). During British colonial rule, Hong Kong was divided into three segments: Hong Kong Island with the capital of Victoria; Kowloon Peninsula and an array of other islands including the largest, Lantau; and the mainland section known as the New Territories. While Hong Kong Island and Kowloon were ceded in perpetuity to Britain in 1842 and 1860, the New Territories were leased by Britain in 1898 for 100 years.

The expiration of that lease in 1997 impelled Britain to negotiate the return of the entire colony to China. Then, Hong Kong, British Crown Colony, became the Xianggang Special Administrative Region, although the name Hong Kong remains in general use. The return of Hong Kong in 1997 (and Macau in 1999) was celebrated throughout China. On July 1, the handover date, Beijing's hand-picked Chief Executive of Hong Kong Special Administrative Region announced, "For the first time in history, we, the people of Hong Kong will shape our own destiny. We will finally be master of our own house."

Hong Kong is governed by a mini-constitution known as the Basic Law. The Basic Law gives Hong Kong a high degree of autonomy except in matters of foreign policy and defense, which are under the control of Beijing. Beijing is committed to preserving Hong Kong's capitalist system and lifestyle for 50 years. Capital is free flowing in and out of the region; Hong Kong is a member of the World Trade Organization, the World Bank, and a variety of other economic and trade structures. Many of the people of Hong Kong resent the fact that Beijing appoints the Chief Executive and other government members. They have demonstrated in significant numbers for true democracy with an elected government. The Beijing government continues to deny these rights to the people of Hong Kong.

As a free port, unencumbered by tariffs or other restrictions, with a magnificent, deep-water harbor, Hong Kong soon became a hub of economic activity in East Asia. It has become one of the world's largest container ports, a giant in banking and finance with its own stock exchange, and a major processing and re-exporting center. Hong Kong is now the eleventh largest trading entity in the world.

China's economic policies in adjacent Guangdong Province have changed the structure of Hong Kong's economy. When Shenzhen SEZ was established,

Figure 11-17

Hong Kong (Xianggang) Region. Here you can see the relationship between Hong Kong and Shenzhen SEZ in Guangdong Province. Most of the businesses in this region are funded by Hong Kong business people. From H. J. de Blij and P. O. Muller, *Geography: Realms, Regions, and Concepts,* 14th Edition, 2010, p. 500. Originally rendered in color. H.J. de Blij and P.O. Muller. Reprinted with permission of John Wiley & Sons, Inc.

thousands of factories moved there from Hong Kong to take advantage of cheap land and labor. Workers poured in, especially from Guangdong and Guangxi-Zhuang Autonomous Region. The former fishing and duck-farming community grew to nearly 4 million, and Shenzhen took on the appearance of high-rise Hong Kong with corporate offices, hotels, department stores, and gambling casinos.

Geographer Chi Kin Leung (1993) found that the subcontracting activities of Hong Kong were widely distributed in the Zhujiang Delta with Shenzhen and Guangzhou two of the major centers. Leung discovered that existing kinship and established business ties were even more important than proximity in forging these links. Another factor was the efforts of Chinese authorities to attract foreign investment into the Delta. An interdependency has arisen, with Hong Kong relying on low-cost labor to maintain its export competitiveness, and the Delta depending on Hong Kong's technological and marketing expertise to procure overseas markets.

The vast majority of Hong Kong's employees work in the service industry. In fact, 91 percent of the city's GDP in 2005 derived from service activities, which are enmeshed in its globalized production network. Less important are Hong Kong's manufacturing activities, which focus primarily on printing and publishing, food and beverage production, textiles, and electronics. The textile and apparel industries are losing employees as the population ages. Most young people are going into the service sector.

The 2007 economic downturn hit Hong Kong hard with unemployment rising to 10 percent. However, the city government with its pro-market policy—"Market leads, government facilitates"—installed stimulus packages to rescue the economy, which currently is in the process of recovery. Even so, Hong Kong's role as *the* bridge linking China and the West is diminishing as other coastal cities rise in importance. For instance, as of 2009, Shanghai had overtaken Hong Kong as China's economic powerhouse.

Population Issues

From the beginning, Mao Zedong believed that China's people were the country's greatest resource, and that to reduce population growth was a capitalist, anticommunist plot. Therefore, initial population efforts were to reduce mortality. A program of sending minimally trained "barefoot doctors" into the countryside, coupled with anti-sterilization and anti-abortion policies, forged this goal. Only in the 1970s, after disastrous famines associated with the Great Leap Forward, did consensus shift from the idealistic view of people as producers to the realistic understanding of people as consumers.

In the early 1970s, China's yearly natural increase rate was about 3 percent. Beginning in 1971, couples were encouraged (often by coercive means) to marry later, have fewer births, and space children farther apart. A more vigorous population control program was pursued after Mao's death in 1976. Families were ordered to have only one child, with severe economic and social consequences for failure to follow this rule. By the mid-1980s, China's natural increase rate was reduced to 1.2 percent. That rate has been further reduced to an officially claimed 0.5 percent. Nevertheless, China's population exceeds 1.3 billion—the largest population in the world! Figure 11–18 shows population concentrations in relatively humid lowlands.

The one-child policy has been most successful in the cities, where it is rigorously enforced (Figure 11-19). Policies vary in rural areas, depending on local needs. Minorities are permitted to have several children. Consequently, both rural populations and minority populations are growing faster than urban and Han populations.

The biggest census in the world was China's 2000 census, which documented a total of 1.27 billion people in November of that year. Census-taking in China is extremely labor intensive. Census workers visit every individual household to fill out or assist in the filling out of census forms. This is necessary because a large portion of the population, especially those living in rural areas, have low levels of education. In November of 2000, 6 million census workers went from door to door to complete the census.

Between 1990 and 2000, the average annual rate of population growth in China was 1.1 percent, the lowest since the founding of the PRC. No doubt, the decline was in large part due to the one-child policy. Currently, the average number of children per woman is 1.6, which is among the lowest in the world. Fertility decline more than offset the effect of a large childbearing cohort, namely those born during the 1960s and early 1970s when no birth control policy was enforced. Demographers predict that by the year 2020 China will have achieved zero population growth but will still have a population of 1.5 billion.

Despite the success of birth control, its ramifications are far-reaching and can be problematic. Fertility decline, in conjunction with increased life expectancy, accelerates the aging of the population. Given shrinking numbers of youngsters and poor social security, the Chinese are increasingly susceptible to lack of care in their old age. In addition, census results confirm the persistence of male preference among Chinese. In 1982, for every 100 girls born, 108.5 boys were born. In 2000, that ratio jumped to 100:119. Official surveys report that the sex ratios of the second- and third-born are considerably higher.

Imbalance of the sexes has already had adverse effects on those in matrimonial age groups, as the number of men seeking wives significantly outnumbers women of similar ages—a phenomenon referred to by demographers as "marriage squeeze." The number of unmarried young men called "bare branches" is predicted to be 30 million by 2020. Furthermore, 45 percent of Chinese women say that they do not want to give up a career to get married. Thus, both the age structure and sex ratio of the population are expected to induce new social problems in China.

China is primarily an agrarian society, but the 2000 census documented a steady increase in the level of urbanization to 36.09 percent. As of 2009, that figure had

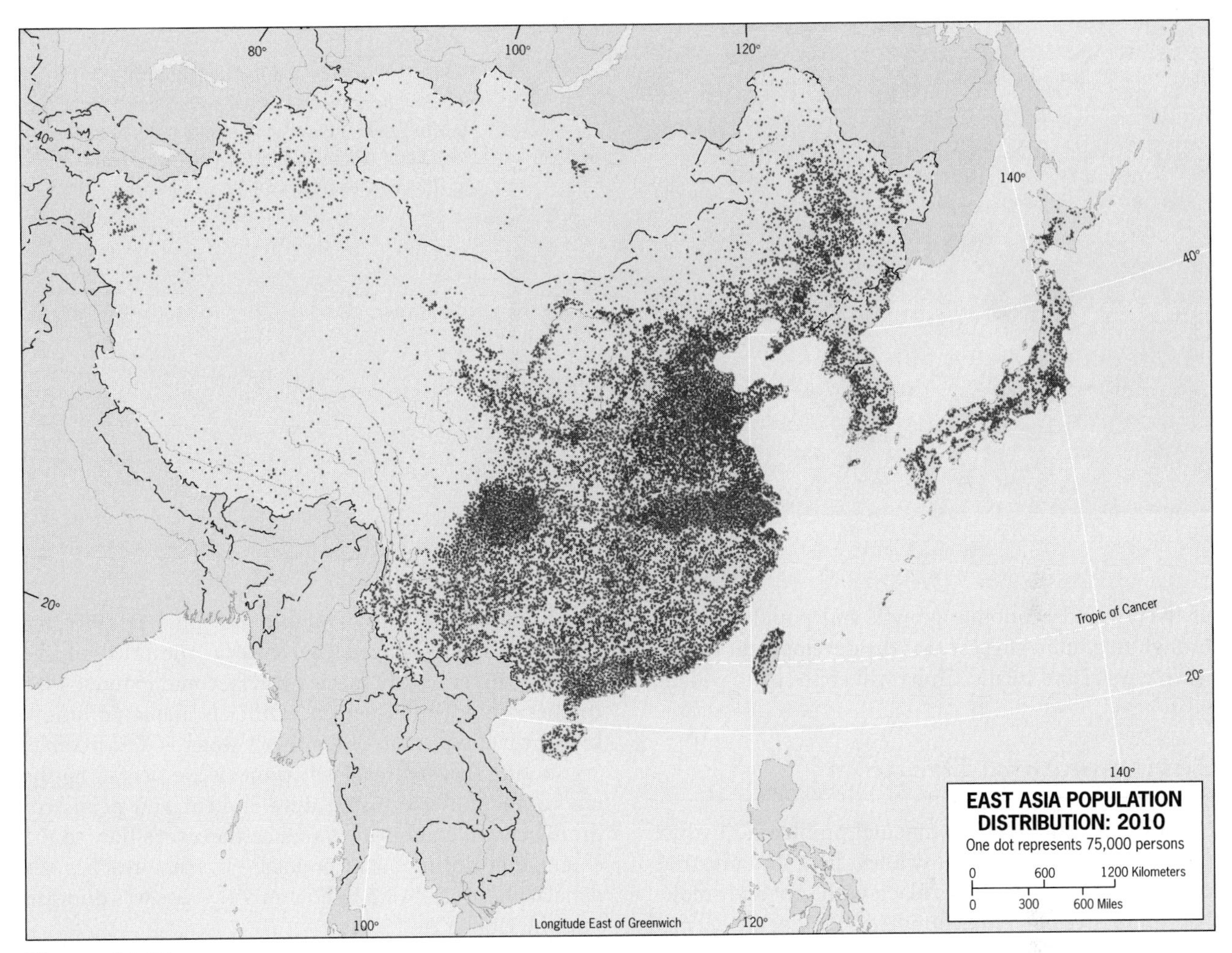

Figure 11-18

Population distribution. What is the relationship between lowland regions and settlement. Which half of the country do most people live in and why is this the case? From H. J. de Blij and P. O. Muller, *Geography: Realms, Regions and Concepts,* 14th edition, 2010, p. 458. Originally rendered in color. Reprinted with permission of John Wiley & Sons, Inc.

risen to 51 percent. Two processes contributed to urbanization. First, a rural area, because of changes such as rural industrialization and intensified interaction with cities, may be redefined as an urban area, thus boosting the size of the urban population. Second, rural–urban migrants directly add to the size of the urban population. Both have happened in China and will continue to elevate the level of urbanization.

Population mobility in China is significantly lower than that in most Western economies. However, the 2000 census documents that mobility has increased significantly. Inter-provincial migration surged in volume from 11.53 million for the 1985–1990 period to 33.24 million in the 1995–2000 period. More than 60 percent of inter-provincial migrants in the latter period originated from inland provinces and moved to coastal provinces, thus exacerbating the unevenness in regional population distribution. Areas with rapid economic growth, such as Beijing, Shanghai, and Guangdong, are especially attractive destinations of migration. The 12 coastal provinces account for less than 14 percent of the country's land area and more than 42 percent of its population. Their

Figure 11-19

I met this woman and her daughter in Guangzhou. When I asked permission to take the photograph, the woman proudly pointed to her child and announced in English: "One child—girl." Photo courtesy of B. A. Weightman.

relatively rapid economic growth and population gain through migration suggest that the development gap between coastal and inland China will continue to widen.

Environmental Dragons

China has a spate of environmental problems, of which it is possible to touch on only a few here. In industrialized China proper, the air is black with soot and smoke. Only about five of China's hundreds of cities are deemed to have clean air. People in the iron and steel city of Baotou have brittle bones and decayed teeth due to decades of hydrogen fluoride being pumped into the air. China has more deaths linked to pollution than any other country.

Up to 30 percent of acid rain in Japan is linked to sulphur dioxide from coal burning in China, where such emissions are expected to exceed those of the entire world by 2035. The country is buiding the equivalent of two coal-burning plants each week. China has now surpassed the United States in carbon dioxide emissions.

Nearly half of China's major river systems are seriously polluted—in part because 80 percent of industrial waste is dumped untreated into them. Shanghai's drinking water is contaminated with oil, ammonia, and nitrogen among other chemicals. Huge quantities of silt and chemicals from intensive farming pour into the Yellow Sea. This silt is laced with heavy metals from industry.

In 1989, the Huang He dumped 751 tons of such minerals as cadmium, zinc, arsenic, and chromium, along with 21,000 tons of oil into the Bohai. In 1997, the government announced that nearly one-third of the "Mother River's" fish species had become extinct. Fish catches are down 40 percent. Officials blame pollution, saying that 66 percent of the river's water is undrinkable, and noting that hydroelectric projects are degrading the environment. Water-using development and persistent drought have combined to reduce the river's flow to the extent that during some periods, no water reaches the delta at all. In 2005, 4.5 billion tons of waste was dumped into the Huang He.

In China's coastal waters, metals in fish have risen dramatically. Oysters have 10 times acceptable chromium content. In many areas, fish have disappeared altogether. The more coastal development and migration occurs, the more marine habitat is lost or destroyed.

Increased urbanization has diverted water needed for irrigation to non-farm use. Aquifers are being depleted. Now more than 400 Chinese cities are short of water. For example, the annual drop in the North China Plain's water table has increased from an average of five feet (1.5 m) in 1993 to 9.8 feet (3 m) in 2003. Deep wells drilled around Beijing now have to reach below the surface more than half a mile (1,000 m) to get fresh water, adding dramatically to the cost of supply. There have been riots in the countryside as farmers contest over water supplies.

An enormous scheme is on the books for a South–North water transfer. For example, water from the Yangzi is being chanelled to the Grand Canal to feed the water needs of the North China Plain. However, pollution has endangered the success of the project. Other plans for pipelines further west have been put on hold.

As aquifers are depleted, the ground subsides. Forty-six of China's cities are sinking. Shanghai is sinking half an inch (1.5 cm) every year. Planners are considering putting limits on the heights of buildings because the weight of skyscrapers contributes to the sinking process.

Sea levels along China's coast are rising. For example, seawater has risen 4.5 inches (11.5 cm) and waters off Tianjin have risen 7.7 inches (20 cm) over the past three decades. This is being caused by global warming and the fact that aquifers are being depleted causing the land to sink. Salt water is leaching into Shanghai's freshwater aquifer.

Deforestation and desertification are also serious problems. Throughout history, China's forests have been decimated through human occupance and farming, the requirements for home construction, and the need for domestic fuel. Forests were also ravaged to make industrial charcoal and to manufacture ships. During the last three decades alone, a quarter of China's forests have vanished. Soil erosion and desertification are direct consequences of deforestation. Desertification and land degradation are wiping out nearly a million acres of grassland a year—an area almost as large as Rhode Island.

One of the most severely eroded regions is the Loess Plateau of north central China, which is the most extensive loess region on the face of the Earth. Wind and water erosion have taken their toll over the centuries but human activities aggravate the situation. Grazing of sheep and goats on semiarid steppe lands removes the already sparse vegetative cover. Wind and water then erode and dissect the region, and thousands of tons of loess blow or flow away.

Unwise use of water for poorly planned or inadequate irrigation schemes also contributes to land degradation. This will probably get worse as most global-warming models predict increasing aridity for countries in China's latitudes.

Efforts have been made to reforest areas, erect wind breaks, and lock the sand in place with layers of straw or clay basins. If successful in soil retention and irrigation, these areas can be reclaimed for such crops as apples and walnuts. Aerial seeding of grasses such as alfalfa (lucerne) has been employed to vegetate bare hills. But salinity is an ongoing hindrance as high potential evaporation rates cause salts to rise to the surface through capillary action. Then, flushing out the salts and special drainage are required, making the whole process very costly.

Sand storms are increasing in frequency. One storm in 2006 dumped 300,000 tons of sand on Beijing. In March 2010, a ferocious storm inundated the region with dust and sand. In Beijing, the sky turned magenta and buildings disappeared in the onslaught. Even Hong Kong—1,240 miles (2,000 ha) away—was enshrouded. People as far away as Taiwan wore masks. South Korea said that this was the worst "yellow dust" haze since 2006. These storms can be expected to increase in frequency and ferocity as desertification continues.

China has the world's largest hydroelectric power potential, yet it is developed to less than 10 percent. One massive project has come to world attention for its controversial impacts. This is the **Three Gorges Dam** on the Yangzi (see Figure 11-15 for its location).

The Chang Jiang (Long River) is the longest river in Asia and third longest in the world (Figure 11-25). Known as the Yangzi to foreigners, the Chinese use that name only for the last 300 or 400 miles (480 or 645 km) of its 3,937-mile (6,300 km) course. Yichang, 1,000 miles (1,600 km) from the sea, is the head of navigation for river steamers. Ocean vessels may navigate the river to Hankou, about 600 miles (1,000 km) from the sea. Between Yichang and Chongqing, at an altitude of 130 feet (40 m), are the spectacular Three Gorges, noted for their geological magnificence and cultural significance.

The river and its tributaries drain 650,000 square miles (1,683,500 km^2) of territory. The entire region is subject to floods. From the Han to the Qing Dynasties (206 BC to 1911), devastating floods occurred once every 10 years. During the last 300 years, dams have been breached more than 60 times and flooding continues to be a problem all along the river. About one-third of China's population lives along its banks.

Beginning in May 2010, China experienced its worst flooding in a decade. The floods and associated landslides affected some 28 provinces and regions throughout the country. More than 1,000 people lost their lives and 12 million had to be evacuated. Millions of acres of farmland were inundated or totally destroyed.

The Three Gorges Dam project is the world's largest water conservation project, including the world's most powerful hydroelectric dam. The normal water level of 574 feet (175 m) will be reached in 2009, with a reservoir covering 408 square miles (1,045 km^2). Its generating capacity will be 18.2 kilowatts with a yearly output capacity of 84.7 billion kilowatt hours. This is equivalent to burning 50 million tons of coal or 25 million tons of crude oil. Should water power actually replace mineral power, massive amounts of carbon dioxide, sulphur dioxide, and dust would be eliminated from the air. The hydroelectric power is intended for use in the energy-deficient eastern and central regions as well as in Sichuan Province.

Shipping is expected to improve markedly and affect Chongqing's status as a *zhixiashi*. In effect, the city would be tied to the Pacific Rim economy. However, its greater fortunes would appear to lie westward with the rapidly developing Sichuan Basin, a densely populated, resource rich, agricultural, and industrial region virtually isolated by topography from the rest of China. Chongqing might evolve as the most important growth pole in China's interior.

Between 1.2 and 1.9 million people have been forced to leave their homes, and some 620,000 acres (248,000 ha) of farmland have been lost. Approximately 8,000 cultural sites and some 1,000 industrial enterprises have been inundated, thereby contributing their pollution to the reservoir.

The Three Gorges Dam is extremely controversial. Forced removal of people is seen by many as a human rights issue. Although China claims that people are treated fairly with appropriate compensation and special privileges, many Chinese and others claim that resettlement is racked with corruption and unfairness. Compensation is inadequate, promised farmland is unproductive, and guaranteed jobs do not exist.

Scientists are also concerned. Some fear that the dam will ultimately collapse. In 1975, 62 dams in Henan Province gave way, killing anywhere from 86,000 to 230,000 individuals. Disease and famine affected over 10 million in the aftermath. A Three Gorges flood would be 40 times as great! Environmentalists claim that the Yangzi River ecosystem will be irreversibly upset. The Chinese alligator, the finless porpoise, and the Chinese sturgeon will be severely endangered or even wiped out. It is believed that the Yangzi white river dolphin is already extinct. Hydrologists postulate that sedimentation will cause sewage backups and floods upstream around Chongqing. Downstream, people worry about a lack of silt for agriculture.

Another aspect worth consideration is the fact that control of the dam by the northern power structure means control over water supply to the south. Water could be used as a weapon against any separatist movements in the south.

The debate over the potential benefits and detriments of the Three Gorges Dam project continues around the world. Both the United States and the World Bank elected to withdraw their support of the project.

A Spark of Light at Rizhao

Buildings in Rizhao, a coastal city of nearly 3 million on the Shandong Peninsula, are covered with solar panels—that is, heat-collectors. Houses use solar water heaters, and traffic lights are powered by solar voltaic cells. More than 60,000 greenhouses employ solar heat. Even the school is heated in winter—unusual in China. This achievement became possible because of dedicated city officials who cooperated with local solar industries in light of a supportive national government policy. Widespread use of solar energy has negated the need for much coal, with the result that Rizhao is among the top 10 of China's cleanest cities. Consequently, the city has drawn foreign investment, new residents, tourists, and university professors who build retirement homes.

Severe Acute Respiratory Syndrome (SARS)

SARS is a highly contagious pneumonia-like disease that infected more than 8,400 people and killed more than 800 worldwide between November 2002 and July 2003. China was identified as the source of the disease and topped the list of SARS hotspots with more than 5,300 cases and approximately 350 deaths during the nine-month period. After the World Health Organization (WHO) issued travel advisories and made concerted efforts to halt the spread, the global outbreak was finally contained in July 2003, though health experts caution that it may be a seasonal disease and can return in cold weather.

It is believed that SARS originated in Guangdong in southern China. Travelers quickly spread the disease to Hong Kong, Taiwan, Vietnam, Singapore, and as far as Canada, South Africa, Australia, and Brazil. Asia was hit the hardest, and Toronto—a city with a large Asian population and frequent travelers from and to Asia—had the most cases outside Asia. The scope and speed of the outbreak, unfortunately, is in large part explained by globalization. As China is increasingly connected to the rest of the world, its influence is no longer limited to global economic affairs but also includes health issues. Densely populated cities such as Hong Kong and Singapore are most vulnerable to the spread of infectious diseases. In Hong Kong, within just a matter of several days, more than 300 residents in one single apartment block were infected, probably via faulty sewage pipes.

The SARS outbreak has been both a challenge and an opportunity for the Chinese Communist Party (CCP). China was severely criticized for covering up the epidemic and underreporting its magnitude. International travel to China dropped sharply, which adversely affected tourism. It became clear that the public health-care system had major flaws. However, the CCP quickly admitted that mistakes had been made and vowed to implement measures to control the spread of the virus. SARS patients were isolated through mandatory quarantines, and schools and businesses were shut. The one-week May Day holiday was slashed to three days in order to discourage travel. The country's health minister and the mayor of Beijing were removed from their posts, which signaled that the Chinese government was serious about enforcing accountability and honest reporting. The CCP's quick response to criticisms suggests that we may well be observing the beginning of a more open society.

AVIAN FLU

Avian flu, more commonly known as "bird flu," is a particularly virulent form of the disease. Originating in Hong Kong in 1997, it attacks primarily waterfowl such as ducks as well as chickens. Cases have also been found among swine and even tigers.

Bird flu struck China in full force in 2005, wiping out thousands of migratory birds and chickens. It then spread to other places, particularly in Southeast Asia. Affected countries slaughtered millions of chickens to counter further diffusion of the disease.

The first human case of bird flu occurred in Vietnam. Since then, 247 people have died, 22 of them in China. The most recent related death in China was in 2009.

Our Western Storehouse: Xizang/Tibet

Tibet, according to geographer Antonia Hussey (1992), "is indeed a special place, at the top of the world, a landscape of light and shadow dancing over towering mountains, gravel plains and lush green valleys." A mysterious place, isolated by massive topographic barriers and profoundly influenced by Tibetan Buddhism, the region is centered on a previously forbidden city—Lhasa, the soul of the nation. But Professor Hussey declares that Tibet is no longer mysterious. "The distinctive landscape of this ancient theocratic state with its Buddhist traditions and institutions has been shattered and replaced with the utilitarian socialist landscape of China and renamed Xizang."

Back in the seventh century, Tibet was a great military power with borders extending far beyond its current ones. In fact, the Chinese paid a tribute of 100,000 rolls of silk to Tibet. By the tenth century Tibet had foresworn the arts of war and had withdrawn from its Chinese and Indian territories.

In 1207, Tibet fell under Mongol domination but maintained an amenable priest–patron relationship with the Mongol court, many of whom had adopted Tibetan Buddhism. During this period, the Dalai Lama system emerged. A specially chosen child would become the Dalai Lama, spiritual head of the Tibetan people. Great monasteries such as Drepung and Potala were built, eventually housing thousands of monks (Figure 11-20).

By the mid-seventeenth century, Mongol power weakened. In 1720, the Qing Dynasty sent a garrison of Chinese soldiers to Lhasa. The next year they proclaimed that Tibet had always been a vassal of China.

By the late nineteenth century, Tibet was caught up in an imperialist power struggle between the British and the Russians. British troops were placed in Tibet, and Tibet agreed to keep other powers out and trade with British India. When Chinese forces entered Tibet in 1910, the Tibetans asked the British for help and Britain refused. Meanwhile in 1911, corrupt China had fallen to the Nationalists. In 1912, the Dalai Lama proclaimed Tibet's independence.

The Tibet of the early twentieth century was dominated by powerful monasteries and feudal landlords. There was no secular education, and the masses lived in dire poverty. The Dalai Lama's attempts at reform were crushed by the traditionalists. The victory of the Communists in China in 1949 had little meaning to most ignorant and isolated Tibetans.

The story for China was different: in 1950, Radio Beijing announced, "The task of the People's Liberation Army for 1950 is to liberate Tibet." On October 7, 84,000 Chinese soldiers entered Tibet. India and Nepal expressed sympathy but did nothing, and Britain again refused to get involved. The 15-year-old Dalai Lama was forced to agree to return Tibet to the "Motherland." The resistance movement headed to the mountains to pursue guerrilla warfare.

Figure 11-20

This is the magnificent Potala Palace in Lhasa. Its 13 stories are filled with narrow corridors and more than 1,000 rooms filled with 200,000 images of Buddha and an array of Tantric gods. The odor of yak butter sculptures and candles permeates everything, and swarms of humanity cluster around the statues to make offerings and pray. The complex occupies 5 square miles (13 km^2) and stands 425 feet (130 m) above the Lhasa River valley. Photo courtesy of B. A. Weightman.

In 1959, Tibetan resistance exploded and the Dalai Lama went into exile in India. Chinese retaliation caused anywhere from 87,000 to 430,000 deaths. Torture and humiliation were ubiquitous. Peasants and nomads, forced into communes, were made to grow wheat instead of their traditional barley. Tens of thousands starved to death in the famines following the Great Leap Forward. Tibet was declared a Xizang Autonomous Region of the People's Republic of China in 1965.

The Cultural Revolution, led by the Red Guards, decimated Tibet. Thousands were killed or sent to labor camps. Monasteries and religious artifacts were destroyed. By 1978, the number of monasteries had plummeted from 2,700 to 8. Holy texts were turned into toilet paper. Only Mao's death in 1976 ended this hideous onslaught.

In the 1980s, the Chinese apologized and proceeded to rebuild monasteries and restore the cultural landscape. However, about 130,000 Tibetans lived in exile and the Dalai Lama, headquartered in Dharamsala, India, began to seek understanding and aid from abroad. Support for Tibet's independence has grown worldwide. However, in the absence of the Dalai Lama, socialist transformation had begun in earnest.

China now sees Tibet as a storehouse of natural resources and is determined to sinicize it. Coal, lithium, uranium, gold, chromium, and tin are only some of the minerals sought by the Chinese. The Qaidam Basin (once part of Tibet) is of special interest for oil.

Chinese settlers continue to stream into Tibet, a region of barely more than 2 million inhabitants. With travel enhanced by the new rail line (mentioned above) and an improved road from Kathmandu to Lhasa, this and other towns are becoming dominated by Han Chinese. Out of some 13,000 shopkeepers in Lhasa, for instance, only 300 are Tibetans.

China is forcing nomads to settle into permanent houses built especially for them. The government says that this will enable them to pool their resources and gain access to education and health care. While nomads are decrying their lack of choice, officials claim that controlling nomadic movements is essential for keeping "the upper hand in our struggle with the Dalai clique."

Tourism is being promoted, and huge hotels and related tourist attractions are being built. Much of traditional Lhasa has disappeared to the typical white tile and blue glass of Chinese new development. The caretaker monks of the Potala Palace have been replaced by Chinese-trained "guides," and the Drepang ("rice heap") monastery on the outskirts of Lhasa, once home to 10,000 monks, now has around 700. Most of the monastery lies in ruins.

Crackdowns on religion continue. As late as 1997, the Chinese authorities announced that religion would have to bow to communism. "Development" has meant that half of Tibet's forests have been felled and more than a quarter of its mineral resources have been extracted since 1959. A fiber-optic cable has been laid between Lhasa and Shigatse, and plans are underway for a rail link with Sichuan.

The year 2008 saw more violence that was triggered by monks from the Drepang monastery who marched in peaceful protest against Chinese oppression. Monks from other monasteries joined the protest and 50 monks were arrested. The march turned violent as mobs of Tibetans rioted and burned and looted Chinese businesses. Hundreds of monks, nuns, and civilians were arrested and many will be subject to torture that is widely used in Tibet. Chinese officials claim that these individuals are "splittists" (separatists) and blame the Dalai Lama for the uprising. Tourism has been curtailed and foreign as well as Hong Kong media have been denied access to Tibet.

It is clear that the Chinese are attempting to marginalize the Tibetans and make Tibet more livable for the Han. Tibet's new name is a metaphor for its relationship with China: Xizang—Our Western Storehouse.

Mongolia: In Range of the Dragon's Breath

Mongolia, a landlocked country of 2.7 million and one of the world's most sparsely populated countries, is three times the size of France and four times the size of California. Here in this remote and arid land, livestock outnumber people 12 to 1. Mongolia, lacking in infrastructure, modern industrial skills, and managerial knowledge, is struggling to develop its economy in light of its 1990 freedom from Soviet control. It was the first country in Asia to experience a bloodless transition from communism to democracy.

THE PHYSICAL LANDSCAPE

Mongolia's physical environment is characterized by extremes as well as fragility. Well-watered mountains in the west and northwest descend to steppe on the eastern plateau. Steppe lands graduate into the Gobi—semidesert and desert (*gobi* means "waterless place" in Mongolian). Rivers are concentrated in northern Mongolia but flow to the Arctic, the Pacific, or into landlocked basins. The Gobi has no perennial rivers, therefore underground wells are important. However, most of these freeze in winter. Soviet-installed pumps have fallen apart, and spare parts are not available. Getting water is problematic in the arid regions. At other times, dry river beds are subject to flash floods.

During the Mesozoic Period, much of the Gobi was swamp and favored by dinosaurs. In the early 1920s, the world's first known dinosaur egg was discovered here. In 1993, a 75-million-year-old fossilized embryo of a meat-eating dinosaur was found in the western Gobi, a region rich in vertebrate fossils.

Soil erosion is common throughout Mongolia. In many areas, soils are only a few inches thick and overlie granite. Dessicating winds easily whip up the earth, thereby filling the air with dust and leaving the ground barren. Overgrazing contributes to the problem.

Mongolia has an extreme continental climate with sharp fluctuations in temperatures both seasonally and diurnally. Temperatures average 15°F (–9°C) in January and 64°F (18°C) in July. Winter is long, lasting from October to May. Ulaanbaator (Ulan Bator) is the coldest capital in the world, with winter temperatures sinking to as low as –50°F (–45°C). Precipitation is limited, varying between 10 and 20 inches (25–51 cm). The Chinese used to refer to Mongolia as *Han-hai*: "dry sea."

HISTORICAL BACKGROUND

Cultural Mongolia is much larger than indicated by current political boundaries. There are many Mongol tribes speaking various dialects, and several of these live outside Mongolia and China's Inner Mongolia. These various warring groups were unified under Genghis Khan in 1206.

Inner and Outer Mongolia are political designations dating from the Manchu Dynasty (1644–1911). In 1622, the tribes of South and East Mongolia allied with China. This is why China claims legitimacy over Nei Mongol. In 1689, the Treaty of Nerchinsk laid down the frontiers between Russia and China. At that time Mongolia was part of China. Two years later, the Mongols agreed to pay a yearly tribute to the emperor known as the "Nine Whites"—eight white horses and one white camel. Mongolia then became known as Outer Mongolia. These historic connections provide rationale for China's notions of hegemony over this region.

Outer Mongolia persisted until 1911, when it became an autonomous monarchy ruled by a religious leader, a Lama-King. However, it soon fell under the influence of Soviet communism and in 1921 was taken over by the Mongolian People's Party. With the help of the Russians, the People's Republic of Mongolia was proclaimed in 1924, and Mongolia became the world's second communist nation. Mongolia remained a Soviet satellite until the collapse of the U.S.S.R. in 1990.

SOVIET IMPACTS

The Soviets' chief aim was to transform pastoral Mongolia into a source of raw materials for Siberian industrial efforts. Concerted attempts were also made to settle both rural dwellers and urban workers out of their traditional *gers* (round felt or canvas dwellings) and into multistoried buildings associated with industrialization and regional development. By 1981, more than half of Mongolia's population was classified as urban as compared to less than 22 percent in 1956.

A major contribution to this rapid urbanization was the expansion of Ulaanbaatar with multistoried suburbs encroaching on *ger* communities. Second, administrative centers of Mongolia's 18 provinces acted as magnets for regional growth. Third, the establishment of new towns mediated the dispersal of industry.

Darkhan is an example of a new town that has burgeoned from a railway station of 2,000 people to an industrial center of some 75,000 by 2007. Located close to the Soviet border, Darkhan's thermal power (based on local coal), as well as its brick works, meat-packing plants, and other industries, easily served Soviet needs. Other centers were developed to exploit and process new resources such as molybdenum and copper. Some administrative centers added manufacturing functions.

Mongolia got its first motor road in 1937 from Ulaanbaatar to Altanbulag on the Soviet Frontier. The Ulan Bator Industrial Combine was also developed in the 1930s, producing carpets, wool garments, felt, and leather goods.

To further effect control, the Communists established state farms and required the permanent settlement of the nomadic herders. Needs were met and subsidized by the state. When the subsidies ended after 1990, the herders returned to their nomadic lifestyle, moving frequently across the steppe in summer and clustering in more sheltered valleys in winter.

Another Soviet goal was to transform Buddhist Mongolia into a secular state. Most of the country's 700 temples were destroyed in the 1930s, and more than 100,000 lamas dispersed or killed. Five hundred truckloads of religious art treasures disappeared into the Soviet Union. Since 1990, monasteries are being rebuilt, and the country's Tibetan-style Buddhism is experiencing a revival.

MODERN MONGOLIA

Mongolia is undergoing its own version of perestroika—*shinechiel,* meaning a "renewal." A free-market economy has replaced state-controlled enterprises. People are studying English as well as their traditional Uyghur-Turkic script, which had been replaced by the Russian Cyrillic alphabet. Trade, once as high as 90 percent with the Soviet Union, has dropped to 50 percent and has increased with China, South Korea, and Japan. Foreign aid is pouring into the country, with Japan as the largest aid donor.

Not all aid projects are successful. One cross-breeding program resulted in Mongol-European cattle that died in the winter cold. A prefabricated mini-mill for steel production was brought from Japan. It was to manufacture steel from scrap metal. However, no one realized that Mongolia, with few cars or machines, possessed virtually no metal. Japan has also introduced windmill, battery, and TV assemblies. People once isolated in their round, felt *gers* on the steppe now can choose from 52 channels, including the Cartoon Network and MTV. What impact do you think this will have on Mongolia's development?

Even with the exploitation of mineral resources such as coal, oil, molybdenum, copper, and uranium, animals and animal products account for half of industrial output and 90 percent of exports. Material infrastructure is either lacking or obsolete, and few Mongols are educated in modern business and industrial management.

A third of Mongols are livestock herders practicing transhumance and living either in their large, permanent winter *gers* or in their smaller, portable summer *gers* (Figure 11-21). Because animal pastures are sparsely covered with low-yield grasses, families must move around 30 times a year, covering distances of 125 to 185 miles (200–300 km).

One of the most important animal products is cashmere from cashmere goats. Until recently, China was buying over half of Mongolia's raw wool, to the detriment of Mongolia's processing industry. Consequently, the Mongolian government banned the sale of raw wool. It is now smuggled into China, whose cashmere products challenge those made in Mongolia.

Development, in the context of transition from rigid control under communism to freedoms presented by democracy, has its downside. Mongolia is short of food, and there are thousands of destitute people, mainly women and children. As of 2004, 36 percent of Mongols live below the official poverty line. In addition, literacy has fallen dramatically and school dropouts have increased. Maternal mortality has doubled. Although Mongolia's medium- to long-term economic prospects are good, social reforms are sorely needed.

Relationships with China are difficult because Mongols do not wish to be dominated by China, as is the case in Nei Mongol where millions of Chinese have settled. China is anxious to intensify relations, however, as it needs lumber, minerals, and animal products as well as markets for its farm produce and low-end manufactures.

Mongolia's trade interactions are strongly tied to historic relationships and geographic proximity. Mongolia ships 75 percent of its exports to China and China supplies 28 percent of its imports. It gets 90 percent of its oil from and exports 38 percent of its products to Russia. In order to avoid falling under the sway of either country, Mongolia is pursuing what it calls a "third neighbor"

Figure 11-21

Mongolian herders live in gers. Gers, like yurts, have a collapsible wood frame and are covered with felt and canvas. A cooking and heating fire occupies the center and there is a hole in the roof for the smoke to escape. Many gers now have a battery-run light bulb of 20 or 40 watts. Photograph courtesy of B. A. Weightman.

policy. This involves remaining on good terms with its giant neighbors but also reaching out to countries such as America and Japan.

Now that Mongolia is out from under Russian control, China is exerting its authority. For example, China has loudly protested visits of the Dalai Lama to Mongolia and has discouraged connections with the Mongols of Nei Mongol. For decades, Mongolia has served as a buffer state between China and the Soviet Union with its strongest links to the U.S.S.R. Today, Mongolia is a buffer state between China and Russia with increasing ties—wanted and unwanted—to China.

Recommended Web Sites

http://afe.easia.columbia.edu/china/geog/maps.htm
Site designed by Dr. Ronald Knapp, Professor of Geography at State University of New York (SUNY). Interactive maps, photos, and information on all aspects of geography of China. Excellent site.

www.asiatradehub.com/china/ports.asp
List of China's ports. Click on port name for details on infrastructure, industry, energy, etc.

www.chinabusinessreview.com/
Magazine of the U.S.-China Business Council.

www.chinadaily.com.cn/english/
The China Daily newspaper in English.

www.china.org.cn
China—questions and answers from the People's Republic.

www.chinapage.com/map.html#satellite
Extensive collection of maps of China and surrounding countries.

www.cia.gov/library/publications/the-world-factbook/geos/mg.html
Information on geography, people, government, and economy of Mongolia.

www.economywatch.com/world_economy/china
Economic profiles, structure, industry, finance, imports/exports by country.

http://factsandetails.com/china.php?itemid=394&catid=10&subcatid=66
Information and photos about environmental problems in China.

www.fao.org/sd/LTdirect/LTanoo31.htm
Good account of the household responsibility system from the U.N. Food and Agriculture Organization.

www.freetibet.org/
Information about Tibet from Tibetan exiles.

www.historyforkids.org/learn/china/
Variety of topics and photos. Excellent site for teachers.

www.internationalrivers.org/en/node/4335
China's dam-building projects.

http://news.xinhuanet.com/english/
PRC news agency.

www.panda.org/about_our_earth/about_freshwater/rivers/
World Wildlife Fund site on natural environments, biomes, habitats, animals, rivers, etc.

www.time.com/time/asia
Time Magazine Asia site.

www.yearbook.gov.hk/2006/en
Information on the history, government, and economy of Hong Kong.

Bibliography Chapter 11: China: Great Dragon Rising

Alden, Curtis. 2007. *China in Africa*. London: Zed Press.

Allen, Thomas B. 1996. "Xinjiang." *National Geographic* 189: 2–43.

Baird, Vanessa, ed. 1995. "Tibet A Cause For Courage." *New Internationalist*, No. 274.

Bandurski, David. 2007. "Pulling the Strings of China's Internet." *Far Eastern Economic Review* 170/10: 18–21.

Beachey, P. 1995. "China's Superdam: The Three Gorges Project." *Environment and Politics* 4/2: 333–336.

Benewick, Robert, and Stephanie Hemelryk Donald. 2009. *The State of China Atlas*. Berkeley: University of California Press.

Benson, Linda, and Ingvar Svanberg. 1998. *China's Last Nomads: The History and Culture of China's Kazaks*. Armonk: M. E. Sharpe.

Blaikie, Piers M., and Joshua S. S. Muldavin. 2004. "Upstream, Downstream, China, India: The Politics of Environment in the Himalayan Regions." *Annals of the American Association of Geographers* 94/3:520–548.

Broadman, Harry G. 2007. *Africa's Silk Road: China's and India's New Economic Frontier.* Washington, D.C.: World Bank.

Brunn, Ole, and Ole Odgaard, eds. 1996. *Mongolia in Transition: Old Patterns, New Challenges*. Richmond, Surrey: Curzon.

Callerman, Thomas E. and Linda G. Sprague. 2007. "All Roads Lead to Beijing." *Far Eastern Economic Review* 170/5: 57–59.

Cannon, Terry, ed. 2000. *China's Economic Growth: The Impact on Regions, Migration and Environment*. New York: St. Martin's Press.

Cannon, Terry, and Alan Jenkins, eds. 1990. *The Geography of Contemporary China: The Impact of Deng Xiaoping's Decade*. New York: Routledge.

Casadei, Maria. 2009. "Remember Kashgar." *Far Eastern Economic Review* (May): 77–79.

Chang, Kuei-sheng. 1961. "The Changing Railroad Pattern in Mainland China." *The Geographical Review* 51: 534–548.

Dai, Q. 1998. *The River Dragon Has Come: The Three Gorges Dam and the Fate of China's Yangtze River and Its People*. Armonk, N.Y.: M. E. Sharpe.

Demick, Barbara, and David Pierson. 2010. "Shortage of Cheap Labor? In China?" *Los Angeles Times*, A8–9.

Edelmann, Frederic, ed. 2008. *In the Chinese City: Perspectives on the Transmutations of an Empire*. Barcelona: ACTAR. Palais de Chaillot, the Cite' de L'architecture & du Patrimonie, and Barcelona Centre for Contemporary Culture.

Fan, C. Cindy. 1996. "Economic Opportunities and Internal Migration: A Case Study of Guangdong Province, China." *Professional Geographer* 48: 28–45.

Fan, C. Cindy. 1995. "Of Belts and Ladders: State Policy and Uneven Regional Development in Post-Mao China." *Annals of the Association of American Geographers* 85: 421–449.

Fan, C. Cindy, and Youqin Huang. 1998. "Waves of Rural Brides: Female Marriage Migration in China." *Annals of the Association of American Geographers* 88: 227–251.

Fan, C. Cindy. 2005. "Interprovincial Migration, Populaton Redistribution, and Regional Development in China: 1990 and 2000 Census." *The Professional Geographer* 57/2: 295–311.

Fang, Cai. 2007. "Pay-Back Time for China's One-Child Policy." *Far Eastern Economic Review* 170/4: 58-61.

Fishman, Ted C. 2006. *China Inc*. New York: Scribner.

Fullen, Michael, and David Mitchell. 1991. "Taming the Shamo Dragon." *Geographical Magazine* 63: 26–29.

Gately, David. 1997. "Growth in China's Grain Imports an Opportunity for Exporting Countries, Not a Threat to World Food Supply." *International Food Policy Research Institute News Release*. Washington, D.C.: IFPRI.

Harney, Alexandra. 2008. "Bye Bye Cheap Labor: Guangdong Exodus." *Far Eastern Economic Review* 171/2: 29–35.

He, Bochuan. 1991. *China on the Edge: The Crisis of Ecology and Development*. San Francisco: China Books.

Hsieh, Chao-min, and Max Lu., eds. 2004. *Changing China: A Geographical Appraisal*. Boulder: Westview Press.

Hsu, Immanuel, C. 1995. The Rise of Modern China, 5th edition. New York: Oxford.

Hsu, Mei-ling. 1994. "The Expansion of the Chinese Urban System, 1953–1990." *Urban Geography* 15: 514–536.

Jacka, Tamara. 1997. *Women's Work in Rural China: Change and Continuity in an Era of Reform*. New York: Cambridge.

Johns, Chris, ed. 2008. "China: Inside the Dragon." *National Geographic*, Washington D.C.: National Geographic Society.

Kuhn, Anthony. 2004. "The Death of Growth at Any Cost." *Far Eastern Economic Review* 1 (April): 28–30.

Kung, James K., and Yi-min Lin. 2007. "The Decline of Township and Village Enterprises in China's Economic Transition." *World Development* 35/4: 569–584.

Kurlantkitz, Joshua. 2006. "China's Charm Offensive in Southeast Asia." *Current History* 105/692: 270–276.

Lampton, David M. 2008. *The Three Faces of Chinese Poser: Might, Money, and Minds*. Berkeley: University of California Press.

Larmer, Brook, 2006. "Manchurian Mandate." *National Geographic* (September): 42–73.

Leppman, Elizabeth J. 2002. "Breakfast in Liaoning Province, China." *Journal of Cultural Geography* 20/1: 77–90.

Leung, Beatrice, and Joseph Cheng, eds. 1997. *Hong Kong SAR: In Pursuit of Domestic and International Order.* Hong Kong: Chinese University Press.

Li, Dun J. 1978. Modern China: From Mandarin to Commissar. New York: Scribner's.

Li, Lianjiang. 2006. "Driven to Protest: China's Rural Unrest." *Current History* (September): 250–254.

Loo, Becky P.Y., and Kai Liu. 2005. "A Geographical Analysis of Potential Railway Load Centers in China." *The Professional Geographer* 57/4: 558–565.

McDonoch, Gary, and Cindy Wong. 2005. *Global Hong Kong*. New York: Routledge.

Magnier, Mark. 2006. "Farmers in China Face Great Wall." *Los Angeles Times*, April 19: A1 and A22.

Mozur, Paul. 2007. "Retaining the Loyaty of Xinjiang's Hans." *Far Eastern Economic Review* 170/10: 27–30.

Ngai, Ppun. 2005. *Made in China: Women Factory Workers in a Global Workplace*. Durham, N.C.: Duke University Press.

Pannell, Clifton, W. 1988. "Regional Shifts in China's Industrial Output." *The Professional Geographer* 40: 19–32.

Pearce, Fred. 1995. "The Biggest Dam in the World." *New Scientist* 28 (February): 25–29.

Percival, Bronson. 2007. The Dragon Looks South: China and Southeast Asia in the New Century. Westport, Conn.: Praeger.

Perkowski, Jack. 2006. "The Coming China Car Boom." *Far Eastern Economic Review* 169/3: 23–26.

Perry, Alex. 2007. "Africa's Oil Dreams." *Time*, June 11: 22–30.

Postel, Sarah. 2007. "China's Unquenchable Thirst." *World Watch*, November/December: 20–21.

Postiglione, Gerard A., ed. 2006. *Education and Social Change in China*. Armonk, N.Y.: M. E. Sharpe.

Restall, Hugo. 2009. "China Enters a Period of Eruptions." *Far Eastern Economic Review* (July/August): 46–8.

Riley, Nancy. "China's Population: New Trends and Challenges." *Population Reference Bureau (Population Bulletin)*, 59/2.

Robertson, Benjamin. 2005. "China: Selling Out the Family Farm." *Far Eastern Economic Review* 168/9: 48–51.

Rotberg, Robert I., ed. 2008. China into Africa: Trade, Aid and Influence. Washington D.C.: Brookings Institute.

Sautman and June Teufal Dreyer, eds. 2006. *Contemporary Tibet: Politics, Development, and Society in a Disputed Region*. Armonk, N.Y.: M. E. Sharpe.

Schmalzer, Sigrid. 2002. "Breeding a Better China: Pigs, Practices and Place in a Chinese County, 1929–1937." *The Geographical Review* 92/1: 1–22.

Smil, Vaclav. 1993. *China's Environmental Crisis: An Inquiry into the Limits of National Development*. Armonk, N.Y.: M. E. Sharpe.

Smil, Vaclav. 2007. "Poor Visibility on China's Air Pollution." *Far Eastern Economic Review* 170/10: 35–39.

Smith, Christopher J. 2000. *China in the Post-Utopian Age*. Boulder, Colo.: Westview.

Starr, S. Frederick. 2007. *Xinjiang: China's Muslim Borderland*, Armonk, N.Y.: M. E. Sharpe

Stevenson-Yang, Anne. "China's Online Mobs: The New Red Guard?" *Far Eastern Economic Review* 169/8: 53–57.

Sun, Jian. 2006. "China: The Next Global Auto Power?" *Far Eastern Economic Review* 169/2: 37–41.

The Economist. 2006. "Battle for Mongolia's Soul.' *The Economist*, December 23: 94–96.

The Economist. 2007. "Missing the Barefoot Doctors." *The Economist*, October 13: 27–29.

The Economist. 2008. "Rushing On by Road, Rail and Air." *The Economist*, February 16: 30–32.

Unger, Jonathan. 2002. *The Transformation of Rural China*. New York: M. E. Sharpe.

Veeck, Gregory. 2005. "From Communism to the World Trade Organization: The Remarkable Transformation of China's Agricultural Sector." *Focus on Geography* 48/3:1–10.

Vriens, Hans. 1997. "Lhasa Lost." *Far Eastern Economic Review* 29 (May): 44–45.

Wang, Dong. 2007. "Prospects of Chinese Beef Production and the Role of Dairy Beef." *Asia Today*, September 26.

Wu, W. 2004. "Cultural Strategies in Shanghai: Regenerating Cosmopolitanism in an Era of Globalization." *Progress in Planning* 61: 159–180.

Yeh, Anthony Gar-on, and Xueqiang Xu. 1996. "Globalization and the Urban System in China." In *Emerging World Cities in Pacific Asia*, eds. Fu-chen Lo and Yue-man Yeung, pp. 219–265. New York: United Nations.

Zich, Arthur. 1997. "China's Three Gorges." *National Geographic* 192: 2–33.

Chapter 12

Japan: Century 21

"Our greatest glory is not in never falling but in rising every time we fall."

CONFUCIUS (CA 551–479 BC)

Rich, highly urbanized, densely populated, and one of the world's largest users of natural resources, Japan is Asia's leading tiger. The Tokyo–Yokohama–Kawasaki urban area is the largest metropolitan region on Earth. Japan consumes the products of more tropical forests than any other nation, a reminder that Japan is virtually devoid of natural resources to exploit.

Dynamic Archipelago

Having achieved full status as an economic tiger, Japan is reckoning with problems facing all developed economies in the twenty-first century. It has been eclipsed by China for its position as the world's second largest economy after the United States. How did Japan reach its pinnacle of economic development? What has gone awry, and how is this affecting economic structure and spatial organization? And what social changes are occurring in the context of economic transformation?

Most Japanese live on 3 percent of the country's land area. Japan's physical landscape comprises mountains thrust up from the Pacific Ocean. Only 25 percent of the land area has level-to-moderate slopes, while 75 percent consists of hills and mountains too steep to be easily cultivated or settled. Consequently, the Japanese are densely crowded on very limited lowland.

Landforms show evidence of tectonic uplift with coastal terraces, and glaciation with ice-carved features in the Japanese Alps. In many places, the mountains descend directly into the sea. Plains are restricted to narrow river floodplains and deltas. The most extensive plains on the main island of Honshu are the Kanto Plain around Tokyo and Yokohama, the Nobi Plain around Nagoya, and the Yamato Plain around Osaka, Kyoto, and Kobe. Needless to say, these areas are very densely populated (Figure 12-1).

As a result of its midlatitude location and the fact that it is surrounded by water, Japan is very humid with a humid continental climate in the north and a humid subtropical climate in the south. Japan is also affected by the monsoon. In summer, warm monsoonal winds from the south bring heavy rainfall and, frequently, typhoons. In winter, cold winds from the northwest carry heavy snows to the Sea of Japan coast and the north. An important dividing line is the 37° line of latitude, which marks the northern limit of double-cropping of rice.

Four main islands dominate the Japanese archipelago: Hokkaido, Honshu, Kyushu, and Shikoku. These and hundreds of smaller islands lie at the junction of three tectonic plates: the Pacific, Philippine, and Eurasian (Figure 11-2). The Pacific Plate dips between the Eurasian and Philippine plates in the Japan trench north of Tokyo. The Philippine Plate is being consumed by the Eurasian Plate in the Nankai trench south of Kyushu and Shikoku. The Pacific Plate drops at a rate of 2.9 inches (7.5 cm) a year.

Japan is the most seismically active country in the world. Such instability results in many earthquakes. In Tokyo, there is an average of three noticeable quakes every month. In 1923 an earthquake registering 8.3 on the Richter scale devastated Tokyo and the Kanto Plain around it. Cooking stoves set fire to the mostly wooden and paper structures. The quake, the fires, and a 36-foot (10.8 m) tsunami claimed 140,000 lives. This was known as the *Great Kanto Quake.*

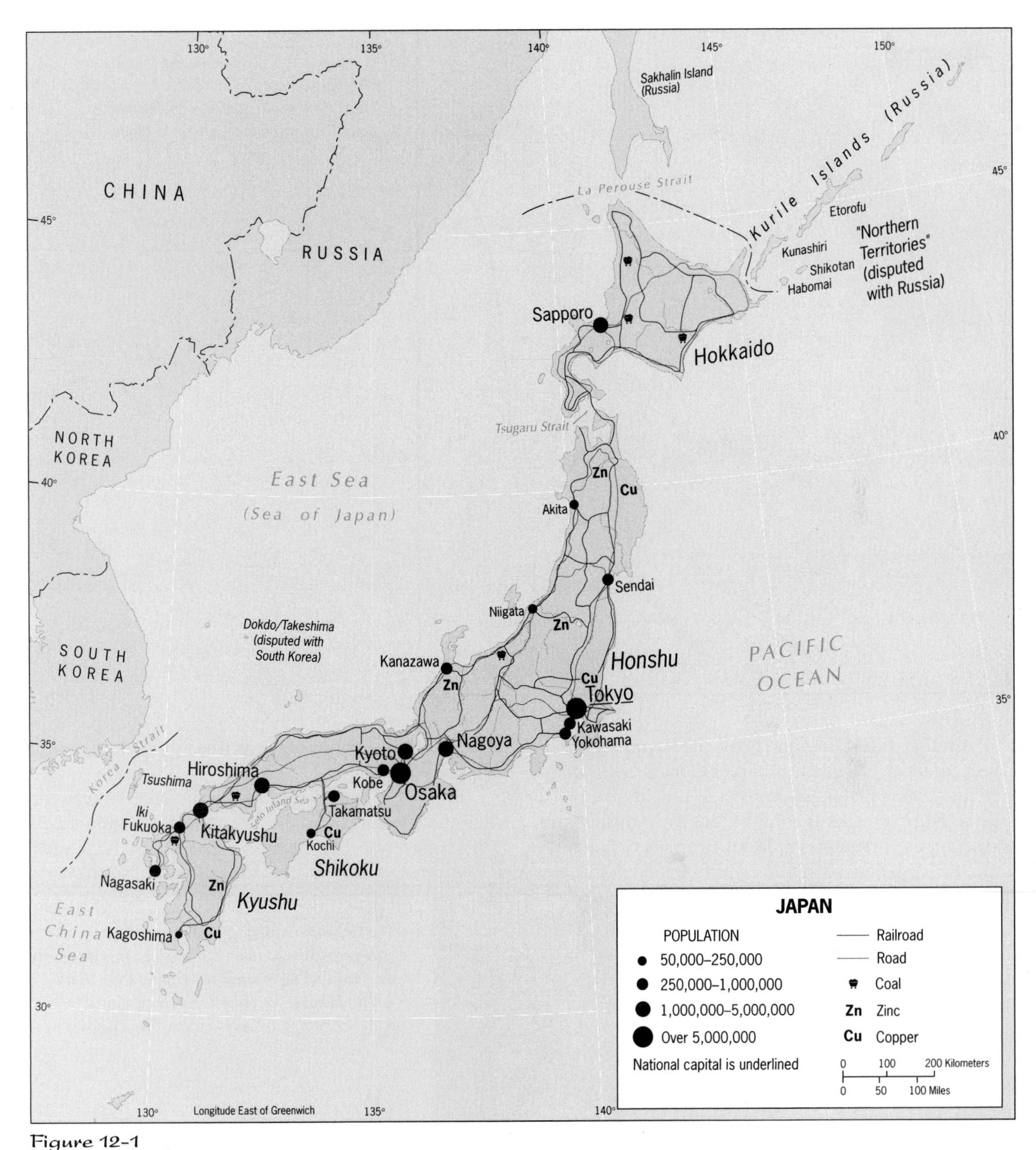

Figure 12-1

Major islands and cities of Japan. From H. J. de Blij and P. O. Muller, *Geography: Realms, Regions and Concepts*, 14th edition, 2010, p. 509. Originally rendered in color. Copyright H. J. de Blij and P. O. Muller. Reprinted with permission of John Wiley & Sons, Inc.

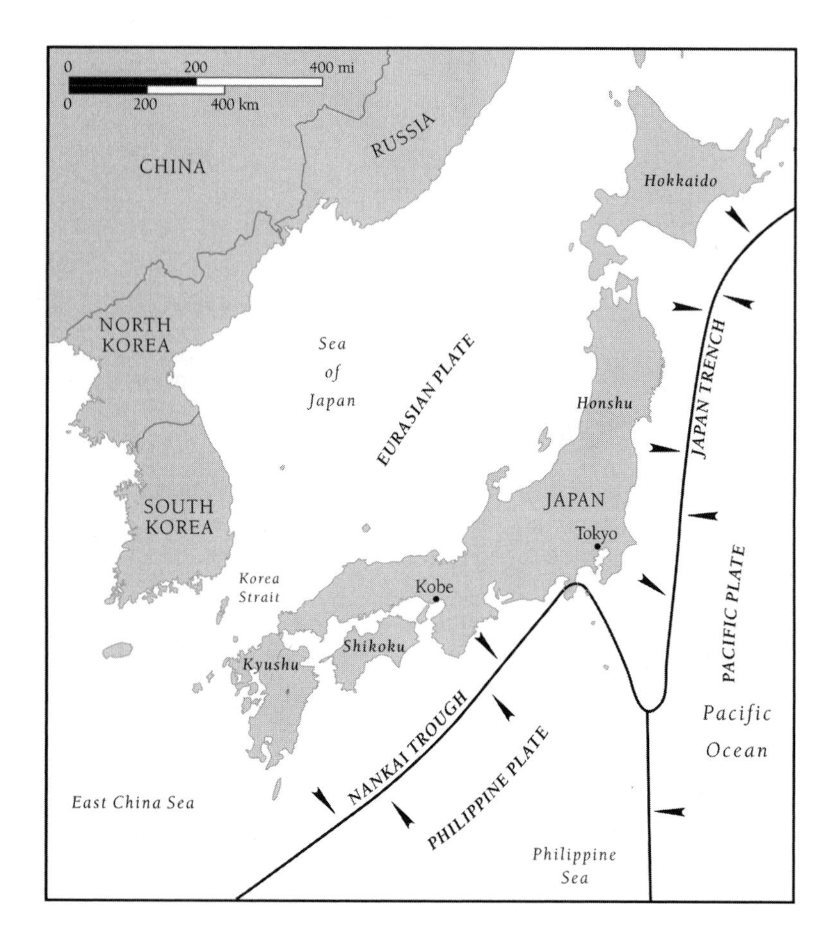

Figure 12-2

Earth in action. Note how the Pacific and Philippine plates push against the Eurasian Plate. This geological movement makes Japan one of the most unstable places on Earth. ©NG Maps.

Another severe tremor shook the Kanto region in 1983, resulting in much property damage. All Japanese know that major quakes strike roughly every 70 years. With acres of landfill to liquify, thousands of skyscrapers to crumble, miles of freeways to collapse, and miles of underground gas lines to rupture, Tokyo is a scary place.

In 1995, another ruinous earthquake struck: this time the epicenter was the city of Kobe (Figure 12-3). This quake was caused by a "strike-slip" (sideways)

Figure 12-3

The Hanshin Expressway fell over as part of the disastrous Kobe Quake. Fortunately, this was early morning and there was little traffic. Even in the midst of chaos, the orderly Japanese waited for lights to change and used crosswalks. Photo courtesy of Risa Palm.

Figure 12-4

Mount Fuji is a dormant volcano towering to 12,385 feet (3,776 m) and is one of Japan's more than 200 volcanoes. This mountain is sacred to the Japanese and is indicative of the country's mountainous terrain. © Royalty-Free/Corbis Images.

motion along a minor fault. Older, concrete buildings collapsed, while those of flexible construction or fitted with shock absorbers sustained less damage. Some buildings still had severe damage, however. Fires raged out of control for days; 5,000 people were killed, and 250,000 were made homeless.

Volcanic activity is another hazard in Japan, which boasts more than 200 volcanoes. Sixty of these have been active in historic time. Mount Fuji, a dormant volcano towering to 12,385 feet (3,776 m), is regarded as sacred by the Japanese (Figure 12-4). Japan is also the world's foremost hot springs country, with close to 20,000 springs and some 2,000 resort spas, many in Kyushu.

Historical Development

ORIGINS

Popular myth says the Japanese and the first emperor are direct descendants of the Sun Goddess Amaterasu. The first emperor took the throne in 660 BC, but recent archeological evidence tells a different story.

About 400 BC, people from the Korean peninsula migrated to the island of Kyushu. Bringing with them rice culture from China, their numbers rose about two percent a year. In a few hundred years, their culture overwhelmed that of the indigenous Jomon hunters and gatherers who had inhabited the region for at least 10,000 years. Using metal tools and the techniques of large-scale rice cultivation, the newcomers established a regional core on the Yamato Plain. Japanese culture diffused from the Kansai Plain northward throughout the islands. Even though this story is based on scientific evidence, it is difficult for many Japanese to accept the fact that they are related to Koreans, whom they long have considered inferior.

The language that originally diffused to Japan was not Korean as we know it today. It is thought that the Japanese language diverged from the Korean language at least 4,000 years ago, as they have few words in common. However, Japanese and Korean both employ common Chinese words.

Japanese Writing

Initially, the Japanese simply borrowed the Chinese character system and called it ***kanji***. However, profound differences between Chinese and Japanese grammar meant that sentences were constructed awkwardly. So the Japanese developed a syllabary known as ***hiragama*** that permitted the expression of words and parts of speech not easily represented by Chinese characters. While there are thousands of Chinese characters, there are only 51 *hiragama* symbols. Each represents a syllable or a combination of a consonant and vowel. A different system of sounds called *katakama* is used for words of foreign origin.

Figure 12-5

Signage in Japan reveals the country's numerous writing systems. Photo courtesy of B. A. Weightman.

> Unlike Chinese, Japanese is easily written in the Roman alphabet. The resultant *romanji* style of writing is very important in advertising and computer use. Today's technology allows the translation of *romanji* into the traditional Japanese meld of *kanji, hiragama,* and *katakama* (Figure 12-5).

OPENING CLOSED DOORS

Japan continued to be influenced by its Asian neighbors for centuries to come. However, as an island-nation it was isolated enough to also develop its own distinctive cultural patterns, including the unique Japanese language and many customs, art forms, and other traditions. Japanese people are justifiably proud of their distinctive culture, and continue to enjoy traditional foods, festivals, theater, music, and other activities, even as they lead modern lifestyles and follow fashions and current trends from countries around the world.

For much of Japanese history, the capital was Kyoto, the seat of imperial rule and repository for many aspects of traditional Japanese culture. It is a beautiful city with spectacular temples and gardens, as well as Nijo Castle and the historic Imperial Palace, all outstanding examples of traditional Japanese architecture and landscaping (Figure 12-6).

After a time, the emperor lost power to Yoritomo, the leader of one of Japan's many clans. Yoritomo was the

Figure 12-6

This is the Nijo castle in Kyoto. It is a landscape symbol of Kyoto's role as a castle town. Photo courtesy of B. A. Weightman. ©Werner Forman/Corbis.

first **shogun**, an all-powerful military leader who rewarded his followers with lands and castle towns from which they ruled the various provinces and often waged war on one another. Eventually, Japan was unified under the rule of the Tokugawa line of shoguns who governed Japan from 1603 to 1867 from a capital city of their own choosing, **Edo**. Edo grew quickly under the shoguns' influence to become one of the biggest cities in the world. It was renamed Tokyo when the Tokugawa shogunate ended and imperial rule was restored.

From 1603 to 1854, Japan was essentially closed to foreigners and Japanese themselves were not permitted to leave the country. This was the shoguns' response to the intrusion by Western powers that was changing other parts of Asia and that threatened Japan. However, some trading was permitted under close scrutiny in the southern port of Nagasaki, far from Japan's heartland. Then, in 1853, an event took place that would change Japan forever. This was the arrival in Edo (Tokyo) Bay of a squadron of armed American ships under the leadership of Commodore Matthew C. Perry. Because of superior military technology, Perry was able to engineer an unequal treaty with Japan that favored the intruders and opened Japan for trade to outsiders.

It became readily clear to the Japanese that they had fallen behind Western nations in technology during their time of isolation. The ensuing crisis hastened the demise of rule by the shoguns and brought back to power the nation's emperor, who in turn moved his seat of authority to then newly named Tokyo.

Britain, an island nation in the Atlantic that had acquired the largest colonial empire in the world, became a model for Japanese modernizers who, under the forward-thinking Meiji emperor and his advisors, led the transformation of Japan. Confucian values were called upon to educate the Japanese and move the nation into the twentieth century. Foreign teachers of everything from languages to technology were imported in great numbers. Japanese also began to study abroad. Japanese industry developed especially quickly, with huge military-industrial conglomerates called ***zaibatsu*** rising to great power. This time of fast-paced, catch-up modernization is referred to as the Meiji Period (1868–1912).

Shintoism

While most Japanese are Buddhist, many follow the ancient religion of **Shinto** for certain aspects of their lives. However, most Japanese combine elements of Shinto with Buddhist practices. A person might have a Shinto wedding and a Buddhist funeral, for example. Shinto began as the worship of spirits in nature and evolved into a belief system focused on the relationship between natural harmony and human existence. In the late 1800s, the Japanese government elevated Shinto into a nationalistic cult that worshipped the emperor as divine. The more salient nationalistic elements of the religion were expunged after World War II.

Shinto is centered on places and nature. Certain mountains, such as Mount Fuji, are considered sacred and are climbed by thousands of people. Major Shinto shrines attract large numbers of pilgrims. The most notable of these is the Ise Shrine south of Nagoya. It represents the cult of the emperor.

EXPANSIONISM AND EMPIRE

Japan's industrialization was accompanied by an expansion of its military might and acquisition of territory (Chapter 1). Flexing its new muscles, Japan launched wars with China in 1894–1895 and with Russia in 1904–1905. In 1895, it annexed Taiwan. In 1920, Japan annexed the whole of Korea, which it ruled until 1945. Manchuria became another Japanese colony in 1931. Such expansion was Japan's response to the colonial ambitions of Europeans and Americans in Asia, and reflected Japan's vision of a Greater East Asia Co-Prosperity Sphere (GEACPS) in which "Asia would be for Asians." Yet Japanese rule was often harsh and oppressive, and was opposed by rebellious patriots in the occupied lands.

Many of Japan's Asian neighbors still harbor resentments against Japan for its time of colonialism, citing specific atrocities that the country has yet to officially acknowledge. The most famous of these is the December 1937 Nanking (Nanjing) Massacre, named for a Chinese city in which up to 300,000 Chinese civilians were killed and 20,000 women raped by Japanese troops. For Koreans, a key unresolved issue is the conscription of thousands of women as sex slaves for Japanese soldiers during the occupation of their country.

Japan's expansionism was brought to a halt in 1941 after its attack on the American naval base of Pearl Harbor in Hawaii. Japan justified the attack by claiming that Americans had interfered with its oil supply lines in Asia. The United States entered World War II and began to push back against Japanese control in Asia. Although Japanese troops managed to advance across a wide area of China and Southeast Asia and the South Pacific, they eventually lost ground in fierce fighting against the Americans and their allies. Japan came within range of allied bombers (based in China and on Pacific islands), which destroyed Tokyo and almost all other Japanese cities. The war in the Pacific was brought to an abrupt end after the Americans dropped atomic bombs on the cities of Hiroshima and Nagasaki in 1945.

PHOENIX FROM THE ASHES

Defeat in the war meant the end of GEACPS but a new beginning for Japan. The country was almost completely devastated by the bombing, and its people were literally starving in the immediate postwar period. Yet Japan's recovery in the ensuing years was the most rapid of any nation in history. Relying on the Confucian values of hard work, loyalty, and cooperative efforts and with American assistance, infrastructure was rebuilt. A democratic form of government was put into place, the *zaibatsu* were broken up and divorced from a reduced military, and land reform gave land to the war-ravaged peasants.

Grounded in a fresh technological base, the government, financial institutions, and industry banded together to build the new Japan. Another form of industrial conglomerate emerged: ***keiretsu***.

Keiretsu

Keiretsu means something close to "link" or "affiliated with." A *keiretsu* is a hierarchical organization of companies. There are "horizontal *keiretsu*" and "vertical *keiretsu*."

Horizontal *keiretsu* revolve around a financial core including a major bank, a giant trading company, and a huge manufacturer. For instance, there are Mitsubishi Bank, Mitsubishi Corporation, and Mitsubishi Heavy Industries. There are also about 30 companies and more than 200 affiliates (10 percent equity ownership) that are part of this horizontal *keiretsu*.

Many vertical *keiretsu* lie within the borders of the major horizontal *keiretsu* such as Mitsubishi, Sumitomo, or Mitsui. Vertical organizations are either in production or distribution. A collection of firms assembles parts for a manufacturer who directs its end product through a distribution network of wholesalers and retailers. Most production *keiretsu* concern automobiles and electronics. Most distribution *keiretsu* are involved in consumer electronics.

A member, but very independent, of the Mitsui Group is the Toyota Motor Group. Toyota was founded in 1894 by the Toyoda clan that had its start in the manufacture of wooden spinning machines. Toyota, with its automobiles, real estate firms, insurance companies, and other interests, is becoming a combination vertical-horizontal *keiretsu*.

A LEGACY OF TENSION

More than half a century after Japan's defeat in World War II, there are still tensions between Japan and the neighboring countries that once comprised Japan's empire. Survivors and their descendants in China and Korea have little inclination to forgive the Japanese for their harsh and oppressive rule. Moreover, the Japanese have yet to atone for their atrocities. Lingering resentments damage international relations to this day.

One example of the bad relations concerns a recent episode of sex tourism by Japanese men to the southern Chinese city of Zhuhai, one of that country's fast-growing SEZs. Such cities attract poor migrants from the Chinese countryside in search of a livelihood and have a flourishing underside of prostitution and other illicit activities. A package tour group of some 400 Japanese men, all employees of an Osaka-based construction company, was entertained there for three days by 500 local prostitutes.

Even though it is not unusual for Japanese males to travel in groups to China and other Asian countries for packaged sex, this particular orgy happened to coincide with an anniversary of Japan's pre-World War II invasion of China. The incident raised particularly strong indignation in the Chinese media, which described the event as an affront to Chinese dignity. It then became the center of a diplomatic flap between the two nations, causing Japanese officials to issue a public apology.

Japan's relations with neighboring Russia are also strained by historical tensions. The two countries could be enormous trading partners, given the proximity and richness of Siberia's natural resources. However, full exchange is held back by a territorial dispute that dates back to the end of World War II. At that time Russia snapped up a chain of small islands northeast of

Hokkaido that had been Japanese territory. Japan claims that these "Northern Territories" are illegally occupied and insists on repatriation as a condition for full economic and diplomatic relations. This issue is a source of frequent protest by Japanese right-wing political groups who hold noisy demonstrations almost daily near the Russian embassy in Tokyo. Their favorite tactic is the use of sound trucks to blast ear-splitting demands that Russia give back the disputed islands.

Population and Society

A comparison of the population pyramid in Chapter 3 (Figure 3-5) with the one below (Figure 12-7) makes it clear that Japan's population of 127.6 million is expected to decline during the twenty-first century. In fact, demographers predict that the country will have 25 percent fewer people by 2020!

Japan has an "old" population. The average life expectancy for women is 86 years, for men 79 years. For the first time in history, Japan has more people over 65 (23 percent) than it does children under age 15 (13 percent). By 2035, the government estimates that more than 30 percent of Japanese will be 65 or older. Japan is preparing to launch a public nursing insurance program, and several companies, such as Sanyo and Mitsubishi, are jumping on the bandwagon to produce senior-oriented products and services.

Young Japanese are putting off marriage and having fewer children than in the past. The average age of marriage has risen to 26.3 for women and 28.5 for men. Furthermore, many women are seeking careers and are less likely to have or even want children.

Women complain about the high cost of education and the dearth of affordable housing. They spurn the physical and emotional burden of raising children in their husband's absence. Japanese men work from dawn until late at night only to come staggering home from the last subway train at midnight. Some call them "7-11 husbands."

In 2008, there were 9.0 births per 1,000 people, compared with 28.1 births per 1,000 people in 1950. The average Japanese mother has about 1.4 children, well below the 2.1 replacement level. Falling birth rates have resulted in decreasing enrolments in elementary and secondary schools and many of them have been forced to close.

Changing population dynamics have serious ramifications throughout both economy and society. The working age group began to decline in 1997. The Japanese government is concerned about its tax base and available work force. It must also face the fact that family structure is changing from extended to nuclear. In 1960, 87 percent of Japanese over 65 lived with adult children. By 1990 this number had dropped to 60 percent and now

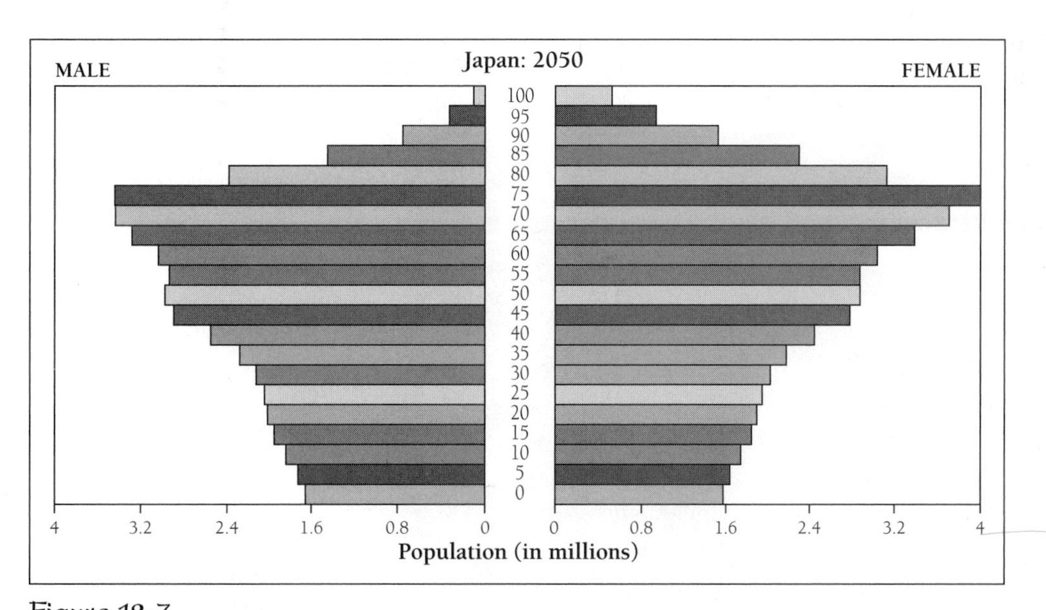

Figure 12-7

Japan's population structure in 2050. Notice how the base of the pyramid gets increasingly smaller. If this trend continues, what will happen to the Japanese population?

Source: U.S. Census Bureau, International Data Base.

(2010) it is only 42 percent. Even so, Japan is still a country where the elderly are revered and taken care of by their family. Moreover, the government must cater to older people not only out of this inherent respect for the elderly but also because they are the most likely to vote.

Out of its concern, Japan's government has established a program of subsidies for each preschool child along with an expanding network of day-care centers that are opened until late at night. Local governments have their own programs. For example, one prefecture in rural western Japan has launched "Operation Stork." This promotes such things as TV commercials and post cards that state: "Get a brother or sister for your child."

The majority of public surveys show that such efforts are having little effect on childbearing patterns. Most women simply do not want to get stuck raising a child by themselves. Many believe that Japanese husbands work long hours and come home late not simply because of work pressures, but because they enjoy the company of their male colleagues after-hours more than what is called "home service."

Feminists in Japan say that deep-rooted male attitudes, steeped in cultural tradition, are changing but all too slowly. They view the low birth rate as a "silent resilience by women" to a male-dominated system.

THE DARK SIDE

A myth of Japan is that it is an ethnically and culturally homogeneous society. In fact, Japan has three classes of minorities: native ethnic groups such as the **Ainu**; foreign ethnic groups such as the Koreans and Chinese; and social outcasts known as ***Burakumin*** or ***Eta***. Ethnic and social minorities make up about 4 percent of the population, roughly 5 million people.

Native to Hokkaido, the Ainu are racially unrelated to the Japanese (Figure 12-8). It is thought that are related to peoples of eastern Russia. Less than 20,000 identify themselves as Ainu. Their culture has been undermined by assimilative pressure; they suffer discrimination in all walks of life, and they live in relative poverty. There are only 10 Ainu speakers left. Media portray them as adorable and folksy and as living close to nature.

Recent public awareness of the Ainu plight has not reduced discrimination, an issue most Japanese do not face. In 2008, Japan's legislature finally passed a resolution that recognizes the Ainu as a people in their own right. According to historian David Howell (1996), recognizing some aspects of Ainu identity may generate pride but this "also shields Japanese society at large from the need to re-examine its basic premises."

There are about 600,000 ethnic Koreans living in Japan, descended from the 2.5 million drafted as soldiers or forced laborers during Japan's 35 years of colonial rule in Korea. Most are classed as resident aliens; until the 1990s, even those Koreans born in Japan were routinely fingerprinted and denied citizenship. About 20,000 Koreans died in the bombing of Hiroshima, but the needs of bomb-diseased survivors are virtually ignored. Discrimination in housing, employment, and education is

Figure 12-8

Here, Ainu community members perform a spiritual ceremony called "Kamui-nomi" at the Charanke festival in Tokyo. ©Scilla Alecci/AFP/Getty Images, Inc.

rampant. Consequently, the majority of Japan's Koreans are trapped in poorly paid, unskilled jobs.

The Chinese number more than 655,000. About 20 percent of these people are descendants of residents who were in Japan before World War II. The rest are temporary residents, guest workers, and migrant laborers. In fact, China is becoming *the* major source for low-cost labor in Japan. Chinese are most visible in "Chinatowns," the largest of which is in Yokohama. While discriminated against, they fare better than the Koreans.

Another minority is the Okinawans from the Ryukyu Islands south of Kyushu. The Ryukyus were not incorporated into Japan until the seventeenth century. Okinawans are similar to Japanese and speak a variant of the language. They are treated as second-class citizens.

There are also foreign residents in Japan. These hail from places such as the Philippines, Thailand, Indonesia, and Vietnam. Coming to Japan for economic opportunities, they are forced into low-end jobs that no Japanese will take. There are also foreigners of Japanese ancestry—the *Nikkei*. Numbering about half a million, these people come from Brazil, Peru, and other parts of South America from where they emigrated in the twentieth century. Now returning to their homeland for work, they look Japanese but cannot speak the language. Moreover, their cultural bent is Latin American. Consequently, they are treated as outsiders.

The largest minority group is the *Burakumin*, who are Japanese descended from under-classes defined in medieval times. They number 2 to 3 million. The *Hinin* (nonpersons) were beggars, prostitutes, entertainers, and the like. The *Eta* (filth) were a subclass of untouchables regarded as spiritually unclean because of their occupations burying the dead or slaughtering animals. Since myths of spiritual and racial purity still abound in Japan, *Burakumin* are shunned and remain segregated in urban ghettoes. Some attempt to quietly integrate into society but they are usually found during background checks for employment and marriage. Then they are rejected.

Many companies, including the major ones, will not hire minorities. Students from the best schools will not work for companies that do. To marry a minority person is unthinkable.

Several minority action groups deal with government agencies to obtain improved conditions, but not all minorities are equally represented. Moreover, some minorities (e.g., the Koreans) eschew full participation in Japanese society. This results in a paradox: sustaining group identity in the face of discrimination and relying on organizational intermediaries to deal with major issues shields Japanese society and individuals from confronting their prejudicial behaviors and practices.

IMMIGRATION DEBATE IN A GLOBAL AGE

In 2006, Japan's population began to decline exponentially. In terms of available workers, this was thought to be unimportant because Japan's business would either be automated or be offshore. However, the demand for low-wage workers for the country's labor-intensive industries has actually increased.

Japan has a history of xenophobia. Government policies and public opinion share a common perspective: they think that the presence of foreigners should be tightly controlled and kept a respectable distance from the routines of Japanese life. However, foreigners are arriving in even greater numbers, some with phony visas and passports. Also, they are bringing their families and forming new households with Japanese nationals, and residing in cities throughout the nation. These migrants include many women who come to work in the "entertainment industry."

The voluntary or forced movement of Asian women to Japan has a long history stretching back more than a 100 years. This movement has become highly structured and highly complex as agents, brokers, and prostitution rings and clubs in Japanese cities have created extensive networks for the recruitment and distribution of foreign women. A 1997 survey of Filipinas coming to Japan revealed that only 10 percent knew that they would be working as prostitutes. The remainder believed that they would be working as domestics or in some other regular job.

For the most part, Japanese women have gained enough power to avoid working in the sex trade. Foreign women have taken their place. Japan's government and most Japanese ignore the exploitation of foreign women, reducing them to "others" not worthy of concern. Japanese men view prostitutes as a "necessary evil" while Japanese women note, "For those [foreign] women, it's just their way of life."

A City that Wants Immigrants

Desperate for workers, some cities aggressively pursue foreign migrants. For example, Hamamatsu, a coastal city of more than 500,000 half way between Tokyo and Osaka, is recruiting workers from

Japanese migrant communities in Peru and Brazil. Hamamatsu's economy relies on vehicle manufacturing in factories including Honda, Yamaha, and Suzuki.

Recognizing that the city would lose its economic base without new residents, it set out to get foreign workers. Officials assumed that these ethnic Japanese would blend easily into society. But the Spanish and Portuguese-speaking immigrants are more Peruvian and Brazilian than they are Japanese. Now there are a number of Spanish and Portuguese newspapers, schools, and community centers. City officials now recognize immigrant holidays, often using them to launch political campaigns.

One of the most hotly debated topics in Japan is whether or not to relax the nation's strict controls on foreign immigration and allow more foreigners to enter. In comparison to other countries, Japan's population is almost entirely ethnically homogeneous, with non-Japanese accounting for less than five percent of the population.

Foreign immigration would diversify the country, as well as help change the age structure of the population by bringing in younger people of childbearing age. Moreover, immigrants are said to be helpful to the economy by providing candidates for difficult, dirty, and dangerous (three words beginning with a "k" sound in Japanese) work that most Japanese avoid. The result would be more taxpayers to provide financial support to Japan's aging society by paying in to the increasingly stressed social security system. However, Japan is not used to hosting immigrants, and anti-foreigner feelings are common.

Immigrants from developing countries in Southeast Asia and Africa face an extra amount of discrimination although it is they that do the unpopular "3K" jobs in the country. "Japan for the Japanese" is a popular attitude. In many cities one sees signs at the doors of bars, bathhouses, and pachinko (Japanese pinball) parlors that read "Japanese only."

Many Japanese unfairly associate increasing social problems in their country such as crime and sexually transmitted diseases with the presence of foreigners. When foreign visitors came to Japan in bigger numbers than usual in the summer of 2002 for the World Cup soccer tournament, the Japanese media was filled with warnings from police and political leaders that more prisons would have to be built to house the "hooligans" and that hospitals should prepare for an overflow in births nine months later. As it turned out, the "hooligan" problem never materialized and there was no spike in births.

Tiger's Transformation

THE FADING OF AGRICULTURAL JAPAN

In 1920, 51 percent of Japan's labor force was in agriculture. Now less than 4 percent is farming. About 90 percent of these individuals are part-time farmers who earn most of their income from non-farm activities. In fact, much of the farm work is done by women and older people; men work other jobs, and young people, who are not interested in farming, move to the cities.

Agriculture accounts for a mere 13 percent of the land area and farms are very small, averaging 4 acres (1.6 ha). Farms in the United States average 500 acres (200 ha). Since agriculture competes with other activities for level land, farmland is very expensive. While there are some terraces, most cropping is done along flat, alluvial plains and in valley bottoms. Climatic factors and limited space call for intensive land use. The growing season varies from 260 days a year in Kyushu to 150 days in Hokkaido. The northern limit for double-cropping is 37° N. latitude.

There are two types of agricultural fields in Japan: irrigated rice-fields called *tambo* and non-irrigated fields called *hatake*. *Hatake* are used primarily to grow vegetables. The primary crop is rice, planted in summer. Rice is rotated with wheat or barley in the double-cropping zone. Tea and fruit such as apples grow on hill slopes.

Much of Japan's agriculture is **market gardening**: producing vegetables and poultry products for the vast urban market. For this purpose, greenhouse agriculture is very important and organic farming is on the increase. Nevertheless, agriculture is highly mechanized and chemical fertilizers and pesticides remain widely used (Figure 12-9). In fact, Japan uses more chemicals per unit of land than any other country. Consequently, crop yields are very high, with rice yields three times the average Asian yield.

Figure 12-9

This rice harvester symbolizes Japan's high degree of mechanization in agriculture despite the small size of its farms. ©Ric Ergenbright/Corbis.

Honorable Rice

O-kome or "honorable rice" figures significantly in Japanese culture. The early indigenous name for Japan was *mizu ho no kuni*, which means: "Land of the water-stalk plant (rice)." The Japanese word *gohan* means "cooked rice" or, more tellingly, "meal." *Asagohan* (breakfast) means "morning rice." In the Shinto religion, rice products are the most sacred of offerings. Rice is also an important offering in Buddhist temples.

According to mythology, rice was intimately associated with the creation of Japan. Supposedly, the sun goddess, Amaterasu, gave grains of rice to one of her descendants—the mythical first emperor Jimmu. His task was to turn the country into a land of rice. Today's Emperor Akihito, who is descended from Jimmu, harvests a small crop of rice on the Imperial Palace grounds.

Rice is a symbol of communality; it is the only dish shared from a single bowl. A famous proverb reveals how rice is imbued with the all-important Japanese concept of humility: "The heavier the head of rice, the deeper it bows."

Agriculture is also becoming urbanized in that many urban areas are dotted with rice paddies, small orchards, and vegetable plots. These can even be found in between and next to shopping malls, parking lots, and factories. About 5 percent of Tokyo's land is being farmed by some 13,000 families.

Many farmers are independently rich after selling their land to industry. However, resistance to parting with land imbued with centuries of family tradition has made acquisition of farmland for urban and other development difficult and expensive. Nevertheless, there are those farmers who hold on to their land, waiting for the highest bidder.

Animal herds are rare other than in Hokkaido, where extensive plains make dairying feasible. An ordinance of 1896 established non-military settlements on Hokkaido, and these were modelled on the township system in the United States. Here, dairy farms look like American farmsteads. Some beef cattle are kept near Kobe. These animals are beer-fed and hand-massaged daily. A steak from Kobe beef cattle can cost hundreds of dollars.

Japan's major source of protein is fish—the country has the world's largest fishing fleet. In addition, **aquaculture** is common, especially around the Inland Sea where shrimp, oysters, and seaweed are cultivated. Fish are also raised in ponds and in ocean-based fish farms.

Agriculture in Japan, especially rice production, faces an array of problems:

- Farmers who are working paddies are ageing and dwindling in number. Of the country's 3 million farmers, 70 percent are age 60 or older.
- Given the current economic crisis, the government has cut back on public works projects in rural areas

that once provided part-time employment to farmers and propped up rural economies.

- Young people want city jobs and small towns are devoid of youth. As one farmer bemoans: "Japanese agriculture has no money, no youth, no future." Abandoned, overgrown plots are a common sight.
- The number of farmers dropped from 12.2 million in 1960 to 2.2 million in 2004.
- Rice production has fallen 20 percent in the last 10 years, and Japan now imports 61 percent of its food. For example, most of the country's soybeans come from the United States and Brazil.
- People are eating less rice. For instance, rice, fish, pickled vegetables, and *miso* soup for breakfast are being replaced by cereal, eggs, and toast. People are eating more meat, bread, and dairy products.
- Traditional meals are being replaced by packaged instant noodles and McDonald's and other fast foods. Changing diets have resulted in Japanese who are taller and heavier than before. In addition, heart disease and cancer are on the rise.
- Agricultural output has declined drastically; the only food items that are produced domestically in sufficient quantities are rice (enough for the new diet), eggs, and vegetables such as onions and cucumbers.
- Domestically produced foods are becoming very expensive. A box of cherries can sell for as much as US$140.
- Government subsidies pay for surplus crops. Rice is six times more expensive to grow in Japan than elsewhere, yet it is a tradition supported by the government, which also prevents imports of cheaper Japonica rice from places such as California.
- The countryside is viewed as a repository of pristine national values such as harmony and community spirit. These emotional ties have helped make agriculture sacred in Japanese politics.
- Pressure from trading partners and global treaties has forced the Japanese government to negotiate trade agreements with a variety of nations such as Mexico, Thailand, and China. Consequently, a growing number of farm imports are being allowed into the country. Stiffest competition comes from China.

In the early 2000s, Japanese farmers were shocked to see cheap Chinese vegetables stocking supermarket shelves. Tomatoes, onions, eggplant, and garlic grown in China can be sold for far less than the same Japanese produce. The Chinese have already garnered a 40 percent share of the shiitake mushroom market. At one point, the government put a tariff on Chinese mushrooms and China retaliated by putting tariffs on Japan's machinery exports.

As family farms close, a new phenomenon is appearing in the countryside—the factory farm. For example, a subsidiary of Daiei Inc., one of Japan's largest supermarkets, has opened a factory farm to produce 20,000 pigs a year to be sold in Daiei markets.

Lettuce is the most commonly factory-farmed vegetable. Grown under artificial light in closed, bacteria-free environments, it is grown hydroponically (in water) and inputs are computer-regulated. One factory can produce 1.6 million heads of lettuce a year!

Japan hopes to boost its agricultural sector by producing and exporting "luxury crops" such as square watermelons, high-quality apples, grapes, pears, persimmons, and green tea. Fruits are exported to Taiwan and large amounts of green tea are sent to Europe.

The government is encouraging farms like this to make the country more competitive in an increasingly globalized marketplace. Currently, it is cutting subsidies to smaller farms in order to encourage mergers into larger corporate entities. Still, there were only a few more than 1,200 factory farms in 2000—a mere 0.34 percent of all farms. The fate of the family farm is yet to be decided.

FOUNDATIONS OF INDUSTRIAL JAPAN

The foundation of modern Japan's economic growth is manufacturing. From the mid-1800s, the economy responded to external trading pressures as well as a desire to become a military power. Factories were built in the coastal zone from Tokyo to Osaka. These specialized in iron and steel production, shipbuilding, and textile manufacturing. In the period prior to World War II, industrial expansion was based on links between business and government, cheap labor, large numbers of small businesses, and an improved infrastructure. Japan became known for its cheap but poor quality goods. This industrial base was devastated in World War II.

Rebuilding and restructuring of Japanese industry and infrastructure began with the infusion of American capital. Following the reconstruction of heavy industries such as steel and chemicals in the 1950s, Japan diversified into shipbuilding, automobile production, and light manufacturing.

Japan's industry has always relied on imported raw materials. In fact, Japan has the most limited natural resource base of any of the world's major industrial powers.

In post-war Japan, raw materials were cheap and, with its own low-cost labor force, this meant that Japan could easily compete in world markets.

Conditions changed in the 1970s with labor shortages, increased oil prices, and a slowdown of global shipping. As Japan now depended on more costly oil for 70 percent of its energy needs, the government decided to reduce the cost of hydroelectric and nuclear power and to subsidize unprofitable coal mines.

By the 1980s, older heavy industries were reaping the effects of overcapacity and competition from emerging tigers such as South Korea. Factory closures led to rising unemployment. Japan reacted to these changes by diversifying into light industries such as electronics, cameras, and appliances. Another venture was robotics—the building of mechanical devices to do the work of humans. Robots became widely used in the auto industry. However, their use has been cut back in recent years because they have proven to be very expensive.

At the end of the 1980s, Japanese corporations began investing more overseas. First, they bought out foreign mineral suppliers and then they built factories in other countries. After major investments in the United States and Britain to establish auto plants, they turned to Southeast Asia where economies were stimulated by the infusion of Japanese capital. In Japan, industries that lost jobs to cheap labor overseas were said to be **hollowed out**.

By the mid-1980s, Japan expanded its investment in cheap labor and assembly-line factories by moving research and design facilities overseas as well. It also built on its relationship with China, especially in China's northeast. By the early twentieth century, most Japanese multinationals such as Sony, Hitachi, and Toshiba and automakers such as Toyota, Honda, Mazda, and Nissan were operating in China. But recent economic crises beginning in the 1990s are proving that all is not well in Japan Inc.

AILING TIGER

Despite Japan's generally high standard of living and great wealth in comparison to other parts of Asia, all is not well in the country. After the unprecedented prosperity of the 1970s and 1980s, the economy went into a painful slump in the 1990s, so much so that the decade is now being called the "Lost Decade."

The more recent crisis, beginning in 2007, has struck Japan hard. Countless companies, including several major banks, well-known supermarket and department store chains, and once prosperous manufacturers, have gone bankrupt. They have been forced to terminate long-term workers just to stay afloat. The factory sector has been particularly hard hit, as more and more employment has shifted to overseas locations where labor is cheaper. Migrant workers can also be employed at low wages to cut costs. Companies in traditional industries such as steel production and shipbuilding have gone bankrupt altogether because of competition from South Korea, China, and Taiwan.

For Japanese employees in general, the promise of a guaranteed job for life has all but disappeared. Many workers face unemployment or moves downward to inferior occupations at lower wages. Likewise, recent college graduates, once accustomed to having a choice of jobs, struggle to find any employment at all and are often forced to settle for part-time work without benefits. Moreover, a new class of young workers called *freetah* has emerged. They eschew the "rat race" of full-time employment and instead go from one part-time job to another in accordance with short-term financial needs and personal inclinations. To lower the cost of living, many *freetah* delay or avoid marriage and live at home with parents long after the usual age of striking out on one's own.

There are many explanations for the malaise, including the troubling assessment that Japan's economy is chronically inefficient with redundant employees and outmoded business practices. A look at virtually any construction site or inside any bank shows the excess of workers, many of whom seem to do nothing but drive up the cost of business. Every road repair site, for instance, seems to have a bevy of traffic control assistants with flashing light batons to wave vehicles and pedestrians around the obstruction, whether they are needed or not, as well as what seems like an inordinately large number of workers and supervisors.

The bad economy is also blamed on Japan's financial institutions because of the many poorly secured loans they made to cronies during the bubble. Many loans were for dubious investments such as speculative real estate. The national government has recently had to bail out several large companies with special loans and tax breaks. This has caused critics to point to Japan's cozy relationships between politicians and businessmen as still another flaw in the nation's economy.

The influence of the construction industry is seen everywhere in Japan. Not only does the country have some of the longest tunnels and bridges in the world, such as the tunnel to Hokkaido and the bridge to Shikoku, but it also has countless other more ordinary tunnels, bridges, and highways built at great cost across the nation's difficult terrain. Similarly, Japan has the world's highest density of dams, which it continues to build even as other countries dismantle some of theirs for sound environmental reasons. Moreover, there is hardly a free-flowing river in the country, as almost all have been reconfigured and realigned in concrete by various kinds of flood control projects. In small towns, where employment is lacking, project funding has been halved in light of the recent economic crisis.

Ashio's Declining Fortunes

A town that is representative of Japan's economic woes is Ashio, a once prosperous copper mining center in the mountains of Honshu north of Tokyo. As mining declined over much of the twentieth century because of cheaper copper from other sources and depletion of the richest ores, the town faced a problem of unemployment and uncertainty about its future. Population dropped from a peak of nearly 40,000 around 1915 to less than 5,000 today, as residents relocated to Tokyo and other cities for jobs, and young people moved away soon after finishing high school because of a lack of opportunities.

The proposed solution for the stagnation was to open Ashio for tourism. Government economic development funds helped with the construction of a museum about copper mining, as well as with construction of new roads and bridges to make the town less isolated. Unfortunately, the planners overestimated the number of people who would travel out of their way, even by improved roads, to learn details about copper. The museum and souvenir stands have few visitors and Ashio remains another depressed town in Japan's countryside.

JAPAN'S AUTO INDUSTRY

Vehicles are Japan's largest export product, with more than 7 million being exported in 2004. The "Big Five" (Toyota, Honda, Mitsubishi, Nissan, and Mazda) account for about 80 percent of the 9 to 10 million vehicles produced each year.

The industry had a slow start with a small domestic market. Moreover, there was competition with U.S. companies producing cars in Japan in the 1920s and 1930s. In the 1950s, the government targeted the car industry as one with great potential. As incomes rose, the domestic market expanded. At the same time, there was a growing market in Europe, where people were demanding small, well-developed, fuel-efficient cars. The industry took off and has reached the point where one-tenth of Japan's workforce is now in car manufacturing.

There is a trend in the worldwide auto industry to integrate national production, thereby making autos international entities. For example, Toyota, Nissan, and Honda manufacture vehicles and parts in China. As of 2000, Toyota had invested more than US$10 million in 18 facilities in the United States that employ more than 27,000 workers. The Camry, for instance, is produced in Kentucky. In fact, more than 63 percent of Toyotas sold in the United States are made somewhere in North America. The "Big Five" are making big profits on overseas sales.

In the recent economic downturn, vehicle production has slowed. This is, in part, because of a decrease in Japan's domestic market. China took over Japan's spot as the world's largest vehicle manufacturer in 2009.

HIGH-TECH REVOLUTION

Even in times of economic crisis, Japan is furthering its high-tech revolution, which began in the post-war years. It is turning to unexpected ways to harness technology and cater to the ever-picky Japanese consumer.

High-tech factories have been built throughout the country (Figure 12-10) in part to disperse populations of overcrowded metropolitan regions such as Tokyo and Osaka. High-tech industries are knowledge-intensive and focus on such things as microelectronics, semiconductors, medical and aerospace technologies, and communication systems. Integrated circuits (microchips and silicon chips) comprise the core of production for everything from microwaves to advanced weaponry. Many of these are manufactured by Fujitsu, Hitachi, and NEC corporations.

The Japanese love gadgets. Sharp has invented a microwave that automatically cooks recipes off the internet. It also offers an 18-minute washing machine that washes people instead of clothes. Also, what about a GPS that people can use to track the whereabouts of their children or elderly parents? Japan is the "gadget-king," and as geographer Pradyumna Karan observes: "There is no country like Japan when it comes to fusing and cross-fertilizing technologies to create something new."

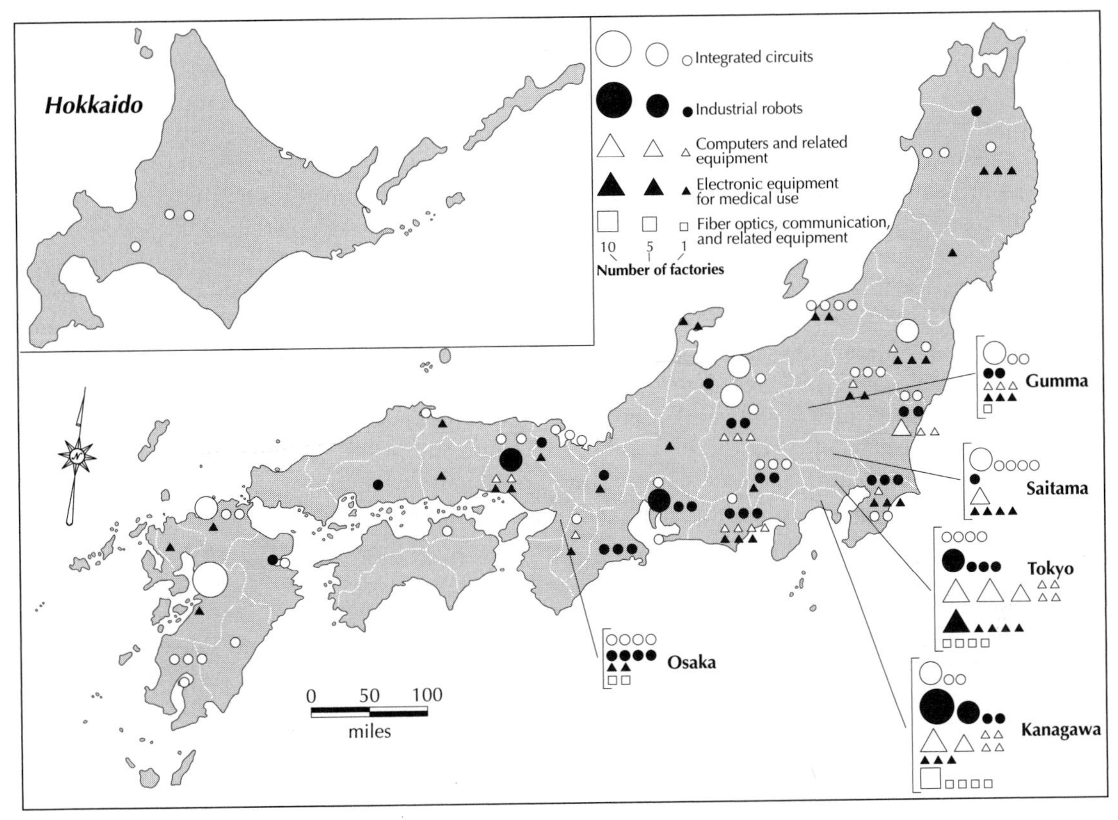

Figure 12-10

Distribution of high-technology factories. Why are there relatively few such places on the outer islands? From P. P. Karan, *Japan in the 21st Century*, 2005, p. 246. Reprinted with permission of the University Press of Kentucky.

Urban Japan

City development in Japan can be traced back to the eighth century, when the first permanent capitals of Heijokyo (now Nara) and Heiankyo (now Kyoto) were established. Soon other cities were built, primarily for political and military purposes. Some, such as Naniwa (now Osaka), were built to meet the needs of travelers. As the central authority of Japan's military government waned in the fifteenth and sixteenth centuries, towns increasingly evolved around castles constructed by regional warlords.

CASTLE TOWNS

The catalyst for urban growth was the fortified **castle town**. One of the most important of these was Osaka, where a grand castle was built in 1583. This served as the nucleus for urban growth. Moreover, government policies stimulated the rise of Osaka and other towns. After the mid-1630s, the prohibition of foreign trade stimulated internal interchange of goods and subsequent town growth. Minor feudal castles were destroyed so that there was only one castle town per province, furthering focus of activity in major centers. These centers became consolidated as civilians were encouraged to move to the more important castle communities.

Castle towns, with their political and economic functions, were typically located on level land near important landscape features. Osaka emerged as the principal business, financial, and manufacturing center in Tokugawa Japan. Cities were also linked by highways that stimulated growth and trade. The most famous was the Tokaido Road running from Osaka, eastward through Nagoya, to Edo (Tokyo), the capital. By the middle of the eighteenth century, Tokyo had one million inhabitants. With rapid post-war development in Tokaido Road cities, an urban-industrial region emerged to be known as the **Tokaido megalopolis**.

Edo was founded in the fifteenth century when a minor feudal lord built a castle on a bluff near the sea. With a natural harbor, hills that could be fortified, and the Kanto Plain behind it for expansion, Edo had an excellent geographic situation.

Tokyo got its real start when the Tokugawa shogun Ieyasu decided to make Edo his capital in the sixteenth century. He and his descendants designed a vast area of palaces, parks, and moats in the heart of the city. This grand plan incorporated an imperial enclosure for the emperor. Reclamation of land from the sea set the pattern of development for many Japanese cities that faced a shortage of level land for expansion and needed good port facilities. By the mid-eighteenth century, Tokyo was one of the largest cities in the world.

Edo's initial growth was based on its function as a political center tied to other places by a network of roads. The rivalry between Tokyo as a cultural and political center and Osaka as a business center lingers to this day. With the restoration of the Meiji emperor in 1868, the royal court was transferred from Kyoto to Edo, which was renamed "Tokyo," meaning "Eastern Capital." With its new political functions and the rapid industrialization and modernization programs undertaken from the 1870s onward, the city experienced profound growth into the twenty-first century.

RAPID URBANIZATION

Despite more than a century of industrialization, Japan did not reach the 50 percent urban mark until after World War II. Between 1950 and 1970, urban population increased to 72 percent, a figure that took the United States many decades to achieve. Since 1970, the proportion of people living in cities has risen more slowly, reaching 78 percent in 2003 and 86 percent in 2010. As the urban population grew dramatically, so did the number and size of cities. While small towns and villages (fewer than 10,000 people) declined sharply, medium and large cities expanded rapidly both in population numbers and spatial extent. This process reflected the country's phenomenal economic growth after World War II.

THE CORE REGION

A distinctive feature of Japan's urban pattern is the concentration of cities in a relatively small area. Almost all of the major cities are found in the core region extending from Nagasaki and Fukuoka in northern Kyushu, along both flanks of the Inland Sea on Shikoku and Honshu, through Nagoya to the Tokyo region (Figure 12-11). The core region, especially the southern coast of Honshu, contains strings of industrial cities such as Hiroshima, which has recovered from the devastation created by the atomic bomb dropped on it at the end of World War II. The Tokaido megalopolis, with 45 percent of the Japanese population, is contained within the core area.

Japan has a classic core-periphery imbalance—the "developed" capital region focused on Tokyo, and the "less-developed" (in a relative sense) regions elsewhere. Since the 1970s, migration and growth have been increasingly toward Tokyo at the expense of other places, a phenomenon called **unipolar concentration**. As the city continues to expand, it drains both capital and people from other areas, many of which are in a state of stagnation. The Osaka region, for example, has not seen the growth of many industries to replace the smokestack ones of steel and shipbuilding. In addition, Osaka businesses continue to relocate to Tokyo. This process stimulates out-migration and lowers personal consumption. To be in the mainstream of modern Japan is to live and work in or near Tokyo.

TECHNO-ARCHIPELAGO

Today, the Capital Tokyo Metropolitan Area (CTMA) houses 31 million people. CTMA's phenomenal growth has paralleled Japan's integration in the global economy. Its geographical location between New York and London is perfectly suited to the online, global, 24-hour, international telecommunication and financial transaction system. As a result, Tokyo was expected to provide an array of businesses and services not directly related to the domestic economy. Other sectors of the economy increasingly demanded specialized services in order to maintain their competitive edge.

The crucial role of research and development was soon recognized. Kansai Science City began to take shape in the 1980s. Kansai Techno-Research Complex includes five cities and three towns in the Kyoto–Osaka–Nara prefectures and is an integrated, multinodal development. This structural format models plans for a new urban Japan intended to possess a multicentered regional and urban system founded on advanced information technology. The system was to reorganize from an assemblage of hierarchical structures to a unicentered, functional network with CTMA as the central node.

The **technopolis** was integral to the master plan that conceived a vast network of science cities across the

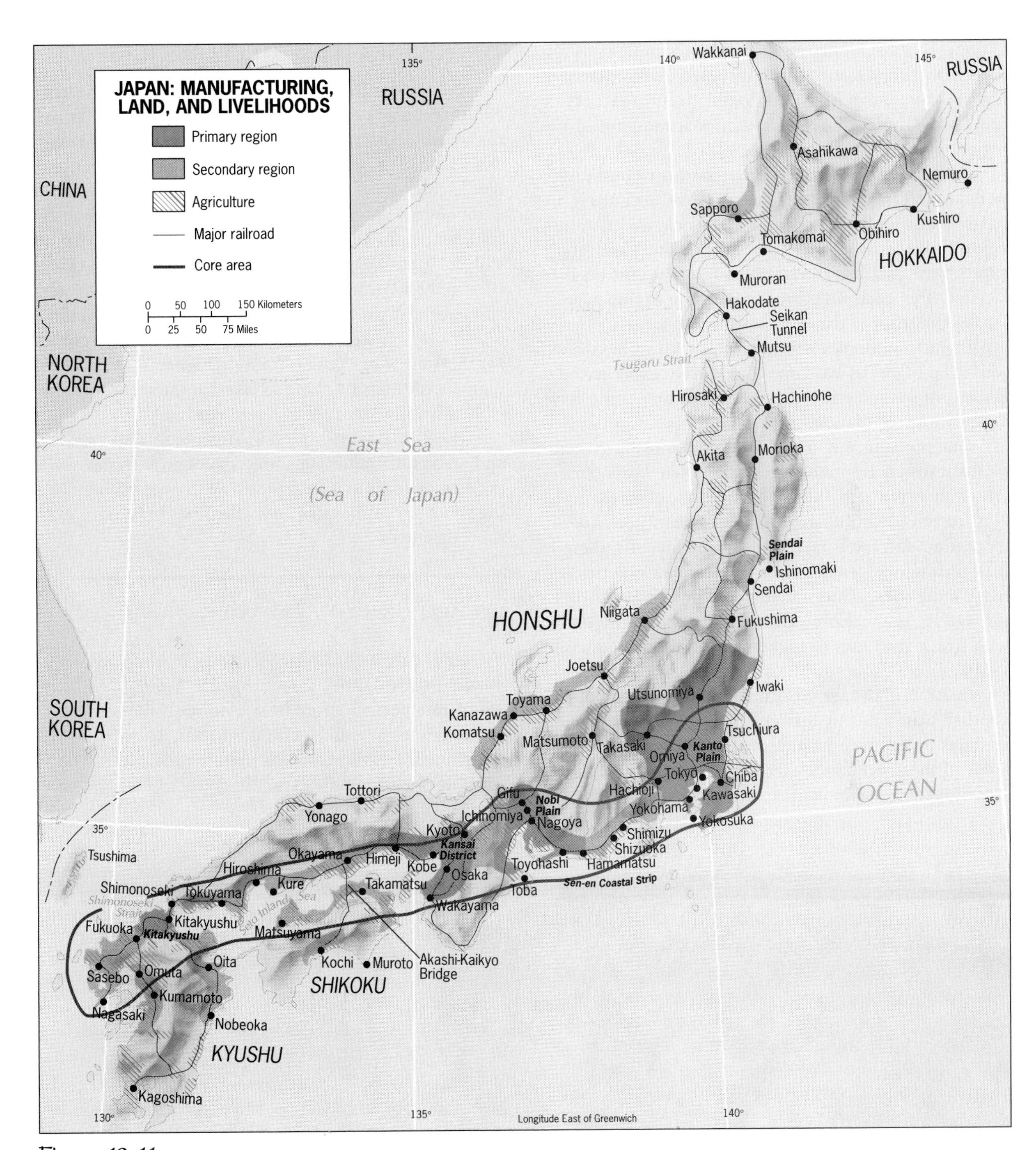

Figure 12-11

Japan's core region and major manufacturing areas. From H. J. de Blij and P. O. Muller, *Geography: Realms, Regions and Concepts*, 14th edition, 2010, p. 510. Originally rendered in color. © H. J. de Blij and P. O. Muller. Reprinted with permission of John Wiley & Sons, Inc.

nation. Fourteen industries such as biotechnology, robotics, software, and space were targeted for development in tandem with research and development and other service functions. These are to fuel additional industry in the region.

Each technopolis centers on a "mother city" and is to be linked by rapid transit to Tokyo, Nagoya, or Osaka by no longer than a one-day return trip. The new undersea tunnel to Hokkaido and bridge to Shikoku along with more expressways and high-speed bullet trains are intended to facilitate this goal. This evolving **Techno-Archipelago** is Japan's landscape of the twenty-first century.

Eight technopoles within reach of Tokyo have fared well. The other 18 have not. Some have experienced a decline in manufacturing employment and some have actually lost population.

One problem is that major corporations have established research laboratories close to their headquarters rather than close to the production site. For example, NEC research facilities are in Kawasaki near Tokyo while its product factory is in Kyushu. Consequently, there is little technology transfer between incoming factories and local industries. Thus far, the program has failed to achieve its original objective to integrate techno-cities with R&D, institutes of higher-learning and production facilities.

In 2002, under the government's "Big-Boned Policy," another plan was put forward to better integrate such facilities to enhance Japan's economic competitiveness. Some of these new integrated facilities, especially in the area of medical technology, are meeting with success.

TOKYO: THE CITY WITH EVERYTHING

Tokyo is famous for its crowding. Perhaps one of the most enduring images of the city is of "commuters' hell" when rush-hour trains and subways are jammed with homeward-bound passengers who, like sardines in a can, are hardly able to move. In the morning the scene is repeated as inbound trains are packed with commuters, with many stations employing white-gloved "pushers" to cram in still more passengers as the train car doors slide closed (Figure 12-12). Many riders commute very long distances and have to endure the crowding for two hours or more.

Tokyo is linked to most other regions of the country by *Shinkansen* or "Bullet Trains" (Figure 12-13). These high-speed trains fly along so fast that the scenery is only a blur. The city ones are also jam-packed.

Tokyo highways and side streets are also crowded and standstill traffic jams are common. At times, even pedestrian zones can be jammed with people complaining about slow sidewalks and difficulties in walking even short distances.

The Yakuza Network

The Yakuza are traditional, organized crime syndicates in Japan. They are arranged in 'families" with a hierarchical structure from a number-one boss through to the minions on the streets. The second-in-command controls areas in individual cities. The whole organization is structured according to

Figure 12-12

Trains in Tokyo are so crowded that people are employed to push people into the cars. ©Paul Chelsey/Stone/Getty Images, Inc.

Figure 12-13

Central Japan Railway Company's N700 series Shinkansen bullet train travels past Mount Fuji. ©Tomohiro Ohsumi/Bloomberg /Getty Images, Inc.

"superior–inferior" or "parent–child" relationships. Any perceived slight on the part of an underling requires him to cut off part or all of a finger. Everyone knows that someone missing a finger is a Yakuza.

The largest family, the *Yamaguchi-gumi* with its 45,000 adherents, has 45 percent of Yakuza membership of at least 100,000. It is divided into 750 clans and has its headquarters in Kobe.

Yakuza often have full-body tattoos that are imprinted into the flesh with sharpened bamboo or steel needles. They have long, slick-backed hair and dress in shiny, tight-fitting suits and pointed-toed shoes. They also like to drive Cadillacs and Lincolns.

The Yakuza syndicate runs such rackets as prostitution, drugs, gambling, human-trafficking, and protection. They are spread across Japan and operate in other Asian countries such as Korea. Deeply involved in the corporate world, they even have connections to American crime organizations.

The police refer to the Yakuza as *boryokundan,* meaning "violence group." Yakuza call themselves *ninko dantai,* meaning "chivalrous organization."

Housing is another key issue in Tokyo. Because of the excessively high cost of land, most people in Tokyo feel that their housing conditions are inadequate: too distant, too small, and too expensive. It is not uncommon for parents and their children to share a single bedroom, or for families to share a housing unit with in-laws, grandparents, or other relatives.

Most young couples can only dream about owning their own home in Tokyo, as increasingly overcrowded apartments in high-rise buildings and small condominium units that are ironically called *manshon* (mansions) are the typical dwelling types. However, recent data show that Tokyo rents are dropping because of the poor economy, and that "empty nesters" and two-income households in particular are increasingly able to afford housing in Tokyo's central wards. A new boom in inner-city apartment construction is a Tokyo variant of the gentrification process (Figure 12-14).

Planners and government officials in Tokyo have tried to address the housing shortage by building large

Figure 12-14

One of the main problems in Tokyo and other large Japanese cities is the long commute, often under very crowded conditions, between home and workplace. As a result, planners are advocating constructing high-quality, high-rise housing, with ample amenities and community facilities in central locations. This photograph is from the upscale Roppongi Hills development near downtown Tokyo. Courtesy Roman Cybriwsky.

residential towns, such as Tama New Town in Tokyo's western suburbs, and by requiring developers of large commercial properties in central wards to include residential buildings as part of their projects. The latter reduces commuting times for residents and helps to repopulate the city core, which had been hollowed out over the years by commercial expansion.

Although living in Tokyo means living with crowding, the city can also be surprisingly comfortable and enjoyable. It is quite clean despite the number of people, and crime rates are among the lowest in the world. While bad things do sometimes happen, for the most part one can go just about anywhere at any time of day or night in Tokyo, without fear of being robbed or assaulted. This is true for both men and women (Figure 12-15).

The city is a paradise for lovers of restaurants, cafes, music pubs, and dance clubs, and a magnet for the best entertainment from around the world, be it classical music or the hottest stars of rock and hip-hop (Figure 12-16). It is also an outstanding place for shopping, keeping pace with all the latest fashions in Europe and North America, as well as setting fashion trends of its own.

It is not surprising that Tokyoites are often on the move. Since dwellings are too small for comfort or for entertaining friends, a very common Tokyo lifestyle involves going out after work or school and going home just to sleep. Friends come together at designated meeting places at train or subway stations and then go out for dinner, shopping, a movie, a concert, or whatever other activity interests them. For those with intimacy on their minds, Tokyo has thousands of inviting "love hotels" that rent fantasy theme rooms by the hour. As much as any city in the world, Tokyo has it all.

Figure 12-15

These Japanese women are dressed in traditional costume. Note that they are on their cell phones. ©Jeremy Woodhouse/Getty Images, Inc.

Figure 12-16

The sport of sumo is one of countless activities available in Tokyo. Sumo, along with baseball, is a national passion in Japan. Wrestlers have to be not only big but also quick and wily. Courtesy Roman Cybriwsky.

Recommended Web Sites

http://factsanddetails.com/japan-.php?itemid=941&catid=24&subcatid=159
Numerous short articles on Japanese agriculture. Eco-farms; high-tech; co-ops; exports/imports; foreign competition.
www.Japan-101.com/
Japan's best Web site, according to *Japazine* magazine. Multiple topics such as population, physical, urban and cultural geography Good source for photos.
www.japantimes.co.jp/news.html
The *Japan Times* newspaper. Latest news in a variety of categories.
www.pbs.org/wgbh/nova/satoyama/rice.html
Nova/PBS online. Excellent article about rice: history, cultivation, genetic engineering, etc.
www.stat.go.jp/english/index.htm
Ministry of Internal Affairs statistics. Demographic, economic, and social data. Statistical Yearbook. Note that information is available in Japanese, Chinese, Korean, English, and Portuguese.
www.tokyofoundation.org
Short articles on government policies, economy, society, and culture.
www.worldwatch.org/ww/ainu
Sources and resources for the Ainus' struggle for recognition.

Bibliography Chapter 12: Japan: Century 21

Aoki, Hidekazu. 2000. "Factory Location in Electronic Products and Regional Production Linkages: The Example of Sony Group." *Human Geography* 57/5: 23–42.
Chang, Iris. 1997. *The Rape of Nanking: The Forgotten Holocaust of World War II*. New York: Basic Books.
Cybriwsky, Roman. 1988. "Shibuya Center, Tokyo." *The Geographical Review* 78: 48–61.
_____. 1998. *Tokyo: The Shogun's City at the Twenty-First Century*. New York: Wiley.
Cybriwsky, Roman, and Aoi Shimuzu. 1993. "Rocks, Rust, and Tourism on Japan's Copper Mountain." *Focus* (American Geographical Society) 43: 22–28.
Diamond, Jared. 1998. "Japanese Roots." *Discover* 19: 86–94.
Douglass, Mike, and Glenda S. Roberts. 2000. *Japan and Global Migration: Foreign Workers and the Advent of a Multicultural Society*. Honolulu: University of Hawaii Press.
Edgington, David, W. 1997. "The Rise of the Yen, "Hollowing Out" and Japan's Troubled Industries." In *Asia Pacific: New Geographies of the Pacific Rim*, eds. R. F. Watters and T. G. McGee, pp. 170–189. Vancouver: UBC.
Fackler, Martin. 2009. "Japan's Rice Farmers Fear Their Future Is Sinking." *The New York Times* March 28, p. A5.
Facklet, Martin. 2004. "The Decline of Japan's Farmers." *Far Eastern Economic Review* 26 (February): 12–15.
Fujita, Kuniko, and Richard Hill, eds. 1993. *Japanese Cities in the World Economy*. Philadelphia: Temple.
Hohmann, Skye. 2008. "The Ainu's Modern Struggle." *World Watch* November/December: 20–23.
Hoon, Shim Jae, et al. 1993. "Rural Exodus: Seeds of Despair." *Far Eastern Economic Review* 4 (March): 20–21.
Howell, David, L. 1996. "Ethnicity and Culture in Contemporary Japan." *Journal of Contemporary History* 31: 171–90.
Karan, Pradyumna. 2005. *Japan in the 21st Century: Environment, Economy and Society*. Louisville: University Press of Kentucky.
Kerr, Alex. 2001. *Dogs and Demons: Tales From the Dark Side of Japan*. New York: Hill and Wang.
Kingston, Jeffrey. 2001. *Japan in Transformation 1952–2000*. London: Longman.
Kinsella, Kevin, and David R. Phillips. 2005. "Global Aging: The Challenge of Success." *Population Bulletin* 60/1. Washington D.C.: Population Reference Bureau.
Lamont-Brown, Raymond. 1995. "Earthquake—The Ultimate Japanese Nightmare." *Contemporary Review* 266: 117–119.
Lie, John. 2001. *Multiethnic Japan*. Cambridge: Harvard University Press.
Miyashita, Kenichi, and David Russell. 1996. *Keiretsu: Inside the Hidden Japanese Conglomerates*. New York: McGraw-Hill.
Okimoto, Daniel L., and Thomas P. Rohlan, eds. 1988. *Inside the Japanese System: Readings on Contemporary Society and Political Economy*. Stanford, Calif.: Stanford University Press.
Rothery, David A. 1995. "Kobe—Anatomy of an Earthquake." *Geographical Magazine* 67: 54–55.
Ruble, Blair A. 2006. Melange Cities: "The Disruption that Immigrants Bring Is Often a Benefit." *Wilson Quarterly*, Summer: 56–59.
Selleck, Yoko. 2001. *Migrant Labour in Japan*. Houndsmills, Hampshire, UK: Pelgrave.
Suganuma, Unryu. 2001. "The Geography of Toyota Motor Manufacturing Company." In *Japan in the Bluegrass*, ed. P. P. Karan, pp. 61–97. Lexington: University Press of Kentucky.
Takahashi, Junjiro, and Noriyuki Sugiura. 1996. "The Japanese Urban System and the Growing Centrality of Tokyo in the Global Economy." In *Emerging World Cities in Pacific Asia*, eds. Fu-chen Lo and Yue-man Yeung, pp. 101–43. New York: United Nations.
The Economist. 2008. "The Ainu: People at Last." *The Economist*, July 12: 54.
Williams, Jack F., and Kam Wing Chan. 2008. "Cities of East Asia." In *Cities of the World*, eds. Stanley D. Brunn, Jack F. Williams, and Donald J. Zeigler, pp. 475–527. New York: Rowman and Littlefield.

Chapter 13

Korea and Taiwan: Tigers Rising

"An oppressive government is more to be feared than a tiger"

CONFUCIUS (CA 551–479 BC)

In this chapter we will discuss two tigers of East Asia: South Korea and Taiwan. We will also talk about North Korea, which is definitely not an economic tiger. As you read on, it will become increasingly evident that the historical evolution of each of these regions is related to developments in China.

Korea: Two Countries, One Nation

PHYSICAL SETTING

Korea is one of the world's oldest known land areas, dating back 1,600 to 2,700 million years. Made of granite and limestone, the peninsula is tilted westward to the Yellow Sea where hundreds of islands dot the coast. Only 20 percent of Korea is level land. The rest consists of low-rising, sharp, and often bare mountain ridges. These ridges never reach above 9,200 feet (2,800 m). The mountains are more concentrated in the north and east. Along the east coast, the Taebaek Range runs like a spine with southwestern spurs that cut the peninsula into narrow valleys and alluvial plains (Figure 13-1).

Korea is bordered by China on the north by the navigable Yalu and Tumen rivers, both of which flow from Paektusan, the highest point in Korea at 9,000 feet (2,744 m). A range of extinct volcanoes leads from Paektusan to the Sea of Japan.

Like the rest of East Asia, Korea has a monsoon climate with warm, humid summers and cold, dry winters. While North Korea is influenced by continental extremes, the south is warmed by the Japan Current. Southwestern Korea and Cheju Island are, in fact, subtropical and have considerably more rainfall than the rest of the peninsula. About half the rainfall comes in June, July, and August, amounting to about 30 to 40 inches (76 to 102 cm) in the south and less in the colder and drier north.

North-south differences are also apparent in resource availability. Although the peninsula is only moderately endowed with mineral resources, the north is favored with substantial deposits of coal, iron, copper, zinc, and other ingredients needed for heavy industry. The north also has extensive hydroelectric power resources, particularly along the Yalu, Tumen, and Taedong rivers. In the case of the Yalu and Tumen, the power is shared with China. In South Korea, minerals are similar to those in the North but, with the exception of graphite and tungsten, exist in smaller quantities and are of lower quality. Consequently, the South relies on imported raw materials for its heavy industrial base.

Only about a sixth of mountainous North Korea is suitable for cropping. Coastal lowlands produce rice, corn, wheat, millet, and soybeans but food sufficiency is a problem. The hills have rich timber reserves, extensive orchards, and livestock grazing.

South Korea, with its population of 48.7 million, is the peninsula's traditional rice bowl, with about 20 percent level land. However, greater productivity is offset by a considerably larger population, more than double that of North Korea's 22.7 million in 2009. South Korea's agriculture is intensive, focused on rice, vegetables, fruit,

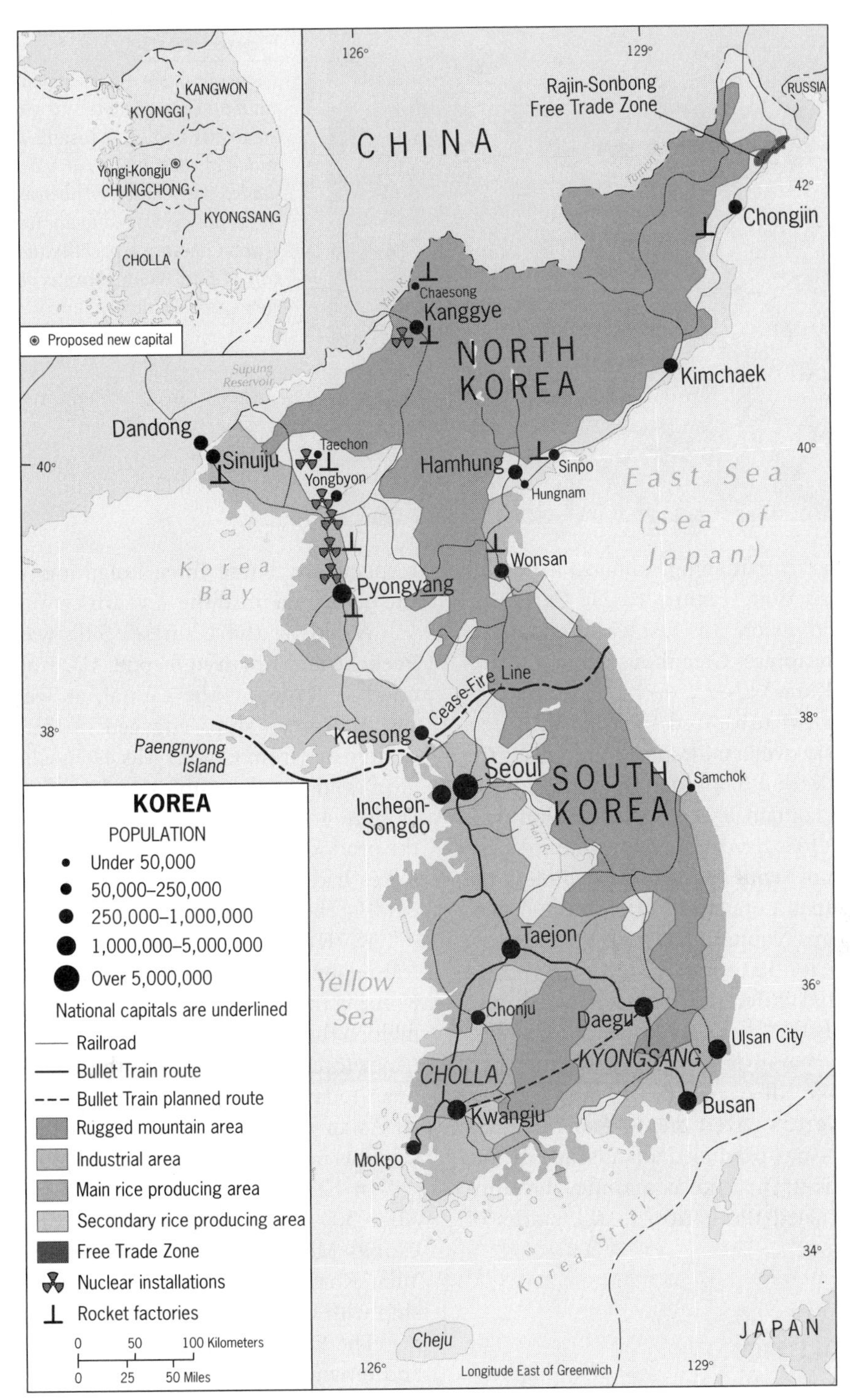

Figure 13-1

Geography of North and South Korea. From H. J. de Blij and P. O. Muller, *Geography: Realms, Regions and Concepts*, 14th Edition, 2010, p. 521. Originally rendered in color. © H. J. de Blij and P. O. Muller. Reprinted with permission of John Wiley & Sons, Inc.

Figure 13-2

These are fields of ginseng outside Kumsan, the main marketing center for ginseng in Korea. Here the root is known as insam. *The soil is a rich mulch of chestnut or oak leaves, and the plants are shaded with thatch or fiber mats. Once used, the soil is not cultivated again for another 10 to 15 years. Ginseng, especially the red variety, has been part of East Asian pharmacopoeia for centuries.*
Photograph courtesy of B. A. Weightman.

and other types of market gardening. Kumsan is notable for its ginseng production (Figure 13-2). Cattle are raised on Cheju Island, where it is just warm enough to grow oranges and pineapples. Greenhouse agriculture is ubiquitous in South Korea.

South Korea is more urbanized than North Korea, which has only one city over a million people. This is the capital of Pyongyang with 2.6 million. The twin cities of Hamhung-Hungnam contain fewer than a million people. In the South, six cities have more than a million people. The conurbation of Seoul and its port Inchon is the largest, with a combined population of 18 million people. Seoul, which means "capital," has been Korea's (now South Korea's) capital for 500 years.

The quintessential difference between North and South is political. North Korea's strict communist system has produced a harsh landscape of rigidity and poverty, isolated from all except China and Russia. Variation in South Korea's government—from authoritarian to democracy—has produced a landscape of free enterprise, modernization, and economic development, which have raised the nation to the status of Tiger in East Asia.

HISTORICAL EVOLUTION

Archaeological evidence suggests that the Korean Peninsula was first settled 5,000 years ago when Tungusic tribes migrated to the area from Manchuria and northern China. Koguryo, the first state, developed around 100 AD. The kingdoms of Paekche in the southwest and Silla in the southeast emerged around 250 and 350 AD, respectively. These three kingdoms vied for control of the peninsula until the seventh century.

Aided by the Chinese, Silla was able to conquer Paekche and Koguryo by 668 AD. Trading centers developed in Kyongju, the capital, as well as in provincial towns. Korean ships engaged in the region's sea trade. Buddhism (from China) was established as the state religion, and lavish temples and tombs were constructed (Figure 13-3). Woodblock printing was developed for the production of religious texts. From this time forward, Korea had an important cultural influence on Japan, including the introduction of Buddhism.

By 918, the kingdom of Koryo emerged in the central region of the peninsula and overthrew Silla in 935. Korea is the westernized version of Koryo. Buddhism remained the state religion but Confucianism emerged as a major force in government ethics and political structure. Even today, Korea is considered the most Confucian of all Asian nations. Check out its flag. Where do these symbols originate?

In 1231, Mongol armies moved into the peninsula. After 30 years of war, Koryo became a vassal state of China's Mongol Yuan Dynasty. Following the fall of the Yuan, Korea's Yi Dynasty continued the tributary relationship with China's Ming Dynasty (Chapter 10).

The Yi moved their capital to Seoul on the Han River and renamed their kingdom Chosun in 1392. Under Yi rule, Confucianism replaced Buddhism as the state belief system. The world's first movable, metal type was developed. More importantly, a Korean lettering system was developed in the fifteenth century. Known as ***Hangul,*** it reduced dependency on Chinese characters and made literacy achievable for the general public.

Figure 13-3

This corner eave of the great Buddhist temple at Pulguksa illustrates the detailed craftsmanship of Korean temple builders. The central dragon is meant to ward off evil. Built in the sixth century, this is one of the oldest surviving Buddhist monasteries in Korea. Photograph courtesy of B. A. Weightman.

Hangul

In the fifteenth century, King Sejong (who ruled from 1419 to 1450) commissioned scholars from the "Hall of Worthies" to develop a Korean alphabet system for those who had not mastered Chinese characters. After analyzing the phonetic structure of the language, they distinguished three separate elements of words: initial, medial, and final sounds. Of these, the medial element is a vowel. The new alphabet consisted of 29 letters—22 consonants and 7 vowels. Consonants and vowels are grouped into a syllabic block form to create a word. This new system, called Hangul, differed radically from the Chinese ideographic system in which a concept is conveyed by each character. King Sejong presented his "correct sounds for the instruction of the people" in 1443 and literacy became widespread.

TANGLING WITH A JAPANESE TIGER

Japanese forces invaded Korea in 1592 and again in 1597. With the help of China, the Koreans fended off the Japanese but were forced to send artisans and scholars to teach Chosun's advanced technology in Japan. The Japanese ceramic and lacquer-ware industries owe their origins to the Koreans. Later, the Japanese were given residence rights in a small area of Busan (Pusan).

The seventeenth century brought Manchu control and subsequent decline. After much struggle, Korea fell to Manchu power in 1637. Significant change was on the horizon. Factional strife within Korea weakened the political structure. Christian ideas, introduced via official tribute missions to China, clashed with Confucianism. New agricultural techniques fostered a commercial economy, despite aristocratic disdain for such activity. Corruption and intrigue hampered orderly administration. By the mid-1800s, Korean politics and the economy were in shambles.

In 1864, the Korean government closed the country to all except the Chinese and the Japanese at Pusan. It was during this period that Korea became known as the Hermit Kingdom. All other foreign interests were spurned until Japan forced Korea to enter into its first foreign commercial treaty in 1876. This period also marked a great divide in the nation's history. Prior history was China-centered. The future would be determined by Japan and the West.

Korea fell prey to Japanese imperialism in the late nineteenth century. The Japanese engineered the murder of the anti-Japanese Korean Queen Min in 1895. The Russo-Japanese War of 1904–1905 saw Japanese troops operating on the Korean peninsula as they moved the war front into Manchuria. Korea became a Japanese protectorate in 1905, and despite anti-Japanese efforts, Japan was able to annex the country in 1910. Japanese rule was highly resented by the Koreans, who viewed themselves as culturally superior. Japan relied on the use of force to transform its colony.

Figure 13-4

A North Korean flag flaps atop a tower at the propaganda village at Gijungdong in the DMZ. This picture is taken from the Dora observation post in Paju almost 33 miles (55 km) from Seoul.
©Kim Kyung-Hoon/Reuters/Landov LLC.

While the traditional, Confucian agrarian order was retained, a modern administration was established, with Japanese functioning in key positions. Both agriculture and industry were structured to meet the needs of Japan and its plans for aggrandizement. Transportation, communication, mining, and other infrastructure were modernized and expanded. Thousands of Koreans were sent to Japan to work in factories. In Japan, the Korean language was banned and Koreans had to take Japanese names. In Korea, resistance movements were brutally crushed, deepening Korean hatred of their Japanese overlords. Japanese rule of Korea did not end until Tokyo surrendered to the Allied Forces in 1945.

NORTH-SOUTH DIVISION

In 1945, the Soviets and the Americans divided the peninsula into two spheres of influence and military occupation at the 38th parallel. Free elections were supposed to follow, but disagreement on this issue ensued in the next two years. In the face of the adamant refusal of the Soviets to subscribe to United Nations–supervised elections, the Republic of Korea (ROK) was established south of the 38th parallel on August 15, 1948. The Soviet response came in September, with the formation of the Democratic People's Republic (DPRK) in the north. The United States withdrew its military forces the following year.

On June 25, 1950, North Korean armed forces attacked South Korea across the 38th parallel. The United Nations responded with the support of sixteen nations, led by the United States with 90 percent of the troops, and the Korean War ensued. The conflict was prolonged by the entry of Chinese troops in support of North Korea in 1950. After three years of fighting, an armistice was signed, with an agreement to find a political solution to the problem of division. The 1953 agreement led to the creation of a Demilitarized Zone (DMZ), which has become heavily militarized on both sides.

The DMZ is an interesting site (Figure 13-4). You can visit the southern flank and look across a "no-man's-land" to a phony "ideal" North Korean village where propaganda billboards tout the superior lifestyle of North Koreans. The route to the DMZ, just 25 miles (40 km) north of Seoul, is marked by military bunkers and bomb-filled overpasses ready to crush any invasion from the North. The North Koreans have made several tunneling attempts into the South, but these have been detected and destroyed.

The DMZ is about 2.5 miles (4.2 km) wide and 155 miles (258 km) long, making it one of the largest unpopulated areas in northeast Asia. Riddled with landmines, tank traps, sensors, and automatic artillery, and festooned with barbed wire, the Korean DMZ is also home to a host of wildlife. Black bears, antelope, roe deer, rare cranes, eagles, and other species now inhabit this otherwise desolate strip of territory. Some environmentalists fear that a reunification of the two Koreas might spell doom for this vibrant ecosystem.

The Korean War devastated the country, especially the North, where systematic bombing eradicated the industrial plants. Some 4 million perished, and millions of refugees remained trapped on the wrong side of the divide, most never to see their families again. Around 1 million refugees fled to the South with the retreating U.N. forces.

Subsequently, the Russians and Chinese aided the North and the Americans supported the South in recovery efforts. Korea was now two countries, each pursuing its own path of development.

North Korea: A Closed Society

SOCIETY AND ECONOMY

The Democratic People's Republic of Korea (DPRK) is one of the world's most closed and rigid societies, if not the most. Under the absolute power of "Great Leader" Kim Il Song, the North was brought from the brink of disaster and turned into an industrial economy based on central planning and state ownership of the means of production. Individual needs are subjected to the will of the state, and ownership is limited to individual and household possessions.

The Great Leader espoused the ***juche* philosophy**, meaning self-reliance or self-sufficiency. Juche is based on the proposition that humans are masters of all things and that they, among all creatures, are endowed with the capability of self-reliance and independence. However, the masses cannot succeed in their struggle for these goals without a supreme leader to give form to their causes and direction to their actions. Kim Il Song was followed by his son, "Dear Leader" Kim Jong Il in 1994.

North Korea has a highly socialized command economy, in which farm land is collectivized and agriculture accounts for 37 percent of the workforce. State-owned industry produces 95 percent of manufactured goods, mainly heavy industrial products and armaments. There is a shortage of arable land, and the country has not achieved self-sufficiency in food production.

North Korea needs at least 6 million tons of feed grain a year to get by. Torrential rains in a badly deforested environment drastically reduced harvests from 1995 to 1998. Aid workers and others who managed to sneak across the Chinese border to the Korean community there reported widespread starvation. Hillsides were stripped of vegetation; people ate bark. There were no cats, dogs, or chickens. It is said that the military and the Communist Party elite remained well fed, due to food aid from China, Russia, Japan, and the United States.

In 1995, peasants were allowed to cultivate private plots, but this resulted in reduction of yields at the collective level. The DPRK estimates that famine claimed at least 220,000 lives between 1995 and 1998. International organizations estimate that 1.5 million North Koreans have perished due to famine-related causes. Although climatic conditions have improved in recent years, the nation is still significantly short of food and lacks the minimal calories required each day to stave off malnutrition. The government has broadcast recipes for boiling grasses.

Data on urbanization in North Korea are notably unreliable. However, according to DPRK information, urbanization has increased from less than 20 percent in 1953 to about 60 percent in 2009. The definition of urban is unclear. Most urbanites live in 23 cities of more than 89,000. Pyongyang has over two million. Hamhung, the second largest city, is about a third the size of Pyongyang. The port of Chongjin is another large, industrial city.

Pyongyang

Pyongyang, North Korea's largest city, is also its national capital. Situated along the Taedong River near the west coast, it controls all commerce moving north and south along North Korea's western side of the peninsula. Leveled to the ground in the Korean War, Pyongyang was rebuilt with wide boulevards and massive, grim government buildings typical of those built in the Soviet Union in the late 1940s and 1950s (Figure 12-5). Drab apartment blocks have few amenities. It is the only city in North Korea to have an underground subway system. Although there are only two lines, the subway is considered a national accomplishment and ranks with political and cultural monuments as a point of pride.

In 2002, North Korea allowed wages and prices to rise, and created incentives whereby factories could turn a profit. Consequently, consumerism has increased, at least in Pyongyang. Billboards advertise a North Korean–made car—the *Huiparam* or "The Whistle." The Whistle is assembled in the port city of Nampo and is made with Fiat parts imported via Italy and South Korea. The enterprise is a joint venture with a South Korean automobile company. However, most North Koreans cannot afford to purchase a car.

Along major streets, sidewalk kiosks are open for business. Locally grown apples and peaches are sold alongside imported pineapples. Clothing,

Figure 13-5

Pyongyang. Tony Waltham/Getty Images.

televisions, and other cheap goods from China, dishwashing soap from Thailand, and beer from Singapore are among the goods that are available. Prices are determined by customer demand and people pay with cash, not with coupons as in the past.

Since the changes were introduced, increasing numbers of cars and bicycles are seen in the streets. People are wearing brighter colored clothes from China instead of the grey and brown Mao suits and peaked caps that they wore before. Some young people even wear baseball caps, and the sight of Mickey Mouse T-shirts is not uncommon.

Economic reforms have at least affected Pyongyang. Their impact in other North Korean cities is unclear. Some experts say that the country will achieve widespread market socialism like China. Others say that these reforms are not enough to generate much change overall. The United States views the changes as insignificant and considers North Korea to be a "rogue state."

Beijing has cast its relationship with North Korea as being "close as lips and teeth." It is in China's interest to prop up the North Korean regime while at the same time effecting its redirection along a more liberal path according to the China model. China wants regional stability. It does not want a floundering country on its northeastern border next to its own Korean population, nor does it want a unified Korea, friendly to the West and housing American troops. China is the DPRK's chief trading partner.

SONGUN: ARMY-FIRST POLICY

North Korea, with its population of a mere 22.7 million, has the fifth largest standing army in the world after China, the United States, Russia, and India. Kim Jong II established *Songun*—Army-First Policy to protect his nuclear program. *Songun* is the overarching state ideology that "puts the army before the working class." It instructs that, "the gun barrel should be placed over the hammer and sickle" (representing workers).

Furthermore, the DPRK has one of the world's largest biological and chemical arsenals with enormous stocks of anthrax, cholera, and plague. It also has eight biochemical production facilities.

More than 1 million troops hover over the half-million South Korean forces. Only the tallest soldiers get stationed at the DMZ. This is true for South Koreans and Americans as well.

Today, many young North Koreans eschew higher education in favor of joining the armed forces. Here they will get good pay and sufficient food. It is estimated that military expenses consume one-quarter to one-third of the entire national budget despite near-famine conditions in many parts of the country.

MISERY AND REALITY

Although 37 percent of North Koreans are engaged in agriculture, the state farms are notoriously inefficient. Hunger and starvation remain persistent problems, with some 40 percent of North Korean children chronically malnourished. Consequently, stunting and other diet-related deficiencies are widespread. Food is rationed.

In 2008, the United States agreed to give food assistance via the World Food Program and NGOs but rejected shipments in 2009. The economy is believed to have grown that year due to more favorable climate conditions and energy assistance from other countries.

In the DPRK, you can be arrested for the most minor infraction. For example, if you don't keep your pictures of Great and Dear Leader perfectly dusted and shined, you can be put into one of 5 to 7 prisons in the gulag. Typically, your entire extended family will be arrested as well under the country's "guilt-by-association system." You are also expected to spy on your friends and neighbors. Listening to foreign radio broadcasts is punishable by death. Public executions are held to remind people to obey their Leader at all times.

There are an estimated 200,000 political prisoners—men, women, children, and elderly. Prisoners are provided starvation-level rations and forced to work long days in brutal conditions. Many face torture or execution for minor offences. More than 400,000 have died in the DPRK gulag over the last 30 years.

North Korean society is stratified according to Communist Party loyalty; level of commitment determines job prospects and access to housing, education, and health care. Subjugation to a tightly structured administration is a familiar mode of rule in Korea, and this reinforces the absolute authoritarianism that North Koreans now experience.

NUCLEAR CRISIS

In 1994, the DPRK suggested that it could build a nuclear bomb by extracting plutonium from one of its test reactors. In response, the United States, South Korea, and other countries agreed to provide the North with oil until the South Koreans completed two nuclear power reactors in the North. The North Koreans were to dismantle their bomb-making program.

In 2002, North Korea admitted that it had not dismantled its bomb-making program. Interested parties, including South Korea, Japan, Russia, and the United States have made several attempts to reason with the North Koreans, but to no avail. U.S. President George W. Bush cut off shipments of oil and other supplies to the DPRK, which responded by evicting the United Nations inspectors who had been monitoring the nuclear program. The situation remains unresolved.

The DPRK also has an arsenal of missiles capable of striking Japan and beyond. It sells these to Middle Eastern countries such as Iran for badly needed foreign exchange. In 1998, while engaged in talks with the United States regarding normalization of relations and food aid, North Korea proposed that Washington pay US$500 million to stop missile sales. This didn't happen, and the DPRK continues to sell arms to anti-American governments around the globe.

North Korea conducted a nuclear missile test in 2006. DPRK's actions have provoked concerns in the region and have revived support for "Star Wars" type missile defense systems in Japan and Taiwan. China is not pleased with this turn of events. Clearly, North Korea's actions are detrimental to any consideration of reunification of North and South.

BEIJING'S HEADACHE

In order to get its way with China, Pyongyang can threaten to destabilize Northeast Asia. It can play the "collapse card" by appearing to be on its last legs, or it can play the "cataclysm card" by preparing to lash out at its perceived "enemies."

It is in China's interest to maintain a stable North Korea. It does not want millions of North Koreans pouring across its 816 mile (1,360 km) border in the northeast to where an estimated 30,000 to 300,000 have already escaped. In fact, China has constructed a fence along the border. China routinely sends refugees back across the border where they are arrested, taken to labor camps, and often executed.

The plight of Korean refugees in China is indeed a sad one. Men are exploited and are lucky if they get a living wage. Women are raped and shipped off to other parts of China as "wives" or prostitutes. In recent years, the Chinese authorities have been conducting house-to-house searches to root out refugees that China calls "defectors." This gives them criminal status. The only way to get across the border and stay in China is to pay expensive bribes.

China continues to ship low-cost food and fuel to North Korea. It also encourages Chinese entrepreneurs to invest there. However, in private talks, China has forewarned the DPRK that it should not count on China to rescue it out of serious trouble. Should China support the DPRK's invasion of South Korea, it would damage its economically beneficial relationships with the rest of Asia and the world. Further, China cannot bring down another communist country without losing face.

China is concerned about North Korea's nuclear program and has stated publically that it is committed to a non-nuclear DPRK. However, its position most likely represents a long-term ideal as opposed to an immediate policy priority.

Travel in Dear Leader's Land

No, you can't go to North Korea on your own. You must apply for a visa; you must be "approved;" you must go on a guided tour. No, you can't walk outside by yourself. You must have two "minders" with you at all times. Each minder watches the other and both watch you. No, you can't take pictures of just anything. You must have permission of your minders to take any photograph. No, you can't eat or shop just anywhere. These decisions are made for you by your minders. No, you can't switch channels on a TV; there is only one channel—the government one. No, you can't say or do anything that might be construed as an "insult" to Dear or Great Leader. You will probably be arrested.

If you do manage to get into the DPRK, you will be provided with entertainment. For instance, you might get to watch 10,000 children dancing in synchronization, dressed as eggs. In this way you will learn to appreciate what Great Leader Kim Il Sung proclaimed: "We must give to all the workers a proper and enough education about the very long history and culture of our country, so they'll appreciate all the splendid traditions and heritage of our nation in all its beautiful morals and customs in the way that they'll carry them forward and develop them according to the new today's socialist life."

The two Koreas had an agreement to allow guided tours for South Koreans to the mutually revered Mt. Kumgang just across the border. South Korea suspended the tours in 2008 after a Seoul housewife was shot dead because she had inadvertently wandered into a poorly marked "closed" military zone. North Korea has threatened an array of retaliatory measures including the abrogation of a joint venture contract in Kaesong, where 42,000 North Koreans work in a Hyundai factory.

South Korea Inc.

South Korea has transformed itself from a poor, agrarian society to one of the world's most highly industrialized nations. With huge support from the United States following the Korean War, South Korea has become one of the newly-emergent, urban-industrial countries linked to the global economic system. Land reform, education, and economic expansion have eliminated the *yangban*, the landlord elite. The growth of a middle class has been accompanied by dramatic population growth and urbanization.

South Korea's governments have ranged from paternalistic democracies to dictatorships. Absolute power was used to promote personal savings and export industries. By the 1980s and 1990s, there was rising prosperity and large trade balances. Freer elections in the mid-1980s led to an expansion of democratic processes in the 1990s. Because of the influential role played by the government, South Korea is said to be a product of **state capitalism**.

STATE CAPITALISM

South Korea's transition from an agricultural to an urban and industrial society followed several phases. At the end of the Korean War, three-quarters of the South Korean people were engaged in agriculture. In post-war reforms, the government took over poorly run large estates and encouraged a new group of small landowners to farm with the aid of imported chemical fertilizer. Productivity rose rapidly and the country became largely self-sufficient in food.

Government and business targeted specific industries for development. Initially, emphasis was placed on light manufactures and textiles. In the 1970s, heavy industries such as iron and steel and chemicals were emphasized. Then the focus was shifted to making automobiles, ships, and electronics.

The lynchpin of industrialization was the ***chaebol***. A *chaebol* is a giant corporate conglomerate. Hyundai, Daewoo, Samsung, and LG (Lucky-Goldstar) are leading *chaebols*. In the early years of development, preferential investment in exports discouraged the production of consumer goods. This policy was reversed in the 1980s. Then labor unions demanded wage increases, and domestic consumption rose. Higher wages saw a decline in South Korea's international competitiveness in labor-intensive manufacturing, and overseas investment for cheap labor sources became essential.

Since 1990, South Korea has become a major investor in China and Southeast Asia, where it has access to cheap labor. Much of this investment is in China's nearby coastal province of Shandong. Initially, the investors were small businesses but now it is the *chaebols* that are establishing small companies in China to produce their components and products.

Chaebols

Chaebols are the equivalent of the Japanese *keiretsu*. Both terms are written with the same Chinese characters meaning "fortune cluster." Owned and managed by family groups, they are enormous conglomerates of diversified businesses, each having hundreds of ancillary firms. Of about 30 such concerns, Samsung is the largest. Samsung manufactures products ranging from airplanes and semiconductors to flour and paper. It also operates an array of service firms such as insurance, hotels, and department stores.

The rise of *chaebols* to economic dominance can be attributed, in part, to Confucian ideals and loyalties. The *chaebol* is like a family with its structured loyalties intensifying toward the family head: the boss. Unlike Japan, where consensus rules, in Korea consensus is what the boss says it is.

In the Hyundai *chaebol*, Confucian spirit became the "Hyundai Spirit." A manual given to new recruits observes: "The indomitable driving force, a religious belief in attaining a goal, and a moral diligence tempered with the frugality of the Hyundai group have materialized as a major part in the heavy industry of Korea."

Another important *chaebol* anchor is the support of state-owned banks and guaranteed loans. Government-infused capital has fostered expansion and diversification. In 2000, the four top *chaebols* accounted for 40 to 45 percent of South Korea's economic output.

More recently, with rapid wage increases under democratic reforms, *chaebols* have become less competitive. Disbanding or selling off numerous unprofitable "family members" (ancillary firms), venturing more into research and development, and procuring cheaper labor overseas are part of recent restructuring efforts. *Chaebols* today do not have the power they once had.

South Korean enterprises are found worldwide. For example, Daewoo, which means "Great Universe," has poured billions of dollars in planet-circling investments including electronics and components, hotels, autos, construction, finance, textiles, and heavy industries. Daewoo's globalization plans include hiring 250,000 foreign workers at 1,000 overseas subsidiaries. Daewoo Motors has plants in such countries as Poland, India, Iran, Uzbekistan, and the Philippines. Daewoo Electronics hopes to capture 10 percent of the world market in TVs, VCRs, PC monitors, refrigerators, washing machines, and microwave ovens.

South Korea is the world's number one shipbuilder. Hyundai, Samsung, and Daewoo are the largest corporations in this sector of the economy. More than 40 percent of global ship orders go to these companies. Some worry about competition from China but such fears are probably ill-founded.

New shipyards in China cannot compete with those in Sough Korea in terms of engineering, technical expertise, or quality. Moreover, China has trouble meeting production deadlines. In addition, China manufactures "simple" ships such as oil tankers and dry-bulk vessels. China is more likely to cut into the Japanese market, as Japan builds similar ships that are more costly due to higher labor costs.

ECONOMIC CRISES AND CHANGE

The Asian financial crisis of 1997 was especially severe in South Korea. Having unlimited access to bank loans, many *chaebols* incurred serious debt. That year, the Korean currency was devalued 50 percent and many subsidiaries went bankrupt. Per capita income declined by 35 percent and 1.6 million people became unemployed. The International Monetary Fund (IMF) came to the rescue with the largest loan ever awarded a single country. Reforms were introduced to reduce the power of the *chaebols* and more loans were given to small and medium-sized companies. State-owned industries are being privatized.

Unemployment continues to be a problem. About five million South Koreans between the ages of 20 and 34 are unemployed or underemployed. Graduates fare the worst in terms of having difficulty finding jobs. In 2002, six out of ten higher-education graduates failed to find employment.

There are two reasons for this situation. The first is demographic. Children of the country's second baby boom—the progeny of the first baby boomers born after the Korean War—are reaching adulthood and pouring into the job market. The second reason is economic. Since the Asian financial crisis of the late 1990s, companies have become more focused on profits and are reducing workforces. Many rely on temporary employees and others with more experience and skills. They have also turned to technology for labor-intensive tasks and are outsourcing jobs to China and Southeast Asia.

Figure 13-6

This is the food floor in the multistory Lotte Department Store in Seoul.
Photo courtesy of B. A. Weightman.

Improvements in the economy will not necessarily translate into positions for younger workers. Take the giant Lotte Department Store as an example (Figure 13-6). Lotte has developed a checkout counter system that replaces workers with personal digital assistants and is planning to install these in its 20 stores nationwide. The company expects to replace at least 70 percent of its workers, mostly in their 20s.

The government is trying to help. In 2004, it spent nearly 500 million dollars in subsidies to companies that hire jobless youth. Job creation programs last between six months and a year and are designed to give young people experience and a chance to improve their job skills. Government ministries have opened up 142,000 positions and hope to stabilize the youth unemployment rate at 5 percent over the next four years.

South Korea has managed to avoid a serious impact of the recent economic downturn experienced by Japan and the United States among a spate of other countries. This is thanks to strong export growth, low interest rates, and major stimulus expenditure by the government. In fact, it was the only developed economy to actually expand in 2009.

The Development of Cheju (Jeju-do)

In 2002, the government announced plans to develop the island of Cheju off the southern coast. The island is very attractive with its relatively mild climate, volcanic mountains, waterfalls, and sandy beaches. It is a traditional honeymoon destination and possesses fruit plantations, fish farms, and abundant seafood. It can get cold and even snow in winter. Consequently, crops such as citrus are grown in protected areas or in heated greenhouses.

Because of its historic spatial isolation, Cheju language and culture evolved differently from the mainland. Cheju has a matriarchal society. It also has women who deep sea dive without scuba gear or oxygen. These *haenyo* gather conch, abalone, and other seafood. Today, all *haenyo* are older than 40 and young women are not interested in pursuing this occupation. Evidently, this is a dying art.

In contrast to modern skyscrapers in the two major cities, village houses are made with rough stones and their thickly thatched roofs are tied down with wide rubber bands to protect them from howling winter winds. Houses are heated from underneath by the *ondol* system. In winter, heat from an outside fire is fanned into channels under the house. The warm air rises to heat the room and the thick straw roof keeps it inside (Figure 13-7).

Cheju is known for its carved basalt figures called *dol Hareubang (harabang)* (Figure 13.7). These mushroom-like statues are called "stone grandfathers." Interestingly, their eyes are more Western than Asian. Their origin is unknown.

Officials who are developing the project boast of South Korea's and Cheju's geographic position between China and Japan. Cheju is two hours flying

Figure 13-7

This "stone grandfather" (harabang) *is hand-carved out of basalt. The origin of* harabangs *is unknown.* Photograph courtesy of B. A. Weightman.

time from Hong Kong, Beijing, and Tokyo. The island is becoming an international business and tourism hub as well as a free-trade area where English is the second language. There are shopping malls, a science and technology park, and several golf courses. It now has a World Cup soccer stadium, a new airport, and many hotels have been built. An eco-friendly naval base is to be installed in Cheju by 2014.

URBANIZATION

South Korea's population more than doubled from 20 million in 1949 to 48.7 million by 2009. A major contributor was the influx of 3.3 million Koreans who had been living in Japan, China, and the Soviet Union. An additional 2 million refugees migrated from North Korea. A population control program launched in 1965 curbed the birth rate, but improved living standards, lowered the death rate, and increased life expectancy. Rural–urban migration depleted the countryside and provoked a major urban housing crisis. Millions of apartment units (flats) have been constructed to house this 82 percent urban society.

Songdo City is a new development under construction on reclaimed land near the port of Incheon on the Yellow Sea. Songdo will be a free economic zone with 80,000 apartments (already rented or sold out), schools, hospitals, extensive office and retail space, and underground train links to Seoul. Plans call for 40 percent of the city to be "green." Computers will be built into every building, office, and street as part of a ubiquitous network linking every person and place. Urban theorist Mike Davis calls Songdo City a spectacular example of "imagineered urbanism," When completed in 2015, it will be the gateway to Northeast Asia. Songdo City is Korea's answer to Shanghai and Dubai. It is billed as the largest real estate development in history.

Urbanization and apartment living have altered Korean traditional family structure. Most Koreans now live in nuclear family settings rather than in extended family systems. However, Confucian values prevail. Although women now have more control over their lives, they are far from equal to men in decision making and economic opportunity. Women are found mostly in service or "nimble-finger" jobs. Corporations are overwhelmingly male enterprises at management levels.

Two-thirds of South Koreans live in cities of 50,000 or more with 47 percent living in and around Seoul. Greater Seoul has sprawled eastward and southward with an agglomeration of new satellite cities (Figure 13-8). Formerly focused on a single central business district (CBD) dating from Japanese times, Seoul has evolved from a monocentric to a multicentric city. Occupation rates are well over 100 percent, reaching even 600 percent in older, more densely populated areas! This means that most flats house two families—one owner and one renter.

Much of downtown Seoul is underground. Miles of shopping arcades lie under the streets. All are connected to a network of subway lines and linked to major department stores and hotels above ground.

Another modern development is the rejuvenation of the Han River. Formerly a trash-filled, dry ditch in summer, its water level is now kept constant by a series of dams. The river is criss-crossed by bridges and lined with parks and high-rise housing developments (Figure 13-9). Whereas 80 percent of raw sewage used to flow into the Han channel, new sewage systems and treatment plants now treat all sewage. Today, the Han River carries sailing and passenger boats.

Figure 13-8

This is a southern expansion of Seoul looking past the Han River and the Olympic Expressway to the suburb of Tongjakgu. More than half of Seoul's population live south of the Han River. Photograph courtesy of B. A. Weightman.

Incheon and Taegu are cities of more than 2 million, while Taejon and Kwangju have more than a million residents apiece. The port of Busan has a population of close to 4 million. With their concentrations of industry, these large cities continue to grow, while medium-size and smaller places tend to stagnate. However, even their rate of growth is decreasing as traditional industries have begun to decline.

With the bulk of industry still concentrated in the capital region of Seoul, the Korean government has devised a **technobelt** system whereby research and industry are to be linked via communication and telecommunication. The key nodes of this system are Seoul and Taedok, a new research and development center near Taejon.

Each belt has designated specializations. For example, the Seoul to Kwangju technobelt will focus on energy and food production. The Seoul–Taedok–Busan technobelt will concentrate on electronics and textiles. Other areas designated for growth (e.g., Masan and Changwon on the south coast) have become engineering centers linked with the Busan–Ulsan heavy industrial region. Ulsan is dominated by Hyundai car manufacturing and shipbuilding.

Figure 13-9

This housing development is on the island of Yeouido by the Han River in Seoul. Photo courtesy of B. A. Weightman.

Overseas Koreans

More than 5 million Koreans live overseas. The largest group resides in China. As many as 2 million live in China's Manchurian region. The strongest concentration is in Yanbian Korean Autonomous Prefecture just north of the Tumen River. Here, Koreans are able to exercise a degree of autonomy and speak Korean in accordance with the rights given to all minority nationalities in China.

Koreans were pioneers in opening up the cold Manchurian region to paddy rice farming. Rice was first grown in Yanbian in 1877, and Koreans are still responsible for producing much of the rice crop in the most northerly province of Heilungjiang.

Large contingents of Koreans also live in the Central Asian republics of Kazakhstan and Uzbekistan and on Sakhalin off the east coast of Siberia. Unofficial estimates range up to 750,000. These populations are important links for South Korea's recent Central Asian investments.

As discussed earlier, the Korean community in Japan makes up 85 percent of its foreign population. Regarded as citizens of the Japanese empire during colonial times, the majority of Koreans continue to be classified as aliens and suffer from various forms of discrimination.

Emigration to the United States began in the early twentieth century. From 1903 to 1905 a group of 7,226 Koreans was allowed to live in Hawaii and work on sugar plantations. Under Japanese pressure, this migration was stopped and many Korean settlers abandoned the plantations and became shopkeepers. Some 2,000 moved to California. This was the seed of a subsequent migration stream to the United States.

The number of Koreans living in the United States has increased from 38,711 in 1970 to more than a million today. Korean-Americans are the seventh largest immigrant group in the country. Over half live in just four states: California, New York, New Jersey, and Virginia.

The majority of Koreans living outside the country are Christian. Christians comprise 20 percent of the Korean population. Most westerners don't realize that Korea is home to the world's five largest mega-churches.

About a third of America's Koreans live in California, mainly in the Los Angeles area. Other significant concentrations are in New York and Chicago.

THE UNITED STATES AND SOUTH KOREA

For the past 50 years, South Korea's foreign policy has been in virtual lock-step with that of the United States. However, in 2002 when President Roh Moo-hyun was elected, he said that his country would no longer "*kowtow*" to the United States.

There has been a wave of nationalism in South Korea triggered by the acquittal of two American soldiers charged in the deaths of two girls killed during a training exercise. "Wrongs will be righted," declared President Roh as he referred to grievances associated with the bilateral treaty that governs the 37,000 American troops stationed in his country. Even North Korea has criticized the treaty because, in essence, it gives extraterritorial rights to the soldiers. In other words, American troops are subject to American and not South Korean laws.

As the relationship between the two countries has changed, the United States has taken steps to downplay its profile. It is closing a large military base in Seoul and has pulled back its troops from the DMZ. However, troops are being maintained in other locations. South Korean conservatives are not happy with these moves as they feel vulnerable to attack from North Korea. Moreover, they do not want to pay for the moves. Only time will tell the ramifications of this situation.

More recently, South Korea has drawn closer to the United States. At the London G-20 Summit in 2009, President Obama called the country: "one of America's closest allies and greatest friends." The G-20 is a group of 20 countries representing some of the world's leading economies. China, Japan, Australia, South Africa, France, and Saudi Arabia are among the 20. The summits are designed to share ideas on global economic and financial issues.

PROSPECTS FOR REUNIFICATION

North and South Korea have entered the twenty-first century as two diametrically opposed systems with divergent paths of development. Both Koreas remain pledged to reunification and occasionally present proposals. A strong emotional commitment to reunification exists among many Koreans, especially those with cross-border family ties (Figure 13-10).

Figure 13-10

Buddha for "Reunification Desire" at Tonguasa Temple near Daegu (Taegu).
Photo courtesy of B. A. Weightman.

Extant differences make reunification problematic. North Koreans are not even an eighth as rich as South Koreans. South Koreans are twice as numerous. North Koreans have virtually no idea of how to function in a modern, technologically advanced society.

North Korean defectors encounter great difficulty in the South, despite being offered generous financial aid, job training, and other assistance. They have trouble dealing with the competitive nature of a market society, handling money, and making wise choices among competing goods. They have discovered that their DPRK education is insufficient given the need for computer skills, grasp of English, and knowledge of Chinese written characters. Furthermore, many of the more than 6,000 defectors suffer psychological problems much more so than do immigrants from other countries.

There are only 24 telephone lines between Seoul and Pyongyang, and these are rarely used. In the 1980s, a visiting DPRK delegation accused the South of rounding up every car in Korea to create traffic jams in Seoul. Even the language has evolved differently, with the South's technological vocabulary unknown in the North.

Raising the North's productivity level to stem a southward flow of migrants would cost billions. Although the South has traded with the North since 1990—food for raw materials—investment has been severely limited by the DPRK government.

South Korea wants to reunify in stages: diplomatic contacts to build trust; investment in special economic zones in the North; and, down the line when the economic gap has been narrowed, full unification. In 2000, an historic event took place. President Kim Dae Jung of South Korea met with President Kim Jong Il of North Korea to reduce tension between the two countries and promote peace and reconciliation. This decision, made by President Kim, became known as the "Sunshine Policy."

As a result of the meeting, both sides have stopped propaganda broadcasts across the border and allowed a limited number of cross-border visits for family reunification. Symbolically, North and South Korean athletes marched together under one flag at the Olympic Games in Sydney, Australia, in 2000 and in Athens, Greece, in 2004. A joint team was also sent to the Beijing Games in 2008 and, for the first time in history, the Olympic torch was brought from Seoul to Pyongyang. More talks are scheduled in the future. Kim Dae Jung has been awarded the Nobel Peace Prize for his efforts toward peace in the region. However, North Korea's recent nuclear program threatens reunification considerations.

Taiwan: Little Tiger and Big Dragon

Small but rich, economically powerful but politically isolated: this is Taiwan, an island just 100 miles (165 km) from the big dragon, mainland China (Figure 13-11). The history of Taiwan is inextricably tied to that of the People's Republic, and its current development remains couched in that relationship. It is one of East Asia's newly

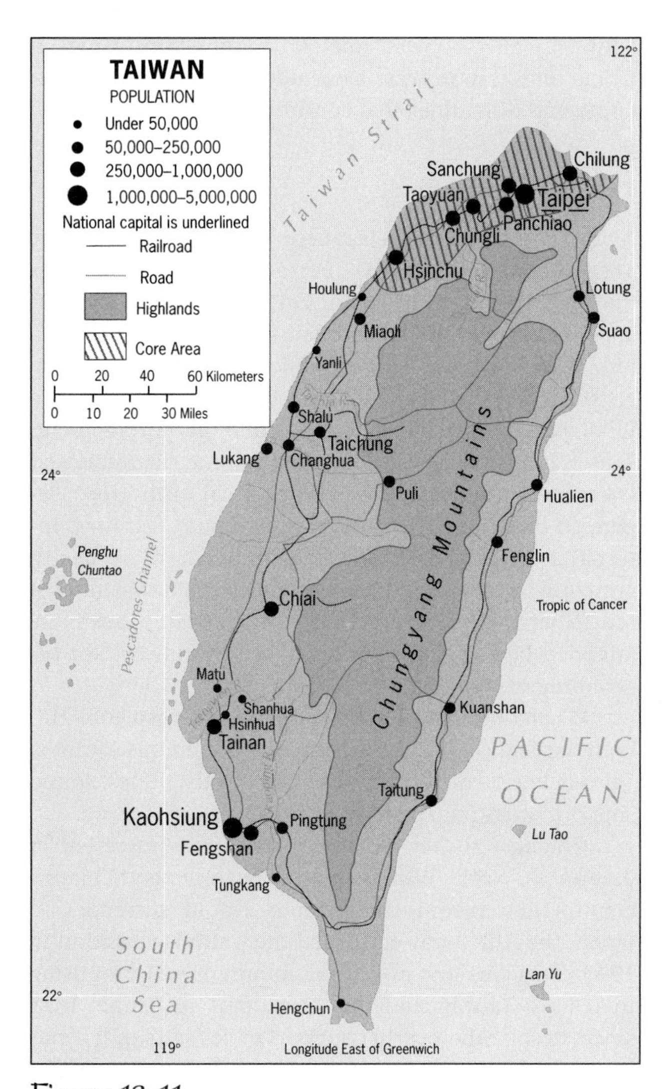

Figure 13-11

Taiwan. From H. J. de Blij and P. O. Muller, *Geography: Realms, Regions and Concepts*, 14th Edition, 2010, p. 505. Originally rendered in color. © H. J. de Blij and P. O. Muller. Reprinted with permission of John Wiley & Sons, Inc.

industrializing "countries" and has achieved tiger status in terms of economic development.

THE PHYSICAL LANDSCAPE

Taiwan incorporates the main island and 64 smaller islands called the Penghu Islands. The Penghu were colonized by the Portuguese, who named them the Pescadores (Fishermen). Here, fringing reefs provide good shelter for ships and there is an important naval base at Makung. Because of their position in the Taiwan Strait, the Penghu Islands are of great strategic importance to Taiwan.

The island of Taiwan can be divided both vertically and horizontally. The vertical division is created by the north-south ridge of the Chungyang Shan (Central Mountains) ranging from 10 to 35 miles (16–56 km) in width. The Chungyang are extremely rugged with more than 30 peaks reaching over 10,000 feet (3,000 m). The highest is Yu Shan at 13,113 feet (3,997 m). River valleys and canyons cut the mountainsides, which are covered with coniferous and mixed forests. With the exception of the Taitung rift valley, cool temperatures, thin soils, and severe erosion hinder cultivation. Still, the mountains are important for forest products and hydroelectric power.

Most people dwell on the western coastal plain, which can be divided horizontally by the Choshui River. To the north are hills and basins, including those of Taipei and Taichung. To the south is a low, flat, alluvial plain. The river also marks the climatic boundary between north and south. In the north, the rainy season is from October to March; in the south it is from May to September. Typhoons are common in July and August. Agricultural patterns differ as well with such crops as rice and tea in the north and soybeans, pineapples, sugarcane, and other tropical crops in the south.

Agricultural output has steadily declined over the years as farmland and employment shifts to urbanization and industry. Taiwan is 78 percent urban with major population clusters in the capital of Taipei, and in the ports of Chilung, Tainan, and Kaohsiung.

FROM *BAO DAO* TO FORMOSA TO TAIWAN

Ancient Chinese maps portray the island of Taiwan as the island of *Bao Dao*. Written accounts referred to it as "a mudball across the sea, not worthy of China." *Bao Dao* was known to Chinese fishermen and pirates but was inhabited by non-Chinese people. These aboriginal people are Malay-Polynesian and are related to the indigenous people of the mountains of northern Luzon, the closest island of the Philippines. There are nine distinct ethnic groups among the aborigines, each with its own language and culture. However, immigration from China pushed these people into the mountains and subsequent discriminatory policies of the Taiwanese government have rendered their cultures almost extinct.

Immigrants from China began arriving in the fourteenth century, a migration stream that continued until World War II. These were Han people from the Fujian coast who spoke the dialect known variously as Amoy, Minnan, or Hoklo. This dialect is called Taiwanese, and these people are known as native Taiwanese. Native Taiwanese are 70 percent of Taiwan's population. In recent years the term Taiwanese has come to be used to mean everyone who lives on the island.

In the sixteenth century, the Portuguese were struck by the beauty of the forested mountains and pristine beaches of the island and called it Ilha Formosa, "the beautiful island." Taiwan was known to the world as Formosa for much of the twentieth century.

Japanese traders also visited the island and even established a settlement near present-day Tainan. They called the island Tai Wan, meaning "Big Bay."

In 1624, the Dutch established forts Zeelandia and Providentia near the by then defunct Japanese settlement at Tainan. Spain established colonies in 1626 but was later ousted by the Dutch. The Dutch established sugar plantations and encouraged increased immigration from China as fieldworkers. The Dutch were thwarted by a Ming general in 1662.

In the seventeenth century, Hakka speakers from Guangdong arrived and settled in the foothills. There, they developed Taiwan's forest products industry including camphor wood and oil. Hakka account for 10 to 15 percent of the population and make up a disproportionate share of Taiwan's industrial workers.

Emigration from China had been illegal since the Ming Dynasty. Even so, people left to avoid taxes and seize new opportunities elsewhere. The Chinese regime regarded them as people who renounced civilization. To China, Taiwan was uncivilized.

In 1684, the Qing Dynasty achieved control of Taiwan for the first time. But the island's population rebelled against them repeatedly over the next 200 years. One official Qing document complained: "Every three years an uprising! Every five years a rebellion!" Chinese immigration continued, and by the late nineteenth century, the island had a population of 2.5 million.

Western traders reappeared after the Opium Wars of the 1840s. Britain and Canada expanded agricultural exports, sent Presbyterian missionaries, and established hospitals and schools.

The safety of foreign traders was a major issue. China denied responsibility for incidents of violence because these were taking place "beyond the boundaries of China." Nevertheless, China made Taiwan a province of China in 1887. These unclear relationships between China and Taiwan were a prelude to decades of uncertainty and difficulties that continue to this day.

JAPANESE RULE

At the end of the Sino-Japanese War (1894–1895), the Treaty of Shimonoseki gave Formosa and the Pescadores (along with Korea) to Japan. On June 1, 1895, a resistance movement proclaimed Asia's first republic: the Republic of Formosa. This republican group was quickly crushed, and the Japanese began their era of harsh rule.

Japan accelerated Taiwan's economic development. This was meant to provide Japan with agricultural and later, industrial products. The colonial authorities expanded electric power, and improved port facilities, internal transportation, and communication. A health system was established, and an educational system promoted Japanese language and culture. Many Taiwanese attended Japanese universities. The groundwork for future progress was laid at this time.

When the Qing Dynasty was overthrown in 1911, both the Chinese Nationalists and the Communists wanted Taiwan liberated from the Japanese. Both groups voiced support for an "independent and democratic" Taiwan.

Chiang Kai-shek became increasingly interested in Taiwan as World War II ensued. Fearing that Chiang's corrupt and incompetent troops would surrender to Japan, the Allied powers issued the Cairo Declaration in 1943. This was one of several communiqués promising to restore Taiwan and the Pescadores to China. Ever since, despite their earlier support of Taiwan's independence, the Nationalists and Communists have cited this declaration to justify their view that "Taiwan is a sacred and inseparable province of China." But the Cairo Declaration was, in fact, a statement and not a legally binding treaty. The island's uncertain status was furthered by the Allied-Japanese peace treaties that left the fate of the island to its inhabitants.

Another element of the island's population consists of immigrants who arrived from the mainland between 1945 and 1949. These are referred to as mainlanders, although their "mainlander" descendants are now born in Taiwan. In fact, more than 90 percent of Taiwanese were born on the island.

THE "ONE CHINA" QUESTION

In 1949, Chiang Kai-shek and his two million KMT loyalists retreated to Taiwan where they set up what they claimed to be the true government of the Republic of China (ROC) in exile. In that same year, Mao Zedong renamed China the People's Republic of China (PRC). Both Mao and Chiang claimed to head the government of "one China," of which Taiwan is a province.

In Taiwan, Mandarin was declared the official language by the KMT government. However, Taiwanese is widely spoken in informal settings, and current demand has resulted in the production of Taiwanese language media.

The United States was entrenched in the Cold War against communism and consequently supported Chiang Kai-shek and the KMT. However, this support met with world realities in 1971 when the ROC lost its seat at the United Nations in favor of the PRC. The seating of the PRC in the world body quickly eroded Taiwan's status as the sole representative of China. In 1972, U.S. President Nixon signed the Shanghai Communiqué by which the United States acknowledged the PRC view that "all Chinese on either side of the Taiwan Strait maintain there is but one China and that Taiwan is a part of China." It gave no credence to the Taiwanese desire for self-determination.

In 1978, President Carter established formal diplomatic relations with the PRC and reiterated the Shanghai Communiqué. The United States also severed formal relations with Taiwan, although economic and cultural offices maintain informal ties. The United States wants the issue resolved peacefully. The PRC says that the issue of reunification is an internal affair.

Regional considerations are also critical. If the PRC controlled both flanks of the Taiwan Strait, it would have a chokehold on the main sea lanes to South Korea and Japan.

As PRC-international relations increased, Taiwan became represented only as "Taipei" in international organizations. Taiwan has no official embassies abroad but allows its trade offices to handle similar functions. In the 1990s, the island began promoting itself as "China Taipei."

Both Taiwan and the mainland cling to their notions of one China. The PRC hopes that the examples of Hong Kong and Macau and their "one country, two systems" policy will stimulate the ROC to formally rejoin the PRC. But many Taiwanese, aware of conditions on the mainland, have little interest in this goal.

The number of islanders who consider themselves "Taiwanese" only (not "Chinese") was 44 percent in 2007, or more than double the 20 percent recorded in 1994, and up from 37 percent in 2000. There is also talk of changing the name PRC to simply "China." The belief that all China should be united seems to be becoming increasingly amorphous.

While the China-Taiwan problem remains unresolved, tensions have eased since Taiwan's President Ma Ying-jeo declared a "diplomatic truce" shortly after he took office in 2008. Using a conciliatory approach, he declared that while Taiwan will not move toward political unification with China, neither will it declare independence. This stand is in line with the views of 75 to 80 percent of Taiwanese who want to preserve the status quo. On the mainland, President Hu Jin-tao has also backed away from an aggressive stance and seems satisfied to increase economic and cultural integration.

The tenuous political situation has not precluded ongoing and intensifying economic relationships between China and Taiwan. In fact, big dragon China is of significant economic consequence to little tiger Taiwan.

RISE OF THE TIGER

Taiwan's achievements as an economic tiger are underpinned by its excellent educational system, in-place infrastructure, and American connections. Initially, KMT land reform gave land to tenant farmers and labor-intensive industries employed thousands. The state confined itself to key sectors such as finance and energy, leaving most industrial opportunities to private entrepreneurs. Farmers now focus on specialty crops such as mushrooms and aquaculture products such as eels. While agriculture is very efficient, the country must import increasing amounts of food. Industry has moved steadily up the value-added ladder from apparel, to footwear, to electronics.

Like Japan, Taiwan must import most of its raw materials to produce finished exports. Consequently, most of the industries are located on the alluvial plain along the west coast where the major ports are located and where raw materials are unloaded. This industrial corridor is anchored by Taipei and its port of Chilung in the north and the port of Kaohsiung in the south. A high-speed rail line between the two cities is under construction.

Taiwan has long met the criteria of a *kotadesasi* (discussed in Chapter 4) through its economic transformation and evolving spatial structure. Growth of population in the four metropolitan regions of Taipei, Taichung, Tainan, and Kaohsiung and the construction of a north–south freeway from Taipei to Kaohsiung have created a metropolitan corridor similar to that between

Tokyo and Osaka. With its multiplex of service, industrial, and agricultural activities, linked by transportation and communication networks, this part of Taiwan exhibits all the features of a *kotadesasi* zone.

Hsinchu Science Park, established in 1980, has become the world's third largest high-tech industrial center with strong links to California's Silicon Valley. Hundreds of U.S.-schooled Taiwanese are now employed at Hsinchu. A second science park is being built near Tainan. Taiwan now leads the world in the production of notebook computers and is among the top three makers of desktop computers and peripherals. However, a small domestic market, increasing labor costs, shortage of land for expansion, and lack of raw materials have compelled structural and spatial changes.

Taiwan has taken advantage of commoditization of the personal computer (PC) industry: the division of the PC market into two sectors—big name-brand marketing and sales and hundreds of no-name component suppliers. Taiwan has garnered the latter—the Original Equipment Manufacture market. Three-quarters of the country's electronic products are sold under another brand name. For example, more than 80 percent of flat-screen displays sold under such names as Acer and Dell are manufactured by Taiwan and Korea.

In this export-driven economy, small and medium-size firms are dominant. Unlike Japan and South Korea where large corporations were beholden to large banks in the financial crisis of 1997, Taiwanese companies borrowed from savings clubs and mutual aid associations. Bad loans were minimal and banks did not collapse. Trade with China is easing Taiwan through the current economic downturn.

A penchant for expansion, combined with cost efficiency, has led Taiwan to extend its operations overseas, where both space and cheap labor are available. More than 80,000 firms have moved offshore—half to Southeast Asia and half to the PRC. For example, Paochen footwear could employ 10,000 workers in Taiwan but employs 120,000 in the PRC. On the mainland, it is the world's largest producer of Nike, Reebok, and other name-brand shoes.

CROSS-STRAIT BAMBOO NETWORK

A **bamboo network** of trade has grown more complex as China has become Taiwan's largest trading partner. In 2009, 30 percent of Taiwan's exports went to China. These include such things as machinery, chemicals, plastics, flat panels, electronics, and a variety of optical, photographic, and other technical instruments. Many exports are parts to be assembled on mainland factories—the final products destined for Japanese, American, and European markets.

Taiwanese businesses have invested an estimated US$150 billion in mainland projects. Along with Hong Kong, it shares half the investment projects in Shanghai's Pudong development project and China's Xiamen SEZ. Some estimates put Taiwan's involvement in China's IT production at 70 percent. In 2009 Taiwan opened up 100 of its industries to mainland investments. The first sectors to be opened will include such industries as automobiles, textiles, plastics, computers, cell phones, hotels, herbal medicines, and wholesale firms. About one million Taiwanese—mostly business executives—now live in China. Half of these live in Shanghai.

Until recently, because of the political situation, most Taiwan–China trade was indirect via Hong Kong. But Hong Kong, even with its SAR status, is now part of the PRC. In 1997, transshipment services commenced between Kaohsiung and Xiamen and Fuzhou in Fujian Province—the first authorized cross-strait shipping links.

Beginning in 2003, direct air links were put in place for the Chinese New Year celebrations. In 2009, the number of direct flights between China and Taiwan were increased from 108 to 270 per week. Furthermore, Taiwan increased its daily quota of visitors from China to 3,000 from 300.

To further ease restrictions, vessels from China or Taiwan are no longer required to transfer cargo at a third port. Instead, they only have to pass through the waters of a third port on their way across the Taiwan Strait. Foreign ships are not allowed to participate in this trade because the PRC considers cross-strait traffic to be domestic. Foreign ships can ply the Hong Kong–Shanghai route but not the Shanghai–Taiwan route.

Taiwan is also creating five Free Trade Zones (FTZs) to enhance the bamboo network and handle the China trade. FTZs now exist in the major ports of Taipei, Kaohsiung, and Taichung, and Keelung Taowan is the country's first aviation FTZ.

As part of its efforts to lure Taiwan into the PRC fold, China has deepened and widened, and increased the number of berths at all its major ports to accommodate large Taiwanese ships. It also gives preferential treatment to Taiwanese investors on the mainland who have built factories from Suzhou to Urumchi. Some fear that Taiwan's increased foreign investment and plant location in the PRC could ultimately hold the country hostage to the whims of the Communist government in Beijing.

Despite intermittent diplomatic friction, the cross-strait relationship has blossomed in recent years. China entered the WTO in 2001 and within a month, Taiwan entered as "Chinese Taipei."

Another symbol of rapprochement is the fact that China did not object to "Chinese Taipei's" participation as an observer at the World Health Assembly, the governing body of the World Health Organization (WHO). This is the first time Taiwan was granted observer status at a United Nations body since it lost its seat to China in 1971.

SOUTHWARD POLICY

Taiwan's **southward policy** involves increasing investment in Southeast Asian countries to balance investments in China. The ASEAN nations, including Vietnam, comprise Taiwan's third most important export destination after China and Japan. Subic Bay in the Philippines, Batam, Indonesia (just south of Singapore), and Vietnam are key targets for investment, construction, and development loans. Southeast Asia supplies Taiwan with raw materials including minerals and forest products.

This investment strategy is intended to upgrade Taiwan's industrial capacity, internationalize its business operations, and maintain its competitive advantage and economic growth in the context of regional and global dynamics. Moreover, strong economic impacts can only strengthen Taiwan's de facto political influence in the region.

Figure 13-12

The Taipei 101 building—the second tallest in the world.
© Reuters/Corbis Images.

THE DOWNSIDE OF DEVELOPMENT

Development has brought a lot of benefits, but many Taiwanese are not happy with its negative side effects. For instance, there is a growing rich-poor gap as money becomes increasingly concentrated in the upper echelons of society. By 2004, the richest 20 percent controlled more than half the wealth in Taiwan. Business is notorious for corruption and nepotism. Gambling, prostitution, and other criminal activities are serious issues. Individualism, materialism, and hedonism are striking blows at traditional Confucian values, to the dismay of the older generation. Americanization is often blamed for these trends.

Environmental quality has sadly deteriorated with unchecked air, ground, and water pollution. Traffic congestion is out of control, and the infrastructure cannot keep up with demands for electricity and power. One good sign is that people are forming issue-oriented public interest groups and that environmental organizations have multiplied. However, environmental and social campaigns have not kept pace with rapid change and the uncontrolled growth of Taiwan's material culture.

One controversial project is the Taipei 101 building. At 1,670 feet (509 m), it was the tallest building in the world until 2010 when it was eclipsed by the Burj Khalifa 2,717 foot (828 m) skyscraper in Dubai. Taipei 101 houses shops, restaurants, and offices (Figure 13-12).

Taiwan is hardly the place to build tall because of the many earthquakes that strike the island each year. A 7.6 quake centered in the middle of the island killed 2,415 people in 1999. The most recent quake was 6.4 in 2010. Furthermore, typhoons with high velocity winds impact the island several times a year. For example, Typhoon Morakot swept the island in August, 2009 causing the worst floods since 1959. More than 100 people were killed. Many Taiwanese believe that Taipei 101 is a dangerous building despite the designers' claim that it is perfectly safe and able to withstand the forces of nature.

In spite of its difficulties, Taiwan will continue to be an economic tiger in Asia. It will increase its economic ties with the mainland, but the future of its political relationship with the PRC, while now at a fragile status quo, remains in question.

Recommended Web Sites

www.adb.ord.org/korea/default.asp
Asian Development Bank site for Korea. "Outlook 2009 Update."
www.asiatimes.com
Articles focusing on various aspects of Asian countries.
www.cfr.org/publication/9223/
Council on Foreign Relations site. China-Taiwan relations. Search by region, country, or key word.
www.cia.gov/library/publications/the-world-factbook/geosks.html
CIA World Fact Book: Korea. Economic information and statistics. Links to North Korea and Taiwan.
www.dynamic-korea.com/contact.php
Official Web site of the Embassy of the ROK in the United States. Covers recent news. Read about the "Korean Wave."
http://end.stat.govtw/mp.asp?mp=5
National statistics for Republic of China (Taiwan).
www.korea-dpr.com/
Official Web site of North Korea. If you want to see real "propaganda," check this one out.
www.korea.net
Official Web site of the ROK. Short articles on current issues.
www.taiwan.com.au
Cultural and business conduit between Taiwan and Australia.

Bibliography Chapter 13: Korea and Taiwan Tigers Rising

Adams, Johnathan. 2009. "Will 'Red Money' Tame Taiwan?" *Far Eastern Economic Review*, June: 37–40.

Altenburger, Engelbert. 2004. Earthquake Hazards in Taiwan—The September 1999 Chichi Earthquake." *Focus* 48/2: 1–8.

Asian Development Bank. 2009. *Asian Development Outlook 2009*. Manila: Asian Development Bank.

Chan, Gerald. 1997. "Taiwan's Economic Growth and Its Southward Policy in Asia." In *Asia Pacific: New Geographies of the Pacific Rim*, eds. R. F. Waters, and T. F. McGee, pp. 206–222. Vancouver: UBC.

Dean, Jason. 2003. "On Top of the World for Now (Taiwan)." *Far Eastern Economic Review*, June 12: 30–35.

Demick, Barbara. 2010. *Nothing to Envy: Ordinary Lives in North Korea*. New York: Random House.

Eberstadt, Nicholas. 2007. *The North Korean Economy: Between Crisis and Catastrophe*. London: Transaction.

Fuller, Douglas, and Eric Thun. 2007. "Engineers to Researchers. *Far Eastern Economic Review*, July/August: 54–58.

Genser, Jared. 2006. "Stop Pyongyang's Autogenocide." *Far Eastern Economic Review* 169/9:15-18.

Garrett, Banning, and Bonnie Glaser. 1995. "Looking Across the Yalu: Chinese Assessments of North Korea." *Asian Survey* 35: 528–45.

Hiebert, Murray. 2003. "Decision Time Looms Over North Korea." *Far Eastern Economic Review*, May 15: 12–13.

Howard, Keith, ed. 1996. *Korea: People, Country and Culture*. London: School of Oriental and African Studies.

Hsieh, Chiao-min. 1964. *Taiwan—Ilha Formosa*. Washington D.C.: Butterworth.

Kang, Chol-Hwan and Pierre Rigolot. 2000. *The Aquariums of Pyongyang: Ten Years in the North Korean Gulag*. New York: Basic Books.

Kaplan, Robert. 2006. "When North Korea Falls." *The Atlantic Monthly* 298/3: 64–73.

Kim, Samuel. 2006. "The Mirage of a United Korea." *Far Eastern Economic Review* 169/9: 9–14.

Lee, Charles. 1997. "Tomorrow the World (South Korea)." *Far Eastern Economic Review*, May 1: 43–45.

Lintner, Bertil. 2007. "A Perilous Escape from Pyongyang." *Far Eastern Economic Review* 170/5: 29–32.

Lintner, Bertil. 2004. "Shop Till You Drop (North Korea)." *Far Eastern Economic Review*, May 13: 14–19.

McNeill, David. 2009. "Songdo City Defies Crisis Odds." *Asia Times*, November 12.

Palka, Eugene, and Francis Galgano. 2004. *North Korea: Geographic Perspectives*. Guilford, Conn.: McGraw Hill/Dushkin.

Scobell, Andrew. 2007. "Beijing's Headache over Kim Jong II." *Far Eastern Economic Review* 178/6: 35-8.

Song, Kimberly. 2004. "Daewoo's Long Road to Recovery." *Far Eastern Economic Review,* May 13: 40–45.

The Economist. 2006. "Free Trade Zones: Taiwan's New Window to the World." *The Economist*, December/January: 42–44.

Tkacik, John. 2005. "North Korea's Bogus Breakthrough." *Far Eastern Economic Review* 168/8: 21–23.

Tsai, H. H. 1996. "Globalization and the Urban System in Taiwan." In *Emerging World Cities in Pacific Asia*, eds. Fu-chen Lo and Yue-ma Yeung, pp. 179–218. New York: United Nations.

Chapter 14

Southeast Asia: Transition among the Nagas

"To understand the present and
anticipate the future, one must know
enough of the past, enough to have a
sense of the history of a people."

LEE KUAN YEW (1980)

"A more highly fragmented part of the globe could scarcely be imagined," said geographer Norton Ginsburg (1972). Southeast Asia, land of the *nagas* (see Chapter 1), is a realm of contrasts in both physical and cultural geography. Fragmented landscapes have formed the backdrop for the establishment of glorious kingdoms and trading states, while hills and valleys have complicated the imposition of boundaries and the demarcation of people and territory.

Separation has enhanced cultural diversity and thwarted regional unity. Nevertheless, several Southeast Asian states have become prominent on the global economic map. Malaysia, Thailand, and Singapore are designated as economic tigers. Indonesia, with its growth rate of four percent (2009) even in the economic crisis, is an emerging tiger. Vietnam boasts of its new label: baby tiger. In contrast, countries such as Myanmar, Laos, and Cambodia lag far behind in both the human condition and regional stability.

Many geographers stress the region's diversity over unity. Geographer Jonathan Rigg (1991) notes, "Southeast Asia represents a residual region: it owes its regional identity not to internal coherence, but to external incoherence." In this chapter, you will learn about the differences, but also the similarities, among the peoples, landscapes, and cultures of this complex region.

Before reading the following discussion, examine Table 14-1, which gives an overview of significant events in Southeast Asia's history.

Physical Landscapes

The difference between land and water is clearly evident. In fact, there are a mainland Southeast Asia and an island, or archipelagic, Southeast Asia (Figure 14-1). The mainland includes Burma (Myanmar), Thailand, Laos, Cambodia, and Vietnam. We will include Malaysia with the insular region as it not only shares the island of Borneo with Indonesia but also is connected to the mainland by the narrow Kra Peninsula. Consequently, Malaysia, Singapore, Indonesia, Brunei, Timor-Leste, and the Philippines form the insular or archipelagic region.

All told, there is four times more water area than there is land area. While civilizations arose in the mainland interior, marine landscapes and coastal locations dominated the fortunes of the island realm. Control of ocean passageways marked the rise and fall of many a power group such as the Majapahits (1298–1500s) whose empire stretched from Java to the Philippines and incorporated the regions of today's Malaysia and southern Thailand. Control of the seas later fixed British-established Singapore as the dominant port of the entire region.

Table 14-1 Time Line of Selected Historic Events in Southeast Asia.

Century	Year	Kingdom or State	Important Events
51st C BC	ca 5000 BC		Ancestor of modern rice cultivation. Pottery.
31st C BC	ca 3000 BC		Domestication of cattle, dogs, pigs, chickens.
21st C BC	ca 2000 BC		Austronesian people arrive in Indonesia and spread throughout archipelago. Domestication of water buffalo. Domestication of red jungle fowl as poultry. Iron industry.
8th C BC	ca 700 BC	Dong S'on culture on mainland Southeast Asia.	Women important in society. Mariners sail Indian Ocean. Bronze. Wet-rice irrigation—Sawah.
3rd C BC	ca 200 BC	Hindu Kingdom in Java and Sumatra. Beginnings of Funan in Cambodia.	
	207 BC	Kingdom of Nam Viet.	
2nd C BC	100 AD		Theravada Buddhism spreads from Sri Lanka throughout region.
	111 BC		China conquers Nam Viet. Vietnamese borrow selectively from Chinese culture—Mahayana Buddhism, Confucianism, and Chinese characters.
6th C AD	ca 500 AD		Hindu influences including concept of God-Kings.
9th C AD	800 AD	Khmer Empire (ca 800–ca 1370).	Angkor Wat capital of Khmer Empire.
	850 AD	Bagan Kingdom in Burma (until 1280s).	Beginnings of Prambanan Hindu Temples on Java.
10th C AD	939 AD	Funan in Cambodia.	Funan extends from Burma to Vietnam. Hinduism in Funan.
11th C AD	1000 AD	First Indian kingdom on Kra Peninsula. Rise of Thai state.	Construction of Borobudur, Buddhist temple on Java.
	1057	Kingdom of Myanmar.	Unification of multiple states.
12th C AD	1136	Kedah Kingdom on Malay Peninsula.	King converts to Islam. Islam spreads throughout the region by sea traders.
	1192	Champa Kingdom in Vietnam until 1471.	
13th C AD	1283	Sukothai Kingdom (Thailand).	Khmer writing system adopted.
	1293	Majapahit Empire in eastern Java until ca 1500.	Expansion of Majapahits throughout region from Malay Peninsula to Moluccas (possibly Papua).
14th C AD	1351	Ayuthaya Kingdom (Thailand).	Great traders.
	1353	Kingdom of Lan Xang (Laos).	Theravada Buddhism made official religion.
15th C AD	1445 AD	Muslim Dynasty in Malacca.	First of many Muslim dynasties in Malay archipelago.
16th C AD	1500		Viets move southward.
	1511		Portuguese take control of Malacca.
	1512		Portuguese begin trading nutmeg and mace from the Spice Islands (Moluccas).
	1521		Spanish at Cebu Island in Philippines.
	1571	Spanish Philippines.	Spanish name the Philippine archipelago after King Phillip II and set up capital at Manila.
	1595		First Dutch expedition reaches Java.

(continued)

Table 14-1 *(Continued)*

Century	Year	Kingdom or State	Important Events
17th C AD	1602		Founding of Dutch East India Company.
	1625		Dutch begin expelling Portuguese from Indonesia.
	1650		Hindu princes and Brahmins leave Java for Bali.
18th C AD	1767		Burmans crush Ayuthaya.
19th C AD	1819	Singapore.	Sir Stamford Raffles in Singapore.
	1824	Netherlands East Indies.	Anglo-Dutch Treaty—British in Malaya and Dutch in Dutch East Indies (Indonesia). British in Burma. Burma becomes a province of British India until 1948.
	1858		French begin colonization of Vietnam.
	1863		French protectorate over Cambodia.
	1885	Burma.	British invade Myanmar and rename it Burma.
	1887	French Indo-China.	Includes Vietnam and Cambodia. Laos incorporated in 1893.
	1896	Thailand.	Anglo-French agreement recognizing independence of Thailand as a buffer state.
	1898	American Philippines.	Spain sells the Philippines to the U.S. for US$20 million.
20th C AD	1927		Rise of Sukarno in Indonesia
	1941	Greater East Asia Co-prosperity Sphere (Japanese control).	Japanese attack Pearl Harbor and the Philippines. Vietminh formed as a guerilla force to liberate Vietnam from the Japanese. Rise of Ho Chi Minh as leader.
	1942		Japan conquers Malaya. Fall of Singapore.
	1945		Japanese surrender to Allied Forces. Independence movements in European colonies.
	1948	Burma.	Independence from British.
	1949	Indonesia.	Dutch concede independence for Indonesia.
	1953	Cambodia.	Cambodia independent from French.
	1954	North and South Vietnam.	Battle of Dien Bien Phu. French troops surrender to Vietnamese, leaving Vietnam divided in North and South.
	1955		U.S. supports an independent South Vietnam.
	1959		Organization of Viet Cong in Vietnam. Subsequent Vietnam War.
	1960	Malaysia.	Formation of Malaysia.
	1975	Democratic Republic of Vietnam. People's Republic of Kampuchea. Lao People's Democratic Republic.	End of Vietnam War.
	1979	Cambodia.	Vietnamese oust Khmer Rouge.
	1988	Myanmar.	SLORC takes over Burma and renames it Myanmar.
	1990	Myanmar.	Aung San Su Kyi wins election in Burma and is subsequently put under house arrest

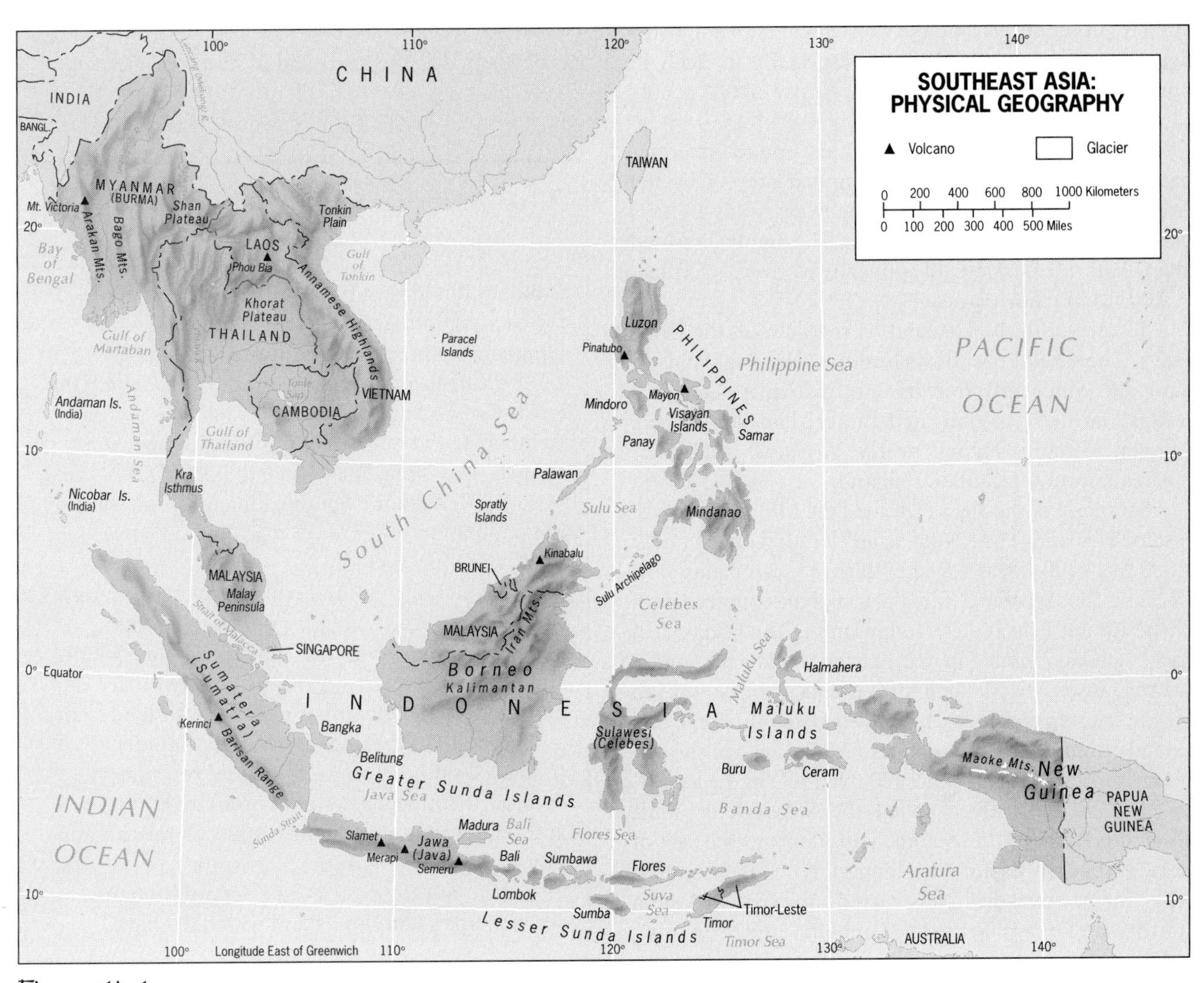

Figure 14-1

Physical features of Southeast Asia. From H. J. de Blij and P. O. Muller, *Geography: Realms, Regions and Concepts*, 14th edition, 2010, p. 534. Originally rendered in color. © H. J. de Blij and P. O. Muller. Reprinted with permission of John Wiley & Sons, Inc.

Nagas in Southeast Asia

Naga, in Sanskrit, means serpent. The ***naga*** is an aquatic symbol that permeates the daily life of people everywhere in the watery realm of Southeast Asia (refer to Figure 1-2). The *naga* appears in art form as a dragon and snake-like creature, and frequently as a mutation of a different being. For instance, Thailand has a *sang*, a cross between a naga and a *singh* (lion). The *naga* form is ubiquitous in Thai and other cosmological representations. The Khmer and numerous other Southeast Asian civilizations are portrayed as emerging from water. Sometimes whole populations have been identified as the *naga* people. In almost everything of significance, the *naga* symbol asserts itself.

Mountain formation remains active and crustal instability is a feature affecting the lives of millions (Chapter 2). The 2004 earthquake off the northwest coast of

Sumatra in Indonesia, and the subsequent tsunami, had a devastating effect in the region, especially in Aceh in northern Sumatra (see Chapter 16). Nearly 80,000 people were killed in Indonesia alone.

Throughout this unstable region, mountains dominate over plains. The hills of Myanmar are cut by the Ayeyarwady (Irrawaddy) River. The Ayeyarwady Delta is used extensively for rice cultivation. The highly dissected Shan Plateau of eastern Myanmar rises over 4,000 feet (1,200 m) in many places.

To the south, in Thailand, is the delta of the Chao Praya. The Chao Praya flows from the Thai highlands to Bangkok. Since this river is confined in a geological trough, its delta is elongated rather than fan-shaped. Alluvial lowlands extend nearly 310 miles (500 km) upstream from the Gulf of Thailand.

Southeast of the Shan Plateau is the Khorat Plateau of eastern Thailand. The Khorat, at an elevation of just under 1,000 feet (300 m), is edged by higher escarpments. The Khorat is thus separated from the rest of the country. South of the Khorat Plateau lies the lowland of the Cambodian Basin with Lake Tonle Sap in the lowest, middle portion. In the wet season, the Mekong River drainage rolls inward to fill the lake up to six times its average depth. Tonle Sap extends from 1,600 square miles (2,580 km^2) in the dry season to 12,400 square miles (20,000 km^2) in the wet.

The Mekong River, called the *Mégôngk* or *Tonle Thom* ("great river") in Cambodian, flows to the east of the Cambodian Basin. The longest river of Southeast Asia, the Mekong has constructed an expansive delta starting at Phnom Penh in Cambodia and extending 175 miles (282 km) into the sea. This vast, alluvial delta has developed as the heartland of southern Vietnam.

The Mekong

The Mekong—"tiger river" or "mother river"—is the twelfth longest river in the world (Figure 14-2). Originating in the Tibetan Highlands of China's Qinghai Province, the river flows 2,600 miles (4,184 km) through southwestern China, along the borders of Myanmar, Laos, and Thailand, across Cambodia, and down through Vietnam, where it sprawls across the 26,000-square-mile (67,340 km^2) Mekong Delta to the South China Sea. One-third of the population of the latter four countries lives in the lower Mekong Basin.

The Mekong cultural realm is overwhelmingly rural. Most of the region is poor, with people engaging primarily in rice cultivation and fishing. However, as foreign aid and investment increase and modern development ensues, both landscapes and lifestyles are being transformed rapidly. Increased population pressure and resource demands, coupled with uncoordinated policies and regulations, are threatening both human and physical environments.

Although the river is an artery of human activity, upstream rapids and waterfalls have

Figure 14-2

I am standing on the Laotian side of the Mekong looking at Thailand. It is January (dry season) and the river level is low. At this time it is quite easy to get across from one country to the other. The story is different in the wet monsoon when the water rises to inundate the shores. The current is fierce and it is dangerous to cross. The building, where the canoes are docked, floats and rises and falls with changing water levels. Photograph courtesy of B. A. Weightman.

prevented any large-scale development for international shipping. Smaller seagoing vessels can travel upstream to Phnom Penh, but travel beyond is limited to local traffic. A bridge connecting Laos to Thailand was constructed in 1994.

In 1992, China initiated plans to open the upper reaches of the river to year-round navigation by large cargo ships. China's primary objective is to facilitate the export of raw materials from its landlocked Yunnan Province to ports in Thailand, Laos, and the rest of Southeast Asia. When fully implemented, the dredging and channeling will more than double the annual shipping capacity of the Mekong to 10 million tons and involve the passage of at least a dozen 500-ton ships daily.

Cooperative efforts to develop the Mekong River basin have been discussed since 1957. Several hydroelectric power, navigation, flood control, and irrigation projects were planned, but war and political upheaval have allowed only some to be implemented. For example, dam construction has made Laos a key exporter of power to Thailand.

In 2000, Thailand, Laos, Burma (Myanmar), and China signed a navigation agreement. Each promised to develop ports and facilitate the passage of ships. However, many plans have been scaled down because of both cost factors and environmental concerns. The first three countries are worried that their markets will be inundated with cheap Chinese goods. Farmers, who plant crops along the riverbank and rely on natural flood waters for irrigation, are concerned that the seasonal flow of water will be disrupted and that large ships will generate waves that will drown their crops.

Blasting rapids to create shipping channels has already damaged the Mekong's environment. Fishers already complain that waves from large ships cause their small boats to capsize and that the fish supply is not as abundant as it used to be. Rapids may impede shipping, but the rocky outcrops underneath harbor more than 1,000 fish species.

Various groups have protested the development plans and projects in progress (Figure 14-3). In June 2002, China announced that it would scale down the blasting and require environmental impact reports. Nevertheless, it has already built three dams on the upper Mekong and a third dam is due for completion in 2012. Villagers downstream are already complaining about the fluctuation in water levels. China appears driven to develop the river regardless of the consequences elsewhere.

The Annamite Ranges run through Vietnam between the Mekong and the South China Sea. In northern Vietnam, the Song Koi (Red) River flows in a trough-like valley to the Gulf of Tonkin, where it forms a 5,460-square-mile (14,000 km^2) delta. The southern Annamites give way to the Mekong Delta.

The shallow sea enclosed by Cambodia, Malaysia, and the island of Borneo is less than 656 feet (200 m) deep, as it is part of the Sunda Shelf (Chapter 2). Borneo is the chief emergent part of the shelf, having several mountains above 6,560 feet (2,000 m). Mount Kinabalu in northern Sabah is the highest peak at 13,451 feet (4,101 m).

Southeast Asia's relief offers little opportunity for human settlement. The region's populations are found

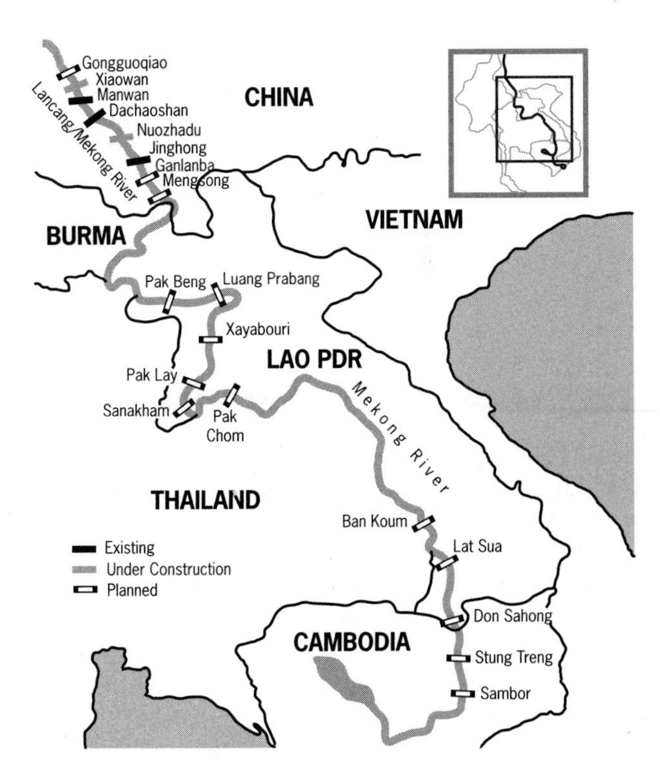

Figure 14-3

Locations of proposed dams on the Mekong. Why are dams constructed in the first place? What are the potential benefits? What are the negative consequences? From International Rivers, Berkeley, California. http://www.internationalrivers.org/en/ node/2275.

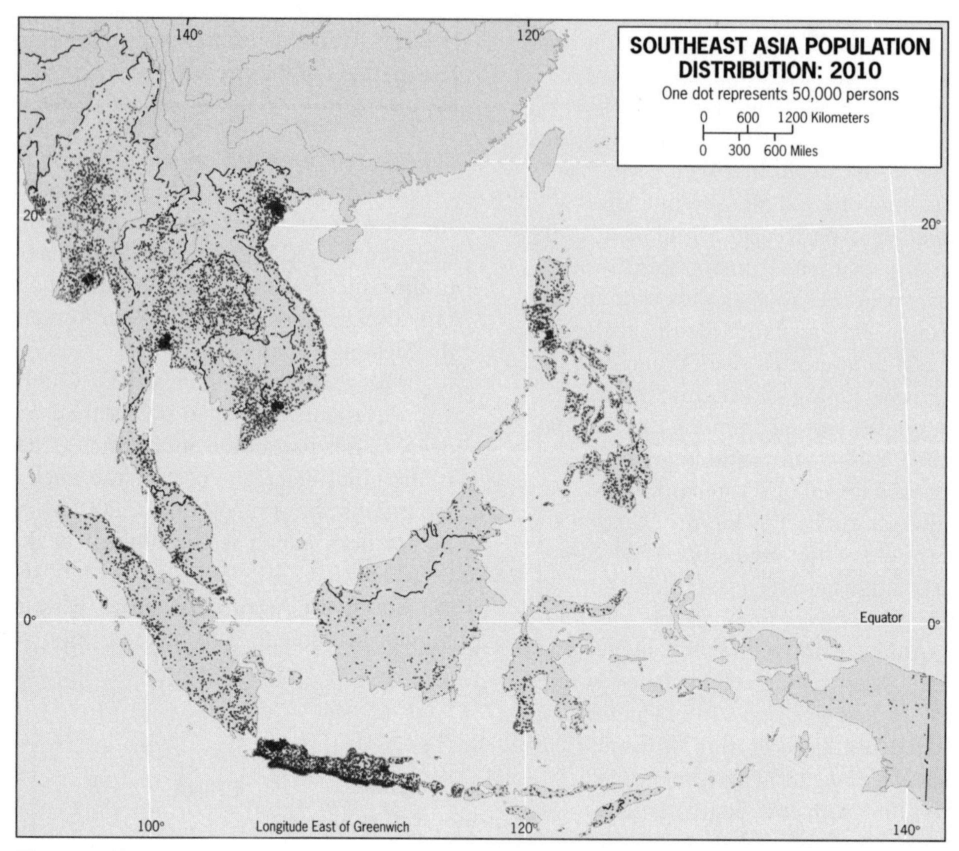

Figure 14-4

Notice how the population lives along waterways and coasts. Why is this the case?

From H. J. deBlij and P. O. Muller, *Geography: Realms, Regions and Concepts*, 14th edition, 2010, p. 535. Originally rendered in color. © H. J. de Blij, P. O. Muller. Reprinted with permission of John Wiley & Sons.

primarily along rivers. Most of the Sunda Shelf, which could have offered livable land, is overlain by the sea. Extensive coastal areas are mangrove swamps. The most densely populated areas in the island realm are Java in Indonesia and Luzon in the Philippines, both of which possess fertile volcanic soils (Figure 14-4).

Even within this tropical realm, straddling the equator, there are considerable temperature variations. Pinang, Malaysia, at 5°21' N averages 80°F (27°C) for 12 months, Hanoi, Vietnam, at 21°04' N might drop to 60°F (16°C) in December or January. Altitude also plays a role. With temperatures falling 3 degrees per 1,000 feet, Southeast Asia had its hill stations comparable to those of the Raj. Java's Bandung and the Malay Peninsula's Cameron Highlands are examples of places where temperatures might be 10 to 20 degrees cooler than in the sweltering lands below.

Precipitation varies both in amount received and its seasonality (Figure 14-5). Sixty inches (152 cm) will support tropical rain forests or monsoon forests. Areas with less than this amount are moisture deficient. Most moisture-deficient areas are on the mainland and on the islands east of Java. Distinct wet and dry seasons increase as one moves away from the equator. In equatorial regions, the wet and dry monsoons are affected by convectional rainfall. Even so, topographic barriers exhibit seasonally humid and dry sides, depending on the monsoon and other wind patterns.

Length of dry season varies as well, lengthening both north and south of the equator. Also, some "dry seasons" are merely less wet ones. For instance, there is no real dry season in Borneo, and in peninsular Malaysia it lasts only a month. But on the mainland the dry season lasts four to five months. Most of the rainfall is in the summer months.

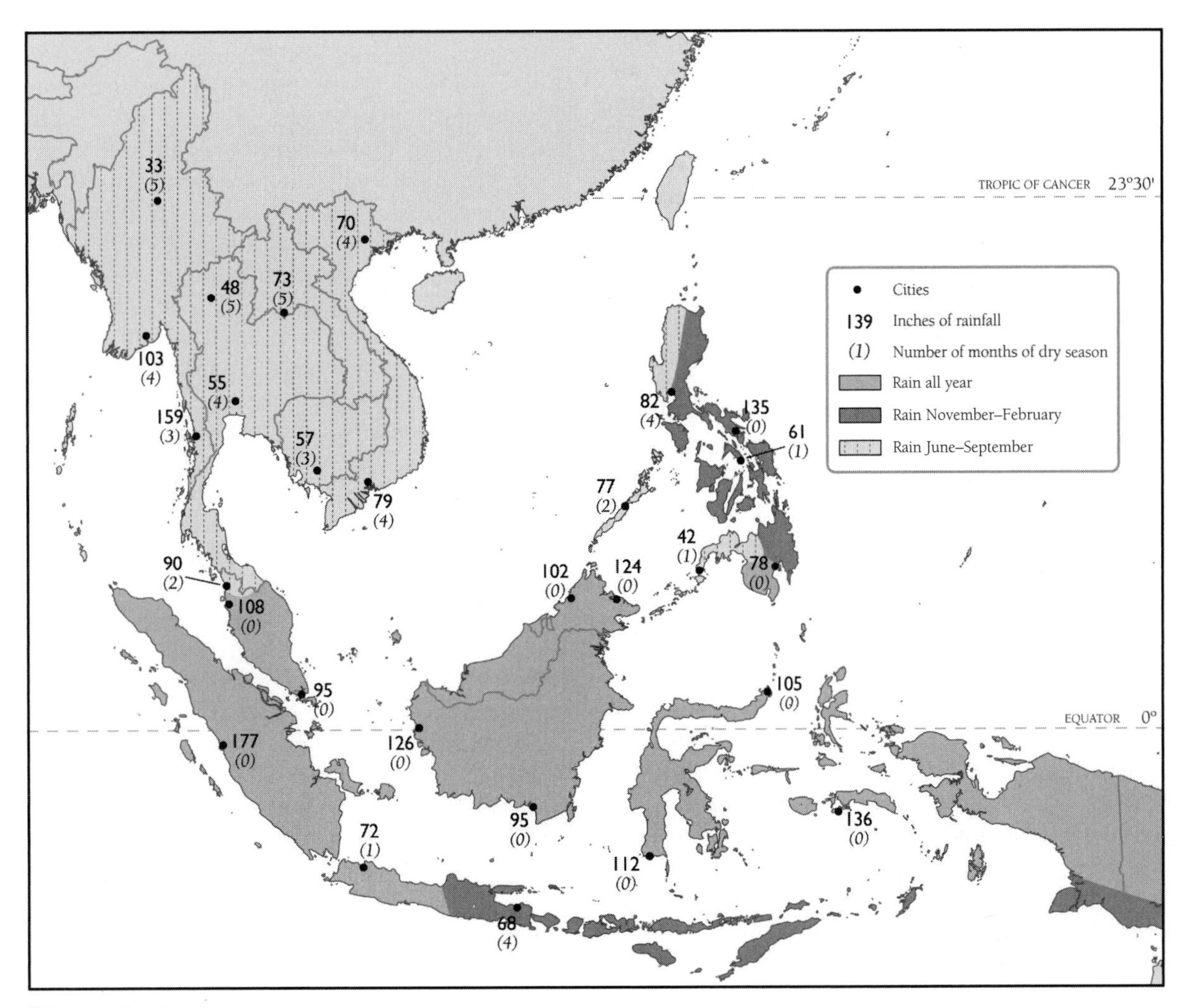

Figure 14-5

Length of dry seasons and precipitation. Note that dry seasons increase in length toward the interior of the mainland.

THE ROLE OF WATER

According to Jonathan Rigg (1991), "Of the environmental elements, water has possibly been the most instrumental in forging Southeast Asia's distinctive personality." Fairly predictable winds, wood supplies located close to shore for shipbuilding, and endless waterways made this region most favorable to maritime activity. Extensive trade existed well before the fourteenth century.

In prehistory, waves of migrants filtered southward from the mainland to the Indonesian Islands and on to the Pacific Islands and Australia. At the beginning of the first century AD, there was an infusion of Hindu and Buddhist culture when Indian sea traders ventured for gold and spices. This Indianization of Southeast Asia peaked with the establishment of empires centered on great complexes such as Srivijaya on Sumatra and Angkor in Cambodia.

In the seventh century, both Arab and Indian ocean traders brought Islam to the island realm although Islam was not widespread until the 1200s. Meanwhile, monks from Sri Lanka disseminated Theravada Buddhism to the mainland.

In the modern era, colonial countries introduced new technology and power into the region and created conditions conducive to the immigration of large numbers of Indians and Chinese. Commercial interaction ensured diffusion of products and ideas and linked the peoples of the region with each other. Interestingly, with the possible exceptions of Vietnam and Burma, no Southeast Asian country established strong ties with either India or China.

The amount and availability of water are crucial to life in Southeast Asia. Virtually all outdoor economic activities are regulated by precipitation. Water is central to rice cultivation. In fact, there is no other staple crop so dependent on its presence. Many regions have rituals surrounding the "gift of water." In Thailand, female cats (the personification of dryness) are carried through the streets and soaked with water while people chant "Hail Nang Maew (Lady Cat) give us rain." In parts of northern Thailand, Bali, Java, Sumatra, and other islands of both Indonesia and the Philippines, elaborate irrigation systems have been constructed to ensure a stable water supply.

Bali's Subak Irrigation

In Bali, rice is a gift of the gods, and the traditional rice-growing cycle commences on a day set according to a religious calendar. Water regulation is an essential aspect of this highly regimented pattern of activities.

Rivers cut through Bali's volcanic mountains to the sea. The Balinese have settled the mountain ridges and covered the slopes with *sawah*—terraced rice paddies (Figure 14-6). Water runs by gravity but must be fairly distributed and regulated according to the needs of farmers and the system. For example, irrigated *sawah* are drained up to three times per year to encourage root development and allow weeding. Fields are drained for harvesting and then re-flooded to allow other plants to grow. Water hyacinth is popular because it can be worked back into the soil or used as pig feed.

Balinese farmers belong to cooperatives called ***subak***. Every farmer who owns or rents more than a quarter of an acre (one-tenth of a hectare) must join. Bali has more than 1,000 *subak*. Each *subak* elects a leader who runs the operation and liaisons with agricultural and other agencies. *Subak* law (*awig-awig*) meetings must be held on one auspicious day each month of the religious calendar. Monies are derived from fees, fines, and land sales.

Cooperation is the key to success. *Subak* members repair and maintain the irrigation system after the harvest in accordance with the amount of water they have received. Those not working pay a fee. The average farmer gives the communal system about 15 days a year. An important aspect of the *subak* is its religious connection. Each keeps a temple, which must be cooperatively maintained by the members.

The *subak* system diffused from Java a thousand years ago. It has survived in Bali because of its isolation as an island and because the *adat* system

Figure 14-6

Irrigated rice terraces (sawah) *in Indonesia. Water is moved by gravity flow.* Photograph courtesy of B. A. Weightman.

of local regulations, stressing personal responsibility and communal consensus, remains strongly entrenched.

It now seems clear that rice was first domesticated in Southeast Asia and diffused from there to China and India, replacing millet as the staple crop. Around 4000 to 3500 BC, a new red-on-tan pottery form appeared in northern Thailand. This is known as Ban Chiang pottery. Shortly thereafter, bronze-making emerged, carried out at the village level.

UPLANDS AND LOWLANDS

From about 1000 BC, major changes began to take place. With the expansion of the Han Chinese in person and influence, small towns began to emerge. In addition, to facilitate trade between India and China, **entrepôts** or transshipment points were established along the coasts of the Malay Peninsula and of what is today southern Vietnam. Entrepôts became centers of diffusion of Indian culture throughout Southeast Asia.

As these events ensued, primitive and tribal people began to lose their dominant positions. The new civilizations centered on the major river valleys and Cambodia's Tonle Sap basin (Chapter 15). These lowlands are separated by sparsely populated mountainous areas—the uplands. Hill people became increasingly differentiated from lowland people, yet they retained part of their cultural realm as "holders of the wild" or "people of the upland fields." Ceremonial exchange relationships validated the bonds between hill people and lowlanders. Tangible trade between the two groups was important as well. Lowlanders also viewed hill people as potential laborers, and slavery was not unusual. Hill tribes often raided each other's settlements and sold the captives as slaves to lowlanders. In some regions such as Vietnam, efforts were made to reduce lowlanders' contact with the hill people to simple tribute payments and restricted trade.

The mountains have acted as barriers between larger settlements. Over time, successive immigrant groups have settled in the lowlands, but more recent and more populous groups have driven earlier, less sophisticated groups into the upland regions.

Lowland areas are inhabited by the dominant ethnic group of each country—Thais in Thailand, Khmers in Cambodia, Javanese in Java, Burmans in Myanmar (Burma), and so forth. Upland, forested areas are inhabited by minority groups such as the Shan in Burma and the Yao and Miao in Thailand.

In these contrasting settings, different ways of life have evolved. Lowlanders primarily practice wet rice cultivation while uplanders focus on some form of dry field agriculture or shifting cultivation. These patterns continue today and affect patterns of development and the resolution of environmental issues.

The Impact of Indian Culture

Mountain barriers channeled the diffusion of Indian culture into Southeast Asia's coastal regions. New ideas, religions, and customs came almost entirely by sea in sailing ships driven by the monsoon across the Bay of Bengal. As early as the sixth century BC Indian traders worked the coasts of Myanmar, the Malay Peninsula, and western Indonesia. The Indians impressed local leaders with lavish gifts, their knowledge of medicine, and their frequent portrayal of themselves as royalty. They also married into Southeast Asian society. Temple forms, styles of art, music, dance, and tales of the Ramayana are woven into the cultural tapestries of most Southeast Asian countries to this day (Figure 14-7). The impact of Indian culture was greatest during the Gupta Empire—India's Golden Age (320–535 AD).

Hinduism and Buddhism were the most salient aspects of Indianization. Missions were sent back and forth to Ceylon (Sri Lanka), the core area of Theravada Buddhism. Mahayana Buddhism was accepted in central Java during the eighth century and in Cambodia's Angkor in the twelfth century. Sanskrit was the first written language of the region. Its related Pali provided the basis for developing local writing systems.

The Hindu concept of kingship, law codes, economic treatises, and other canon were absorbed by Southeast Asian rulers. However, the caste system and inferior standing of women were not adopted. Although there is a patriarchal veneer to present-day Southeast Asian society, couples tend to live with the wife's parents, or at least relate most closely to them. Women have much more social and economic power than in other Asian societies.

The earliest Indian states emerged along the coasts, with their nuclei at river mouths. The first of these, called Funan by the Chinese, became an imperial power in the third century AD. This was followed by the Khmer kingdom in Cambodia. The kingdom of Champa (1192–1471) was established along the east coast of Vietnam. Another early sea power to accept Indian influence was Srivijaya. Its capital at Palembang on Sumatra became a center of Mahayana Buddhism, as was central Java at the time. The

Figure 14-7

Evidence of early Hindu kingdoms in Java, Indonesia, these temples at Prambanan have suffered earthquake damage. The 232 temples were built between 900 and 930 AD. Like Borobudur, they represent the cosmic Mount Meru. The largest three temples are dedicated to Brahma, Shiva, and Vishnu. Prambanan is the largest Hindu temple complex in Indonesia. Photograph courtesy of B. A. Weightman.

temple complex of Borobudur (built in the eighth century) stands as a testament to the strength of Buddhism (Figure 14-8). All of Southeast Asia, with the exception of the Tonkin region of northern Vietnam and the Philippines, was impacted by Indian culture.

The Chinese Influence

The earliest contact with China can be traced to the Qin Dynasty (221–207 BC). Chinese influence diffused overland; the region mainly affected was Vietnam. In fact, northern Vietnam remained a Chinese territory until it broke away in 907 AD. China referred to the region of the southern seas as **Nanyang**. Its cultural impact was the direct result of colonization and conquest. Chinese administration style, writing system, and Confucian classics were all superimposed on indigenous culture. Many Chinese cultural influences have persisted in Vietnam.

By the sixteenth century, Chinese trading communities were permanently established throughout the region. However, large-scale Chinese migration did not take place until the turmoil of China's nineteenth century, when thousands of Chinese migrated from southern China to work in mines, on plantations, and in other European-operated enterprises. Others came as farmers,

Figure 14-8

Built by Buddhists on Java in the eighth century at the height of the Sailendra Dynasty, Borobudur's construction took about 75 years. It comprises 1,600,000 volcanic, andesite stones and is ornamented with at least 1,500 bas reliefs and 500 statues of Buddha. No mortar was used in its construction. The large base platform was added at a later date and hides a panel of reliefs known as the "hidden foot." Borobudur stands as a stupa, a replica of the cosmic Mount Meru, center of the universe, and as a mandala to assist meditation. The complex was covered by tropical vegetation until it was discovered by the Dutch in the 17th century. Photograph courtesy of B. A. Weightman.

fishers, traders, retailers, or skilled craftsmen. Increasing numbers of Chinese came to play a critical role in the region's commerce.

With the extant plethora of ethnic groups and the influx of Europeans, Indians, and Chinese, each functioning in its own realm of expertise, Southeast Asia became what is referred to as a **plural society.**

Overseas Chinese in Southeast Asia

Overseas Chinese are prominent in Southeast Asia far beyond their numbers (Figure 14-9). For example, in Malaysia ethnic Chinese make up 29 percent of the population but control 69 percent of share capital by investing in businesses. In Thailand, ethnic Chinese make up 10 percent of the population, yet they control 81 percent of listed firms by market capitalization. In the Philippines, the Chinese comprise 2 percent of the population but control half of the share market, and in Indonesia 3.5 percent ethnic Chinese control more than 70 percent of the listed firms. Clearly, the Chinese have clout, disproportionate to their numbers, in Southeast Asian economies. In Singapore, Chinese comprise 75 percent of the total population and control 90 percent of the economy. Singapore is the only country in Southeast Asia where the economy is dominated by the Chinese.

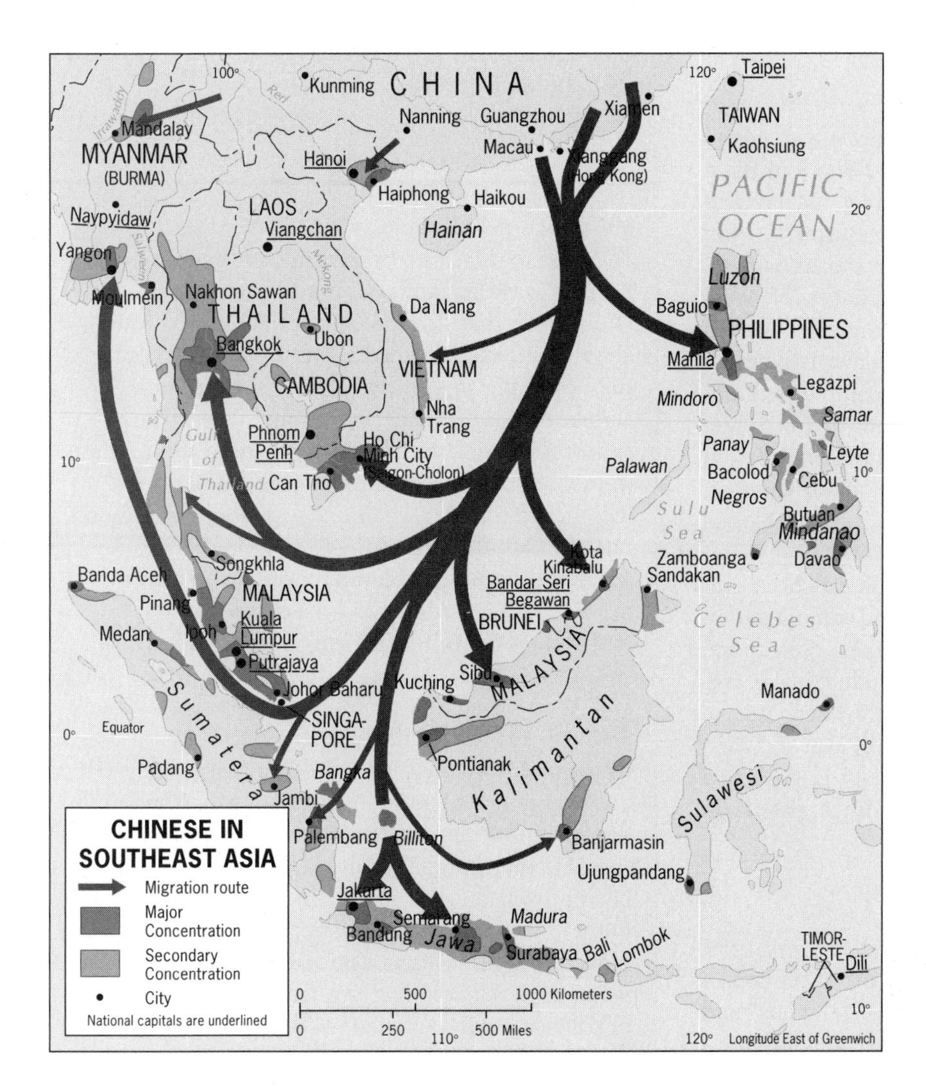

Figure 14-9

Chinese migration flows in Southeast Asia. Historically, Chinese emigrated from the southern regions of China. Most are concentrated in urban areas where they own businesses. From H. J. de Blij and P. O. Muller, *Geography: Realms, Regions and Concepts*, 14th edition, 2010, p. 538. Originally rendered in color.

Southeast Asia's Chinese have a reputation of being rich and influential but loyal to their own group. They are said to do business with each other and hire their kin rather than the local people. Suspicion and tension are rife in several countries such as Indonesia and Malaysia. The least tension is in Thailand and Singapore.

Chinese family businesses take advantage of their kin networks and regional connections to stay afloat in the worst of times. This situation leads to more jealousy and even hatred among non-Chinese. Chinese have frequently been the scapegoats for any sort of trouble. Thousands have been killed, raped, and tortured and their homes and their businesses destroyed. The last attack against the Chinese was in Indonesia during the economic crisis of 1997.

Linguistic Pluralism

Hundreds of distinct languages and dialects are spoken in Southeast Asia. These can be classified into four main language families: Malayo-Polynesian (e.g., Malay and Javanese); Sino-Tibetan (e.g., Thai, Burmese, and Cantonese); Austro-Asiatic (e.g., Khmer and Annamese, or Vietnamese); and the various languages of West Papua. Although the insular region's languages harken mainly to the Malayan division of the Malayo-Polynesian group, the mainland's languages form a complicated mosaic. There are multiple Sino-Tibetan and Austro-Asiatic languages and dialects spoken throughout the region. For example, there are at least six versions of Chinese. Hill regions, with their isolated pockets, have developed a myriad of dialects.

Since the eighteenth century, English has been employed widely in the fields of education and administration in many areas such as the Philippines, Singapore, Malaysia, Thailand, and Myanmar. In Indochina (Cambodia, Laos, and Vietnam), French was the *lingua franca*. Dutch served Indonesia until after World War II, when English came into favor. The prevalence of English as a means of general communication is readily apparent on signage in many places.

Religious Pluralism

The mainland is largely Buddhist with the exception of the Vietnamese lowlands, where "Chinese religions" and Roman Catholicism are practiced. Chinese religions are a mixture of Mahayana Buddhism, Daoism, and ancestor worship, a social practice affiliated with Confucianism. The island world is largely Islamic with the exception of Bali, which retained Hinduism after the last Hindu dynasty in eastern Java had been overthrown in the seventeenth century.

Christians predominate in the Philippines except on the island of Mindanao, where Islam prevails, and in Timor-Leste, where Christianity remains a vestige of Portuguese colonialism. The Dutch introduced Protestantism in Indonesia in the sixteenth century and the Jesuit Francis Xavier worked in the Moluccas (Maluku) from 1546 to 1547. Christian communities remain in northern Sulawesi and in some islands of Maluku (Figure 14-10).

Islamic fundamentalism is on the rise in some areas and is sometimes associated with acts of terrorism. Indonesia, southern Thailand, and Mindanao have been the most affected by this religious revival. (See Chapters 15 and 16 for more details).

Angkor Wat

Angkor was the capital of the Khmer empire, a civilization that flourished from about 800 until about 1370 and included Lower Burma (Myanmar), the upper Malay Peninsula, central Thailand, Cambodia, and southern Vietnam. A mixture of sandstone, brick, and laterite was used to build an immense complex of temples (the Khmer word *wat* means temple), each one a symbolic representation of Hindu or Buddhist cosmology. The whole of the Angkor complex covers an area of more than 15.5 miles (25 km) east to west and almost 6.2 miles (10 km) north to south (Figure 14-11).

At one time, the complex supported a million people sprawled over an area the size of Los Angeles. But by the fifteenth century a prolonged drought punctuated by intense monsoons had destroyed the city's water-preservation infrastructure. Furthermore, Angkor was too far inland to benefit from the maritime trade that was becoming more important in the region. Then the capital was moved to Phnom Penh near the coast. Also, the Siamese were pushing down from the north and northwest, taking over large sections of the Khmer

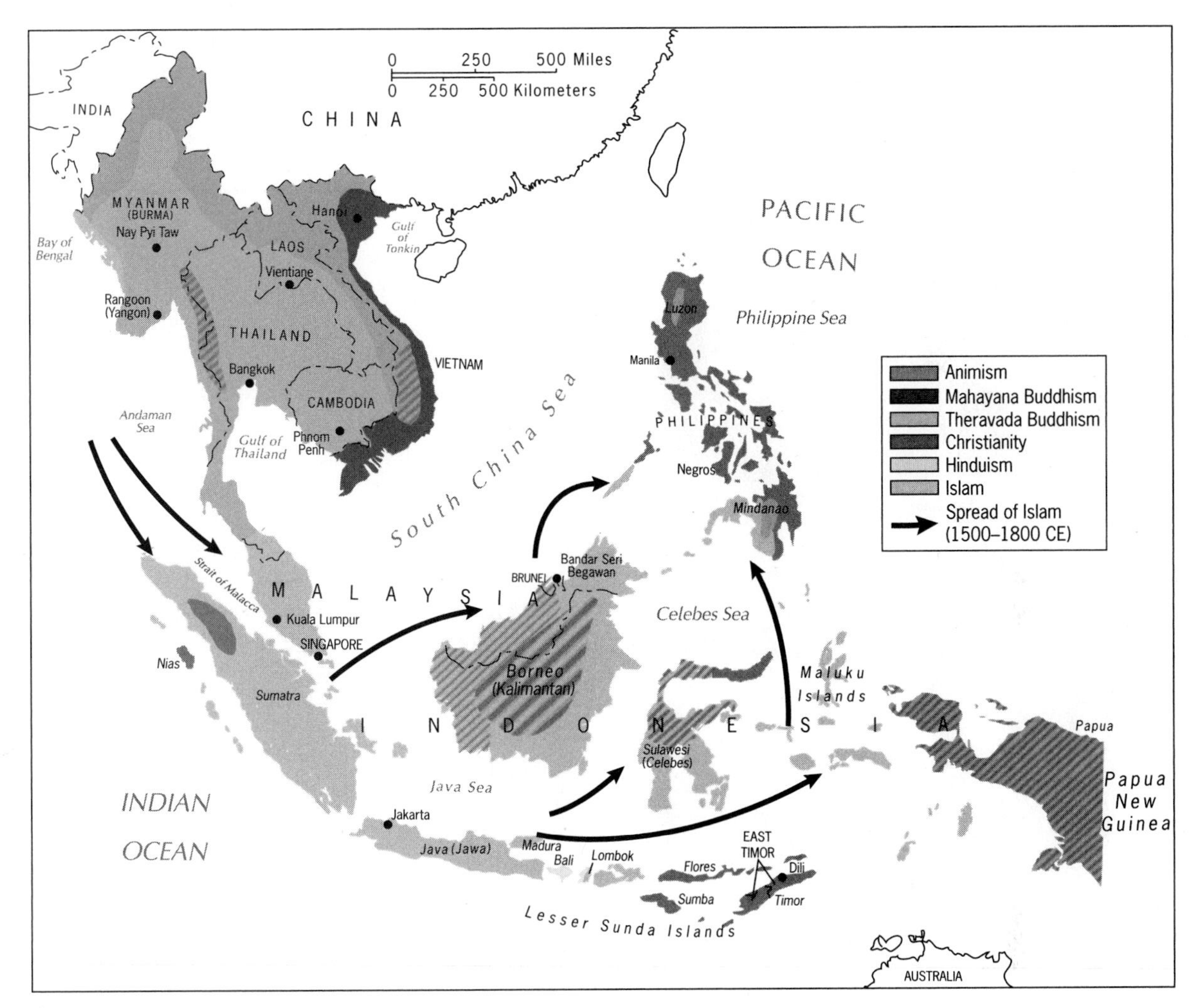

Figure 14-10

Because of animist, Buddhist, Muslim, and Christian influences, Southeast Asia has a complex religious landscape. From Les Rowntree, Martin Lewis, Marie Price, and William Wyckoff, *Diversity Amid Globalization*, 4th edition, 2009, p. 597. Originally rendered in color. Reprinted with permission of Pearson-Prentice Hall.

Empire. Angkor was captured by the Siamese in 1431. It eventually was covered by tropical vegetation and remained buried until it was rediscovered by French anthropologists in the 1860s.

Angkor Wat, dedicated to the Hindu god Vishnu, is just one temple. Erected as the funerary memorial of King Suryavarman II in the twelfth century, it covers 1 square mile (2.5 km^2) and is one of the largest religious edifices in the world. Angkor Thom is slightly larger. It is entered by four axially aligned gateways that define a square almost 0.62 mile (1 km) across.

Angkor has largely escaped the ravages of time but it is not unscathed. Certainly the jungle has taken its toll. Trees have split walls, and lichens etch away illustrious carvings. French and Indian archaeologists have cleared the forest from much of the monument but they can do nothing about the robbers who have absconded with carvings, statue heads, and even whole statues. More recently, Cambodia's war has resulted in further damage to the complex in the form of bullet holes, smashed artifacts, and the carving of political slogans into bas reliefs.

Figure 14-11

The main Buddhist wat at Angkor is reflected in a moat surrounding the complex. This temple, built of laterite, covers one square mile (2.5 km^2) and is one of the largest edifices in the world.
Photograph courtesy of B. A. Weightman.

Fortunately, international assistance is supporting archaeologists and the local people in their restoration efforts. Treasure hunting raids still occur, however, and certain sections can only be visited in the company of armed guards.

The Coming of the Europeans

European political contacts with Southeast Asia commenced with the Portuguese conquest of Malacca in 1511. Still, there was no major European political or cultural impress on the region until well into the nineteenth century. The Philippines is an exception in that European penetration here was greater than anywhere else in Southeast Asia. Indeed, there were a string of fortified ports in the region (Chapter 1), but even the Dutch, who controlled the western third of Java centered on Batavia (Jakarta), were unaware of Borobudur, perhaps the greatest Buddhist monument in the world.

The Spice Islands

Long known as the Moluccas and now Maluku, the Spice Islands are a volcanic mini-archipelago between Sulawesi and Papua north of Timor-Leste. Ambon, Banda, and Ceram are only 3 of the 1,000 islands. Some of the islands were known for their spices, especially nutmeg and cloves. Spices, important for preserving and flavoring meat, were traded for Indian cotton, Chinese porcelain, and Javanese rice for at least 2,000 years before the Europeans came on the scene.

The arrival of the Portuguese in 1511 initiated decades of conflict among the local sultans and the competing colonial powers. The Dutch established settlements in 1599 and by 1667 had completed their conquest of the islands. Their first treaty with Ambon gave the Dutch monopoly over all the cloves. The Dutch East India Company, founded in 1602, put an end to individual profiteering by ousting all other traders by force and intrigue. At one point, the company eliminated the entire population of Banda.

To guarantee control and keep prices high, the Dutch destroyed all the nutmeg and clove plants except in those areas approved for cultivation. Even the number of trees was restricted. Cloves were removed from all of the islands except Ambon. This was a particularly cruel move because the indigenous people planted a clove tree at the birth of a child and to destroy the tree portended ill-will for the child. The penalty for cultivating nonregulated trees was death.

Sago palms were eliminated from Ceram, known as the "mother island" because of its supply of this food staple. The area soon became uninhabitable. Today, it is said that Ceram is populated by ghosts.

Restrictive policies were enforced until 1824. By that time the Mollucans had been reduced to subsistence existence. The company's monopoly was broken when a Frenchman smuggled nutmeg and clove seedlings to Zanzibar. The spice trade also declined with the availability of fresh meat and changing dietary habits in Europe.

The nineteenth century ushered in the Industrial Revolution, creating increased demand for raw materials and markets. The Caribbean, focus of eighteenth-century trade, fell into decline with the abolition of slavery and the slave-labor system. The introduction of steamships and the opening of the Suez Canal in 1869 shortened the journey to Asia, and increasing interest in China put Southeast Asia once more at the crossroads of activity. European missionary zeal led to additional ventures to the region. From about 1825 to 1870, most of Southeast Asia came formally under European control (Figure 1-14).

EUROPEAN IMPRESS

The various powers had their own agendas, which diversified their influences on people and landscapes. In 1896, Britain and France decided to preserve Siam (Thailand) as a buffer state to protect their valuable colonial dependencies of Burma (Myanmar) and Indochina. Britain treated Burma as an extension of its Indian empire and pursued a policy of development of resources and building the economy. France's doctrinaire policy of assimilation of Indochina into France left most of that region undeveloped. The Dutch commercial exploitation policies left most of Indonesia undeveloped and unprepared, economically or politically, for self-government. In the Philippines, Spanish rule gave way to American control in 1898. American anticolonial attitudes assured that country's direction toward self-rule.

The Dutch Cultivation (Culture) System

The Dutch cultivation (culture) system (*cultuur stelsel*) in Indonesia exemplifies the penetration of capitalism and commercialization into the landscape. First established by the Dutch East India Company, the system required that every farm household allocate one-fifth of its land to the production of a cash or export crop. The system worked well in the sugar-growing areas of Java but proved burdensome where land was badly needed for food crops. The system was exploited by local officials who received commissions for the products they delivered to authorities. When rice was added to the list of export crops in the 1840s, famine struck Java causing widespread starvation and emigration.

The Dutch relinquished their monopoly control to private capital after 1877 but compulsory labor was still required. Farmers had to pay substantial taxes and were unable to get reasonable-rate credit. The powerful position of European and Chinese entrepreneurs added to the exploitation and disruption of traditional Indonesian society.

The British, who ruled Burma from Calcutta, brought in Indian moneylender castes to bank and fund the expansion of rice cultivation in the Ayeyarwady (Irrawaddy) Delta. Indians often foreclosed on the Burmese, who were hit especially hard during the world depression. Anti-Indian sentiments in Burma derive from the tactics employed by the British.

Economic concerns also stimulated population movements. For instance, the cultivation of new crops and the opening up of new lands attracted masses of migrants. The rapid expansion of rice cultivation into the Mekong, Ayeyarwady, and Chao Praya deltas are testament to increased global demand for rice in the late nineteenth century. In lower Burma, where the rice area burgeoned from 988,000 acres (400,000 ha) to 9,880,000 acres (4 million ha), populations increased from 1.5 million in 1855 to 8 million in 1930.

Populations were also drawn to other commodity regions such as rubber and tin. Rubber was planted over large areas of Malaya, Sumatra, and Java, literally transforming these landscapes. Malay and Chinese smallholders accounted for 39 percent of the rubber industry by 1939, yet it was the capital and labor-intensive plantations that dominated production. As Malays were not keen on wage labor, companies imported large numbers of indentured laborers—Indians and Ceylonese (from Sri Lanka). More than 700,000 Indian workers entered Malaya between 1907 and 1917. The tin industry in Malaya, Indonesia, and Thailand was dominated by Chinese mining enterprises employing cheap Chinese immigrant labor. Between 1890 and 1899, almost 165,000 Chinese laborers a year were entering the Straits Settlement of Singapore and Penang alone.

The influx of Europeans, Indians, and Chinese created plural societies; each group occupied its own residential quarters but amalgamated in the marketplace. Immigrants concentrated in urban places and zones of commercial agriculture or mining.

Figure 14-12

This photo shows the traditional Chinese/European shop-houses that once dominated Kuala Lumpur. These stand in sharp contrast to today's modern architecture. Photograph courtesy of B. A. Weightman.

Cities were now founded on commerce and, with the exception of Bangkok, were dominated by external forces. Cities such as Jakarta, Manila, and Kuala Lumpur were largely foreign places, alienated from the circumstances of the rural masses that made up 90 percent of the population (Figure 14-12).

Transportation (river, road, and rail) was immensely improved to serve the colonial presence. Routes linked raw materials with coastal ports, which served as entrepôts for outgoing and incoming goods. This selective transport development meant that many areas were devoid of communication systems, isolated from the larger economy. Dual economies emerged, only to be exacerbated as development increased. The rural–urban divide widened as education and health care were made available to urban elites. Only the Philippines had widespread educational facilities.

POLITICAL PATTERNS

Politically, territorial units had little relevance to historical and ethnic circumstances. Territorial forms were derived from European considerations: the need to access raw materials, the need to avoid conflict, and administrative conveniences. Boundaries did not coincide with settlement patterns of ethnic groups that were often divided by boundary constructs. One of the few instances where boundaries were drawn according to ethnic differences is between parts of Laos and Vietnam. Villages where the majority of houses were built on piles with entrances facing east were assigned to Laos. If the majority of houses were built on the ground facing south, that village went to Vietnam.

Burma and Thailand each have long tails trailing down the Malay Peninsula. Laos is landlocked, and what eventually became Malaysia is fragmented between the mainland and island realms.

Although colonials frequently turned internal hostilities to advantage, no attempts were made to integrate territorially-based ethnic identities into unified wholes. In particular, Chinese and Indian migrants remained largely unassimilated and resented. Indians were expelled from Myanmar (Burma) after independence.

Portuguese, British, and Dutch introduced changes that shredded the fabric of indigenous societies; the net result was the creation of social systems that could not survive without the paternalism of colonial governments. France, Spain, and the United States also played roles in the colonial period but their activities did not as directly alter regional systems.

Rising Nationalism and Independence

Geographer Jonathan Rigg (1991) suggests, "The uneven pattern of economic development, and the subordinate position of the local populations, engendered a great deal of bitterness. In addition, the paternalistic attitude of many colonial administrators and their occasionally harsh treatment of the local inhabitants created a reservoir of resentment that nationalist leaders were only too happy to exploit in order to achieve their aims."

Nationalistic leaders emerged because certain elites had been given the opportunity to study abroad. These individuals, such as Sukarno of Indonesia, Aung San of Burma, and Ho Chi Minh of Vietnam, spearheaded independence movements. New ideologies were framed in the context of the Boxer Rebellion in China (1899), the Chinese Revolution of 1911, and the rise of Japan and its defeat of Russia in 1905, events that proved that the Western powers were not invincible.

After the Japanese invasion of 1941, Japan promoted "Asia for the Asians." It also encouraged emerging nationalist movements and even transferred some powers to local leaders such as the sultans of the Malay States. Although the Japanese occupation proved to be no better than that of the former colonial powers, the period between 1941 and 1945 was important in laying the groundwork for subsequent independence.

After the defeat of Japan in World War II, the colonial powers returned to discover they had to negotiate with well-organized and informed nationalists with popular support. Not all efforts were peaceful, with much acrimony between the negotiating parties. The Philippines gained independence in 1946, Indonesia in 1949, and Malaya, along with British North Borneo and Sarawak (today's Malaysia), in 1957. The process in some of the other states was prolonged. For example, Singapore did not gain independence until 1963 and Brunei in 1984. (Figure 1-14). Whatever the case, the colonial period left what Rigg (1991) calls "an institutional and ideological legacy" of Westernization that became superimposed on or integrated with former modes of governance.

A Bitter Battle: French Indochina

French Indochina—Cambodia, Laos, and Vietnam—became nominally independent in 1949, although political and economic power was retained by France. Various nationalistic leaders including Ho Chi Minh resisted the French with aid from Communist China and the Soviet Union. A guerrilla war ensued in Vietnam, and the French were finally defeated at the Battle of Dien Bien Phu in 1954.

The United States, panicked over the further spread of communism, stepped into what became known in America as the "Vietnam War" and in Vietnam as the "American War." The American-supported non-Communist government of South Vietnam waged battle against the Communist forces of North Vietnam. After years of terrible conflict, loss of life, and increasing corruption in South Vietnam, public opinion in America forced U.S. withdrawal in 1973. The war finally ended in 1975. Divided Vietnam was then unified by the Communists.

The Vietnam/American war was a human and environmental tragedy. More than 4.5 million died in Indochina, including more than 58,000 Americans. Bombs, napalm, and chemical defoliants destroyed millions of acres of forests and cropland as well as people's health. Thousands have been killed or maimed by unexploded ordnance throughout Vietnam, Cambodia, and Laos. As you will see in the next chapter, the ramifications of this war were severe, widespread, and ongoing.

Changing Urban and Rural Worlds

Although there were a number of very large cities in 1900, most of mainland and insular Southeast Asia remained sparsely populated. However, the twentieth and twenty-first centuries witnessed unprecedented population growth in concert with economic, social, and political transformation (Figure 14-13).

Historically, settlement patterns were geared to coasts and rivers and based on seaborne interchange. Inland transportation was difficult until the road and railroad building of the 1800s.

Settlement has been tied to rice cultivation, which has, with Green Revolution improvements, supported ever-larger populations. Centralized policies, in Angkor and the Red River Delta, were required to command and control the expansion of rice production. Given the fractured terrain, for most of history, small societies and local economies dominated the Southeast Asian world. This situation changed in the late nineteenth century as the tentacles of European imperialism reached beyond port cities into the interior. Large-scale plantations, mines, and administrative cities were constructed in any potentially profitable niche.

With the growth of colonial enterprises, urban centers including traditional entrepôts and new colonial cities exhibited spurts of growth. By 1910, there were eleven Southeast Asian cities of more than 100,000 people: Mandalay, Rangoon (Yangon), Bangkok, Hanoi, Saigon, Georgetown

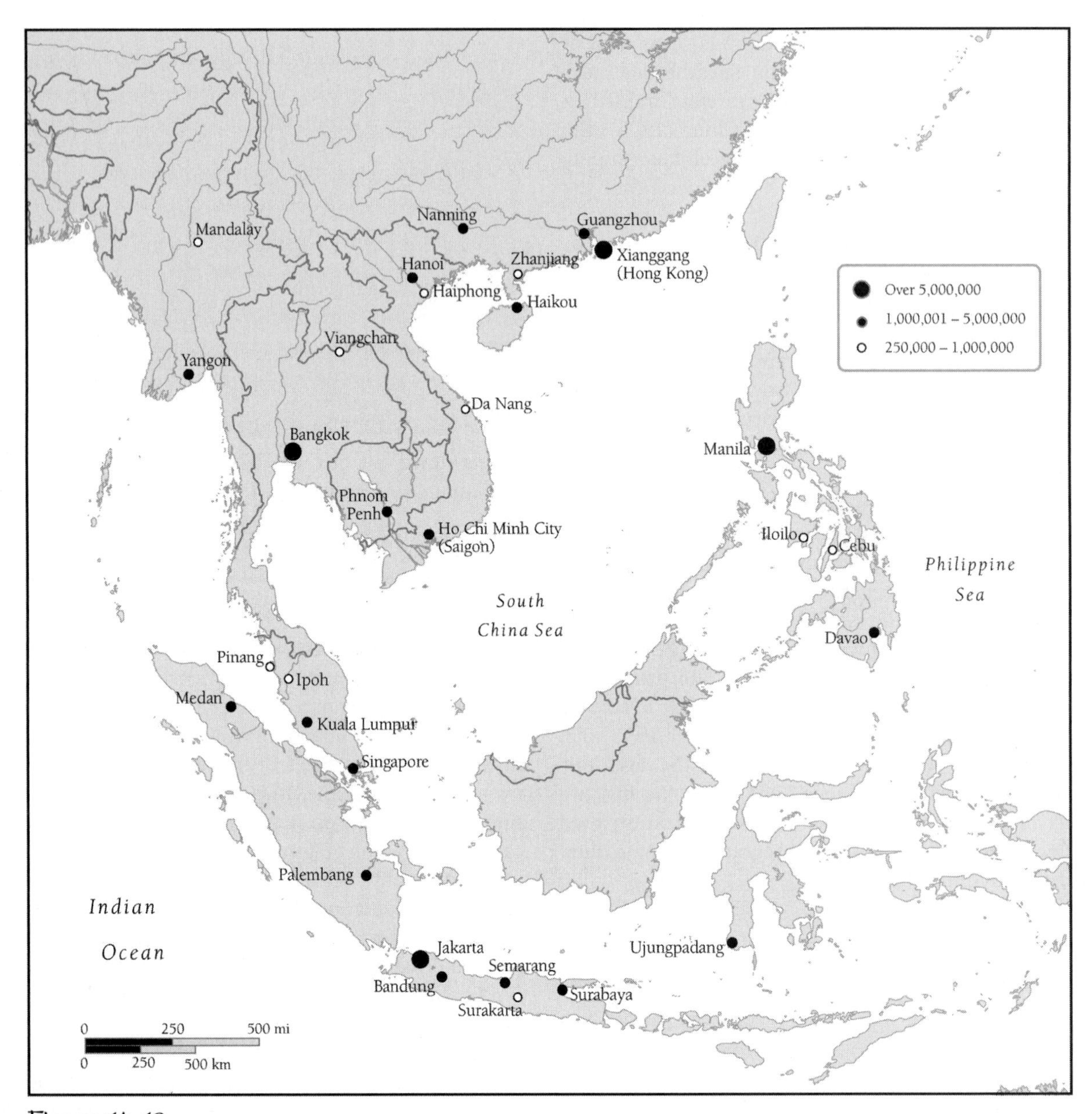

Figure 14-13

Note that the majority of cities in Southeast Asia are either coastal or on major rivers. Why is this?

(Pinang), Singapore, Batavia (Jakarta), Surakarta (Solo), Surabaja, and Manila. These cities were primarily administrative and commercial centers without large industrial bases. Moreover, the colonial economy did little to stimulate economic development beyond the export sector.

The rural sector exhibited three categories. The largest was the peasantry tied to a subsistence economy. Second was the commercialized peasantry who participated in the growing market economy based on urban needs. Third was the enclave economies of plantations and mines opened up with Western and Chinese capital and imported wage labor.

A demographic picture of Southeast Asia at the turn of the twentieth century reveals two primary settlement patterns. At one extreme were the dense populations engaged in irrigated rice cultivation. Java, with its population of 30 million in 1901, stands as an example. At the other end of the spectrum were vast insular and mainland areas sparsely settled by shifting cultivators. Around 1900, irrigated rice began diffusing, and several areas such as the remaining frontiers of Java, lower Burma, central Siam (Thailand), and the Mekong Delta transitioned from low to high population densities. The first three decades of the twentieth century continued this pattern of opening up the region to export industries and tightening the links between the colonies and the imperial powers overseas.

The end of the Japanese interregnum in 1945 ushered in a period of transition to independence. During this time, international trade withered, urban and industrial employment fell, and many were forced to revert to a subsistence existence. Nevertheless, there were factors of continuity as well as those of change.

The transition to national independence increased population mobility, only part of which was rural to urban migration. Many moved from overcrowded rural areas to less crowded ones. Others moved to cities as squatters, and still others populated the first suburban and even exurban areas outside large cities.

Out-migration of Western colonial administrators and business people—along with ethnic Chinese from several areas, and all Indians from Burma, plus Vietnamese from Cambodia—were other significant movements. Plantation and commercial rice regions were the areas mainly affected by these out-migrations. Many foreign managers and overseers were replaced by indigenous people.

METROPOLITAN EXPANSION

In Southeast Asia, it is the largest metropolitan areas that are growing much faster than other cities. For example, Kuala Lumpur has grown at almost double the rate of Singapore in the post-war period, reflecting its position as the capital of the new state of Malaysia. Once French Indochina split into Vietnam, Laos, and Cambodia, Phnom Penh grew rapidly in contrast to the former Indochina capital of Hanoi. Primate cities such as Bangkok, Kuala Lumpur, Jakarta, Manila, and Ho Chi Minh City (Saigon) continue to dominate the urban hierarchy.

Urbanization in general is on the increase. For example, the proportion of Indonesians living in urban areas of more than one million rose from 32.6 percent in 1971 to 43 percent in 2009. In the Philippines, the Manila Metropolitan Region has expanded from 1.5 million in 1950 to more than 11.5 million in 2010.

Many of these increases are related to processes of the transactional revolution, whereby flows of people, commodities, capital, and information are no longer hampered by the limits of space. While cities continue to grow, they do so over larger areas. This is called **region-based urbanization**, and it produces **extended metropolitan regions (EMRs).**

It is in the areas adjacent to currently defined urban areas that urbanization is occurring most rapidly, and this is reflected in the growing proportion of populations engaged in nonagricultural activities (refer to *kotadesasi* settlements in Chapter 4). EMRs have two main components: the central city and the metropolitan region containing adjacent urban settlements. With the exception of Singapore, it is these outer rings that are experiencing the most rapid growth. Meanwhile, city cores have maintained substantial shares of metropolitan populations. In fact, city populations continue to grow with increasing density and congestion (Figure 14-14). EMRs also continue to attract rural–urban migrants. At the social level these areas are extraordinarily heterogeneous with multiple types of land use existing side by side. You will learn more about these trends and individual cities in the next two chapters.

NEW RURAL DYNAMICS

There have always been market dependencies and interactions between rural and urban worlds, but the rural world of today is different in many ways. While village may describe life for the majority of Southeast Asians, millions now spend some of their time in urban areas both living and working and millions more spend time working in urban-type activities that have come to the countryside.

There is an assumption that people who live in villages are farmers (i.e., that "rural" is synonymous with "agriculture"). But this is not necessarily the case. Urban residents from rural areas often still identify themselves as villagers and farmers. In other words, domicile does not always mean "home" and rural and urban are frequently melded.

Recent studies of village economies consistently indicate that between 30 and 50 percent of total household income is derived from non-farm or off-farm sources. Many households have different family members participating in either rural or urban livelihoods, making it problematic to classify the household as rural or urban.

The transport revolution has been instrumental in expanding networks of contact and communication. The

Figure 14-14

More than 10 percent of Thailand's population lives in Bangkok, the capital city. Bangkok is also known as Krung Thep, meaning "City of Angels." Here, the crowded old city is surrounded by new office and apartment towers. Hundreds of people pour into the city every day seeking employment. Densities are as high as 5,800 people per square mile (3,600 per km^2), and an estimated 400 new vehicles are registered every day. Bangkok is one of the most congested cities in the world.
Photograph courtesy of B. A. Weightman.

construction of new roads, the easing of restrictions on travel, and the advent of the bicycle, motor scooter, and other cheap forms of transport encourage rural and urban interchange in terms of both job transfers and industrial location (Figure 14-15). This increases economic opportunities and supplements incomes. There is a definite link between poor infrastructure and high rates of poverty.

The merging of rural and urban worlds has had other impacts. The use of chemical fertilizers, wage labor, and mechanization has increased. More emphasis is placed on raising animal stock and cash crops for the urban market. More people engage in non-farm activities on the farm, such as mat and basket weaving, pottery, and furniture making. Off-season, full-time employment in factories or public sector jobs has increased, and there is increasing migration for employment beyond local areas.

There are also age and gender considerations. Agriculture remains the dominant occupation among older age groups. Younger groups are more likely to engage in wage labor, which points to the availability of such opportunities in or near rural areas. A second change is the growing importance of female employment. Rising female labor is pronounced in export-oriented activities such as the garment industry, electronics, and footwear. Declining fertility, later marriages, education, and increased

Figure 14-15

I met these two students in Viangchan, Laos. Both were very proud of their new motorbike.
Photograph courtesy of B. A. Weightman.

mobility have fostered conditions for women to work beyond the confines of the village.

Rural and urban changes function in tandem. There is more to rural life than agriculture, and at any time in the year different members of different households may be engaged in rural- or urban-type occupations. Although better accessibility allows rural dwellers to live and work in cities, governments are encouraging factories and other businesses to locate in non-urban settings where a pool of workers is ready to take up non-farm employment.

WOMEN'S CHANGING ROLES

Southeast Asian women are in a relatively better position than those in South or East Asia. Part of this is due to historical circumstance. Traditional kinship was traced through *both* maternal and paternal lineages. A daughter was not a financial burden because of the widespread practice of paying a bride price. A married couple typically lived with or near the wife's parents. Women had important roles in religious rituals. Finally, their labor was essential in agriculture and they dominated local markets. However, the position of women changed over time.

The rise of centralized states and the spread of imported philosophies and religions—Confucianism, Daoism, Buddhism, Islam, and Christianity—increasingly privileged men and subordinated women. Even so, these influences, most notable among the elite, were always moderated by local traditions.

The position of women was altered further under the auspices of colonialism. In some areas, women were recruited as cheap wage laborers on rubber, tea, and other plantations and in processing factories. At the village level, colonial authorities strengthened the role of males as "head of the household" and re-formed customary laws that gave women considerable autonomy. Even in Siam (Thailand), the only non-colonized country, legal codification strengthened patrilineality. These developments encouraged a preference for sons rather than daughters. Nonetheless, women were still influential in community life, at times even leading anti-colonial rebellions. Increasing female literacy, especially in the Philippines, and exposure to Western feminism encouraged elite women to confront issues of gender inequality.

As nationalism increased in the late nineteenth century, men concentrated on independence but educated women focused on such issues as polygamy, divorce, domestic abuse, and the responsibility of fathers. However, men avowed that women should wait to address their issues until after independence was achieved. Despite women's involvement with independence movements as fighters, strike organizers, spies and couriers, and outspoken journalists, they were seen as auxiliary, not equal partners.

Theoretically, the newly independent states that emerged in the 15 years following WW II were committed to gender equality, but this has rarely translated into reality. The numbers of women in government have increased, but only in the Philippines has female representation risen above 10 percent. Moreover, women in government usually find themselves marginalized in the shadow of men. Women who have risen to head-of-state positions as in the Philippines or Indonesia have done so as the daughters or wives of politically powerful men. Notably, they have not been promoters of women's issues as this might alienate them from their male supporters.

Greater female involvement in politics is impeded by prevailing societal attitudes that consider women as wives and mothers. Gender stereotypes are reinforced in school textbooks and are sometimes encouraged by religious teachings. For example, Buddhists still believe that rebirth as a woman rather than a man indicates that less merit was accrued in past lives. Southeast Asian Islam has been relatively tolerant, but over the past 20 years, there has been more stress on "correct dress" and "appropriate behavior" in public. Although all countries, with the exceptions of Laos and Cambodia, have signed the "Convention on the Elimination of All Forms of Discrimination against Women" and have made advances in gender equality, it is difficult to change the preference for sons, especially in Vietnam with its strong Confucian heritage.

Fortunately, people do hold to the idea that women can earn and control their own income. However, they are paid less than men for the same work and are typically hired in subservient positions. As Southeast Asian countries have become export-driven, women are vital in factory work. Many educated women find skilled employment abroad as in the case of nurses from the Philippines. With the exception of Cambodia and Laos, the numbers of women progressing to post-secondary training is rising. In some countries such as Brunei, Malaysia, Thailand, and the Philippines, there are more female graduates than males. Many of these graduates become involved in female-oriented NGOs.

Despite the region's economic, political, and cultural diversity, Southeast Asian countries generally fare

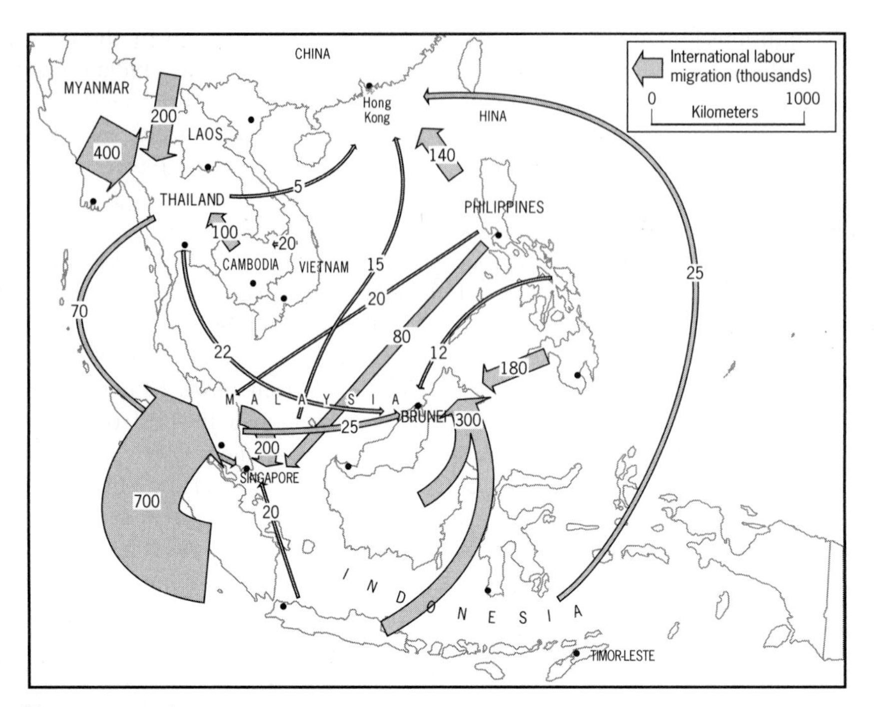

Figure 14-16

Selected flows of intra-regional labor migration. Extra-regional flows such as Thais and Filipinos to Japan are not included. From Jonathan Rigg, *Southeast Asia: The Human Landscape of Modernization and Development*. 2003 (Reprinted 2006), Figure 4.1, p. 163. Reprinted with permission of Routledge.

well in measures of human development. The heritage of relatively favorable gender relations and the resilience and pragmatism of local societies suggest that the status of most Southeast Asian women can only improve.

LABOR MIGRATION

According to geographer Jonathan Rigg (2006), "Southeast Asia is rapidly becoming a regional human resource economy." This is because borders are becoming more permeable; fewer legal and physical barriers allow people to cross into countries with labor shortages such as Malaysia, Singapore, and Brunei. Wage differentials between countries and between different sectors within countries encourage worker migration (Figure 14-16).

Migrant laborers include both men and women. For instance, both enter Thailand from Myanmar, Cambodia, Laos, and China to work in the construction and sex industry. Thai agricultural workers cross into Malaysia to harvest rice. Malaysian men and women take the causeway to Singapore to work in a variety of occupations there.

Unfortunately, foreign workers are exploited. Even legal immigrants are paid less than locals for the same work. They are housed in miserable dormitories or live in shacks with few services. They work long hours and can be fired for the tiniest infraction. This is especially true for women, who are also subject to sexual harassment. Illegal immigrants fare even worse.

One study of Cambodian workers in Thailand shows that they are engaged primarily in low-skilled, low-paid, and often hazardous work. They are paid 30 to 50 percent less than Thai workers and are required to work longer hours (10 or more hours a day). They are often swindled and subject to arbitrary arrest and detention. Female workers are regularly harassed. Yet they still come because even these conditions are, at least economically, better than what they experience at home.

REGIONAL INEQUALITIES

Even though Southeast Asian economies are growing apace, regional inequities persist. Poverty studies in various countries indicate that poor people have several things in common. For example, poverty is more intense among rural, farming populations in peripheral regions. Poor people are more likely to be landless, tenant farmers or day laborers. Subsistence farmers are more likely to be deprived than are commercial farmers. Those with larger families or with young household heads, particularly if they are uneducated, are more likely to have to do without. Finally, poor families have little or no access to health care and education or credit and modern technology. We will discuss some specific examples in the following chapters.

When we speak of poverty, it is important to note that the poor are not necessarily concerned with income, resources, and consumption. Their version of deprivation is more likely to concern vulnerability, stability, regular employment, dependency, and self-respect. Lack of appropriate shelter is also a factor. Most people would like to have a permanent, brick dwelling with a tiled roof as opposed to a woven, fiber walled and roofed house that requires constant expenditure on repairs.

Population Dynamics

Prior to the intensification of the European role, Southeast Asia was a comparatively healthy region. However, by the nineteenth century, conditions had deteriorated. Western advances in health provision and medical techniques had spurred population growth, meaning that increased concentration and contact along with the influx of new groups reduced public health effectiveness. Epidemics became a problem. Paradoxically, populations increased, with growth rates climbing from 0.2 percent to well over 1 percent in many areas.

The Spread of Disease

When the rain forest was cleared over vast expanses to plant rubber trees, water bodies were exposed to sunlight. This caused the *Anopheles maculatus* to breed. This mosquito is one of the most dangerous epidemic vectors of malaria in the world. Epidemics were so severe that plantation owners had to carry Tamil workers off in droves and replace them with new ones. These deadly mosquitoes continue to plague settlement schemes, logging operations, agricultural clearance, and the like.

In another case, the British accidentally introduced the *Aedes aegypti,* another mosquito originally from Africa. As a super vector of dangerous viruses, it caused dengue fever to spread explosively in urban areas. Today, it is responsible for the emergence and spread of the deadly dengue hemorrhagic fever.

Interregional connections and land conversion policies of the Europeans changed the biotic ecology of diseases, especially vectored ones. Millions of victims have suffered or died as a result.

This growth is thought to be attributable to colonial activities. For instance, the spread of Christianity in the Philippines increased birth rates considerably. The expansion of settled agriculture and general peace provided an atmosphere of stability and security in which to raise families. Improvements in agriculture meant improvement in nutrition, encouraging longer life expectancies and lower mortality rates. By the 1960s, fertility rates had increased dramatically. For instance, Indonesia's total fertility rate (TFR) was 5.67 and Thailand's TFR was 6.4. However, with family planning (begun in 1971 in Indonesia) and social pressure, these rates had fallen to 2.5 and 1.8, respectively, by 2009. The achievements of these two countries in reducing fertility are to be noted for their noncoercive nature (unlike China).

From 1950 to 2009, population growth was impressive. Thailand's population increased from 20 million to 67.8 million. In the Philippines, population increased from 19.9 million to 92.2 million. Indonesia's population grew from 83 million to 242.3 million.

At the beginning of the twenty-first century, Southeast Asia's population reached more than half a billion. In 1900, the population was a mere 80 million. Even with birth control programs on the part of such countries as Thailand, Indonesia, and Malaysia, population dynamics are uneven. On the one hand, richer countries such as Singapore and Thailand have low fertility rates. On the other hand, poorer countries such as Cambodia and Laos have high fertility rates: 3.0 and 3.5, respectively. Further, different groups within countries have different rates. For example, Malaysian women of Chinese ancestry (24 percent of the population) tend to have about two children each, while Malays and other indigenous groups average close to three children each. In Southeast Asia, the use of birth control is at 60 percent, with contraceptive use rising with educational level. A major problem is access to contraception, as in remote areas of Vietnam and the Philippines.

Environmental Concerns

The most severe environmental problem in Southeast Asia is deforestation. While slash-and-burn cultivation, clearing for migrant settlement, and the development of tree crop plantations are important causes, perhaps the most important is commercial logging. The forests of Southeast Asia have high commercial value in addition to high yields. Much of today's imported wood products originate in these forests.

Commercial logging is a lucrative activity. The entire process has involved considerable corruption, especially in Indonesia, Malaysia, the Philippines, and Thailand. Much forest is logged illegally. In other cases, timber exports are under-invoiced and the logs smuggled overseas. Most logs from the Philippines are exported illegally. Finally, timber concessions are granted for political or money favors. Who has access to and control of forests is one of the most salient aspects of their use today.

Southeast Asian sources of tropical timber products are concentrated in the Philippines, the Malaysian states of Sarawak and Sabah, and Indonesia's outer islands. The primary destination of logs and related products is Japan, which has been the world's largest importer of tropical timber since the 1960s. Forest devastation has been so great that today only Sarawak continues to export raw logs legally.

Timber extraction in Southeast Asia is related to Japan's investment and development plans in the region. For example, Japan's early role in the timber industry in Kalimantan began with technical assistance in exchange for timber export agreements. After 1967, most of the major Japanese trading companies such as Mitsubishi, Mitsui, and Sumitomo were involved in loaning money for infrastructural development, including access roads into timber areas. In 1971, Indonesia replaced the Philippines as Japan's first source of raw logs, which were largely processed into plywood in Japan.

By 1980, after rampant forest destruction, Indonesia shifted its efforts to develop its own forest products industry. It also began a phased ban on the export of raw logs, which was fully implemented by 1988. Most Japanese companies withdrew from the Indonesian timber industry, but Japan remains the largest importer of Indonesian plywood. This example shows the relationship between one country's political and economic activities and environmental consequences in another.

Although several countries have taken steps to control wholesale forest destruction, illegal logging continues. Deforestation, with its accompanying soil erosion and habitat elimination (Chapter 2), continues as one of the major issues in the region.

Conflicts at Sea

Territorial conflicts have permeated the history of Southeast Asia. Given this ocean-bound world, it is no surprise that some conflicts would be over ocean spaces. Control over ocean space becomes critical when natural resources are involved, as in the case of the Paracel and Spratly islands where fishing rights and oil potential drive countries to occupy islands, reefs, rocky outcrops, and anything else sticking out of the ocean.

DISPUTED ISLANDS

The Paracels are disputed islands and reefs approximately 16 degrees north of the equator and about 233 miles (375 km) east of Da Nang in Vietnam. They were a part of French Indochina until the Japanese held them in World War II. Although these islands are closer to Vietnam than they are to China, the Chinese took the islands from Vietnam in 1974, claiming they were always a part of China. The Paracels are also claimed by Vietnam and Taiwan. The interest centers on potential oil reserves beneath the ocean.

The Spratly Islands are another group of a hundred or so islands and reefs in the South China Sea lying about two-thirds of the way from southern Vietnam to the Philippines. They were used as a submarine base by the Japanese in World War II. Their natural resources are fish, guano, and undetermined natural gas and oil potential. They are also situated near several primary shipping lanes. About 45 of the islands are claimed and occupied by China, Malaysia, the Philippines, Taiwan, and Vietnam (Figure 14-17). In 1984, Brunei established an exclusive fishing zone in the southern Spratlys.

In 1994, the American company Exxon signed a multibillion dollar deal with Indonesia to develop the

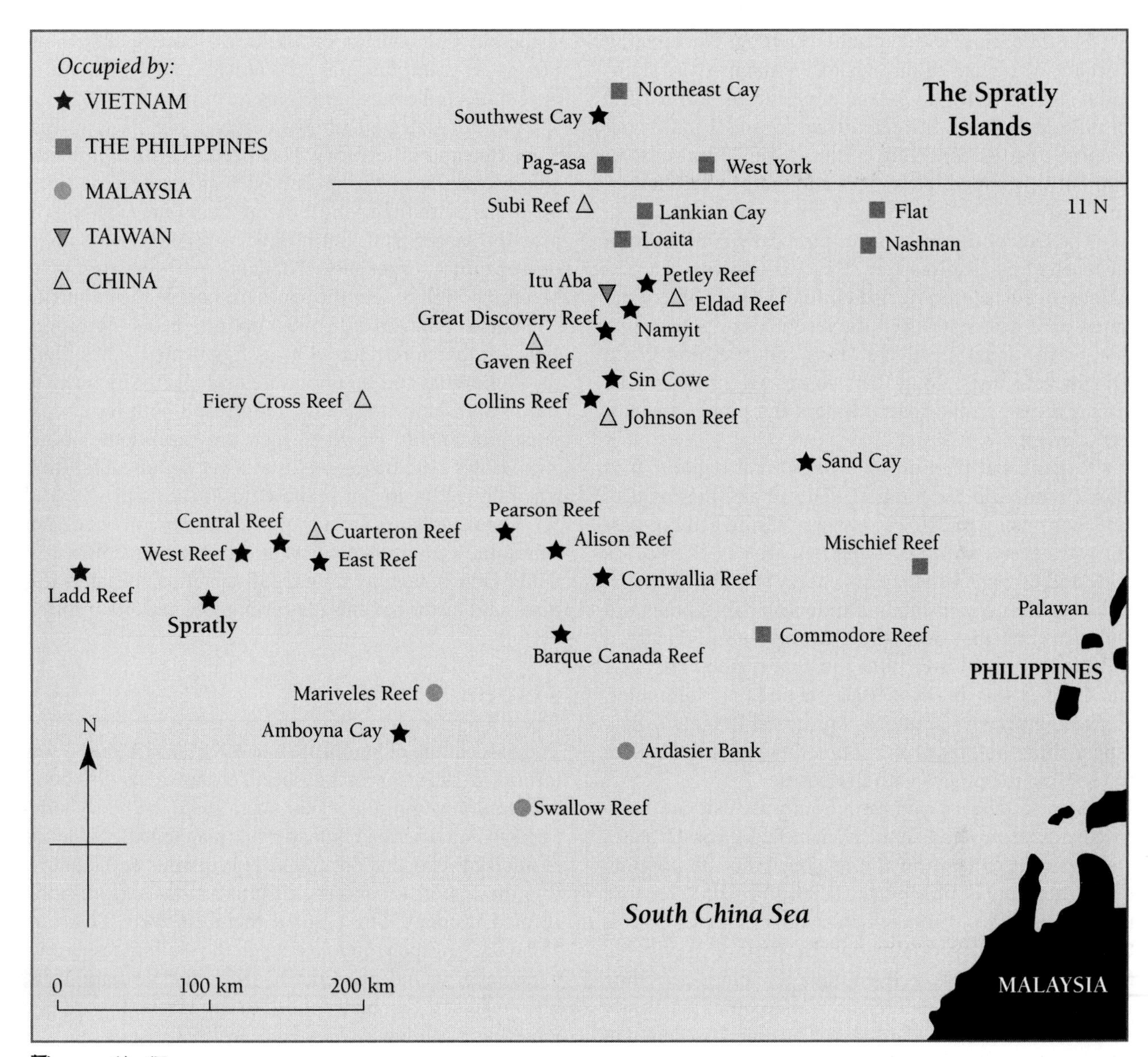

Figure 14-17

Note how various countries have occupied islands of the Spratlys. China claims them all.

Natuna gas fields in the southern Spratlys, an area partly claimed by China. There were clashes between China and the Philippines in 1995 because China occupied one island within 124 miles (200 km) of the Philippine island of Palawan. That same year, Indonesia expressed concern over China's claim to the natural gas fields. Some have suggested joint development of this region, but no multilateral agreements on dividing or exploiting the region have yet been reached.

PIRACY

Traders have been harassed in the Southeast Asian waters for centuries. The nomadic seafaring Bugis from Sulawesi were once infamous as pirates and continue to ply goods throughout Indonesia. But piracy is not just a thing of the past. Pirates continue to attack from the Straits of Malacca to the Celebes Sea. Cargo ships, fishing vessels, passenger liners, and refugee boats are all victims.

The situation was particularly tough during the Vietnam War when thousands of Vietnamese and Cambodians were trying to escape persecution. From 1980 to 1986 alone it is estimated that around 1,500 "boat people" (escapees) were killed, and hundreds of women were raped, abducted, and sold to brothels in Thailand.

The Law of the Sea has put piratical victims beyond the reach of international law. Piracy is only piracy when attacks occur within jurisdictional waters of coastal states, and most states lack the resources to police their vast marine areas effectively. Many countries now claim straight base lines along their coasts within which the marine areas become coastal waters and pirates are subject to arrest.

Commercial piracy occurs in several regions. It is most common in the Strait of Malacca and the Strait of Singapore. Some 6,000 vessels pass through these passageways every month carrying a quarter of the world's trade and 80 percent of Japan's energy supplies. Such attacks irritate relations among Indonesia, Singapore, and Malaysia. Endemic overfishing in the Gulf of Thailand and poverty also foster piracy in that region. The Sulu and Celebes seas, between Malaysia and the Philippines, are frequent scenes of attacks on coastal freighters, passenger ships, and fishermen. Some freighters have taken to repelling the pirates with fire hoses.

Because modern pirates in Southeast Asia are land-based and not rovers like the ancient Bugis, coastal states are expected to take on the responsibility of stopping them. Incidents of piracy have fallen since 2000 because more ships have security systems and air and sea patrols have increased, especially on the part of Indonesia, Malaysia, and Singapore. Even India has joined patrolling operations. Even so, this region still accounts for nearly 40 percent of the world's piracy. Thus far, it seems that maritime frontiers are very difficult to police effectively.

Shapes of Countries

You have probably noticed that Southeast Asian countries take on many different shapes, some rather odd. Country shapes are significant in that they may affect a country's ability to consolidate its territory and control trans-border interactions. The most desirable shape is circular, because that provides the shortest possible border in relation to territory and allows equal access to all places from the center. Although this implies stability, stability is not necessarily the case. Cambodia has a relatively circular or compact shape but Cambodia does not have a history of stability. Elongated countries such as Vietnam and Laos tend to have unity and control problems.

Prorupt states are almost compact but have at least one extension of territory. This makes control problematic, especially in the prorupted regions. Thailand and Myanmar, with their long territorial tails, are examples of prorupt states and both have problems with their proruptions—especially Thailand with its minority Muslim population in the southern part of the country.

Some countries are broken up into pieces. Archipelagoes are therefore referred to as fragmented states. Both the Philippines and Indonesia are examples of fragmented states. Both have trouble with unity and both have separatist movements. However, there are exceptions. Brunei and Malaysia are fragmented states yet do not suffer from instability. Why do you suppose this is the case?

Country shapes provide grist for discussion but do not determine a country's political or economic circumstances. Other factors such as resource distribution, physical features, and historic events play even more important roles.

Asean

The **Association of Southeast Asian Nations (ASEAN)** was formed in 1967 to promote social, economic, and political cooperation within the region. After 1945, the newly independent countries had little to bind them together in terms of interstate cooperation, and since economic development was the foremost concern, an international organization seemed prudent. The original members were Thailand, Malaysia, Singapore, Indonesia, and the Philippines. Brunei joined after its independence in 1984, and Vietnam, Laos, Myanmar, and Cambodia joined in the 1990s. Timor-Leste joined after gaining independence in 2002.

ASEAN's main goals are to promote and integrate economic development, encourage social and cultural progress, and guarantee regional peace and stability. Actually, member countries trade more with outside nations than with each other, as so many of them produce the same things. Moreover, international trade activity is dominated by Singapore, Malaysia, Indonesia, and Thailand. Intra-ASEAN trade is also dominated by Singapore. Minerals and fuels account for more than half the volume of exchange.

Three obstacles stand in the way of increased intra-ASEAN trade. The economies of the members are competitive rather than complementary, which supports trade outside the organization. Second, non-tariff barriers that are employed to protect domestic industries serve as barriers to intraregional exchange. Third, the productivity of countries such as Vietnam, with its low level of development, or Brunei, with its small size, cannot be expected to match the volume and diversity of output of more developed countries such as Singapore or Thailand.

Facing competition from India and China, ASEAN is being forced to confront its failure to realize a long-held goal: integration into a single market that is attractive to foreign investors. Singapore and Thailand are pushing for change and Brunei and Malaysia are somewhat supportive. However, the newer members—Cambodia, Laos, Burma, Vietnam, and Timor-Leste—are reluctant to open their economies fully because they fear they cannot compete with the stronger countries. Some have backslid from original commitments to cut tariffs by 60 percent. The backsliders include Malaysia with its car industry, Indonesia with rice and sugar, and the Philippines with rice and petrochemicals.

India and China sent representatives to the ASEAN summit in Bali, Indonesia in 2003. They boasted about their investments in Southeast Asia and pointed out the importance of their economic roles in the region. These countries want to forge links with ASEAN in a larger economic community.

ASEAN now has what it calls "dialogue partners." ASEAN +3 adds China, Japan, and South Korea. 2ASEAN refers to Australia and New Zealand. Dialogue partners participate in discussions not only on economic issues but also on political issues such as democracy in Myanmar, the spread of bird flu, and the insurgency in southern Thailand.

Pursuing consolidation, ASEAN leaders have endorsed a blueprint that envisions a single common market and a single production base with the free flow of goods, services, investment, and capital. The year 2020 is the target date, but Singapore and Thailand want to accomplish it sooner. ASEAN leaders have agreed to remove non-tariff barriers, and harmonize customs procedures and product standards. Targets have been set for 100 priority sectors ranging from wood products and cars to air travel and tourism.

These economic agreements might change the face of ASEAN, which until recently has been most effective in the political arena, curbing hostilities and acting as a forum to solve numerous disputes. Most importantly, a strong regional identity has been created among its members. Geographer Antonia Hussey (1991) wrote, "The sense of identity and cooperation espoused by ASEAN is not found elsewhere in the third world." However, countries have been too hesitant in criticizing blatant human rights violations among their members—such as in Cambodia in the past and in Myanmar (Burma) today.

Economic Crises of the 1990s and 2000s

Rapid economic growth in the non-Communist ASEAN countries came to a rapid halt in the late 1990s when overlending by banks caused a major financial crisis. Thailand, the Philippines, Malaysia, and Indonesia were forced to devalue their currencies in 1997. The problems were caused by high levels of foreign borrowing, government budget deficits, large bank loans to glutted property markets, and slower than expected economic growth. Japan, in its own economic crisis, slowed investment and technical transfers to Southeast Asia.

Thailand suffered most as the crisis led to a change of government and a new constitution. Austerity measures were imposed. Malaysia and Indonesia, long ruled by authoritarian governments, found it more difficult to admit mistakes and launch reforms and restructuring. President Suharto of Indonesia was forced to resign, leaving the country's economic and social life in chaos. Indonesia has had the hardest time recovering from this 1997 crisis, partly because a backlash against the Chinese led to their withdrawal in large numbers together with their money. Indonesia also had to absorb more than a million workers who were deported from Malaysia during the crisis there.

By the early 2000s, China presented new challenges with increased competition and increased markets. The redirection of foreign direct investment to China slowed Southeast Asia's recovery from the 1997 crisis. Moreover, Chinese goods compete in textiles (Vietnam and Indonesia) and higher-value products (Thailand and Malaysia). ASEAN knows that it must attract investors, lower trade barriers to open up markets, and upgrade domestic worker skills to counter China's impact, but inevitably it will face severe competition from China in many areas in the years to come.

The more recent economic crisis that ensued in 2007 hit Southeast Asia's economies as exports fell and the influx of foreign capital slumped sharply. However, the crisis was abated when governments provided stimulus packages to banks and businesses and cash transfers to the public. Moreover, those economies with large numbers of workers overseas, such as Indonesia and the Philippines, were cushioned by cash remittances.

Recommended Web Sites

http://aero-comlab.stanford.edu/jameson'sworld_history/A_short_History_of_South_East_Asia1.pdf
Short history of Southeast Asia from early states such as Funan to 1970.

www.africanwater.org/mekong_river.htm
Maps, photos, and discussion of water agreements on the Mekong. Maps and diagrams of water flows of Tonle Sap. Statistics on countries along the Mekong.

www.art-and-archaeology.com/seasia/map1.html
Maps of historical kingdoms of Southeast Asia from National Geographic.

www.aseansec.org/index2008.htm
Official site of the Association of Southeast Asian Nations (ASEAN).

www.asianamerican.net/southeast.html
Links to numerous sites on countries, religion, leaders, etc.

www.asiarice.org/sections/riceheritage/riceheritage.html
Excellent site about rice culture, heritage, ceremonies, rituals, and ceremonies in Southeast Asia.

www.asiasociety.org/
Home page of the Asia Society. Numerous short articles on topics such as women in Asia, Asian religions, history, economics, politics, etc., by country.

www.bing.com/images/search?q=mekong+river&FORM_IGRE#
Hundreds of outstanding photos of the Mekong and life along the river.

www.en.wipedia.org/wiki/Architecture_of_Penang
Excellent photos and short discussion of colonial, Chinese, and other architecture of Penang, Malaysia.

www.en.wikipedia.org/wiki/Chinese_in_Southeast_Asia#Southeast_Asia
Good account of the Chinese in Southeast Asia.

www.fsmitha.com/h2/c23t.htm
Short account of independence movements in Southeast Asia.

www.learner.org/courses/worldhistory/support/reading_10_1.pdf
Readings on "Riverine and Island Empires." Funan, Khmer, Srivijaya, and Majapahit. Annenberg Foundation.

www.mrcmekong.org/
Mekong River Commission agreement on sustainable development of the Mekong River Basin. Photos and links to various related articles and reports.

www.neworldencyclopedia.org/entry/Mjahapit
History of Indianized Majapahit Kingdom from 1293 to about 1500. Links to related sites.

www.orientalarchitecture.com/index.php
Excellent site with maps, diagrams, and photos of temples, stupas, and other ancient structures. Information and virtual tours of Asian art and architecture by country and city.

Bibliography Chapter 14: Southeast Asia: Transition among the Nagas

Andaya, Barbara Watson, 2008. "Women in Southeast Asia." New York: The Asia Society.

Brierley, Joanna Hall. 1994. *Spices: The Story of Indonesia's Spice Trade*. New York: Oxford.

Dauvergne, Peter. 1997. *Shadows in the Forest: Japan and the Politics of Timber in Southeast Asia*. Cambridge: MIT.

Drake, Christine. 1989. *National Integration in Indonesia*. Honolulu: University of Hawaii.

Duchin, Dian. 1986. *The Subak Agriculture of Bali*. The New York Botanical Garden 10: 6–10.

Dutt, Ashok K., ed. 1985. *Southeast Asia: Realm of Contrasts*. Boulder, Colo.: Westview.

Dzurek, Daniel J. 1989/90. "Piracy in Southeast Asia." *Oceanus* 32: 65–70.

Ginsburg, Norton. 1972. "The Political Dimension." In *Focus on Southeast Asia,* ed. Alice Taylor, pp. 3–12. New York: Praeger.

Hirschman, Charles. 1994. "Population and Society in Twentieth-Century Southeast Asia." *Journal of Southeast Asian Studies* 25: 381–416.

Hussey, Antonia. 1991. "Regional Development and Cooperation through ASEAN." *The Geographical Review* 81: 87–98.

Jumsai, Sumet. 1997. *Naga: Cultural Origins in Siam and the West Pacific*. Bangkok: Chalermnit Press.

Keyes, Charles F. 1995. *The Golden Peninsula: Culture and Adaptation in Mainland Southeast Asia*. Honolulu: University of Hawaii.

Kummer, David M. "The Physical Environment." In *Southeast Asia: Diversity and Development,* ed. Thomas R. Leinbach and Richard Ulack, pp. 6–34. Upper Saddle River, N.J.: Prentice Hall.

Ma, Laurence J. 1985. "Cultural Diversity." In *Southeast Asia: Realm of Contrasts,* ed. Ashok K. Dutt, pp. 53–64. Boulder, Colo.: Westview.

Mabbet, Ian, and David Chandler. 1996. *The Khmers*. Oxford: Blackwell.

Mastny, Lisa. 2003. "Messing With the Mekong." *World Watch* (November/December): 21–28.

McBeth, John. 2003. "ASEAN Summit: Taking the Helm." *Far Eastern Economic Review* (October): 38–39.

McCawley, Tom. 2004. "Sea of Trouble (Piracy)." *Far Eastern Economic Review* (May): 50–52.

McCloud, Donald G. 1995. *Southeast Asia: Tradition and Modernity in the Contemporary World*. Boulder, Colo.: Westview.

McGee, T. G. 1995. "Metrofitting the Emerging Mega-Urban Regions of ASEAN: An Overview." In *The Mega-Urban Regions of Southeast Asia,* ed. T. G. McGee and Ira M. Robinson, pp. 3–26. Vancouver: UBC.

McGee, T. G. 1967. *The Southeast Asian City*. London: G. F. Bell.

Michell, George, and John Fritz. 1993. "Angkor Regained." *Geographical* 65: 33–38.

Minority Rights Group. 1992. *The Chinese of Southeast Asia*. Manchester: Free Press.

Noble, Allen G. 1985. "The Physical Environment." In *Southeast Asia: Realm of Contrasts,* eds. Ashok K. Dutt, pp. 36–51. Boulder, Colo.: Westview.

Rigg, Jonathan. 1991. *Southeast Asia: A Region in Transition*. London: Cambridge.

Rigg, Jonathan. 2003. *Southeast Asia: The Human Landscape of Modernization and Development*. 2nd edition. New York: Routledge.

Sanderson, Warren C., and Jee-ping Tan. 1995. *Population in Asia*. Washington, D.C.: World Bank.

Sar Desai, D. R. 1989. *Southeast Asia: Past and Present*. Boulder, Colo.: Westview.

Shaffer, Lynda Norene. 1996. *Maritime Southeast Asia to 1500*. London: M. E. Sharpe.

Vatikiotis, Michael, and Prangtip Daorueng. 1998. "Survival Tactics (Overseas Chinese)." *Far Eastern Economic Review* 26 (February): 42–45.]

Chapter 15

Mainland Southeast Asia: Turmoil and Peace

"The waves of human passion rise and fall; worldly things are as transitory as passing clouds."

CAO BA-QUAT (1809–1854)

In this chapter we will learn about the five countries of mainland Southeast Asia: Burma/Myanmar, Thailand, Vietnam, Laos, and Cambodia (Figure 15-1). As we will discover, the history of each unfolds in context of the history of the others. Landscapes, with their impress of rice cultivation, Buddhism, and battlegrounds, are remarkably similar. Yet each country exhibits its own character in terms of ethnicity, language, political system, level of development, and various other cultural aspects (Figure 15-2). The complexity of mainland Southeast Asia makes it a geographic conundrum leaving us with more questions than answers.

Burma/Myanmar: Paradise Lost

Imagine living in a country where your universities are closed most of the time. Where you cannot stay at a relative's or friend's house without government permission. Where you voted for democracy but your party is outlawed and your leader is under house arrest. Where barbed wire barricades are stacked on streets ready to block any march or protest. Where people meet in secret to tell you about persecution of minorities and slave labor. This is Myanmar today—oppressive for most within and despised by most without.

Officially, the country is called Myanmar (Figure 15-3), but many people refer to it as Burma. The country's original name was Myanmar, but this was anglicized as "Burma" by the British who invaded in 1885. In 1989, an unelected military dictatorship changed the name back to Myanmar and this is the name recognized by the United Nations. However, there is a "democracy movement" in Myanmar that insists on the name "Burma." Those who support this movement and who do not recognize the military dictatorship as a legitimate government support the name "Burma" as well. The problem with "Burma" is that it refers to the majority Burman people (68 percent) and does not represent the many other ethnic groups that live in the country. Some think that Myanmar/Burma should have a new name, one that represents all the people. We will use Burma and Myanmar interchangeably in the text.

The 135 officially recognized ethnic groups of Myanmar are categorized into eight "major national ethnic races" by region. For example, the Shan include 33 ethnic groups speaking numerous languages in four different language families (Figure 15-4). Aside from the 8 categories, there are many other unrecognized groups such as the Burmese Indians and Chinese, and the Rohingyas.

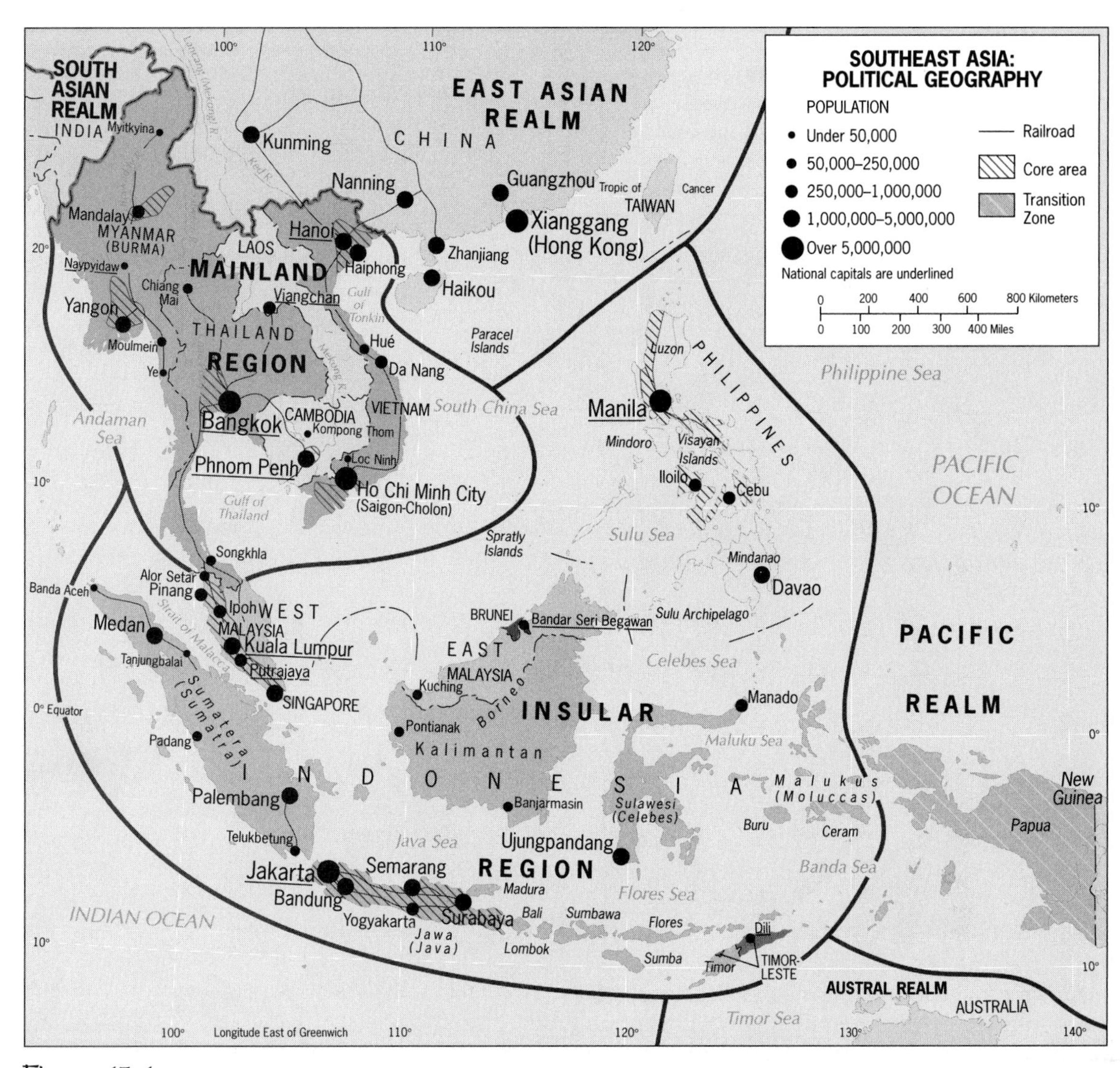

Figure 15-1

Defining the realm: mainland and insular Southeast Asia. Note the locations of the core regions.
From H. J. de Blij and P. O. Muller, *Geography: Realms, Regions and Concepts*, 14th edition, 2010, p. 532. Originally rendered in color. © H. J. de Blij and P. O. Muller. Reprinted with permission of John Wiley & Sons, Inc.

The Rohingyas

The some 730,000 Rohingyas are Sunni Muslims of northern Rakhine (Arakan) State. They have been persecuted since the Myanmar army launched operation Nagamin (Dragon King) in 1978 whereby they killed, raped, and plundered Rohingya communities. Most Rohingyas are denied Myanmar citizenship. They are subject to extortion, arbitrary detention and torture, land confiscation, and forced labor. To avoid further atrocities, around 200,000 fled to Bangladesh, but the Bangladeshi government withdrew its support of them in 2005. Now, thousands have escaped to Thailand where they live in refugee camps. Unfortunately, the Thais don't want them either. In 2009 and 2010, there were several cases of Rohingyas being towed out to sea by the Thai army and cut adrift to die.

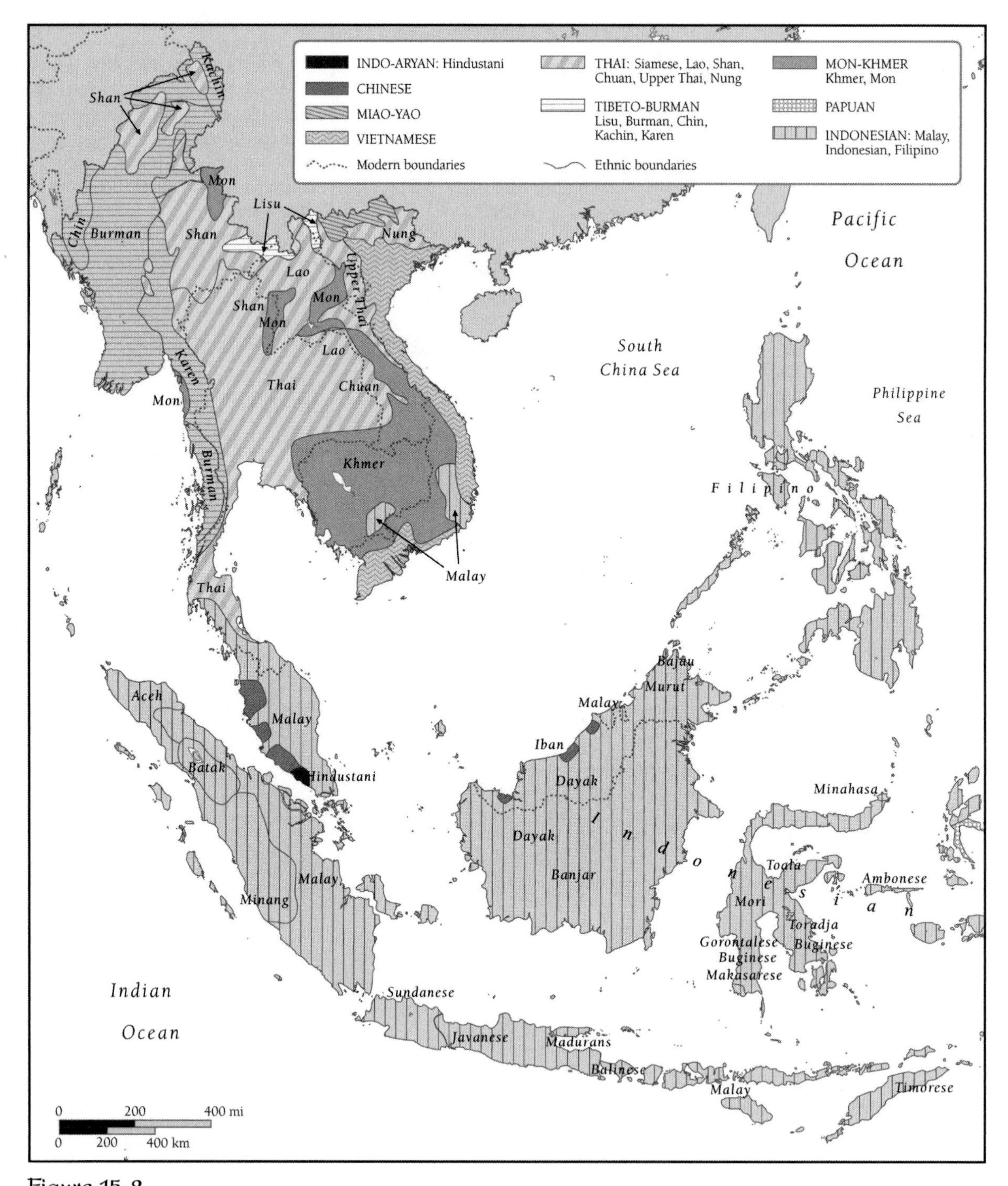

Figure 15-2

This region is a dramatic ethnic-cultural complex. Note the presence of Chinese and Indian communities.

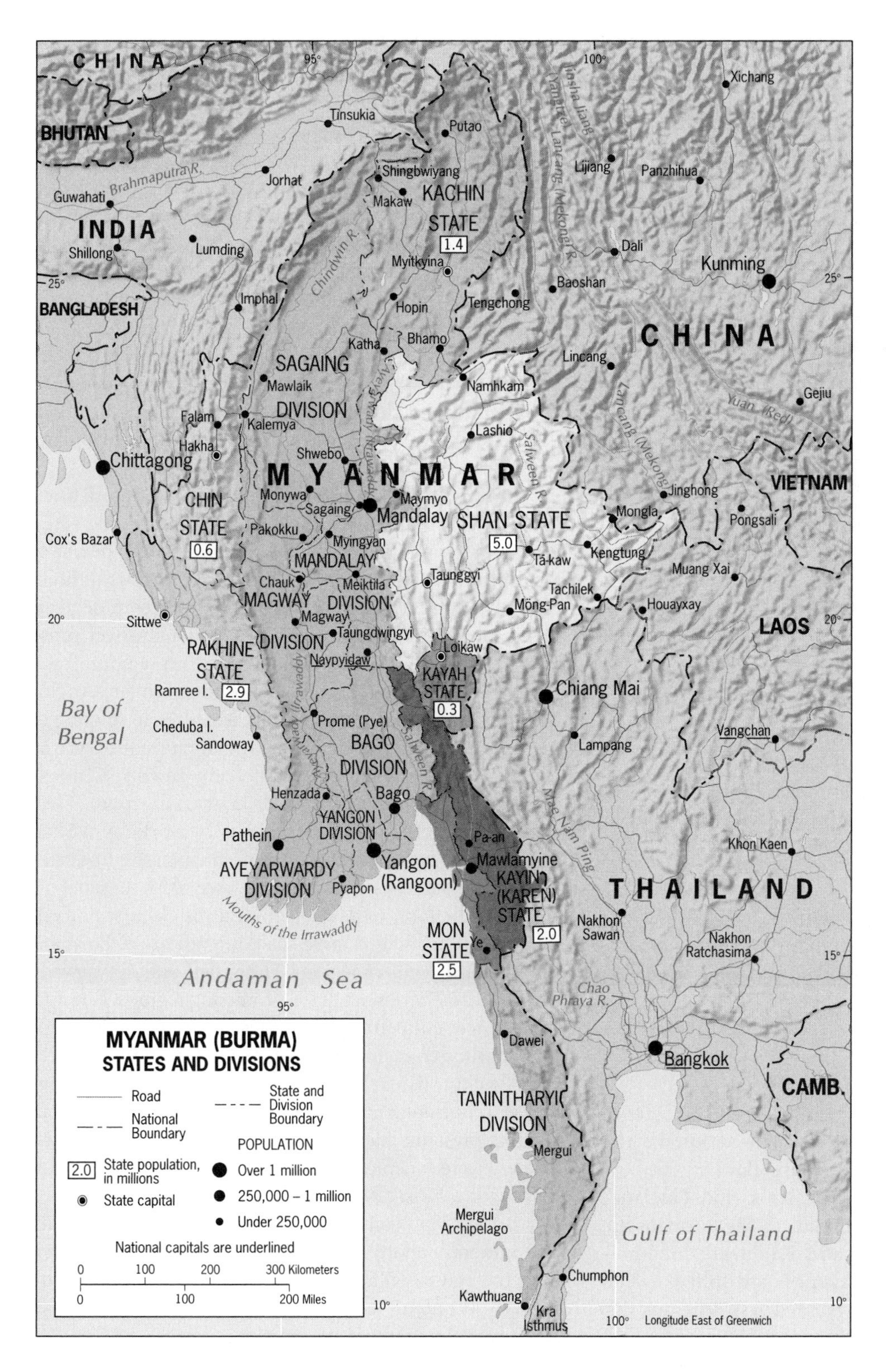

Figure 15-3

Myanmar (Burma) and its core areas. Note the importance of rivers in the location of urban places and core areas. From H. J. de Blij and P. O. Muller, *Geography: Realms, Regions and Concepts*, 14th edition, 2010, p. 558. Originally rendered in color. H. J. de Blij and P. O. Muller. Reprinted with permission of John Wiley & Sons Inc.

Figure 15-4

The Shan, estimated at 6 million, are one of four major Theravada Buddhist ethnic groups in Burma. They live mainly in the Shan State in the northern part of the country.
Photograph courtesy of B. A. Weightman.

In 1948, Burma was the world's leading rice exporter. After years of government mismanagement, Burma has slumped into being the poorest country in Southeast Asia. How did this happen?

POLITICAL BACKGROUND

Burma is the largest and most ethnically diverse state in mainland Southeast Asia. The Burmans dominate the alluvial plains and cities. Tribal groups, flanking the borders of India, Bangladesh, China, Laos, and Thailand, have long resisted Burman dominance. The Burmans are mainly Buddhist. The Shan and Karen are predominantly Christian, and the Arakanese are split between Christian and Muslim (Rohingyas). These religious differences have deepened the rifts among Myanmar's ethnic groups.

Burma has always been concerned about the ambitions of its neighbors. War, competition, and animosity between Burma and Thailand are centuries old. The Mongols once invaded Burma, and the Chinese are still seeking access to the Indian Ocean. Myanmar's economy was once dominated by Indian migrants. Thus, historic acrimony shapes the country's attitudes to outside powers.

The British East India Company, headquartered in Calcutta, envisioned Burma as a buffer zone. Clashes in Arakan (Rakhine State) adjacent to Bengal mushroomed into the first Anglo-Burmese War from 1824 to 1826. Superior arms gave victory to the British, and Burma was forced to cede territory along the Bay of Bengal. A second conflict erupted in 1850, whereby the British acquired more territory in lower Burma. In 1885, the Burman king was exiled to Calcutta (even the throne was removed), and Britain annexed Burma in 1886.

Ruling from Calcutta, the British imposed alien Indian forms of government. Lower Myanmar—the Burman region—fell under direct rule, and the power of local elites was squashed. In Upper Myanmar, indirect rule left the hill tribes pretty much to their own devices. Consequently, the minorities developed a stronger sense of identity in opposition to the British-run region.

British colonial rule was founded on an Anglo-Burmese and Indian elite. Indians were free to migrate to Burma. By 1931, about 7 percent of the country was Indian, and Rangoon was 53 percent Indian. Chinese were recruited from British Malaya and Singapore. They soon added to the population diversity, working in Shan state mines and operating small businesses and rice mills. Shan state is the largest province of Myanmar.

British efforts to develop the Burma delta regions in the 1850s encouraged large-scale migrations of Burmans from the arid northern regions. The Sittang and Ayeyarwady deltas became prominent rice-producing regions. The area under rice cultivation expanded from 800,000 acres (1.9 million ha) in 1852 to around 6 million acres (14.8 million ha) in 1901. The associated population increase went from 1 million to 4 million. By the 1920s, population pressure had become a major problem: many farmers were indebted, and tensions among Burmans, Indians, and Chinese were spiraling into violence.

Britain was content to have Burma as a rice, teak, and mineral exporter. There was no attempt to industrialize the country. This plural society rested singularly on primary economic activities. Moreover, economic position was coincident with ethnicity.

Economic development under colonial rule was accompanied by the spread of Western education and the rise of nationalist leaders. These were not only dominated by ethnic Burmans but also were strongly anti-Chinese and anti-Indian. The fact that Buddhism was at the core of Burman life further alienated the non-Burman minorities, especially those who were Christian.

Three of the prominent nationalists and independence seekers were Aung San, U Nu, and Ne Win. Aung San was assassinated in 1947, U Nu became the first prime minister in 1948, and Ne Win led a coup in 1962 placing the military in power. In 2010, the military was still in power.

The anti-Western Japanese occupation in the 1940s was welcomed by many Burmans, seduced by the Greater East Asia Co-Prosperity Sphere idea of "Asia for Asians." However, reality did not meet the vision, and the disillusioned Burman nationalists worked toward independence. Burma was terribly damaged in World War II, with most of its infrastructure destroyed.

After the war, independence came quickly (in 1948). Minorities found themselves under a government relentless in its efforts to Burmanize them. A Karen nationalist movement launched several insurrections. These continue into the 2000s. Burma was a democratic state between 1948 and 1962, but its failings were critical. A failing economy, regional insurrections, social urban and rural unrest, and rampant corruption led to Ne Win's overthrow of U Nu. Political and ethnic minority leaders were arrested and opposition from university students and Buddhist monks was ruthlessly suppressed. The Westernized, often Anglo-Burman elite that had run the country under colonial rule and through the 1950s fled, along with large contingents of Indians and Chinese.

The Demise of Teak

To finance his campaign against U Nu, Ne Win's junta sold off much of Burma's teak. An estimated 80 percent of the world's teak is found within Burma's 900-mile (1,500 km) boundary with Thailand. Thailand is hungry for logs since it placed a moratorium on domestic logging in 1988. Scores of logging and fishing concessions were sold to Thai firms, along with the rights to gems, minerals, and oil to foreign concerns. Leaders of the Karen and other minorities have been and are still being displaced by these concessions and associated road building. Even so, they, too, are selling off teak to finance their insurgencies against the Burmans. However, tribal peoples use elephants for logging, and elephants do not require roads. Consequently, their operations are less damaging than the mechanized ones that call for forest clearance for road development.

Burma is losing at least 2 million acres (800,000 ha) of forest cover a year. Forests have been virtually eliminated in coastal areas and in the Ayerawaddy and Sittang valleys.

In 1974 the country was partitioned into administrative divisions: seven minority states and seven Burman divisions based on Western notions of nation and state. The regions were not designed for any level of autonomy but rather to facilitate further centralization. Delineations are not consistent with ethnic or residential integrity; groups and villages are splintered and mixed up. This aggravates the existing tensions among different linguistic and religious groups. Moreover, as geographer Curtis Thomson (1995) explains, traditional mapping with bounded concepts "contrasts sharply with the ephemerally bordered, hierarchically centered territories of mainland Southeast Asia." Boundary definitions remain an important focus of rebel movements.

In 1988, a series of protests and strikes called for an end to military rule. In response, a new organization, the State Law and Order Restoration Council (SLORC), took control and killed thousands of protestors. SLORC decided to hold elections in 1990. Surprisingly, Aung San Suu Kyi, daughter of the nationalist hero Aung San, returned from London to lead the National League for Democracy (NLD). She was placed under house arrest in 1989. Nevertheless, the NLD won the election convincingly. However, SLORC declared the elections null and void and arrested NLD members. Aung San Suu Kyi was put under house arrest without a telephone, cell phone, or computer. She was awarded the Nobel Peace Prize in 1991. SLORC, now known as the State Peace and Development Council (SPDC), held an election in November, 2010 which it won by fraudulent means. Anug San Suu Kyi has since been released but forbidden from political activity. How long she will stay free remains to be seen.

Rules of SLORC/SPDC

The rules of SLORC are known as the "People's Desire." They are printed in daily newspapers and on billboards and include such orders as: Oppose those relying on external elements acting as stooges; Oppose those trying to jeopardize stability of the state and programs of the state; Oppose foreign nations interfering in internal affairs of the state; and Crush all internal and external destructive elements.

ABODE OF KINGS

In 2005, the military junta moved the capital Yangon to Naypyidaw (Abode of Kings) in the middle of nowhere and began to build from scratch. New government buildings, residential and commercial zones, and a military sector are set to be completed by 2012. At this point rutted, jam-packed roads funnel into eight-lane highways going into the capital. However, few outsiders are permitted to go there.

The world wondered why this move was made. The government explains that Yangon was too crowded and lacked space for expansion. However, speculation has it that the paranoid junta wanted to get away from Yangon, which might be attacked from the sea. A land base is easier to defend. The locals believe that the decision was based on the portents of fortune-tellers, who are consulted for almost every move people make.

THE SAFFRON REVOLUTION

In 2007, Burma experienced the largest protest it had seen in 20 years. Thousands of students, political activists including women, and monks marched in 25 cities to protest rising consumer prices and government oppression. The protest is called the "Saffron Revolution" in reference to the saffron-colored robes worn by most monks in Southeast Asia (even though Burmese monks wear red).

The army brutalized and killed hundreds of the marchers and threw many of their bodies into the jungle. A Japanese reporter was murdered in the street. Both India and China maintained their policies of non-interference. Numerous other countries in the European Union along with the United States, Canada, and Australia exercised sanctions against Myanmar. Meanwhile, protests continue to be forbidden, and any inkling of dissent is quickly quashed.

SCRAPING THE RICE BOWL

Burma was once known as the "Rice Bowl" of Southeast Asia. Rice production has dwindled during decades of socialist rule.

Rice trade is one of the regime's most important sources of foreign revenue. The government monopolizes rice exports. It buys rice from farmers at low prices, sells it on the world market, and absconds with the profits. Quotas are set for each cultivated area—15 to 20 percent of each farmer's crop is forcibly purchased for less than half the market price. Many farmers cannot make ends meet and wind up losing their land to the government or to other, richer farmers.

There are 7,000 "agricultural supervisors" posted around the country to encourage harder work. Signs such as "Lazy People Lose Their Land" and "Be Confident in Your Ability" are intended to stimulate productivity.

Rice production fell dramatically in 2009 due to the impact of Cyclone Nargis and a late monsoon. Those farmers who did have rice to sell were hurt by lowered prices related to the global economic crisis. There is very little government support, and most farmers have no cash for inputs.

In border regions, quotas and inadequate technology have compelled many farmers to leave their land altogether and migrate to Thailand. Others have been forced off their land by the military. About a million Burmans now work in factories and farms in Thailand.

Pests are a serious problem. Insects devour rice because storage techniques are poor. Moreover, more double-cropping with fast-growing seeds means that much rice is harvested in the rainy season. Without pesticides and adequate storage facilities, the wet grain is even more susceptible to pests.

"BAMBOO DEATH"

Chin State highlands are being devastated by a plague of forest rats that have devoured nearly 80 percent of the region's crops. The catalyst for the rat infestation is an ecological phenomenon known as *mautam,* which translates as "bamboo death." Once every 50 years, the *Melocanna baccifera* species of bamboo goes into flower and its fruits attract rats and increase their fertility. An exponentially expanded rat population scourges all fields gobbling up rice, maize, and sesame crops. Then they move on to tea plantations and tamarind orchards. The last cycle in 1958–59 led to the deaths of some 15 thousand people in Chin State and the neighboring Indian states of Mizoram and Manipur.

While the Indian government has introduced a number of programs to stop the plague, the Myanmar government has done relatively little. Cat-owning farmers were lucky until they realized that if the cats killed the rats, they would have no food at all. They needed the rats for food since all their other foodcrops were decimated. Moreover, if they poisoned the rats, they would be deadly to eat.

Thousands of Chin farmers have given up and migrated to border areas where food can be purchased. Others simply leave to scavenge for grass, roots, and bark in less-populated areas. However, it is a social taboo—a disgrace—for Chin to abandon their villages and communities. Many leave during the night.

Recently the military junta has allowed the World Food Program and a limited number of other organizations to generate "work-for-food" projects. Relief from across the Indian border is absolutely forbidden. Any assistance is given clandestinely, often by Christian groups. The Chin, fearing reprisals, are afraid to talk about the deteriorating situation because this would draw attention to the failure of the government to manage the crisis.

Figure 15-5

Inle Lake. This man is tending his gardens from his dugout boat. Note that the paddle is operated by hand and foot. Increasing numbers of motorboats are displacing the polers. Photograph courtesy of B. A. Weightman.

Inle Lake's Floating Agriculture

Inle Lake, located in one of the Shan states, is Burma's largest lake. The lake is home to vast areas of floating gardens anchored in place with long bamboo poles driven into the lake bottom (Figure 15-5). Adjacent houses are built on stilts and stand in 6 to 10 feet (2 to 3 m) of water.

The gardens are strips roughly 82 feet (25 m) long and 3 feet (1 m) wide. They are cut from floating vegetation at the lake's edge and towed to the garden site. Sediment from the lake bottom is scooped up and placed on top of these mats to provide topsoil for the planting of vegetables. Plants, especially tomatoes, with their roots close to the water, absorb nutrients from the lake, which is a repository for all the residential waste in addition to runoff from the surrounding hills. The gardens are tended by farmers who move from strip to strip in dugout canoes. Much of the produce is sold at markets around the lake. Fish farming is also practiced, with the floating gardens acting as nurseries.

AUTHORITARIAN DEVELOPMENT

To make way for development projects such as roads and tourist facilities, SPDC has forcibly moved people off their land. Tens of thousands have been moved to satellite towns around Yangon (called "Rangoon" by the British and during the period before SLORC) and Mandalay. The entire population of Bagan was displaced to eliminate dwellings from among the temple complexes. The displacement was part of the government's tourism promotion. When the government wanted to widen a road in Mandalay, it simply sliced the fronts off all the shops lining the route.

Most displaced people work in the cities but find transport costly in both time and money. Many have moved to squatter settlements closer to their place of employment. In any case, community and family relationships and village economic systems are broken or strained.

The country has a serious shortage of electricity. In Mandalay, power is rationed. Many homes have electricity on alternate days, and factories can only operate part time. However, government-supported tourist hotels always have power.

The government relies on forced labor for public works and eco-tourism development. In 1996, when people protested, at least 300 villages were destroyed and many people were killed. There are numerous accounts of people "worked to death." Many eco-tourism projects are in minority regions, as are other developments such as new roads and oil pipeline construction. These projects are controlled by the SPDC military, whose forces have doubled since 1988. Recruitment of 13- to 15-year-olds has boosted numbers to 400,000.

Cyclone Nargis

In May 2008, Burma was hit by the worst natural disaster in its history: Cyclone Nargis. With winds of more than 135 mph (215 km/h) the southern part of the country was devastated. A 12-foot (3.6 m) storm surge worsened the situation, which affected 2.4 million people. The death toll was at least 78,000 and a million people were rendered homeless. The government was condemned around the world for its unprecedented action of preventing assistance by international agencies including the United Nations. By the time aid got into the region, it was too late for many who drowned, starved to death, or died from injuries or disease.

THE YADANA GAS FIELDS

In 1992, Burma signed a contract with the French company Total S.A. to develop the Yadana gas reserves in the Andaman Sea and build a pipeline to Thailand where the gas would be sold. Total invited other companies to participate in the project. The American company UNOCAL was attracted to Burma because of its cheap labor and by the fact that it was an entry point into Southeast Asian markets. While UNOCAL was concerned about Burma's human rights violations, it regarded the government as stable. Despite the risks, UNOCAL decided to invest in the project. In the face of criticism from abroad, it argued that "engagement" rather than "isolation" was the way to engender social and political change in Burma.

UNOCAL constructed the 256 mile (412 km) pipeline. Most of the pipeline lies under the ocean, but the final 40 miles cross southern Burma through a region inhabited by the Karen, a minority ethnic group hostile to the Myanmar government. Land clearing began in 1993 and the pipeline was completed in 1998. Greenpeace, Human Rights Watch, and Amnesty International reported that the Myanmar army had used forced labor and brutalized the Karen "in order to provide security" for UNOCAL workers and their equipment. A UNOCAL investigation team concluded that "egregious human rights violations" had occurred.

Work on the project continued, and commercial gas production from Yadana ensued in 2000. UNOCAL has taken steps to improve the lives of the Karen. New medical programs have reduced infant mortality and new schools have raised educational levels. By 2004, the project was delivering gas to Thailand, enabling it to deliver clean-burning natural gas to fuel its electric plants instead of dirtier fuel oil. Revenues from sales yield several hundred million dollars a year to Myanmar's military government.

In 1996, 15 Karen filed a class-action suit against UNOCAL in a U.S. federal court. They claimed that they and their families had been subjected to forced relocation and labor, torture, murder, and rape on the Yadana project. The right of foreigners to seek compensation in U.S. courts was upheld by the Supreme Court in 2004. The suit was settled in 2005, with UNOCAL agreeing to pay compensation and ensure the safety of the Karen. The exact terms of the settlement remain undisclosed.

THE CHINA CONNECTION

The Mekong River and the Burma Road from Kunming, Yunnan in China, to Mandalay now function as major arteries for the influx of Chinese into Burma and elsewhere in Southeast Asia. Thousands of Yunnanese have poured into Mandalay, where they now dominate commercial life.

China refers to the region as the "Great Golden Peninsula." Trade between Yunnan and Burma increased from US$15 million to well over US$800 million annually between 1984 and 1994. Trade has been growing at a rate of 10 percent a year since then. Manufactured goods that once came from Thailand now come from China. In addition, the Burma Road has become the main route for timber into China and arms shipments from China. It is also a major conduit for drugs.

Figure 15-6

This woman is pounding little sheets of gold. These are used in Buddhist temples where they are plastered onto statues of the Buddha as offerings. Photography courtesy of B. A. Weightman.

The Chinese are anxious to gain connections to the Indian Ocean. It is surmised that China already has access to Andaman Sea islands for spy operations. Fishing boats with sophisticated monitoring equipment have been detected in the area as well. India has protested to Naypyidaw, and Indonesia has expressed concerns over potential Chinese threats to commerce in the Strait of Malacca.

LANDSCAPE OF REPRESSION

For decades a tyrannical military regime has kept Burma in political and economic turmoil. Infrastructure is pathetic at best, and thousands of Burmans and minorities are unemployed, engaged in forced labor, or working under oppressive conditions.

I have a vivid memory of Mandalay. I went to a place where they make gold paper to be used for Buddhist temple offerings. Perched at the edge of a muddy lane by a stream was a thatched, wooden structure with a cellar carved out of the earth. The cellar was filled with the stream's murky water. Floating on the water was a platform, and on the platform squatted four women. The only light was that which crept past a tarpaulin shielding the pit's doorway from the sun's intensity. It was 87°F (30°C). Both heat and humidity were tortuous. Each woman had a substantial piece of wood in each hand with which she rhythmically beat the small stack of gold sheets at her feet (Figure 15-6). The pounding resonated in the enclosure; the noise was deafening. These women work in this manner for eight hours a day, seven days a week. Think about it. This could be your life!

Unfortunately, Myanmar has been shoved so far into life's cellar that even a change in government will take years to drag the country out of its current state of oppression, corruption, and impoverishment.

Drugs and the Golden Triangle

In 1987, the United Nations gave Myanmar "Least Developed Nation" status. However, there are two economies: one legal and the other illegal. The illegal economy focuses on opium, of which Myanmar is one of the world's major producers.

The production of illicit narcotics has more than doubled over the past few years in the Burmese section of the Golden Triangle, a place where the borders of Burma, Laos, and Thailand intersect. The area under opium poppy cultivation increased from 228,000 acres (92,300 ha) in 1987 to 380,542 acres (154,000 ha) in 1995 (Figure 15-7). As of 2003, poppy acreage had been reduced by half, but more intensive efforts on the remaining land means that opium production has not declined. Heroin is derived from opium. The potential heroin output rose from 54 tons in 1987 to 166 tons in 1995 and to 825 tons in 2002. Close to 400,000 households earn the bulk of their income from opium production.

For years the drug trade was controlled by the infamous Khun Sa, but in the 1980s he lost ground to better equipped heroin barons near the Yunnan frontier. Khun Sa surrendered to the Burmese government in 1996 and moved with his heroin-generated fortune to Yangon, where he lives in luxury. Much of his money was invested in perfectly legal businesses such as hotels, real estate, supermarkets, and construction, generating a mini-boom in Myanmar's economy.

Figure 15-7

Opium poppies in the Golden Triangle in northern Laos. Photograph courtesy of B. A. Weightman.

Khun Sa's turf has been taken over by the United Wa State Army, a splinter group of the Burmese Communist Party in northern Shan state. This group maintains a cease fire with the Yangon regime and has formed a militia to protect both their interests. The United Wa Army comprises an ethnic minority of about 20,000. Their lingua franca is Chinese. Having links to opium poppy growers in the region as well as in China, they control the largest narcotic trafficking organization in Southeast Asia. China provides the group with arms.

While drug "kingpins" operate heroin and methamphetamine factories, they also build infrastructure for the local people to ensure their support. Drug earnings have funded roads, schools, and clinics in the otherwise poverty-stricken region. For these residents caught in the cycle of poverty and repression that is modern Burma and largely isolated from the outside world, drug money has become the primary engine of state-building in the Wa "state."

Since the 1980s, a new string of refineries has been established in the Shan state close to the Chinese drug market and with links to new markets abroad. One new market is Taiwan, which in 1989 was virtually drug free. Now drug abuse is a major concern. Drug use is increasing in Vietnam, Bangladesh, and India. There are more than 700,000 Burmese heroin addicts, 70 percent of whom have HIV. There are at least half a million addicts in Thailand, and 14 percent of Malaysian teenagers are now using hard drugs. China counts at least 250,000 addicts, with Yunnan having the highest rate of addiction.

Heroin and other drugs from Burma reach international markets via land and sea. Most drugs travel overland via Yunnan or Guangxi in China to ports along the eastern seaboard. Drugs are carried by people of all ages who have little other means to earn a living. At times when the Chinese government is cracking down on this illegal trade, the drugs are shipped by fishing trawlers to the Chinese ports.

For years, Thailand was the main conduit for drugs, but stepped-up law enforcement and poppy production alternative programs have altered this pattern. Cambodia and Laos have emerged as major transit routes. Cambodia is also a key center for money laundering. Drug laws are vague in Cambodia, and marijuana is freely available in markets. Cambodia is also a major shipper of marijuana. In addition, after years of war, Cambodia has become a major source of arms for the Golden Triangle's drug lords. Cambodia's deep-water ports at Koh Kong and Kompong Som, with their hundreds of fishing trawlers, make it easy for drug runners to smuggle the drugs to larger ships in the international distribution web. Cambodia handles about 15 percent of Burma's heroin production. In Laos, drug smuggling is protected by the Mountainous Area Development Corporation, a company controlled by the Laotian army.

Another new trail is from Burma and northern Thailand via northern Laos into Vietnam via Hanoi and the port of Haiphong. Danang is another distribution point. Vietnam, like its neighbors, has a long history of opium smoking among ethnic minorities in mountainous areas. Vietnam provides an alternative, and a less-policed shipping route. Addiction is becoming problematic in Hanoi among wealthy teens and the disadvantaged.

Profits are huge. A poppy farmer earns US$330 for 7,000 grams of raw opium, which is then converted into a 700-gram heroin brick, worth US$4,250, at a Burmese refinery. In Sydney, the brick will be worth US$53,050. In New York, the brick will be valued at US$80,000 and will be converted into 28,000 packets of adulterated heroin with a total retail value of US$280,000.

Clearly, drug production in the Golden Triangle has widespread international implications. Chinese-organized crime families from Hong Kong, China, Taiwan, and Singapore show great interest in Cambodia. Various African groups ferry drugs to Nigeria by sea and then on to Europe. The Tamils deal in arms and heroin to finance their war in Sri Lanka.

Democracy alone will not alleviate the drug crisis in northern Burma. Even with a new government, present and past drug lords are unlikely to listen to calls for reconciliation and stop their lucrative activities. In the absence of a comprehensive policy, the fracturing of northern Burma into small, well-armed, and violent drug-trafficking states will continue to create instability in the region and serious drug problems abroad.

Thailand: Tiger in the Metropolis

Like an elephant with its trunk drooping down the Malay Peninsula, Thailand sits in the heartland of mainland Southeast Asia. Because the north–south distance stretches 16 latitudinal degrees, Thailand's climate exhibits great diversity. Mountains in the north give way to limestone-encrusted tropical islands in the south. Rivers and tributaries drain into the Gulf of Thailand via the Chao Praya Delta near Bangkok (Figure 15-8).

The country is divided into four main zones: the fertile central region dominated by the Chao Praya; the drought and flood-prone, poor, northeast plateau; the rugged northern region dominated by mountains and fertile valleys; and the southern peninsular region characterized by rain forest.

Thailand's population is relatively homogeneous. It is unique in Southeast Asia in that it avoided the disruptions of Western colonial rule and the upheavals of decolonization. After World War II, which produced minimal effects on the landscape, Thailand pursued capitalist development. By the 1990s it was deemed an NIC and now is one of the tigers of Asia. A brief look at Thailand's history will set the scene for the discussion of this Bangkok-centered economy.

THE TAIS

The Tais were the principal ancestors of the Thai, Lao, and Shan peoples. Tai is a cultural and linguistic term used to denote the various Tai people who became differentiated into a larger number of separate identities. Over many centuries, the Tai migrated southward from western China.

The Tai were wet-rice farmers organized in *muang*—one or more villages under a chieftain. By the thirteenth century, further expansion and alliances transformed *muang* into kingdoms that adapted many of the beliefs of peoples they came into contact with. For example, the Tai probably adopted Theravada Buddhism from the Mon people of central Thailand and the Burmans at Pagan. Angkor served as a blueprint for Tai state builders. From Angkor came the concept of society as a divinely ordained hierarchy under a ruler incarnate of a Hindu deity or Buddhist *boddhisattva*. Angkor also provided a model for administering large and scattered populations as well as a range of arts and technologies. Cultural borrowings increased as a result of Tai attacks on Angkor in the fourteenth and fifteenth centuries.

SUKOTHAI

Meanwhile, in the thirteenth century the Tai kingdom of Sukothai had arisen. Modern Thais regard Sukothai as the birthplace of the Thai nation. The people of Sukothai adopted the name "Thai" to distinguish themselves from other Tai-speaking people. The Khmer system of writing was adopted and modified for a Tai written language. This is the predecessor of today's Thai writing system. Ultimately, Sukothai faded. In 1351 a new kingdom emerged to the south, Ayuthaya, which eventually became known as Siam.

AYUTHAYA

Ayuthaya lasted until 1757, prospering from its strategic position only 43 miles (70 km) up the Chao Praya from the sea. Commanding the vast Chao Praya plain, Ayuthaya provided rice for export. Ayuthaya was seen by the Europeans as a great trading power. However, Ayuthaya had to face a challenge from the Burmans, who

Figure 15-8

Thailand has many cities and towns but the core area focuses on the primate city of Bangkok. Why do you think this is the case? From H. J. de Blij and P. O. Muller, *Geography: Realms, Regions and Concepts*, 14th edition, 2010, p. 555. Originally rendered in color. © H. J. de Blij and P. O. Muller. Reprinted with permission of John Wiley & Sons, Inc.

crushed the empire by 1767. Ayuthaya's ruling class was decimated. Thousands of people and their wealth were taken away. The city, which had been one of the greatest and wealthiest in Asia, was burned and vast tracts subjected to a scorched earth policy.

THE RISE OF BANGKOK

By 1785, new Thai rulers had emerged to smash the Burmese and establish new kingdoms. These eventually consolidated into an area larger than Ayuthaya had ever known. The new empire included all of mainland Southeast Asia excluding Burma and Vietnam. It also included some of the northern Malay states. A new capital was established at Bangkok in 1782.

In 1896 an Anglo-French agreement recognized Thailand as an independent buffer state between British Burma and French Indochina. Not experiencing any intense Western pressures, Bangkok remained traditional until the mid-nineteenth century. At this time, many Thai nobles studied Western languages, science, and mathematics. Bangkok was prepared to orient itself toward the West. In 1855, the Bowring Treaty was signed with Britain, giving Britain extraterritorial rights. Treaties with other countries followed.

Eventually the Thai empire was stripped of some of its holdings. It had to abandon its Cambodian and Laotian territory to the French and the Malay states to the British. In 1896, a treaty between England and France guaranteed the independence of most of the territory that today forms Thailand. Siam became Thailand in 1939, although it was still referred to as Siam during World War II.

Modernization proceeded with the building of railways and Western-style education for royal and upper-class children. Rice exports boomed, as did teak and rubber. But there was no significant industrialization. Western and Chinese interests controlled the economy.

The government stimulated Chinese assimilation into Thai society through business partnerships and intermarriage. The Chinese also adopted the Thai language, education, and culture.

European predominance in Southeast Asia was challenged by the Japanese, who invaded Thailand in 1941. When Western forces collapsed in the area, Thailand nominally allied itself with Japan, although effectively, it was an occupied country.

The defeat of Japan in 1945 was followed by an era of close relations with the United States in an alignment against the spread of communism in the region. Thailand has served as a refugee center for hundreds of thousands fleeing the conflicts of former French Indochina.

CHANGE, FAITH, AND ORDER

America's involvement in Vietnam was a powerful factor in opening up Thai society. In the 1970s, international tourism struck the country in a big way and labor migration began in earnest. This stimulated consumer culture and the emergence of a plethora of "lifestyles." What does this mean for Thai culture?

The Thais are confident that future developments will be adapted to the Thai way of life. One important aspect of Thai culture is order. Thais like a well-regulated life with a bent for aestheticism and appreciation for harmonious social order. Implicit in this is respect for interpersonal relationships and hierarchy in concert with ensuing rights and obligations. Even with modernization and social transformation, the expectation that society should be orderly remains.

The role of the *sangha* in the knowledge of Theravada—original teachings—is predominant. The monks are fed twice a day by the common people before noon. This gives people a chance to acquire merit. Almost every Thai male earns merit for his family by going into the monkhood as a novice for at least a rainy season. Traditionally, few Thai women would consider marrying a man who had not spent time thinking about ethics, learning to read, meditating on the Four Noble Truths, or focusing on how to be an upstanding human being. Even in the whirling commercial world of Bangkok, the role of monks and *wats* and the belief system of Theravada Buddhism are central.

In Thai thought, Buddhism encompasses virtue and wisdom, which can liberate one from the common order of life. It transcends and relates to the ultimate order of morality and goodness symbolized by the home, the mother, and the female symbols of Mother Earth and Mother Rice. Buddha and these female entities constitute the domain of moral goodness.

Next comes the less tenuous order of the realm of supernatural powers outside the home. This area of *saksit* is the focus of religious preoccupation, with rules for dealing with it.

Beyond the area of *saksit* lies the area of chaos represented by wickedness and immorality. Forces of goodness can intermingle with forces of evil. Pure order can descend into chaos; Buddha can emerge into evil; pure virtue can become deeply immoral; and safety can transform into danger.

Figure 15-9

This is a phrapbuum, *a Thai spirit house. Thais believe that if it is well-treated and respected, it will protect homes and businesses.* Photograph courtesy of B. A. Weightman.

Each Thai house and business possesses a spirit shrine of the *phrapbuum*—"lord of the place" (Figure 15-9). If respected, well-treated, and given offerings, the spirit shrine will protect and care for the place. If things do not go well, it is not the fault of the *phrapbuum;* the cause is always elsewhere.

Just as individual households have lords of the place, so do temples, villages, and provinces where the lord resides at the city pillar. These rulers have no power outside their local territories and must be vested with appropriate rites and offerings.

What's a Wat?

Wat is a Thai word from a Pali-Sanskrit word meaning "dwelling for pupils and ascetics." A *wat* is a Buddhist compound where men can be ordained as monks or women as nuns (Figure 15-10). Virtually every village has a *wat*, while towns and cities may have several. Each *wat* has a number of buildings such as a *bot* where ordinations are held; a *sala* for community meetings and lectures; a *tripitaka* or library for Buddhist scriptures; and various *chedis* (mounds or monuments) and *stupas* (large pagodas) housing the ashes of worshippers. *Wats* are multifunctional and are fundamental in villages. They operate schools, clinics, community centers, old age homes, and so forth. The typical *wat* is a locus of social activity. Especially important are the temple fairs that take place on auspicious dates such as Buddha's birthday.

SUSTAINABLE AGRICULTURE

Sixty percent of Thailand's exports are agricultural. Thailand ranks sixth in the world in rice production but number one in rice exports—35 percent of the global market. It ranks third in tapioca (manioc), fourth in sugar, and sixth in coconut production. It is the world's largest producer of natural rubber. Other exports include corn, jute, pineapple, palm oil, and canned or frozen shrimp, tuna, and chicken. Chickens are grown American-style in large, industrial feed houses, and they are exported frozen. Pigs are also raised in industrial fashion. One of the risks of this kind of industrial food production is the outbreak of epidemics such as Japanese encephalitis. Freshwater aquaculture is also important. Thailand exports mainly to the United States, China, Japan, and Australia.

About 42 percent of Thailand's workforce is engaged in agriculture, but the sector is fraught with problems. One of these involves the availability of water, which dwindles in the dry season. Some worry that decreased flow in the Chao Praya will draw salt water from the Gulf of Thailand to foul waterworks north of Bangkok. Climatic change has resulted in less rainfall. Water shortage is also attributed to deforestation. Thailand has lost 50 percent of its forests over the last 30 years, and there is now a moratorium on logging. The country is procuring its wood from Burma and Laos.

Farmers are used to expanding space rather than intensifying inputs to increase production. This is no longer possible. The industrial boom has sent harvest wages soaring, and many people have left the farm for work in industry.

Rising competition is another problem. Sugar, the only crop subsidized by the government, faces stiff

Figure 15-10

Wat Pho, originating in the sixteenth century, is the oldest and largest wat in Bangkok. It is also the earliest center of public education. Photograph courtesy of B. A. Weightman.

competition in the newly established ASEAN Free Trade Area. Still, Thailand is the world's number two sugar exporter after Brazil. Vietnamese rice exports, especially lower grades, are challenging thousands of Thai subsistence farmers. Chicken and shrimp exporters face competition from China. The rubber industry is threatened by cheaper labor in Indonesia. Thailand is especially vulnerable to world market prices as 70 percent of its land is planted in only six commodities: rice, sugar, rubber, maize, cassava, and sorghum.

Some changes are taking place. Improvements are being made in mechanization and intensification, along with land consolidation and better merchandising. A dairy industry is taking off as cows from New Zealand and the Netherlands are being crossbred with other breeds to withstand Thailand's climate. Agribusiness firms are entering into new areas such as fresh and processed fruit and cut flowers. The country has become a center of tissue culture; laboratories commercially mass-produce plant clones for customers around the world. Further, Thai firms have been exporting agribusiness technology to other developing countries (e.g., China, Indonesia, and Vietnam).

Export farming is practiced by large operators who typically have mono-crop operations. Recently, the government has been promoting sustainable agriculture for small farmers with 6 to 7 acres (2.5–3 ha) of land. These farmers make up half of all farmers in the country. There are four types of sustainable cropping being practiced in various areas of Thailand.

- Integrated production: Farmers raise at least two crops in the same field. For example, paddy might be combined with fish, or pigs with vegetables. Farmers are being encouraged to select high-value crops such as flowers and to maintain soil fertility with animal, instead of chemical, fertilizer.
- Organic farming: This is a system whereby only organic fertilizer and herb-based pesticides are used. This is designed to meet the increasing demand for organic foods. Urban populations are already buying such crops as organic rice and vegetables.
- Natural farming: With this methodology, there is no tillage and no application of fertilizers, herbicides, or insecticides. This is considered the ultimate form of sustainable farming. A few natural farms are operating in northern Thailand. The government is conducting further research on the suitability of this method in other regions.
- Agro-forestry: This format involves combining agriculture with reforestation. For example, fruit trees, coconut palms, and banana plants can be grown in the same area and animals can forage underneath.

New Theory Farming

New Theory Farming is a form of land and agricultural management introduced by King Bhumiphol Adulyadej in 1993. In the initial stage of this community-based economy, farmers divide their 6-acre (2.4 ha) plot of land into three 30 percent and one 10 percent portion allotted to rice, field crop or fruit, water supply, and residence. This first stage should enable farmers to become self-sufficient and to pay off their debts.

In subsequent stages, farmers join to form cooperatives and then link up with financial and energy sources such as commercial banks and oil companies. In the third stage, farmers will manage their own mills and co-ops and take on joint ventures.

In the first year of the project, 1998, thousands of villages participated in various aspects of the program. Community markets, networks, water source development, agricultural service centers, and civil society activities were put into place. More than 5,000 families have participated in the New Theory self-sufficiency project with positive results. It is the hope of the Thai government that these programs will help farmers pay off their debts and ease themselves out of poverty.

TO BE THE HUB OF EVERYTHING

Thailand is driving ahead to be Southeast Asia's vehicle hub—the "Detroit of Asia." Car and truck manufacturing are on the rise since several international manufacturers such as Toyota, Isuzu, Mitsubishi, Mazda, Ford, and GM have established themselves in Thailand. The Thai auto industry is booming with investments, with Toyota and Isuzu leading the market shares. These two companies also dominate the pickup truck market, and Thailand is second only to the United States in pickup production. Other companies such as general Motors, Ford, and an array of Japanese companies are in on the act. Not only does Thailand build vehicles but also it manufactures 80 percent of their component parts. The country now has 700 automobile factories.

Thailand also plans to be the region's aviation and tourism hub. It has redesigned and expanded its main Suvarnabhumi airport and has introduced an open skies policy to allow all travelers to visit the array of historic and cultural sites in Thailand as well as in surrounding countries. An airport-rail link is under construction to facilitate passengers getting into Bangkok. Chiang Mai's airport is being expanded, and thousands of new hotel rooms are being built to develop this northern hub for tourism in the tribal regions.

Since 1987, tourism has become a leading earner of foreign exchange, outdistancing even textiles. Ecotourism has increased substantially among the northern hill tribes. Sex tourism, with its accompanying risk of AIDS, continues in Bangkok and some of the southern resorts such as Pattaya. Unfortunately, many of the hill tribe traditions have been compromised to suit tourist desires, leaving substantial numbers of indigenous peoples culturally marginalized. Now, much of hill tribe life is "staged" for the tourists. The timing of work or festivals is directed to fit in with tourist schedules. Traditional lives are disrupted as the tribes are reduced to being "objects" of tourist attention.

Thailand is gaining a name for itself in medical tourism. Thai hospitals have an excellent reputation for quality surgeons and high nurse-to-patient ratios. Moreover, surgery costs considerably less in Thailand. People come from around the world for everything from heart transplants to knee replacements to cosmetic surgery. Customers buy packages that include luxury tours and hotels in addition to their surgery.

Bangkok has designs on being the region's fashion hub as well. Thai textiles, with their ancient and complex patterns, are unique and in great demand by designers. The government initiated the "Bangkok Fashion City Project" in 2004 with international fashion shows to tempt foreign buyers and to promote tourism.

Thailand is pursuing plans to replace Singapore as Asia's fuel-trading and transport hub. This includes the construction of an energy land bridge between the Indian Ocean and the Gulf of Thailand. In January 2004, a state-owned energy company opened a new petroleum trading center at Sriracha, a deep-water port on Thailand's eastern seaboard just southeast of Bangkok near Chon Buri. The government plans to build a "Strategic Energy Land Bridge," which will include two deep-sea oil terminals (the one at Sriracha and another at Laem Chabang just north of Pattaya) and a pipeline for transporting oil across the southern Thai isthmus from the Andaman Sea to the Gulf of Thailand.

This project is of particular interest to the oil-importing countries of China, South Korea, and Japan. Almost 70 percent of East Asia's oil currently passes in tankers through the Strait of Malacca, where piracy remains a concern. Some analysts speculate that China is concerned about a potential American blockade of the Strait of Malacca that would cut off China's oil imports, although this is highly unlikely. It is estimated that the 22 million to 25 million barrels of crude oil that East Asia imports every day will more than double by 2020 led by China's growing appetite for fuel. Oil security is a top priority for East Asia.

An especially grand plan is to cut a canal through the Isthmus of Kra to shorten shipping routes between Europe and East Asia. Investors from Australia and Japan have expressed interest in the project. Some economists argue that the relatively small savings for shippers will not be enough to recoup the estimated cost of $US35 billion, but at least the canal would allow ships to avoid the dangerous Strait of Malacca.

Thailand also hopes to be the regional hub for export of tropical fruits. Thai tropical fruits are unique with such varieties as rambutan, star fruit, mangosteen, and durian all being marketed overseas. Middle-range to high-end North American and European grocery stores and supermarkets are now selling these delicious fruits.

These grandiose schemes due for completion by 2020 are, in large part, in response to China's emergence as a rival low-cost manufacturer and a magnet for foreign investment. China's cheap manufactured goods are eroding key Thai export markets in the United States and Japan. The Thai economy has dropped dramatically since the 2007 crisis. The economy declined 2.3 percent in 2009. It is hoped that the new developments will attract more foreign investment and subsequent employment.

The 1980s and 1990s witnessed a surge of foreign investment from such countries as Japan, South Korea, and Taiwan where labor costs had risen. Thailand had numerous attractions for relocating industrial operations from these countries. Unskilled, trainable, and cheap labor was available. It had a decent infrastructure of roads, airlines, and port facilities, as well as sufficient power and water supplies for industry. Tax incentives, infrastructure provision, and other favorable policies encouraged foreign investment. Bureaucratic procedures were streamlined. Moreover, there was political stability and lack of racial conflicts in most of the country. Finally, there was no adversity toward Chinese business, an important consideration for firms coming from Hong Kong and Taiwan.

Much foreign capital has flowed toward primary industries such as rubber, tiger prawns, livestock, nonferrous metals, and marble. It has also gone to resource-based industries such as processed foods, palm oil, sugar products, wood-furniture parts, rubber gloves, and paper products.

Manufactured goods have become increasingly important as a source of foreign revenue. Textiles and footwear, food products, rubber, automobiles, computers, and electronics lead the way, accounting for 75 percent of the country's GDP. Most of Thailand's exports go to the United States, China, and Japan.

Thailand now has 10 export-processing zones in various locations such as Chon Buri, Ayuddhya, and Bangkok. In these zones, raw materials and other inputs may be imported duty-free.

The country remains an attractive destination for foreign involvement. Japan, the major investor, and other countries view Thailand as the hub of a golden pentagon comprising Thailand, Myanmar, and Indochina. Thailand is seen as an ideal site to establish satellite companies into a surrounding hinterland of cheap labor and a potential market of more than 120 million consumers. Thai companies, offshoots of joint ventures with the Japanese, are already trading in Vietnam and have a strong presence in Laos, where they dominate extractive industries.

Unfortunately, during the current economic crisis, many factories have either had to lay off workers or close down completely. The first groups to be let go have been foreign workers and women. Government stimulus packages are easing the blow for some.

More recent events do not bode well for Thailand. A military coup in 2006 ousted then Prime Minister Thaksin who was accused of being "thuggish, greedy and corrupt." However, the coup did not ruin him as a political force and served to unleash a wider backlash against the economic, social, and military elite that perceives itself as the agent of a divine hierarchy descended from revered King Bhumibol. This elite, represented by current Prime Minister Abhisit, has little sympathy or interest in the aspirations of rural and working class voters.

Many of those who benefited from Thaksin's populist policies, especially in the areas of health care and education, feel disenfranchised under the present government. In 2010, they launched massive protests demanding new elections. Wearing red shirts, they stormed and occupied several areas in the capital as well as in towns in the poor northern and northeastern areas of the country. In the wake of ensuing violence, more than 80 were killed and over 100 injured. Moreover, Bangkok was paralyzed and the tourism industry in ruins. Protesters were finally overwhelmed by the Thai military.

The rise of the "red-shirts" (supported by the exiled Mr. Thaksin) has revealed the growing social and economic injustices that permeate once-peaceful Thai society. At this point, Thailand's future is uncertain.

Ya Baa

Coming from clandestine laboratories in Myanmar and the Golden Triangle, a yearly supply of more than 800 million methamphetamine pills have supplanted heroin as the drug of choice in Thailand. The number of drug offenders in Thailand rose from just over 10,000 in 1992 to 2.4 million in 2010, thereby swamping prisons and courts. The highest rates of drug use are found among 15 to 24-year-olds.

Most of the drugs are manufactured in Myanmar, particularly in the states controlled by the United Wa State Army. This pro-Yangon ethnic group operates at least 80 permanent and countless mobile production labs near the Thai border and is said to be in collusion with the Yangon military government.

Ya Baa, or "crazy medicine" as it is known in Thailand, has broad usage across all segments of society. Construction workers, sailors, farmers, and taxi drivers take the drug to help them through long and arduous days. Students use it to stay awake to study and get through exam periods. Nightclub devotees take it to dance the night away. Children as young as six become hooked and hospitals treat more Ya Baa addicts than alcoholics. The worst symptoms, after years of addiction, are paranoia and hallucinations leading to uncontrolled violence. Cures take several months and relapse is common.

In 2003, former Prime Minister Thaksin decided on all-out war against drug trafficking. More than 50,000 drug dealers were arrested including 2,750 large-scale retailers. Forty million pills were seized as well. Anti-drug programs have succeeded in 82,500 villages that are currently drug free. Nevertheless, about 5 percent of Thais still abuse methamphetamines—Ya Baa.

SOUTHERN INSURGENCY

In the five southernmost provinces of Thailand, 85 percent of the inhabitants are Muslim. These are not Thai but rather ethnic Malays with ties to the population of Malaysia to the south. For more than a century, the southern provinces have had stronger ties with Malaysia than with Bangkok 1,000 km (600 mi) away. With the rise of Islamic fundamentalism and incidents of terrorism in recent years, southern Thailand has come to international attention. Amnesty International says that this is the third most significant Muslim insurgency in the world after Afghanistan-Pakistan and Iraq.

Violence has flickered in the south ever since Thailand's 1902 annexation of the region, which used to be an independent sultanate. Its efforts to acculturate this Malay-speaking Muslim area into its Thai-speaking Buddhist society have failed. Many Muslims feel that they are being assimilated out of existence. Violence has increased over the years with more than 1,700 Buddhists being killed by suspected militants from 2004 to 2006, for example. The Thai army seized control of the region in 2006 but the violence has not been quelled. The death toll on both sides now stands at more than 4,000.

The insurgents apparently have links to Southeast Asia's Jemaah Islamiah, a group linked to the 2002 nightclub bombing in Bali, Indonesia, that killed 202 people. However, the strength of these ties is debatable.

Thai authorities see the south's network of Muslim religious schools as a key source of recruits for the insurgency. Each month about 100 sectarian attacks take place in southern Thailand, down from 200 a month in 2007. The decrease is attributed to the fact that Thailand is employing more local militias to maintain control.

Thai police have arrested numerous suspects and are watching this region very closely. The government is particularly concerned about the Andaman Sea coast. This is an important tourist area and one that might become a target of terrorism directed against non-Muslim tourists. The Thailand–Malaysia border is another area of concern. This area is extremely porous. Muslim Malays can cross into Thailand with ease.

TIGER IN A METROPOLIS: BANGKOK

Why is Thailand a tiger in a metropolis? Although only 16 percent of Thais live in Bangkok, it produces more than half of the GDP. Bangkok is mainland Southeast Asia's largest city and is the economic pulse of the region. At 34 times the size of Thailand's second largest city, Nakhon Ratchasima, it is a textbook example of a primate city.

A city of more than 10 million, the historic and current primacy of Bangkok relates to its excellent location. It has been the capital since its founding in 1792 and lies in the rich Chao Praya rice region. The government is highly centralized, the city is the spiritual center of the nation, and it is the residence of the highly esteemed royal family.

Bangkok is a rapidly industrializing city. Economies of agglomeration are supported by a significant informal sector. Urbanization has spread from the Bangkok Metropolitan Area (BMA) into the Bangkok Metropolitan Region (BMR), which comprises five provinces circumscribing the city. The BMR contains 15 million people. Transportation, water, and utility costs are lower in Bangkok than elsewhere in the country. Labor is more productive because of the higher educational levels and greater experience of workers. The city attracts labor-intensive, export activities. Smaller firms are concentrated in the BMA where land is more expensive. Larger firms prefer the cheaper land of the BMR. More than a third of Thailand's manufacturing enterprises are located in the BMR.

Ribbon development has extruded onto agricultural land. Rice-producing and other agricultural areas are being displaced by industry, housing estates, and recreational facilities such as golf courses. Thailand has more golf courses than any other Southeast Asian country. Although development is clustered in the BMR, numerous benefits flow from this wealth, including upgraded health and education systems and improved social and physical infrastructure throughout the country.

Rapid industrialization has taken its toll on the environment. Air and water pollution are severe and traffic congestion is among the worst in Asia. These seemingly unmanageable problems are constraints on development. Yet construction continues as profits are made in the development process rather than from the final product. Many construction projects are joint ventures with Japanese firms.

Aside from air and water pollution and inadequate waste disposal services, Bangkok is also sinking. As a consequence of overdrawing of well water, the city is experiencing land subsidence. It is sinking at a rate of 4 inches (10 cm) a year. In fact, some areas have subsided more than 3 feet (.9 m) since the 1950s. Global warming will only exacerbate the problem.

The population of Bangkok fluctuates daily and seasonally. An estimated 1 million people commute into the city every day. In February, a slack agricultural period, many workers move to the city in search of temporary jobs in the informal sector of the economy.

In Bangkok, an estimated 4 million cars and 1.5 million motorcycles are in the city every day. The number of cars is rising by 1,000 a month. Nearly half the traffic police are undergoing treatment for respiratory disease. A commuter spends an average of 1,026 hours—44 days a year—in traffic (Figure 15-11). Average speeds are around 6 miles (10 km) an hour. Gas stations sell Comfort 100, a portable toilet for the motorist stuck in traffic jams.

Traffic congestion is only getting worse because of the Thai's new obsession with owning a car. Car ownership has become an enormous status symbol even among the lower middle-class who can barely make their payments.

Road surfaces account for only 8.5 percent of urban space; they should make up 20 percent. A sky train and mass transit systems including an underground train

Figure 15-11

Bangkok is on every Top 10 Worst Traffic Jams list. Even with new freeways and public transportation, more cars than ever pour into the city every day. Increasing numbers of people can afford cars, which adds to the mayhem. ©Thomas Ma/EPN/NewsCom.

Figure 15-12

Bangkok is famous for its khlongs. *Many people live in stilt-houses along these canals, and floating markets are held in many areas.* Photograph courtesy of B. A. Weightman.

have been installed, but they only reach limited areas of the city. Motorcycles, buses, taxis, and trucks comprise the bulk of the traffic. Buses, which account for a mere one percent of the traffic, contribute more than half of the pollution.

Even though masses of vehicles such as buses continue to belch toxic, black smoke, Bangkok has cleaned up its act to a considerable degree. Officials convinced oil companies to produce cleaner fuel, used higher taxes to phase out the once-ubiquitous two-stroke motorcycles, and converted all taxis to run on clean-burning, liquefied petroleum gas. Having no emissions standards before 1992, Thailand now has controls based on European standards. Bangkok, once an icon for smog, has emerged as a role-model for Asia's pollution-choked capitals, boasting considerably cleaner air than Jakarta, New Delhi, Beijing, and Shanghai.

Bangkok is called the Venice of Asia because of its *khlong* (canal) system linking various parts of the city with the major artery—the Chao Praya (Figure 15-12). Now many of the *khlongs* have been paved over and quicker boat traffic has given way to slower vehicle traffic.

With development so concentrated in the BMR, the question is this: Is Thailand a newly industrializing country or is Bangkok a newly industrializing city? It does appear that Thailand is more of a metropolitan tiger than a national one.

Vietnam: Baby Tiger

"Two baskets of rice separated by a bamboo pole." This is Vietnam—the densely populated Red River Delta in the north strung by 1,000 miles (1,610 km) of mountains and a narrow coastal strip to the watery world of the Mekong Delta in the south. On these two rich, alluvial plains live two-thirds of the Vietnamese population, now numbering 87.3 million (Figure 15-13).

Eighty percent of the people in Vietnam are ethnic Vietnamese. In addition, there are at least 50 minority groups. Some minorities are identifiable by their religious affiliation. Roman Catholics are the largest group. At independence in 1954, there were about 2 million. When Vietnam was divided later into two zones, about two-thirds of the 900,000 who left the north were Catholic. Another group is the Cao Dai, with its headquarters in Tay Ninh Province near Ho Chi Minh City. Cao Dai means "high tower" and is a mix of Buddhist, Christian, Confucian, and Taoist beliefs.

Although the population is dominated by the ethnic Vietnamese, the presence of these other groups has made it difficult for any government to create a centralized and highly integrated society. Moreover, regional differences are strong among the northerners, southerners, and the central Vietnamese. Currently, the Communist regime is attempting to reduce these dichotomies by integrating southerners into the more regimented system of the north. Moreover, modernization is moderating differences throughout the country.

ORIGINS AND CHINESE INFLUENCE

It is possible that the ancestors of the Vietnamese originated in southern China, but they first appear in history as *homo erectus* in the Red River Delta (Tonkin region) at the same time as Peking and Java Man, about 500,000 years ago. Available evidence supports the fact that they had mastered primitive agricultural techniques as early

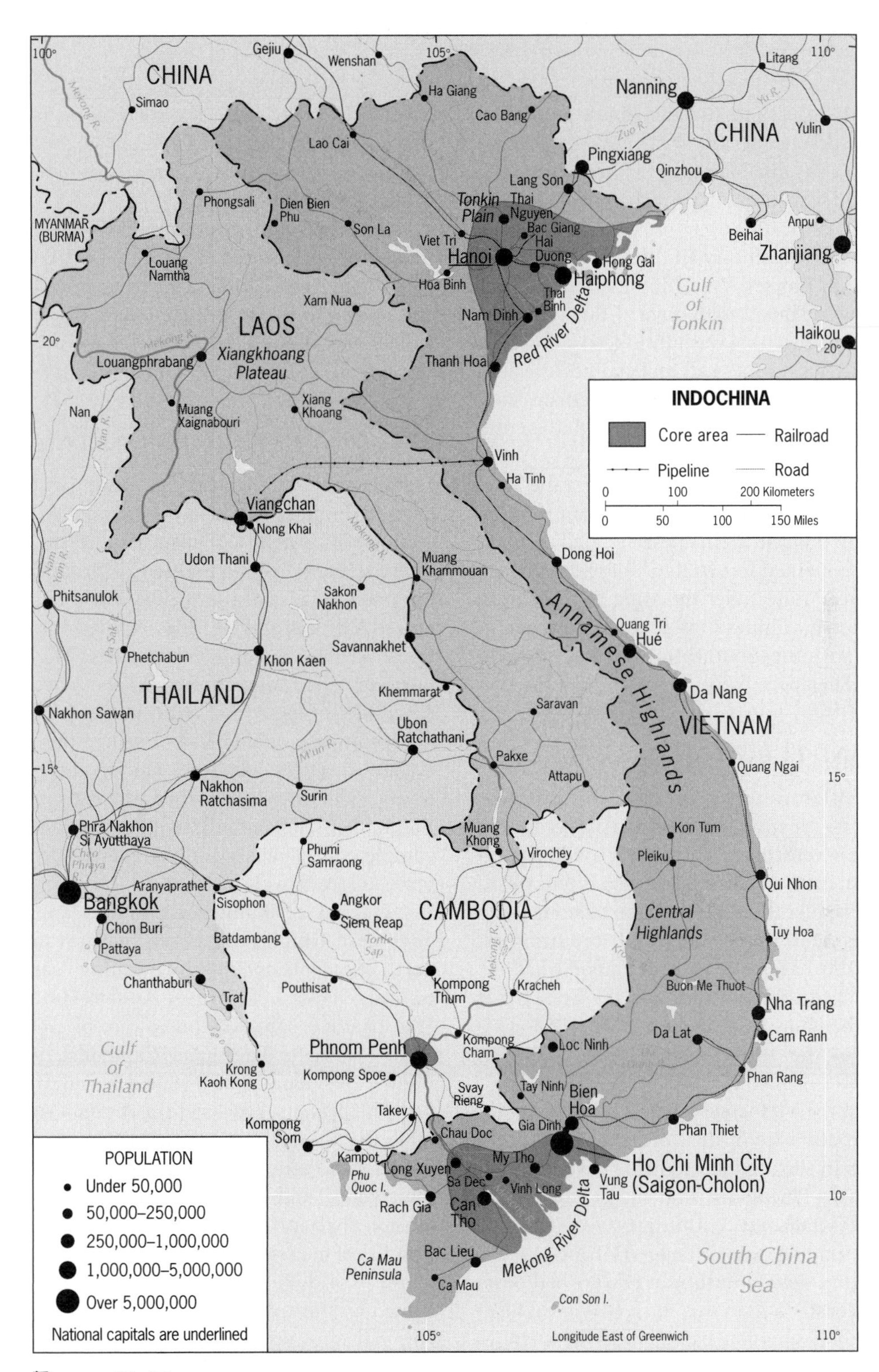

Figure 15-13

Vietnam is part of a region known as Indochina, including Laos and Cambodia. All were French colonies. Note how the Annamese Cordillera dominates the physical geography of Vietnam. With rice-producing core areas at either end of the country, Vietnam is indeed "two baskets of rice separated by a bamboo pole. From H. J. de Blij and P. O. Muller, *Geography: Realms, Regions and Concepts*, 14th edition, 2010, p. 548. Originally rendered in color. © H. J. de Blij and P. O. Muller. Reprinted with permission of John Wiley & Sons, Inc.

as 9,000 years ago. If this is so, the Vietnamese are among the first people to practice settled agriculture. By the second millenium BC, the inhabitants of the Red River delta had created an advanced civilization based on foreign trade and wet rice cultivation.

Then in the second century BC, the region was absorbed into the Han Empire, the Chinese calling it Nan-Yueh or Nan-Viet. Thus began over 1,000 years of Chinese rule. Chinese writing was introduced along with the Confucian classics, Chinese art and architecture, and Mahayana Buddhism. This differentiated Vietnam from the rest of mainland Southeast Asia with its Indian and Theravada Buddhist ideals.

The peak of Chinese influence was during the Tang Dynasty, whose rulers termed Vietnam Annam, or "the pacified south." In 939 the Vietnamese asserted their independence and remained free from Chinese dominance except for a 20-year stint under the Ming Dynasty from 1407 to 1428. Today, Chinese influence is stronger in Vietnam than in any other Southeast Asian country, with the exception of Singapore.

EXPANSION AND EUROPEAN INFLUENCE

Vietnam was constantly under pressure to find more farm land for its growing population. Between the eleventh and the seventeenth centuries, it gradually extinguished the kingdom of Champa in what is today central Vietnam. Subsequently, it acquired the Mekong Delta from the Khmers. During the nineteenth century Vietnam would probably have gotten all of Cambodia had the Thais not challenged their advance and the French not established Cambodia as a "protectorate" in 1863. This was the origin of the anti-Vietnamese sentiments in Cambodia today.

Expansion of the Vietnamese state southward coincided with the appearance of the Portuguese and other European traders in the region. Vietnam's first exposure was in 1535, when a Portuguese ship entered the port of Da Nang on the central coast. Within a few years the Portuguese set up a trading post at Faifo (Hoi An). By the seventeenth century several European powers had posts along Vietnam's coast.

The first Catholic mission arrived in 1615, when Jesuits established a mission at Hoi An. Another mission was set up at Thang Long (Hanoi). The Jesuit priest Alexander de Rhodes organized a society to train missionaries to propagate the faith. However, he was known more for his transliteration of the Vietnamese spoken language into the Roman alphabet. Known as quoc ngu or "national language," it did not come into general use immediately but was used to translate the Bible into Vietnamese. Consequently, thousands of new converts were gained.

Missionary activities eventually antagonized the authorities, who feared a subversion of Confucianism. Christianity was ultimately barred from both the north and south, and European missionaries were either expelled or executed. In 1697, the French closed down their small trading post at Hoi An.

FRANCE AND THE *MISSION CIVILISATRICE*

French interests changed the history of the region forever. In France, religious organizations demanded protection for their colleagues overseas. Commercial interests feared the expansion of Britain in Burma and the potential loss of the "China market." The French attacked Vietnam and in 1862 acquired three provinces in the south, the opening of port cities, and the freedom to propagate their religion. In 1867, French units seized the remainder of the south and transformed the region into the colony of Cochin China. Further, they assumed Vietnamese rights in Cambodia and turned it into a protectorate. The French finished their conquest of all of Vietnam in the mid-1880s. Shortly thereafter, France completed its "balcony on the Pacific" by making Laos a protectorate as well.

France's civilizing mission was applied to an Indochinese Union comprising five territories or states: north Vietnam, or Tonkin (meaning "eastern capital"); central Vietnam, known as Annam (from the Chinese "the pacified south"); the colony of Cochin China in South Vietnam; Laos; and Cambodia. A hierarchy of French "residents" was installed to run the country.

The primary goals were economic—to exploit natural resources and open up markets for goods produced at home. The French also emphasized the moral aspect of their deeds, and the French citizenry believed that this *mission civilisatrice* (civilizing mission) would bring the benefits of modern civilization to the "primitive peoples" of the world. In fact, the civilizing mission was subsumed by the greater goal of commercial profit. A small colonial elite was educated in Western ways, but the masses of people had only rudimentary exposure to Western culture.

With all its faults, there was economic progress under the French regime. A vigorous commercial and manufacturing system emerged, although it was dominated by foreigners including Chinese and Indians. Coal mining

Figure 15-14

Women sifting and cleaning rice in the Mekong Delta. Photograph courtesy of B. A. Weightman.

north of Haiphong was boosted; Vietnam remains a coal exporting country to this day. Bridges were built over major waterways, and railroads were laid from Saigon to Hanoi and on to Kunming in China. It is said that Vietnam, under the French, had the best system of paved roads in all of Asia.

The French also expanded agricultural land. In the Mekong Delta, they drained marshlands and constructed canals, putting thousands of acres under cultivation for the first time (Figure 15-14). The development of the Mekong Delta enabled Vietnam to become one of the world's leading rice exporters by the 1920s. The French also established coffee, tea, and rubber plantations along the hill slopes toward Cambodia. However, a lot of the Mekong Delta lands were held by absentee landlords who charged exorbitant rents for the land. Many farmers lacked security as sharecroppers on these vast estates and suffered from poor living and working conditions. Further, the commercialization of agriculture forced thousands into the cities looking for jobs.

By the 1930s, there was substantial opposition to the French. However, there was also a new class of native "collaborators" working in tandem with the colonial regime. These French-speaking individuals wore French clothing, ate French food, and lived in French-style houses. They were the middlemen between the traditional masses and the foreign regime and to outsiders served as evidence of the "success" of *mission civilisatrice*. Nevertheless, Vietnam was a cauldron of social unrest.

THE ROAD TO COMMUNISM

The collapse of Confucian government and the triumph of Western colonial aims put traditional Vietnamese beliefs into question. What path should be taken into the future? Some opted for Western-style democracy; some looked to China (itself in turmoil) for a restoration of Confucian values. Others became interested in the new Indochina Communist Party (ICP) founded in 1927 by young idealists including Ho Chi Minh. In fact, however, the most imposing popular movements were Cao Dai, with over a million adherents by the 1930s, and a Buddhist sect known as Hoa Hao, also popular in the south. Christianity had also grown, claiming 10 percent of the population.

The Japanese entered French Indochina in 1940 and set up a nominal government under the emperor Bao Dai. Wartime disruptions in the economy and disastrous weather created terrible famine in Tonkin and Annam. More than 1 million people died. When the war ended, there was a power vacuum in Vietnam, and the stage was set for the August revolution of the Viet Minh, a front organization of the International Communist Party.

The Viet Minh took control of most of northern and central Vietnam. They were less successful in the south, where they were opposed by politicians, business people, and religious groups. Ho Chi Minh set up his government in Hanoi. Meanwhile, by late 1945, the French had regained control of the south. War ensued, with the French pitting their conventional tactics against the guerrilla tactics of the Viet Minh. With assistance from China and the Soviet Union, Ho Chi Minh's forces were able to subdue the French at Dien Bien Phu in 1954. The power of the Viet Minh ended French efforts to hold Vietnam.

PARTITION AND WAR

As Dien Bien Phu fell, the great powers were meeting in Geneva to plan Vietnam's future. The decision was made to divide the country at the seventeenth parallel, with the northern Democratic Republic of Vietnam (DRV) to be ruled by Ho Chi Minh and the southern Republic of Vietnam to be run by Bao Dai. Elections were to be held in 1956 to reunite the country. These never took place. Ultimately, the Americans gave support to Ngo Dinh Diem, an administrator in the Bao Dai government, and Bao Dai left the country.

While President Diem spent his aid money on military acquisitions, the north pressed ahead with socialism and agricultural collectivization. Even though the "rich" suffered, popular support for Ho Chi Minh's government remained high. By 1959, the National Liberation Front (NLF) or the Viet Cong had organized to oppose Diem's oppressive and ineffective rule. By the early 1960s, the Viet Cong were in control of vast areas of the southern countryside, which compelled the Americans to increase their aid to Diem. By 1963, there were 17,500 American military advisors in Vietnam.

By now Diem and his followers had antagonized all groups in South Vietnam. Buddhist monks were burning themselves alive to protest his regime. Ultimately Diem was assassinated. Several years of unstable government followed, but meanwhile the Americans decided to confront the NLF. The United States then began to bomb targets in both the north and the south and, in 1965, landed troops in the south. Thus began the Vietnam or American War (Chapter 14), which did not end until 1975, with a combined DRV/NLF victory.

The environmental damage of the war on Vietnam was appalling. From 1965 to 1971, Indochina, an area slightly larger than Texas, received twice as many bombs as were dropped in all the combat theaters of World War II combined. In a typical B-52 mission, seven planes delivered 756,500 pounds of bombs in a pattern that saturated nearly 1,000 acres. The B-52s alone created about 100,000 new craters a month. The area had as many as 26 million bomb craters. These craters filled with weedy grasses and became breeding grounds for mosquitoes. About 10 percent of the agricultural land of southern Vietnam had to be abandoned. In addition, thousands of tons of chemical defoliants were dropped. Vast tracts of forests, rubber plantations, and mangroves were obliterated or seriously damaged. In these areas, there was a significantly higher rate of pregnancies that resulted in stillbirths and congenitally abnormal babies. The effects of the wartime bombing are still felt today.

Thousands fled the country to refugee camps in adjacent countries, especially Thailand. The first migrants used a map torn out of a school atlas to navigate the Gulf of Thailand. More than a million refugees have been resettled since 1975. Of those who remained in refugee camps in Southeast Asia, most were "screened out" as economic rather than political refugees. For them, there was no hope of resettlement. These people were repatriated to Vietnam voluntarily or involuntarily.

AFTERMATH OF WAR

Since the 1950s, the northerners had lived under an austere form of socialism that reemphasized the traditional regard for social hierarchy and community obligation. The southerners had been introduced to a quasi-capitalistic consumer economy and the trappings of American popular culture. In 1975, the north moved swiftly to subordinate the south to direct rule from Hanoi. In 1976, the country was renamed the Socialist Republic of Vietnam; Saigon became Ho Chi Minh City. Socialization of the southern economy proceeded quickly over the next two years.

However, things did not go as well as planned. There was an acute economic crisis, worsened by floods and other natural disasters in 1977 and 1978. Ambitious industrial targets were not achieved, rice production diminished, and food had to be rationed. In 1978, Vietnam attacked Cambodia to oust the genocidal, anti-Vietnamese regime of Pol Pot. In the 1980s, thousands of Chinese and Sino-Vietnamese fled as boat people. These refugees went to Australia, Canada, the United States, France, and many other countries.

Doi Moi

A hard-line communist approach was held in Vietnam until 1985 when reforms in the Soviet Union gave impetus to reforms in Vietnam. In 1986, the Vietnam Communist Party (VCP) instituted the policy of ***doi moi***, meaning "renovation." Measures were introduced to encourage private-sector production, make prices more flexible, and give more discretion to managers on the use of their profits. This was followed by the introduction of a peasant agricultural contract system in 1981. The system allowed farmers to keep anything over their set quotas. State enterprises were permitted to sell production exceeding their quotas on the open market.

Reforms were given further impetus with the collapse of the Soviet Union and the European Eastern Bloc's abandonment of socialism. (This resulted in a 30 percent cut in Vietnam's budget.) Support for reform was also bolstered by China's new policy of market socialism. Today, Vietnam is a hybrid of a one-party government, faltering state enterprises, and burgeoning free enterprise. Vietnam joined the

WTO in 2007 and this has provided the country an anchor to the global market and reinforced the domestic economic reform process.

Figure 15-15

I came across this scene in the Red River Delta in northern Vietnam. The two women are operating a bucket to lift water from a lower to higher level in the rice paddies.
Photograph courtesy of B. A. Weightman.

TODAY'S ECONOMY

Vietnam is anchored by two cities: the capital of Hanoi in the north and Ho Chi Minh City (Saigon) in the south. Hanoi, a quiet city of bicycles in 1995, is now a bustling city of motorcycles and automobiles. Hanoi dominates the Tonkin Basin and its productive agricultural hinterland. Agricultural methods remain traditional in many areas—dependent on human and animal labor (Figure 15-15). The port of Haiphong, a mere 60 miles (97 km) away, is linked to Hanoi by a new four-lane highway that passes through a developing *kotadesasi* landscape.

Ho Chi Minh City (Saigon) and Cholon (Chinatown), with its 8.5 million people, is an urban agglomeration on the Saigon River. It contains about 10 percent of the country's population. The name Saigon is commonly used. Unlike Hanoi, Saigon can be accessed by ocean-going vessels, and a Special Economic Zone has been developed downstream from the port. Saigon is a bustling city of every form of transportation imaginable, with traditional districts selling a wide array of consumer goods. High-rises punctuate the skyline everywhere. Modernizing Saigon is distant in form and fact from the more traditional capital of Hanoi. With the lifting of a United States embargo in 1994, foreign investors began investing in Vietnam. Modern high-rises and construction cranes mark the landscapes of both Saigon and Hanoi.

Now about 52 percent of the labor force works in the primary sector in agriculture, forestry, and fisheries. The primary sector includes such products as rice, coffee, rubber, cotton, tea, peanuts, and fish and animal products. Vietnam is the world's second largest coffee exporter after Brazil. Land is being taken from hill people in large swaths and planted with coffee and other export crops such as pepper.

Mining is important as is the manufacture of cement and fertilizers. Food processing and the textile and apparel industries, which rely on cheap female labor from the countryside, are also significant sectors of the economy. Some 30 percent of Vietnam's exports are destined for the United States and Japan.

Deep poverty has declined significantly, and Vietnam is working to create jobs to meet the challenge of a labor force that is growing by more than a million people every year. Fortunately, the country has attracted new investments in electronic components assembly. The Saigon Hi-Tech Park has emerged as a technology hub bringing together foreign investors with domestic companies and research and training facilities. The goal is to replicate the Taiwanese and Chinese strategy of moving beyond mere assembly to link up foreign hi-tech investors with their domestic suppliers such as Intel, and Japan's Nidec have invested in the Park. Another such high-tech park is being developed on the outskirts of Hanoi.

Meanwhile, the global recession has hurt the country's export-oriented economy with GDP growing less than the 7 percent per annum average achieved during the last decade. Exports fell 10 percent in 2009, prompting the government to use stimulus spending and a subsidized

Figure 15-16

Halong Bay is famous for its karst topography. It is also Vietnam's premier shrimp and crayfish producing region. Photograph courtesy of B. A. Weightman.

lending program to support the economy during the crisis. While foreign investment has declined overall, foreign donors have promised US$8 billion for new development assistance in 2010.

FISH, SHIPS, AND TOURISTS

In 2007, Vietnam published its "Ocean Strategy 2020." Offshore oil and gas production along with beach-front tourism are additional ways by which the country hopes to make the most of its (3,200 km) coastline. Global demand for farmed fish and large ships is on the rise. However, Vietnam still does not have a deep-water port and infrastructure is still lacking. Even so, the government intends that by 2020, fish, ships, and tourism and related services will account for more than half of the GDP as compared to 15 percent in 2005.

Although Vietnam has at least 50 varieties of food fish available in its seas, rivers, and canals, almost half of its seafood production is shrimp. Its chief shrimp-producing area is Halong Bay east of Hanoi (Figure 15-16).

Vietnam has some 200 shipyards that build every type of craft from fishing vessels and container ships to oil tankers. Plans are to be using at least 75 percent local inputs by 2015. Vietnam expects to become one of the world's top five shipbuilders by 2020.

Beach resorts have been developed all along Vietnam's magnificent coast. The most well known is the new complex on China Beach near Danang—a site that became famous during the Vietnam War. Tourism industry experts project that 100 million Chinese will travel overseas by 2020. Many of them will join Japanese, Taiwanese, and Korean vacationers at Vietnam's new golf courses and beach resorts.

If Vietnam can weather the global downturn, it will be able to live up to its reputation as "Asia's Baby Tiger."

Laos: Land of a Million Elephants

The Lao People's Democratic Republic, now often referred to as simply *Lao* or *Laos,* is a mountainous, landlocked, sparsely populated country sharing borders with Thailand, Myanmar, China, Vietnam, and Cambodia (refer to Figure 15-13). Laos, as we know it today, is a creation of the French, but much of its history has involved conflict with its surrounding countries.

LAN XANG

There are no early documentary records of early Lao history, but we know that the region had links to Chiang Mai in the eleventh century and that it was a satellite of the Khmer empire in the twelfth century. The first kingdom of Laos emerged in 1353. This was known as **Lan Xang**: "a million elephants."

Lan Xang stretched from China to Cambodia and from the Khorat Plateau in present Thailand to the Annamite Mountains in the east. Louang Phrabang, on the Mekong, became the capital of Lan Xang, and the king made Theravada Buddhism the official religion. A census in 1376 indicates that there were 300,000 Thais living in the kingdom. Muang administrative districts were organized, and these lasted until abolished by the Communists in 1975.

In the sixteenth century, Vientiane (Viangchan) on the Mekong became prominent as a religious and trading center. Lan Xang weathered wars with the Vietnamese, the Thais, and eventually the Burmese. Nevertheless, Vientiane reached its zenith in the seventeenth century.

The first European visitors reported on the city's imposing buildings and religious structures. For lack of a royal heir, the kingdom broke up in 1694 and Vientiane and Louang Phrabang became centers of rival states. The south fell under Thai patronage.

FRENCH TAKEOVER

The nineteenth century was a period of devastation for a Laos caught in regional rivalries. The surviving kingdom at Louang Phrabang acknowledged the overlordship of both the Vietnamese and the Thais, but was really within the Thai orbit of control.

The French takeover between the 1860s and 1885 resulted from several factors. They perceived the Mekong as a route to China, although this was an incorrect assumption, as the Mekong is broken by impassable rapids. They feared Thai interests in their territories. They were concerned about the intentions of armed bands of renegade Chinese who were attacking northern Laos and Vietnam. Also, they were worried about British encroachment from Burma. By 1885, the French controlled the Vietnamese emperor's claims to overlordship of the Lao territories. Still, Laos remained the center of rivalries among the surrounding states in addition to the competing colonial powers.

The French's attitude to Laos was one of "benign neglect." Few French actually went there, and Vietnamese were placed in administrative positions. Trade was left to the Vietnamese and growing numbers of Chinese. Most people in the region continued as subsistence farmers, the lowlanders growing rice and the uplanders pursuing slash-and-burn cultivation. The colony's most important products were tin, mined by the Vietnamese, and opium grown by the Hmong and other mountain dwellers. The French had a monopoly on the opium trade throughout Indochina.

Prior to World War II, modernization was extremely limited. A mere 8,050 miles (5,000 km) of mostly unpaved roads eased communication, but most services, including health care, were limited to the towns, and no Western-style education was available beyond the primary level. However, a small Vietnamese or French-educated elite did emerge in Viangchan, and by the 1940s it became the core of a Lao nationalist movement.

THE WAR YEARS

Nationalist politicization was a feature of the war years in Laos. After 1945, there were attempts at independence, but the French recaptured Laos in 1946. In the late 1940s, Lao guerrilla groups developed along the mountainous Lao–Vietnam border, aided by Viet Minh know-how and supplies. Meanwhile, the French declared the king in Louang Phrabang king of all Laos, held elections for a national assembly, and declared the country independent in 1949. Nevertheless, the guerrilla movement expanded, known by then as the Pathet Lao. By 1954, it controlled large areas of northern Laos.

In 1954, the same year as the French were defeated at Dien Bien Phu, a conference was held in Geneva to settle Indochina's disputes. The Pathet Lao were not invited. By 1959, guerrilla warfare was in full swing, with the United States involved in supporting the Royal Lao Forces. At the same time, the CIA was clandestinely forming links with the opium-growing Hmong, whom they assisted in selling drugs to new markets in Southeast Asia.

Unfortunately, Laos became deeply involved in the Vietnam War. About one-third of the bombs dropped in Indochina fell on Laos, especially on the Plain of Jars, part of the route of the Ho Chi Minh Trail. Thousands of unexploded bombs still mar the landscape of northern Laos. When I was in Laos in 1998, seven children were killed by an unexploded bomb. It will take many years for these bombs to be removed. Meanwhile, the people collect bomb casings, aircraft fuel tanks, and other materials and incorporate them as practical items in their landscape (Figure 15-17).

By 1972, the Pathet Lao, backed by the well-armed DRV, was gaining ground. In 1973, the two sides reached an "Agreement on the Restoration of Peace and Reconciliation in Laos." When, in 1975, communist organizations toppled the power structures in Saigon and Phnom Penh, the Pathet Lao supported a "popular revolution." The king abdicated. A total of 300,000 Lao—10 percent of the population—left the country between 1973 and 1975. These were mostly middle-class people. Many ended up in the United States and France. Then, in December 1975, a National Congress of Peoples Representatives voted for the establishment of the Lao People's Democratic Republic.

The Plain of Jars

The Plain of Jars is a plateau 3,280 feet (1,000 m) high, covering an area of 390 square miles (1,000 km^2). It is also known as *Plaine des Jarres*. Upon this bomb-crater-pocked plateau lie more

Figure 15-17

In Laos, a bomb casing makes a perfect container for a herb garden. Photograph courtesy of B. A. Weightman.

than 300 stone jars ranging from 3 feet (1 m) to 6 feet (2.5 m) high and around 3 feet (1 m) in diameter. Each 2,000-year-old jar equals the weight of three small cars (Figure 15-18).

Generations of archaeologists, geographers, and others have pondered the origin and use of the jars. Were they used to brew *Lau-lao*, a wicked local drink? Perhaps they were used to store grain or water. The most likely theory holds that they were burial urns with the larger jars being for the elite.

Tools, bronze, and ceramic objects have been found in the jars, indicating that they were used by a sophisticated society. But no one can relate them to any previous civilization. The stone that they are hewn from is not found locally.

During the war, the Pathet Lao set up their headquarters in a cave near the jars and the region was heavily bombed. Miraculously, the majority of jars remain unscathed.

RURAL SOCIETY

Over 70 percent of Laos is rural. Most people live in villages ranging from 10 to 200 households, or up to approximately 1,200 persons. Towns are mainly administrative and market centers. Most district centers have little more than a middle school and a few officials. Some places have minimally supplied clinics. Since 1975, the government has spent a lot of effort at national unification and instilling a sense of belonging to a Lao state.

One priority has been to improve infrastructure. However, both mountainous topography and torrential rains make access, work, and maintenance problematic. Lao's low population density means that the cost per capita for road construction is often prohibitive. War damage such as bomb craters and the presence of mines and cluster bombs inhibit most development.

Lowland Society

Lao Loum (Laotians of the valleys) have been the dominant group since the founding of Lan Xang in the fourteenth century. The *Lao Loum* make up half of the population of 6.3 million. They live in stilt houses constructed of wood, bamboo, and thatch. Progressive village houses sport a corrugated metal roof. Steamed glutinous rice, or "sticky rice," is the staple food. Eaten with the fingers, it is dipped into a soup or stewed dish.

Lao Loum live in stable villages along rivers or streams and engage primarily in paddy rice cultivation. Where terrain is not level, they practice swidden cultivation. Other crops are cotton, tobacco, or sugarcane, but these are grown for personal use. Villagers keep chickens, pigs, and draft animals. Hunting, gathering, and fishing supplement subsistence activities.

Household tasks are usually divided by gender, but divisions are not rigid. For instance, both men and

Figure 15-18

The Plain of Jars in Laos is an amazing place. I'm sitting on one of the jars to give you an idea of the scale of these objects. Their origin remains a mystery. Photograph courtesy of B. A. Weightman.

women cut and carry firewood. Women and children carry water, tend gardens, look after small animals, and are the main marketers for produce. Men look after and market draft animals. In field work, men plow and harrow the rice paddies, women plant seedlings, and both sexes transplant, weed, and harvest the rice.

Lowland Lao are mostly Buddhist. Most villages have a *wat,* which serves as a religious and social center. The *wat* also serves as the village meeting house. *Wats* are populated by two to six monks and a few novices, depending on the size of the village.

Midland Society

Lao Theung (Laotians of the mountain slopes) comprise about a quarter of the population and consist of 37 different ethnic groups. They speak languages of the Austro-Asiatic family. Many are mutually incomprehensible. Characterized as swidden farmers, the people of the mountain slopes are semi-migratory, moving their villages on occasion as swidden plots are exhausted. Many of these villagers were disrupted in the war, and many have moved closer to the lowlands. Traditionally, however, lands were managed in a way to maintain sedentary communities.

Lao Theung villages are generally smaller than *Lao Loum* villages. Some are divided into two segments with a men's common house in the middle. Villages of stilt houses are situated near a stream to access water that may be diverted via bamboo aqueducts.

Swidden rice economy is the basis for most communities. Both regular and glutinous rice are cultivated. Corn, cassava, and wild tubers supplement this diet. Hunting and fishing are also important.

Gender role differentiation is much greater among the hill people. Men clear and burn the swidden fields and punch holes for seeds. Women follow, dropping seeds in the holes and covering them with topsoil. Both sexes weed, but this is regarded primarily as women's work. Harvesting is a joint activity. In the house women cook, gather wood and water, and look after children. Men weave baskets, repair farm tools, hunt for small game, and are much more likely than women to manage finances and engage in trade. Where villages have schools, both boys and girls attend but girls have significantly higher dropout rates.

Upland Society

Lao Sung (Laotians of the mountain tops) include six ethnic groups. The Hmong are among the most numerous. They also live in Vietnam and Thailand. The *Lao Sung* are among the most recent migrants to Laos, having initiated a series of moves from China in the nineteenth century. All *Lao Sung* villages are located in the north. Only the Hmong live as far south as Viangchan.

Lao Sung typically reside above 3,280 feet (1,000 m) on steep slopes and mountaintops. As swidden cultivators, most groups are semi-migratory. Yet some villages have remained sedentary for over 100 years. Hmong houses are built directly on the ground, with walls of vertical wood planks and roofs of thatch or split bamboo.

Hmong swidden farming is based on white (non-glutinous) rice, corn, and a variety of tubers, vegetables, and squash. They also raise pigs and chickens in large numbers. Opium poppies, a cold-season crop, are planted after the corn harvest.

Opium, the sap drawn from the poppies, is a low-bulk, high-value item that is easy to transport. The Laotian government has outlawed opium production and instigated a program of alternative crops. Nevertheless, the practice continues in the remoter regions.

Hmong gender roles are strongly differentiated. Task divisions are similar to those of the *Lao Theung*.

Most *Lao Sung* are animists, but some have converted to Christianity after contact with missionaries. Most believe that illness is caused by spirits, and shamans play a powerful role in the communities. Shamans are "chosen" by the spirits and may be either male or female.

Lao Sung swidden cultivation is endangered. In the late 1980s, the Thai government imposed a ban on logging, so Thailand began looking to surrounding countries for timber. Logging in Laos is causing deforestation and soil erosion. However, the practice is to blame this on the swidden farmers. The government wants to see the disappearance of swidden farming early in the twenty-first century and the resettlement of Lao Sung in the lowlands.

URBAN SOCIETY

With a population of 680,000 in 2009, Viangchan is the only city of any significant size. Other cities, such as Louang Phrabang, have a mere 20,000 people, and Pakse has only 70,000. Laos is 27 percent urbanized, including district centers of 2,000 to 3,000 people. The fundamental village nature of Laotian society is evident in the cities, where many residents journey out to their fields to farm on a daily basis. Very little rural-to-urban migration occurs as there is not much to migrate to.

NEW THINKING

In recent years, the government has shied away from the centrally planned economy and embarked upon a policy of *Chin Thanakan Mai*, or New Thinking. The hammer and sickle were removed from the state crest in 1991 and a series of economic reforms was instigated. With the demise of handouts from Moscow, Viangchan has turned to the West for investment. At first the pace of change was *koi koi bai*—"slowly, slowly." But in 1991, everything became privatized. Limited tourism is being promoted. Nongovernmental organizations are ubiquitous. Thai television, with its materialist message, is beamed across the Mekong. Likewise, more consumer goods, such as children's toys and motor scooters, are appearing in the markets of at least the largest cities. Foreign investment, mainly Japanese, is evident in the capital. Many Thai businessmen have appeared on the scene as well and new businesses are opening every day.

The average annual inflation rate was about 11.5 percent from 1985 to 1989 after which a massive increase in money supply and devaluation of the *kip* compounded by a reduction in foreign exchange forced the rate up to 52 percent. The government tightened monetary policy in 1990 and restricted credit to unprofitable state enterprises. Lower food prices as a result of good food harvests also helped to slow inflation to about 20 percent. Inflation has declined slowly and now stands at 6.8 percent (Figure 15-19).

Infrastructure is still very poor. Only a few of the larger towns have municipal water systems. However, electrification has now reached cities and towns and many rural areas. Still, most roads are not paved, and many are impassable in the wet season. The Friendship Bridge (built by the Australians in 1994) links Laos and Thailand by road across the Mekong. With upgrading of roads in Laos, soon it will be possible to drive all the way from Singapore to Beijing!

Many Lao see their country's future as a transport hub linking northeast Thailand with the port of Da Nang in Vietnam, as well as southern China with Thailand, and, from north to south, Malaysia and Singapore. Many fear the effect of rapid development and influences of the surrounding states.

Although economic growth has reduced official poverty rates from 46 percent in 1992 to 26 percent in 2009, Laos remains one of the poorest countries in Asia. More than 75 percent of the population lives on less than two dollars a day. Subsistence agriculture, mostly rice cultivation, accounts for about 40 percent of GDP and 70 percent of employment. Even worse, inequality between various parts of the country is increasing. Development has been concentrated in Viangchan and other towns on the Mekong River plain, adjacent to Thailand, while remote provinces in the mountainous north have seen minimal change.

Even though Laos remains on the United Nations' list of "Least Developed Countries," the economy has benefitted from high foreign investment in hydropower, mining, and construction. Laos gained normal trade

Figure 15-19

In periods of high inflation, money becomes close to worthless. When I crossed the border into Laos from Thailand in 2002, I went to a bank in Pakse to exchange dollars for kip. I was stunned to see stacks of kip holding the door open. People were bringing in wheelbarrows of black plastic bags filled with money. Heaps of bills were being counted and tied into bundles to be piled on the floor. Only a few of my dollars bought me several stacks of kip, which I had to carry in my backpack.
Photo courtesy of B. A. Weightman.

relations with the United States in 2004 and is taking steps required to join the WTO. The country also launched an effort to ensure the collection of taxes in 2009 as the global economic meltdown reduced revenues from mining projects. Simplified investment procedures and expanded bank credits for small farmers and entrepreneurs are expected to improve Lao's economic prospects. The World Bank predicts that the country's goal of graduating from the UN Development Program's list of least-developed countries by 2020 is achievable.

Cambodia: Beyond the Killing Fields

Cambodia is famous for two things. It is highly noted for the wonders of Angkor and it is highly notorious for the genocide that took place from 1975 to 1978. What happened in this small and compact country of Southeast Asia?

This land of the Khmer people is the fourth smallest nation in the region after Timor-Leste, Brunei, and Singapore. About 5 percent of Cambodia is rivers and lakes. The plains of Cambodia are ringed by the last foothills of the Himalayas, the Cardamom range in the west, the Elephant Mountains to the southwest, and the lengthy Dangrek chain to the north. The eastern frontier with Vietnam runs up into the high and inaccessible Moi Mountains, which are crossed by only one highway. Rising out of the plains are ranges of hills called *phnom*. These were worshipped and shelter numerous settlements. The Mekong crosses the Cambodian Plain and south of Phnom Penh, splits into the Bassac to the southwest and the lower Mekong to the east. Inland, 340 miles (550 km) from the sea, the Mekong is still navigable year round in Cambodia.

Many thousands of years ago, Cambodia lay under the waters of the Pacific Ocean. The action of the Mekong River filled in the submerged land to form a fertile plain. This plain, watered by rivers and tributaries, formed natural irrigation systems. Where the land was not uniform, large lakes such as Tonle Sap were created. These lakes are relics of the former ocean gulf. The civilization of Cambodia developed on this moist plain.

The history of Cambodia is one of rivalries between outside forces and internal power struggles. Although the country has experienced periods of peace, the peace has often been enforced. This is the situation today.

THE EARLY PERIOD

By the early centuries of the Christian era, the Khmer and the related Mon peoples occupied a broad swath of land stretching from Burma to Vietnam. This was Funan, which flourished from the third to the seventh centuries. Early records also mention the states of "water Chenla" in the Mekong Delta and "land Chenla" further inland. Chenla brought down Funan.

The consolidation of Khmer society was clearer by the ninth century when Jayavarman II ruled over

Angkor and established the concept of *Devaraj*: the god-king. Jayavarman II built temples at widely spaced sites across Cambodia. Eventually, Angkor, near Siem Reap, became the core region with an irrigated area of 13,585 acres (33,500 ha) and a large population. Then, King Suryavarman II initiated the construction of Angkor Wat, the largest religious complex in the world. The complex was enlarged and enhanced by subsequent rulers (Figure 14-11).

The focus of Khmer society slowly changed. Varieties of Buddhism had long existed with Hindu cults, but during the thirteenth century, Theravada Buddhism won general allegiance. Now rulers demonstrated their power through the building of temples, schools, and monasteries.

In the 1440s, Angkor was abandoned in favor of capital sites in the region of Phnom Penh. The abandonment could have been the result of the silting up of the irrigation system or even malaria. The rise of the Thai state of Ayuthaya was probably the main reason. The capital sites were closer to the sea and the booming sea trade of the fifteenth century.

Until the late sixteenth century, the Khmer kingdom appears to have been as strong as its rivals Ayuthaya and Lan Xang. Thereafter, however, the Khmers became embroiled in regional machinations of the Thais and the Vietnamese. Vietnamese and Chinese began to dominate Cambodian ports. Lively trade ensued even as Cambodia was caught in the pincers of Thai and Vietnamese expansionism. Repeated incursions left the Thais dominant in the eighteenth century.

In the nineteenth century, the Vietnamese rose to the fore and Vietnamese people were encouraged to colonize the region. Vietnamese language, law, and modes of dress were imposed. A countrywide rebellion broke out in 1840; for five years Thais, Vietnamese, and Cambodian factions fought an inconclusive war.

COMING OF THE FRENCH

The French, having established their colony of Cochin China by 1862, were interested in Cambodia. In 1863 France signed a "treaty of protection" with the king that placed a French resident at Phnom Penh. This gave the French control of foreign policy and opened the door to commercial interests. However, the French soon realized that there was little economic return to be had from Cambodia and so concentrated on Cochin China.

By 1904, the French, with the intent of expanding their presence in Southeast Asia and countering British interests, achieved complete authority over the protectorate. In 1921, the population of Cambodia was assessed at 2.5 million. A Chinese rice export business had developed with rice purchased from the Khmer farmers. However, Cambodian rice was regarded as inferior to the rice of Cochin China so did not fare as well in the market. There were also small Chinese timber and pepper industries and French-financed rubber plantations using Vietnamese labor. Maize, kapok, and fish were gleaned from the region of Tonle Sap. The Mekong remained Cambodia's main trade artery but the port of Saigon handled the shipping.

Around 95 percent of Cambodians survived as subsistence farmers. They were characterized by the French, Vietnamese, and others as "lazy," "ignorant," "fatalistic," and "child-like." Western observers condemned them as a "decadent race." As the interlopers took in their profits, no industries of consequence were developed. Towns remained small, with Phnom Penh housing only 100,000 by 1930. Towns were dominated by aliens, and the first Khmer language newspaper only appeared in 1938.

WORLD WAR II AND INDEPENDENCE

During World War II, the French reached an agreement with Japan to retain control of Indochina in return for allowing free movement of Japanese troops. In order to deflect Cambodian fascination with Japanese rhetoric, the French stimulated a nationalist movement. They tried to institute a new form of writing as in Vietnam but the Buddhist community rebelled against what they saw as an attack on Cambodian culture and heritage. When the war ended, Cambodia had no clear nationalist group as existed elsewhere in Southeast Asia. The country drifted and the French returned to power.

While keeping control of key elements of government, the French permitted the formation of political parties. By the 1950s however, lack of progress and cooperation produced acute strains. In 1951, the Khmer People's Revolutionary Party (KPRP) was founded and began to organize guerrilla operations in outlying areas. Meanwhile, Prince Sihanouk toured France and the United States demanding independence. The French agreed and granted it in 1953.

SIHANOUK, WAR, AND REVOLUTION

Sihanouk, who came to power under rigged elections, promoted "Buddhist socialism." He was intolerant of opposition and forced hundreds of "dissidents" to disappear.

He greatly expanded education but expressed little interest in economic affairs. The Cambodian economy went into decline. Meanwhile, the Vietnam conflict was omnipresent and Sihanouk permitted the northern Viet Minh and the southern Viet Cong to use Cambodian territory. He rejected the United States and opened relations with China. This eliminated United States aid and angered conservatives. Communist-led revolts over government seizures of rice were brutally crushed. In 1970, while Sihanouk was overseas, the National Assembly withdrew its support and installed Lon Nol as the head of the new government of the "Khmer Republic."

The anti-Sihanouk coup polarized the Cambodian population. The United States supported Lon Nol and his totally inept government. Aid programs funded by the U.S. fostered gross corruption in both government and military. At the same time, the United States was carpet bombing Cambodia with more than half a million tons of bombs as part of its offensive against North Vietnam. About 500,000 soldiers and civilians were killed. The Communist insurgency was slowed by this devastation but not defeated. The Lon Nol regime ultimately collapsed and the insurgents took Phnom Penh in April 1975.

The Khmer Rouge (Red Khmer) set up the state of Democratic Kampuchea. This name derived from Kambuja, meaning "the sons of Kambu"—the ascetic who married a celestial nymph and founded the kingdom of Chenla, forerunner to the great Khmer empire. Their communist leadership was made explicit in 1977 when they announced the existence of the Communist Party of Kampuchea (CPK), which had been founded in 1968. A Paris-educated school teacher rose to take the helm. His name was Saloth Sar, later known as Pol Pot. One of his first acts was to declare 1975 as the year zero.

The Killing Fields

In April 1975 the insurgent Khmer Rouge marched triumphantly into Phnom Penh. The city was filled with refugees and its population had jumped from 600,000 to more than 2 million. Pol Pot immediately imposed his radical Maoist-style agrarian society onto the Cambodian people. Within four days, most of the inhabitants of Phnom Penh had been forcibly relocated to the countryside. Later, hundreds of thousands of people were forced from the southeast of the country to the northwest.

Pol Pot's goal was pure socialism. Everything modern was to be eliminated. A typical slogan was "We will burn the old grass and new will grow." Money, newspapers, education, and technology were outlawed. Intellectuals were killed. This was to show the strength of the people who, "though bare handed . . . can do everything."

Food was scarce under Pol Pot's inefficient system of farming, and administration was built on fear, torture, and execution. Little was known of the Khmer Rouge atrocities until a few refugees trickled over the Thai border. During the Khmer Rouge reign of terror, more than a million people died; some estimates are as high as 2 to 3 million. The Khmer Rouge turned Cambodia into a terrifying work camp in which families were abolished and murder was used as a tool of discipline. Pregnant women were disembowled, babies were torn apart limb from limb, people were buried alive or bashed to death with axe handles. Of 64,000 Buddhist monks, only 2,000 survived. Half of the Cambodian Vietnamese were eliminated, along with thousands from other groups. The CPK even turned upon itself, purging those who had received training in Vietnam.

In 1978, Vietnamese troops entered Cambodia. The Khmer Rouge collapsed before them, abandoning their centers of power. The Vietnamese took Phnom Penh in 1979 but failed to capture Pol Pot and his close comrades. The Vietnamese quickly established the People's Republic of Kampuchea (PRK) with a Cambodian government. This included Hun Sen, who became premier in 1985 and won in elections held much later in 1998.

The country suffered a devastating famine, and it was not until the mid-1980s that subsistence farming reached an even keel and shops and markets began to function normally. Meanwhile, the PRK became an international pariah supported by only a few countries, including the Soviet Union. In 1989, the Vietnamese withdrew from Cambodia. Kampuchea was renamed the State of Cambodia and committed itself to a private enterprise economy and the restoration of Buddhism.

The United Nations became involved in Cambodia's internal affairs and elections were held in 1993. Multiple parties contested and a coalition government was

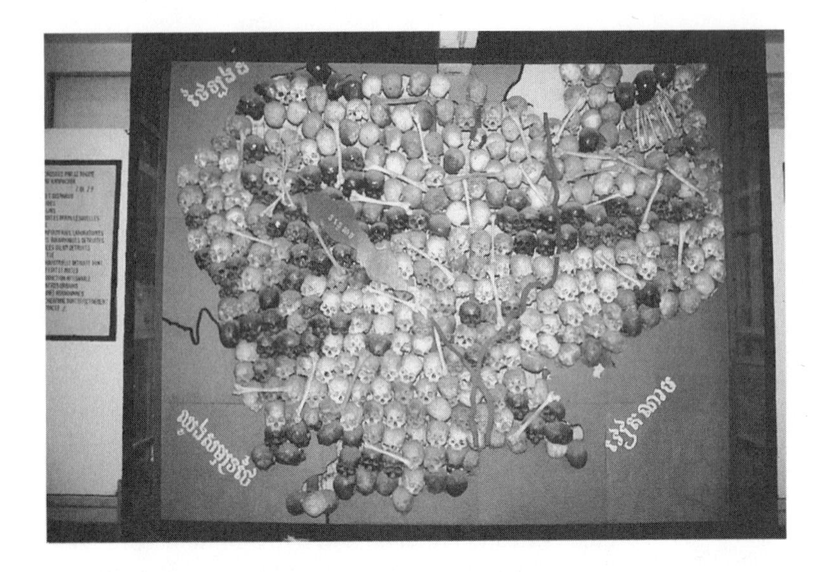

Figure 15-20

This map made with real human skulls symbolizes the millions of people who died at the hands of the Khmer Rouge during their reign of terror in Cambodia. Photograph courtesy of B. A. Weightman.

formed. King Sihanouk was reinstalled, although he later died of cancer. The coalition government was heavily financed by foreign aid, but internal dissension and corruption meant that little progress was made toward rebuilding the shattered economy and infrastructure. Phnom Penh University opened in 1988. Only 5,000 teachers had survived of the original 20,000 in 1975. A sex ratio imbalance was evident in employment. For example, 70 percent of workers in the textile and tobacco industries are women, mostly widows of pre-1975 workers. A former school in Phnom Penh, where 20,000 died, has been turned into a museum to Pol Pot's genocide (Figure 15-20).

In disputed elections in 1998, Hun Sen became the head of government. Pol Pot committed suicide in the same year, and other Khmer Rouge leaders have either defected or live clandestinely in various parts of the country.

Death Awaits

Because of the influx of military hardware to various factions during the 1980s, Cambodia remains one of the most heavily land-mined places in the world today. At that time, here were more than 3 million mines, mostly concentrated in the northwest province of Battembang, a Khmer Rouge stronghold. There have been more than 40,000 land mine victims since the end of the insurgency. It is estimated that in 2004, more than 5 million people in some 6 thousand villages were at risk. From 2000 to 2005, there were at least 850 deaths or serious injuries from land mines each year. With the help of international agencies, mine clearing is proving successful. Currently there are fewer than 300 new amputees a year. One in every 236 Cambodians is an amputee.

CAMBODIA TODAY

Cambodia continues as a very poor country although conditions are improving. Now only 26 percent of children under age five are malnourished as compared to 40 percent only five years ago. Nevertheless, infant mortality rates remain high at 62 per 1,000 live births. More than 90 percent of births in rural areas are at home and attended by traditional midwives. Only 13 percent of primary school students complete grade five in five years. Only 19 percent of students in upper secondary school are female. The main reason given for the fact that girls do not attend much more than primary school is that "they are needed at home." Of the rural population, only 26 percent have access to safe water. A mere 6 percent of rural people have access to improved toilet facilities. One-third of Cambodians live on less than US$0.50 a day. Clearly, Cambodia remains one of the least developed countries in the world.

Rice production accounts for nearly 88 percent of the available cultivated area. Farms tend to be very small—around 2.5 acres (1 ha). A farm of this size supports a family of five. Rice is supplemented by fruit and vegetable cultivation. Although rice yields are low, Cambodia became self-sufficient in rice in 1996 and had a surplus to export.

Fishing is important as it provides the main source of protein in the Cambodian diet. Fisheries are both inland and marine, and fish are also raised in ponds, pens, and cages where water is available around the year. Some frozen fish are exported.

Poor food distribution and soaring food prices have led the International Food Policy Research Institute to list Cambodia as a country having an "alarming" level of hunger in 2008. A recent impact study concluded than 1.7 million Cambodians are experiencing food insecurity and this number is expected to increase to 2.8 million—a fifth of the population.

Cambodia's rubber plantations are second only to timber as a source of export earnings. Natural rubber is Cambodia's biggest export to South Korea. Expansion of rubber production by Vietnamese investors is planned for the eastern provinces. Cambodia should be able to produce 600,000 tons of rubber annually.

Teak and rosewood are among Cambodia's most valuable assets. However, more than 70 percent of Cambodia is deforested and the government has taken measures to stop timber exports. Illegal cutting and exporting continues despite the prohibitions. In 2003, the government expelled Global Witness, an environmental group that was monitoring efforts to combat illegal logging. As a consequence, aid to the country's forestry program from the International Monetary Fund and the World Bank has been stalled amid concerns over high-level government involvement in illegal logging.

Cambodia is rich in minerals including iron, bauxite, gold and precious stones such as sapphires and rubies. It has oil in the Gulf of Thailand, which is now being exploited by both domestic and foreign concerns.

Foreign aid and investment is critical to Cambodia's development. Light industry is more developed than heavy industry. The more developed industries include household goods, textiles, soft drinks, alcohol, tires, pharmaceuticals, and cigarettes. The extent of foreign involvement can be demonstrated by the textile industry. Of 165 companies, Hong Kong runs 45, Taiwan 25, Cambodia 18, and China 17. The rest are divided up among South Korea, Malaysia, Indonesia, Singapore, France, Britain, and the United States.

After decades of civil war and isolation, Cambodia has become a favorite destination for foreign tourists. With most wanting to experience Angkor Wat, some two million foreigners visited the country in 2008. The largest numbers of tourists come from Vietnam and South Korea. Tourism is Cambodia's second largest earner of foreign exchange after textile exports.

DANGEROUS LIAISONS

On the more sinister side, tourists also include thousands of pedophiles who take advantage of Cambodia's weak law enforcement. Prostitution and drug abuse have also led to the highest HIV/AIDS infection rate in Asia. According the United Nations, 160,000 Cambodians between the ages of 15 and 49 have HIV. Some 30,000 children under the age of 5 have lost their parents to AIDS.

AIDS so far has killed more than 200,000 people since it was discovered in the country in 1991. Another 200,000 people are expected to develop the disease in the next 10 years. This will overwhelm Cambodia's grossly inadequate health-care system. Moreover, the economy will be impacted because most of the sick and dying will be young people—potential workers. The epidemic, which has slowed slightly in recent years with increased intervention on the part of the United Nations and other organizations, is still referred to as "the new killing fields."

Recommended Web Sites

www.aseanaffairs.com/
News on current political and economic issues in Southeast Asia by country.

http://books.sipri.org/files/PP/SIPRIPP20.pdf
2007 report on the conflict in southern Thailand by the Stockholm International Peace Research Institute.

www.business-in-asia.com/
Information on business by country including maps and charts.

www.cambcomm.org.uk/ff.html#
Christian site with a wealth of information on Cambodia.

www.cambodia.org/
Official site with historical and current information on government, economy, etc. Link to Phnom Penh Post.

http://earthobservatory.nasa.gov/IOTD/view.php?id=8723
Satellite photos of the impact of Cyclone Nargis on Burma.

www.foodsecurityatlas.org
Information on a variety of topics such as food security, education, and literacy by country.

www.maplandia.com/asia/
Interactive maps for any region in Asia. Link to Google Earth satellite maps.

www.maritimeshows.com/vietnam/Vietnam_Country_Information-270907.pdf
2009 report on shipbuilding, ports, and logistics in Vietnam.

www.mizzima.com/
Site founded by Burmese exiles. Specializes on recent and past events in Burma. Photo gallery.

http://thailand.prb.go.th/ebook/king/new_theory/.html

Details on King Bhumiphol's New Theory Agriculture. Examples, charts, and photos.

http://Thailand.prd.go.th/ebook/review/indez.php?bookID=5

Inside Thailand Review. Information on government development programs.

http://theasiamag.com/countries/southeast.asia?page=1

Human interest stories, perspectives, and photos by country.

http://uclibraries.colorado.edu/govpubs/for/laos.htm

University of Colorado at Boulder's Web site for information about Laos. Numerous links to articles, reports, and databases from official and non-official sources. Excellent site.

Bibliography Chapter 15: Mainland Southeast Asia: Turmoil and Peace

Aye, Henri-Andre'. 2009. *The Shan Conundrum in Burma.* Charleston, S.C.: BookSurge Publications.

Chin, Ko-lin. 2009. *The Golden Triangle: Inside Southeast Asia's Drug Trade*. Ithaca, N.Y.: Cornell University Press.

Crispin, Shawn, W. 2003. "Thailand: Big Risks in Big Plans." *Far Eastern Economic Review* (February 27): 16–17.

Crispin, Shawn, W. 2004. "Thailand: Pipe of Prosperity." *Far Eastern Economic Review* (January 19): 12–16.

DeWeerdt, Sarah. 2008. "War and the Environment." *World Watch* (January/February): 14–21.

Duiker, William, J. 1995. *Vietnam: Revolution in Transition.* Boulder, Colo.: Westview.

Freeman, Donald B. 1996. "Doi Moi Policy and the Small-Enterprise Boom in Ho Chi Minh City, Vietnam." *The Geographical Review* 86: 178–197.

Huke, Robert E. 1998. "Myanmar Visited." *Focus* 45: 18–25.

Hussey, Antonia. 1993. "Rapid Industrialization in Thailand, 1986–1991." *The Geographical Review* 83: 14–28.

Jackson, Karl D., ed. 1989. *Cambodia 1975–1978: Rendezvous with Death*. Princeton, N.J.: Princeton University Press.

Jacobs, Jeffrey. 2002. "The Mekong River Commission: Transboundary Water Resources Planning and Regional Security." *The Geographical Journal* 168/4: 354–364.

Lintner, Bertil. 1994. "Burma: Enter the Dragon." *Far Eastern Economic Review* (December 22): 22–23.

Lintner, Bertil. 1997. "Drugs, Insurgency and Counterinsurgency in Burma." In *Burma: Myanmar in the Twenty-First Century*, ed. John J. Brandon, pp. 207–245. Bangkok: Open Society Institute.

Lintner, Bertil. 1998. "The Dream Merchants (Narcotics)." *Far Eastern Economic Review* (April 16): 26–27.

Lintner, Bertil. 2003. "Cambodia: Dangerous Liaisons." *Far Eastern Economic Review* (March 13): 47.

Lintner, Bertil. 2003. "Laos: Aid Development." *Far Eastern Economic Review* (July 3): 45.

Ma, Lawrence, and Carolyn Cartier, eds. 2003. *The Chinese Diaspora: Space, Place, Mobility, and Identity*. Lanham, Md.: Rowman and Littlefield.

Mabbett, Ian, and David Chandler. 1995. *The Khmers*. Cambridge: Blackwell.

Maguire, Mark. 2010. "Southern Thailand Mired in Violence." *The Los Angeles Times* (April 4): A3.

Montlake, Simon. 2007. "Thailand's Exposed Southern Flank." *Far Eastern Economic Review* 170/4: 35–38.

Mulder, Niels. 1996. *Inside Thai Society: Interpretations of Everyday Life*. Kuala Lumpur: Pepin.

Norman, Colin. 1983. "Vietnam's Herbicide Legacy." *Science* 219: 1196–97.

Osbourne, Milton. 2000. *The Mekong: Turbulent Past, Uncertain Future*. New York: Grove Press.

Penrose, Jago, Jonathan Pincus, and Scott Cheshier. 2007. "Vietnam: Beyond Fish and Ships." *Far Eastern Economic Review* 170/7: 43–46.

Perry, Peter John. 2007. *Myanmar (Burma) Since 1962: The Failure of Development*. Burlington, Vt.: Ashgate.

Reardon, Douglas. 2004. "The Qualms of a Geographer in a Pariah State." *Focus* 48: 7–11.

Rigg, Jonathan. 1995. "In the Fields There is Dust." *Geography* 80: 23–32.

Rigg, Jonathan. 1995. "Managing Dependency in a Reforming Economy: the Lao PDR." *Contemporary Southeast Asia* 17: 147–172.

Ryan, John C. 1990. "War and Teaks in Burma." *World Watch* 3: 8–9.

Savada, Andrea, ed. 1995. *Laos: A Country Study.* Washington, D.C.: U.S. Government.

Savage, Victor R., et al. 2004. "The Singapore River Thematic Zone: Sustainable Tourism in

Shawcross, William. 1994. *Cambodia's New Deal*. Washington, D.C.: Carnegie.

Thomson, Curtis N. 1995. "Political Stability and Minority Groups in Burma." *The Geographical Review* 85: 269–285.

Tyner, James. 2008. "Cities of Southeast Asia." In *Cities of the World*, eds. S. D. Brunn, Maureen Hays-Mitchell, and D. J. Zeigler, pp. 429–473. New York: Rowman and Littlefield.

Wellner, Pamela. 1994. "A Pipeline Killing Field: Exploitation of Burma's Natural Gas." *The Ecologist* 24: 189–193.

Chapter 16

Insular Southeast Asia

"There is not one state truly alive if it is not as if a cauldron burns and boils in its representative body, and if there is no clash of convictions in it."

SUKARNO (1970)

Welcome to the peninsular and island realm of Southeast Asia (see Figure 15-1). With six nations including Malaysia, the Sultanate of Brunei, Singapore, Indonesia, Timor-Leste, and the Philippines, this is one of Asia's most complex regions. Like the rest of Southeast Asia, the region's history is bound up with ocean interactions, foreign intervention, and struggles for identity in the modern world. In this chapter, starting with Malaysia, you will read some of the most intriguing stories of geographic interrelationships in the world today.

Malaysia 2020

The course of Malaysia's history has been influenced by its strategic position, which made it a natural meeting place for traders from both east and west. The lush tropical forests and abundance of water sustained numerous small, self-supporting communities. However, unlike mainland Southeast Asia, the presence of mountains and the absence of broad plains precluded the development of elaborate systems of water control so fundamental to mainland societies. Sea trade bred contacts from the outside, and this was the foundation of Malaysia's history.

Whence "Malay"?

Many Malay areas were colonized by Sumatrans long ago. It is conceivable that the word *Melayu* derived from the *Sungei Melayu* (Melayu River) in Sumatra. Then there is the possibility that it came from a Tamil word *malai*, meaning hill.

The geographer Ptolemy called the Malay Peninsula *Aurea Chersonesus*, or "The Golden Chersonese." This was the fabled land of gold.

During the Portuguese and Dutch colonial periods, and prior to the British colonial period, the whole peninsula was labeled Malacca. There was very little mapping of the peninsula until the early nineteenth century. As you will read shortly, many name changes would occur before reaching the name "Malaysia."

EARLY SETTLEMENT

The earliest of the current inhabitants of Malaysia are the ***Orang Asli***, meaning "Original People," who inhabited the Malay peninsula (Figure 16-1). Other groups include the Dayak and Penan of Sarawak and the Rungus of Sabah, people whose nomadic way of life has been almost completely destroyed in the wake of development.

The next arrivals were the Malays, the Proto-Malays establishing themselves around 1000 BC. These were followed by other immigrants, including the Deutero-Malays, streaming down the peninsula from mainland regions, including China. Many migrated even further to the island realms of Indonesia and elsewhere. The peninsular Malays had their closest links with the Malays of

Figure 16-1

This man is an Orang Asli, one of the original inhabitants of the Malay Peninsula. He is carving a sago palm, one of the most useful palms in the forest. The carbohydrate-rich pith is dried as "pearl sugar" and is eaten with fish or fruit. The palm provides fiber for rope, matting, or cloth; the leaves can be used for thatch; and the sap ferments into a liquor. Photograph courtesy of B. A. Weightman.

Sumatra, and for centuries the Strait of Malacca served as a corridor for interaction between the two regions.

Together with the *Orang Asli,* all these people share a common culture and make up the ***Bumiputera*** *(Bumiputra)*—"Sons of the Soil." Cultural characteristics were rooted in an agrarian-maritime economy and were reflected in village society, where leadership was largely consensual and attitudes were impregnated by a belief in an all-pervasive spiritual world. Although this culture came to be overlaid by Hinduism, Islam, and Western ideologies, elements of the basic culture persist.

Trade with China and India began around the first century BC, although Hindu and Buddhist elements from India were more influential in terms of impacts on language, literature, and social customs. Numerous temples were constructed throughout the region. For the greater part of this time, Malaysia was subject to either Javanese or Sumatran power structures.

The Peranakan

By 1400, Malacca had become an important outpost for Chinese traders. They arrived from November to March on the northeast monsoon and departed in June with the southwest monsoon. Malacca's sultans even paid tribute to the Ming court. Many Chinese settled in Malacca, especially around *Bukit Cina*—Chinese Hill. Today, *Bukit Cina* is a historical park, housing the world's largest traditional Chinese cemetery.

Subsequent generations of Straits Chinese became known as Peranakan. The Peranakans were known for their business acumen and lavish lifestyles. They built classic townhouses and developed all the trappings of "high society". They also developed their own cuisine, a blend of Malay and Chinese, which is noted for its meticulous preparation and pungent flavors.

Blending easily into British society, the Peranakan remained divorced from other Chinese. Many became doctors, lawyers, and entered public office. In Pinang, they were known as the "Queen's Chinese."

The Hindu-Buddhist period of Malaysian history ended with the arrival of Islam. Introduced primarily by Indian and Arab traders, Islam became a major influence after 1400 with the conversion of formerly Hindu rulers of Malacca. Islam diffused along the trade routes into the rest of the insular region. The Sultanate of Malacca and the Kingdom of Brunei became major proponents of the faith.

Both Malacca and Brunei were shattered by the influx of Europeans. The Portuguese, Dutch, and Spanish made their presence felt through force of arms. Despite their technological superiority, the Europeans had limited influence until the coming of the British at the end of the eighteenth century and their introduction of the resources and organization of the Industrial Revolution.

MAKING MALAYSIA

Under the British, Pulau (Malay for "island") Pinang (1786), Singapore (1819), and Malacca (1824) became known collectively as the Straits Settlements. Garnering other Malay states under their wing of control, the Straits Settlements were later expanded to include the island of Labuan off the northern coast of Brunei. On the peninsula, a system of Federated Malay States was established with the capital at Kuala Lumpur. Several other peninsular states were given British advisors and became known as the Unfederated Malay States. Sarawak on the island of Borneo became a British protectorate ruled by the Brooke family. Sabah also became a protectorate ruled by the Chartered Company of British North Borneo.

This confusing arrangement satisfied few and opposition frequently turned violent. In 1948, the British established the Federation of Malaya, including all nine Malay states of the peninsula (Figure 16-2). They also committed to the region's independence. Local elections were held in 1951 and federal elections in 1955. The British released their sovereignty in 1957.

The idea of the formation of Malaysia was introduced in 1961. The new nation was to include the peninsular states, the Borneo states, and Singapore. The name "Malaysia" is derived from "Malay"—the people, "si"—Singapore, and "a"—Malaya. Elections were held shortly thereafter, with only Brunei declining to join. Tension was apparent with regard to Singapore, however. Its large contingent of Chinese (with their potential political power aligned with Chinese elsewhere in Malaysia) concerned the Malay majority. Singapore was pressured out of the union in 1965 (Figure 16-3).

TIN AND RUBBER

At the beginning of the nineteenth century the population of peninsular Malaysia was no more than 300,000. Most were Malay rice cultivators. A heavily forested terrain deterred the British, who were more interested in their ventures in Pulau Pinang and Singapore. However, the success of the British entrepreneurs stimulated migrations from India and China. The Indians and Chinese would eventually become intermediaries between the British and the Malays.

Chinese miners revolutionized tin mining, which had previously been on a very small scale. Given a low-cost means of extraction and the location of tin-bearing gravels along the western foot of the Main Range, a mere 25 to 30 miles (40 to 50 km) from the coast, production surged. In the 1890s, a "tin rush" ensued in the three western states of Perak, Selangor, and Negeri Sembilan.

Relations between the Chinese and Malays were not good, so the British decided to provide a system of law and order in the region. This took the form of an elaborate infrastructure of road, rail, and port facilities. Smelting and exporting of tin were centered at Pinang and Singapore.

In the early 1900s, an increasingly motorized world was demanding rubber. The climate of Malaysia was perfect for the raising of this commodity. Consequently, new rubber estates were established within the area already possessing an infrastructure. Indian rubber tappers were brought to work the trees, and the rubber industry boomed until the Great Depression of 1930. When the global economy recovered, so did rubber. Malaysia remains the world's largest rubber producer.

The peninsula's development was severely unbalanced. A rubber and tin belt occupied the southwest. It covered only a quarter of the land area but accounted for 85 percent of its economic activity. Here, too, were the 10 largest towns and a large but diverse population of Chinese, Malays, and Indians (Figure 16-3). The east was largely undeveloped and contained a much lower density of predominantly Malay peasants. When Malaysia became independent, a major goal was to rectify this imbalance.

EAST MALAYSIA

East Malaysia consists of Sarawak and Sabah (Figure 16-4). Prior to the economic intervention of the Europeans, northern Borneo's trading economy was dominated by indigenous, unprocessed, and luxury products. Jungle products such as bezoar stones (concretions from the stomachs of animals such as goats), hornbill beaks, camphor, and birds' nests were the most important. Other products included rattan, bamboo, other woods, and gutta percha (a resin from the gutta tree used to make cable coverings and glue).

This was an economy separate from the world market. Traditional networks of collecting and marketing channeled jungle products to coastal ports where Chinese or Malay traders sold them on the regional market (Figure 16-5).

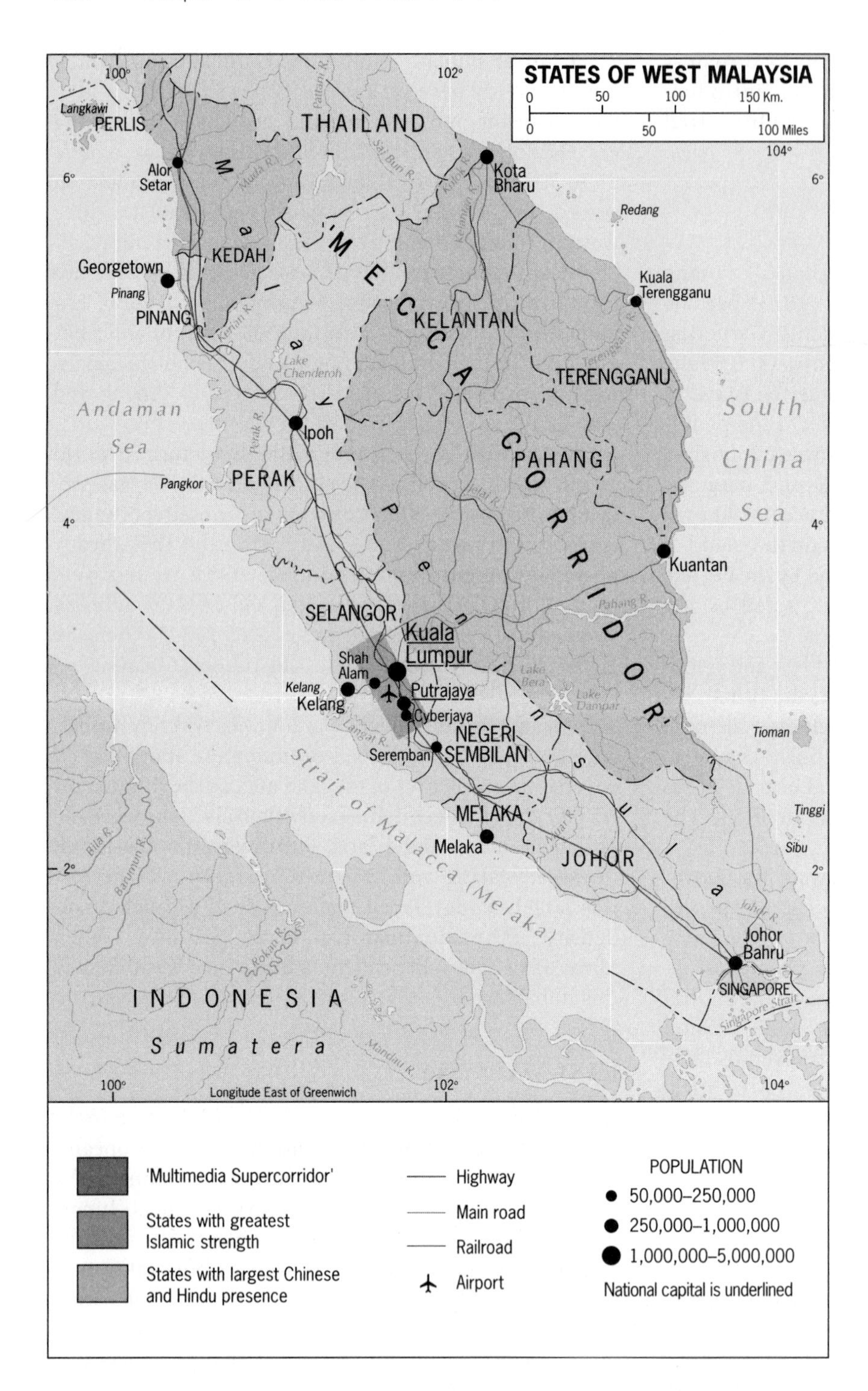

Figure 16-2

Malaysia. Like Indonesia and the Philippines, Malaysia is a fragmented country. From H. J. de Blij and P. O. Muller. *Geography: Realms, Regions and Concepts*, 14th edition, 2010, p. 561. Originally rendered in color. © H. J. de Blij, P. O. Muller. Reprinted with permission of John Wiley and Sons.

By the late nineteenth century, the East Malaysian economies had begun to shift toward meeting the needs of world markets. Production systems were increasingly controlled from abroad as plantation farming, rubber, timber, and oil took precedence over jungle products. Indigenous groups, such as the Penan and the Dayak (engaged directly in accessing these products), fell increasingly under the influence of others, particularly the Chinese. Javanese workers were also recruited to work in the region. However, as jungle produce continued to de-

Figure 16-3

Sri Mahamariamman, dating from 1873, is the oldest Hindu temple in Kuala Lumpur. The Raja Gopuram (tower) was erected in 1968 in South Indian Style. The gopuram is 75 feet (22.9 m) tall. Photograph courtesy of B. A. Weightman.

liver significant income, indigenous people clung to those activities while outsiders involved themselves in tobacco, pepper, and other new production enterprises. Oil was very important in British North Borneo, where it was the stable backbone of revenue.

Pre-colonial and colonial trading economies had become intermeshed. Whatever the case, the focus was on environmental exploitation. Native populations became ever more marginalized as commodity trade increasingly fulfilled the requirements of the growing world economy. This heritage affects the developmental patterns of East Malaysia and Brunei to this day.

ETHNICITY AND DEVELOPMENT

Malaysia has a plural society and development policies have favored indigenous Malays. However, the relationship between ethnicity and status is difficult to define as there are not viable definitions of ethnicity. Ethnic pluralism is a historical construct. For example, in Malaysia the British perceived the Malay as farmers unwilling to do wage labor and the Chinese and Indians inherently suited for mining and plantation work. Thus, Malays were shoved to the margins of development by both British practice and policy. As geographers Mark Cleary and Brian Shaw (1994) observed: "Ethnic pluralism in Malaysia might be more accurately represented as a series of overlapping Malay, Chinese or Indian ethnic dimensions (language, business practice, occupation, religion) rather than lines of rigid demarcation. They represent socially constructed rather than 'natural' difference."

Just as economic and infrastructural factors were unbalanced, so was population distribution. By the late 1960s, some 15 percent of the Malay population was classified as urban. About 45 percent of Chinese and 35 percent of Indians were classified as urban. Malays in the cities were heavily concentrated in government employment. Very few participated in the commercial and industrial sectors.

Modern development in Malaysia has been defined by a New Economic Policy (NEP) and subsequent "Malaysia Plans." A fundamental goal of the NEP was to restructure Malay society by raising incomes for all groups and eliminating the association of ethnicity and economic function. This meant an increase in the production of export commodities such as timber and oil as well as export-oriented industrialization. Explicit within the NEP was the intention to create a Malay (*Bumiputera*) business community that would own or manage at least 30 percent of all these activities by 1990. This goal date has been reset to 2020.

Even though most Malays are not as rich as Chinese business people, they have seen major increases in prosperity over the past two decades. While the Chinese dominate business in Kuala Lumpur and Pinang, established Malay businessmen have become a major force in many areas. Moreover, Malays dominate the government. Government efforts to encourage *Bumiputera* entrepreneurs have spawned a lively and increasing influence of Malay business at all levels of the economy.

THE RISE AND FALL OF ISLAMISM

Malaya's headlong rush into development generated a backlash among conservative Muslims. Consequently, in 2001, Islamists won political victories in two states: tin-producing Kelantan and energy-rich but socially poor Terengganu. Fundamentalist officials imposed strict

Figure 16-4

Sarawak and Sabah on the island of Borneo are both part of Malaysia. Note how the Sultanate of Brunei is also a fragmented state. Kalimantan is a part of Indonesia.

Figure 16-5

Here, the Iban people of Sarawak are selling "jungle products," which they have collected from the forest. The fruit in the foreground is "rambutan." The Iban are traditional shifting cultivators, but many have moved to coastal towns to participate in the oil, lumber, and plantation economy. They make up nearly one-third of the population of Sarawak. Photograph courtesy of B. A. Weightman.

Islamic laws in both states. Malaysians in general discussed the notion of the country being divided into two corridors: the "Mecca Corridor" in the east and the "Multimedia Supercorridor" in the west. Subsequently, new elections brought down the two Islamic state governments along with the autocratic head of state, Mahathir bin Mohamad. With the appointment of his more moderate successor, Abdullah Badawi, Islamic fervor abated, and calls for *jihad* in Malaysia faded into the background.

DEVELOPMENT SCHEMES

One widely acclaimed land development scheme was the Federal Land Development Authority (FELDA) settlement plan. Commencing with a 2,470 to 4,940 acre (1,000 to 2,000 ha) block of land, contractors were brought in to clear the jungle, plant the main crop—rubber in the early years and palm oil, cocoa, and coffee in later years—and build basic infrastructure such as roads, houses, and schools. Settlers were selected, moved in, and employed as salaried workers until the crop began to produce. Each family was given title to a plot of land once the mortgage was paid off. By 1987 there were 422 FELDA schemes in all states except Pinang.

Today, FELDA schemes are more likely intensively run "agro-businesses" than traditional Malaysian farming communities. Moreover, the schemes increasingly are becoming part of the Western, mechanized, uniform production system. FELDA plans were cut to zero in the Sixth Malaysia Plan of 1991.

The Ninth Malaysia Plan of 2006 promotes a shift from those mono-crop, low-technology, and small-scale operations that still exist, to integrated, high-tech, and large-scale enterprises. Emphasis on palm oil and rubber is to be reduced while the production of fruit and vegetables is to be promoted. The plan also proposes to intensify land use by integrating more food crops with plantation crops. Furthermore, the government is encouraging urban horticulture and agro-technology parks. Land settlement is now seen as the concern of the private sector in the context of regional planning frameworks.

Rural Poverty

For all its impressive growth rates and higher per capita incomes, Malaysia ranks among the most unequal societies in Southeast Asia in terms of income distribution. While it is true that absolute poverty levels have declined—from 29 percent in 1980 to 6 percent in 2000—critics complain that these figures are misleading. The poverty line is at US$134 a month and this cannot keep up with the rising costs of living. The World Bank reports that the trend of inequality reduction has reversed itself since 1990. Another report that measures Malaysia's Gini Index—where 0 indicates perfect equality and 1.0 indicates perfect inequality—finds that the country's 0.49 score is one of the highest in the region. Indonesia is at 0.34 and Thailand is at 0.42.

According to another research study, intra-ethnic inequality appears to be higher among Malays. Two decades of affirmative action policies did increase the stake in the economy of the *Bumiputeras* and led to the emergence of a Malay middle class. However, a huge amount of this stake is held by state-backed investment agencies holding shares in trust.

Another disparity concerns the rural-urban divide. Rural incomes are much lower than urban incomes. Many of the poor Malay households can be found in Kelantan, Terengannu, Kedah, and Perlis states. It is in this "heartland" that support for conservative Islam is the strongest. In East Malaysia, indigenous groups in the interior of Sarawak and Sabah are very poor as are the *Orang Asli* of peninsular Malaysia.

The Malaysian government is in the process of developing two new regions in order to accelerate economic growth and raise incomes. The first, launched in 2007, is the Northern Corridor Economic Region (*Korridor Utara*) in Perlis, Kedah, Pulau Pinang, and the north of Perak. Pinang, with its 1.6 million inhabitants, has become Malaysia's "Silicon Island," attracting both foreign and domestic industrial concerns. Chinese make up 42 percent and Malays comprise 40 percent of the population. Pinang is part of an economic "Growth Triangle" that incorporates Sumatra and Thailand.

The second focus will be on *Iskandar Malaysia*, a development zone in southern Johor with links to Singapore. These projects are expected to be completed by

2025 by the end of Malaysia's Tenth Year Plan. These plans have six featured goals:

- Agriculture: Transformation from traditional practices to a "new agriculture" to improve the productivity of land.
- Manufacturing: To decentralize Pinang and balance production across the region in order to provide more employment opportunity.
- Tourism: Significantly diversify the tertiary sector and accelerate the move into more modern service industries.
- Infrastructure: Build new infrastructure to facilitate the above developments with the intention of providing integrated services and logistics such as irrigation, utilities, and transportation.
- Human Capital: Improve educational facilities and teaching resources to properly train people in the appropriate science and technology to compete in the global marketplace.
- Environment: Protect and conserve natural environments in the course of development with the goal of sustainability.

A Multimedia Super Corridor (MSC) was launched in 1996. This 15 by 50 km (9 by 30 mi) corridor stretches from the twin Petronas Towers in Kuala Lumpur southward to the international airport and includes the new cities of Cyberjaya and Putrajaya. Kuala Lumpur and the airport are joined by a high-speed rail system that opened in 2002. There is a multimedia university in Cyberjaya, and Putrajaya is designed to be the new administrative capital of Malaysia with a paperless, electronic government.

The MSC is served by a fiberoptic/coaxial cable network with direct links to ASEAN, Japan, the United States, and the European Union. Firms locating in the MSC enjoy tax and investment incentives and are allowed to hire foreign "knowledge workers." The "knowledge workers" will live in "smart homes" in which they can shop, entertain themselves, and further their education online. As of 2004, over 1,000 high-tech firms had located in the MSC.

A POPULATION ANOMALY

Malaysia's population growth is relatively high, given its level of development. It is also relatively higher than the average rate for Southeast Asia. Malays have significantly higher total fertility rates (TFRs) than the Chinese and the Indian populations. The national TFR in 2009 was 2.6, of which Malays had a two-thirds contribution.

High Malay TFRs can be related to Islamic fundamentalism and associated decline in the use of contraception. Furthermore, population policy in 1984 promoted five-child families. Malay families receive preferential treatment in university education, government employment, and the like. Malaysia 2020 has pronounced a goal of 70 million people as the optimum population size. This would substantially enlarge the consumer market in light of industrialization ambitions.

The structure of Malaysian population has been altered, with Malays now forming a dominant majority. Many Chinese, seeing no future for their children, left for Singapore, Canada, and Australia. Malays are now 58 percent of the population as compared to 56 percent in 1970. The Chinese make up 25 percent and the Indians 7 percent. Multiple non-Malay ethnic groups, mostly in East Malaysia, make up the final 12 percent.

In East Malaysia, populations are concentrated along the coast and in surrounding urban centers. This contrasts with the sparsely populated interior where settlements are along river arteries. Sabah's population has been boosted by the in-migration of nearly half a million Filipinos. Many are illegal residents and every few years there are crackdowns, arrests, and deportations. Every year, there are 4,000 to 6,000 Filipinos deported for immigration offenses. Filipinos work in the banking, construction, engineering, and medical industries. Some 20,000 Filipinas work as nurses, nannies, and maids.

In 2010, an estimated 2.2 million Indonesians lived and worked in Malaysia. After the financial crisis, about one million of these people were deported. In 2009, an agreement was signed between Malaysia and Indonesia that would allow Indonesians to work in the country but they would have to pay six months wages to the government for the privilege. Moreover, they have no rights to decent working hours, fair wages, or reasonable living conditions.

Many Indonesian women work as maids in Malaysia. When a law was passed to allow them one day off a week there was an outpouring of anger. One maid was beaten to death by her employer.

URBANIZATION

Migration has taken place in two ways. In the 1960s and 1970s, government efforts were concentrated in redistributing rural populations to other rural areas via land development schemes. The 1980s witnessed a dramatic change, with the largest share of migration being rural to urban. The rural sector was less able to absorb more

Figure 16-6

Kuala Lumpur. The colonial era, Moorish-style Sultan Abdul Samad building, housing city hall and the Supreme Court, contrasts sharply with the modern office and bank buildings nearby. The tall structure to the right is the Islamic Bank.
Photograph courtesy of B. A. Weightman.

people due to a fall in commodity prices, and the NEP pointedly tried to reduce rural poverty and dissociate Malays from strictly agricultural pursuits.

The result has been the cultural diversification of urban places once dominated by Chinese. More than 50 percent of Malays now live in urban areas. Moreover, the cultural landscapes of the largest urban centers have been noticeably "Islamicized" in recent years. More mosques, the high-rise Islamic Bank, Islamic archways over roads leading into the cities, and the presence of more women wearing head coverings are some of the more noticeable landscape elements (Figure 16-6).

Urban population distribution is still unbalanced. The west coast states accounted for more than 60 percent of the country's urban population in 2000. Historical inertia plays a role in this phenomenon. In East Malaysia, rates of urbanization have been rapid, yet the percentage of urban dwellers is relatively low. This can be attributed to the absence of an urban-based manufacturing economy.

The paramount urban conurbation is that of Kuala Lumpur and the Kelang Valley. Because of the domination of Singapore and Pinang in the colonial period, Kuala Lumpur's population exceeded 100,000 only by 1930. Kuala Lumpur was designated a federal capital in 1974. Rapid growth ensued in the 1970s under NEP policies, and today the city with 1.6 million residents is the economic, political, and cultural core of the country.

The city incorporates a number of satellite cities in the state of Selangor, including an east–west corridor between it and the port of Kelang. Petaling Jaya was established in 1952 as a "new town," and by 1991, Petaling Jaya and Kelang were the nation's fourth and fifth most populous urban centers. In the 1970s, a free trade zone was established at Sungei Way. Further west is Shah Alam, the newly planned Selangor state capital that houses one of Southeast Asia's largest mosques (Figure 16-7). All these are joined by superhighways.

Kuala Lumpur is a modern landscape with a mix of modern and colonial buildings, Chinese shops, shopping malls, a fabulous Moorish-style railroad station (to become a hotel), and the 88-story, 1,483 feet (452 m) Petronas "Twin Towers"—the world's tallest twin-tower buildings (Figure 16-8).

Housing ranges from squatter settlements (fewer than in most of Asia) to condominiums, to suburban homes, to mansions. Because of a housing shortage, the government has built high-rise public housing (Figure 16-9). Squatters do not like these structures because it removes them from ground level, where they can garden and watch their children play. Moreover, it is difficult to operate a home-based industry in a high-rise. In addition, these projects are frequently distant from employment and services.

Since this area does not have vast expanses of paddy, it does not fit the *kotadesasi* model exactly, although it has some elements of it. *Kampungs* (Malay villages) and agricultural land do exist in the interstices of urban places, highways, industrial parks, and the like (Figure 16-10). Many original inhabitants supplement their income with off-farm jobs.

Figure 16-7

Selangor's new capital building at Shah Alam is part of the Kuala Lumpur-Kelang development corridor. Photograph courtesy of B. A. Weightman.

Figure 16-8

These are the Petronas Towers. They were the tallest buildings in the world until Taipei 101 was erected. However, they remain the tallest twin towers in the world. Petronas is Malaysia's state oil company. © Goh Seng Chong/Bloomberg/Getty Images, Inc.

A MULTI-SECTOR ECONOMY

Malaysia's national economy is now anchored by high-value manufacturing, including electronics for export and textiles, steel, and auto assembly for domestic consumption. In the 1980s, the main catalyst in economic growth was an outward-looking industrialization strategy. Rapid industrialization was encouraged by free trade zones and other strategies.

The manufacturing sector is the largest generator of employment. The Ninth and Tenth Malaysia Plans (2006 and 2010) point to placing more emphasis on diversified manufacturing for export to foreign markets. The plans also stress the need for the acquisition of more knowledge-intensive industries. Malaysia is also attempting to lessen its dependence on its state oil producer—Petronas, which supplies 40 percent of the government revenue.

Pinang, with its 1.6 million inhabitants, has become Malaysia's "Silicon Island," attracting both foreign and domestic industrial concerns. Chinese make up 42 percent and Malays comprise 40 percent of the population. Pinang is part of an economic "Growth Triangle" that incorporates Sumatra and Thailand.

The Proton Saga

The Proton Saga automobile company is one of several government-owned public enterprises managed by Malays. Its assembly plant is in Shah Alam. A

Figure 16-9

This public housing project in Kuala Lumpur is in Chinatown. Photograph courtesy of B. A. Weightman.

joint venture with Japan's Mitsubishi, Proton Saga produces a car replete with the star and crescent symbols of Islam. Originally, there was only one model; now there are several with different names. One new model is the Proton Iswara, named after a native butterfly.

The first exports of Protons were to Singapore and the United Kingdom. In 2008, a successor was released for export to China, India, Australia, and other countries in Southeast Asia. In 2009, an electric Proton was demonstrated at a car fair.

In 1994, another car called the Kancil (after a Malaysian mouse-deer) was introduced. Small and fuel efficient, the Kancil is often bought by beginning drivers and is used by driving schools. Proton owners often buy one as a second car. The Kancil is sold as the Perodua Viva in Indonesia and as the Daihatsu Ceria in Great Britain.

DEFORESTATION AND DEVASTATION

Deforestation is one of Malaysia's most pressing environmental problems. From its origins in early swidden agriculture and colonial plantation, mining, and timber operations, deforestation has reached a zenith with land development schemes and logging over the past two

Figure 16-10

This is a typical kampong house. Usually built of wood with a tin roof, they are often raised above the ground. This provides air circulation, keeps the floor dry in the monsoon, discourages snakes and other unwanted beasts from coming into the house, and can be used for storage. Curtains, while providing privacy, allow breezes to flow through the house in the hot and humid climate. Photograph courtesy of B. A. Weightman.

Figure 16-11

I was on my way to the Mulu bat and swallow caves when I saw this boat loaded with logs sailing toward a coastal lumber mill. The river water was dark, reddish brown from the tropical soils and debris that are continually washing into the river as the region is deforested and eroded. Photograph courtesy of B. A. Weightman.

decades. Although devastated, forest cover is greater in East Malaysia than in West Malaysia.

One-quarter of the deforestation is attributable to logging. I recall driving from Kuching, Sarawak, to Brunei along a beautiful, modern road, lined with barren hillsides, stacks of enormous logs, and fires burning off debris. The route is punctuated by Malay *kampungs*, squeezed on the edges of coastal towns that are dominated by Chinese shops and fronted with docks piled high with logs and cut lumber. Rivers run brown with eroded material from clear-cut plots deep in the once magnificent rain forest (Figure 16-11). I thought of the polluted and debris-ridden water, the damage to fish, and the resultant flash floods. All along the route, signs announced the coming of palm oil, pepper, or other estate enterprises, along with new housing and industrial sites.

As I looked upon government-built, tin-roofed long houses lining parts of the highway, I thought of the once-proud Dayak hunters and gatherers whose lives have been disrupted and homes displaced (Figure 16-12). I also thought of the Penan, another group of hunters and gatherers who unsuccessfully protested the removal of their forest home. Many of these displaced people now work for the logging operations or on plantations.

Figure 16-12

An Iban longhouse along the main highway in Sarawak. As deforestation and modernization ensue, many Iban are forced to seek work in the coastal towns. Although they retain their nuclear family structures and unified housing, the influx of income enables their acquisition of such luxuries as metal roofing and televisions. Photograph courtesy of B. A. Weightman.

A striking note is that women play a leading role in direct action protests. While most men are forced into the cash economy, women are left behind to care for villages and families. Women often work part-time in local industries but still must maintain their subsistence agricultural plots. With their diverse experiences and interests, women are driven to participate in protest events against the logging companies.

A sign of the times is a cultural center near Kuching, which has model villages of all the major tribes of the region. This has been developed for tourists. If you want to see the old ways of doing things, you can go there. Traditional ways of life in Malaysia are sadly disappearing in the wake of development. Both environments and peoples are being devastated.

Brunei: Micro-State

Brunei Darussalam—the country's official name—means "Abode of Peace." Territorially, Brunei is about the size of Luxembourg, a mere 3,577 square miles (5,769 km^2). In 1981, the government purchased a cattle ranch in Australia's Northern Territory. The ranch is larger than Brunei. Brunei is a fragmented state, and people go from one part to the other by boats referred to as "flying coffins." These long-boats are totally enclosed. Passengers are entertained by programs of *Kung Fu* genre films.

In the early 1900s, Brunei was known for its exports of *cutch* (or gambier), a dye produced from mangrove trees and used for dying leather. Today, the country is known for its oil, with Royal Dutch Shell as the major company. Oil accounts for half of Brunei's GDP and 90 percent of its ports. So much wealth derives from oil that Bruneians are the wealthiest people in Asia after the Japanese. Brunei is sometimes called the "Shellfare State," as its citizens do not pay income tax, and enjoy free education, medical care, and old-age pensions. Bruneians seldom do manual labor; this is done by Filipino, Thai, and Indonesian immigrants. More than half of Brunei citizens hold government jobs. Most Malays are well-housed, although there is a long-lived stilt village in the primate city capital of Bandar Seri Begawan (Figure 16-13). This more traditional village has electricity and running water.

The Sultan of Brunei is one of the world's richest men, reputedly earning $100 a minute from oil. He lives an extravagant lifestyle and has no problem reconciling his lavish spending and partying with the strictures of Islam. Rock star concerts and alcohol consumption in the palace are generally ignored by the populace. After all, "he is the Sultan." He also has a team of polo horses outfitted at-the-ready every morning whether he is going to play or not. To ensure his safety, the Sultan maintains a contingent of (British) Gurkha troops in addition to his own army.

Figure 16-13

Note the contrast between Kampung Ayer and the Omar Ali Saifuddin Mosque in Bandar Seri Begawan, the capital of Brunei. Kampung Ayer is 400 years old and is the world's largest stilt village, housing around 30,000 people, mostly Malays. In 1987, it was declared a national monument and remains Brunei's most popular tourist attraction. Photograph courtesy of B. A. Weightman.

As of the 2001 census, Malays made up 66 percent of the total population of Brunei. Most of the rest are Chinese and a handful of indigenous groups. About 11 percent are Asian and Western expatriates (primarily British and Australian) connected to the gas and oil industries.

Aside from oil, other primary sectors are not well developed. Brunei is one part of the island of Borneo where most of the forests remain intact. Oil and gas are concentrated at Seria and Kuala Belait. Production began at an onshore field at Seria in 1932. Now most production is from offshore fields that continue to be explored by Total and Shell. Reserves of oil and natural gas are expected to last for 40 years. There is little diversification of the oil industry. Downstream processing takes place in facilities at Miri and Bintulu in Sarawak. In terms of patterns of development, Brunei has emerged as a tertiary-based economy with a small primary sector and no secondary sector.

Brunei must import about three-quarters of its food needs. Modern grocery stores are dominated by *halal* foods—foods that are processed according to Islamic law—but the main ones have small "non-*halal*" sections for the benefit of the expatriate community. McDonald's and other fast-food chains are prominent in Bandar.

Singapore: The "Intelligent Island"

Geography is central to Singapore's evolution. This 240-square-mile (600 km^2) island, holding close to 5 million people, is situated at the tip of the Malay Peninsula, separated from the mainland by a narrow stretch of shallow water. It is a pivotal, highly developed island in the Strait of Malacca (Figure 16-14). This island city-state's success is contingent on its strategic location, as well as its relationships with its closest neighbors—Malaysia and Indonesia.

ORIGINS

The British sought strategic ports to counter Dutch power in insular Southeast Asia. They acquired Pinang in 1786, and, in 1819 the British flag was hoisted over Singapore by Sir Stamford Raffles. At that time Singapore, with no more than a thousand people, was a small fishing village known as Temasek (Sea Town) and later as Singapura (Lion City). Control of Singapore meant that the British East India Company (BEIC) controlled both the northern and southern entries to the Strait of Malacca.

By the 1830s, Singapore had become the major trading port in Southeast Asia. It was challenged by both Manila and Batavia (Jakarta), but the island had certain advantages. Most ships trading between China, India, and Europe had to pass Singapore. Unlike its competitors, Singapore was a free port. Furthermore, it was enmeshed in the powerful British commercial and industrial empire.

Tin played a major role in Singapore's development. Tin miners in Malaya and Thailand imported their supplies from Singapore and used the port to ship their tin to the world. The island attracted traders of all kinds: British, Indians, Malays, Arabs, and Chinese.

The Chinese were the labor force upon which Singapore was built, and the city was the conduit for thousands of Chinese laborers going to Malaya and the Netherlands East Indies. Most Chinese came to Singapore as poor indentured laborers hoping to make a fortune and send money to their ancestral home. The forced opening of the Treaty Ports in China and the annexation of Hong Kong in 1842 (Chapter 10) accelerated the migration of Chinese—not only to Singapore, but worldwide. Singapore remains a predominantly Chinese country with a 78 percent Chinese population.

By the late nineteenth century, Singapore had become a major transshipment point as well as a commercial and financial center. Here, products from Southeast Asia, such as tin and rubber, were received, packaged, and re-exported. The largest commercial firms were British. Growing numbers of Chinese-owned enterprises were linked to the business web of the **Chinese Diaspora**. There was little manufacturing in Singapore before 1960; three-quarters of the population were involved in the service sector.

There was also a significant Indian contingent, fluctuating from 6 to 12 percent of the population. The minority was large enough to form its own community known as "Little India." The Indian community was far from united, however. There were regional, religious, and caste divisions. During colonial times, thousands were imported from India as convict laborers to build Singapore's infrastructure. Free Indians were primarily merchants or worked in public employment as teachers, policemen, and the like. There were strong links between the Indian communities of Singapore and Malaya. By the time of World War II, Singapore was a multiracial, multireligious, multilingual society governed by a British elite.

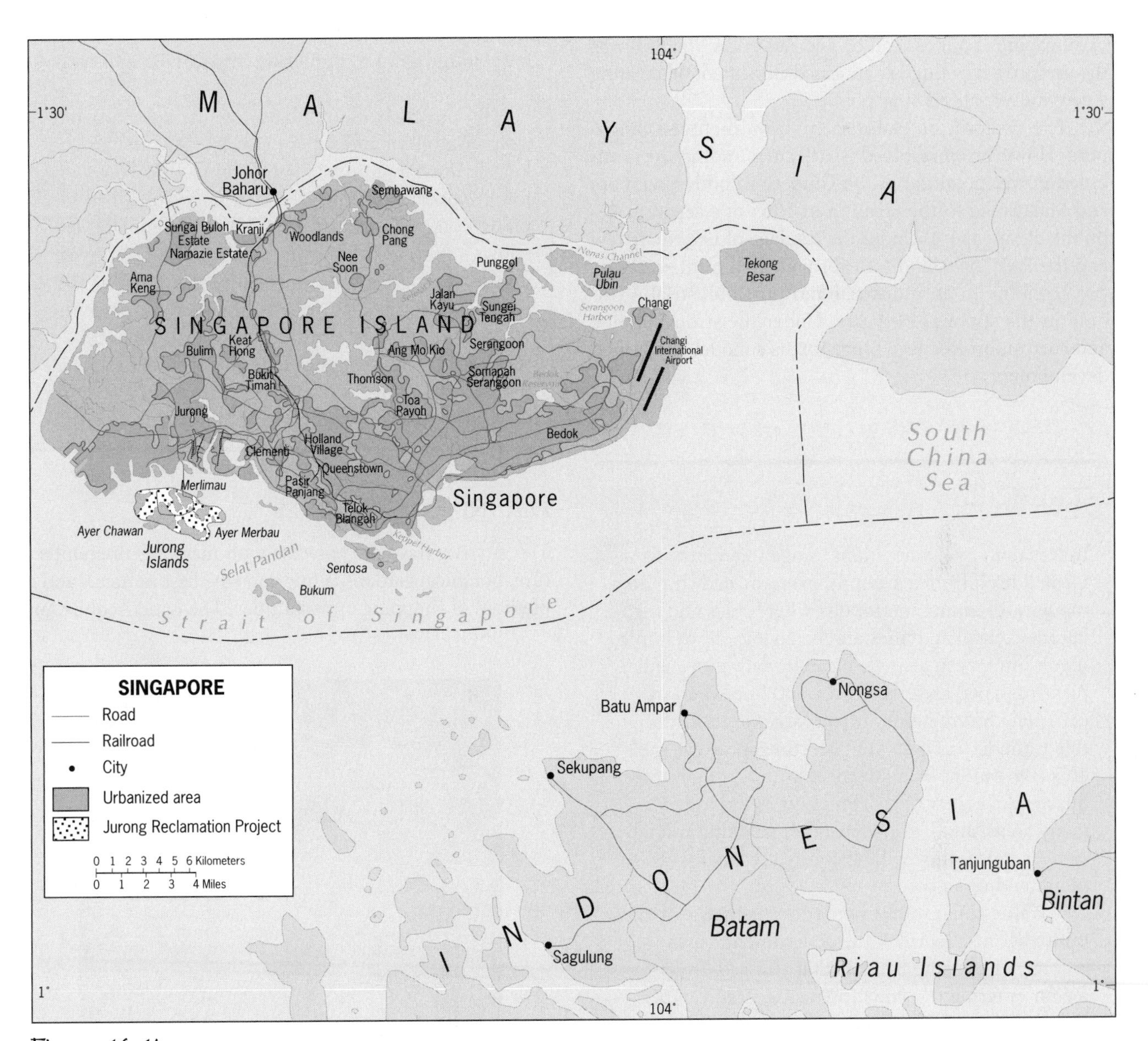

Figure 16-14

Singapore. Note how the urbanized area is oriented toward the sea. From H. J. de Blij and P. O. Muller, *Geography: Realms, Regions and Concepts*, 14th edition, p. 563. Originally rendered in color. © H. J. de Blij and P. O. Muller. Reprinted with permission of John Wiley & Sons, Inc.

Sea-oriented Singapore was ill-prepared for a land invasion and fell to the Japanese in February 1942. All communities suffered greatly at the hands of the Japanese during World War II, a fact that has contributed to a legacy of distrust of Japan.

After the war, the British reoccupied the island. Although its vision for Malaya was independence, it looked upon Singapore quite differently. The British feared the Chinese majority in light of the Communist takeover in China in 1949 and the insurgence of communism in the region. It was believed that an independent Singapore would come under Communist control.

Limited self-government was introduced in 1955. In 1959, the People's Action Party (PAP) began its dominance of Singapore's politics. Led by Lee Kuan Yew, the PAP was the party of a new English-educated elite intent on creating a new, industrialized Singapore.

The creation of Malaysia in 1963 included Singapore. However, unresolved issues surrounding the combined power potential of the Chinese in both Singapore and Malaya led to the creation in 1965 of a separate Republic of Singapore. Under the tutelage of Lee Kuan Yew and the PAP, the island transformed itself from an entrepôt economy to an industrial and post-industrial economy in the space of 30 years. Under the strong, stable, and corruption-free PAP, Singapore is a model of planned development.

The Rules

In less than two generations, Singapore has transformed itself from a swampy, overcrowded slum to a modern, gleaming metropolis where rules and regulations control activities and behavior. For example, if you litter, even a single cigarette butt or a small piece of paper, there is a US$1,000 fine. Vendors can't sell chewing gum. If you forget to flush the toilet, there is a US$500 fine. A Singapore car can't cross the border from Malaysia with less than three-quarters of a tank of gas. Smoking is forbidden in many areas, and if you are caught smoking Indonesian cigarettes they will be confiscated and you might end up in court.

Traffic congestion in the central city is at a minimum because in 1975, the government began to enforce a Licensing Scheme that charged drivers a fee for entering the downtown area. Traffic experienced a 45 percent reduction and vehicle crashes fell 25 percent. In 1998, an Electronic Road Pricing System was installed. Required transponders were installed in vehicles free of charge. Consequently, traffic was reduced by an additional 15 percent. This has helped Singapore to maintain ideal traffic speeds of 30–40 mph (46–64 km) per hour on expressways. Moreover, 65 percent of commuters use the elaborate network of public transportation.

The air quality in Singapore meets the health standards of the Environmental Protection Agency of the United States. For instance, reduced traffic in the central business district has led to an 176,400 pound (80,000 kg) reduction in CO_2 emissions and a 22 pound (48 kg) reduction in particulate matter (soot).

Buses and taxis are ubiquitous, and the city has a super-modern and super-clean subway system. People must line up neatly to avoid a fine. Also, food, especially durian (a very smelly fruit), is forbidden on public transport. There are signs in the subway stations to this effect (Figure 16-15). If you decide to ignore a rule, you will be chastised by the people around you or arrested if police are present.

Singapore is recognized as the cleanest and most orderly city in Asia. This achievement largely derives from Singaporeans' acceptance of behavioral rules and great pride in their pristine environment.

PLANNING OFFENSIVES

The city is a traveler's paradise with magnificent architecture, botanical gardens, a fine zoo (the best in Asia), and a plethora of shopping opportunities. The Land Acquisition

Figure 16-15

This is only one of multiple signs in a Singapore Metro station. Rules of behavior are posted everywhere in the city. Photograph courtesy of B. A. Weightman.

Act of 1967 allowed the government to acquire land anywhere, making serious urban planning possible. Before then, the city was plagued with a soaring population and housing shortages. The PAP set up a Housing Development Board (HDB), which cleared slums and renewed or built new housing along the Singapore River. Hundreds of Chinese shops (with living quarters above) gave way to multi-unit apartment blocks. In the case of older housing, residents paid only 10 to 20 percent of the cost of refurbishing. The government paid the rest.

The HDB then proceeded to build new towns, complete with high-rise apartment blocks, town centers, markets, schools, and light industries to provide jobs. With government loans, apartments (flats) are sold far below free market prices and now house about 90 percent of the city's population.

Population planning is also part of the grand plan. Prior to the 1980s, strict family size controls and incentives were introduced, with a resultant drop in fertility rates, especially among the Chinese majority. Suddenly, the government realized that there were not going to be enough workers for the growing economy. In 1987, new policies were introduced encouraging people to have more children. Nevertheless, Singapore's TFR of 1.3 is one of the lowest in Asia. While Malays and Indians have higher fertility rates, the Chinese remain the majority at 75 percent of the population. Indians comprise 9 percent and the Malays 14 percent. The remaining 2 percent are Europeans and other non-Asians.

Many more plans are in the making, especially for acquiring water. Singapore has developed what is called a "Four Tap" policy on water. This includes:

- Imported Water: For decades, Singapore has imported water from Malaysia, which has accounted for half of the city's supply. As of 2009, this dependence had been reduced to 40 percent. However, the two water agreements that bring this water to Singapore are set to expire by 2011 and 2061. Currently, the two countries are engaged in a debate over the price of water.
- Rainfall: Singapore's first reservoir was built by the British in 1867. Since then, numerous reservoirs and catchment areas have been installed. By 2001, there were more than 40 reservoirs, treatment works and water catchment areas. Two additional reservoirs are under construction.
- Recycled Water: NEWater is the brand name given to reclaimed water produced by the city's public utilities. This is, in fact, purified sewage. Currently, there are five NEWater factories that can meet 30 percent of Singapore's water needs.
- Desalinization: In 2005, Singapore opened Sing Spring—its first desalination plant. The plant can produce 30 million gallons of water (113,562 m^3) a day. The plant also produces bottled water known as *Desal H2O*. Sing Spring is one of the largest such enterprises in the world and is meeting 10 percent of the city's water requirements.

CYBERCITY

Already one of the world's largest hubs for shipping and commerce, this economic tiger is staking its future on becoming a global nerve center for media and communications. The National Computer Board's report titled "IT2000: Vision of an Intelligent Island" proposes an infrastructure that includes both a fiber-optic network and the information services that require this form of digital communication. Singapore is second to Japan in terms of the number of computers it manufactures. In addition, the island makes more than half of the world's hard disc drives and almost all the sound cards for personal computers.

A new emphasis is being placed on biomedicine. The island wants to make the biomedical sciences, which include the research and development of pharmaceuticals, biotechnology, medical equipment, and health-care products and services, the "fourth pillar" of its economy. The other three pillars are electronics, engineering, and chemicals.

Several pharmaceutical companies such as GlaxoSmithKline and Pfizer and Merck & Co. have opted to locate in Singapore. In 2006, GlaxoSmithKline announced that it was building another plant to produce pediatric vaccines—the first such facility in Asia.

Singapore's largely corruption-free government, skilled workforce, and advanced and efficient infrastructure have attracted investments from more than 3,000 multinational corporations (MNCs) from the United States, Japan, and Europe. Foreign firms are found in virtually every sector of the economy, and account for more than two-thirds of manufacturing output and direct export sales.

More than 7,000 MNCs operate out of Singapore, employing 60 percent of the island's workforce and producing 80 percent of its exports. Companies regard Singapore's infrastructure as superb. It has the world's busiest container port, busier than even Hong Kong or Rotterdam, state-of-the-art utilities, and one of the world's most advanced telecommunication networks. It should be noted that Shanghai's port recently surpassed Singapore's in terms of tonnage handled and Singapore was devastated. There is a word that sums up the Singaporean existential condition and that is *kiasu*—"afraid to lose."

CREATING A HINTERLAND

Building an external economy is considered a national imperative. Foreigners, especially Indonesians, already make up 20 percent of the labor force, and an overseas push is the only place for expansion. Many companies have already relocated in such countries as China, India, Thailand, Vietnam, Russia, Hungary, and Belgium. China is Singapore's top location for investment. For instance, 22 Singapore companies are constructing a huge industrial park near Suzhou (west of Shanghai). The rapidly growing economy of India, especially the high-technology sector, is also becoming an expanding source of foreign investment.

In 2007, 46 percent of Singapore's investment was in Asian countries, primarily in China, Malaysia, Indonesia, and Thailand. The rest is spread around the world in regions including Latin America, North America, Australia, and the EU. Most investments are in the financial, insurance, and manufacturing sectors. Cheaper labor in some of these regions is not the only reason for these investments. Singapore is hoping to tap into large potential markets as well. The country has made free-trade agreements with 11 countries plus several more in the EU.

Figure 16-16

Crocodiles, snakes, and other reptiles are raised in Singapore. Their meat is eaten and their skins are used to make fancy boots, belts, and handbags. Photograph courtesy of B. A. Weightman.

Singapore's goal is for greater interdependence with its neighbors, especially with Malaysia and Indonesia. Johor at the tip of the adjacent Malay Peninsula (The Iskander Development Region) and Indonesia's Riau Islands, linking Singapore to Sumatra, have been integrated as the **Sijori Growth Triangle**. "Sijori" refers to **Si**ngapore, **Jo**hor, and **Ri**au. This combines the competitive strength of the three areas to make the subregion more attractive to regional and international investors. It links the infrastructure, capital, and expertise of Singapore with the natural and labor resources and the abundance of land of Johor and Riau.

Riau's Bantam island is being developed with industrial estates while Bulan island is being converted into agribusiness properties on which pigs, chickens, shrimp, and orchids are raised. Pig production supplies more than 50 percent of Singapore's pork and the chickens are destined for Kentucky Fried Chicken outlets. Crocodiles and snakes are also raised for luxury clothing items (Figure 16-16).The newest frontier is the island of Bintan, which is being developed as a tourist mecca.

Singapore's growth triangle is neither complete nor proven entirely successful. While these areas are not yet fully integrated, it is still argued that there is more to be gained from cooperation than from competition. This growth triangle is still in the making. The lack of formalization is partly because the growth triangle means different things to each of the participants. According to geographers Scott Macleod and T. G. McGee (1996), "In many ways the strength of the Growth Triangle as a concept lies in its vagueness; it can mean all things to all sides."

FINDING AN IDENTITY

One way in which Singapore is forging a sense of place is through heritage tourism. Certain districts of the city are being refurbished to display their cultural origins. "Little India" and "Chinatown" are examples of this phenomenon. However, localism is combined with globalism, as many buildings now host international business chains

Figure 16-17

This former Chinese shop-house along the Singapore River has been renovated and turned into a trendy Peranakan restaurant. It is only one of many restaurants along the waterfront.
Photograph courtesy of B. A. Weightman.

such as upscale shops and restaurants (Figure 16-17). In advertising, the city is being portrayed as having, "its head in the future and its soul in the past." As geographer T. C. Chang (1999) points out, "On the conceptual front, the power relations between the 'global' and the 'local' in Singapore demonstrate that 'globalization' and 'localization' are mutually constitutive." These relations are revealed in the city's carefully orchestrated heritage landscapes.

Language, as the context for thought, is vital to forging identity. In the 1960s, the PAP stressed the importance of Malay and English. As Singapore's prosperity grew and the economy became more internationalized, the PAP shifted its emphasis to Mandarin and English at the expense of Malay and other Chinese dialects. Today, there are four official languages: English, Malay, Mandarin, and Tamil. English is the language of education and government. Singapore is the only country in Asia to have English as an official language.

Singlish

Singlish is a bewildering mixture of English, Mandarin, Chinese dialects, and Malay. There is concern that the rise of this dialect will demote the official languages, which are intended to internationalize the state. The artistic community has embraced the dialect enthusiastically, and Singlish is even popularized on television shows. An example is: "Don't pray pray *lah*," which means "Don't fool around." Many people feel that Singlish is their own personal language for everyday communication and "*So shiok*" ("That's great"). New words and expressions are invented almost on a daily basis, fueled by the Internet experience. One of the latest terms is "You blur like *sotang lah*." This translates as, "You're dumber than squid man." As the debate over Singlish and English rages, it is clear that Singlish is the language spoken on the street.

In the twenty-first century Singapore will be among the few prosperous countries in the world. Its geographic situation may lessen in importance as it becomes enmeshed in the global economy's computer revolution. In regional terms, Singapore regards itself as the economic engine of the region. The growth triangle is evidence of this view. Nevertheless, Singapore worries about its image and is constantly advocating campaigns extolling what it means to be Singaporean. Its 1990s drive to reveal historic aspects is only part of this continuing search for national identity. However the future is charted, it will be in the context of "Okay can!"

Indonesia: Integration Versus Disintegration

With more than 17,000 islands, Indonesia is the world's largest archipelago (Figure 16-18). With a population of more than 240 million, it is the fourth most populous

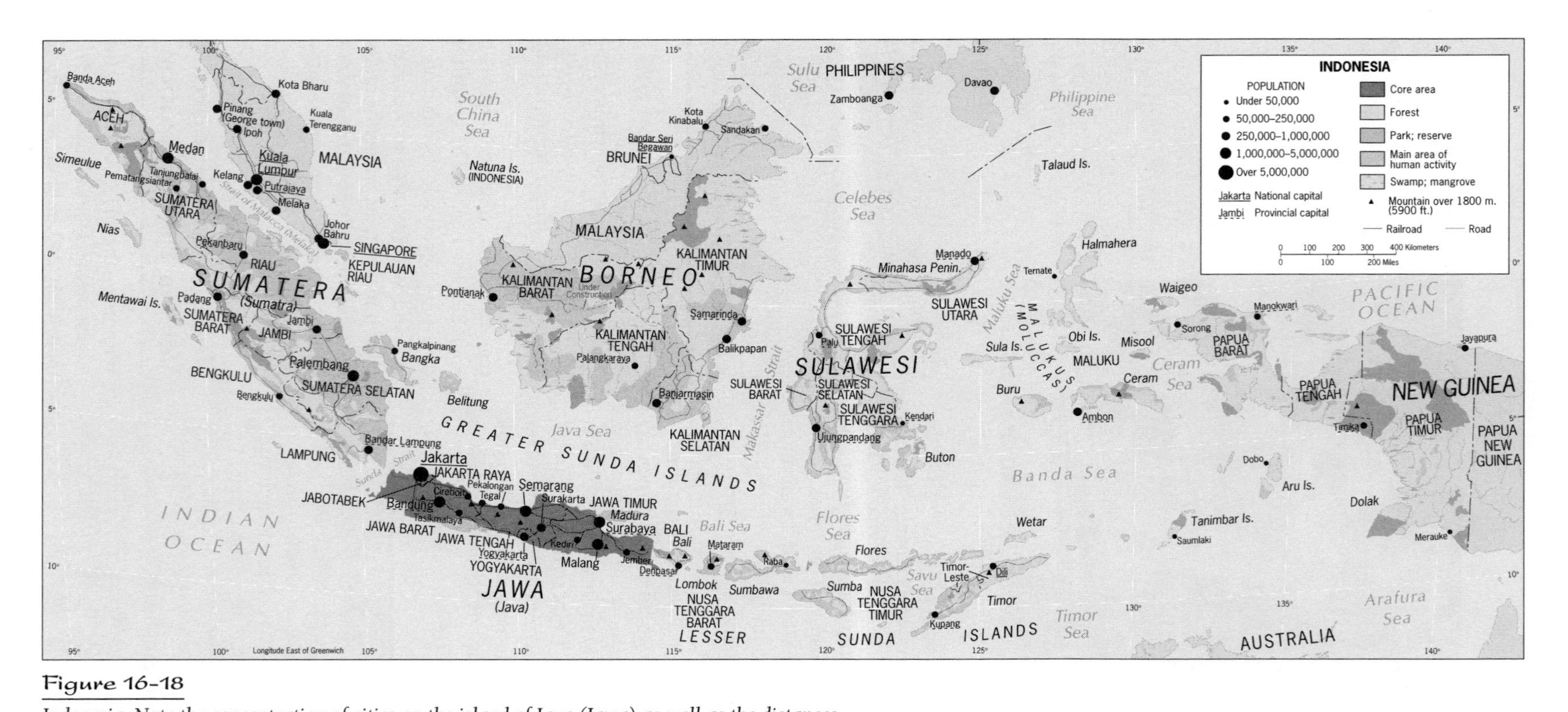

Figure 16-18

Indonesia. Note the concentration of cities on the island of Java (Jawa) as well as the distances between Java and the outer islands. From H. J. de Blij and P. O. Muller,. *Geography: Realms, Regions and Concepts*, 14th edition, 2010, pp. 558–559. Originally rendered in color. © H. J. de Blij and P. O. Muller. Reprinted with permission of John Wiley & Sons, Inc.

country in the world. There are more than 350 ethnic groups speaking an estimated 583 languages and dialects. Indonesia also has the world's largest Muslim population.

Indonesia's geography is just as varied. There are vast expanses of tropical swamps in Kalimantan (part of the island of Borneo) and Sumatra and glaciers in the central mountains of Papua (the western part of the island of New Guinea formerly known as Irian Jaya). There are more than 300 volcanoes, 200 of which have been active in historical time. Indonesia's motto *Bhinneka Tunggal Ika*—"Unity in Diversity"—is fitting indeed.

Because of Indonesia's cultural diversity, most people first develop a regional identity, only learning the national language *Bahasa Indonesia* when they attend school, and with it a national identity. The Indonesian government is fully aware that national unity and national cultural identity must be created and sustained.

Indonesia can be divided into the inner islands and outer islands. The inner islands are Java (Jawa), Bali, and Madura. The outer islands are the rest, such as Sumatra (Sumatera), Kalimantan, Sulawesi, and the islands of Maluku (Figure 16-19). The political, historical, and economic heart of the country is the islands of Java and Bali. These two islands contain 67 percent of the population but only 7 percent of the land area.

Figure 16-19

This woman is a Dayak who lives in the interior of eastern Kalimantan. The Dayaks were once head-hunters. Their traditional way of life in the jungle is rapidly changing with resource exploitation, deforestation, influx of migrants, and government impress from Java. Note the filter cigarette. Clove cigarettes are commonly smoked in Indonesia. Photograph courtesy of B. A. Weightman.

THE EARLY YEARS

As you already know, Southeast Asia lies across the great trading routes from India to China. Chinese, Indian, and Arab traders were common sights in the ports that dotted the region.

In the Indonesian archipelago, there were two types of states. First were the coastal states located at river mouths with their secure harbors, which were dependent on regional and international trade. Second were the inland states. These were based on agricultural production derived from the rich volcanic soils and alluvial plains. The most prominent states were in central and eastern Java and Bali.

The earliest kingdoms were Hindu/Buddhist states. The religions were adopted by local rulers who were attracted by religious and philosophical ideas as well as court rituals. Yogyakarta is the center of this history and culture. Within 48 miles (30 km) of this city are the great monuments of Borobudur (Buddhist) and Prambanan (Hindu) (Figures 14-7 and 14-8). The island of Bali remains predominantly Hindu.

Founded in 650 AD, Sri Vijaya was one of several sea-oriented states. Its capital was at Palembang in southeastern Sumatra, and its location gave it access to the region's two strategic waterways: the Strait of Malacca and the Sunda Strait. Palembang provided an excellent port for ships riding the monsoon winds from India to trade with China. The city became a center of Buddhist learning and had more than 1,000 Buddhist monks in its monasteries. Sri Vijaya expanded its territory so that it could monopolize trade in the region. At its peak, the empire controlled parts of southern Thailand, peninsular Malaysia, western Java, and parts of western Borneo. Sri Vijaya ruled the region's sea lanes and trade until it was defeated by the Javanese in 1290.

Muslim traders arrived as early as the seventh century but the Islamicization of Indonesia began in the thirteenth century with the conversion of the ruler of Aceh in northwestern Sumatra. The nature of Islam in Indonesia varies greatly, ranging from strict adherence among the Acehnese to elsewhere where people have a more relaxed attitude to the faith.

When the Europeans arrived in the middle of the sixteenth century, they found a series of well-established states across the archipelago. The major ones were at Aceh, Solo (Surakarta), and Yogyakarta in central Java, Bali, the Moluccas (Maluku), and Makassar (Ujungpandang). There was a constant flow of people and goods across the region. Traders used Malay as their means of communication. The Indonesian and Malay languages are similar in the same way that British and American English are similar.

THE NETHERLANDS EAST INDIES

The Netherlands United East India Company (VOC) began to trade with local kingdoms in the early seventeenth century. In 1619, they attacked Jayakarta (Jakarta), then a major fort in the Javan kingdom of Banten. Jayakarta was renamed Batavia. It remained the capital of the Netherlands East Indies until independence in 1945, when it was renamed Jakarta.

Total control of Java was not achieved until 1756 after the bloody Java War whereby the Dutch defeated the Yogyakarta sultanate. Several years later, in 1796, the Dutch Crown took over from the corrupt Company and ruled what by then was the largest state in the archipelago. The Dutch extended their control gradually through the nineteenth century. Sumatra and eastern Indonesia came under Dutch rule and, with the destruction of the Balinese kingdom in 1905 and Aceh in 1911, the colony of the Netherlands East Indies was complete.

The Dutch maintained a centralized state. Power was vested in Batavia, and the colonial government kept a close eye on religious leaders. Any stirrings of resistance were quickly quashed. The Dutch promoted a Western, secular elite by educating the children of the pre-colonial elites and making every effort to prevent the formation of an Islamic elite.

Regional trading networks were gradually destroyed. External trade became the preserve of European companies and internal trade became dominated by the Chinese, who were encouraged to migrate from southern China.

Javanese agriculture was transformed by the Dutch cultivation (culture) system (Chapter 14). Islanders were compelled to produce designated crops, and by the end of the nineteenth century Java was the world's largest sugar producer. Village Java's subsistence economies faded to commercialization, and cities and towns emerged to handle the export trade. By the beginning of the twentieth century, most Javanese were tenant farmers, sharecroppers, or wage laborers.

In Sumatra, vast areas of virgin forest were removed to make room for tobacco and rubber plantations. When oil was discovered in the 1920s, it became the nucleus of what would become the Royal Dutch Shell Oil Company. Much of the labor was Chinese, and although they never made up a large portion of the colony's population, they became dominant in local trade and urban commerce.

The economic transformation of Indonesia led to increased urbanization, and by 1910 Java's cities were unable to cope with the flow of rural-to-urban migration. Urban places were crowded, without piped water or sanitation systems, and people were suffering from malaria and other waterborne diseases. Living conditions for most Indonesians worsened into the 1970s.

The Dutch introduced Western education to provide a needed pool of skilled labor, but entry to these schools was very difficult for ordinary Indonesians. The best schools used Dutch as the medium of instruction. Graduates entered the professions or even attended universities in the Netherlands. Most who benefited were children of the elite or government officials. By the end of the Dutch colonial era, literacy rates were lower than in any other European colony in Asia, with the exception of the Portuguese colony of East Timor (Timor-Leste).

NATIONALISM AND SUKARNO

The term *Indonesia* was first used in the early 1920s. By 1928 there were enough educated youth with determination to create a modern Indonesia free from Dutch rule. Although several political parties emerged, the most prominent nationalist was **Sukarno**, who charmed the masses with his brilliant oratory. (Many Indonesians go by only one name.) Sukarno succeeded in spreading the idea of freedom across all levels of society. His most important contribution was the popularization of nationalist ideology. An important theme was that people should set aside their ethnic and religious differences to unite in opposition to colonial rule. It was generally agreed that Indonesia should be a secular state. However, none of this pleased the Dutch, who expelled or jailed many nationalist leaders including Sukarno.

THE JAPANESE AND INDEPENDENCE

The Japanese occupied Indonesia in March 1942 and quickly alienated the Indonesians by requiring all males to work for the war effort and treating people with disdain and brutality. However, they removed the Dutch from administrative functions and replaced them with Japanese and Indonesians. Thousands of Dutch were thrown into concentration camps where many perished from torture, disease, and starvation.

Certain Japanese policies eventually served the cause of freedom for Indonesia. The Japanese prohibited the use of the Dutch language and encouraged the use of Japanese and Indonesian. Various military training programs were initiated for young people. Nationalist leaders, including Sukarno, were freed from jail. As the Japanese realized they were losing the war, they promoted moves toward independence. The Republic of Indonesia was declared on August 17, 1945, in Jakarta. The Dutch rejected this notion and a four-year guerrilla and diplomatic war ensued. An agreement was reached to end Dutch colonial rule in 1949.

AFTER THE REVOLUTION

The revolution witnessed the rise of a strong army intent on taking the initiative in leading the country to national unity and modernization. The army continues to play a strong role in Indonesian politics to this day. It insists on its *dwifungsi*, "dual function," to defend the nation from external threat and internal subversion and to promulgate economic development.

Despite the perceived power of the army, it did not come into full power until after 1965. Numerous political parties vied for control of Indonesia, and by 1960 competition was intense. Many were concerned about the influence of the Indonesian Communist Party (PKI). Moreover, in 1965, there were rumors that Sukarno was terminally ill and that the PKI was preparing a coup. On September 30, a group of army officers overthrew the government and took power under the leadership of General Suharto, who proclaimed a government of the "New Order."

Over the next six months, the army vigorously sought out PKI members. At least 400,000 were killed, mostly in Java and Bali. Many of the victims were Chinese. The PKI was destroyed.

Pancasila

With the demise of the PKI, there arose a fear of the power of Islam. The consensus political system has embraced what is called ***Pancasila*** democracy. *Pancasila* is the five principles adopted by Sukarno in 1945 as the basis for Indonesian ideology and development: belief in one God, national unity, humanitarianism, democracy based on consensus and representation, and social justice. Its vagueness allows for many interpretations. All students, members of the civil service, and the armed forces must pass exams based on the principles of *Pancasila*.

Despite its great mineral, timber, and agricultural wealth, Indonesia remains a relatively poor country with a GDP per capita of US$4,000 which ranks it 155th in the world. Eighteen percent of Indonesians live below the official poverty line. An estimated 8 percent of its workers were unemployed in 2009. Millions are underemployed.

The mid-1997 financial crisis dealt the country a devastating blow, with Indonesia's currency falling in value lower than in any other Asian country. This economic crisis generated political turmoil that saw the fall of General Suharto in 1998. This was followed by a period of *reformasi*, "reform" that encouraged liberalization. Elections were held in 1999 and a new and more liberal government was put into place. Also exacerbated by the economic turmoil were Indonesian-Chinese tensions and criticisms regarding the country's population distribution and transmigration program. After 1997, thousands of Chinese fled the country, taking more than one billion dollars of investment with them. Peaceful elections were held in 2004, with a new president—Susilo Yudhoyono—defeating Sukarno's daughter, Megawati Sukarnoputri.

Indonesia has weathered the current global financial crisis quite well because of its reliance on domestic consumption as the driver of economic growth. The government is using fiscal stimulus to counter the effects of the crisis and delivers cash transfers to poor families. However, the country still struggles with poverty and unemployment, inadequate infrastructure, corruption, a complex regulatory environment, and unbalanced resource distribution among regions.

Figure 16-20

These residents of Jakarta have one and two children, respectively. Smaller family size reflects the success of family planning in Indonesia, as well as modernizing trends in terms of the status of women and growing preference for nuclear families. Photograph courtesy of B. A. Weightman.

POPULATION DYNAMICS

In the case of Indonesia, economic growth has far outpaced population growth. Women of the 1990s had only half as many children as women did in the 1960s. In part, this is due to the changing status of women and the decline of the extended family (Figure 16-20). However, strong family planning programs have been in place since 1971.

The program was started in Java and Bali and was progressively extended to the larger outer islands, where it gained support from local governments as well as Muslim leaders. The program involved hospitals and family planning services, a battery of field workers, and village leaders. It focused on both reducing family size and improving quality of life.

In the last quarter century, there has been nearly a 50 percent decline in fertility. Mortality levels have also been reduced, a function of education, improved access to health care, and better child nutrition. Life expectancy has risen but still is not at the same level as in many other Asian countries such as China, Malaysia, and Vietnam. Maternal mortality remains a problem. Furthermore, there has been an increase in chronic and degenerative diseases such as cancer and heart disease.

Water and sanitation remain problematic in Indonesia. More than 100 million people lack access to safe water and more than 70 percent rely on water from potentially contaminated sources. About 30 percent of water delivered by water companies is contaminated with *E. coli* or fecal coliform and other pathogens.

Population access to improved sanitation in rural areas has remained stagnant at around 38 percent. More than 40 percent of rural households use unsanitary open pits or defecate in fields or water bodies. "Toilet" shacks with simple holes on the floor that hang over or float on rivers are common sights. Urban sanitation is the least well addressed, with disposal and treatment of sewage available for less than 2 percent of the population.

Transmigrasi

Population distribution has long been a concern of Indonesian governments. Even the Dutch had colonization programs to provide pepper estates with labor. People were moved from the crowded island of Java to work on plantations in Sumatra, for example. ***Transmigrasi*** started up again in 1969 with the first five-year plan of development. In the next three decades, some 6.5 million people were moved to try to correct population imbalances among the islands. The proportion of Indonesians living in Java declined from 66 percent at the time of independence to nearly 60 percent in 1995. This relatively small change is due to the fact that a number of people from the outer islands moved to Java. Java continues to be one of the most densely populated islands in the world.

The aims of *transmigrasi* were to relieve population pressure in crowded areas, to open up new lands, and to increase the presence of Javanese in sensitive areas such as Aceh, Papua, and Kalimantan. People with farming experience were usually selected for the program, but it was rife with problems.

Lack of funds to prepare areas and assist settlers was one problem. Another was the unsuitability of the land to be developed. For instance, attempts were made to grow wet rice on sandy soil. Some migrants to Sumatra ended up in mangrove swamps. In many cases there were cultural clashes between the immigrants and the original settlers. For example, pig farmers from Bali insulted the sensitivities of Muslims—who regard pigs as unclean—in parts of Sulawesi.

From 1997, there were violent outbreaks between tribal groups and settlers who were cutting down forests for palm oil plantings. More problems arose as illegal loggers and miners chopped down more sections of rain forest. Many migrants tried to move back to their place of origin but no longer had homes or land to return to.

In the 1930s, thousands of Madurese (from the island of Madura) moved to central Kalimantan, which is a Dayak region. Over the years, the Madurese have taken control of mining, logging, and other industries in Dayak territory. In 2001 violence broke out in a town called Sempit and 500 Madurese were killed. Many were decapitated by the former head-hunters. Some 100,000 Madurese were displaced from their homes.

The *transmigrasi* program is blamed for internecine strife and environmental degredation. Coming under much criticism, the government has scaled down the program and currently relocates only about 15,000 families a year.

URBANIZATION

The dominant population movement in Indonesia today is rural-to-urban migration. The proportion of Indonesians living in urban areas has doubled during the last 25 years. Indonesia was 43 percent urban in 2009. Moreover, many people living in rural areas actually work in urban areas. Improved transportation has fostered commuting on a daily basis. Other people spend part of the year working in the city while maintaining a family in a rural community. These are **circular migrants**. Both commuting and circular migration have increased with expanded transportation systems, advances in education, changes in the role of women, and greater industrial development.

Jakarta is Indonesia's mega-city, a sprawling metropolis of 10 million situated on the north coast of Java. Jakarta originated as a fortified port town called Sunda Kelapa. As with many primate cities in Southeast Asia, urban growth has spread into the hinterland. In recognition of this sprawl, the government began to refer to the entire urban region as **JABOTABEK** in the mid-1970s. JABOTABEK stands for **Ja**karta and the contiguous administrative units of **Bo**gor, **Ta**ngerang, and **Bek**asi. With a combined population of 23 million, JABOTABEK is the largest urban agglomeration in Southeast Asia.

JABOTABEK

This is the fastest growing area in Indonesia and is becoming a single, integrated urban region with Bandung. JABOTABEK and Bandung consist of nine administrative units. In 2010 the combined population of this region was nearly 30 million people. JABOTABEK is home to over 10 percent of Indonesia's entire population and 25 percent of its urban population. Increased spatial unification is expressed in socioeconomic corridors of activities including agriculture, residences, and retail trade.

An important part of this mega-urban region is housing development. New housing developments have been stimulated by the high cost of city-center land; the lack of affordable housing for middle-income families in the main cities; increasing employment opportunities in urban fringe areas; improved transportation facilities; and an increase in the social and economic-facilities provided by real estate developments outside the main cities.

Housing development is in the form of large-scale subdivisions. Towns associated with industrial developments are also on the drawing board. Much of this development is associated with foreign investment. Both government and private enterprises are participating in development. However, the government is losing control to private enterprises, which dominate construction projects. The result is the emergence of numerous problems.

Prime agricultural land has been lost to developers. Farmers are reluctant not to sell as demand increases and prices rise. Water supply has become critical—both groundwater and surface water are in short supply. Seawater is encroaching inland as groundwater is depleted. Mangrove habitats are giving way to shrimp farms and grand tourism schemes. Displacement of local people is a serious problem as well. While local communities are the last to be told about forthcoming development, they are the first to be removed. So far, the government has not adequately addressed these and other problems of rapid growth in the JABOTABEK region of Java.

Jakarta remains Indonesia's most important metropolitan area. It is both the national capital and the principal administrative and commercial center of the archipelago. The greater part of foreign investment goes to Jakarta and focuses primarily on manufacturing, construction, and service sectors. The city remains an important manufacturing center, but current and future growth does and will derive from tertiary and quaternary activities.

In the past decades, the central core of the city has been transformed from residential to higher-intensity commercial and office land use. There are also many new high-rise apartments there. However, Jakarta reflects the urban woes characteristic of many primate cities. There is a lack of adequate public housing. Traffic congestion, air and water pollution, and sewage disposal present additional problems. The government recognizes that these issues must be addressed as the metropolitan population continues to increase.

Figure 16-21

Here in a Dayak village in Kalimantan, a chief wears traditional garb specifically for the benefit of tourists. Tourism is an important source of Indonesia's foreign exchange. Photograph courtesy of B. A. Weightman.

ECONOMIC TRANSFORMATION

The Sukarno years were ones of stagnation. Under General Suharto, five-year plans were initiated and development ensued. Beginning in 1969, economic development actually outpaced population growth. From 1970 to the early 1980s, economic growth was sustained by global increases in oil prices. However, with the fall in oil and gas prices during the 1980s, the government was forced to shift its strategy. It proceeded to adopt a market-oriented approach.

Liberalization of controls and processes of deregulation have allowed the economy to become more diversified, and growth picked up again. Now much less reliant on its oil and gas sector, Indonesia has expanded its apparel and footwear, food processing, rubber and wood products, and chemical industries. Indonesia also has the world's largest gold mine and the second largest copper mine, both in Papua. Nevertheless, less than 20 percent of workers are employed in industry. Much of the manufacturing growth is in and around Jakarta and Surabaya. Nevertheless, in terms of numbers of firms, increases were greater in the outer islands such as Sulawesi. Tourism has fallen off dramatically since September 11, 2001 (Figure 16-21).

Unemployment is concentrated in the young, educated group. On the other hand, underemployment is estimated to be as high as 40 percent of all workers. Another important aspect not to be ignored is the informal sector, which absorbs labor and redistributes income within the larger economic scheme of things. Women dominate the informal sector—68 percent of participants.

Although agriculture has decreased as a contributor to Indonesia's GNP, rice is still an important commodity. Paddy land has expanded in most regions, but a substantial increase in rice production has been achieved through Green Revolution technology—-better hybrid seeds, more fertilizer and pesticides, and better use of irrigation—rather than with increased labor. In the 1990s, the agricultural workforce shrank to less than half the overall workforce for the first time.

Change in village life is endemic. In the traditional way, all villagers participated in rice cultivation for a share of the harvest. But with commercialization, agricultural work was put on a pay basis, which broke the patron–client relationships between landed and landless peasants.

Another change concerns the number of young women who seek employment in factories. Women are paid less than men, partly because they do not have access to jobs done by men. Moreover, they are usually less educated than men as they are expected to be home helpers until at least the age of 15. Generational shifts are apparent in context of female employment in Indonesia:

- Ages 16–30: Many women seek employment in large-scale manufacturing enterprises, often geared to export. Unmarried women with no responsibilities (assumed) who are willing to work long hours for low wages are preferred by employers.
- Ages 31–40: Often young mothers work in smaller-scale, home or locally-based manufacturing enterprises typically geared to the domestic market. Hours are regular or flexible and can be worked around child-care responsibilities.
- Ages 41–50: Women subcontract with firms and conduct their work at home. Highly flexible from the company perspective and more convenient for women with more than one child.
- Ages 51+: Home work, self-employment, or family labor. Older women have a difficult time gaining employment and usually are paid the lowest wages.

Factories are supposed to adhere to regulations regarding minimum wages and working conditions for both women and men, but few are registered with the Government Ministry that oversees these matters. Bribery of inspectors to overlook poor conditions is commonplace. Protests and strikes draw police oppression.

Although the Indonesian government increasingly appears to be sensitive to the unacceptable situation for most factory workers, it has little room to maneuver given the global reality of the marketplace. Having embraced foreign-investment-driven, export-oriented growth, it is aware that improvements in wages and work environments will raise costs and impel investors to move to lower-wage areas such as China or Vietnam.

The Chainsaw Massacre

Indonesia is home to 10 percent of the world's rain forests. Products derived from these forests such as rattan, plywood, sawn timber, and wood furniture are its most profitable export after oil and gas. The biological diversity of these forests is astonishing—more than 10,000 species of trees, 1,500 varieties of birds, and 500 types of mammals.

One well-known endangered animal is the orangutan—"man of the forest." This animal is native to the forests of Sumatra and Borneo. They are largely vegetarian, so destruction of the forest is destruction of their habitat and food source. Adult orangutans are frequently shot and their offspring are taken as pets or to be sold to zoos. Conservation efforts exist in both Malaysia and Indonesia, but these are small in relation to the actual devastation of these and other precious animals.

Of 356 million acres (144 million ha), 74 million acres (30 million ha) are designated as protected; 49 million acres (20 million ha) have been set aside for national parks and wildlife preserves; 77 million acres (31 million ha) are "conversion forests" for settlers; and the remaining 158 million acres (64 million ha) are earmarked for logging. However, millions of acres not so designated have already been deforested or converted for use by small land owners. Others claim that the logging sections are not always suitable for logging.

Indonesian loggers are granted 20-year concessions to log in areas ranging in size from 49,000 acres (20,000 ha) to 247 million acres (100 million ha). Many of these concessions are connected to Chinese or military interests. Selective cutting is supposed to be practiced in order to ensure a regenerative growth to allow another cutting in 35 years. But the realities are often different.

The World Bank claims that Indonesia's forests are disappearing at the rate of 247 million acres (100 million ha) a year (Figure 16-22). About a quarter of that amount is attributable to

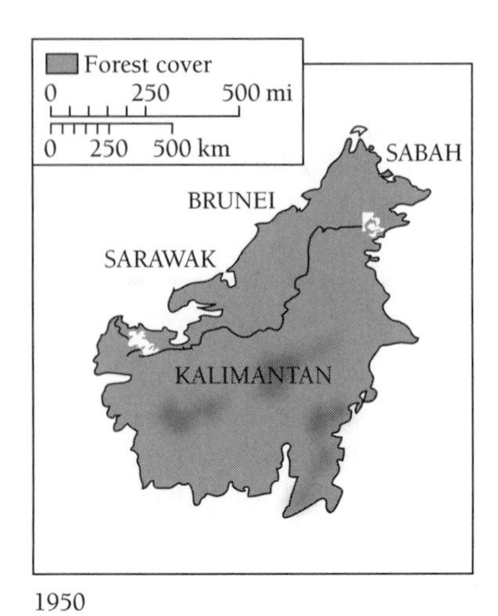

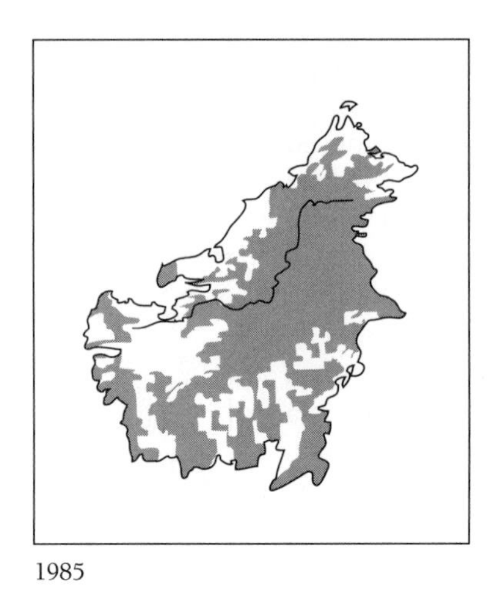

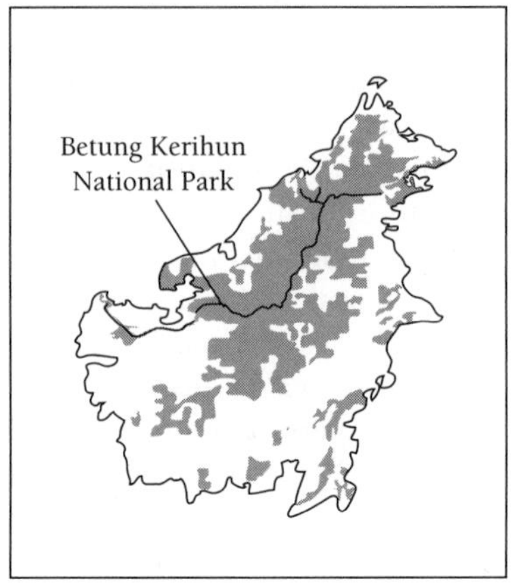

1950 1985 2020 (Projected)

Figure 16-22

This map sequence demonstrates the rapid rate at which Borneo's forests are disappearing. Think of the hundreds of plant and animal species that will become extinct with this rampant destruction. From George White, Joseph P. Diamond, Elizabeth Chacko, and Michael Bradshaw, *Essentials of World Regional Geography*, 2nd edition, 2011, Figure 5-10, p. 148. Originally rendered in color. Printed with permission of McGraw-Hill.

poor logging practices and over-logging. A variety of development projects such as palm oil plantations have also contributed to forest losses. Illegal logging is rampant in all designated forest environments. Many experts have said that Indonesian logging practices are not conducive to sustaining its forests.

Fire has been a major destructive force. As of 1997, according to official estimates, fire had swept through more than 198,000 acres (80,000 ha) in Kalimantan, Sumatra, Sulawesi, and West Papua. In 1982 to 1983 a blaze eradicated more than 9 million acres (3.6 million ha). Vast areas of Kalimantan, with its underground coal deposits, still burn. A blackened wasteland is the result (see Figure 5-8). The blazes reach temperatures that roast the subsoil and kill off root systems and microorganisms to a depth of 6.6 feet (2 m). You can walk on this smoldering landscape and have the soles of your shoes melt. This is what happened to me! Forest fires in Indonesia often affect the air as far north as mainland Southeast Asia and the Philippines, where visibility has been cut to 50 feet (15 m) and the sun seems to disappear. Satellite photos reveal that many of these fires are deliberately set by logging and plantation companies.

Fires have often been blamed on shifting cultivators who set fire to their plots before planting them. However, this claim is misleading because there are very few shifting cultivators left in Indonesia. Moreover, through generations of fire use and management, they are skilled fire users who rarely produce uncontrollable fires. To fault these poor and powerless agriculturalists is convenient for those authorities who want to steer the blame away from deeper underlying causes that are embedded in a nexus of influences linking powerful political and commercial interests. The Kalimantan peat project provides an illustration.

The Kalimantan peat swamp project dates from 1995 when the Indonesian government decided to develop 2.5 million acres of peat swamp (1 million ha) in central Kalimantan to grow rice and settle 1.5 million migrants from Java. By 1997, it had become a logging free-for-all as illegal logging operations, in connivance with transmigration program managers, stripped the area of its valuable timber. Fire was used to clear the land. Out of control, it became one of the largest conflagrations in all of Borneo.

The prospect of work drew thousands of new migrants into the region, while the livelihoods of the local Dayaks were destroyed by the environmental consequences of land clearance, canal digging, and firing. Even the rice that was supposed to grow in the devastated land never materialized. As new fires broke out in 1999, the government pronounced the failure of the project and blamed small-farmers for the debacle.

NATIONAL INTEGRATION

National integration is of primary concern in Indonesia. After Suharto's downfall in 1998, there was a power vacuum and a fear of disintegration. In many outlying areas, local autonomy or independence is seen as the panacea for long-felt ills, and *reformasi* has allowed long-seething discord to surface.

In northern Sumatra, there have been violent clashes over land rights between farmers and security forces protecting the land rights of plantation owners. In southern Sumatra, there has been fighting between the locals and *transmigrasi*. On the island of Ambon in Maluku, Christians have clashed with Muslims. Christians and Javanese Muslims have also clashed on the island of Sulawesi.

Since 1998, the government has recognized the need for decentralization and more autonomy for Indonesia's different regions. Autonomy would allow outlying regions to make their own arrangements with foreign investors and set their own development priorities. This could spur voluntary migration from overcrowded Java. On the other hand, these areas are typically rich in natural resources, and the revenue they generate is needed by the central government in Jakarta. Autonomy is not likely for most areas.

Take Kalimantan as an example. This province is rich in coal resources, most of which lie under primary forest. Indonesia's oil is expected to run out in the next 20 or 30 years and coal will be an essential fuel. What will happen to the people and the environment of Kalimantan? Kalimantan will remain an essential part of a developing Indonesia.

Two very important regions of dissent are Aceh and Papua. Aceh, in northern Sumatra, and Papua, in the western half of New Guinea, both have had long-running separatist movements.

ACEH

About 4.2 million Acehnese live in the northern Sumatran province of Aceh. The region is important for oil, gas, gold, tin, and coal. It also produces coffee, pepper, rice, tobacco, rubber, and timber.

Aceh was where Islam first gained a foothold in Indonesia, and it was the last part of the archipelago to fall to Dutch rule. After a difficult colonial rule lasting 69 years, the Dutch withdrew from Aceh in 1942. Aceh was persuaded (bribed) to become a part of an independent Indonesia in 1949. Since then, many Javanese have been moved to Aceh in an attempt to integrate the region into the larger Indonesia. Strict Acehnese Muslims resent the more religiously relaxed and alien Javanese.

Complaining that Indonesia was sucking the resources out of Aceh without fair revenue return, the Aceh-Sumatra National Liberation Front was formed in 1976. The movement is also known as *Gerakan Aceh Merdeka* or "Free Aceh Movement." In that same year, Aceh declared itself independent. Since then, thousands of Acehnese were killed by the Indonesian army. Thousands more simply disappeared. There are more than 16,000 recorded orphans in Aceh. Furthermore, about 15,000 Javanese migrants were driven out of areas where they had lived since the 1970s.

Human rights were systematically violated in Aceh by the Indonesian army. The people were forbidden to use their own language in written form. There was no freedom of expression. Workers who worked for foreign companies at below-subsistence wages were prevented from forming trade unions. Hundreds of schools and other public buildings were burned and 70,000 refugees driven into impoverished camps along the northern coast. Nevertheless, the rebels fought the Indonesian army to a stalemate, a situation that would probably have continued had it not been for a dramatic turn of events in 2004.

The epicenter of the 2004 undersea earthquake in the Indian Ocean was near the far northern coast of Sumatra, and Aceh was directly in the path of the most powerful tsunami. Entire towns and villages were swept away and hundreds of thousands died. The capital city of Banda Aceh was devastated.

Subsequent international relief efforts opened Aceh to the outside world as it had never been before. Suddenly, the Indonesian army and the rebels found themselves working together in salvage and rescue missions rather than warfare. The upshot was a truce and an agreement whereby the army would withdraw and the rebels would drop their demand for independence.

The 2005 peace agreement between Jakarta and the Free Aceh Movement will provide the province with upward of US$3 billion a year in special disbursements from oil and gas fields now controlled by the central government. The development community has seeded a broad return of family businesses and helped spawn new service companies such as automobile rentals and Internet providers. Now hundreds of new motorbikes clog the streets where bulldozers once plowed piles of post-tsunami debris. Talented Achenese who fled the conflict are also returning to win construction contracts.

In 2003, Aceh became the first of Indonesia's 33 provinces to adopt Sharia Islamic law. The issue is not so much the law itself, but rather how it is implemented. The Acehnese now face harsh punishments for even minor infractions. There are public floggings at mosques for drinking alcohol, gambling or having premarital sex. In addition, a radical "vice and virtue" patrol that includes women has become feared and despised. Moderates, who constitute most of Indonesia's Muslim community, are worried that the secular 1945 Constitution will not be enough to hold the extremists at bay in other parts of the country where Islamism is spreading.

PAPUA

In Papua, whose name was changed from West Irian Jaya in 2007, there is a 13,500-foot (4,050 m) mountain thrusting out of the crocodile- and malaria-infested swamps of the glacier-capped Sundirman Range. Here is a hunk of rock called Grasberg that holds the world's richest copper and gold deposits. The Grasberg (estimated worth of US$50 billion) is owned by the mining company of Freeport-McMoRan Copper & Gold of Louisiana. The mine was one of Suharto's first development projects and is Indonesia's single largest source of tax revenue.

Since the Freeport mine opened, the lives of the indigenous inhabitants have been radically altered. Tribes have received little if any compensation for their lands. Indonesian law requires indigenous peoples to relinquish customary rights over their land in the national interest. After riots in 1996, the mine implemented a scheme to parcel out 1 percent of its revenues to the 20,000-strong Melanesian population. However, the scheme collapsed due to corruption and mismanagement among provincial officials.

The Freeport mine has provided 18,000 jobs, but only 1,500 of these are filled by Papuans, and only 400 are filled by local people. In addition, locals are paid only one-seventh as much as other employees. There have been intertribal conflicts. Alcoholism and prostitution abound in the nearby town.

On the mountain's eastern border lies the Lorentz National Park, a 6-million-acre (2.5 million ha) expanse whose biotic richness is incomparable. The park's ecological integrity is endangered. Lava flows containing the mineral deposits also exist in the national park. The mountains are rich in timber and oil and gas deposits lie offshore. How is this area to be protected from exploitation?

Freeport has one of the world's worst environmental records. It dumps tailings and other chemical and toxic wastes into local rivers, and the mining operations leave behind a bare patch large enough to be viewed from space. Tailings devastate lowland vegetation by choking the life out of the plants. Trees are killed by tailings-laden floodwaters. Ninety-seven percent of ore ends up as tailings, which contain arsenic, lead, mercury, and other dangerous metals. These have killed fish, poisoned sago forests (a traditional food source), and made the water unsafe to drink.

Some scientists claim that the vegetation will regenerate on its own. Experiments are taking place to find out what crops might grow on the tailings. World pressure has forced Freeport to put more resources into environmental considerations.

Papua is an Indonesian province but it lies in the Pacific geographic realm. Along with the ethnic Papuans, about 200,000 Indonesians from other parts of the archipelago, especially Java, inhabit the province. Most of the non-Papuans are concentrated in the capital of Jayapura (a non-Papuan name). Local opposition against Indonesia's rule has continued for decades. For example, the *Organisasi Papua Merdeka* (Free Papua Movement or FPM) held rallies in Jayapura in 1999 and in 2000 demanded independence. They went so far as to display a Papuan flag—the Morning Star. Later that year, Indonesian troops killed demonstrators on Papua's southern coast.

Papua has seen the introduction of a series of measures aimed at closing down the political space that opened up in the post-Suharto period. Papua is a state of de facto military operations that are responsible for the displacement, disappearance, and death of thousands. The division of Papua into three provinces in 2003 is perceived by Papuans as an effort to fragment their internal control and concentrate the power of "security forces" within each region. Papuan animosity toward other Indonesians and the Indonesian government is deepening, and there is no solution in sight.

RADICAL ISLAM

For centuries, the Indonesian version of Islam was a moderate and tolerant one. Many Indonesians were "nominal" Muslims holding only loosely to Islam's precepts. In recent years, however, radical Islamic movements have arisen. As the West's war on terror reached into Indonesia, the government was pressed to investigate possible al-Qaeda links. Terrorists responded in 2002 by bombing a nightclub in Bali. More than 200 people, mostly vacationing Australians, were killed. A Marriott hotel and the Australian Embassy in Jakarta were bombed in 2003.

An organization called Jemaah Islamiah is said to be responsible for these and other terrorist acts. Jemaah Islamiah adheres to the Wahabi sect of Islam, a strictly fundamentalist form of the faith imported from Saudi Arabia. This and other fundamentalist groups want *sharia* applied as the law of the land, as it already is in Aceh. Mosques and religious schools promote hatred of the United States and Israel. More than 30 people linked to Jemaah Islamiah have been convicted and sentenced to prison for their part in the Bali and Marriott bombings.

The group's suspected leader, Abu Bakar Bashir, was arrested but released in 2006 after serving two years in prison. In 2007, he gave a sermon in which he referred to Bali tourists as "worms, snakes and maggots" with specific reference to the "immorality of Australian infidels." He has also claimed that the Bali bombings were the result of a "micro-bomb" planted by the CIA in concert with "Australia and the Jews." The actual Bali bombers were executed by firing squad in 2008. While denying his connection with Jemaah Islamiah, Bashir is forming a new group—Jemaah Ansharut Tauhid—"partisans of the oneness of God," in order to continue his *jihad* for a *sharia*-run Islamic state.

Once regarded as a haven for moderate Muslims, Indonesia is in the throes of a struggle between moderates and conservatives. The moderates seek pluralism and accommodation, a form of Islam that steers away from the sort of orthodoxy that discriminates against women and non-Muslims, and embraces democracy. The conservatives want Islam to control all aspects of all citizens' lives. However, it should be noted that Islamic political parties have been losing influence at the national level.

Orthodox Islam appeals to the poor and also to the middle class, many of whom have a Western education. The transition to democracy since 1998 has done little to improve the lives of the vast majority of Indonesians. The government claims to have stabilized the economy and secured an annual growth of almost four percent. But critics point to rising prices and rampant corruption. At the same time, political freedom has allowed proponents of conservative Islam to promote their views.

The outcome of the struggle has profound implications, not just for security, but also for economic progress and political stability in the region. Should the conservatives prevail, it could provoke conflict with the country's sizeable non-Muslim community (more than 20 million Christians, for example), produce a foreign policy more critical of the West, and an investment environment less friendly to non-Muslim investors. It would also ensure a steady stream of potential recruits for radical groups like Jemaah Islamiah.

PROGNOSIS

As Indonesia's resources are increasingly eroded and as population growth continues, it will be difficult, if not impossible, to reign in the outer resource regions to the will of the industrial core of Java. It also will be difficult, if not impossible, to suppress the many self-determination movements across this vast region. It also will be problematic to control the progress of conservative Islam. These are only three of many problems facing the current government and President Susilo Bambang Yudhoyono.

Timor-Leste: New Nation

Timor is the easternmost of the lesser Sunda Islands. The eastern part of Timor was a Portuguese colony until it was overrun by Indonesia in 1975 and annexed in 1976. Subsequently, East Timor became the scene of a bitter struggle for independence. Tens of thousands of Timorese were killed by the Indonesian army. The infrastructure was destroyed and thousands of people were dislocated. Foreign intervention, led by Australia (once Indonesia's only supporter of its claim to East Timor), brought some semblance of order. East Timor became the independent nation of Timor-Leste in 2002.

As Figure 16-23 reveals, sovereignty has created geopolitical complications. The political entity of Timor-Leste comprises a main territory containing the capital of Dili, along with a small exclave on the north coast of Indonesian West Timor. This exclave is called Ocussi. Although there is a road from Ocussi to the main territory, relations between Indonesia and Timor-Leste may not allow this road to be used, so that the two parts of the state would have to be connected by boat traffic only.

A second problem has to do with oil. South of Timor-Leste is an area called the Timor Gap that contains major oil and gas reserves. While Indonesia ruled East Timor,

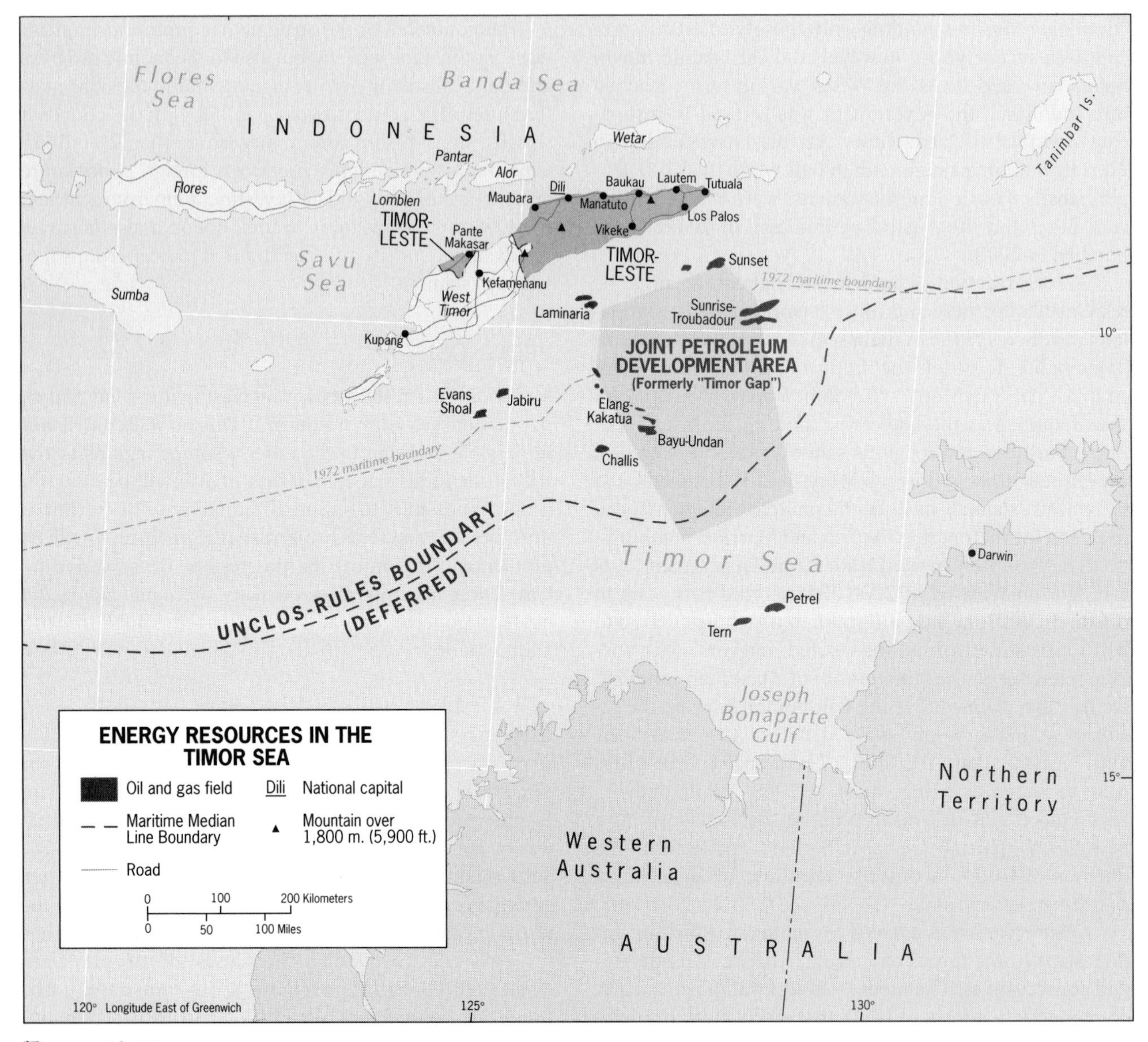

Figure 16-23

Timor-Leste and the Timor Gap. From H. J. de Blij and P. O. Muller, *Geography: Realms, Regions and Concepts*, 14th edition, 2010, p. 572. Originally rendered in color. © H. J. de Blij and P. O. Muller. Reprinted with permission of John Wiley & Sons, Inc.

Indonesia and Australia agreed on a maritime boundary greatly favoring Australia in 1972. Once East Timor became independent, it wanted to get rid of this boundary and replace it with one based on United Nations recommendations. The United Nations boundary would be midway between the island and Australia and give the bulk of the resources to Timor-Leste. Australia rejected this proposal. However, in 2004, the two countries hammered out a proposal to develop the Greater Sunrise gas field. Revenues will be shared with Timor-Leste gaining US$3.5 billion in tax and royalty payments. The problem of the maritime boundary has yet to be solved.

East Timor is one of Asia's poorest countries with 42 percent of its people living below the official poverty line as of 2004. The population is growing at a rate of 3.1 percent. The total fertility rate is 6.5, and 45 percent of the population is under 15 years of age.

In 1999, anti-independence militias, along with the Indonesian army, destroyed the country's entire infrastructure. Homes, schools, irrigation and water supply systems, and the electrical grid must be rebuilt from scratch. As of 2009, 90 percent of Timorese are involved in agriculture and coffee is the main export. More than half of those employed work in the service sector. Industry accounted for only 13 percent of the economy. Unemployment and underemployment are widespread—as much as 50 percent. However, there is the possibility of new jobs in the production of coffee and vanilla and in the oil industry.

The development of oil and gas resources in offshore waters has boosted government revenues. However, as there are no production facilities on the island, few jobs have been created. Gas is piped to Australia. In 2005, the government created a Petroleum Fund to preserve the oil and gas revenues for future generations.

In 2008, the government resettled tens of thousands of an estimated 100,000 internally displaced people. Problems with the development of Timor-Leste are extremely challenging. As the government has not yet determined how best to use oil and gas revenue to create employment and reduce poverty, it remains heavily reliant on international assistance.

The Philippines: Pearl of the Orient Seas

The Philippine archipelago began to take shape about 50 million years ago, the result of volcanic eruptions and the buckling of the Earth's crust. The archipelago, comprising more than 7,000 islands, stretches 1,100 miles (1,770 km), forming a land chain between the Pacific Ocean and the South China Sea. Only 154 of these islands exceed 5 square miles (14 km^2) in area, and 11 of them contain about 95 percent of the total population. Luzon and Mindanao are the largest islands. These bracket a regional grouping of islands, known as the Visayas (Figure 16-24).

Rainfall is variable because of the mountainous nature of the islands and shifting wind directions. On Luzon, for instance, rainfall ranges from 35 to 216 inches (84–549 cm) a year. Manila is wettest during the southwest monsoon, which delivers 82 inches (208 cm) from June to November. The islands also straddle the typhoon belt and experience about 15 cyclonic storms yearly, which are increasingly violent and destructive. Active volcanoes and earthquakes can also wreak havoc.

In June 1991, Mount Pinatubo erupted and devastated an area of 154 square miles (400 km^2) with volcanic ash. Fortunately, predictions allowed precautions to be taken, and only 250 people died in the event.

Knowledge of the Philippines' prehistory is sketchy, but Hominid remains estimated to be 30,000 years old have been discovered on Palawan. Malay-related people of Mongoloid descent are predominant today. Other groups are Negritos of obscure origin. Chinese trace their origins back to traders of the tenth century. Five centuries later, Islam from Borneo and the Sulu Islands diffused as far as Mindanao, only to be stopped by the arrival of the Spanish Christians in the sixteenth century.

Philippine society is complex, with at least 60 ethnic groups among the Malay majority. Chinese, Americans, other Asians, and indigenous people are also part of the population mix. Several languages are spoken, with **Tagalog**—the language of the people around Manila Bay—being the official language. However, English is the lingua franca throughout the islands.

THE SPANISH PHILIPPINES

Ferdinand Magellan claimed the Philippines for Spain in 1521. Magellan came upon the islands during his circumnavigation of the world. He never completed that journey, however, and was murdered in the Philippines. Only one of his ships managed to complete the entire voyage.

Spain controlled the Philippines for the next 377 years, with the familiar twin goals of spreading the Christian faith and setting up trade. The first settlement was on the island of Cebu in 1565, but headquarters were moved to Manila in 1571. The islands were named after the Spanish Crown Prince Felipe.

The economic base of Spanish occupation of the islands was the galleon trade between Macao on the south China coast, Manila, and Acapulco in Mexico. Great galleons left Acapulco for Manila laden with silver. In Manila the silver was exchanged for silk brought from Portuguese Macao. The silk was then taken to Acapulco and on to Europe.

In the seventeenth and eighteenth centuries, Manila became a rival of the major colonial port of Batavia (Jakarta). By this time Manila was in many ways a Chinese city. Chinese traders and ***mestizos*** (Chinese-Filipino

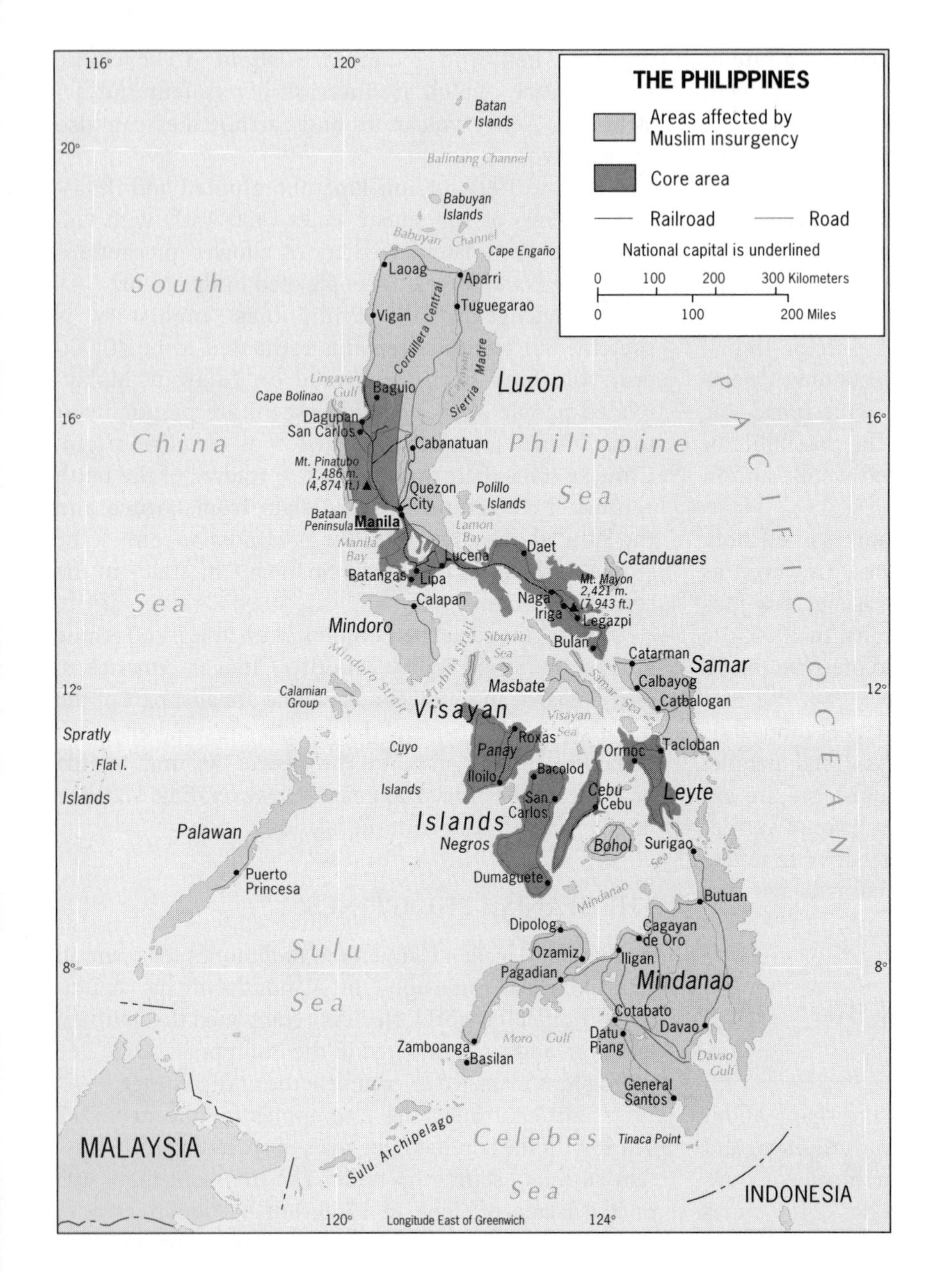

Figure 16-24

The Philippines. Note the focus of activity on Luzon and the Visayan Islands. From H. J. de Blij and P. O. Muller, *Geography: Realms, Regions and Concepts*, 14th edition, 2010, p. 574. Originally rendered in color. © H. J. de Blij and P. O. Muller. Reprinted with permission of John Wiley & Sons, Inc.

or Mexican-Filipino mix) organized the entrepôt's trade and soon became the most influential political and economic force in the Philippines.

Power was vested in Luzon in the north. Efforts to control the Muslim-controlled Sulu archipelago and Mindanao were repeatedly rebuffed by the Islamic sultanate. These regions were not incorporated into the Philippines until the American takeover at the end of the nineteenth century.

Spanish control was based on a close relationship between church and state. Jesuit, Dominican, and Colombian friars were sent into the countryside to proselytize and at the same time establish plantations. Today, more than 80 percent of Filipinos are Roman Catholic. Eventually, a state-controlled tobacco monopoly was established in northern Luzon with local people forced to provide the labor.

As the galleon trading monopoly came to a close, the islands were open to private investors and traders. In 1785, the Spanish established the Royal Philippine Company to invest in export crops such as sugar, coffee, indigo, and pepper. By exporting these crops, the Philippines came into direct competition with the Netherlands East Indies. Furthermore, British and American interests had become prominent. By the nineteenth century, British-American export houses dominated the export economy. The volume of trade increased 15 times between 1825 and 1875, with major exports being sugar, coffee, tobacco, and abaca—the strongest vegetable fiber obtained from trees in the banana family.

The incorporation of the Philippines into the world economy had two chief results. First, it saw the rise of Philippine nationalism, as well as the emergence of a modern nation state. Second, it created regional economic, political, and social forces that would ultimately weaken the country. Powerful regional elites became the critical social forces of the twentieth century.

Philippine nationalism was the earliest of its kind in Southeast Asia but it had little impact on the other colonies. The Chinese *mestizos* dominated ownership of export plantations. The *haciendas* (large estates) developed by powerful regional families were worked by tenant farmers. The landowners became richer and the tenants became poorer. From this base arose the powerful landed families that controlled politics (a situation that lasted even into the 1990s). This elite educated their children in Spanish schools, seminaries, and universities. These Spanish-educated children, known as *ilustrados,* were influenced by liberalized policies in Spain. They were anti-clerical and demanded separation of church and state. As their demands were ignored, they began to call themselves Filipinos. The name Filipino became a symbol of nationalism. In 1896, a rebellion erupted.

THE AMERICAN PHILIPPINES

At the same time Spain was dealing with open rebellion in the Philippines, it was at war in Central America and Cuba. U.S. intervention in Cuba resulted in the Spanish-American War. The Americans sailed to Manila, smashed the Spanish fleet, and laid siege to Manila. Philippine opportunists seized the opportunity to declare independence in 1898. The Filipinos were the first people in Asia to fight their colonial oppressors and declare a nation-state. However, the Americans were not done with the Philippines.

In 1899, war broke out between America and the Philippines. Hostilities ended in 1901, when the Americans effectively bought off the *ilustrado* elite, promising to maintain their wealth and power in return for cooperation with American colonial rule.

The United States occupied the Philippines with a "democratic mission" of leading the Philippines into self-rule. The Tydings-McDuffie Act of 1934 promised independence in ten years. The Americans stressed the development of health, education, and democratic processes. In 1941, the literacy rate was five times that of Indonesia. In 1935, a Democratic Republic was established with Manuel Quezo as its first president. Government leaders were from the landed elite. While espousing complete independence, their political and economic interests were well served by American rule.

World War II interrupted the Philippines' march to independence. Fighting in the Philippines was more intense than elsewhere in Southeast Asia. However, Japanese slogans such as "Japan, the light of Asia" had little attraction among Filipinos, and resistance to the Japanese was strong. Filipino nationalists welcomed the American liberators, who promised independence by 1946.

INDEPENDENT PHILIPPINES

Independence changed very little in that the landed elite maintained control. Patron–client relations and vested links between rich landowners and poor peasants have continued to underpin the political process. Although rebellions have been crushed, rural discontent and unrest remain problems.

In the 1950s and 1960s, a new industrial class emerged. Some were from the landed elite who were diversifying their capital. Others were professional traders who had merged interests with American or local Chinese businessmen. By the 1960s, the Philippines was the most successful manufacturing country in Southeast Asia. Agribusiness, real estate, and banking were the basis of several multinational conglomerates. Unfortunately, subsequent government policies spoiled the record and the economy became stagnant.

During the rule of Ferdinand Marcos, who proclaimed martial law in 1972, per capita income fell, foreign indebtedness rose to serious levels, and the middle class protested corruption and the inability of the government to handle the country's social ills. Moreover, Marcos, his friends and relatives, and the military practiced crony capitalism—absorbing private businesses and state enterprises—on an unprecedented scale. Family conglomerates controlled 78 percent of all corporate wealth in the country. Land reform and the end of rural poverty remained sheer rhetoric.

By the 1980s, the nation was fragmenting. The Moro Nationalist Liberation Front (MNLF) had 50,000 to

60,000 guerrillas fighting for an independent Muslim state in Mindanao. Half the Philippines' army was fighting that battle. Most of the others were fighting communist-controlled guerrilla attacks elsewhere in the country. Opposition to Marcos became strident. He called an election in 1986 and lost to Corazon Aquino, wife of assassinated popular politician Benigno Aquino. Marcos ultimately fled the Philippines.

Although the government returned to its 1970s' democratic form, little changed in the social structure. The landed elite retained their riches, land reform was minimal, and political loyalties remained personal rather than institutional. However, there was an impressive economic turnaround.

ECONOMIC PROGRESS

Ferdinand Ramos took power in 1992, and by the mid-decade the Philippines had reached growth levels approaching those of other ASEAN countries. However, the country has a long way to go. Although the agricultural sector has remained relatively constant, with 17 percent of the total economy, it employs 34 percent of the labor force. Industry and manufacturing account for 15 percent of employment while services account for 51 percent. But still, a third of the population is unemployed or underemployed.

Subic Bay

In 1992, under pressure from the Philippine government and people, the United States gave up the Subic Bay Naval Base and Clark Air Base, both located on the island of Luzon. Subic Bay alone had employed 32,000 local people and indirectly created 200,000 jobs. However, it also imported some negative aspects of American culture, such as sex work and rampant consumerism. Soon Subic Bay was turned into a free-trade zone, and 200 companies—including many from Taiwan—pledged investments in the zone. In 1995, the American company Federal Express established Subic Bay as its air hub for overnight parcel deliveries in the region.

Many consider the Subic Bay Freeport zone's greatest asset, in terms of its potential, to be its deep-water harbor. In 2007, South Korea's Hanjin Heavy Industries Corporation, one of the largest shipbuilding companies in the world, began its container-ship building operations employing 15,000 workers. In addition to the shipyard, more than 600 companies have located in the Subic Bay Freeport Zone.

Subic is divided into four key investment areas: the Central Business District, Subic Gateway, Subic Bay Industrial Park, and the Subic Techno Park. These areas house manufacturing-related businesses, electronic communications and computer industries, warehousing and transshipment operations, financial and educational facilities, and a vibrant tourism, resort, and lodging industry.

The traditional industrial base was the processing of agricultural products, but this was altered in the 1990s with a policy of import substitution. For example, the Philippines' largest company, the San Miguel Corporation (of beer fame) diversified into food and soft-drink processing and invested in coconut products, banking, and other commodities.

Recent emphasis has been placed on export-oriented industrialization. Export-processing zones (EPZs) are located near Manila, Cebu, and in Baguio. In addition to Subic's free port, Clark Air Base is now a special economic zone. Numerous private enterprises have been established in other areas, mainly in Metro Manila. Today, the leading exports of the Philippines are electronics and apparel. Gateway Electronics has developed a business park, and numerous chip manufacturers have gravitated to the site. Texas Instruments, Motorola, and Philips have also established industrial sites near Manila.

Manufacturing is highly concentrated in Metro Manila. Its metropolitan area accounts for more than half the total Philippine manufacturing employment. The next largest centers are Cebu and Davao.

Cebu is a world center for the manufacture of rattan products. Nearby locations such as Mactan Island—where Ferdinand Magellan was killed during the first circumnavigation of the globe—include an international airport and an export-processing zone. Mactan has also been developed for tourism, with fine hotels and beach resorts. One-third of all foreign tourists, mostly Japanese and Taiwanese, visit Cebu and Mactan. The Japanese are the major foreign investors in the EPZ. One of the zone's first factories was Timex. Now, 80 percent of all Timex watches are made in the Mactan plant. Other companies produce semiconductors, cameras, furniture, apparel, and so forth.

Another important manufacturing center is General Santos City on the southern coast of Mindanao. The major industries here are food processing, especially of fish.

It is the largest producer of tuna, most of which is exported to Japan as sashimi. The agricultural region surrounding the city accounts for 30 percent of the nation's maize production, 10 percent of its coconuts and copra, 20 percent of its rice, 99 percent of its asparagus, and 40 percent of its pineapples. One of the largest employers is the Dole Corporation, involved in everything from pineapples and bananas to vegetables.

Another economic scheme links Brunei, Indonesia, Malaysia, and the Philippines into an East Asian Growth Area (BIMP-EAGA or EAGA). EAGA is developing agro-industry focusing on products such as marine products, coconuts, palm oil, poultry, and high-value tropical fruits and vegetables. It also specializes in the production of *halal* food for the region's Muslim population. EAGA is a leading eco-tourism destination. It has pristine rain forests, coral reefs, and is ethnically and culturally diverse. Many World Heritage Sites are in EAGA such as Mount Kinabalu and the Mulu Caves in Sarawak. A new international airport, expanded harbor facilities, and a new EPZ are part of the future of this new economic growth triangle.

OVERSEAS FILIPINOS

Foreign exchange revenue inflows are an important element of Philippine earnings. With millions of workers deployed in over 190 countries and territories, billions of dollars come into the country in the form of remittances every year.

The first significant outflow of Philippine workers was to Hawaii to work on plantations. Others have worked in U.S. defense industries in Guam and Vietnam. These ***balikbayans*** (overseas Filipinos) remain important sources of revenue for the Philippines. More than 90 percent (around 1.4 million) live in the United States, Canada, and Australia. Many go to work in the Middle Eastern oil fields of Saudi Arabia, Kuwait, and the United Arab Emirates.

Adjacent Asian countries also offer opportunities for employment, but these are for low-paid domestics and "entertainers." Geographer James A. Tyner (1996) has found that the marketing and recruiting strategies of the Philippine government, especially in promotional literature, present Filipinos as loyal, disciplined, and obedient and that promotional materials draw attention to gender-associated roles and occupations. For example, women are portrayed as nurses and men as construction workers. Tyner notes that, "Global marketing of workers is organized around specific gendered assumptions of male versus female occupations. The use of gender by government and private institutions as an organizing device provides a gendered context in which migrants and their households must subsequently operate." Gendering of jobs is also most pronounced in Japanese firms. Unfortunately, many Filipino foreign workers find themselves turning to prostitution to pay labor recruitment fees, and otherwise suffer abuse, unable to save up enough money to go home.

ONGOING PROBLEMS

The Philippines has many problems to overcome. Environmental damage is severe, with diminishing forests and damaged coral reefs. Little has changed for the poor in either urban or rural areas. Manila is marred by vast squatter settlements and lack of adequate housing. Wealthy families still control the land and tenant farmers remain impoverished. Income differentials are huge—some of the largest in Asia—and the middle class remains small. According to the WB, nearly 40 percent of Filipinos still live in poverty.

Another pressing problem focuses on the Philippines' small Muslim population that is concentrated in the southeastern part of the archipelago. Muslims feel that they are marginalized in this predominantly Christian country. Over the past three decades, several Muslim organizations, including the Moro National Liberation Front and the Moro Islamic Liberation Front, have promoted the Muslim cause with tactics ranging from peaceful negotiation to violence. Basilan Island has become the base for the extreme Abu Sayyaf, which receives support through al-Qaeda. The United States has sent troops to help the Filipinos pursue these terrorists. Deadly bombings in southern towns and even in Manila have forced the Philippine government to spend a large share of its budget on crushing the insurrection. In addition, a communist uprising in 2004 required a response in difficult, remote mountain terrain where military operations are very expensive.

Conclusion

Clearly, the countries of insular Southeast Asia are moving along the path toward modernization. Singapore has achieved this end and is well entrenched as a fully developed country in the world of transnationals and cyberspace. However, it is now challenged to preserve its historic landscapes. Malaysia is challenged by the problem of uneven development exacerbated by its fragmented nature. Environmental destruction is taking its toll, especially in Sarawak and Sabah, where traditional lifestyles are being destroyed. Both Indonesia and the Philippines are fraught with problems deriving from their fragmented nature and their ethnic and religious complexities. Intergroup rivalries are detrimental to a smooth transition to more even economic development in addition to political freedom and self-determination for all concerned. Furthermore, environmental problems overshadow all forms of progress.

Recommended Web Sites

http://app.www.sg/
Official "Gateway to Singapore." History, geography, Year Book, etc.

www.bruneipress.com.bn/brunei/brunei.html
Good site for information on various aspects of Brunei. However, statistics conflict with those of other sources such as Population Reference Bureau and the CIA Factbook.

www.cyborlink.com/besite/indonesia.htm
Very interesting site on Indonesian customs and etiquette. Links to related resources.

www.dbkl.gov.my/pskl2020/english/index.htm
The Kuala Lumpur plan for 2020.

www.economywatch.com/world_economy/
News on economies by country.

http://en.wikipedia.org/wiki/Colony_of.Sarawak
Good site for information on Sarawak—its history, cultural and social geography. Look at Wikipedia for Sabah as well.

www.icg.org/home/index
International Crisis Group reports and briefings by country. Check on Timor-Leste, for example.

www.indonesia.go.id/en/
National portal for the Republic of Indonesia. Geography, history, politics, economy, etc. Current news articles.

www.rsc.ox.ac.uk/PDFs/RSCworkingpaper2.pdf
Oxford University Refugee Study Center research papers on conflict and displacement in Papua, Indonesia (2007).

www.sbma.com/index.html
Official site for Subic Bay Freeport.

www.singstat.gov.sg/
Statistics Singapore. Excellent collection of statistics, reports, and surveys; economic, demographic with charts and graphs.

www.state.gov/r/pa/ei/bgn/2700.htm
U.S. Department of State. Excellent account of Brunei. Select links for information on other countries in Asia.

www.straitstimes.com/
The *Straits Times*, major regional newspaper out of Singapore.

www.theodora.com/wfb/
Countries of the world: geography, history, economy, etc.

Bibliography Chapter 16: Insular Southeast Asia

Airriess, Christopher. 2000. Malaysia and Brunei. In *Southeast Asia: Diversity and Development*, eds. Thomas R. Leinbach and Richard Ulack, pp. 341–378. Upper Saddle River, N.J.: Prentice-Hall.

Aspinall, Edward. 2009. *Islam and Nation: Separatism and Rebellion in Aceh, Indonesia*. Stanford, Calif.: Stanford University Press.

Augustya, Heather. 2007. "A Burning Issue: Palm Oil Shows Promise as a Biofuel, but the Environmental Coast of Production Can Be High." *World Watch* (July/August): 22–25.

Bakshian, Douglas. 2007. "Winding Down the Mindanao War." *Far Eastern Economic Review* 170/10: 47–49.

Bresnan, John, ed. 2005. *Indonesia: The Great Transition*. New York: Rowman and Littlefield.

Brooks, Oakley. 2007. "The Rebirth of Aceh." *Far Eastern Economic Review* 170/9: 30–42.

Bunnell, Tim. 2004. *Malaysia, Modernity and the Multimedia Corridor*. New York: Routledge.

Cartier, Carolyn L. 1993. "Creating Historic Open Space in Melaka." *The Geographical Review* 83: 359–373.

Chang, T. C. 1999. "Local Uniqueness in the Global Village: Heritage Tourism in Singapore." *The Professional Geographer* 51: 91–103.

Cleary, M. C. 1996. "Indigenous Trade and European Economic Intervention in North-West Borneo c. 1860–1930." *Modern Asian Studies* 30: 301–324.

Cleary, Mark, and Brian Shaw. 1994. "Ethnicity, Development and the New Economic Policy: The Experience of Malaysia, 1971–1990." *Pacific Viewpoint* 35: 83–107.

De Ayala, Jaime Augusto. 2006. "The Service Heart of Asia." *Far Eastern Economic Review* 169/9: 55–57.

Dharmapatni, Ida Ayu Indira, and Tommy Firman. 1995. "Problems and Challenges of Mega-Urban Regions in Indonesia: The Case of Jabotabek and the Bandung Metropolitan Area." In *The Mega-Urban Regions of Southeast Asia,* eds. T. G. McGee and Ira M. Robinson, pp. 296–314. Vancouver: UBC.

Dhume, Sadanand, 2007. "Step Up the Fight Against Islamism." *Far Eastern Economic Review* (July/August): 6–13.

Drake, Christine. 1989. *National Integration in Indonesia.* Honolulu: University of Hawaii Press.

Drake, Christine. 1992. "National Integration in China and Indonesia." *The Geographical Review* 82: 295–312.

Finkel, Michael. 2009. "Facing Down the Fanatics." *National Geographic* October: 76–99.

Ford, Michael and Lyn Parker, eds. 2008. *Women and Work in Indonesia.* New York: Routledge.

Freeman, Donald B. 2003. *The Straits of Malacca; Gateway or Gauntlet?* Montreal: McGill-Queen's University Press.

Heyzer, Noeleen. 1992. "Rainforest Management and Indigenous Livelihoods: A Malaysian Case Study." *Development* 4: 14–17.

Hugo, Graeme. 2000. "Indonesia." In *Southeast Asia: Diversity and Development*, eds. Thomas R. Leinbach and Richard Ulack, pp. 304–340. Upper Saddle River, N.J.: Prentice-Hall.

Jacobson, Mark. 2010. "The Singapore Solution." *National Geographic* (January): 133–149.

Jardine, Matthew. 1995. *East Timor: Genocide in Paradise.* Tucson, Ariz.: Odonian.

Jayasankaran, S. 2004. "Brunei: Buoyed By Oil." *Far Eastern Economic Review* (August 12): 41.

Lepawsky, Josh. 2005. "Digital Aspirations: Malaysia and the Multimedia Super Corrridor." *Focus on Geography* 48/3: 10–18.

Lee, Boon Thong. 1995. "Challenges of Superinduced Development: The Mega-Urban Region of the Kuala Lumpur-Klang Valley." In *The Mega-Urban Regions of Southeast Asia,* eds. T. G. McGee and Ira M. Robinson, pp. 315–327. Vancouver: UBC.

Leinbach, Thomas R., et al. 1992. "Employment Behavior and the Family in Indonesian Transmigration." *Annals of the Association of American Geographers* 82: 23–47.

Lintner, Bertil. 2009. "Inside the Papuan Resistance." *Far Eastern Economic Review* (July/August): 53–57.

Macleod, Scott, and T. G. McGee. 1996. "The Singapore-Johore-Riau Growth Triangle: An Emerging Extended Metropolitan Region." In *Emerging World Cities in Pacific Asia*, eds. Fu-chen Lo and Yue-man Yeung, pp. 417–464. New York: United Nations.

Mapes, Timothy. 2003. "No Solution in Sight for Aceh." *Far Eastern Economic Review* (September 25): 21–22.

McBeth, John. 2004. "East Timor: Not Yet Ready To Go It Alone." *Far Eastern Economic Review* (May 20): 12–14.

McKay, Steven C. 2006. *Satanic Mills of Silicon Islands: The Politics of High-Tech Production in the Philippines.* Ithaca, N.Y.: Cornell University Press.

Montlake, Simon. 2008. "Race Politics Hobbles Malaysia." *Far Eastern Economic Review* 171/2: 36–39.

Netto, Anil. 2004. "Rural Malaysia Mired in Poverty." *World Press Review* (January): 14.

Parsonage, James. 1992. "Southeast Asia's "Growth Triangle": A Subregional Response to Global Transformation." *International Journal of Urban and Regional Research* 16: 307–317.

Renner, M., and Zoe Chafe. 2007. "Aceh: Peacemaking after the Tsunami." *World Watch*, Washington, D.C.: 20–25.

Rigg, Jonathan. 2006. *Southeast Asia: The Human Landscape of Modernization and Development.* New York: Routledge.

Robinson, Kathryn M. 2009. *Gender, Islam and Democracy in Indonesia.* New York: Routledge.

Rushford, Greg. 2009. "Clan Warfare Hobbles the Philippines." *Far Eastern Economic Review* (July/August): 42–45.

Sharma, Gouri. 2009. "Timor-Leste: The Road to Recovery." *New Internationalist* (October): 26.

Smith, Paul J., ed. 2005. *Terrorism and Violence in Southeast Asia.* Armonk, N.Y.: M. E. Sharpe.

Sutton, Keith, and Amriah Buang. 1995. "A New Role for Malaysia's FELDA: From Land Settlement Agency to Plantation Company." *Geography* 80: 125–137.

Taylor, D. M., et al. 1994. "The Degradation of Rainforests in Sarawak, East Malaysia, and Its Implications for Future Management Policies." *Geoforum* 25: 351–369.

The Economist. 2009. "A Golden Chance: A Special Report on Indonesia." *The Economist* (September 12): 1–16.

Tyner, James A. 1996. "The Gendering of Philippine International Labor Migration." *Professional Geographer* 48: 405–416.

Ulack, Richard. 2000. "The Philippines." In *Southeast Asia: Diversity and Development*, eds. T. R. Leinbach and R. Ulack, pp. 408–33. Upper Saddle River, N.J.: Prentice-Hall.

Vatikiotis, Michael. 2003. "The Struggle for Islam." *Far Eastern Economic Review* (December 11): 54–58.

Glossary

Activity space The cumulative space employed in the activities of daily life.

Adi Granth The Sikh holy book containing the teachings Guru Nanak, the founder of Sikhism.

Adivasi Means "original inhabitant." Applied to scheduled tribes in India.

Aeolian Referring to wind.

Age of Exploration The period of European exploration of the world beginning in the fifteenth century.

Age of Imperialism (1870–1914) The period during which competing European powers acquired, controlled, and exploited colonies.

Ainu Aboriginal people of Japan. Now confined mainly on Hokkaido.

Allah In Islam, "God."

Alluvial soils Fertile, water-deposited soil.

Altitudinal zonation The variation in plant communities caused by climatic change associated with increasing altitude.

Amenity-migration The movement of people to places offering amenities such as rest and recreation or a more comfortable climate.

Americanization The promotion and acceptance of American culture.

Ammonite A flat, spiral fossil shell from the ancient Tethys Sea.

Analects The ideas of the Chinese philosopher and teacher Confucius as recorded by his students.

Annam Chinese word meaning "the pacified south." The Chinese name for Vietnam during the Tang Dynasty. During the period of French rule, one of the five territories of the Indochinese Union.

Anti-globalization movement Argument that corporations seek to maximize profits at the expense of fair compensation and treatment of workers as well as environmental conservation and integrity of governments.

Aquaculture The cultivation of sea products such as shrimp, oysters, seaweed, and pearls.

Aquifer A permeable subsurface rock layer that can store, transmit, and supply water.

Archipelagoes Collections of islands. Japan, the Philippines, and Indonesia are all archipelagoes.

Arithmetic density The number of people existing per square mile or kilometer of land in a particular country.

Artesian basin An aquifer in which water is held under pressure. Drilling a well would allow water to flow freely to the surface without artificial pumping.

Aryans Relatively light-skinned nomadic herders who invaded India from Central Asia around 1500 BC. The Aryans spoke Sanskrit.

Asia-Pacific Similar to the economic concept of the Pacific Rim. The countries of the Asia Pacific region trade primarily with one another.

Asia Pacific Economic Cooperation (APEC) A regional trading block similar to the European Union (EU) or the North American Free Trade Association (NAFTA).

ASEAN Association of Southeast Asian Nations.

Asoka The most important Mauryan ruler in India. Asoka adopted Buddhism as the State religion.

Aung San Suu Kyi Leader of the National League for Democracy (NLD) in Burma. She was under house arrest from July 1989 to November 2010.

Azad Kashmir Revolutionary government in northern Kashmir.

Bagasse Sugarcane fiber.

Bamboo network Trade network between China and Taiwan.

Bangla The Bangladeshi name for Bengali, the language of Bangladesh.

Balikbayans Overseas Filipinos. They are important sources of revenue for the Philippines.

Bao Dao The name for Taiwan as portrayed on ancient Chinese maps.

Barani Dry, non-irrigated regions of Pakistan.

Bari The basic spatial unit in a Bangladeshi village consisting of six houses built around a square or rectangular courtyard.

Base population The actual population from which growth or decline occurs.

Batavia The Dutch name for Jakarta, the capital of Indonesia. Batavia was renamed Jakarta in 1945.

Benazir Bhutto The youngest and first female leader of a Muslim state. Served as prime minister of Pakistan from 1988 to 1990 and again from 1993 to 1997. Assassinated in 2009.

Bible The holy scriptures of Christians and Jews. Christians accept the whole Bible while Jews adhere only to the Old Testament.

Bilateral aid Government to government transfers of cash, low-interest loans, food, and the like.

Biodiversity The range of living organisms defined in terms of genes, species, and ecosystems.

Biological oxygen demand (BOD) The measure of the amount of oxygen required to support microorganisms that break down organic wastes such as those found in raw sewage.

Biome Major territorial ecosystem.

Biotech (BT) crops Genetically engineered seeds/crops.

Bodhisattva In Buddhism, a divine being such as Avalokitesvara in Sanskrit, who postpones entry into *nirvana* in order to save other beings; meaning "the lord who looks in every direction."

Bollywood India's movie industry located in Mumbai (Bombay).

Brahman In Hinduism, the one ultimate reality.

Brahmins Hindu priests in India's ancient social hierarchy. In most recent times, any one of this class.

Brain drain The process of educated people in many poor countries leaving for jobs in richer countries.

Bride price Payment made by the groom's family to the bride's family to compensate for loss of labor and fertility by the bride's kin group.

Buddha In Sanskrit, the enlightened one. The name of Siddhartha Gautama upon his enlightenment.

Buffer state A country situated between two other opposing countries. Nepal and Bhutan are buffer states.

Bumiputera Means "sons of the soil." The Malay peoples of Malaysia.

Bungalow A low-lying, thick-walled, one-story house found in the hill stations of northern India.

Burakumin Social outcasts in Japan.

Burghers Descendants of Portuguese or Dutch in Sri Lanka.

Buttresses Giant, supportive blade-like structures of a shallow-rooted tree.

Canopy The upper layer of the rain forest. The forest's biomass.

Cao Dai Vietnamese word meaning "high tower." A Vietnamese religion that is a mixture of Buddhism, Christianity, Confucianism, and Daoism.

Carbon offset programs Governments agree to invest in actions and projects that will reduce or even eliminate greenhouse-gas emissions.

Cardamom A highly valued spice commonly used in South Asian cuisine.

Caste system The division of Hindu society according to social ranking and occupation.

Castle town In early Japan, a fortified town along a transportation corridor.

Chaebol A giant, corporate conglomerate in South Korea.

Chain migration Where the mover is part of a migrant flow from a common origin to a prepared destination.

Chars In Bangladesh, the name given to small, sandy islands formed by rivers as they sway between valley walls.

Chawl Slum in Mumbai (Bombay).

Ch'i The vital force or cosmic breath that gives life to all things. The most important force of *feng shui*.

Chiang Kai-shek The leader of the Nationalist forces during China's civil war and World War II. Chiang opposed Mao Zedong and the Communists and was defeated by them in 1949. He fled to Taiwan in 1949 and ruled the island as the Republic of China until his death in 1975.

China Frontier The rugged, arid, and sparsely populated environment of western China.

China Proper The relatively low relief, humid, and densely populated environment of eastern China.

Chinese Diaspora Dispersal of Chinese from the mainland to all parts of Southeast Asia.

Chipko Movement A grass roots, women's movement to save trees in the Indian Himalayas.

Christianity A monotheistic faith proclaiming the idea of a single deity, God, and Jesus Christ, the manifestation of God on Earth.

Cilao Portuguese name for Ceylon (Sri Lanka).

Cinchona Source of quinine extracted from tree bark; it was the only treatment for malaria in historic times.

Circular migrant A person who lives and works in the city for part of the year and spends the remainder of the year in his or her village.

Civil line British civilian district socially and spatially separated from Indian society.

Cloud forests In the Himalayas, forests existing in fog and mist at elevations between 8,000 and 13,000 feet (2,400–39,000 m).

Cochin China A colony in southern Vietnam. One of the five territories of France's Indochinese Union.

Collectivization The end to land ownership in China as instituted by Mao Zedong in 1950. Land was worked collectively.

Colonizers' model The idea that civilization and progress diffuse from Europe to the rest of a culturally barren world.

Commoditization The division of a particular market into two sectors: big-name brand marketing and sales, and hundreds of no-name component suppliers. The garnering of the no-name component sector is known as the Original Equipment Manufacture market.

Confucius A Chinese teacher of ethics, ritual, and philosophy.

Congress of Vienna 1815 A meeting at which European boundaries were defined and Europe was demarcated from Asia by the Ural Mountains, through the Caspian and Black seas, and along the eastern end of the Mediterranean Sea to the African continent.

Congress Party The Indian National Congress was formed in 1865 to provide a forum for nationalist sentiments. The Congress Party was run by *brahmin* Hindus.

Continentality The size of a landmass.

Continental drift A process that took place millions of years ago in which a continent known as Pangaea broke into pieces forming separate continents.

Convectional rainfall A feature of equatorial regions in which precipitation is brought about by the immediate rising, cooling, and condensing of moisture-laden air.

Coral reefs Biological entities formed in seawater by small marine animals living in symbiotic relationships with algae. Coral reefs occur along continental shelves and around islands.

Coral Triangle "Amazon of the Sea." Extensive region of coral reefs that fringes six countries in Asia.

Core-periphery The concept that there are powerful core regions of development surrounded by less powerful regions of lesser development. The core-periphery concept is applicable in any context and at any scale.

Core region Focal point/area of economic and related activities.

Cultural convergence Takes place when lifeways and belief systems blend in a landscape.

Cyberspace The sphere of information technology.

Cyclone A low-pressure center. Can be associated with cyclonic storms having high-velocity winds and copious precipitation.

Dalits The name used by India's untouchables in reference to themselves. It means "oppressed."

Dao de Jing (*Tao-te-ching*) Means "classic of the way." A book of philosophy concerned with the *Dao*, or the way the universe works.

Dasht Expanses of black pebbles common to Baluchistan, Pakistan.

Deciduous forests These consist of trees that lose all their leaves in a cold or dry season.

Decolonization A nation's process of gaining of its independence from a colonial power.

Demographic transition model Sequential changes in population growth or decline beginning with high births and high deaths and ending with low births and low deaths.

Deng Xiaoping Ruler of China from 1976 to 1989. A colleague of Mao Zedong and survivor of the Long March. He began to open up China to a market economy.

Dependency ratio Ratio of percentage of a population under 15 and over 65 who are considered to be dependent to the percentage of the remaining people who are considered to be working.

Dependency theory The idea that core regions can grow only through exploitation of their dependent peripheries.

Deposition The depositing of material through the process of erosion.

Deserts Regions receiving less than 10 inches (254 mm) of precipitation a year.

Developing countries (DCs) Countries that are beginning the industrialization process. India and Indonesia are examples.

Dien Bien Phu The battle in 1954 between the Communist Viet Minh and the French. The French loss effectively ended their efforts to hold Vietnam.

Diffusion The spatial spread of a phenomenon.

Digital divide Argument that possession and use of electronic devices such as cell phones occur primarily in rich countries that are mostly in the Northern Hemisphere.

Dipterocarps Trees with two-winged seeds that dominate the upper canopy of the rain forest.

Doab The level and fertile land between the Ganga and Yamuna rivers. It means "two rivers." Also, any such land between two rivers.

Doi Moi Vietnamese word meaning "renovation." Reform policies instituted by the Vietnamese Communist Party in 1986.

Domestication The process of learning how to grow plants and manage animals for food and related products. The invention of agriculture.

Dowry Payment in cash and kind by bride's family to groom's family.

Dowry deaths The death of a new bride at the hands of her husband's family who are displeased over the size of the young woman's dowry and harass or kill her.

Dragon and tiger economies Newly industrializing economies of East and Southeast Asia.

Dravidians Dark-skinned people of the Indus Civilization in India.

Drignam Namzha The traditional values and etiquette policy launched by the king of Bhutan in order to salvage Bhutanese culture.

Drukpas The Bhote tribe of northern Bhutan, they were originally from Tibet. They revere the king of Bhutan as their spiritual leader.

Drukyul Original name of Bhutan.

Dual economy The economy characterized by the existence of two levels of economic activity; a formal sector functioning at a large scale in the realm of national and international trade, and an informal sector, functioning at a smaller scale at local levels.

Dutch cultivation (culture) system During the colonial period, the Dutch requirement that every Indonesian farm household allocate one-fifth of its land to the production of a cash or export crop.

Dwifungsi Means "dual function." Refers to the Indonesian army's dual role of defending the country and promoting economic development.

Dynasties In ancient China, periods of rule by one family.

Dynastic cycle The rise and fall of a particular dynasty in China. A dynasty would begin as virtuous, peaceful, and productive but over time would become corrupt, warring, and a victim of natural disasters, thereby losing the Mandate of Heaven.

Dzongs Fortified monasteries in Bhutan.

Economic Cooperation Organization (ECO) Established by Pakistan, Iran, and Turkey in 1985. Now includes Afghanistan and the Central Asian Republics.

Economic development zones (EDZs) Part of Deng Xiaoping's new economic policies. Areas of China designated as special zones for foreign investment.

Ecumene The inhabited world.

Edo Traditional name for Tokyo.

Eightfold Path The Buddhist way to overcome desire and achieve the ultimate state of nirvana.

Emergents Trees rising above the rain forest canopy.

Endemic Plants or animals occurring in a given area only.

Endogamy The practice of marrying within one's caste or *jati* (occupation).

Entrepôt A transshipment point, usually a port city such as Malacca.

Environmental criticality Environmental degradation associated with socioeconomic deterioration such as the kind that takes place when forests are destroyed.

Environmental determinism The notion that physical environmental forces determine human behavior.

Environmental refugee A person who has lost his or her home and land due to environmental disaster or deliberately planned environmental change.

Environmental transformation Changes in the physical environment wrought by values, attitudes, policies, and practices.

Ephemerals Desert plants that bloom sporadically in response to rain.

Epiphytes Plants (such as orchids and ferns) physically but not nutritionally supported by other plants. These are found in the middle level of the rain forest canopy.

Erosion The wearing away of the Earth's surface by the action of wind and water.

Estate Tamils Indian, Tamil laborers imported by the British to Sri Lanka to work on plantations or estates.

Ethnocentrism Judging other countries or cultures by one's own which is considered superior.

Eurasia The geographical landmass comprising Europe and Asia.

Eurocentric A worldview that puts Europe at the central and superior position relative to the rest of the world.

Exclusive Economic Zone (EEZ) The 200-nautical-mile zone belonging to coastal states as defined by the 1982 Law of the Sea Treaty issued by the United Nations.

Extended Metropolitan Regions (EMRs) Regions that have evolved through the continual growth of cities over a large area.

Extraterritoriality A consular jurisdiction whereby a foreigner is subject to the laws of his or her own country.

Exudates Resins (such as copals and dammars) that ooze out of a tree.

Factories Trading posts established by the Dutch and British in South and Southeast Asia.

Fa-hsien The first Chinese Buddhist pilgrim to visit India. His journey took place from 399 to 412 AD.

FELDA Federal Land Development Authority in Malaysia. FELDA develops schemes to open up jungle and virgin forests to settlement and agriculture.

Female infanticide The deliberate killing or neglect of female children.

Feminization of agriculture Women representing a larger proportion of laborers than men in the agricultural sector in Asia.

Feminization of labor In Asian economic development zones, women being the majority of workers.

Feng shui Chinese term meaning "wind and water." Forces of nature that must be dealt with appropriately in behavior and landscape design.

Filial piety Confucian concept of hierarchical loyalties of ruler to heaven; subject to ruler; son to father; younger brother to older brother; and wife to husband.

Filipinos People of the Philippines.

Five Pillars of Islam Rules for proper adherence to Islam.

Floating population The movement of more than a million individuals in China from city to city seeking work opportunities.

Fluvial River-related.

Flying Geese Model Developed by a Japanese economist in the 1930s. Describes industrial life cycles in newly industrializing countries.

Foot-binding In China during the dynasties, the practice of binding women's feet with the toes folded under the arch. Known as the Golden Lotus, bound feet were considered sensual. The practice began among the upper classes and court circles of the Southern Song Dynasty (1127–1278) and continued into the early 1900s.

Food insecurity Not having enough food to lead a healthy and productive life.

Forward-thrust capital A new capital city built in a frontier region to decentralize population from elsewhere and serve as a growth pole. Islamabad, in Pakistan is an example.

Four Noble Truths The Buddhist notion that the world is filled with suffering which is caused by desire and that desire can be overcome by following a set of rules.

Frontline state A state that serves as a barrier to the progress of an undesirable political system.

Functional interactions Relationships between people and the natural world that are expressed in political, social, and economic landscapes.

Ganga Ma Mother Ganges. A river in India that is considered sacred by Hindus.

Gender A culturally constructed social institution. Gender roles are expected and assumed in cultural contexts and society is organized around gender.

Gender Development Index (GDI) Statistic that indicates the well-being of women in a particular society.

Genghis Khan Lived 1155–1227. A military genius who amassed his nomadic hordes into the world's most formidable cavalry. Occupied northern China, Mongolia, and the Far East.

Geographic realms Based on physical and cultural characteristics, the largest units into which we can divide the ecumene, or the inhabited world.

Geomancer An expert in *feng shui*. Often called upon in building orientation and design.

Gini Index Measures relative income inequality within a country.

Glaciation The action of ice in the process of erosion and deposition.

Global feminization of poverty The fact that increasing numbers of women are becoming poor and that the -majority of the world's poor are women.

Globalism The concept that countries are linked in -various ways on a global basis.

Globalization The process by which regions become increasingly linked with global capitalist enterprises and economic systems.

Golden Quadrilateral India's nationwide four lane expressway linking major nodes of its urban system.

Grameen Bank A non-government organization in Bangladesh. The bank offers small loans to the poor, mostly women, to get a business started.

Great Leap Forward The communization of China's workers beginning in 1958. Agricultural cooperatives were reorganized into People's communes holding several cooperatives and thousands of individuals. Communization was expanded to include urban and industrial regions.

Great Proletarian Cultural Revolution A period of giant social, political, and economic upheaval in China. Beginning in 1966, an estimated 100 million people were targeted for reeducation and forcible movement to different regions of the country.

Greater East Asia Co-Prosperity Sphere (GEACPS) Japan's ambition to lend an empire composed of Asian countries minus European imperialists. The plan was halted when the United States was drawn into World War II by the Japanese attack on Pearl Harbor in 1941.

Greenhouse gases Gases such as CO_2 that are released into the atmosphere by clear-cutting of the Earth's forests.

Green Revolution The use of high-yielding, pest-resistant seeds in agriculture.

Growing season The number of consecutive, frost-free days in a year.

Growth triangle A triangle of economic growth regions often involving more than one country.

Growth pole A center designated to attract industry and development.

Gun-salute states Indian princely states and estates where the number of gun salutes—21, 15, 10—indicated the importance of the ruler.

Guptas Rulers of India who united northern India around 320 AD. Art, mathematics, science, and literature flowered during the Gupta period, also known as India's Golden Age.

Gurkhas British-trained Nepalese troops.

Guru A spiritual advisor in Hinduism.

Guru Nanak Founder of Sikhism in the fifteenth century.

Hajj A pilgrimage to the holy cities of Mecca and Medina. One of the Five Pillars of Islam.

Hamun Dry lakes or *playas* found in Baluchistan, Pakistan.

Hangul Korean lettering system developed in the fifteenth century under King Sejong.

Hartal A labor strike in Bangladesh.

Hearth areas Places of origin of cultural phenomena.

Hill stations Communities at higher elevations used by the British as refuges from the summer heat of the lowlands. During the Raj, they were socially and spatially segregated from Indian areas.

Hindutva "Hinduness," a desire to make India a country where Hindu principles prevail.

Hiragama Japanese syllabary that permits the expression of words and parts of speech not easily represented by Chinese characters.

Historic inertia A phenomenon in which people continue to live in and migrate to places where other people already live and opportunities are perceived to exist. As a consequence, densely populated areas are most likely to continue to be densely populated.

Ho Chi Minh The founder of the Indochina Communist Party. Leader of North Vietnam from 1954 until his death in 1969.

Hollowed-out industry In Japan, an industry that has lost its competitive edge due to rising labor costs and is forced to seek cheaper labor elsewhere, typically overseas.

Honor killings Killing of females who are perceived to have dishonored the family in Muslim societies.

Household responsibility system A return to family farming in China whereby farmers are to meet government quotas but have the remainder of the crops to keep or sell on the open market.

***Hsia-fang* (downward transfer) movement.** The transfer of millions of youths and others to the countryside during China's Cultural Revolution.

Hsuan-tsang Chinese pilgrim who spent 13 years in India from 630 to 643 AD. He furthered the spread of Buddhism in China.

Hu Jintao General Secretary of the Chinese Communist Party since 2002. Also president of the People's Republic of China since 2003.

Hukou Household registration system in China. Every Chinese citizen is registered as nonagricultural or agricultural. Registration includes one's particular geographic location, and it is very difficult to alter one's classification.

Human Development Index (HDI) A composite index that measures a country's average achievements in health, knowledge and a decent standard of living

Human trafficking Trafficking of girls (rarely boys) within and between countries for purposes of sex or servitude.

Hydrological Referring to water.

Hyperurbanization Extremely rapid city growth.

Ilustrados Spanish-educated descendants of powerful landed families in the Philippines.

Imam Islamic religious leader.

Indentured servitude A state in which poor farmers who borrow from landlords, become victims of usurious interest charges, and remain indebted for the rest of their lives.

India's Golden Age A period during the fourth to sixth centuries when art, science, and mathematics flourished under the Guptas.

India's Silicon Valley A region of high-tech industries centered on Bangalore.

Indus Waters Treaty In 1960, divided the tributaries of the Indus River between India and Pakistan.

Industrial crops Non-food crops such as cotton and rubber.

Infant mortality Death rate of infants 0 to 1 year old per 1,000 births.

Infidels According to Muslims, people who do not believe in Islam.

Intermediate technology Technology that lies between the most primitive and the most modern.

Intertropical Convergence Zone (ITCZ) The equatorial low-pressure trough that dominates the central area of the Indian Ocean.

JABOTABEK The fastest growing area in Indonesia. Comprises the metropolitan regions of Jakarta, Bogor, Tangerang, and Bekasi.

Jainism A religion focusing on non violence against any life form. Founded in the sixth century by Mahavira.

***Jajmani* system** Punjabi custom for Harijan laborers to work for their Jat landlords in exchange for grain and fodder. The system is both an economic system and a social contract rooted in caste relationships.

Jarkhand Name for a new independent state in the Jamshedpur industrial region in India. Jarkhand came into existence in 2000.

Jati A subdivision of *varna* defined by and named by occupation.

Jawaharlal Nehru Leader of the Congress Party and the first prime minister of India.

Jesus Christ Believed by Christians to be the Son of God and the manifestation of God on Earth.

***Juche* philosophy** Kim Il Song's ideas of self-reliance and self-sufficiency for the people of North Korea.

Jungle Dense vegetation of the tropics that occurs where light penetrates the rain forest canopy.

Jute A fiber crop used to make sacking, twine, ropes, carpet, car interiors, lining for asphalt roads, and many other products. One of the major crops in Bangladesh.

Kamikaze Japanese term meaning "the divine wind." A typhoon was responsible for the defeat of Kublai Khan in his attack on Japan in 1281.

Kampuchea Cambodia's name under the Communist Khmer Rouge and Pol Pot.

Kampung A village in Malaysia and Indonesia.

Kanji The Japanese writing system that was borrowed from the Chinese character system.

Kansai Science City Research and development center focusing on Kyoto, Osaka, and Nara in Japan.

Karez Underground irrigation systems consisting of hand-dug, gravity-flow tunnels. *Karez* are found in northwest China and in southwest Asia. The tunnels carry water from aquifers to surface channels and ponds.

Karma In Hinduism, the accumulation of all of a person's deeds in a lifetime.

Karst topography Pitted and lumpy limestone formations caused by chemical weathering.

Katchi abadis Squatter settlements on the periphery of Karachi, Pakistan.

Keiretsu Japan's modern industrial conglomerates. A hierarchical organization of companies.

Khalistan Name for the independent state desired by many Sikhs in Punjab.

Kharif In South Asia, crops (such as rice) sown after the onset of the wet monsoon. *Kharif* crops are harvested in autumn.

King Harsha Controlled an empire from Punjab to the Narmada River during the 600s AD. The last north Indian ruler to encroach south of the Narmada for 600 years.

Koran The holy book of Islam; from the Arabic word *qur'an*, meaning "recitation."

Korean Wave III Korean popular culture.

Koryo A Korean kingdom that emerged in 918 AD in the central region of the peninsula. Origin of the name "Korea."

Kotadesasi From the Indonesian words *kota* (town) and *desa* (village) meaning the coincidence of urban and rural activity in the same spatial territory.

Kshatriyas Members of the warrior class in India's caste system.

Kublai Khan Grandson of the Mongol warrior Genghis Khan. Kublai Khan dominated all of China, Korea, and northern Vietnam. The Mongols ruled China as the Yuan Dynasty, 1279–1368.

Lan Xang A Laotian word meaning "a million elephants." The first kingdom of Laos that emerged in 1353.

Landscape The world that we see and experience.

Lao Loum Laotian term meaning "Laotians of the valley." *Lao Loum* make up one-half the population of Laos.

Lao Sung Laotian term meaning "Laotians of the mountain tops." *Lao Sung* are among the most recent migrants to Laos.

Lao Theung Laotian term meaning "Laotians of the mountain slopes." *Lao Theung* make up about a quarter of the population of Laos.

Laozi In Chinese, "the old one." Laozi was a contemporary of Confucius. The philosopher most closely associated with Daoism, the study and contemplation of the way the universe works.

Laterite Rock-like, lower layer of tropical soils formed by the presence of aluminum and iron oxides.

Lianas Woody climbers or vines found in the middle level of the rain forest canopy.

Life expectancy How long one can expect to live from birth.

Link-language A language used to allow communication between speakers of other languages. English is a link-language in Sri Lanka.

Literacy rate Percentage of males and females who can read and write at least at an elementary level.

Loess Fine, windblown soil lacking horizontal stratification. Loess can stand in vertical cliffs.

Long March The name given to the 6,000-mile-long trek (9,660 km) by the Communist Army in 1934 and 1935, as it fled the forces of the Nationalists. The Long March was led by Mao Zedong and took the communists from southern China to Yenan in the loess hills of Shensi Province.

Madrassas Islamic religious schools.

Maharaja A word meaning "great ruler." Maharajas ruled various states in India following the breakdown of the Mughal empire.

Mahavira Founder of Jainism.

Mahayana Buddhism The school of Buddhism practiced in China, Korea, Japan, and Vietnam.
Mainlanders Those who moved to Taiwan from the Chinese mainland between 1945 and 1949. They are associated with Chiang Kai-shek and the Nationalist forces.
Malesia Malaysia and Indonesia.
Manchukuo Japanese puppet state in Manchuria established in 1931.
Mandate of Heaven In dynastic China, the belief that Heaven bestowed the right for Chinese rulers to rule. The mandate demanded virtue and the integrity of proper relationships.
Mangal A mangrove community of tropical and subtropical coasts.
Mao Zedong Leader of the Chinese Communist Party whose forces won the revolution and control of China in 1949. Mao ruled China until his death in 1976.
Mappae Mundi Religiously-inspired maps of the world that portrayed the Earth as flat.
Mariculture Aquaculture—the practice of raising fish, shellfish and other marine products in various water bodies.
Market gardening The production of fresh agricultural products for urban markets.
Maternal mortality The number of women who die in child birth for every 100,000 live births.
Mecca Place of origin of Islam and Islam's holiest city.
Medina Associated with Mecca as place of origin of Islam.
Mega-cities Giant cities often incorporating several cities in a cluster.
Mega-urban regions Urban agglomerations that extend beyond metropolitan boundaries.
Mercantilism An economic system whereby raw materials are extracted from a colony and brought to the colonial power for manufacture. The manufactured goods are shipped back to the colony for purchase.
Mestizos In the Philippines, people of Chinese-Filipino or Mexican-Filipino descent.
Microcredit Small loans given to poor people, usually women, to start and maintain a business enterprise.
Mission civilisatrice French term meaning "civilizing mission." The French belief that colonialism would bring the benefits of modern European civilization to the primitive peoples of the world.
Mohandas Gandhi He believed in non violent protest and was a key figure in gaining India's independence.
Monotheistic Having one god.
Monsoon A climatic system based on intense seasonal changes in pressure and wind patterns. A realm affected by the monsoon has wet and dry seasons.
Monsoon Asia Those parts of Asia under the influence of the monsoon climatic system.
Mosque Place of worship for Muslims.
Mother Theresa (1910–1997) Christian who founded the Missionaries of Charity in Kolkhata, India.
Mughals Ruled India from 1497 to 1858.
Muhajirs Urdu-speaking, post-partition, Muslim migrants to Pakistan from India.
Muhammad The prophet of Islam.
Muhammad Ali Jinnah Leader of the Muslim League and first prime minister of Pakistan.
Mujibur Rahman (Mujib) Leader of the Awami League in Pakistan.
Multilateral aid Aid deriving from international organizations such as the United Nations, World Bank, or Organization for Economic Cooperation and Development.
Multinational enterprise An economic, social, or political entity with ties to several nations.
Multiregional hypothesis Proposes that humankind originated in several locations simultaneously.
Mullah Islamic religious leader.
Mura Traditional rural village in Japan.
Muslims Those who submit to God; adherents of Islam.
Myth of Asia The fact that there is no cultural or historical entity that equals Asia.
Naga Sanskrit word meaning "serpent." The *naga* is an aquatic symbol that appears in art form as a dragon, snake, or other bodily form. Part of cosmologies in Southeast Asia.
Nanyang China's name for the southern sea regions (Southeast Asia) in early times.
Native Taiwanese Immigrants from southeastern China who arrived from the fourteenth century into the twentieth century. They speak the Taiwanese dialect and make up about 70 percent of Taiwan's population.
Natural boundary A boundary demarcated by physical features such as mountain ridges or rivers.
Natural vegetation Undisturbed, mature plant communities.
Naxalites Far left-wing, radical communists in India who support Maoist ideology.
Neo-Malthusian Concept that while fertility rates are important, it is absolute numbers of people that count.
New Theory Farming In Thailand, a form of land and agricultural management employed as a means to introduce community-based economies.
New Thinking A policy aimed at liberalizing the centrally planned economy of Laos.
Newly industrializing countries (NICs) Countries that have recently begun to industrialize. Used in reference to several East Asian countries such as South Korea and Taiwan.
Nirvana In Buddhism, a state of desirelessness and peacefulness.
Non-Governmental Organizations (NGOs) Non-governmental aid organizations. NGOs have an agenda, be it political, humanitarian, or religious.
North-South dichotomy The difference between the rich, industrialized countries of the Northern Hemisphere and the poorer, developing countries of the Southern Hemisphere.
Nucleated Clustered, as in human settlements.
Nutritional density The number of people existing per square mile or kilometer of cultivated land.

Occident The West.

Offshoring The substitution of less costly foreign labor for domestic labor.

Oracle bones bones, especially scapulae, which are heated and their cracks interpreted by ancient Chinese rulers and sages. Also inscribed with questions and interpretations.

Orang Asli Means "original man." Earliest of the current inhabitants of the Malay Peninsula.

Orient The East.

Orographic precipitation Precipitation brought about by the presence of a topographic barrier that forces a moisture-carrying air mass to rise, cool, and condense.

Oustees Uprooted persons who flee from war and violence, oppression, disasters, economic stagnation, and human-induced environmental displacement. About half of all oustees are landless or sharecroppers.

Out of Africa Theory that postulates that humankind originated in Africa.

Overseas Chinese Chinese—both individuals and corporations—living outside of mainland China.

Pacific Rim An economic concept referring to all countries situated around the Pacific Ocean. These countries' primary trading partners are each other.

Pancasila The five principles adopted by Sukarno in 1945 as the basis for Indonesian ideology and development.

Pan-gu In Chinese mythology, a primeval being whose body parts transmuted into various parts of the cosmos.

Pat Clay desert in the Sind province of Pakistan.

Pathet Lao Communist guerrilla fighters in Laos.

Patriarchy Type of society in which family unit structure focuses on males who exert authority over the rest of the family members.

Pax Mongolica A period of peace and prosperity when China was ruled by the Mongols from 1280 to 1300.

People's communes A collection of cooperatives housing thousands of people organized under China's Great Leap Forward of 1958.

Per Capita Income (PCI) Statistic that divides a country's gross national or domestic product among all of that country's population.

Percentage urbanized The percentage of a country's population living in cities.

Peripheral regions Poor regions that are remote from the center of economic and other activity in core regions.

Phrapbuum Thai word meaning "lord of the place." It refers to the spirit shrine exhibited in houses, office buildings, and other private and commercial establishments in Thailand.

Pinyin A simplified character and transliteration system for the Chinese language. Pinyin closely approximates the sound of the word in the Mandarin dialect of Chinese.

Plantation Large, commercial, agricultural enterprise that typically focuses on one crop such as tea or palm oil.

Plate tectonics A theory that postulates that rigid plates in the upper portion of the lithosphere float on the more plastic upper layer of the Earth's mantle. Movement along the plates' fracture zones cause seismic activity.

Playas Salt-encrusted dry lakes found in Baluchistan, Pakistan.

Plural society A society in which two or more culture groups live adjacent to each other in the same country without mixing.

Political ecologist One who investigates the inner workings of the "development" process. [As in Chapter 5]

Pol Pot Leader of the Khmer Rouge in Cambodia from 1975 to 1978. Responsible for the death of millions.

Polytheistic Having many gods.

Popular culture Usually refers to American popular culture that is diffusing via transnationals and media into all corners of the world. The popular cultures of Japan and South Korea are also diffusing elsewhere, especially in Asia.

Population composition The age and sex structure of a particular population.

Population migration The internal or international movement of people from one place to another on a temporary or permanent basis.

Population pyramids Pyramidal graphs showing the percentage of a country's males and females in various age cohorts.

Post-development theory Holds that pigeonholing countries into Western-contrived categories is pejorative and associated with old colonialist paternalism and racism.

Power relationships The fact that men and women are accorded unequal roles in virtually every society.

Primate city A nation's largest city, within the context of urban primacy.

Purchasing Power Parity (PPP) A measure of income based on gross domestic product that accounts for price differentials among countries and translates local currencies into international dollars.

Purdah The Muslim custom of secluding women from unrelated men.

Quality of life The state of human well-being.

Quoc Ngu Means "national language." Transliteration of Vietnamese spoken language into the Roman alphabet.

Rabi In South Asia, crops (such as wheat) sown at the end of the wet monsoon. *Rabi* crops are harvested in spring.

Rainshadow The protected, leeward side of a topographic barrier which is the dry side in the process of orographic precipitation.

Raj A word meaning "the rule." The *Raj* refers to the period of British rule in India.

Ramadan A month of daytime fasting; one of the Five Pillars of Islam.

Reformasi A period of economic reform and liberalization in Indonesia following the end of General Suharto's rule in 1998.

Regional state A natural economic zone shaped by the global economy in which it participates. Guangdong Province in China is an example.

Regionalism Focusing on the local rather than the global.

Region-based urbanization The continual growth of cities over large areas.

Reincarnation In Hinduism, the notion that a person is reborn into another cycle of existence depending on the *Karma,* or sum of ideals, of the previous life.

Religion Set of sacred beliefs and practices.

Religious revivalism In Islam, the resurgence of religious fundamentalism. In the early 2000s, characterized as extreme hostility to the West and Western culture.

Remittances The transfer of funds by a foreign worker to her home country.

Replacement fertility rate Rate at which a population is growing in order to replace itself in the future.

Restructuring The rapid change in local, national, and international social, political, and economic systems wrought from the dynamics of global interaction.

Rich-poor gap The difference between rich and poor between parts of the world, but also between countries and within individual nations.

Saint Thomas Introduced Christianity to southwestern India in 52 AD.

Sangha Monastic orders of Theravada Buddhism.

SARS Severe acute respiratory syndrome, a highly contagious pneumonia-like disease that infected thousands of people between November 2002 and July 2003. China was the source of the disease.

Sati In early India, the practice of burning a widow alive on her husband's funeral pyre. *Sati* was outlawed by the British in 1829 but continued to exist for many years.

Sawah Terraced rice paddies in Indonesia.

Scheduled castes and tribes Also called *adivasis,* original inhabitants of India such as the Nagas and Bondas.

Sawah Terraced rice paddies in Indonesia.

Semi-deciduous Refers to trees that lose some of their leaves in the dry months of the monsoon.

Semi-peripheral region A region that acts like a core to a peripheral region and like a periphery to a core region.

Sepoy British-trained Indian soldier.

Serendip Arab name for Ceylon (Sri Lanka).

Sex cities Cities that are known for their prostitution, sex clubs and sex tourism.

Sex ratio The numerical ratio between men and women in a population.

Sex tourism National and international tours that are geared to sexual experiences.

Sex-selective abortion The deliberate abortion of males or females (typically females) after sex determination of the unborn fetus.

Shadow economy Illegal economic activities, such as black market enterprises, that are unregulated, untaxed, and unseen.

Sharia The sacred law of Islam.

Shatter-zone A region of multiple, independent cultural groups often generated by fragmented terrain.

Sherpas Herders and traders of Nepal's Himalayas who are now dependent on tourism. Many serve as guides for climbing expeditions.

Shifting cultivation A migratory, field rotation system of agriculture found mainly in forested regions with relatively sparse populations.

Shiites A breakaway Islamic group with roots in Persia (Iran).

Shinto Traditional animist and nature-centered religion of Japan.

Shoguns Military leaders who ruled Japan from the medieval period to 1868.

Sikh A follower of Sikhism.

Sikhism Belief system founded in fifteenth century by Guru Nanak in India. Sikhs believe that all people are the children of God. They also oppose the caste system.

Simhala Dvipa Island of the Lion Tribe (Sri Lanka).

Sinhala Language of the Aryan, Buddhist, Sinhalese in Sri Lanka.

Sinicize To make Chinese.

Sinocentric A world view that places China at the central and superior position relative to the rest of the world.

Slash-and-burn method The cutting down of all but the largest trees in an area and the burning of the debris to clear the land for planting.

Social construction The construction of landscapes by internal and external political, social, and economic forces.

Sodic land Land where crusts of salt or other alkalis prevent nutrients and water from going below the surface. Reclamation of sodic lands is a goal of the Indian government.

South Asian Association for Regional Cooperation (SAARC) Formed to disassociate political conflicts from economic cooperation. It includes Bangladesh, Bhutan, India, the Maldives, Nepal, Pakistan, and Sri Lanka.

Southward policy Taiwan's policy of increasing investment in Southeast Asian countries to offset investments in China.

Spatial duality The differences between the landscapes of core and peripheral regions.

Special economic zones (SEZs) Zones set aside by China's government to attract foreign technology and investment.

Species diversity Variation in plant or animal species.

Squatter settlements Makeshift settlements in and around cities; associated with rural to urban migration.

State capitalism A system in which the government plays a strong role in economic development such as in South Korea.

Steppe Short grasslands.

Storm surge Water pushed toward shore by the force of the winds swirling around the storm. When combined with a tide in an increasingly narrow channel across low, flat terrain, a storm surge can cause an increase in water level of 15 feet or more.

Subak Agricultural cooperative in Bali.

Sudras Anyone belonging to the class of serfs in India's caste system.

Sukarno Indonesian nationalist who ruled from 1945 until 1965.

Sun Yat-sen Asia's first Asian nationalist. Proclaimed a Chinese Republic in 1912.

Sunderbans An area of ever-shifting mangrove, mudflats, islands, and distributary channels found in the Indus and Brahmaputra delta in the Bay of Bengal.

Sunna The way of life prescribed in Islam, based on the teachings of Muhammed and the Koran.

Supply chain The division of products into specialized dissemblies to drive down costs, improve quality and reduce the time it takes to get the product to market.

Sustainable development Development that sustains environmental integrity.

Swidden Shifting cultivation in Southeast Asia.

Tagalog The official language of the Philippines.

Taiping Rebellion An uprising against the corrupt and ineffective Manchu rulers of China. It began in 1850 and ended in 1864 with the loss of 20 million lives.

Tais Principle ancestors of the Thai, Lao, and Shan peoples. *Tai* is a cultural and linguistic term used to denote the various *Tai* people who became differentiated into a larger number of separate identities.

Taiwanese Everyone who lives on the island of Taiwan.

Tamil Language of the Dravidian Hindus in Sri Lanka and the state of Tamil Nadu in India.

Tamil Eelam The proposed name for an independent state that was being fought for until 2010 by the Tamil Tigers in northern Sri Lanka.

Tamil Tigers A militant Tamil group of northern Sri Lankans who were agitating for an independent state called Tamil Eelam. They were defeated by the Sri Lankan army in 2010.

Tantric Buddhism Also called *Vajrayana* Buddhism; it developed in north India in the seventh century and spread to the Himalayas where it was influenced by Tibetan shamanism. Official religion of Bhutan and Sikkim.

Techno-Archipelago A series of technopolises linked by information and communication channels.

Technobelt Zone selected for research and development and linked by communication and telecommunication.

Technopoles Places where information and technology disseminate from. In India, Bangalore's "silicon valley" is an example of a technopole.

Technopolis A network of science cities centering on a "mother city" and connected to a major metropolis.

Technoscapes Vast, intertwined and overlapping channels of communication produced by flows of technology, hardware, and software disseminated by transnational or supranational corporations, organizations, and agencies.

Terai A marshy plain in southern Nepal situated 900 feet (270 m) above sea level.

Theravada Buddhism The school of Buddhism practiced in Sri Lanka and most of Southeast Asia. It stresses the responsibility of the individual for his or her own salvation through the accumulation of merit by good behavior and religious activity.

38th parallel The line of latitude that divides North and South Korea into two nations. The parallel division was instituted in 1945 by the Soviets and Americans.

Three Gorges Dam World's largest water conservation project on the Yangzi River in China. Three Gorges has the world's most powerful hydroelectric dam and improved shipping from the east to the interior. However, thousands of acres of farmland were lost and countless numbers of residents were forcibly relocated. In addition, numerous important natural habitats and cultural sites were destroyed.

Tied aid The linking of foreign aid to structural adjustment or specific activities or purchases.

Tienanmen Square Large parade square in Beijing. Site of ruthlessly crushed "democracy" protests in 1989.

Time-space convergence More space can be covered in less time due to improvements in transportation and communication.

Timor Gap The disputed ocean and seabed between Australia and Indonesia.

Tokaido megalopolis Urbanized, industrialized region from Tokyo to Osaka. Named for the old Tokaido Road.

Tonkin Region of the Red River delta in northern Vietnam. Under French colonial rule, Tonkin was one of the five territories of the Indochinese Union.

Topographic barrier An area of raised relief (such as the Himalayas) that prevents the passage of wind or other phenomena.

Topography Landforms such as mountains and valleys.

Toponyms Place names.

Total Fertility Rate (TFR) The number of children a woman would have if age-specific birth rates remained constant throughout her childbearing years—age 15 to 45 or 15 to 49.

Township and village enterprises Economic enterprises established in China's rural areas to provide employment and discourage out-migration.

Trans-border development region An economic development zone in adjacent political units.

Transgenic plants Human-engineered plants.

Transmigrasi People who move from crowded areas of Indonesia to open up and populate frontier regions.

Transmigration of souls In Hinduism, the changing of people's life status through rebirth according to the merits of their behavior.

Transnational enterprise An economic, social, or political entity with ties to several nations.

Tree line The line above which the growing season and the soil depth do not support trees.

Tribal areas Mountainous, tribal, and fractious regions, especially those in the western boundary of South Asia, where tribal chiefs and religions leaders rule.

Tropical cyclone A violent storm created by a deep, circular, low-pressure cell. Called a *typhoon* in the Pacific.

Tsunamis Seismic sea waves caused by earthquakes.

Tundra A subarctic environment above the tree line that is a zone of low-lying flora, mosses, and lichens.

Typhoon In the Pacific, a violent storm created by a deep, circular, low-pressure cell. Widely known as a tropical cyclone.

Underemployment Western notion regarding the underutilization of labor.

Unipolar concentration A situation arising from the migration to and growth of one city such as Tokyo at the expense of other places.

Untouchables Non-caste individuals associated with "polluting" occupations in India.

Upanishads A work of Hindu literature, from the Sanskrit phrase meaning "sitting down near" one's spiritual advisor.

Upazilas A subdistrict in Bangladesh created in the 1980s.

Urban agglomerations Clusters of cities.

Urban primacy A situation that exists when the population of a nation's largest city is greater than the combined populations of its next three largest cities.

Vaishyas Peasant farmers in India's caste system.

Value chain A chain of activities in a production system in which each activity adds value to the end product.

Varnas "*Varna*" is a word meaning "color." In Hinduism, classes of people such as priests or warriors.

Vedas The earliest known Hindu literature, from the Sanskrit word *veda*, meaning "knowledge."

Veddas Aboriginals of Sri Lanka.

Vegeculture The cultivation of tuber and tree crops such as taro and bananas. Vegeculture preceded seed cultivation.

Vertical integration Exists when one company or conglomerate controls all stages of a production process.

Viet Cong National Liberation Front in southern Vietnam. Opposed to the American-supported South Vietnamese government.

Viet Minh The front organization of the International Communist Party in Vietnam. After World War II, the Viet Minh took control of most of northern and central Vietnam.

Wallace Line The line of demarcation between the distinct zoogeographic regimes of Malaysia and Indonesia.

We–they syndrome Ethnocentric peoples (we) have a sense of superiority to other cultures (they) that are considered inferior.

White revolution The transformation of India's dairy industry to produce more milk.

Wind-chill factor The role that wind plays in reducing temperature.

World cities Cities at the top of a global urban hierarchy such as London, Tokyo, or New York.

World systems theory Proposed by Immanuel Wallerstein. Dichotomy of capital and labor as competing agents in the accumulation of capital in context of globalization. Economic interchange takes place on unequal terms between cores and peripheries on a global scale.

Xenophobia Fear of foreigners.

Xerophytic Drought-resistant.

Yang With *yin*, one of two opposing yet complementary forces that govern the universe.

Yi jing (I-ching) A Chinese literary work in which hexagrams depict the harmonies of complementary opposites.

Yin With *yang*, one of two opposing yet complementary forces that govern the universe.

Zaibatsu Japanese military-industrial conglomerates. They were disbanded after World War II.

Zeylan Dutch name for Ceylon (Sri Lanka).

Index

Note: *A reference with an italicized number refers to figures, tables, boxes, and captions.*

C

G

H

L

N

Q

R

T

U

V

W